Ball · Kinderlehrer · Podio-Guidugli · Slemrod (Eds.)
Evolving Phase Interfaces in Solids

Springer
Berlin
Heidelberg
New York
Barcelona
Hong Kong
London
Milan
Paris
Singapore
Tokyo

Morton E. Gurtin

J. M. Ball D. Kinderlehrer
P. Podio-Guidugli M. Slemrod (Eds.)

Fundamental Contributions to the Continuum Theory of Evolving Phase Interfaces in Solids

A Collection of Reprints of 14 Seminal Papers
Dedicated to
Morton E. Gurtin
on the Occasion of His Sixty-Fifth Birthday

With an Introduction by Eliot Fried

Springer

Professor John M. Ball
Mathematical Institute
24-29 St. Giles'
Oxford OX1 3LB, United Kingdom

Professor David Kinderlehrer
Department of Mathematics
Carnegie-Mellon University
Pittsburgh, PA 15213-3890, USA

Professor Paulo Podio-Guidugli
Dipartimento di Ingegneria Civile
Universitá di Roma "Tor Vergata"
I-00133 Roma, Italy

Professor Marshall Slemrod
Center for Mathematical Sciences
University of Wisconsin
1308 West Dayton St.
Madison, WI 53715-1149, USA

ISBN1-13: 978-3-642-64188-6 e-ISBN-13: 978-3-642-59938-5
DOI: 10.1007/978-3-642-59938-5

Library of Congress Cataloging-in-Publication Data

Fundamental contributions to the continuum theory of evolving phase interfaces in solids: a collection of reprints of 14 seminal papers, dedicated to Morton E. Gurtin on the occasion of his sixty-fifth birthday/J. M. Ball ... [et al.] (eds.); with an introduction by Eliot Fried.
p. cm. Includes bibliographical references. ISBN 3-540-64683-3 (acid-free paper)
1. Solids--Surfaces--Congresses. 2. Phase transformations (Statistical physics)--Congresses. 3. Crystals--Congresses. 4. Molecular structure--Congresses. 5. Gurtin, Morton E. I. Gurtin, Morton E. II. Ball, J.M. (John MacLeod), 1948--QC176.8.S8F86 1999 530.4'17-dc21 98-41920

Softcover reprint of the hardcover in 1st edition 1999

Production Editor: Goldener Schnitt R. Kusche, Sinzheim, Germany

Cover design: *design & production* GmbH, Heidelberg

SPIN: 10481363 55/3144 - 5 4 3 2 1 0

Preface

A traditional way to honor distinguished scientists is to combine collections of papers solicited from friendly colleagues into dedicatory volumes. To honor our friend and colleague Mort Gurtin on the occasion of his sixty-fifth birthday, we followed a surer path to produce a work of intrinsic and lasting scientific value: We collected papers that we deemed seminal in the field of evolving phase interfaces in solids, a field to which Mort Gurtin himself has made fundamental contributions. Our failure for lack of space to include in this volume every paper of major significance is mitigated by the magisterial introduction prepared by Eliot Fried, which assesses the contributions of numerous works.

We hope that this collection will prove useful and stimulating to both researchers and students in this exciting field.

August 1998

John M. Ball
David Kinderlehrer
Paulo Podio-Guidugli
Marshall Slemrod

Contents

Contents

III. Papers on Mathematics

Introduction: Fifty Years of Research on Evolving Phase Interfaces

Eliot Fried

Department of Theoretical and Applied Mathematics, University of Illinois at Urbana-Champaign, Urbana, Illinois 61801, USA

A material body most often consists of one or more regions, termed phases, within which the chemical composition and the crystal or molecular structure do not vary with position. A phase transition is a process in which a region occupied by one phase grows at the expense of a region occupied by another phase, with the essential physical activity occuring at the interfaces separating phases. Processes of this type can be used to control the scale and distribution of microstructural features that are associated with different phases and, thus, granted an understanding of the various influences that such features may exert on macroscopic properties, to create materials with optimally determined properties. To realize such a program requires a physically faithful mathematical framework with the capacity to describe and predict the evolution of phase interfaces.[1] During the last fifty years, materials scientists, continuum physicists, and mathematicians have made substantial progress toward establishing a broadly applicable continuum-level theory of this nature.[2] This theory is based on the approach of Gibbs (1878: *1*), who treated phase interfaces as surfaces across which bulk material properties may suffer discontinuities and, as a means to account for localized interactions between phases, endowed these surfaces with excess fields. Here, some of the relevant developments are summarized, with emphasis being placed

[1] The following passage, due to Smith (1948: *1*), shows that the need for such a theory was apparent prior to the midpoint of this century: "The art of metallography is mature and the forms in which various micro-constituents appear are well known. Investigations almost without end have disclosed the importance of the exact manner of distribution of phases on the physical properties and usefulness of an alloy. Surprisingly, however, relatively little attention has been paid to the forces that are responsible for the particular and varied spatial arrangements of grains and phases that are observed. Like the anatomist, the metallurgist has been more concerned with form and function than with origins."

[2] That continuum physics can provide a useful framework for the study of evolving microstructure is clear from the following statement, written by Herring (1953: *2*) in support of the suitability of macroscopic concepts for the study of surface science: "These concepts, besides being simpler and neater to work with than those of atomistic theory, have the great advantage that they often lead to soundly based conclusions which are independent of any detailed assumptions regarding the atomistic mechanisms of the phenomena being studied. Of course, this does not mean that the user of macroscopic concepts should think no atomistic thoughts! It merely means that he should wherever possible use such thoughts as a guide to his intuition in formulating macroscopic laws and deciding the range of their validity."

on the contributions made by those researchers whose works are reprinted in this volume.[3]

Works in Materials Science

A major impetus to the development of a framework capable of characterizing microstructural evolution was Herring's (1951: ***2***) paper on the sintering of powders. Herring considered powders that consist of monatomic crystalline grains and allowed for atomic diffusion through the migration of point defects (i.e., vacancies and/or interstitials), with his primary goal an explanation of the manner in which atomic diffusion is influenced by localized interactions across junctions separating grains. To describe diffusion, Herring introduced densities and chemical potentials, for atoms and point defects, as well as bulk and interfacial diffusive fluxes. He then observed that the bulk diffusive flux should be proportional to the gradient of the diffusion potential,[4] which, in the context that he considered, is the surfeit of the chemical potential of the atoms over that of the point defects.[5] Significantly, Herring recognized that the description of equilibrium requires a condition that determines the diffusion potential at an interface,[6] a condition that he derived by computing the change in the total energy of the system that arises on altering the interfacial configuration. In recognition of the crystalline nature of the grains, Herring was careful to account for orientation dependence in the response function that determines the interfacial energy density.[7] From his derivation, it is clear that Herring's expression for the interfacial value of the diffusion potential is both an interface condition that is supplemental to and distinct from the standard statements of force balance and mass balance that arise from variations of either the deformation or the arrangement of atoms and point defects, respectively.[8] Not content to consider only situations where the

---

[3] When papers that appear in this volume are cited, a bold-face italic reference number (e.g., ***1***) is used; otherwise, a light-face italic reference number (e.g., *1*) is used.

[4] The term "diffusion potential" was introduced by Larché & Cahn (1978: *1*).

[5] Herring also noted that the diffusive flux ***within*** grain boundaries should be proportional to the interfacial gradient of the diffusion potential.

[6] Herring worked under the hypothesis of "local surface–volume equilibrium," which implies the continuity of the diffusion potential.

[7] That the interfacial energy density of a crystalline solid should generally depend on orientation goes back at least as far as Gibbs (1878: *1*). See, also, Curie (1885: *1*) and Wulff (1901: *1*).

[8] The additional condition derived by Herring can be interpreted as a generalization, that encompasses orientation dependence in the response function for the interfacial energy density, of what is often called the "Gibbs-Thomson condition" (cf., e.g., Volmer (1939: *1*) and Porter & Easterling (1981: *6*)). The full set of interface conditions that ensue on varying the total energy of the system constitute Weierstrass–Erdmann corner conditions (cf. Weierstrass (1927: *1*) and Erdmann (1877: *1*)).

interfaces between grains are smooth, Herring obtained an expression for the diffusion potential at a faceted interfaces[9] as well as an equilibrium condition at a junction where three grains meet.

Concurrently with Herring's work on sintering, Eshelby (1951: *1*) published the first in a series of investigations, including (1956: *1*; 1970: *1*; 1975: *2*), aimed toward a reconciliation of classical elasticity theory with the solid-state theory of lattice defects.[10] Eshelby worked from the hypothesis that an appropriate continuum analog for a defective crystal is an elastic solid in a state of stress induced not by externally applied loads (i.e., body forces and surface tractions) but, rather, by the presence of material imperfections (i.e., localized nonuniformities). Granted this, his primary goal was to derive expressions for the forces that act on material imperfections. To do so, Eshelby relied, from the outset, on a constitutive relation giving the energy density as a function of the strain and observed that, with the applied forces held constant, the total energy of a materially imperfect elastic solid is a function of the parameters that specify the configurations (i.e., positions and shapes) of the imperfections. Interpreting these parameters as generalized coordinates of the sort introduced by Lagrange (1788: *1*) in his treatment of analytical mechanics, Eshelby then defined the generalized force acting on an imperfection as the negative of the variational derivative of the total energy with respect to the parameter that determines the configuration of the imperfection and noted that, in equilibrium, each such generalized force must vanish. (From this vantage point, Herring's calculation of the diffusion potential at a grain interface can be reinterpreted as a determination of the generalized force associated with the interfacial configuration.) In his 1951 paper, Eshelby derived expressions for the forces that act on vacancies, interstitial atoms, dislocations, and (three-dimensional) inclusions; in his 1956 paper, Eshelby revisited the topics that he addressed in his 1951 paper and also considered surface defects, calculating the forces that act on cracks and also on grain and twin boundaries. Attention, in both of these early works, is restricted to linearly elastic solids. Later in his career, Eshelby (1970: *1*; 1975: *2*) extended his approach to deal with nonlinearly elastic solids and with dynamical effects. Except in his work on fracture, Eshelby did not associate excess fields with imperfections. Of central importance in Eshelby's work is a conservation law involving a field that he called the "energy–momentum tensor"[11] and will be referred to here as the "Eshelby tensor." Indeed, to de-

[9] More than twenty years before Herring's work on sintering, Volmer (1939: *1*) presented the version of this expression that holds if all facets are planar.

[10] Eshelby attributed the earliest contributions in this vein to the materials scientists Koehler (1941: *1*), Leibfried (1949: *1*), and Bilby (1950: *1*), but also drew attention to works by Burton (1892: *1*) and Larmour (1897: *1*) concerning the passage of a planetary body through the "æther," which might be likened to the diffusion of a defect in a crystalline solid.

[11] Prompted by analogy to electromagnetic theory, Eshelby (1951: *1*) first referred to this tensor as the " Maxwell tensor of elasticity." In their studies of mixtures,

rive expressions for the generalized forces that act on point, line, and surface defects, Eshelby repeatedly exploited this conservation law.[12] Insofar as a relation determining the bulk energy density through a response function depending on the strain was a key ingredient in Eshelby's approach, his results, like those of Herring, hinge on a provision of constitutive information.

Although Eshelby (1951: *1*; 1970: *1*) discussed the generalization of his approach to account for inertia and suggested a suitable dynamical version of the Eshelby tensor, he never arrived at laws governing the evolution of the defects that he considered.[13] However, in the same year that Eshelby's second major paper on imperfections in linearly elastic solids appeared, Mullins (1956: *2*) presented a theory that results in a dynamical equation for grain growth.[14] In this work, Mullins was motivated largely by a desire to improve on a soap-bubble model of grain growth due to Smith (1952: *3*),[15] a model in which, by the Young–Laplace relation,[16] the pressure jump across a surface separating two bubbles and the curvature of that surface cause gas to pass through the surface in such a way that the larger bubble grows at the expense of the smaller. In pursuit of this goal, Mullins recognized that,

---

Bowen (1967: *1*; 1969: *1*) and Truesdell (1969: *3*) derived, for each species, a "chemical potential tensor" analogous to the energy-momentum tensor. (In that the results of Bowen and Truesdell are expressed in terms of variables associated with the deformed state, the connection between their chemical potential tensors and the energy-momentum tensor is not immediately recognizable. However, a straightforward calculation remedies any apparent discrepancy.) In their work on phase transitions, Grinfeld (1981: *3*) and Truskinovsky (1983: *9*) referred to the energy-momentum tensor as the "chemical potential tensor," while Gurtin (1994: *2*; 1995: *1*) employed the terms "Eshelby tensor" and "configurationalstress tensor." The Eshelby tensor actually has two alternative forms, one based on deformation and the other based on displacement; but these are, in a certain precise sense, equivalent.

[12] The conservation law in question issues as a consequence of Noether's (1918: *1*) theorem concerning the invariance of functionals (see also Günther (1962: *2*) and Knowles & Sternberg (1972: *1*)).

[13] The following remarks, from Eshelby's 1956 paper, indicate that he was aware that such dynamical laws must involve the generalized forces that, on variational grounds, he otherwise equated to zero in equilibrium and, moreover, must be distinct from the standard balance laws of continuum physics: "In nonequilibrium problems the generalized forces, being derivatives of the free energy, are the driving forces which provide the raw material for a kinetic calculation of the rate of approach to equilibrium by arguments outside the scope of a continuum theory." Eshelby (1953: *1*; 1956: *1*; 1969: *2*; 1970: *1*) did consider problems involving propagating dislocations and cracks. However, his emphasis in these works was to determine expressions for the energy released when these defects propagate at a given constant velocity.

[14] Before the work of Mullins, an equation governing the evolution of a spherical grain was proposed, on an ad-hoc basis, by Burke & Turnbull (1952: *2*). That equation is consistent with the general evolution equation derived by Mullins.

[15] Prior to Smith's work, Bragg & Nye (1947: *1*) performed analog experiments for grain growth using soap bubbles.

[16] Cf. Young (1805: *1*) and Laplace (1806: *1*).

whereas the high mobility of gas molecules leads to a uniform pressure distribution within bubbles and, ultimately, to the Young-Laplace relation, the significantly lower mobility of atoms within a crystal grain allows for no such uniformity of pressure. Accordingly, he noted that the evolution of a grain boundary should be governed solely by local conditions, and with reference to the experimental results of Beck (1952: *1*)[17] – which showed that, when annealed, grain boundaries in a recrystallized alloy evolve toward their centers of curvature – inferred that the evolution of a grain boundary should be described geometrically by an equation relating the interfacial scalar normal velocity and the mean curvature.[18]

The work of Mullins accounted for neither the diffusion of atoms within grains nor the deformation of grains, which are potentially important physical effects. Bearing this in mind, one might expect that the evolution equation proposed by Mullins should express an appropriately simplified dynamical version of the interface condition that Herring arrived at by stipulating that the total energy be stationary with respect to variations of the interface configuration. Indeed, this is the case. In addition to his evolution equation, Mullins derived an important relation that reckons the rate at which grain-boundary area decays. In particular, this relation demonstrates that, given sufficient time, all but one of the grains comprising a polycrystal will, in principle, shrink and disappear, leaving a single crystal.

In an investigation of the stability of the shape of a spherical precipitate undergoing diffusion-controlled growth into a supersaturated matrix, a situation in which diffusion in the matrix *cannot* be ignored, Mullins & Sekerka (1963: *3*) were led to consider an extension of the equations developed by Mullins (1960: *2*), who, to study the development, by the mechanism of volume diffusion, of a grain boundary groove on a interface separating a solid phase and a saturated fluid phase, introduced an additional interface condition wherein the concentration at the interface differs from its equilibrium value by an amount proportional to the interfacial mean curvature.[19] Drawing on the techniques of hydrodynamic stability analysis,[20] Mullins & Sekerka (1963: *3*) then derived a dispersion relation governing the growth rate of a small-amplitude disturbance of given wavelength in terms of that wavelength. Consistent with experimental predictions, this relation identifies a critical radius above which the spherical precipitate grows and below which it shrinks. Mullins & Sekerka also developed an analogy between the diffusion of atomic species and that of heat, and modified their analysis to

[17] See, also, Harker & Parker (1945: *1*).

[18] As Smith did for his soap-bubble model, Mullins restricted his attention to the description of grain-boundary evolution in two spatial dimensions.

[19] Provided that the interfacial energy density is constant, this condition arises either by linearizing Herring's (1951: *2*) expression for the diffusion potential at the interface or by choosing a constitutive description in which the diffusion potential and concentration are linearly related.

[20] Cf., e.g., Chandrasekhar (1961: *1*).

discuss the stability of a solidifying spherical particle in an undercooled melt. Shortly after this work appeared, Mullins & Sekerka (1964: *2*) studied the linear stability of a state where a planar solidification front separates solid from melt in a dilute binary alloy, and, hence, considered a situation in which atomic diffusion and heat diffusion are coupled.

In another direction, Larché & Cahn (1985: *3*) provided an overview of their contributions, spanning more than a decade,[21] to the theory of thermochemical equilibrium in stressed multiphase solids. In doing so, they sought to resolve a number of controversies. The most significant of these concerned the problem of describing atomic diffusion in conjunction with deformation. Classically, a crystalline solid is thought of as a spatially periodic arrangement of atoms. The crystal lattice is a subsidiary object defined by the positions of atoms in space. With such a description, however, a distinction between atomic diffusion and deformation is problematic, because both processes seem to involve the motion of atoms. To resolve this difficulty, Larché & Cahn introduced the notion of a "network." For a crystalline solid, the network is a set of sites upon which substitutional atomic species may reside and diffuse, while interstitial species reside and diffuse on sites not belonging to the network. Most importantly, deformations are associated not with the motion of atoms but, rather, with distortions of the network. In addition to their fundamental contributions, Larché & Cahn used a variational approach to develop the full set of interfacial equilibrium conditions, treating both the coherent and the incoherent cases, but ignoring effects associated with interfacial energy and stress. Among these interface conditions is one derived, in the manner of Herring and Eshelby, by considering variations of the interface configuration. Larché & Cahn also investigated diffusional creep, a matter explored originally by Herring in his work on sintering, demonstrating the importance of nonlinear terms that Herring overlooked, and illustrated the manner in which strain exerts a long-range influence on bulk diffusion.

Works in Continuum Physics

Somewhat prior to Herring's work on sintering, researchers in continuum physics began a reexamination of the foundation of their discipline. This development originated in response to the needs of rheologists, who found the classical theories of linearly elastic solids and linearly viscous fluids to be inadequate for describing the behavior of polymeric materials subjected to large strains or strain rates.[22] However, inasmuch as the classical linear theories for solid and fluid response describe only a portion of the range of

[21] Cf. Larché & Cahn (1973: *1*; 1978: *4*, *5*; 1982: *2*) and Cahn & Larché (1983: *1*).

[22] Truesdell (1952: *4*; 1953: *4*) put this activity into context and provided an overview of the rheological literature up to 1953 (with particular emphasis on those fundamental contributions made after 1945). See, also, Coleman, Markovitz & Noll (1966: *1*).

behavior observable in almost any material, the far-reaching significance of a clarification of the conceptual basis for continuum physics was manifest from the inception of these foundational investigations. What emerged from this undertaking was a framework in which a careful distinction is maintained between the role of the basic laws of mechanics and thermodynamics, which govern large classes of materials, and that of constitutive equations, which characterize particular classes of materials.[23] In particular, the basic laws are stated in integral form so as to apply equally well in regions where the fields vary smoothly, in which case they yield field equations, and at surfaces where the fields may suffer discontinuities, in which case they yield jump conditions.[24] Further, constitutive equations are required to be properly invariant under changes of observer, to embody any underlying material symmetry,[25] and, moreover, to be thermodynamically consistent, in the sense that they satisfy the local entropy production inequality in all processes.[26]

Except for the work of Varley & Day (1966: *2*), which investigated whether an isotropic nonlinearly thermoelastic solid can, at uniform temperature, uniform pressure, and zero deviatoric stress, sustain two (or more) equilibrium states that differ other than by a uniform dilation, and of Gurtin & Williams (1967: *2*), which discussed the constitutive description of elastic solids that possess more than one phase, studies pertaining to phase transitions seem to be absent from the early literature of continuum physics. However, a paper published by Ericksen (1975: *1*) launched a new era, focusing attention on the topic of phase transitions as one that could be addressed effectively from a continuum perspective, and creating circumstances that would bring workers in continuum physics into contact with the works of Herring, Mullins, Mullins & Sekerka, and Larché & Cahn and into renewed contact, from a different point of view, with the work of Eshelby. In particular, Ericksen's paper showed that, if smoothness requirements are relaxed to allow for strain discontinuities, and if the underlying energy density possesses multiple (local) minima, then the theory of nonlinear elasticity provides a context in which it is possible to model coherent heterophase equilibria of the sort that are observed in materials that undergo diffusionless (i.e., martensitic) phase transitions. This insight inspired a series of pa-

[23] Early expositions of this distinction were given by Truesdell (1960: *3*) and Truesdell & Toupin (1960: *4*).

[24] Truesdell (1960: *3*) and Truesdell & Toupin (1960: *4*) provided complete discussions of the integral statements of the balance laws.

[25] Authoritative discourses on observer invariance and material symmetry were supplied by Truesdell & Noll (1965: *1*).

[26] Coleman & Noll (1963: *2*) fashioned the first mathematically sound approach to using the entropy production inequality to obtain restrictions on constitutive equations (see also Coleman & Mizel (1963: *1*; 1964: *1*)). Green & Adkins (1960: *1*) made an early attempt to develop such an approach. However, in doing so, they failed to include a heat supply term in their statement of energy balance and, hence, to allow for arbitrary variations of the independent constitutive fields without violation of that balance.

pers, including those of Ericksen (1978: *2*; 1981: *2*), Fosdick & MacSithigh (1983: *3*), Gurtin (1983: *5*), Gurtin & Temam (1981: *4*), James (1979: *3*; 1981: *5*; 1986: *4, 5*), Parry (1980: *8*), and Pitteri (1985: *5*), that explored the kinematics and energetics of coherent heterophase solid states.[27] These works culminated in an approach in which direct variational methods are employed to explore the fine structure often present in coherent heterophase equilibria, an approach exemplified by the paper of Ball & James (1987: *2*).[28]

Coincident with the appearance of Ericksen's paper, Knowles & Sternberg (1975: *3*) published a complementary work that reveals the link between a loss of ellipticity in the partial differential equations of elastostatics, a nonconvexity of the energy density, and an existence of equilibria where the deformation is continuous but the deformation gradient may suffer discontinuities across surfaces.[29] Subsequently, on noting the analogy between a loss of ellipticity in the equilibrium equations for an elastic solid with nonconvex energy density and a change of type in the equations governing the steady irrotational flow of an inviscid compressible fluid, Knowles & Sternberg (1978: *3*) introduced an admissibility criterion expressing the requirement that the interfacial dissipation be nonnegative during any quasistatic process involving a pair of homogeneously strained regions separated by a planar interface.[30]

While the condition derived by Knowles & Sternberg is the direct counterpart of the classical gas dynamical requirement that the entropy of a particle

---

[27] While Varley & Day did not require their equilibria to be energy minimizers, their attention to the construction of heterophase states free from deviatoric stress foreshadowed, by over two decades, the approach taken in these later works. Further, the treatment of the symmetry of elastic phases presented by Gurtin & Williams, although somewhat lacking, prefigured, to some extent, the constitutive description relied upon in these works.

[28] See also Ball & James (1992: *1*), Bauman & Phillips (1990: *2*), Bhattacharya (1991: *2*; 1992: *2*), Chipot & Kinderlehrer (1988: *2*), Fonseca (1989: *3*), and Fosdick & Zhang (1993: *2*).

[29] In this work, Knowles & Sternberg obtained conditions for the loss of ellipticity in the plane-strain equilibrium equations that ensue on taking the elastic energy density to be isotropic and compressible and of the particular form suggested by the experiments of Blatz & Ko (1962: *1*). Conditions for the loss of ellipticity in progressively more general situations were obtained by Knowles & Sternberg (1977: *3*), Abeyaratne (1980: *1*), Zee & Sternberg (1983: *10*), Simpson & Spector (1983: *8*), and Rosakis (1990: *7*) (who treated three-dimensional deformations of compressible anisotropic materials). Interestingly, Varley & Day, in presenting the solution to a boundary-value problem involving the torsion of a circular cylinder about its axis, noted changes of type, from elliptic to hyperbolic, in the equilibrium equations, and, hence, were the first to expose the connection between the ability of a material to support a heterophase state and a loss of ellipticity in the partial differential equations that govern its equilibrium.

[30] Knowles (1979: *2*) extended the formulation of Knowles & Sternberg to account for situations where the strain field is not necessarily piecewise homogeneous and the interface not necessarily planar and also showed how that formulation arises from the framework of thermoelasticity.

may not decrease on crossing a shock, it is *not* generally sufficient to resolve the lack of uniqueness that arises in problems involving propagating phase interfaces in materials described by nonconvex energy densities.[31] From our vantage point, this development is not astonishing. Granted a monotonic constitutive equation determining the pressure–density relation in a gas, the linear momentum balance (which typically is imposed in the form of the Rankine–Hugoniot relation[32]) and dissipation imbalance determine the magnitude and sign of the shock velocity. However, to determine the motion of an interface during a process involving the growth of one phase at the expense of another generally requires information supplemental to that furnished by the linear momentum balance, the dissipation imbalance, and the stress–strain relation, information that reflects the physical mechanisms governing the underlying transition kinetics. Indeed, that a continuum-level description of a diffusionless solid–solid phase transition might require closure by the provision of such additional information is at least suggested by the works of Herring, Eshelby, Mullins, Mullins & Sekerka, and Larché & Cahn.

Using the statements of thermomechanical balance local to a coherently propagating interface between two solid phases and the appropriate compatibility conditions, Heidug & Lehner (1985: *2*) showed that the entropy production, per unit interfacial area, associated with the growth of one phase at the expense of another is determined by the product of the scalar normal velocity of the interface with the jump, across the interface, of the normal component of the Eshelby tensor.[33] By its lack of any reliance on constitutive assumptions, this deduction constituted a major generalization of one due to Knowles (1979: *2*). Heidug & Lehner interpreted their result within the context of the thermodynamics of irreversible processes,[34] identified the interfacial scalar normal velocity as a "generalized flux" and the jump in the normal component of the Eshelby tensor as a "generalized affinity," and suggested the introduction of an additional interfacial relation, intended to describe the kinetics of the phase transition, a relation in which the flux is determined constitutively in terms of the affinity.

Unfortunately, the paper of Heidug & Lehner, which appeared in the geophysical literature, has been overlooked by workers in both materials science and continuum physics. Were this not the case, it is certain that this work would have been considered carefully for a position among the articles reprinted in this volume. Formulations equivalent, in essence, to that of Heidug & Lehner were advanced independently by Truskinovsky (1987: *5*) and

---

[31] That thermodynamic admissibility does not single out a unique solution in such problems was illustrated first for quasistatic processes by Abeyaratne (1981: *1*), and later for dynamical processes by Abeyaratne & Knowles (1991: *1*). See, also, the remarks of James (1980: *6*), Shearer (1982: *4*; 1983: *6*), Slemrod (1983: *7*; 1984: *5*), and Truskinovsky (1993: *8*).

[32] Cf. Rankine (1870: *1*) and Hugoniot (1887: *1*).

[33] Heidug & Lehner ignored both inertia and interfacial excess fields.

[34] Cf., e.g., Eckart (1940: *1, 2*), Meixner (1943: *1*), and Prigogine (1949: *2*).

Abeyaratne & Knowles (1990: *1*).[35] In particular, of fundamental importance in both these works is the view that the need for supplemental closure information reflects a constitutive deficiency. Concerning the resolution of this deficiency, Abeyaratne & Knowles introduced some terminology that will be useful here, referring to the jump, across the interface, of the normal component of the dynamical Eshelby tensor as the "driving traction" and to a constitutive equation between the interfacial scalar normal velocity and that quantity as a "kinetic relation."[36]

Not long after the 1975 papers of Ericksen and Knowles & Sternberg, Fernandez-Diaz & Williams (1979: *1*) provided a continuum theory for the study of liquid–solid phase transitions that accounts for bulk and interfacial energy, entropy, and heat flux but ignores deformation and atomic diffusion. In a tacit acknowledgement that, as in the case of diffusionless solid–solid transitions, the basic principles and standard constitutive equations are not generally sufficient to determine the evolution of a phase interface, Fernandez-Diaz & Williams suggested, as a closure device, an additional relation whereby the bulk energy jump across a nonplanar interface differs from that across a planar interface by an amount proportional to the interfacial mean curvature.[37] The work of Fernandez-Diaz & Williams contained no explicit discussion of entropy production. As a remedy, Gurtin (1986: *3*) developed a theory in which the bulk entropy production must be nonnegative (as is traditional) but the interfacial entropy production must vanish. The latter requirement leads to a supplemental interface condition equivalent to that of Fernandez-Diaz & Williams and, as such, clarifies the constitutive

[35] In contrast to Heidug & Lehner, Abeyaratne & Knowles and Truskinovsky accounted for inertia.

[36] The notion of a kinetic relation was suggested by Abeyaratne & Knowles (1987: *1*; 1988: *1*) in works concerned with one-dimensional quasistatic motions of two-phase nonlinearly elastic media. In the second of these papers, Abeyaratne & Knowles also observed that, in addition to a kinetic relation, which determines the motion of a phase interface once that interface has been generated, a fully closed theory for solid–solid phase transitions should also require the imposition of a "nucleation criterion" which regulates the process by which phase interfaces are created.

[37] The relation of Fernandez-Diaz & Williams is the counterpart, appropriate to heat diffusion, of the "Gibbs–Thomson condition" which is used in the context of atomic diffusion (cf. Footnote ??). This relation includes as specializations both the classical requirement, imposed first by Lamé & Clapeyron (1831: *1*) and later by Stefan (1889: *1*), that the interfacial temperature coincide with the melting temperature and a modification of that requirement, which is often referred to as the "Gibbs–Thomson condition" (cf., e.g., Woodruff (1973: *2*), Glicksman (1977: *1*), and Langer (1980: *7*)), where the difference between the interfacial temperature and the melting temperature is proportional to the interfacial mean curvature. The latter condition arises by either linearizing the relation of Fernandez-Diaz & Williams or by choosing a constitutive description in which the specific heat is constant.

basis of that condition.[38] Shortly thereafter, in a work that extended the thermodynamic approach of Coleman & Noll (1963: *2*) to a setting involving propagating phase interfaces endowed with interfacial excess fields,[39] Gurtin (1988: *3*) developed a more general theory that embraces the possibility of interfacial entropy production. In so doing, Gurtin allowed for a full range of transition kinetics and arrived at a closure structure entirely analogous, except that it accounts for interfacial excess fields and ignores deformational effects, to that proposed by Abeyaratne & Knowles (1990: *1*), Heidug & Lehner (1985: *2*), and Truskinovsky (1987: *5*).

In a major departure from the approaches described above, Gurtin (1988: *4*) suggested that the need for supplemental information that renders interfacial motion determinate in continuum theories for phase transitions reflects a deficiency at a level more fundamental than the constitutive. Here, Gurtin's reasoning was based on a novel interpretation of the additional equilibrium conditions that Herring, Eshelby, and Larché & Cahn derived by requiring that the total energy be stationary with respect to configurational variations of the interface. Specifically, Gurtin observed that it might be more appropriate to view this condition as a statement of *force balance*. The forces entailed are not, however, of the traditional deformational variety, for, they act not over material particles, but, rather, over points of a nonmaterial entity – the phase interface.[40] Moreover, as the work of Larché & Cahn demonstrates explicitly, this additional equilibrium condition is distinct from the standard force balance, which arises from variations of the deformation. When attention is restricted to a description of equilibrium states characterized by a variational principle, Gurtin's insight concerning the status of this additional interface condition as a force balance may appear to be of dubious value. Gurtin, however, sought to develop a framework capable of describing dynamical processes involving dissipation, where a variational approach seems inapt.[41] Viewing the additional condition under discussion as

[38] Within the framework of Abeyaratne & Knowles (1990: *1*), the equivalent kinetic relation merely imposes the requirement that the driving traction vanish.

[39] Previously, Moeckel (1975: *4*) and Murdoch (1976: *1*) extended the method of Coleman & Noll (1963: *2*) to obtain thermodynamic restrictions on constitutive equations associated with fields defined on *material* surfaces.

[40] Indeed, from Eshelby's (1956: *1*) viewpoint, these forces are *generalized.*

[41] A popular procedure for extending a variational description of equilibrium to a dynamical setting involves the introduction of a "gradient-flow" (or "relaxation") law. In the setting at hand, such a law would require that the variational derivative, with respect to the interfacial configuration, of the energy functional describing equilibrium be proportional to the interfacial scalar normal velocity. The relevant factor of proportionality is signed by the requirement that the total energy decay monotonically in time; otherwise, the properties of that factor are determined constitutively. Hence, the statement of a gradient-flow law is contingent on the provision of constitutive equations governing the bulk and interfacial densities of energy and the kinetics of energy decay, and is, therefore, constitutive. Moreover, since the rate terms that it yields are generally dissipa-

a balance pointed the way toward an alternative approach to such a theory, and led Gurtin to propose a formulation where, in place of supplying additional constitutive information, closure is achieved by: (*i*) an introduction of additional primitive fields, consisting of an interfacial configurational stress and interfacial internal and external configurational force densities;[42] (*ii*) an appropriate reckoning, in the global statement of energy balance, of the power expended by these fields; and (*iii*) an imposition of an interfacial configurational force balance by which these fields are constrained.[43] Substitution, into the configurational force balance, of the thermodynamically restricted constitutive equations for the interfacial configurational stress and internal interfacial configurational force yields a vectorial evolution equation for the interface. The component of that equation in the direction of the interfacial normal field is identical to the equation derived earlier by Gurtin (1988: *3*). This equation holds at points where the interface is smooth. On the other hand, the full vectorial equation must be imposed at points where the interfacial normal jumps, and, hence, governs the evolution of facets.[44]

Shortly after this work of Gurtin, Leo & Sekerka (1989: *7*) extended the framework of Larché & Cahn (1985: *3*) to account for the effects of interfacial excess fields.[45] Leo & Sekerka, like Larché & Cahn, restricted attention to the description of equilibrium and used variational techniques to do so. They treated crystal–melt and crystal–crystal interfaces, considering, in the latter case, both coherent and "greased" crystal–crystal interfaces, and derived interface conditions that express both the standard force balance and,

tive and since it generally leads to parabolic partial differential equations, the gradient-flow approach has only a limited range of applicability.

[42] Gurtin (1988: *4*) did not employ the adjective "configurational" when referring to these additional interfacial fields: in place of the terms "interfacial configurational stress," "interfacial internal configurational force density," and "interfacial external configurational force density," he used the terms "capillary stress," "interaction," and "capillary supply." Subsequently, Gurtin & Struthers (1990: *4*) replaced the term "capillary stress" with "accretive surface stress" (see, also, Gurtin (1993: *4*) and Gurtin & Voorhees (1993: *5*)). The usage here is consistent with that followed by Cermelli & Gurtin (1994: *1*) and Gurtin (1994: *2*; 1995: *1*). Given Eshelby's definition of the generalized force acting on a material imperfection as the variational derivative of the total energy of the system with respect to the parameter that determines the *configuration* of that imperfection, this usage is not inappropriate. Further, it is not without precedent. Indeed, motivated by the works of Reed (1953: *3*) and Roitburd (1971: *1*), Eshelby (1980: *3, 4*) adopted the term "configurational" during the latter stages of his career. Other materials scientists who have used the term "configurational" to refer to forces associated with material imperfections include Régnier (1978: *6*).

[43] This law was referred to as the "capillary force balance" by Gurtin (1988: *4*) and as the "accretive force balance" by Gurtin & Struthers (1990: *4*).

[44] An equilibrium version of the tangential component of the interfacial configurational force balance has yet to be derived from the variational perspective.

[45] See also Alexander & Johnson (1985: *1*; 1986: *1*), Cahn (1980: *2*), Cahn & Larché (1982: *1*), and Mullins (1982: *3*; 1984: *4*).

in Gurtin's terminology, the normal component of the interfacial configurational force balance.[46]

During the same year that the paper of Leo & Sekerka appeared, Angenent & Gurtin (1989: *1*) published a detailed study of a specialized version of Gurtin's (1988: *4*) theory that arises on assuming that both bulk phases are "perfect conductors" and restricting attention to interface evolution in two spatial dimensions. Under these conditions, the theory of Gurtin (1988: *4*) yields a single evolution equation that relates the interfacial scalar normal velocity to the interfacial curvature. In view of the connection that Mullins & Sekerka (1963: *3*) developed between atomic diffusion and heat diffusion,[47] it appears that the perfect conductivity limit of Angenent & Gurtin corresponds to a large bulk mobility limit in the context of mass diffusion and, hence, that the situation considered by Angenent & Gurtin is analogous to that considered by Mullins (1956: *2*) in his theory for grain-boundary evolution. However, since Angenent & Gurtin allowed for orientation dependence in the constitutive equations for the interfacial energy density and the interfacial mobility, their results are more germane to crystal growth than to grain-boundary evolution.[48] Angenent & Gurtin considered the situation where the interfacial energy density is determined by a convex function of orientation, in which case the aforementioned evolution equation is parabolic with a structure similar to that of the evolution equation of Mullins (1956: *2*). Here, Angenent & Gurtin obtained results that, by specifying whether a crystal of given characteristic diameter grows or dissolves as a function of the applied undercooling, generalize the decay relation of Mullins (1956: *2*). Angenent & Gurtin also treated the case where the interfacial energy density is determined by a nonconvex (and perhaps nonsmooth) function of orientation, in which case the underlying evolution equation may become backward parabolic for certain interfacial orientations. This leads to results concerning the structure and stability of propagating facets.

Although Gurtin's approach, which is based on the introduction of additional interfacial configurational fields and an associated balance, was first developed within the context of a theory for transitions between the liquid and solid states, he recognized from the outset that it should be applicable to other types of phase transitions. Indeed, shortly after the work of Angenent & Gurtin appeared, Gurtin & Struthers (1990: *4*) presented a theory for diffusionless solid–solid and liquid–solid phase transitions that embraces both thermal and deformational effects. Thereafter, theories that account for coupled deformation and atomic diffusion were developed by Gurtin & Voorhees (1993: *5*), Gurtin (1993: *4*), and Cermelli & Gurtin (1994: *1*): while Gurtin & Voorhees and Gurtin concerned themselves only with situations where the

[46] Leo & Sekerka referred to the second of these as an "energy balance."

[47] See also Davì & Gurtin (1990: *3*).

[48] Mullins (1956: *2*) argues that it is generally quite sound to assume that the grain-boundary energy density and grain-boundary mobility are constant.

deformation remains coherent, Cermelli & Gurtin treated incoherent deformations; further, whereas Gurtin & Voorhees restricted their attention to circumstances where a single independent atomic species diffuses, the bulk material is linearly elastic, and the interfacial energy density is determined constitutively by a relation that does not involve the interfacial strain, Gurtin and Cermelli & Gurtin considered a multiplicity of atomic species,[49] permitted the bulk material to be nonlinearly elastic, and allowed the response function for the interfacial energy density to depend on the interfacial strain.

The statement of configurational force balance that Gurtin (1988: *4*) introduced, which served as a central ingredient in the subsequent papers of Angenent & Gurtin (1989: *1*), Gurtin & Struthers (1990: *4*), and Gurtin & Voorhees (1993: *5*), involves only interfacial fields and, thus, is of an exclusively *interfacial* character. Since the attention of Gurtin and his co-workers in this collection of papers was focused exclusively on phenomena that involve (propagating) interfaces, the exclusively interfacial character of the configurational force balance may seem perfectly natural. If, however, the configurational force balance is to be regarded as basic, with a status equal to that of the well-established laws of mechanics and thermodynamics, should it not also include bulk terms and thus possess a structure consistent with that of these other laws?

Motivated, at least in part, by the foregoing question, Gurtin (1993: *4*) began to explore the consequences of introducing bulk counterparts of the interfacial configurational fields that enter his earlier works. In this paper, Gurtin first developed a simple purely mechanical theory for coherent diffusionless solid-state phase transitions: the only configurational fields that appear in this theory are a bulk stress and an external interfacial force density; aside from these, traditional deformational fields – consisting of a bulk stress, a bulk external force density, and an interfacial external force density – also appear. In this setting, Gurtin recovered the isothermal specialization of the frameworks developed by Abeyaratne & Knowles (1990: *1*), Heidug & Lehner (1985: *2*), and Truskinovsky (1987: *5*), showing, in particular,

[49] Following Larché & Cahn (1985: *3*), Gurtin treated vacancies as a mobile substitutional species, required that all lattice sites be occupied, and removed the lattice constraint by expressing the density of vacancies in terms of the remaining substitutional densities. What emerges under these circumstances is a framework equivalent to that of Mullins & Sekerka (1985: *4*), who, by positing the existence of a defect mechanism that accommodates an excess or deficiency of substitutional atoms, arrived at a description of multicomponent solid-state diffusion involving only *unconstrained* atomic species. In light of these observations, the situation considered by Gurtin & Voorhees can be interpreted as describing two substitutional species whose atoms are constrained to diffuse on a lattice that is free from vacant sites, or a single substitutional species whose atoms are constrained to diffuse on a lattice with vacant sites, or a single interstitial species whose atoms diffuse in the absence of substitutional species. Cermelli & Gurtin allowed for the possibility that a moving incoherent phase interface may, even in the absence of external supplies, generate vacancies, interstitial atoms, and dislocations.

that, given constitutive equations consistent with the dissipation imbalance,[50] his normal configurational force balance yields an interface condition that is identical to the kinetic relation of Abeyaratne & Knowles. Next, Gurtin generalized this theory to account for interfacial energy density, interfacial configurational stress, and interfacial deformational stress. This generalization leads to a recovery of the framework developed by Gurtin & Struthers (1990: *4*), as that framework specializes under isothermal conditions. In this context, Gurtin proved that, given constitutive equations consistent with the dissipation imbalance, the tangential component of the interfacial configurational stress tensor coincides with the interfacial Eshelby tensor.[51]

Granted the aforementioned result, one might expect that a similar identification should hold between the bulk configurational stress tensor and the bulk Eshelby tensor. However, in the approach taken by Gurtin (1993: *4*), the bulk configurational stress tensor is left indeterminate and, hence, unrelated to the bulk Eshelby tensor. Not long after the appearance of his 1993 paper, Gurtin (1994: *2*; 1995: *1*) arrived at an approach that resolves this discrepancy. What is remarkable about this approach is that it not only yields the coincidence of the bulk configurational stress tensor with the bulk Eshelby tensor and reaffirms Gurtin's previously derived result concerning the tangential component of the interfacial configurational stress tensor and the interfacial Eshelby tensor but, also, that it does so solely as a consequence of an invariance argument, without recourse to constitutive assumptions. The postulate underlying that argument stems from the consideration of "evolving referential subregions" (i.e., regions that evolve, in time, within the region that the body is associated with in a given fixed reference state).[52] To define an evolving referential subregion for some sufficiently small time interval, it suffices to supply a time-dependent parametrization of its boundary. For the given time interval, the time-rate of that parametrization defines a velocity field for the boundary of the evolving referential subregion. In general, such a velocity field possesses components both normal to and tangential to the boundary, with only the former being invariant with respect to reparametrization of that boundary. Gurtin observed that the total power expended on an evolving referential region must account for the motion of its boundary and, most significantly, stipulated the bulk configurational traction to be

[50] In the purely mechanical situation considered by Gurtin, the dissipation imbalance enforces the second law of thermodynamics.

[51] Prior to the 1993 paper of Gurtin, the interfacial counterpart of the Eshelby tensor did not appear in the literature. According to Gurtin (1993: *4*; 1995: *1*), it was P. Podio-Guidugli who first recognized the central importance of the interfacial Eshelby tensor.

[52] Gurtin (1994: *2*; 1995: *1*) referred to such regions as "evolving control volumes." However, since, for the typical U.S.-trained recipient of an undergraduate degree in the mechanical sciences, the term "control volume" connotes a *spatial* region, which may, depending on context, be either fixed or evolving, the terminology used here seems somewhat less susceptible to confusion.

the power conjugate to the velocity of the boundary.[53] By requiring that the total power expended on an evolving referential subregion be invariant with respect to reparametrization of its boundary and that the dissipation imbalance hold for all such subregions, Gurtin proved that the bulk configurational stress tensor must coincide with the bulk Eshelby tensor. Further, by a completely analogous approach, he rederived his earlier, constitutively based, result concerning the equivalence of the tangential component of the interfacial configurational stress and the interfacial Eshelby tensor.

The basic fields that enter the 1994 and 1995 papers of Gurtin consist of those introduced previously in his 1993 paper in conjunction with densities of bulk and interfacial *internal* configurational force. From Gurtin's vantage point, the bulk internal configurational force density is defined by the bulk configurational force balance; that is, granted the Eshelby tensor, this balance is solved to determine the bulk internal configurational force density. In this sense, the bulk configurational force balance is generally tautological. In particular, for an elastic material, this balance is equivalent to the linear momentum balance. For an inhomogeneous elastic material, the bulk internal configurational force density is simply the negative of the gradient, with the strainheld fixed, of the elastic energy density with respect to position. Evidently, then, this force density must vanish for a homogeneous elastic material. On the other hand, the interfacial internal configurational force density is, as a rule, a constitutively specified object that characterizes the drag force on a phase interface and, hence, determines the kinetics of interface motion and, also, the rate at which energy is dissipated during such motion.[54]

Once the notion of an evolving referential subregion and the allied invariance postulate are understood, Gurtin's (1994: *2*; 1995: *1*) approach provides a framework that permits easy derivations of final governing equations. More important, however, is the insight that this approach yields by identifying the bulk and interfacial configurational stress tensors with the bulk and interfacial Eshelby tensors, by specifying the manner in which configurational forces expend power, and by divulging the essential characteristics of internal configurational forces, forces that are not taken into account in classical theories. Further, because Gurtin's approach does not rely on particular constitutive assumptions, his results apply to a broad range of materials, including those that respond *inelastically* in bulk. Finally, while the focus here is on sharp-interface theories for phase transitions, Gurtin's framework can be applied readily to phenomena involving other types of material imperfections, leading to evolution equations for point defects, line defects, cracks, and three-dimensional inclusions.[55]

---

[53] In essence, this postulate defines the bulk configurational stress.

[54] This dissipative force density yields the rate term that would appear otherwise in a formulation based on a gradient-flow law (cf. Footnote 41).

[55] Gurtin & Podio-Guidugli (1996: *1*) have applied this framework to dynamical fracture.

Works in Mathematics

The theories discussed here generally lead to initial-boundary-value problems for the bulk and interfacial fields and the evolving surface separating the two phases. Problems of this type can be viewed as broad generalizations of the classical Stefan problem for solidification, a problem posed by Lamé & Clapeyron (1831: *1*) and later by Stefan (1889: *1*) and which only now, after a period of intensive effort, is considered well understood.[56] Since very few of the methods that developed in connection with the Stefan problem carry over to the problems that arise in the theories discussed here, mathematicians have found in these problems an area fertile for research.

Quite naturally, the majority of the mathematical analyses undertaken thus far concerns problems of pure interface motion. For the most part, these efforts have focused on the special class of problems that arises when the interfacial energy density and mobility are independent of the interfacial orientation, as in the case of Mullins' (1956: ***2***) equation describing grain-boundary evolution.[57] The definitive work in this vein is due to Evans & Spruck (1991: ***5***), who were motivated by the desire to provide analytically based justification for numerical methods for the study of interface propagation that were developed by Osher & Sethian (1988: *7*) and Sethian (1985: *6*; 1990: *8*), methods in which a surface is described as a level set of some continuous function and, therefore, are naturally well suited to handle topological splitting and merging. On the basis of a constructive argument that relies on the notions of viscosity sub- and super-solutions and the existence of a maximum principle for the underlying partial differential equation,[58] Evans & Spruck established the existence of a unique solution to the problem of surface propagation by mean curvature. Concurrent with the appearance of this work, Chen, Giga & Goto (1991: ***3***) considered an anisotropic generalization of the problem of surface propagation by mean curvature. In particular, when the response function that determines the surface energy density depends convexly on the surface orientation, Chen, Giga & Goto established global existence and uniqueness for the solution to the problem of anisotropic motion

[56] Rubenšteĭn (1971: *2*) and Meirmanov (1992: *6*) provided comprehensive overviews of results concerning the one-dimensional and multidimensional Stefan problems, respectively.

[57] Brakke (1978: *1*) used geometric measure theory to study the multidimensional generalization of Mullins' equation – a generalization under which the boundary of a region in (n-dimensional) space evolves at a rate proportional to its mean curvature. Further results concerning this problem were provided by Ecker & Huisken (1989: *2*; 1991: *4*), Gage (1983: *4*; 1984 *2*), Gerhardt (1980: *5*), Grayson (1987: *4*; 1989: *4*), and Huisken (1984: *3*), while the planar problem posed by Mullins was studied by Epstein & Weinstein (1987: *3*), Gage & Hamilton (1986: *2*), and Grayson (1989: *5*).

[58] See, for example, Crandall, Evans & Lions (1984: *1*), Crandall, Ishii & Lions (1992: *4*), Crandall & Lions (1983: *2*), Ishii (1989: *6*), Jensen (1988: *5*), and Jensen, Lions & Souganidis (1988: *6*).

by curvature. In doing so, Chen, Giga & Goto, like Evans & Spruck, relied on methods involving viscosity solutions. Subsequent to the aforementioned works, Barles, Soner & Souganidis (1993: *1*), Evans, Soner & Souganidis (1992: *5*), and Soner (1993: *7*) have also made significant contributions to the understanding of surface propagation by curvature and its generalizations.

Mathematical work on problems related to solidification has also commenced. Luckhaus (1990: *5*) constructed a family of solutions, globally valid in time, to the modification of the Stefan problem that results by requiring that the interfacial temperature differ from the melting temperature by the product of the interfacial energy density with the interfacial mean curvature. To obtain these solutions, Luckhaus considered a time-implicit discretization of the initial-boundary-value problem, expressed the temperature field in terms of (what amounts to) the characteristic function of the liquid phase, determined that characteristic function as the solution to a variational problem, and, finally, showed that a solution of the discretized problem converges to one of the original problem. The primary mathematical innovation in this work is a technique that delivers estimates for fractional time derivatives of the solutions to the discretized problem. In undertaking this work, Luckhaus sought to overcome the difficulties encountered by Visintin (1989: *8*), who constructed weak solutions in which the interface separating the solid and liquid phases develops into a three-dimensional layer. In closing this paper, Luckhaus also suggests a method of extending his approach to deal with nucleation.

Subsequent to the work of Luckhaus, the more difficult problem in which interfacial energy and transition kinetics are both taken into account was considered by Soner (1995: *2*),[59] who showed that a weak solution, globally valid in time, can be obtained as a limit of a solution to the phase-field equations for solidification,[60] thereby providing not only existence for a physically important problem, but, in addition, establishing the phase-field theory as a valid approximation.

The mathematical results that have been obtained so far, with their careful formulations and abstract language, are valuable for the development of the physical theory because: (*i*) they ensure that the mathematical models of the physical processes meet the most primitive of requirements, namely, that their equations admit solutions; and (*ii*) they give promise of a mathematical theory capable of predicting complicated behavior. While these concern special problems, it is hoped that they represent a first step toward an un-

[59] Cf. Chen & Reitich (1992: *3*) and Radkevich (1993: *6*), who established short-time existence and uniqueness of classical solutions to this problem.

[60] These equations, which consist of a heat equation coupled linearly to a Ginzberg–Landau equation, can be derived from various perspectives (cf. Alt & Pawlow (1996: *1*), Fried & Gurtin (1993: *3*), Hohenburg & Halperin (1977: *2*), Penrose & Fife (1990: *6*)), Truskinovsky (1993: *8*), and Wang, Sekerka, Wheeler, Murray, Coriell, Braun & McFadden (1993: *9*)).

derstanding that will encompass the full range of thermal, diffusional, and deformational effects considered in the theories discussed here.

Theories that describe evolving material structures such as phase interfaces lie at the forefronts of both continuum physics and mathematics. Continuum physics provides powerful tools with which to develop such theories; but, as with all scientific endeavors, it is only when the irrelevant features of the conventional theories are discarded and the essential nonstandard concepts discovered that any real progress is made. Further, theories that describe phase transitions generally involve a delicate balance between destabilizing effects arising from the nonconvexity of the bulk material description and the stabilizing capillary and viscous influences exerted by interfacial energy and transition kinetics, respectively, and, because of this, provide new and challenging problems for mathematical and numerical analysts, problems that require, and often yield, new tools and methods and, occasionally, new mathematical theories.

Acknowledgements. The author is grateful to S. S. Antman, D. E. Carlson, P. Cermelli, and H. S. Sellers, who carefully read and suggested improvements to earlier versions of this Introduction. The author also acknowledges support from the program in Physical Mathematics and Applied Analysis of the U. S. Air Force Office of Scientific Research under grant F49620-96-1-0260, the program in Mathematical, Information, and Computational Sciences of the U. S. Depatment of Energy under grant 96-DOE-F-1682, and the program in Mechanics and Materials of U. S. National Science Foundation under grant MSS-93-09082.

Bibliography

1788 *1.* J. L. Lagrange, *Mécanique Analytique*, Desaint, Paris, 1788. Translated as *Analytical Mechanics*, Kluwer, Dordrecht, 1997.

1805 *1.* T. Young, An essay on the cohesion of fluids, *Philosophical Transactions of the Royal Society of London* **95**, 65–87.

1806 *1.* P. S. Laplace, *Méchanique Céleste, Supplement au Xe Livre*, Impresse Imperiale, Paris. Translated as *Celestial Mechanics, Volume IV*, Chelsea, New York, 1966.

1831 *1.* G. Lamé & B. D. Clapeyron, Mémoire sur la solidification par refroidissement d'un globe liquide, *Annales de Chimie et de Physique* **47**, 250–256.

1870 *1.* W. J. M. Rankine, On the thermodynamic theory of waves of finite longitudinal disturbance, *Philosophical Transactions of the Royal Society of London* **160**, 277–288.

1877 *1.* G. Erdmann, Über unstetige Lösungen in der Variationsrechnung, *Journal für die reine und angewandte Mathematik* **82**, 21–30.

1878 *1.* J. W. Gibbs, On the equilibrium of heterogeneous substances, *Transactions of the Connecticut Academy of Arts and Sciences* **3**, 108–248. Reprinted in *The Scientific Papers of J. Willard Gibbs*, vol. 1, Dover, New York, 1961.

1885 *1.* M. P. Curie, Sur la formation des critaux et sur les constantes capillaires de leurs différentes faces, *Bulletin de la Societé Mineralogique de France* **8**, 145–150.

1887 *1.* H. Hugoniot, Mémoire sur la propagation du movement dans un fluid indéfini, *Journal de Mathématique Pures et Appliquées* **3**, 477–492 and **4**, 153–167.

1889 *1.* J. Stefan, Über einige Probleme der Theorie der Wärmeleitung, *Sitzungsberichte der Kaiserlichen Akademie der Wissenschaften in Wien, Mathematisch-Naturwissenschaftliche Classe* **98**, 473–484.

1892 *1.* C. V. Burton, A theory concerning the constitution of matter, *Philosophical Magazine* **33**, 191–204.

1897 *1.* J. Larmor, A dynamical theory of the electric and luminiferous medium – III. Relations with material media, *Philosophical Transactions of the Royal Society of London A* **190**, 205–300.

1901 *1.* G. Wulff, Zur Frage der Geschwindigkeit des Wachsthums und der Auflösung der Krystallflächen, *Zeitschrift für Krystallographie und Mineralogie* **34**, 499–530.

1918 *1.* E. Noether, Invariante Variationsprobleme, *Nachrichten von der Gesellschaft der Wissenschaften zu Göttingen, Mathematische-Physikalische Klasse* **2**, 235–257. Translated in *Transport Theory and Statistical Physics* **1** (1971), 186–207.

1927 *1.* K. Weierstrass, *Mathematische Werke, VII*, Mayer and Müller, Berlin.

1939 *1.* M. Volmer, *Kinetik der Phasenbildung*, T. Steinkopff, Dresden.

1940 *1.* C. Eckart, The thermodynamics of irreversible process, I. The simple fluid, *Physical Review* **58**, 267–269.

2. C. Eckart, The thermodynamics of irreversible process, II. Fluid mixtures, *Physical Review* **58**, 269–275.

1941 *1.* J. S. Koehler, On the dislocation theory of plastic deformation, *Physical Review* **60**, 397–410.

1943 *1.* J. Meixner, Zur Thermodynamik der irreversiblen Prozesse in Gasen mit chemische reagierenden, dissoziierenden und anregbaren Komponenten, *Annalen der Physik* **43**, 244–270.

2. J. Meixner, Zur Thermodynamik der irreversiblen Prozesse, *Zeitschrift für Physikalische Chemie B* **53**, 235–263.

1945 *1.* D. Harker & E. R. Parker, Grain shape and grain growth, *Transactions of the American Society for Metals* **34**, 156–195.

1947 *1.* L. Bragg & J. F. Nye, A dynamical model of a crystal structure, *Proceedings of the Royal Society of London A* **190**, 474–481.

1948 *1.* C. S. Smith, Grain, phases, and interfaces: An interpretation of microstructure, *Transactions of the American Institute of Mining and Metallurgical Engineers* **175**, 15–51.

1949 *1.* G. Leibfried, Über die auf eine Versetzung wirkenden Kräfte, *Zeitschrift für Physik* **126**, 781–789.

2. I. Prigogine, Le domaine de la validité de la thermodynamique des phénomènes irreversibles, *Physica* **15**, 272–284.

1950 *1.* B. A. Bilby, On the interactions of dislocations and solute atoms, *Proceedings of the Physical Society A* **63**, 191–200.

1951 *1.* J. D. Eshelby, The force on an elastic singularity, *Philosophical Transactions of the Royal Society of London A* **244**, 87–112.

2. C. Herring, Surface tension as a motivation for sintering, in *The Physics of Powder Metallurgy* (W. E. Kingston, Ed.), McGraw-Hill, New York.

1952 *1.* P. A. Beck, Interface migration in recrystallization, in *Metal Interfaces*, American Society for Metals, Cleveland.

2. J. E. Burke & D. Turnbull, Structure of crystal boundaries, in *Progress in Metal Physics 3* (B. Chalmers, Ed.), Pergamon, New York.

3. C. S. Smith, Grain shapes and other metallurgical applications of topology, in *Metal Interfaces*, American Society for Metals, Cleveland.

4. C. Truesdell, The mechanical foundations of elasticity and fluid dynamics, *Journal of Rational Mechanics and Analysis* **1**, 125–300.

1953 *1.* J. D. Eshelby, The equation of motion of a dislocation, *Physical Review* **90**, 248–255.

2. C. Herring, The use of classical macroscopic concepts in surface-energy problems, in *Structure and Properties of Solid Surfaces* (R. Gomer & C. S. Smith, Eds.), University of Chicago, Chicago.

3. W. T. Read, *Dislocations in Crystals*, Mc-Graw Hill, New York.

4. C. Truesdell, Corrections and additions to "The mechanical foundations of elasticity and fluid dynamics," *Journal of Rational Mechanics and Analysis* **2**, 593–616.

1956 *1.* J. D. Eshelby, The continuum theory of lattice defects, in *Progress in Solid State Physics 3* (F. Seitz & D. Turnbull, Eds.), Academic Press, 79–144.

2. W. W. Mullins, Two-dimensional motion of idealized grain boundaries, *Journal of Applied Physics* **27**, 900–904.

1960 *1.* A. E. Green & J. E. Adkins, *Large Elastic Deformations*, Clarendon Press, Oxford.

2. W. W. Mullins, Grain boundary grooving by surface diffusion, *Transactions of the American Institute of Mining, Metallurgical and Petroleum Engineers* **218**, 354–361.

3. C. Truesdell, *Principles of Continuum Mechanics*, Socony Mobil Oil Company, Dallas.

4. C. Truesdell & R. A. Toupin, *The Classical Field Theories*, in (*Handbuch der Physik III/1* (S. Flügge, Ed.)), Springer-Verlag, Berlin.

1961 *1.* S. Chandrasekhar, *Hydrodynamic and Hydromagnetic Stability*, Dover, New York.

1962 *1.* P. M. Blatz & W. L. Ko, Application of finite elastic theory to the deformation of rubbery materials, *Transactions of the Society of Rheology* **6**, 223–251.

2. W. Günther, Über einige Randintegrale der Elastomechanik, *Abhandlungen der Braunschweigischen wissenschaftliche Gesellschaft* **14**, 54–72.

1963 *1.* B. D. Coleman & V. J. Mizel, Thermodynamics and departures from Fourier's law of heat conduction, *Archive for Rational Mechanics and Analysis* **13**, 245–261.

2. B. D. Coleman & W. Noll, The thermodynamics of elastic materials with heat conduction and viscosity, *Archive for Rational Mechanics and Analysis* **13**, 167–178.

3. W. W. Mullins & R. F. Sekerka, Morphological stability of a particle growing by diffusion or heat flow, *Journal of Applied Physics* **34**, 323–329.

1964 *1.* B. D. Coleman & V. J. Mizel, Existence of caloric equations of state in thermodynamics, *Journal of Chemical Physics* **40**, 1116–1125.

2. W. W. Mullins & R. F. Sekerka, Stability of a planar interface during solidification of a binary alloy, *Journal of Applied Physics* **35**, 444–451.

1965 *1.* C. Truesdell & W. Noll, *The Non-Linear Field Theories of Mechanics*, in (*Handbuch der Physik III/3* (S. Flügge, Ed.)), Springer-Verlag, Berlin.

1966 *1.* B. D. Coleman, H. Markovitz & W. Noll, *Viscometric Flows of Non-Newtonian Fluids*, Springer-Verlag, Berlin, 1966.

2. E. Varley & A. Day, Equilibrium phases of elastic materials at uniform temperature and pressure, *Archive for Rational Mechanics and Analysis* **22**, 253–269.

1967 *1.* R. M. Bowen, Toward a thermodynamics and mechanics of mixtures, *Archive for Rational Mechanics and Analysis* **24**, 370–403.

2. M. E. Gurtin & W. O. Williams, Phases of elastic materials, *Zeitschrift für angewandte Mathematik und Physik* **13**, 132–135.

1969 *1.* R. M. Bowen, The thermochemistry of elastic materials with diffusion, *Archive for Rational Mechanics and Analysis* **34**, 97–127.

2. J. D. Eshelby, The elastic field of a crack extending non-uniformly under general anti-plane loading, *Journal of the Mechanics and Physics of Solids* **17**, 177–189.

3. C. Truesdell, *Rational Thermodynamics*, McGraw-Hill, New York.

1970 *1.* J. D. Eshelby, Energy relations and the energy–momentum tensor in continuum mechanics, in *Inelastic Behavior of Solids* (M. F. Kanninen, W. F. Alder, A. R. Rosenfield & R. I. Jaffe, Eds.), McGraw-Hill, New York.

1971 *1.* A. I. Roĭtburd, Equilibrium of crystals formed in the solid phase, *Soviet Physics Doklady* **16**, 305–308.

2. L. I. Rubenšteĭn, *The Stefan Problem*, American Mathematical Society, Providence.

1972 *1.* J. K. Knowles & E. Sternberg, On a class of conservation laws in linearized and finite elastostatics, *Archive for Rational Mechanics and Analysis* **44**, 187–211.

1973 *1.* F. C. Larché & J. W. Cahn, A linear theory of thermochemical equilibrium of solids under stress, *Acta Metallurgica* **21**, 1051–1063.

2. D. P. Woodruff, *The Solid–Liquid Interface*, Cambridge University Press, London.

1975 *1.* J. L. Ericksen, Equilibrium of bars, *Journal of Elasticity* **5**, 191–202.

2. J. D. Eshelby, The elastic energy–momentum tensor, *Journal of Elasticity* **5**, 321–335.

3. J. K. Knowles & E. Sternberg, On the ellipticity of the equations of nonlinear elastostatics for a special material, *Journal of Elasticity* **5**, 341–362.

4. G. P. Moeckel, Thermodynamics of an interface, *Archive for Rational Mechanics and Analysis* **57**, 255–280.

1976 1. A. I. Murdoch, A thermodynamical theory of elastic material surfaces, *Quarterly Journal of Mechanics and Applied Mathematics* **29**, 245–275.

1977 1. M. E. Glicksman, Capillary phenomena during solidification, *Journal of Crystal Growth* **42**, 347–356.

2. P. C. Hohenberg & B. I. Halperin, Theory of dynamic critical phenomena, *Reviews of Modern Physics* **49**, 435–479.

3. J. K. Knowles & E. Sternberg, On the failure of ellipticity of the equations for finite elastostatic plane strain, *Archive for Rational Mechanics and Analysis* **63**, 321–336.

1978 1. K. A. Brakke, *The Motion of a Surface by its Mean Curvature*, Princeton University Press, Princeton.

2. J. L. Ericksen, On the symmetry and stability of thermoelastic solids, *Journal of Applied Mechanics* **45**, 740–744.

3. J. K. Knowles & E. Sternberg, On the failure of ellipticity and the emergence of discontinuous deformation gradients in plane finite elastostatics, *Journal of Elasticity* **8**, 329–380.

4. F. C. Larché & J. W. Cahn, A nonlinear theory of thermochemical equilibrium of solids under stress, *Acta Metallurgica* **26**, 53–60.

5. F. C. Larché & J. W. Cahn, Thermochemical equilibrium of multiphase solids under stress, *Acta Metallurgica* **26**, 1579–1589.

6. P. Régnier, Surface tension and surface energies, in *Handbook of Surfaces and Interfaces, Volume 2* (L. Dobrzynski, Ed.), Garland STMP Press, New York.

1979 1. J. Fernandez-Diaz & W. O. Williams, A generalized Stefan condition, *Zeitschrift für angewandte Mathematik und Physik* **30**, 749–755.

2. J. K. Knowles, On the dissipation associated with equilibrium shocks in finite elasticity, *Journal of Elasticity* **9**, 131–158.

3. R. D. James, Co-existent phases in the one-dimensional static theory of elastic bars, *Archive for Rational Mechanics and Analysis* **72**, 99–140.

1980 1. R. Abeyaratne, Discontinuous deformation gradients in plane finite elastostatics of incompressible materials, *Journal of Elasticity* **10**, 255–293.

2. J. W. Cahn, Surface stress and the equilibrium of small crystals – I. The case of the isotropic surface, *Acta Metallurgica* **28**, 1333–1338.

3. J. D. Eshelby, The energy–momentum tensor of complex continua, in *Continuum Models of Discrete Systems (CMDS3)*, (E. Kröner & K.-H. Anthony, Eds.), University of Waterloo Press, Waterloo.

4. J. D. Eshelby, The force on a disclination in a liquid crystal, *Philosophical Magazine A* **42**, 359–367.

5. C. Gerhardt, Evolutionary surfaces of prescribed mean curvature, *Journal of Differential Equations* **36**, 139–172.

6. R. D. James, The propagation of phase boundries in elastic bars, *Archive for Rational Mechanics and Analysis* **73**, 125–158.

7. J. S. Langer, Instabilities and pattern formation in crystal growth, *Reviews of Modern Physics* **52**, 1–28.

8. G. P. Parry, Twinning in nonlinearly elastic monatomic crystals, *International Journal of Solids and Structures* **16**, 43–80.

1981 1. R. Abeyaratne, Discontinuous deformation gradients in finite twisting of an incompressible elastic tube, *Journal of Elasticity* **11**, 43–80.

2. J. L. Ericksen, Continuous martensitic transitions in thermoelastic solids, *Journal of Thermal Stresses* **4**, 107–119.

3. M. A. Grinfeld, On heterogeneous equilibrium of non-linear elastic phases and chemical potential tensors, *Letters in Applied Engineering Science* **19**, 1031–1039.

4. M. E. Gurtin & R. Temam, On the anti-plane shear problem in finite elasticity, *Journal of Elasticity* **11**, 197–206.

5. R. D. James, Finite deformation by mechanical twinning, *Archive for Rational Mechanics and Analysis* **77**, 143–176.

6. D. A. Porter & K. E. Easterling, *Phase Transformations in Metals and Alloys*, Chapman & Hall, London.

1982 1. J. W. Cahn & F. C. Larché, Surface stress and the equilibrium of small crystals – II. Solid particles embedded in a solid matrix, *Acta Metallurgica* **30**, 51–56.

2. F. C. Larché & J. W. Cahn, The effect of self-stress on diffusion in solids, *Acta Metallurgica* **30**, 1835–1845.

3. W. W. Mullins, The thermodynamics of critical phases with curved interfaces: specific case of interfacial isotropy and hydrostatic pressure, in *Proceedings of an International Conference on Solid→Solid Phase Transformations* (H. I. Aaronson, D. E. Laughlin, R. F. Sekerka & C. M. Wayman, Eds.), American Institute of Mining, Metallurgical and Petroleum Engineers, New York.

4. M. Shearer, The Riemann problem for a class of conservation laws of mixed type, *Journal of Differential Equations* **46**, 426–443.

1983 1. J. W. Cahn & F. C. Larché, An invariant formulation of multicomponent diffusion in crystals, *Scripta Metallurgica* **17**, 927–932.

2. M. G. Crandall & P.-L. Lions, Viscosity solutions of Hamilton–Jacobi equations, *Transactions of the American Mathematical Society* **277**, 1–42.

3. R. L. Fosdick & Ġ. MacSithigh, Helical shear of an elastic, circular tube with a non-convex stored energy, *Archive for Rational Mechanics and Analysis* **84**, 31–53.

4. M. E. Gage, An isoperimetric inequality with applications to curve shortening, *Duke Mathematical Journal* **50**, 1225–1229.

5. M. E. Gurtin, Two-phase deformations of elastic solids, *Archive for Rational Mechanics and Analysis* **84**, 1–29.

6. M. Shearer, Admissibility criteria for shock wave solutions of a system of conservation laws of mixed type, *Proceedings of the Royal Society of Edinburgh A* **93**, 233–244.

7. M. Slemrod, Admissibility criteria for propagating phase boundaries in a van der Waals fluid, *Archive for Rational Mechanics and Analysis* **81**, 301–315.

8. H. C. Simpson & S. J. Spector, On copositive matrices and strong ellipticity for isotropic elastic materials, *Archive for Rational Mechanics and Analysis* **84**, 55–68.

9. L. M. Truskinovsky, On the chemical potential tensor, *Geokhimiya* **12**, 1730–1744.

10. L. Zee & E. Sternberg, Ordinary and strong ellipticity in the equilibrium theory of incompressible hyperelastic solids, *Archive for Rational Mechanics and Analysis* **83**, 53–90.

1984 1. M. G. Crandall, L. C. Evans & P.-L. Lions, Some properties of viscosity solutions of Hamilton–Jacobi equations, *Transactions of the American Mathematical Society* **282**, 487–502.

2. M. E. Gage, Curve shortening makes convex curves circular, *Inventiones Mathematicae* **76**, 357–364.

3. G. Huisken, Flow by mean curvature of convex surfaces into spheres, *Journal of Differential Geometry* **20**, 237–266.

4. W. W. Mullins, Thermodynamic equilibrium of a crystalline sphere in a fluid, *Journal of Chemical Physics* **81**, 1436–1442.

5. M. Slemrod, Dynamics of first order phase transitions, in *Phase Transformations and Material Instabilities in Solids* (M. E. Gurtin, Ed.), Academic Press, Orlando.

1985 1. J. I. D. Alexander & W. C. Johnson, Thermomechanical equilibrium in solid–fluid systems with curved interfaces, *Journal of Applied Physics* **58**, 816–824.

2. W. Heidug & F. K. Lehner, Thermodynamics of coherent phase transformations in nonhydrostatically stressed solids, *Pure and Applied Geophysics* **123**, 91–98.

3. F. C. Larché & J. W. Cahn, The interactions of composition and stress in crystalline solids, *Acta Metallurgica* **33**, 331–357.

4. W. W. Mullins & R. F. Sekerka, On the thermodynamics of crystalline solids, *Journal of Chemical Physics* **82**, 5192–5202.

5. M. Pitteri, On the kinematics of mechanical twinning in crystals, *Archive for Rational Mechanics and Analysis* **88**, 25–57.

6. J. A. Sethian, Curvature and the evolution of fronts, *Communications in Mathematical Physics* **101**, 487–495.

1986 1. J. I. D. Alexander & W. C. Johnson, Interface conditions for thermomechanical equilibrium in two-phase crystals, *Journal of Applied Physics* **59**, 2735–2746.

2. M. Gage & R. S. Hamilton, The heat equations shrinking convex curves, *Journal of Differential Geometry* **23**, 69–96.

3. M. E. Gurtin, On the two-phase Stefan problem with interfacial energy and entropy, *Archive for Rational Mechanics and Analysis* **96**, 199–241.

4. R. D. James, Phase transformations and non-elliptic free energy functions, in *New Perspectives in Thermodynamics* (J. Serrin, Ed.), Springer-Verlag, Berlin.

5. R. D. James, Displacive phase transformations in solids, *Journal of the Mechanics and Physics of Solids* **34**, 359–394.

1987 1. R. Abeyaratne & J. K. Knowles, Non-elliptic elastic materials and the modeling of dissipative mechanical behavior: an example, *Journal of Elasticity* **18**, 227–278.

2. J. M. Ball & R. D. James, Fine phase mixtures as minimizers of energy, *Archive for Rational Mechanics and Analysis* **100**, 13–52.

3. C. L. Epstein & M. I. Weinstein, A stable manifold theorem for the curve shortening equation, *Communications on Pure and Applied Mathematics* **40**, 119–139.

4. M. A. Grayson, The heat equation shrinks embedded plane curves to round points, *Journal of Differential Geometry* **26**, 285–314.

5. L. M. Truskinovsky, Dynamics of nonequilibrium phase boundaries in a heat conducting non-linearly elastic medium, *Journal of Applied Mathematics and Mechanics (PMM)* **51**, 777–784.

1988 1. R. Abeyaratne & J. K. Knowles, On the dissipative response due to discontinuous strains in bars of unstable elastic material, *International Journal of Solids and Structures* **24**, 1021–1044.

2. M. Chipot & D. Kinderlehrer, Equilibrium configurations of crystals, *Archive for Rational Mechanics and Analysis* **103**, 237–278.

3. M. E. Gurtin, Toward a nonequilibrium thermodynamics of two phase materials, *Archive for Rational Mechanics and Analysis* **100**, 275–312.

4. M. E. Gurtin, Multiphase thermomechanics with interfacial structure 1. Heat conduction and the capillary balance law, *Archive for Rational Mechanics and Analysis* **104**, 195–221.

5. R. Jensen, The maximum principle for viscosity solutions of fully nonlinear second order partial differential equations, *Archive for Rational Mechanics and Analysis* **101**, 350–362.

6. R. Jensen, P.-L. Lions & P. E. Souganidis, A uniqueness result for viscosity solutions of second order fully nonlinear partial differential equations, *Proceedings of the American Mathematical Society* **102**, 975–978.

7. S. Osher & J. Sethian, Fronts propagating with curvature dependent speed: algorithms based on Hamilton–Jacobi formulations, *Journal of Computational Physics* **79**, 12–49.

1989 1. S. Angenent & M. E. Gurtin, Multiphase thermomechanics with interfacial structure 2. Evolution of an isothermal interface, *Archive for Rational Mechanics and Analysis* **108**, 323–391.

2. K. Ecker & G. Huisken, Mean curvature evolution of entire graphs, *Annals of Mathematics* **130**, 453–471.

3. I. Fonseca, Variational methods for elastic crystals, *Archive for Rational Mechanics and Analysis* **107**, 195–224.

4. M. A. Grayson, The shape of a figure-eight under the curve shortening flow, *Inventiones Mathematicae* **96**, 177–180.

5. M. A. Grayson, A short note on the evolution of a surface by its mean curvature, *Duke Mathematical Journal* **58**, 555–558.

6. H. Ishii, On uniqueness and existence of viscosity solutions of fully nonlinear second order elliptic PDE's, *Communications on Pure and Applied Mathematics* **42**, 15–45.

7. P. Leo, & R. F. Sekerka, The effect of surface stress on crystal–melt and crystal–crystal equilibrium, *Acta Metallurgica* **37**, 3119–3138.

8. A. Visintin, Stefan problem with surface tension, in *Mathematical Models for Phase Change Problems* (J. F. Rodrigues, Ed.), Birkhäuser Verlag, Basel.

1990 1. R. Abeyaratne & J. K. Knowles, On the driving traction acting on a surface of strain discontinuity in a continuum, *Journal of the Mechanics and Physics of Solids* **38**, 345–360.

2. P. Bauman & D. Phillips, A nonconvex variational problem related to change of phase, *Applied Mathematics and Optimization* **21**, 113–138.

3. F. Davì & M. E. Gurtin, On the motion of a phase interface by surface diffusion, *Zeitschrift für angewandte Mathematik und Physik* **41**, 782–811.

4. M. E. Gurtin & A. Struthers, Multiphase thermomechanics with interfacial structure 3. Evolving phase boundaries in the presence of bulk deformation, *Archive for Rational Mechanics and Analysis* **112**, 97–160.

5. S. Luckhaus, Solutions for the two-phase Stefan problem with the Gibbs–Thomson law for the melting temperature, *European Journal of Applied Mathematics* **1**, 101–111.

6. O. Penrose & P. C. Fife, Thermodynamically consistent models of phase-field type for the kinetics of phase transitions, *Physica D* **43**, 44–62.

7. P. Rosakis, Ellipticity and deformations with discontinuous gradients in finite elastostatics, *Archive for Rational Mechanics and Analysis* **109**, 1–37.

8. J. A. Sethian, Recent numerical algorithms for hypersurfaces moving with curvature dependent speed: Hamilton-Jacobi equations and conservation laws, *Journal of Differential Geometry* **31**, 131–161.

1991 1. R. Abeyaratne & J. K. Knowles, Kinetic relations and the propagation of phase boundaries in solids, *Archive for Rational Mechanics and Analysis* **114**, 119–154.

2. K. Bhattacharya, Wedge-like microstructure in martensite, *Acta Metallurgica* **39**, 2431–2444.

3. Y.-G. Chen, Y. Giga & S. Goto, Uniqueness and existence of viscosity solutions of generalized mean curvature flow equations, *Journal of Differential Geometry* **33**, 749–786.

4. K. Ecker & G. Huisken, Interior estimates for hypersurfaces moving by mean curvature, *Inventiones Mathematicae* **105**, 547–569.

5. L. C. Evans & J. Spruck, Motion of level sets by mean curvature, *Journal of Differential Geometry* **33**, 635–681.

1992 *1.* J. M. Ball & R. D. James, Proposed experimental tests of a theory of fine structure and the two well problem, *Philosophical Transactions of the Royal Society of London A* **338**, 389–450.

2. K. Bhattacharya, Self-accommodation in martensite, *Archive for Rational Mechanics and Analysis* **120**, 201–244.

3. X. Chen & F. Reitich, Local existence and uniqueness of solutions of the Stefan problem with surface tension and kinetic undercooling, *Journal of Mathematical Analysis and Applications* **164**, 350–362.

4. M. G. Crandall, H. Ishii & P.-L. Lions, User's guide to viscosity solutions of second order partial differential equations, *Bulliten of the American Mathematical Society* **27**, 1–67.

5. L. C. Evans, H. M. Soner & P. E. Souganidis, Phase transitions and generalized motion by mean curvature, *Communications on Pure and Applied Mathematics* **45**, 1097–1123.

6. A. M. Meirmanov, *The Stefan Problem*, de Gruyter, Berlin.

1993 *1.* G. Barles, H. M. Soner & P. E. Souganidis, Front propagation and phase field theory, *SIAM Journal on Control and Optimization* **31**, 439–469.

2. R. L. Fosdick & Y. Zhang, The torsion problem for a non-convex stored energy function, *Archive for Rational Mechanics and Analysis* **122**, 291–322.

3. E. Fried & M. E. Gurtin, Continuum theory of thermally induced phase transitions based on an order parameter, *Physica D* **68**, 326–343.

4. M. E. Gurtin, The dynamics of solid–solid phase transitions 1. Coherent interfaces, *Archive for Rational Mechanics and Analysis* **123**, 305–335.

5. M. E. Gurtin & P. W. Voorhees, The continuum mechanics of coherent two-phase elastic solids with mass transport, *Proceedings of the Royal Society of London A* **440**, 323–343.

6. E. V. Radkevich, On conditions for the existence of a classical solution of the modified Stefan problem (the Gibbs-Thomson law), *Russian Academy of Sciences Sbornik Mathematics* **75**, 221–246.

7. H. M. Soner, Motion of a set by the curvature of its boundary, *Journal of Differential Equations* **101**, 313–372.

8. L. M. Truskinovsky, Kinks versus shocks, in *Shock Induced Transitions and Phase Structures in General Media* (J. E. Dunn, R. L. Fosdick & M. Slemrod, Eds.), Springer-Verlag, Berlin.

9. S.-L. Wang, R. F. Sekerka, A. A. Wheeler, B. T. Murray, S. R. Coriell, R. J. Braun & G. B. McFadden, Thermodynamically-consistent phase-field models for solidification, *Physica D* **69**, 189–200.

1994 *1.* P. Cermelli & M. E. Gurtin, The dynamics of solid–solid phase transitions 2. Incoherent interfaces, *Archive for Rational Mechanics and Analysis* **127**, 41–99.

2. M. E. Gurtin, The characterization of configurational forces, *Archive for Rational Mechanics and Analysis* **126**, 387–394.

1995 *1.* M. E. Gurtin, The nature of configurational forces, *Archive for Rational Mechanics and Analysis* **131**, 67–100.

2. H. M. Soner, Convergence of the phase-field equations to the Mullins–Sekereka problem with kinetic undercooling, *Archive for Rational Mechanics and Analysis* **131**, 139–197.

1996 *1.* H. W. Alt & I. Pawlow, On the entropy principle of phase transition models with a conserved order parameter, *Advances in Mathematical Sciences and Applications* **6**, 291–376.

2. M. E. Gurtin & P. Podio-Guidugli, Configurational forces and the basic laws for crack propagation, *Journal of the Mechanics and Physics of Solids* **44**, 905–927.

I

Papers on Materials Science

Chapter 8

SURFACE TENSION AS A MOTIVATION FOR SINTERING

By CONYERS HERRING
Bell Telephone Laboratories, Murray Hill, N. J.

The author would like to express his indebtedness to a number of his colleagues, in particular to Dr. G. C. Kuczynski, who first aroused his interest in the sintering problem, to W. T. Read for guidance in the calculation of energies of dislocation arrays, and to Dr. J. K. Galt for valuable discussions of dislocation theory.

INTRODUCTION

The sintering together of powder particles into a dense solid mass at temperatures below the melting point of the particles is a process whose rate and end result are known to be influenced by many factors, *e.g.*, particle size, distribution of particle sizes, compacting pressure, temperature of sintering, surrounding atmosphere, gas dissolved in the particles, etc. Because of the many factors involved, it is difficult to draw from practical metallurgical results any reliable conclusions regarding the detailed laws governing the processes occurring. Recently, however, interest has been growing in attempts to isolate the physical processes likely to be important and to study them in experiments designed for unambiguous interpretation. Such experiments have a twofold interest, in that they not only provide clues toward the elucidation of more complicated metallurgical phenomena but also throw light on some fundamental fields of solid-state physics. This chapter undertakes to provide a broad and logically precise formulation of certain physical laws which underly the interpretation of some of the simplest experiments of this type.

Consider first the motivation for sintering, *i.e.*, the source of the lowering of free energy which occurs in this as in every other irreversible process. There seem to be two possible motivations: pressure and surface tension. Two particles which are being pressed together by an external force can yield to the force by growing together. This may be important in the earliest stages of sintering of compacts which have been compressed beforehand and may be important throughout the whole sintering process when pressure is applied simultaneously with heating. In other cases surface tension is probably the principal motivation: the growing together of particles and the shrinking of cavities decrease the total surface area and therefore decrease the total surface free energy.

There seem to be four possible types of mechanism by which the transport of matter involved in sintering may take place: plastic or viscous flow, evaporation and condensation, volume diffusion, and surface migration. Basic to any theory of sintering is a knowledge of the laws by which the rates of these four mechanisms are determined by the configuration of the particles being sintered through the motivations of surface tension and stress. Most of the sections to follow will be devoted to a phenomenological theory of the way in which the rates of the transport mechanisms are influenced by surface-tension effects; the motivation of diffusion by pressure has been treated elsewhere from a similar viewpoint.[1,*] Now any theory couched in phenomenological terms must be based on some broad physical assumptions which make it unnecessary to consider all the atomic details of the processes which it treats. In the present case the critical assumption is that the solid surface being considered interchanges atoms sufficiently rapidly with the interior of the solid, or with the vapor, so that one can speak of a local surface-volume equilibrium. As will be explained in the concluding section, there is reason to question this assumption, and its validity for one substance will not ensure its validity for another. But it is hard to see how the laws relating transport rates to surface tension could take a simple and tractable form if it is not fulfilled; moreover, there is experimental evidence, to be discussed below, that the assumption is valid in a number of cases. It therefore seems worth while to develop the consequences of this assumption in full detail, in the hope that experiments suggested by the theory will bear it out.

It will be assumed throughout that the substance under consideration is crystalline, and for simplicity it will be assumed that only a single kind of atom or molecule is present, so that we need not be concerned with gradients of chemical composition and osmotic pressures resulting from unequal diffusion rates of different components. On the other hand, careful attention will be given to the variation of surface tension with crystallographic orientation of the surface and to the fact that the relation of surface tension to stress is much less simple for a solid than for a liquid.

As most of the theoretical material to follow is, because of its generality, rather abstract, a few orienting remarks regarding numerical values and physical mechanisms may be in order. The surface tensions of solid metals are undoubtedly of the same order as those of molten metals, *i.e.*, of the order of 10^3 ergs per sq cm. The lowering of free energy accom-

*Superior numbers refer to bibliography listings at the end of this chapter.

panying the complete sintering of a mass initially composed of particles say 10μ in diameter is therefore of the order of 1 gram-cal per mole, a rather small quantity on the scale of chemical reactions. If the particles were composed of a glassy substance instead of a crystalline one, the interior of each particle would be under a pressure due to its surface tension, of magnitude $2\pi R\gamma/\pi R^2 = 2\gamma/R$, where γ is the surface tension and R is the radius. For the 10μ particles mentioned, with $\gamma \backsim 10^3$ ergs per sq cm, this gives a pressure of 4 atm; tensions one or two orders of magnitude higher than this might be expected near the neck where two particles have started to sinter together, since the curvature in this region is concave outward and much sharper. As we shall see later, the relation between surface tension and stress is less simple for crystals than for liquids and glasses; however, the orders of magnitude can be expected to be similar. Finally we may note that according to the Thomson-Gibbs equation,[2] the vapor pressure of the droplet of liquid or glass which we have been considering should exceed the equilibrium vapor pressure p_0 over a flat surface by the ratio

$$\frac{p_r}{p_0} = \exp\left(\frac{2\gamma\Omega_0}{RkT}\right)$$

where Ω_0 is the atomic or molecular volume. With the numerical values used above and with $\Omega_0 = 10^{-23}$ cc, $T = 1000^\circ$K, this ratio is 1.0003. In the neck region between two particles the vapor pressure should be lower than the equilibrium value and by a more sizable factor.

The first few sections will be devoted to a calculation of the boundary values assumed by a certain chemical potential $(\mu - \mu_h)$ immediately beneath the various portions of the surface of a crystalline particle. This quantity can be visualized, though somewhat inexactly, as measuring the concentration of lattice defects—interstitial atoms or lattice vacancies—in the crystal. Its variation from point to point therefore motivates self-diffusion, and it can also be shown to determine the local variations in equilibrium vapor pressure. It is easy to see that the concentration of lattice defects beneath a portion of the surface of a crystal is going to be influenced by surface-tension effects if the surface is curved. For example, if the surface is convex, its area can be reduced by reducing the volume of the region beneath it, and this can be done by decreasing the concentration of lattice vacancies or increasing the concentration of interstitial atoms. Since decreasing the surface area decreases the surface contribution to the free energy, it is reasonable to expect that x_h, the concentration of lattice defects as defined in the following section, will be below normal in the given region. If neighboring regions of the surface are concave, it is natural to expect that the resulting gradient in x_h will give rise to diffusion currents transporting matter from the convex to the concave regions.

It is just such diffusion currents which, according to the theory and experiments of G. C. Kuczynski,[3] seem to play the major role in the sintering of many metallic powders.

Later sections of this chapter will be concerned primarily with the boundary conditions satisfied by the diffusion currents at grain boundaries and take up the question of the relation of surface tension to stress and its relation to plastic flow. Here the phenomenological treatment gives way to a detailed discussion of the motion of dislocations, and arguments are given which give some ground for hope that slip and creep within any single crystal play a negligible role in the sintering of single-component systems, at least in most cases.* The final section discusses the likelihood of fulfillment of some of the basic assumptions of the theory and possible experimental tests of them.

CHEMICAL POTENTIALS FOR A PURE SUBSTANCE

As is the case with many of the laws of physics and physical chemistry, the laws relating diffusion and vapor pressure to surface tension, stress, etc., take a particularly simple form when expressed in terms of chemical potentials. However, it is important in the present applications to distinguish between different ways in which the chemical potential may be defined, and so this section will be devoted to a brief clarification of these definitions.

Consider first the case of a crystal which is perfect except for occasional interstitial atoms, lattice vacancies, or impurity atoms, *i.e.*, a crystal free from dislocations and grain boundaries. Such a crystal will be referred to as "quasi-ideal." What are the properties of such a substance which can vary from one internal point to another? The temperature, the state of strain, and, for other than pure elements, the chemical composition can vary. Even for a pure element, however, there is another degree of freedom, namely, the "concentration of lattice defects" x_h, defined as the difference between the number n_L of lattice sites in unit volume and the number n of atoms per unit volume, divided by n_L. We shall adopt the convention of reckoning this quantity positive when $n_L > n$, *i.e.*, when there are more holes than interstitial atoms, and negative when the reverse is the case. The absolute numbers of lattice defects and atoms in a region of given size will be denoted, respectively, by N_h and N. In the absence of more complicated flaws such as dislocations, it is clearly im-

*In this connection it should be pointed out that the theory advanced by J. Frenkel, *J. Phys.*, vol. 9, p. 385 (1945), is based on an altogether erroneous conception of the nature of creep in crystals. In this theory Frenkel attributes sintering to a viscous flow caused by the presence of lattice vacancies. However, as has been shown by F. R. N. Nabarro, "Report of a Conference on the Strength of Solids," p. 75, Physical Society, London, 1948, lattice vacancies cannot, in a single crystal, give rise to a flow describable by a viscosity.

possible for N_h to change in any other way than by passage of lattice defects across the boundary of the region. If we assume local equilibrium between vacancies and interstitial atoms, it will therefore be meaningful to define the Helmhotz free energy F of a region in the interior of a quasi-ideal crystal as a function of temperature, volume, and state of strain, N, and N_h, or equivalently to define the Gibbs free energy G as a function of temperature, stress, N, and N_h. The chemical potentials μ of the atoms and μ_h of the holes or lattice defects can then be defined by

$$\mu = \left(\frac{\partial F}{\partial N}\right)_{N_h, T, v, \text{strain}} = \left(\frac{\partial G}{\partial N}\right)_{N_h, T, \text{stress}} \tag{1}$$

$$\mu_h = \left(\frac{\partial F}{\partial N_h}\right)_{N, T, v, \text{strain}} = \left(\frac{\partial G}{\partial N_h}\right)_{N, T, \text{stress}} \tag{2}$$

and the Gibbs-Duhem relation

$$G = \mu_h N_h + \mu N$$

will be satisfied. If the concentration of lattice defects at the point in question is in equilibrium with the infinite reservoir of holes outside the surface of the specimen, then of course the free energy must be a minimum with respect to changes in N_h. This implies $\mu_h = 0$ if the crystal is free from stress and provided we can ignore effects associated with surface free energy, an assumption which will be legitimate if the curvature of the outer surface of the specimen is sufficiently small. The value of μ_h when curvature effects are important will be derived in Eqs. (10) to (18).

We must now ask how the situation will be modified by the presence of dislocations in the specimen. To begin with, the quantity N_h which measures the difference between the number of lattice points and the number of atoms in the specimen becomes somewhat indefinite when dislocations are present, since it is hard to say what constitutes a lattice point. However, one can still define N and N_h, and hence μ and μ_h, for elements of volume which do not contain dislocation lines; the significant difference between real and quasi-ideal crystals is that for the latter N_h can change only by migration of lattice defects to or from the surface, a process which may take some time, while for the former it is possible for lattice defects to be created or annihilated at a dislocation by motion of the dislocation in a direction not lying in its slip plane. If widely separated dislocations are to be in equilibrium with the surrounding crystal with regard to displacements of this sort and if the crystal is free from stress, clearly the value of μ_h in the surrounding crystal must be zero. If the concentration of lattice defects is different from that corresponding to $\mu_h = 0$, each dislocation will be effectively subjected to a force tending to move it normally to its slip plane. It is conceivable that in a sufficiently small particle such forces could cause all the dislocations to drift out of the crystal; alternatively, it may be that in responding to these forces the various dislocation lines will get tangled up with each other, the force on each dislocation due to μ_h being balanced by forces due to the stress fields of other dislocations. In either case a stable situation with $\mu_h = 0$ results. A more detailed analysis of this question can be made along the lines of the discussion given below regarding the role of plastic flow in sintering. This gives plausibility to the hypothesis that, in most cases of sintering motivated by surface tension, dislocations will not "anchor" μ_h at the value zero.

It is useful for future reference to note here a few simple thermodynamic relations. If as independent variables describing the local state of the crystal we take the temperature T, the pressure p, the shearing stresses σ_i, and the concentration x_h of lattice defects, we have

$$\left(\frac{\partial \mu}{\partial p}\right)_{x_h, T} = \left(\frac{\partial^2 G}{\partial p \, \partial N}\right)_{N_h, T} = \left(\frac{\partial v}{\partial N}\right)_{N_h, p, T} = \Omega_0 \tag{3}$$

where Ω_0 is the atomic volume. Similarly,

$$\left(\frac{\partial \mu_h}{\partial p}\right)_{x_h, T} = \left(\frac{\partial v}{\partial N_h}\right)_{N, p, T} \tag{4}$$

The latter quantity may be either greater or less than Ω_0, according to whether the atoms neighboring the vacancy prefer to recede from it* or crowd into it. In addition to these relations we have the fact that the variation of μ and μ_h with x_h is primarily that due to the entropy of mixing of the atoms and the lattice defects. Since $x_h \ll 1$, we have from the usual theory of dilute solutions,†

$$\left(\frac{\partial \mu_h}{\partial x_h}\right)_{p, T} = \frac{kT}{x_h} + O(1) \tag{5}$$

$$\left(\frac{\partial \mu}{\partial x_h}\right)_{p, T} = -kT + O(x_h) \tag{6}$$

provided, as is normally the case, the concentration of one of the two types of lattice defect, *viz.*, vacancies and interstitial atoms, greatly exceeds that of the other.

*Calculations which have been made for single-ion vacancies in ionic crystals by N. F. Mott and M. J. Littleton, *Trans. Faraday Soc.*, vol. 34, p. 485 (1938), show that this can occur. See also F. Seitz, *Rev. Mod. Phys.*, vol. 18, p. 384 (1946), especially pp. 397*ff*.

†Ref. 2, pp. 128–129.

RELATION OF THE CHEMICAL POTENTIALS TO DIFFUSION

A transport of matter by diffusion from one region of an isothermal crystal to another will occur only if this transport results in a lowering of the total free energy. In a quasi-ideal crystal this transport can take place only by migration of interstitial atoms or lattice vacancies. Neither of these processes changes the number of lattice points in either region; in other words, the number of atoms N and the number of lattice defects N_h, as defined in the preceding section, change by equal and opposite amounts. The change in the free energy of either region is therefore equal to the number of atoms entering or leaving it multiplied by $(\mu - \mu_h)$, and in order that the total free energy be lowered by transferring lattice defects between neighboring regions, there must be a gradient of $(\mu - \mu_h)$. Thus, if the diffusive flux $\mathbf{J}$ (atoms per unit area per unit time) is expressible as a linear form in the gradients of the quantities specifying local state, it must, for a cubic crystal, take the form of proportionality to $\nabla(\mu - \mu_h)$. The factor of proportionality is easily evaluated by computing $\mathbf{J}$ for the special case where x_h varies with position while p is constant. Using Eqs. (5) and (6) and the usual definition of the self-diffusion coefficient D, we find

$$\mathbf{J} = -\frac{Dn_L}{kT}\nabla(\mu - \mu_h) \tag{7}$$

where $n_L = 1/\Omega_0$ is the number of lattice sites per unit volume.* If the crystal is of lower than cubic symmetry, the D in Eq. (7) must of course be replaced by a tensor.

Note that the quantity $\nabla(\mu - \mu_h)$ occurring in Eq. (7) is affected both by gradients of x_h and by gradients of pressure p. Thus either of these can supply a motivation for diffusion.

The discussion thus far has been concerned with volume diffusion. It is easy to show, however, that an equation similar to Eq. (7) must hold for surface diffusion also. For consider a process in which a number dN of atoms are extracted from a region just beneath the surface at position 1,

*The quantity D occurring in Eq. (7) is not rigorously identical with the self-diffusion coefficient as measured by radioactive tracer techniques, although for our present purposes the distinction can usually be ignored. The difference arises from two sources. If an appreciable contribution to the radioactive diffusion coefficient arises from a process where neighboring atoms in the lattice simultaneously exchange positions, there will be no corresponding contribution to the D occurring in Eq. (7), since processes of this sort involve no net flow of atoms. Also, if diffusion is predominantly due to the migration of lattice vacancies, the D occurring in Eq. (7) can be shown to be greater than that measured by the tracer technique, by an amount of the order of 10 per cent. This is because a tracer atom has a preponderant chance of making two successive jumps in opposite directions. For interstitial diffusion, however, the two diffusion coefficients should be exactly the same.

pass by surface migration to a neighboring position 2 on the surface, and at that place are inserted into the underlying lattice again as interstitial atoms or by filling up lattice vacancies. The first and third of these steps involve no change in free energy, to the first order in dN, provided the concentration of mobile atoms on the surface at each of the two positions is in equilibrium with the concentration of lattice defects immediately beneath the surface. Therefore the total change of free energy in the process, which is $[(\mu - \mu_h)_2 - (\mu - \mu_h)_1]\,dN$, occurs in the surface diffusion stage, and we can express the flux of atoms migrating over the surface as the product of a "surface diffusion coefficient" by the gradient of $(\mu - \mu_h)$ in a direction parallel to the surface. Since most faces of even cubic crystals will show preferred directions for surface diffusion, the mathematical expression of this law will take the form

$$j_\alpha = -\frac{\nu}{kT}\sum_\beta \Delta_{\alpha\beta}\frac{\partial(\mu - \mu_h)}{\partial \xi_\beta} \tag{8}$$

where $\mathbf{j}$ is the surface flux of atoms (atoms crossing unit length per unit time), α, β take on two values corresponding to the two perpendicular directions in the plane of the surface, ξ_1, ξ_2 are coordinates in these two directions, ν is the number of surface atoms per unit area, and the tensor $\Delta_{\alpha\beta}$ may be defined by this equation as the "surface diffusion coefficient."

RELATION OF THE CHEMICAL POTENTIALS TO VAPOR PRESSURE AND EVAPORATION

It is easy to relate the equilibrium vapor pressure over any portion of the surface of a crystal to the value of $(\mu - \mu_h)$ in the volume just beneath it. For, from the requirement that the free energy be stationary with respect to a virtual change in which dN atoms are removed from the vapor and added to this region of the crystal (with a corresponding decrease in the number of lattice vacancies or increase in the number of interstitial atoms), we get

$$\mu_v = \mu - \mu_h$$

where μ_v is the chemical potential of the atoms in the vapor phase. Since the vapor pressure p_v is proportional to $\exp(\mu_v/kT)$, we can write

$$p_v = p_0 \exp\left(\frac{\mu - \mu_h - \mu_0}{kT}\right) \sim p_0\left(1 + \frac{\mu - \mu_h - \mu_0}{kT}\right) \tag{9}$$

if $(\mu - \mu_h - \mu_0)$ is small, where p_0 is the equilibrium vapor pressure corresponding to some standard value μ_0 of $(\mu - \mu_h)$, which may conveniently be taken as the value pertaining to a large flat surface at zero pressure.

According to the usual detailed balancing argument, the rate of evaporation from any element of surface is proportional to the equilibrium vapor pressure over this element, multiplied by a "sticking coefficient" α, which measures the fraction of the atoms impinging on this element of surface from the vapor phase which stick and become incorporated in the lattice, rather than reevaporating while still in mobile positions. If α were known or could be assumed unity for all crystallographic orientations of the surface and if the range of travel of a mobile surface atom were sufficiently small, the relative rates of evaporation from different portions of a crystalline particle could be computed from Eq. (9). However, the range of surface migration may be comparable with, or larger than, the dimensions of the particle,* and in this case changes in the configuration of the given particle and its neighbors will, if transport by evaporation is appreciable at all, be compounded in a complicated way out of surface diffusion and evaporation.

CHEMICAL POTENTIALS BENEATH A SMOOTHLY CURVED SURFACE

The "surface tension" γ ($\mathbf{n}$) of the type of crystal face normal to the unit vector $\mathbf{n}$ is defined as the increase of free energy when the exposed area of this type of crystal face is increased by unit amount, keeping constant the internal state of the specimen and the areas of the other types of faces. Thus the free energy of a crystalline body is the sum of a volume term and a surface term, the latter being of the form $\int \gamma(\mathbf{n})dS$. The use of the term "surface tension" for γ is apt to be misleading, since, as will be shown later, γ is not a correct measure of the state of stress of the surface layers of the crystal, as it would be for a liquid. The term "specific surface free energy" would thus describe γ more correctly; however, to economize on syllables we shall continue to use the term "surface tension."

As was explained in the introduction, it is obvious qualitatively that the value of $(\mu - \mu_h)$ immediately beneath a curved element of surface will be influenced by surface-tension effects. The most obvious way to get a quantitative expression for $(\mu - \mu_h)$ is to make use of the requirement that the free energy of the crystal be a minimum with respect to any infinitesimal virtual change in which the local shape of the surface is altered by removing atoms from the interior and placing them on the surface, or vice versa. As will presently be shown, this requirement determines the value of μ_h immediately beneath each element of the surface if γ is a twice differentiable function of the surface orientation, but not otherwise. Now the polar plot of γ as a function of the direction of the surface normal $\mathbf{n}$ will actually have conspicuous cusped minima in the directions corresponding to the simplest crystallographic planes, and may be expected to

*See discussion of evaporation smoothing and related effects in C. Herring and M. H. Nichols, *Rev. Mod. Phys.*, vol. 21, p. 185, Appendix III (1949).

have, if only atomically smooth surfaces are considered, a number of less and less conspicuous cusps in the directions corresponding to crystal planes of higher and higher indices. If for each direction of **n** we plot not the surface tension γ of an atomically smooth surface, but the value γ' going with whatever hill-and-valley microstructure minimizes the surface free energy for the given orientation of the microscopic surface,[16] the polar plot will consist of portions of spheres and have cusps only where different spheres intersect. It is thus to be expected that smoothly curved surface regions will sometimes be encountered where γ is a twice differentiable function of orientation, but that other cases will exist where the surface of the specimen consists partly or wholly of plane facets whose orientations correspond to cusps of the polar plot of γ. We shall therefore give two types of argument, devoting this section to the former case, the next section to the latter.

Consider therefore a smoothly curved surface with a twice differentiable γ, as shown in Fig. 8–1. Let us try to compute the change in surface free

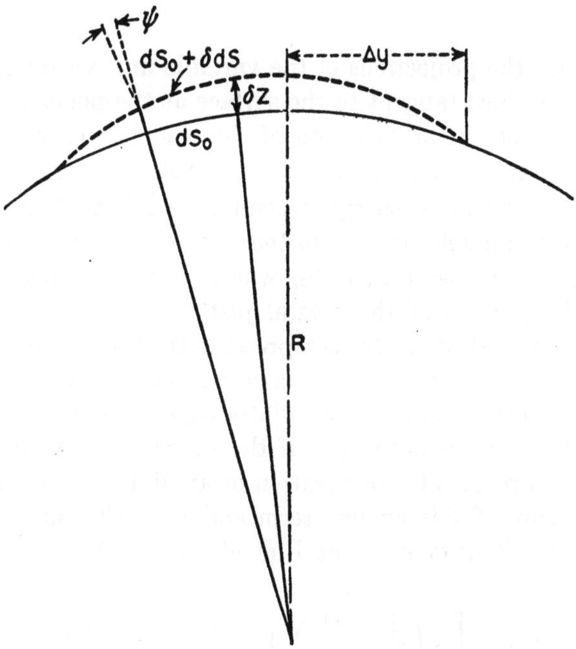

FIG. 8–1. Infinitesimal hump formed by building up a curved surface. Full curve: original surface; dashed curve: built-up surface.

energy accompanying the building up of an infinitesimal hump on the surface, as shown by the dotted line. This change is, to the first order in the thickness of the hump,

$$\delta \int \gamma \, dS = \int \delta\gamma \, dS_0 + \int \gamma \, \delta dS \qquad (10)$$

where dS_0 is the area of an element of the original surface and $dS_0 + \delta dS$ the area of the element of the final surface which is traced out by a family of normals to the original surface erected around the boundary of dS_0. The value of $\delta\gamma$ in the first integrand is determined by the difference in orientation, relative to the crystal axes, between the surface elements dS and dS_0, and by the directional dependence of γ. The value of δdS is easily expressed in terms of the local thickness of the hump and the principal radii of curvature, R_1 and R_2, of the original surface. These calculations, the details of which are given in Appendix 1, yield, with neglect of quantities of the second order in the volume δv of the hump,

$$\int \gamma\, \delta dS = \gamma \left(\frac{1}{R_1} + \frac{1}{R_2}\right) \delta v \tag{11}$$

$$\int \delta\gamma\, dS_0 = \left(\frac{\partial^2\gamma}{\partial n_x{}^2} \cdot \frac{1}{R_1} + \frac{\partial^2\gamma}{\partial n_y{}^2} \cdot \frac{1}{R_2}\right) \delta v \tag{12}$$

where n_x, n_y are the projections of the variable unit vector $\mathbf{n}$, on which γ depends, onto a plane tangent to the surface at the point in question, the x axis being chosen in the direction of the principal curvature $1/R_1$, the y axis in that of $1/R_2$. Insertion of Eqs. (11) and (12) into Eq. (10) gives the change in surface free energy accompanying formation of the hump.

The change in the volume term in the free energy, due to the formation of the hump, is $-p\,\delta v + \mu_h\,\delta v/\Omega_0$, where p is the mean hydrostatic pressure in the portion of the crystal just beneath the surface element being considered, and where $\delta v/\Omega_0$ represents the number of lattice vacancies which must be introduced into this portion of the crystal or the number of interstitial atoms which must be withdrawn, in order to supply the atoms from which the hump is built up. If the shape of the surface is to be in equilibrium with respect to the creation or annihilation of small humps in this way, the sum of this volume term and Eq. (10) must vanish to the first order in δv. This gives, using Eqs. (11) and (12),

$$\mu_h = p\Omega_0 - \left[\gamma\left(\frac{1}{R_1} + \frac{1}{R_2}\right) + \frac{\partial^2\gamma}{\partial n_x{}^2} \cdot \frac{1}{R_1} + \frac{\partial^2\gamma}{\partial n_y{}^2} \cdot \frac{1}{R_2}\right] \Omega_0 \tag{13}$$

The quantity μ is much easier to compute. If the concentration x_h of lattice defects is always $\ll 1$, Eq. (6) shows that variations in μ due to variations in x_h will be negligible compared with the corresponding variations in μ_h given by Eq. (5), and using Eq. (3) we can write

$$\mu = \mu_0 + p\Omega_0 \tag{14}$$

where μ_0 is the constant reference value already used in Eq. (9). Combining Eqs. (13) and (14) gives for the potential motivating diffusion or evaporation

$$\mu - \mu_h - \mu_0 = \left[\gamma\left(\frac{1}{R_1} + \frac{1}{R_2}\right) + \frac{\partial^2\gamma}{\partial n_x^2} \cdot \frac{1}{R_1} + \frac{\partial^2\gamma}{\partial n_y^2} \cdot \frac{1}{R_2}\right] \Omega_0 \qquad (15)$$

This should be applicable whenever γ is a twice differentiable function of $\mathbf{n}$, provided local equilibrium between the surface and the volume immediately beneath it is achieved.

Note that, with γ of the order of 10^3 ergs per sq cm, Ω_0 of the order of 10^{-23} cc, Eq. (15) does not become comparable with kT unless the R's are smaller than a few tens of angstroms. Thus over the range of R's of most interest in sintering problems, *i.e.*, from a fraction of a micron to many microns, Eq. (15) is $\ll kT$, so that the concentration of lattice defects departs only slightly from its standard value. It is therefore quite legitimate to treat the D occurring in Eq. (7) as a constant independent of $(\mu - \mu_h)$.

CHEMICAL POTENTIALS BENEATH A FACETED SURFACE

It is now obvious mathematically why a different treatment must be used for a surface spot whose normal lies in a direction for which the polar plot of the surface tension γ has a cusp: at such a spot the second derivatives are infinite, and Eq. (15) is infinite if the surface is curved, indeterminate if it is flat. In fact, if one were to erect a hump of the sort we have been considering on an initially plane surface whose orientation corresponded to a cusped minimum in the polar plot of γ, the ratio of the term $\int \delta\gamma \, dS_0$ to the other terms in the free energy change would approach infinity as the radius Δy of the hump approached zero. This means that the first-order change in the free energy due to the creation of any hump or hollow would always be an increase; such a surface facet is thermodynamically stable with respect to all infinitesimal local deformations of the surface regardless of the value of μ_h beneath it, though for any given value of $(\mu - \mu_h - \mu_0)$ there will exist a critical radius Δy such that the facet is unstable with respect to formation of humps or hollows of radius larger than Δy. In such cases other means must be used to determine μ_h.

One requirement which may reasonably be imposed is that the total free energy be stationary with respect to any infinitesimal displacement of the plane surface parallel to itself, the atoms added to or subtracted from each element of surface being supplied by changing the concentration of lattice defects in the volume immediately beneath that element of surface. A parallel displacement of this sort is illustrated in Fig. 8-2. The change in surface free energy which it entails can easily be computed in terms of the surface tension γ_0 of the facet in question (AB in the figure) and the surface tensions γ_1 of the surrounding facets such as AA' and BB'. Thus

$$\delta \int \gamma \, dS = \int (\gamma_1 \, \delta z \csc \theta - \gamma_0 \delta z \cot \theta) \, dl \qquad (16)$$

where the integration on the right is over the perimeter of the facet AB, and where θ is the dihedral angle between this facet and the bounding surface, whose surface tension γ_1 will in general be different on different portions of the perimeter. If the material required to build up each portion of this slab is taken from the volume immediately beneath that portion, the change in volume free energy is

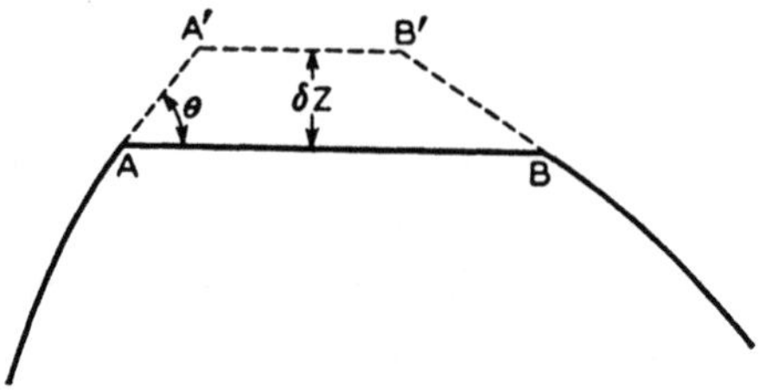

FIG. 8–2. Change in form of a crystal surface due to adding an infinitesimal layer of constant thickness δz to a plane facet. Full lines: original surface; dashed lines: built-up surface.

$$\overline{\mu}_h \frac{A \, \delta z}{\Omega_0} - \overline{p} \, A \, \delta z \qquad (17)$$

where A = area of facet AB

$\overline{\mu}_h$ = area average of μ_h immediately beneath the various points of the facet

$\overline{p}$ = similar average of the pressure

Equating the sum of Eqs. (16) and (17) to zero and combining with $\overline{\mu} = \mu_0 + \overline{p}\Omega_0$ we get

$$\overline{\mu} - \overline{\mu}_h - \mu_0 = \frac{\Omega_0}{A} \int (\gamma_1 \csc \theta - \gamma_0 \cot \theta) \, dl \qquad (18)$$

This is the value which the area average of $(\mu - \mu_h - \mu_0)$ just beneath a plane facet of a crystal must have if volume and surface are in equilibrium with each other.

Knowledge of this area average for all plane facets of a specimen may at first seem far short of knowledge of $(\mu - \mu_h)$ as a function of position over each facet. However, in the quasi-steady state of diffusion the distribution of $(\mu - \mu_h)$ over each facet can be shown to be completely determined by the requirement that the diffusive flux into or away from the surface must be uniform over each facet in order that the facet shall grow by parallel displacement only. (We assume for simplicity that there is no significant transport of matter by means other than volume diffusion; if there is a comparable transport by surface diffusion or by evaporation and condensation, the normal component of volume-diffusive flux need not be constant over a facet.) Since in the quasi-steady state of diffusion in a cubic crystal $\nabla^2(\mu - \mu_h) = 0$, the problem of determining the distribution of $(\mu - \mu_h)$ in a polyhedral specimen is exactly the same as the problem of determining the velocity potential in a fluid confined by a polyhedral container each of whose facets is acted upon by a given normal force, the facets being all rigid and constrained to move by parallel displacement only. It is physically obvious from this analogy that the solution is unique; although we shall not attempt to do so here, it is quite easy to construct a mathematical proof of this uniqueness for a specimen bounded in part by facets on which Eq. (18) holds, in part by curved regions where Eq. (15) applies. Thus, whenever the underlying assumption of local surface-volume equilibrium is fulfilled, the diffusion currents in the interior of any single crystal specimen can be computed by solving a potential theory problem with the boundary conditions Eqs. (15) and (18). It will be shown later how a similar calculation can be made for a specimen containing grain boundaries.

The right of Eq. (18) is easily seen to be equal to $p_\gamma \, \Omega_0$, where p_γ is the normal pressure on the facet which would be required to hold a movable rigid partition of the same size and shape in equilibrium against forces of amount γ_1 per unit length exerted at its boundary in the plane of each neighboring surface, combined with a force of amount γ_0 per unit length exerted in its own plane. From this mechanical analogy it is obvious that Eq. (18) should go over into Eq. (15) when the angles between neighboring facets become very small. It can in fact be shown that the only difference between Eqs. (18) and (15) is that the curvatures occurring in Eq. (15) are replaced by second differences of the inclination of the surface, and the second derivatives by second differences of γ.

INTERFACIAL EQUILIBRIA AT GRAIN BOUNDARIES

We have seen that, provided surface-volume equilibrium obtains, the equations of the two preceding sections suffice to determine the distribution of diffusion currents in any single crystal specimen. When two or more crystals are present separated by grain boundaries, the solution of the diffusion problem obviously requires in addition a knowledge of the boundary condition satisfied by $(\mu - \mu_h)$ on each grain boundary. Moreover, we shall see that Eq. (18) needs some modification when the facet in question is adjoined by a grain boundary. This section and the next will be devoted to these two topics and to the general formulation of the equilibrium condition at the line of intersection of three interfaces.

Before embarking on these questions, however, it should be pointed out that grain boundaries are notoriously prone to accumulate impurity atoms and that their behavior may be seriously influenced by such accumulations. Three cases may be distinguished: the ideal case in which no impurities are present; the case where a small to moderate concentration of impurities is localized at the boundary, so that motion of the boundary cannot take place any more rapidly than diffusion of these impurities can keep up with it; and the extreme case where the layer of impurities is so thick that it constitutes a serious barrier to the diffusion of lattice defects from one side of the boundary to the other. In the present elementary discussion we shall be concerned principally with the first case, although some of the considerations to be given will also apply to the second.

Next, a few comments are in order regarding the factors determining the locations of the grain boundaries in a specimen. In order that the type of phenomenological theory considered in the previous sections be applicable to situations involving grain boundaries, the temperature must be high enough so that all interfaces are in thermal equilibrium with the adjoining volume elements. This means that, under any conditions where the present methods are applicable, the grain boundaries must have a configuration which minimizes the interfacial free energy, or at least makes it stationary with respect to infinitesimal shifts of the grain boundaries. Since motion of an ideal grain boundary involves only local rearrangements and no long-range transport, this configuration will probably be reached very quickly if the grain boundary is ideal, but may or may not be reached if the boundary contains impurities. The problem of finding this configuration is similar in principle to the problem of finding the equilibrium shape of a crystal bounded by vacuum; its general solution presumably involves an extension of the method used by Wulff[4] for the latter problem. If it is true, however, that the interfacial free energy of a grain boundary is almost independent of its orientation,* the problem simplifies greatly, at least when only a very few grain boundaries are present in the specimen.

Where the grain boundaries meet each other or meet the outer surface of the specimen, the three interfacial tensions must satisfy an equilibrium condition. If any one of the three surface tensions involved depends appreciably on the orientation of its surface, the familiar force triangle which applies along the line of intersection of three fluid interfaces must be replaced by a more general condition, derived in Appendix 2, namely,

$$\sum_{i=1}^{3} \left(\gamma_i \mathbf{t}_i + \frac{\partial \gamma_i}{\partial \mathbf{t}_i} \right) = 0 \tag{19}$$

where γ_1, γ_2, γ_3 are the three surface tensions, $\mathbf{t}_i$ is the vector in the plane of the ith surface, normal to the line of intersection of the surfaces and pointing away from this line, and $\partial\gamma_i/\partial\mathbf{t}_i$ is accordingly a vector perpendicular to $\mathbf{t}_i$ and to the line of intersection. If the latter terms vanish, Eq. (19) reduces to the familiar force triangle. The terms in $\partial\gamma_i/\partial\mathbf{t}_i$ imply that each interface strives to contract with a tension γ_i and to

*Evidence suggesting that this is the case has been summarized by C. S. Smith, *Metals Technol.*, vol. 15, *Tech. Pub.* 2387 (1948).

rotate to an orientation of lower γ_i with an effort measured by a torque $\partial\gamma_i/\partial t_i$ per unit area. If the orientation of one of the boundaries corresponds to a cusp in the polar plot of its γ, the quantity $\partial\gamma/\partial t$ is of course indeterminate, and the boundary angles will be in equilibrium if Eq. (19) can be satisfied by any value of $\partial\gamma/\partial t$ lying between the values on the two sides of the cusp. In such case, the boundary in question will be a plane facet; any boundary whose orientation is not that of a cusp in the polar plot of its γ will in general be smoothly curved.

CHEMICAL POTENTIALS AT GRAIN BOUNDARIES

Whether a grain boundary is ideal or not, it is probably safe to assume that at sintering temperatures local equilibrium prevails with respect to exchange of lattice defects across the grain boundary, unless of course there is such a huge aggregation of impurities at the grain boundary as to form a barrier layer against diffusion. Thus we have as one boundary condition that $(\mu - \mu_h)$ must be continuous across all grain boundaries. To make the potential problem soluble, this must be supplemented by another boundary condition, *e.g.*, one concerning the magnitude of the discontinuity in the normal derivative of $(\mu - \mu_h)$. The latter is determined by the fact that matter cannot be created or annihilated at a grain boundary. In those cases where the two adjoining grains are not undergoing any translatory or rotational motion relative to each other, this of course means, for cubic crystals, that the normal derivative of $(\mu - \mu_h)$ must be continuous across the boundary. In more general cases the boundary condition for cubic crystal is

$$\mathbf{n} \cdot \nabla(\mu - \mu_h)_1 - \mathbf{n} \cdot \nabla(\mu - \mu_h)_2 = a + \mathbf{b} \cdot \mathbf{r} \tag{20}$$

since a relative translational velocity $\mathbf{v}$ and a relative angular velocity ω give a normal velocity $\mathbf{n}\cdot\mathbf{v} + \mathbf{n}\cdot\omega \times \mathbf{r} = \mathbf{n}\cdot\mathbf{v} + \mathbf{n} \times \omega\cdot\mathbf{r}$ at a point of the boundary whose position vector is $\mathbf{r}$. In general, the coefficients a and $\mathbf{b}$ will not be known a priori; however, it can easily be shown that $(\mu - \mu_h)$ is uniquely determined everywhere if, in addition to Eq. (20) (with a and $\mathbf{b}$ unknown) and the other boundary conditions already mentioned, the surface average and first moment of $(\mu - \mu_h)$ are known over the boundaries separating each pair of crystal grains. The latter quantities can be determined, at least in simple cases, from the condition that the total free energy be stationary with respect to all infinitesimal rigid motions of the grains relative to each other, separation and interpenetration at grain boundaries being avoided by changing the local concentration of lattice defects. If no external forces are applied, this condition can be formulated in terms of the various surface and grain-boundary tensions; in the presence of external forces, additional terms representing the work done by these forces must be added. These are important in the theory of "diffusional viscosity" of a polycrystalline solid.[1,5]

To illustrate the principles just outlined, let us try to make a rough estimate of the effect of a grain boundary on the rate at which two particles sinter together. If each of them is an approximately spherical single crystal and the two are separated by a plane grain boundary across the narrowest part of the neck, a cross section of the neck region may be expected to look as in Fig. 8–3(*a*). It will in general have a sharp angle where the grain boundary meets the surface. The angles α_1 and α_2 are of course determined by Eq. (19). If the surface tensions of the two crystals vary with orientation, these angles will in general be unequal and will vary from point to point around the edge of the neck; in such case the grain boundary will probably be a twisted surface, rather than a plane. For simplicity, let us assume in the present example that the external surface tension is isotropic, so that these complications do not occur. To determine the mean value of $(\mu - \mu_h)$ over the grain boundary, consider an infinitesimal rigid displacement which brings the two grains a distance δz toward each other, as shown in Fig. 8–3(*b*). To avoid interpenetration,

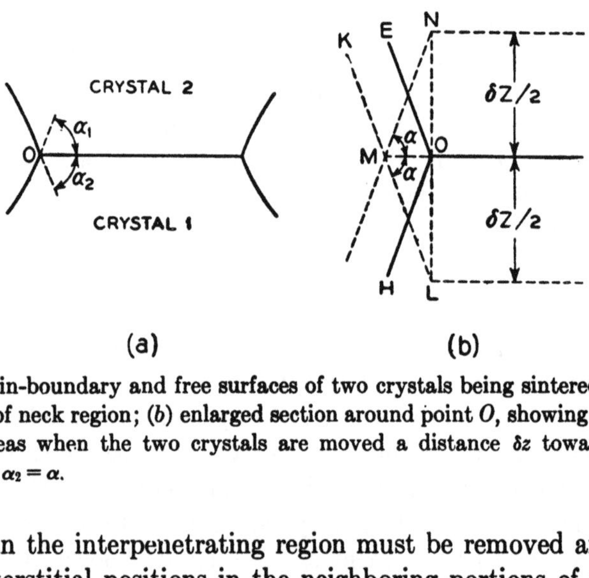

FIG. 8–3. Grain-boundary and free surfaces of two crystals being sintered together: (*a*) cross section of neck region; (*b*) enlarged section around point *O*, showing changes in the interfacial areas when the two crystals are moved a distance δz toward each other, assuming $\alpha = \alpha_2 = \alpha$.

the atoms in the interpenetrating region must be removed and placed in holes or interstitial positions in the neighboring portions of the crystals. This gives a change of volume free energy

$$\delta F_v = \frac{-A\ \delta z}{\Omega_0}\bar{\mu}_h + \bar{p}\,A\ \delta z \tag{21}$$

where A = area of grain boundary
Ω_0 = atomic volume
$\bar{\mu}_h$ and $\bar{p}$ = area averages of μ_h and the pressure over the grain boundary

To compute the change in surface free energy, note that the surface of each crystal is shortened by the amount

$$MN = \frac{\delta z \csc \alpha}{2}$$

and the grain boundary is lengthened by

$$MO = \frac{\delta z \cot \alpha}{2}$$

Thus the change in surface free energy is

$$\delta F_s = \pi r(-2\gamma \csc \alpha + \gamma_b \cot \alpha)\delta z$$

where γ = surface tension
γ_b = grain boundary tension
r = radius of neck

For the present case, Eq. (19) gives

$$\gamma_b = 2\gamma \cos \alpha$$

whence,

$$\delta F_s = -2\pi r \gamma \sin \alpha \, \delta z \tag{22}$$

Equating the sum of Eqs. (21) and (22) to zero gives

$$\bar{\mu} - \bar{\mu}_h - \mu_0 = \bar{p}\Omega_0 - \bar{\mu}_h = \frac{2\gamma\Omega_0}{r} \sin \alpha \tag{23}$$

We can now estimate whether or not the presence of the grain boundary makes the neck grow significantly faster than it would grow if the whole specimen were a single crystal. The normal flux of atoms away from the grain boundary into either crystal will by Eq. (7) have the order of magnitude

$$J_n = \frac{Dn_L}{kT} \mathbf{n} \cdot \nabla(\mu - \mu_h) \backsimeq \frac{Dn_L}{kT} \frac{\bar{\mu} - \mu_h - \mu_0}{2r}$$

The rate of approach of the two particles is $dz/dt = 2J_n\Omega_0$. Multiplying this by $(\cot \alpha)/2 = \overline{MO}/\delta z$ in Fig. 8–3 gives the contribution of the translatory motion of the grains to the growth of the radius of the neck. Using Eq. (23) and the relation $n_L\Omega_0 = 1$, we find

$$\text{Contribution to } \frac{dr}{dt} \backsimeq \frac{D\gamma\Omega_0}{kTr^2} \cos \alpha \tag{24}$$

This may be compared with the expression

$$\frac{dr}{dt} \approx \frac{8D\gamma\Omega_0 R^2}{kTr^4} \tag{25}$$

derived by G. C. Kuczynski[3] with similar assumptions for the case where the grain boundary is ignored and where $r \ll$ the original particle radius R. We see that Eq. (24) $\ll$ Eq. (25) for this case, as one expects. However, there may be some applications, such as the shortening of wires under their own surface tension,[6] where the relative motion dz/dt of two crystals normal to their grain boundary may itself be of primary importance. This is a motion which would not occur at all in the absence of a grain boundary. It has been shown elsewhere[1] that the relative motions of neighboring grains in a long wire, due to diffusion currents of the sort we have just been discussing, may in fact be capable of accounting quantitatively for the observations of H. Udin, A. J. Shaler, and J. Wulff[6] and of other earlier experimenters who measured the contraction of heated foils.[7]

In dealing with problems where a plane surface of a crystal meets a grain boundary, the boundary conditions satisfied by $(\mu - \mu_h)$ on the plane facet of the outer surface of a crystal may have to be computed in a slightly more complicated way than previously. There $(\bar{\mu} - \bar{\mu}_h)$ for the facet was computed by equating to zero the first-order change in free energy in an infinitesimal displacement of the facet parallel to itself, the planes of all other boundaries being kept fixed. In the present case this is not legitimate. For example, in Fig. 8–3(*b*) displacing the boundary of crystal 2 alone will give the boundary *KMOH*; the portion *MO* of this is not of the same nature as the grain boundary, and its contribution to the free energy will not in general be equal and opposite to the contribution which would result if the displacement were in the opposite direction. Clearly what must be done is to find a new equilibrium grain boundary ending on the line where *KL* intersects *HO*, and to compute the difference between its free energy and that of the original grain boundary. This will obviously be difficult in most cases. However, when it is known from symmetry or otherwise that the plane of the grain boundary will not change, the boundary conditions on $(\mu - \mu_h)$ can be determined in the following way: Take as one boundary condition the fact that the normal fluxes on the two planes *EO* and *HO* must be in such ratio as to keep the intersection of these planes on the prolongation *MO* of the grain boundary. Take as the other boundary condition the requirement that the first-order change in free energy vanish when both planes are displaced simultaneously by distances in this ratio. This fixes the value of a linear combination of $(\bar{\mu} - \bar{\mu}_h)_1$ and $(\bar{\mu} - \bar{\mu}_h)_2$, and this linear combination plus the condition just enunciated for the normal derivatives makes the potential problem determinate.

RELATION OF SURFACE TENSION TO STRESS

It has been shown in the preceding sections how to calculate the rates at which volume diffusion, surface diffusion, and evaporation and condensation contribute to the change of shape of a single crystal particle, although unfortunately the rates of the latter two mechanisms depend upon quantities whose numerical values have rarely or never been determined, *viz.*, the surface diffusion coefficients and the sticking coefficients. It remains to consider whether or not surface tension can also motivate changes in shape by the mechanism of plastic flow. It will be shown in this section that, although the interior of a crystalline particle is normally in a state of stress induced by the presence of the surface, this stress cannot be predicted from a knowledge merely of the specific surface free energies γ of the various crystal surfaces, as can be done for liquids. In addition, however, arguments will be presented later which make it seem likely that these stresses, or for that matter any other influence, may be unable to cause plastic deformation to take place in a crystal not acted on by external forces. If the premises of these arguments can be experimentally verified, plastic flow can be ruled out as a mechanism for many of the types of changes which occur in sintering.

One might at first suppose that the interior of a small crystalline particle, like that of a small drop of liquid, would be in a state of stress because of surface-tension forces, and that this stress could be computed by considering a small arbitrary deformation of the particle and equating the change in elastic free energy to the product of the specific surface free energy by the change in area of the surface. This stress, if it existed, could in principle be measured by the strain it would produce in the lattice, which would cause a slight change in the lattice parameters. However, it was pointed out long ago by J. W. Gibbs[8] that no such simple relation between stress and surface tension is to be expected for crystals. Gibbs' argument is that for crystals there is a distinction between the work involved in forming a surface and the work involved in stretching it, a distinction which does not exist for liquids. To put the argument on an atomic basis, we may note that the principal cause of surface tension is the fact that surface atoms are bound by fewer neighbors than internal atoms; surface tension is therefore mainly a measure of the change of free energy with a change in the number of atoms in the surface layer. The difference between a liquid and an ideal solid lies in the fact that for a liquid drop a compression of the interior implies, on the average, a decrease in the number of atoms on the surface, whereas for an ideal crystalline particle the state of strain and the number of surface atoms are entirely unrelated. In the latter case, a statistical preference for configurations with smaller numbers of surface atoms will not of itself imply any preference for compressed states.

These statements by no means imply the absence of a "surface stress"; they merely show that it is not the same as what we have been calling the "surface tension" γ. We may give a precise definition of surface stress as follows: Consider any plane *NN* normal to the surface *SS* of a crystal, as shown in Fig. 8–4. Let $\mathbf{F}_{BA}$ be the total force, per unit length normal to the plane of the figure, which the matter in region *B* on one side of *NN* exerts on that in region *A* on the other side, where region *A* is bounded in depth by some plane *TT* parallel to the surface, while region *B* is unbounded in depth. To make the definition of $\mathbf{F}_{BA}$ free from any possible dependence on the exact depth of x of *TT*, we may specify that it be an average over a

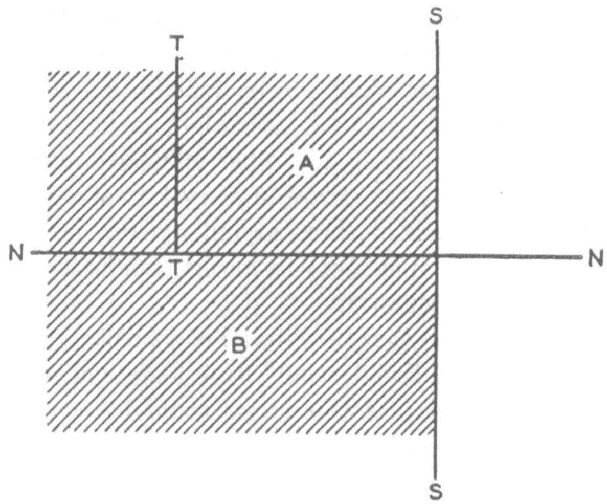

FIG. 8–4. Regions of a crystal near a free surface *SS*.

distribution of values of x covering a range which is macroscopically negligible but still large compared with atomic dimensions. As the depth x is varied, the changes in the μth component of $\mathbf{F}_{BA}$ will be the same as those in the integral

$$\int_0^x \sum_\nu p_{\mu\nu}\, q_\nu\, dx$$

where $p_{\mu\nu}(x)$ is the stress tensor in the interior at depth x and $\mathbf{q}$ is the unit normal to the plane *NN*. However, $\mathbf{F}_{BA}$ will not usually be equal to this integral in total value, and we may call the difference between these two quantities the "surface stress" transmitted across the line of intersection of *NN* and the surface *SS*. By arguments which are the two-dimensional analogue of those used to establish the fact that stresses in three dimensions are described by a tensor, it can be shown that, given any line lying in the surface and having unit normal **q** in the plane of the surface, the μ component of surface stress transmitted across it is given by an expression of the form

$$\sum_{\nu} g_{\mu\nu}\, q_{\nu} \text{ (force per unit length)} \tag{26}$$

where $g_{\mu\nu}$ ($\mu = x, y, z, \nu = x, y$) is a "surface stress tensor" dependent on the properties of the surface but independent of the azimuth of **q** in the surface. For surfaces whose orientations are sufficiently symmetrically disposed with respect to the crystal lattice, the force Eq. (26) will have no component normal to the surface, and we can set the components $g_{\nu z} = 0$, so that $g_{\mu\nu}$ reduces to an ordinary tensor, in two dimensions. If the surface is normal to a three-, four-, or sixfold axis of symmetry, we may expect $g_{\mu\nu}$ to be a multiple of the unit matrix, *i.e.*, of the form $g\delta_{\mu\nu}$ where $\delta_{\mu\nu} = 0$ if $\mu \neq \nu$, 1 if $\mu = \nu$.

The numerical values of the various components of the surface stress tensor $g_{\mu\nu}$ can be computed directly by summing up the forces which the various atoms of the crystal exert on one another and by finding the unbalance in these forces due to the presence of a free surface. The calculations are not difficult if one assumes an idealized crystal model for which the total energy is a sum of interaction energies of pairs of atoms. In many cases, though not all, such calculations yield the result that the surface stress is compressive, rather than tensile.[9]

We shall omit a discussion of the details of such calculations, instructive though these details are, and shall pass on to establish a thermodynamic relationship between the surface stress tensor $g_{\mu\nu}$ and the surface tension γ, a relationship which shows why the two are different for a crystal though the same for a liquid, and why the components of $g_{\mu\nu}$ can be expected to be of the same order of magnitude as the surface tension γ. Consider a crystalline specimen in the form of a slab of thickness d. Choose the z axis normal to the boundary planes of the slab. The stress in the deep interior of the slab will be described by the stress tensor*

$$p_{\mu\nu} = -\frac{2g_{\mu\nu}}{d} \quad (\mu = x, y, z, \ \nu = x, y) \tag{27}$$

The free energy of the slab must be stationary with respect to any virtual distortion, *e.g.*, a homogeneous strain described by the strain tensor $u_{\mu\nu}$. If A is the area of either face of the slab, the change of volume free energy in such a strain is

$$Ad\sum p_{\mu\nu}u_{\mu\nu} = -2A\sum g_{\mu\nu}u_{\mu\nu} \tag{28}$$

*We assume for simplicity that $g_{\mu\nu}$ is the same for the top and bottom faces of the crystal. This will certainly be the case for any crystal class with a center of symmetry.

while the change in the surface free energy $2\gamma A$ is

$$2A \sum \frac{\partial \gamma}{\partial u_{\mu\nu}} u_{\mu\nu} + 2\gamma \sum \frac{\partial A}{\partial u_{\mu\nu}} u_{\mu\nu} \tag{29}$$

Now for μ, $\nu = x, y$, $\partial A/\partial u_{\mu\nu} = A\delta_{\mu\nu}$, so adding Eqs. (28) and (29) and noting that, since the $u_{\mu\nu}$ are arbitrary, the coefficient of each $u_{\mu\nu}$ in the sum must vanish,

$$g_{\mu\nu} = \gamma\delta_{\mu\nu} + \frac{\partial \gamma}{\partial u_{\mu\nu}} \ (\mu = x, y, z, \ \nu = x, y)^* \tag{30}$$

Note that there is no equation of the type Eq. (30) for $\mu = \nu = z$, since for the free slab under consideration $p_{zz} = 0$ and $\partial\gamma/\partial u_{zz} = 0$.

For a solid, the last term of Eq. (30) can be sizable, but for a liquid it will vanish, because straining a liquid slab merely results in the transfer of liquid from the volume to the surface, or vice versa, without change in the state of the surface.

One may expect the value of $\partial\gamma/\partial u_{\mu\nu}$ for a crystal to be, very roughly, of the same order as γ itself or a little larger. We may therefore expect the components of $g_{\mu\nu}$, whether they be positive or negative in sign, to be also of this order of magnitude, *i.e.*, of the order of 10^3 dynes per cm for common metals.

RELATION OF SURFACE STRESSES TO DISLOCATIONS AND PLASTIC FLOW

The discussion given in the preceding section has been rather abstract, and it may therefore be helpful to pause at this point to discuss a typical physical situation, with some typical numerical values. Consider a specimen of a cubic crystal in the form of a cube with faces in the principal crystal planes and, say, 10^{-3} cm on a side. Suppose that for the cube faces $g_{\mu\nu} = g\delta_{\mu\nu}$ with $g = -1{,}000$ dynes per cm (compressive surface stress). Then in the absence of plastic flow the crystal will be elastically deformed in the same way that a cube of laboratory size would be deformed if a spreading stress of this magnitude were applied to each element of its surface. These spreading stresses all over the surface will be equivalent in their action to forces, of magnitude $g\sqrt{2}$ per unit length, applied diagonally outward along all the edges of the cube, as shown in Fig. 8–5.

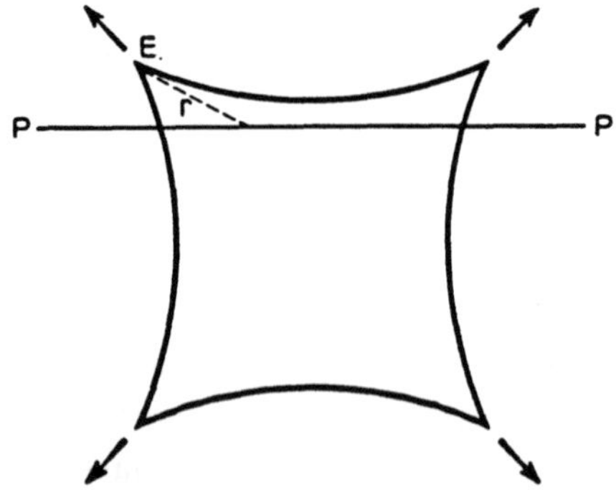

FIG. 8–5. Exaggerated illustration of the deformation of a crystalline cube by a compressive surface stress.

*Since this was written, a similar derivation of the scalar form of Eq. (30) has been published by R. Shuttleworth, *Proc. Phys. Soc. A*, vol. 63, p. 444 (1950).

These forces will pull the middle of each edge diagonally outward by a distance which can be estimated as about ½ to ¾ A, if the crystal is assumed to behave like an isotropic elastic medium with a shear constant of 5×10^{11} dynes per sq cm and a Poisson's ration of ⅓, values approximately characteristic of copper at room temperature. This displacement* is very small on the scale of the specimen. However, the shearing stress in the specimen at a distance r from an edge will be approximately g/r and this can become quite large for small r; it reaches 100 grams per sq mm for the present example when $r = 10^{-4}$ cm. One may therefore legitimately ask whether these stresses, which in sufficiently close proximity to the edges certainly exceed the macroscopic yield stress, will cause slip or creep and thus relieve the surface stress.

It is not hard to see that the surface stress could in fact be completely canceled, as far as long-range effects are concerned, by the presence of a suitable distribution of dislocations a short distance beneath the surface. For example, we have assumed a negative g for the faces of the cube just discussed, so that the surface layers would like to expand if they were not restrained from doing so by the lattice underneath; this desire to expand could be gratified, however, by introducing a distribution of edge-type dislocations having slip vectors parallel to the surface, and with one layer of atoms less on the side toward the surface than on the side toward the interior. Such a distribution is shown in Fig. 8–6. To get an isotropic stress cancellation two such distributions at right angles might be assumed.

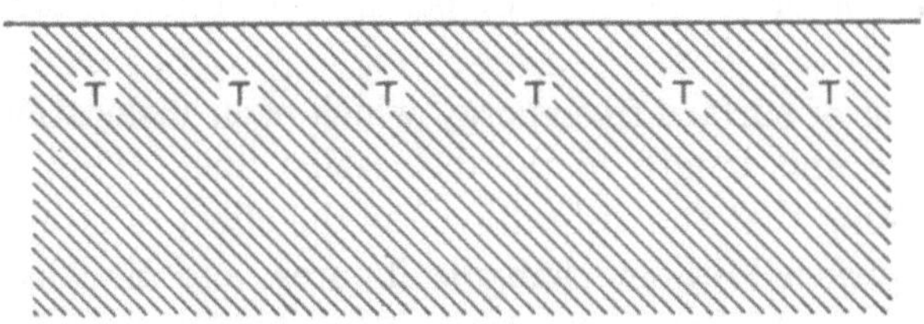

FIG. 8–6. Illustration of how a compressive surface stress could be relieved by a row of dislocations just beneath the surface. Each dislocation is represented by a T-like symbol, the "head" representing the slip plane of the dislocation and the "handle" being drawn on that side of the slip plane which has the extra plane of atoms.

However, work must be done to create such a distribution of dislocations, and it may or may not happen that the total free energy of the crystal will be lowered by introducing dislocations of this sort. We shall try to make a rough estimate of the conditions which must be satisfied for the configura-

*The displacement resulting from application of a force along an infinitely sharp edge is proportional to the logarithm of the distance from the edge; in the present physical situation the displacement of the edge must obviously be given by using as the argument of the logarithm a quantity of the order of the ratio of the radius of the specimen to the lattice constant.

tion with dislocations to be energetically favored. The elastic continuum theory of dislocation energies suggests that the energy of a pair of dislocations of opposite sign separated by a distance z normal to their slip planes is of the form[10]

$$\frac{Ga^2}{2\pi(1-\sigma)} \ln \frac{z}{z_0} \tag{31}$$

per unit length of the dislocation lines, where G is the shear modulus of the material, σ is Poisson's ratio, a is the length of the slip vector of the dislocation (essentially the lattice constant of the crystal), and z_0 is a length of the order of magnitude of a. Now it is known that a dislocation is attracted to a free surface by a force which is very nearly the same as the force which a dislocation of opposite sign in the "image" position would exert on it[10]; since the amount of spreading or shrinking of the surface which a given type of dislocation produces is independent of the distance of the dislocation from the surface, this suggests that the free energy of any array of dislocations decreases monotonically as the dislocations are moved closer and closer to the surface. Detailed calculations for a uniformly spaced row of dislocations have been found to confirm this statement. Now we can hardly speak of a dislocation as such unless its depth below the surface is at least one or two times the lattice constant a, so that its distance z from its image is several times a, making $\ln(z/z_0)$ in Eq. (31) of the order of unity. Thus the volume free energy must increase by at least an amount of order $Ga^2/4\pi(1 - \sigma)$ per unit length for each dislocation introduced.

The lowering of surface free energy which each dislocation produces is $|g - \gamma|\ a$ per unit length of the dislocation line. This is most easily verified by noting that the dislocation has the effect of stretching or compressing the surface without changing its area, and that by Eq. (30) the derivative of γ with respect to strain is essentially $(g - \gamma)$. Thus the introduction of dislocations at a depth sufficient to make the use of the term "dislocation" legitimate will result in a lowering of the total free energy, if and only if

$$|g - \gamma| > \frac{BGa}{4\pi(1 - \sigma)} \tag{32}$$

where B is a factor of the order of unity. With the numerical values used in the above example, *viz.*, $G = 5 \times 10^{11}$ dynes per sq cm, $\sigma = 1/3$, and $a = 2.5 \times 10^{-8}$ cm, the right of Eq. (32) is $1.5 \times 10^3 B$ dynes per cm. This is just of the order of magnitude which one expects for $|g - \gamma|$ for a perfect crystal; it thus seems likely that Eq. (32) will be satisfied in some cases, especially when g is negative (compressive), but not in others.

For a liquid, one should replace the right of Eq. (32) by zero, since there is no work required to produce the analogue of a dislocation in a liquid. This shows that, for this case, $g = \gamma$.

It has been mentioned above that the free energy of a dislocation array increases monotonically with the distance of the dislocations from the surface. This suggests that, when Eq. (32) is satisfied or comes near to being satisfied, the atomic arrangement of lowest free energy will be neither that of a perfect lattice nor that of a lattice with dislocations embedded several lattice spacings deep, but rather an arrangement with a somewhat disorderly surface which might be described, very loosely, as an arrangement with dislocations right at the surface, or even a little outside it. Figure 8–7 shows a conceivable surface of this type. It can be verified that for certain conceivable types of interatomic forces such

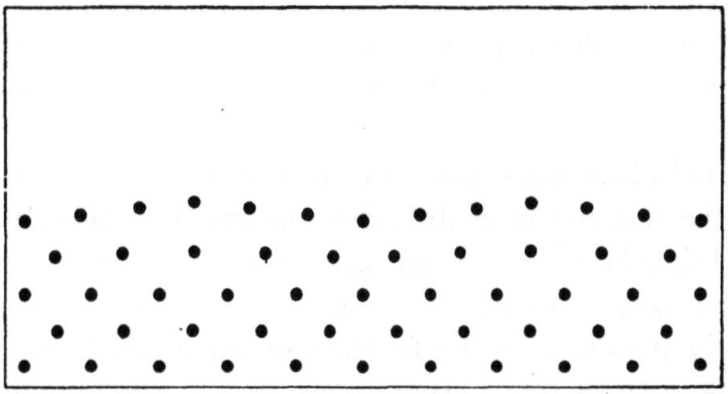

FIG. 8–7. Illustration of a possible "surface disorder" for the (11) surface of a plane quadratic lattice. The surface layer contains six-fifths as many atoms as the deeper layers, yet the nearest neighbor distances are only slightly disturbed.

surface arrangements do in fact give lower energies than a perfect lattice.* When one considers that most metal surfaces studied experimentally are contaminated with foreign atoms, it seems even more likely that in a certain fraction of cases a surface disorder of this sort is present.

The surface stress, as defined above, will be modified by the occurrence of any such surface disorder, presumably in such direction as to make $g_{\mu\nu}$ approximate more closely to $\gamma\delta_{\mu\nu}$ than in the absence of disorder. However, there is no reason to suppose that these two quantities will be exactly equal even in the presence of surface disorder.

We are now in a position to discuss the effect of surface stresses as a

*In this connection, it has been shown by I. N. Stranski and K. Molière, *Z. Phys.*, vol. 124, pp. 421, 429 (1948) that, when the ionic polarizabilities are sufficiently large, the energy of an ionic crystal can be lowered by bunching the surface ions into groups, *e.g.*, alternate strips of closer and wider spacings. This is doubtless an illustration of the phenomenon being discussed here, for the case $g > \gamma$.

motivation for creep or slip, *i.e.*, their ability to move the dislocations initially present in the crystal, and the possibility of their creating new dislocations. Now if a crystalline particle had an initially perfect lattice, its surface stresses would undoubtedly not be able to initiate any plastic flow, since to do so would require the performance of too large an amount of work to create each dislocation and separate it from the surface or from a dislocation of opposite sign. However, it is to be expected that real crystals will contain a certain number of dislocations initially, and that these can move gradually under very small shearing stresses,* or, if the stresses are sufficiently large and extensive, can reproduce their kind in such way as to give rise to a slip which will continue indefinitely as long as the stress is maintained.[11] Processes of the former type will certainly take place under the action of surface stresses, as long as a supply of mobile dislocations remains; processes of the latter type can in principle occur under the motivation of surface tension if the surface tension and surface stress are sufficiently large compared with the energy necessary to form new dislocations, a condition which may however often fail to be fulfilled.

If the original dislocations move out of the crystal without reproducing themselves, the number of mobile dislocations will constantly decrease, and there will thus be a limit to the amount of deformation of the specimen which can be produced by this mechanism, set by the number of dislocations originally present and the fraction of these which can be made to move. In many types of sintering processes, *e.g.*, the growth of the neck between two particles with dimensions of the order of a few microns, the total number of dislocations originally present will be insufficient, even if all of them are mobile, to produce any major change of shape if the dislocations work out of the crystal gradually without reproducing themselves. Moreover, even in experiments where the initial number of dislocations would be sufficient to produce the changes which are observed, such as the slight contraction of wires or foils under their own surface tension,[6,7,†] it is likely that only a small fraction of the dislocations will be able to move under the influence of the small stresses involved. That some such inhibitions to the notion of dislocations must exist is suggested by the fact that dislocations persist in samples subjected to prolonged annealing. It is interesting to note that, if the surface stress is compressive, a specimen might undergo a slight deformation in the direction *opposite* to that required to decrease its surface area, before creep ceases.

*For a discussion of the relation of dislocations to transient creep, see N. F. Mott and F. R. N. Nabarro, "Report of a Conference on the Strength of Solids," p. 1, Physical Society, London, 1948.

†The author has argued in reference 1, however, that the contraction observed in these experiments is due to diffusion within the small crystal grains, rather than to plastic flow.

The arguments just given suggest that the most likely way in which plastic flow can contribute to sintering changes, at least when major changes of shape are involved, is by processes in which the dislocations in working themselves out of the crystal produce new dislocations to take their place. This regeneration of dislocations must be an essentially dynamic process, in which the kinetic energy associated with motion of the original dislocation makes possible the surmounting of the potential energy barrier associated with the early stages of creation of the new dislocation.* As F. C. Frank[11] has shown, this can occur when a rapidly moving dislocation meets an obstruction in the interior, or when it meets the free surface.

In the former case, the old dislocation continues to exist after the new one has been born, and so the process ought to be energetically impossible unless the total work which the shearing stresses have done on the old dislocation exceeds the work required to create the new one. Now if the stress moving a dislocation arises solely from the presence of a free surface, it cannot be expected to do an amount of work on a dislocation of length L greater than something of the order of gLa. Although the stress itself may have very high values in small regions such as the neighborhoods of the edges of the cube of Fig. 8–5, these regions will always be so small that the energy gained by a dislocation in moving through one of them will not exceed the bound mentioned. The work required to create a new dislocation of length L is of the order of $Ga^2L \ln (z/z_0)/4\pi(1 - \sigma)$, where z is of the order of the distance of the new dislocation from the surface or from others of opposite sign. If $\ln(z/z_0)$ is of the order of 5 or 10, the latter work may well exceed the former [see the numerical values mentioned in connection with Eq. (32)], and regeneration of dislocations by this method will not occur.

Regeneration or "reflection" of dislocations at the surface of the crystal will be easier, since the self-energy of the old dislocation which is destroyed may be available to help in creating the new one. However, this process by itself cannot lead to an unlimited plastic deformation unless the inbound dislocation can travel all the way across the specimen to another surface without getting stuck en route. In many cases, *e.g.*, the formation of a neck between two particles, this can probably not occur.

Thus it is possible and even likely that plastic flow may often fail to occur to a significant degree under the motivation of surface-tension forces in small-particle systems, even when local stresses exist which exceed the macroscopic yield stress.

*Since this was written, it has been shown by F. C. Frank and W. T. Read, Jr., *Phys. Rev.*, vol. 79, p. 722, (1950), that regeneration of dislocations by quasistatic processes is possible and even probable. However, the arguments of this section apply just as well to this type of regeneration as to the dynamic type.

CONCLUDING REMARKS

The arguments presented in the preceding two sections make it seem likely that in many cases, perhaps most, of sintering of a pure substance in the absence of external stresses, the contribution of plastic flow within the crystals to the sintering is negligible. If the substance being sintered contains two or more constituents capable of diffusing, however, this simple state of affairs is not likely to persist, since in this case osmotic pressures will usually be set up which can easily exceed the yield stress of the material.* Now it seems likely that, when plastic flow does not occur, volume diffusion will often be the predominant mechanism of transport, vapor phase transport being ordinarily smaller because of the low vapor pressures of solids and surface migration being unimportant unless the ratio of surface to volume is greater than that usually encountered in sintering experiments. This expectation is borne out by the experiments of G. C. Kuczynski[3] on copper and silver. In such cases the course of the sintering process under ideal conditions, *i.e.*, when effects of such things as occluded gas or grain-boundary impurities are unimportant, can in principle be computed from the equations† of previous sections, provided the assumption of surface-volume equilibrium is satisfied and provided the diffusion coefficient and the surface and grain-boundary tensions are known. The diffusion coefficient is easily measured, and the interfacial tensions, though rarely measured to date, can be determined in various ways. (It must be remembered, however, that the latter may be sensitive to contamination.) When evaporation or surface migration is important, the rates of transport, though governed by the same chemical potential as volume diffusion, depend on quantities which are less easily determined, *viz.*, sticking coefficients or surface diffusion tensors.

From what has just been said, it is evident that before theoretical predictions on the course of sintering can confidently be made in the manner suggested, two questions must be answered in the affirmative: Does local surface-volume equilibrium exist under the conditions involved in any sintering experiment? Is the contribution from plastic flow negligible? The former question is closely related to the much discussed problem of the mechanism of crystal growth, and it will be worth while to make a few comments on this relationship before discussing the information on this question which can be derived from past and future experiments. The question regarding plastic flow has already been the subject of considerable theoretical speculation in the previous section, so only its decision by existing and prospective experiments need be discussed here.

*Plastic flow due to such osmotic pressures has been observed on a macroscopic scale by A. D. Smigelskas and E. O. Kirkendall, *Trans. AIME*, vol. 171, p. 130 (1947).

†Some of these equations have been given only for the case of cubic crystals, *i.e.*, the case where the diffusion coefficient is isotropic; however, the generalization to crystals of lower symmetry is easily made.

Local surface-volume equilibrium will obtain if lattice vacancies and interstitial atoms in the region just beneath the surface can migrate to and from the surface readily, so that if one were to cause the concentration of lattice defects beneath the surface to depart from its equilibrium value, even by a slight amount, the mean position of the surface would shift to restore the equilibrium and would do so in a time small compared with the time scale of the sintering process being considered. A necessary, and probably also sufficient, condition for this to be so is that the surface, whose exact configuration may of course be undergoing rapid thermal fluctuations, should spend at least an appreciable part of the time in configurations for which the change in free energy due to adding an atom to the surface and the change due to subtracting an atom are numerically very nearly equal. If this is not the case, the chemical potential ($\mu - \mu_h$) just beneath the surface, or the vapor pressure outside it, can vary over a finite range without any corresponding change in the surface configuration taking place in a reasonable length of time.

For an example, consider the case of an atomically smooth plane surface. An atom added to such a surface would have no neighbors in its own plane and would therefore be bound with an energy considerably less than μ, the free energy per atom of the bulk material, while removing an atom from such a plane would create a hole in the surface and require the expenditure of an energy considerably greater than μ. It was accordingly suggested long ago by J. W. Gibbs* and later by M. Volmer[12] and others that, in a supersaturated vapor, the growth of a crystal in directions normal to facets of this type takes place in jumps. The facet remains smooth for long periods at a time, until by chance isolated atoms deposited on the surface from the vapor acquire somewhere a sufficiently high density to form a two-dimensional nucleus big enough so that taking further atoms from the vapor phase and adding them to the nucleus will lower the overall free energy. After this, the deposition is rapid as long as the new layer is incomplete; when the new layer of atoms has been completed, a period of waiting for another nucleation process ensues. Numerical estimates[13] suggest that quite an appreciable supersaturation of the vapor would be required in such a case before any growth at all would take place on the time scale of ordinary experiments. By the same token, a fairly appreciable departure of $(\bar{\mu} - \bar{\mu}_h)$ from the value given by Eq. (18) would be required before the facet could grow or recede by diffusion.

This difficulty in establishing surface-volume equilibrium does not exist at a portion of a smoothly curved surface where the orientation is not that of one of the crystallographically simple planes, since on such a region of the surface there will be an abundance of sites at which an atom can be added or removed with a free energy change very nearly equal to μ. Thus

*Reference 8, pp. 324-325.

there seems less cause to worry about the validity of Eq. (15) than that of Eq. (18). Even for a facet which is macroscopically plane and of low indices, however, it seems quite likely that the two-dimensional nucleation process described above may be unnecessary, and that atoms can normally be added to or subtracted from the surface with a change in free energy differing only slightly from μ. One way in which this can be ensured has been suggested by F. C. Frank,[14] who points out that if any screw dislocations end on the facet in question, growth without nucleation can occur. Although the simplest form of this theory still seems to require a slight supersaturation before growth can occur, this supersaturation is far less than that required by the nucleation theory, and it is not unreasonable to believe that in some cases Eq. (18) should hold. Even in the absence of dislocations, however, surface-volume equilibrium on a plane facet may still be possible, *e.g.*, through the effect of foreign atoms adsorbed on the surface, or by virtue of the disordered arrangement of surface atoms which the considerations of a previous section suggest may sometimes be the most stable configuration for even a clean crystal surface.

Let us turn now to experiments bearing on the questions of surface-volume equilibrium and plastic flow. The results of G. C. Kuczynski[3] on the growth of the welded neck between a spherical particle and a flat plate are, as far as they go, quite encouraging on both counts, in that they agree with the predictions of a theory based on Eq. (15), with neglect of anisotrophy of the surface tension, and on the neglect of plastic flow. The experiments of H. Udin, A. J. Shaler, and J. Wulff,[6] and of B. H. Alexander, G. C. Kuczynski, and M. H. Dawson,[6] on the shrinking of polycrystalline wires under the influence of surface tension, are also encouraging, for the same reason.[1] However, none of these involved recognizable plane facets, and in the second case the achievement of surface-volume equilibrium may have been facilitated by the presence of grain boundaries.

If a repetition of the experiment of Udin, Shaler, and Wulff using single crystal wires should result in no observable shrinkage of the unloaded wires, the case against the occurrence of plastic flow motivated by surface tension would be greatly strengthened. However, the occurrence of shrinking in such an experiment would not necessarily imply that plastic flow is capable of contributing to major changes in shape when two particles are sintered together, since, as explained in the preceding section, the only process involved might be one of slow motion of dislocations without creation of new dislocations.

It might be possible to obtain fairly conclusive evidence on both questions by comparing the sintering times required, at a given temperature, to produce geometrically similar changes in two systems which are identical except for a difference in linear scale. It has been shown elsewhere[17] that, if only viscous flow contributes to the sintering, the time required goes as the first power of the linear dimension; if the assumption of surface-volume equilibrium is fulfilled, the time goes as the second power of the linear dimension if only evaporation and condensation contribute; as the third power if only volume diffusion is active; and as the fourth power if surface migration is the only transport mechanism. If creep or slip contributes, the behavior may be more complicated, since the relation of stress to strain rate will not usually be one of simple proportionality. If two-dimensional nucleation is required for growth or recession normal to a plane facet, such growth or dissolution will probably not take place at all, and if it does it will occur at a rate which varies with linear dimension in a way quite different from that assumed in deriving the scaling laws just mentioned. Thus, for example, if the sintering time required to develop a certain value of the ratio of neck radius to particle radius, in Kuczynski's experiment of sintering a sphere to a block, were found to be accurately proportional to the cube of the particle radius under conditions where volume diffusion was presumed to be the principal method of transport, it would probably be safe to conclude that surface-volume equilibrium prevailed and that there was no plastic flow. Similar conclusions might be drawn in other types of experiments and under conditions where other modes of transport predominate, since in a comparison of times in geometrically similar experiments many unknown geometrical and physical factors cancel out, including such things as sticking coefficients and surface diffusion tensors. Similarity of crystal orientations, etc., would of course be prerequisite to a completely unambiguous interpretation of the results of such experiments.

Since failure of surface-volume equilibrium is, as we have seen, more likely for plane surface facets than for smoothly curved regions, a direct experimental check on the rate of growth or recession along the normal to such a facet would be worth while. Such a check might be made by measuring the times required for an initially rounded single crystal particle, *e.g.*, a sphere, to develop plane facets of various sizes in the directions corresponding to surfaces of low surface tension. If these times obeyed the scaling law just described, one could conclude that surface-volume equilibrium prevailed. Moreover, if the process were one of volume diffusion, the absolute rates of growth of the facets could be compared with the predictions of an approximate calculation which can easily be made on the basis of Eq. (18).

Another, less fundamental, assumption which underlies the present theory of sintering by volume diffusion is that the diffusion coefficient is a predictable characteristic of the substance being used; *i.e.*, it is not highly structure-sensitive. While this assumption seems to agree with existing experimental data on diffusion, it is conceivable that dislocation lines might in some cases provide channels of easy diffusion, of sufficient importance to make the rate of self-diffusion depend significantly on the concentration of dislocations. The author knows of no systematic attempt to find out experimentally whether this effect exists.

It would be very interesting to know to what extent the surface tensions of simple crystalline substances depend on the crystallographic orientations of their surfaces, since it is much less easy to measure the surface tension of a single crystal facet than the average over a number of surfaces, *e.g.*, the average over the circumference of a wire, as in the experiment of Udin, Shaler, and Wulff.[6] One type of observation which may give information on this point is measurement of the angle at the groove formed where a grain boundary meets a free surface. Since the grain-boundary tension may be expected to be quite small compared with the surface tension, this angle, if it corresponds to an interfacial equilibrium, must be quite shallow unless the surface tensions depend significantly on orientation. R. Shuttleworth[15] has commented that observed grooves seem to be deeper than one expects (neglecting the orientational dependence of γ), and this suggests that the terms in $\partial\gamma/\partial t$ in Eq. (19) may be appreciable, and that they can be approximately evaluated by studies of such grooves.

APPENDIX 1. CHANGE OF SURFACE FREE ENERGY DUE TO ERECTING A HUMP

Referring to Fig. 8–1, if δz is the thickness of the hump over dS_0 and ψ the angle between the tangent to the new surface and the tangent to the old,

$$\delta\, dS = (\sec\psi - 1)dS_0 + \delta z\left(\frac{1}{R_1} + \frac{1}{R_2}\right) dS_0$$

where R_1 and R_2 are the principal radii of curvative at dS_0. Now $(\sec\psi - 1)$ is $O(\psi^2)$ and ψ is $O(\delta z_{\max}/\Delta y)$ where Δy is the radius of the hump in the tangential direction; for any given Δy the quantity $(\sec\psi - 1)$ is thus of the second order in the volume Δy of the hump and may be neglected. If Δy is small, R_1 and R_2 may be taken as constant over the area of the hump, and so to the first order

$$\int\gamma\,\delta\, dS = \gamma\left(\frac{1}{R_1} + \frac{1}{R_2}\right)\delta v \tag{33}$$

In the first term of Eq. (10) we have

$$\delta\gamma = \sum_{\mu}\frac{\partial\gamma}{\partial n_\mu}\,\delta n_\mu + O(\delta n^2) \tag{34}$$

where δn_μ is the μ component of the vector difference between the unit normal to the new surface and the unit normal to the old. The length of $\delta \mathbf{n}$ is thus $2\sin\frac{1}{2}\psi \backsim \psi$, and so by the argument of the preceding paragraph the $O(\delta n^2)$ can be neglected. It is necessary, however, to take account of the fact that both factors in the first term of Eq. (34) vary with position over the surface of the hump. Thus

$$\frac{\partial \gamma}{\partial n_\mu} = \left(\frac{\partial \gamma}{\partial n_\mu}\right)_P + \sum_\nu \left(\frac{\partial^2 \gamma}{\partial n_\mu \partial n_\nu}\right)_P \Delta n_\nu + O(\Delta n^2) \tag{35}$$

where the subscript P refers to some reference point near the center of the hump and where $\Delta\mathbf{n}$ is the vector difference between the normal to the *original* surface at the point to which Eq. (35) refers and the normal at the reference point. The term in $O(\Delta n^2)$ can be neglected if the radius Δy of the hump is $\ll$ the radii of curvature. Thus

$$\int \delta\gamma\, dS_0 = \sum_\mu \left(\frac{\partial \gamma}{\partial n_\mu}\right)_P \int \delta n_\mu\, dS_0 + \sum_{\mu,\nu} \left(\frac{\partial^2 \gamma}{\partial n_\mu \partial n_\nu}\right)_P \int \delta n_\mu\, \Delta n_\nu\, dS_0 \tag{36}$$

Now $\delta n_\mu = -(\partial/\partial x_\mu)\delta z$ where $\delta z(x, y)$ is the thickness of the hump. Inserting this in Eq. (36), the first term on the right vanishes, and the integral in the second is

$$-\int \frac{\partial\, \delta z}{\partial\, x_\mu} \Delta n_\nu\, dS_0 = \int \delta z \frac{\partial\, \Delta n_\nu}{\partial\, x_\mu} dS_0 \cong \left(\frac{\partial\, \Delta n}{\partial x_\mu}\right)_P \delta v$$

if Δy is small. If the x and y axes are chosen in the directions of the principal curvatures of the surface,

$$\frac{\partial \Delta n_x}{\partial x} = \frac{1}{R_1}, \qquad \frac{\partial \Delta n_y}{\partial y} = \frac{1}{R_2}, \qquad \frac{\partial \Delta n_y}{\partial x} = \frac{\partial \Delta n_x}{\partial y} = 0$$

and Eq. (36) becomes finally

$$\int \delta\gamma\, dS_0 = \left(\frac{\partial^2 \gamma}{\partial n_x^2} \cdot \frac{1}{R_1} + \frac{\partial^2 \gamma}{\partial n_y^2} \cdot \frac{1}{R_2}\right) \delta v \tag{37}$$

APPENDIX 2. EQUILIBRIUM AT AN INTERSECTION OF THREE INTERFACES

The full lines in Fig. 8–8 represent three interfaces intersecting in a line normal to the plane of the drawing at O. Consider the change in surface free energy accompanying an infinitesimal parallel displacement of this intersection line, say from O to P in the plane of interface 1. Suppose that in this displacement interface 2 acquires an angle at B and interface 3 at C, as shown by the dotted lines; choose BP and $CP \gg OP$ but still infinitesimal. Per unit length L normal to the paper, the first-order change of surface free energy is

$$\frac{\delta F_s}{L} = (\gamma_1 - \gamma_2 \cos\alpha_2 - \gamma_3 \cos\alpha_3) \cdot \overline{OP} + \overline{BP}\frac{\partial \gamma_2}{\partial \alpha_2}\delta\alpha_2 + \overline{CP}\frac{\partial \gamma_3}{\partial \alpha_3}\delta\alpha_3 \tag{38}$$

where $\delta\alpha_2$ is the difference between the supplement of angle BPO and α_2, and $\delta\alpha_3$ is similarly defined. In terms of the lengths

$$\delta\alpha_2 = \overline{OP}\,\frac{\sin\alpha_2}{\overline{BP}}, \qquad \delta\alpha_3 = \overline{OP}\,\frac{\sin\alpha_3}{\overline{CP}}$$

Inserting these in Eq. (38) and setting $\delta F_s = 0$ as the equilibrium condition gives

$$\gamma_1 - \gamma_2\cos\alpha_2 - \gamma_3\cos\alpha_3 + \sin\alpha_2\,\frac{\partial\gamma_2}{\partial\alpha_2} + \sin\alpha_3\,\frac{\partial\gamma_3}{\partial\alpha_3} = 0$$

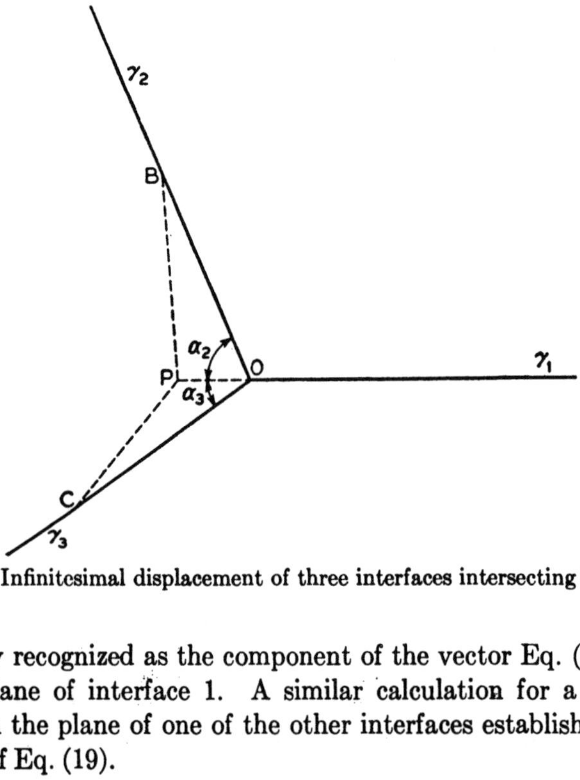

FIG. 8–8. Infinitesimal displacement of three interfaces intersecting in a line.

This is easily recognized as the component of the vector Eq. (19) resolved along the plane of interface 1. A similar calculation for a virtual displacement in the plane of one of the other interfaces establishes the other component of Eq. (19).

References

1. HERRING, CONYERS: *J. Applied Phys.*, vol. 21, p. 437 (1950).
2. GUGGENHEIM, E. A.: "Modern Thermodynamics by the Methods of Willard Gibbs," p. 172, Methuen & Co., Ltd., London, 1933.
3. KUCZYNSKI, G. C.: *J. Metals*, vol. 1, p. 169 (1949).
4. WULFF, G.: *Z. Krist.*, vol. 34, p. 449 (1901); LIEBMANN, H.: *Z. Krist.*, vol. 53, p. 171 (1914); v. LAUE, M.: *Z. Krist.*, vol. 105, p. 124 (1943).
5. NABARRO, F. R. N.: "Report of a Conference on Strength of Solids," p. 75, Physical Society, London, 1948.
6. UDIN, H., A. J. SHALER, and J. WULFF: *J. Metals*, vol. 1, p. 186 (1949); ALEXANDER, B. H., G. C. KUCZYNSKI, and M. H. DAWSON, Chap. 11 of this book.

7. CHAPMAN, J. C., and H. L. PORTER: *Proc. Roy. Soc. (London) A*, vol. 83, p. 65 (1909); SAWAI, I., and M. NISHIDA: *Z. allgem. Chem.*, vol. 190, p. 375 (1930); TAMMANN, G., and W. BOEHME: *Ann. phys.*, vol. 12, p. 820 (1932).
8. GIBBS, J. W.: "Collected Works I," p. 315, Longmans, Green & Co., Inc., New York, 1928.
9. LENNARD-JONES, J. E.: *Z. Krist.*, vol. 75, p. 215 (1930); OROWAN, E.: *Z. Krist*, vol. 79, p. 573 (1932); VOLMER, M.: *Die Physik*, vol. 1, p. 3 (1933).
10. KOEHLER, J. S.: *Phys. Rev.*, vol. 60, p. 397 (1941).
11. FRANK, F. C.: "Report of a Conference on Strength of Solids," p. 46, Physical Society, London, 1948.
12. VOLMER, M.: "Kinetik der Phasenbildung," Kap. 3C, T. Steinkopf, Leipzig, 1939.
13. CABRERA, N., and W. K. BURTON: *Discussions of the Faraday Society*, No. 5, p. 40 (1949).
14. FRANK, F. C.: *Discussions of the Faraday Society*, No. 5, p. 48 (1949).
15. SHUTTLEWORTH, R.: *Metallurgia*, vol. 38, p. 125 (1948).
16. HERRING, CONYERS: To be submitted to *Phys. Rev.*
17. HERRING, CONYERS: *J. Applied Phys.*, vol. 21, p. 301 (1950).

DISCUSSION

Summarized by Frederick Seitz, Technical Chairman

Chairman Frederick Seitz (Univ. Illinois) opened the discussion with the observation that there is considerable evidence that the degree of perfection and speed of growth are influenced by impurities as well and that some work has been done by Egli at the Naval Research Laboratory on the effect of lead in the growth of alkali halides, which indicates that more rapid growth is obtained with lead present. Thus there may be factors other than those considered by F. C. Frank which influence growth and determine places where nucleation occurs.

A. G. Smekal (University of Graz, Austria) then noted that it may be interesting to know what the magnitude of the stress would be for which the first plastic deformations take place. He mentioned experiments of this kind he had made on rock salt some time ago, where it was found that the first very well-defined sheets parallel to the rock-salt plates were observed at 50 grams per sq mm. He described another sensitive experiment made on the birefringence in which there was something lower than 20 grams per sq mm when the first irreversible changes in the optical effect appeared.

A. J. Shaler then pointed out that there are two items in the literature concerning that change of shape of initial spherical crystals. He mentioned one concerning rock salt carried out by Lukirski and another by Daniel and his associates on tungsten points carried out by M.I.T. in 1949.

Herring replied that he was familiar with this work and that Lukirski's experiments were made on the large specimens on which surface-tension effects would be very slight, and that he was not sure the considerations made here would be applicable. Herring added that, in regard to the field emission patterns, there had been quite a number of experiments reported in the literature not only by Daniel but also by Benjamin and Jenkins in England and by Mueller and Haefer in Germany. Herring remarked that these results had all been very puzzling. Benjamin and Jenkins had suggested that the rounded tungsten points of the order of 1 micron in radius acquired a polyhedral shape under the influence of surface tension. However, concluded Herring, the work of other investigators seems to show that this effect, if present, is not nearly so pronounced as one might expect.

Reprinted from JOURNAL OF APPLIED PHYSICS, Vol. 27, No. 8, 900–904, August, 1956

Two-Dimensional Motion of Idealized Grain Boundaries

W. W. MULLINS*
Westinghouse Research Laboratories, Pittsburgh, Pennsylvania
(Received March 2, 1956)

To represent ideal grain boundary motion in two dimensions, a rule of motion of plane curves is considered whereby any given point of a curve moves toward its center of curvature with a speed that is proportional to the curvature. A general theorem is deduced concerning the change of area enclosed by such a curve. Three families of curves are found that obey the curvature rule of motion while undergoing the shape preserving transformations of uniform magnification, translation, and rotation respectively. Pieces of these curves represent the steady shapes of idealized grain boundaries under certain symmetrical conditions.

I. INTRODUCTION

It has been shown by Beck[1] that the grain boundaries of a recrystallized metal, when annealed, migrate toward their centers of curvature. The concomitant reduction in the area of the boundaries, all having positive free energies when referred to an equivalent amount of crystal, provides the driving force for this motion. Thus, it is easily shown[2] that a boundary of mean curvature K and free energy per unit area σ is urged toward its nearest center of curvature with a pressure given by $p=K\sigma$. Such pressures and the motions they produce are now recognized to be responsible for normal grain growth.[3]

Smoluchowski[4] and Turnbull[5] have shown that a pressure p, of the type discussed, produces an unbalance in the fluxes of atoms crossing a boundary. This, in turn, causes the boundary to move in the direction favored by the pressure with a speed S given by $S=pM=K\sigma M$, where $M=Ae^{-q/kT}$ is the speed per unit pressure or mobility. In the ideal case of a pure metal whose boundaries do not undergo any progressive change in their structure or mode of motion, M and σ remain unchanged during isothermal annealing. Hence, the boundary speed at any point will be proportional to the mean curvature at that point. The observed boundary behavior departs from this ideal case because of the effects of impurities, strains, and possibly changes in the boundary density.[6] Nevertheless, the mathematical implications of the proportionality of speed to curvature provide a useful standard of comparison.

In the case of a soap froth, whose well-known analogy to a metal polycrystal has been recently discussed and extended by Smith,[3] the positive free energy of the films causes a pressure difference on the two sides of a curved film. This drives gas through the film causing it to move toward its center of curvature in a manner similar to that of grain boundaries. There is, however, this important difference between the two cases: within each cell of a soap froth, the possibility of a rapid mass flow of air maintains a uniform pressure which in turn causes each film to have a constant mean curvature; within a metal grain there is no possibility of a rapid mass flow and its associated uniformity of pressure so that the motion of any portion of a boundary is governed by local conditions only. Thus the problem of grain boundary motion, according to the curvature rule, is a problem in differential geometry. We will confine the discussion to the two-dimensional case of plane curves which correspond to grain boundaries in sheets.

II. THE CURVATURE RULE AND THE AREA THEOREM

Consider an arbitrary curve[7] given by $r(\theta,t)$ where r and θ are the polar coordinates and t is the time. In defining the curvature K, choose a definite sense of traversal along the curve and regard the tangent as directed in that sense. Referring to Fig. 1, we denote by s the arc length along the curve, by β the angle measured in a counterclockwise sense between the positive x axis and the directed tangent, and by ψ the angle measured in a counterclockwise sense between the polar radius vector and the directed tangent. Assume that any point of the curve moves toward its center of curvature with a speed S given by $S=kK$, where k is a constant equal to the product $M\sigma$. In order to derive the differential equation governing the motion of the curve, we consider the two successive positions which it assumes (shown in Fig. 1) at times t and $t+\Delta t$. We see from Fig. 1 that $\Delta r \sin\psi=-kK\Delta t$. If we divide by Δt, take the limit, and use the standard expressions from the theory of plane curves[8] $K=\partial\beta/\partial s$, $\sin\psi=r(\partial\theta/\partial s)$ we obtain

$$\frac{\partial r}{\partial t}=-k\frac{K}{\sin\psi}=-k\frac{1}{r}\frac{\partial\beta}{\partial\theta}. \tag{1}$$

* Research Engineer, Metallurgy Department, Westinghouse Research Laboratories, Beulah Road, Pittsburgh 35, Pennsylvania.

[1] P. A. Beck, in *Metal Interfaces* (American Society for Testing Materials, Cleveland, 1952), p. 208.

[2] N. K. Adam, *The Physics and Chemistry of Surfaces* (Oxford University Press, London, 1941), third edition.

[3] C. S. Smith, in *Metal Interfaces* (American Society for Testing Materials, Cleveland, 1952), p. 65.

[4] R. Smoluchoski, Phys. Rev. 83, 69 (1951).

[5] D. Turnbull, J. Metals 3, 661 (1951).

[6] W. W. Mullins (to be published).

[7] We assume that the curve satisfies the necessary mathematical requirements, such as continuity of curvature, in order to have consistent behavior according to the purposed rule of motion.

[8] R. Courant, *Differential and Integral Calculus* (Blackie and Son Ltd., London, 1934).

Note that the sign of $\partial\beta/\partial\theta$ is independent of the sense in which the curve is traversed. If we substitute into Eq. (1) the standard expressions for K and $\sin\psi$ in terms of the variables r, $r'=\partial r/\partial\theta$, and $r''=\partial^2 r/\partial\theta^2$ we obtain

$$\frac{\partial r}{\partial t}=-k\frac{r^2+2r'^2-rr''}{\pm r(r^2+r'^2)}, \tag{2}$$

where the sign of the denominator is the same as that of $\partial\beta/\partial\theta$. The solutions $r(\theta,t)$ of this nonlinear partial differential equation describe various one-parameter families of curves, each family being determined by a different initial curve $r(\theta,0)$. As the curve of a given family moves according to the curvature rule, it will successively coincide with all other curves of its family that have a greater value of the parameter t. We will discuss three such families of curves later on. Both for that purpose and for the discussion which follows, we will usually find Eq. (1) a more perspicuous form to use than Eq. (2). It is assumed that k is isotropic.

Let us consider a closed curve of arbitrary shape which moves according to Eq. (1). The area A, enclosed by the curve, is given by $A=\frac{1}{2}\oint r^2 d\theta$ where the integral is taken in a counterclockwise sense around the curve. Using Eq. (1), we obtain for the rate of change of the enclosed area

$$\frac{dA}{dt}=\oint\frac{\partial r}{\partial t}r d\theta=-k\oint\frac{\partial\beta}{\partial\theta}d\theta=-k\oint d\beta=-2\pi k. \tag{3}$$

Therefore the enclosed area decreases at a constant rate. This implies that two curves of different shape but enclosing equal areas will disappear at the same instant. Note that these results are true even if k, instead of being constant, is a function of β, (with a period of 2π).

In order to represent the motion of a two-dimensional array of grain boundaries, we consider a network of arbitrary curves dividing the plane into polygon-like cells (see Fig. 2). Let each vertex be the terminus of three such curves meeting at equal angles of $2\pi/3$ radians. If these curves move according to Eq. (1), all with the same constant k, one may arrive at the analogue of von Neumann's result for a two-dimensional soap froth[9] in which the rate of area loss or gain of a given cell is determined solely by the number of its sides.

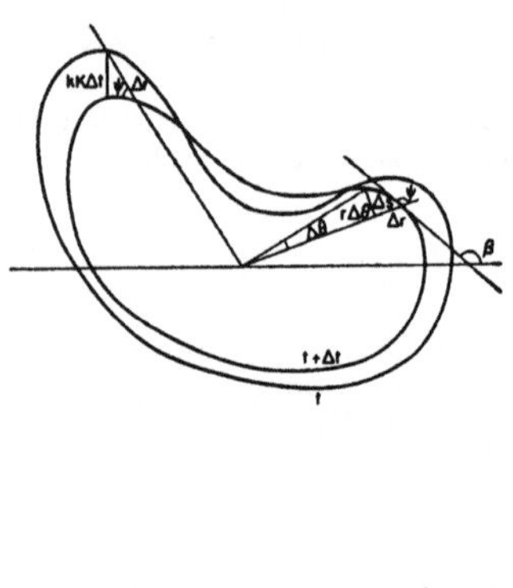

FIG. 1. Right section illustrates definitions; left section derivation of the curvature rule.

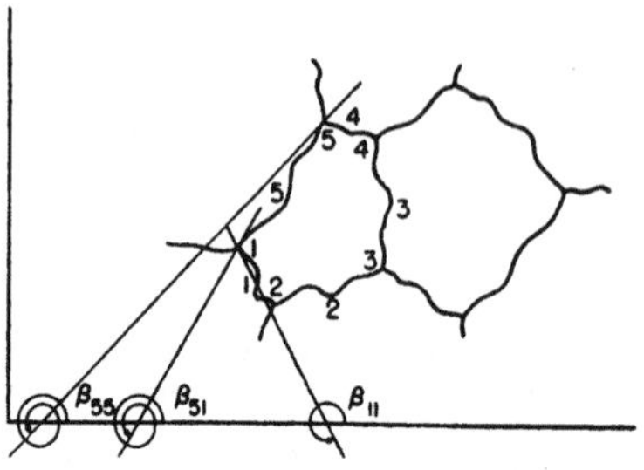

FIG. 2. A network moving according to the curvature rule.

Consider a cell with n sides labeled 1, 2, $3\cdots n$ in a counterclockwise order. Label each vertex with the same number as the side following it (see Fig. 2). Let β_{ij} be the angle between the x axis and the tangent to the ith side at the jth vertex. Then from Eq. (3), we find the rate of change of area of the cell to be

$$\begin{aligned}\frac{dA}{dt}&=-k\oint d\beta=k[(\beta_{11}-\beta_{12})+(\beta_{22}-\beta_{23})\\&\quad+(\beta_{33}-\beta_{34})+\cdots(\beta_{nn}-\beta_{n1})]\\&=k[(\beta_{22}-\beta_{12})+(\beta_{33}-\beta_{23})+\cdots(\beta_{11}-\beta_{n1})],\end{aligned}$$

where the angles are regrouped from pairs on the sides to pairs about the vertices. Since the included angle between intersecting curves at a vertex is $2\pi/3$, each of the first $n-1$ terms equals $\pi-(2\pi/3)=\pi/3$; and the last term equals $\pi/3-2\pi$ because of the revolution of the tangent in circumscribing the cell. Hence,

$$\frac{dA}{dt}=k[(n-1)\pi/3-2\pi+\pi/3]=k\frac{\pi}{3}(n-6). \tag{4}$$

Therefore, as in the soap froth case, a six-sided cell has a constant area, a seven-sided cell gains area as fast as a five-sided cell loses it, etc. The results concern only the continuous changes of boundary positions and do not describe such things as the disappearance of a cell and the resulting sudden change in the number of sides of the neighboring cells.

There is a possible question, in this proof, of the consistency of the curvature rule of motion with the rule of equal intersection angles at the vertices. Since, however, the tendency of the curves to shorten themselves underlies both rules, their mutual consistency seems plausible.

III. SPECIAL SOLUTIONS

Let us now look for some special solutions of Eq. (1), namely those functions describing curves which preserve their shape while obeying Eq. (1) (e.g., a

[9] J. von Neumann, in *Metal Interfaces* (American Society for Testing Materials, Cleveland, 1952), p. 108.

circle). The three types of transformations that preserve shape are uniform magnification, translation, and rotation. Therefore, we try to obtain three solutions, each of which is written in a form guaranteeing invariance under one of these transformations.

A. Invariance under Magnification

In order to find a curve obeying Eq. (1), that moves into a succession of configuration generated by magnifying the original curve from some center (i.e., the polar origin), we need a solution of the form $r(\theta,t) = R(\theta)T(t)$. We see from Eqs. (2) and (1) that $r(\partial r/\partial t)$ and therefore $\partial\beta/\partial\theta$ is homogeneous of degree zero in r and its angle derivatives. Therefore we obtain by substitution of $r=RT$ into Eq. (1) the expression

$$R\frac{dT}{dt}=-k\frac{1}{RT}\frac{d\beta}{d\theta},$$

where β is now a function of θ only. Multiplying through by T/R we separate the variables, obtaining

$$T\frac{dT}{dt}=-k\frac{1}{R^2}\frac{d\beta}{d\theta}=c, \tag{5}$$

where c is a constant.

The time solution of Eq. (5) is $T(t)=[T^2(0)+2ct]^{\frac{1}{2}}$, where the positive square root is taken because both polar radius vectors r and R must be positive. Evidently, for $c>0$ one finds curves which pull away from the origin of magnification ($R=0$), and for $c<0$ curves which shrink toward the origin and disappear at $t=-T^2(0)/2c$. The size of the curves depends parabolically upon the time.

Although a general analytic expression for the spatial solution to Eq. (5) cannot be obtained, one may plot the curves directly from the equation

$$\frac{d\beta}{d\theta}=-\frac{c}{k}R^2 \tag{6}$$

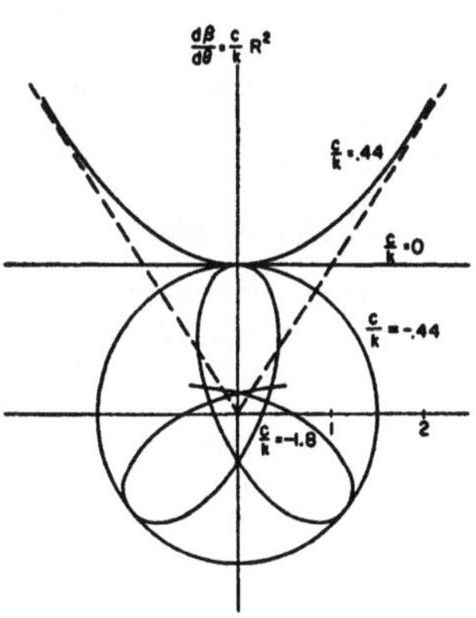

FIG. 3. Curves which undergo uniform magnification while obeying curvature rule.

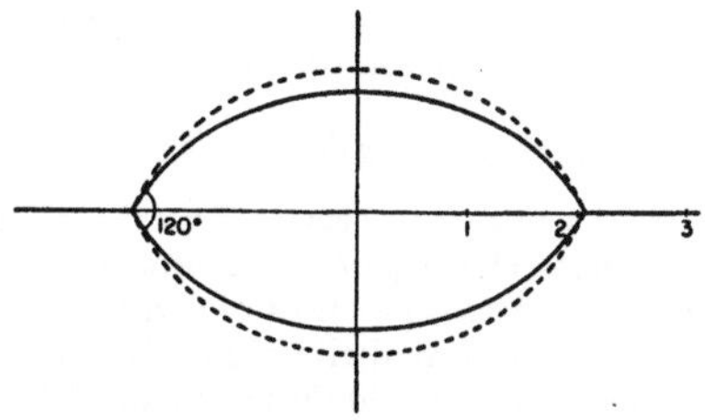

FIG. 4. Steady shape of two sided grain, plotted from $d\beta/d\theta = -0.36R^2$.

or the equivalent version

$$\beta_2-\beta_1=-\frac{c}{k}\int_{\theta_1}^{\theta_2} R^2 d\theta.$$

This latter form states that the tangent turns through angles proportional to the area subtended by the corresponding radius vectors.

Figure 3 illustrates the kind of curves obtained for different values of c/k. For convenience, all curves are shown passing with a zero slope through the fixed point (0, 3/2). When $c>0$, Eq. (6) gives a hyperbola-like curve having asymptotes that pass through the origin. The necessity for these asymptotes is shown by the following argument: a pair of rays from the origin that forms a V, symmetric about the y axis, subtends an increasing area of the curve as the angle of the V increases; unless the V forms a pair of asymptotes at some position, one has the contradiction, from the integral version of Eq. (6), that an unlimited subtended area cannot cause the two tangents to turn through the finite angles required to become parallel to the rays of the V. The hyperbola-like curves represent the shape of a boundary meeting the edge of a sample (y axis) at right angles and having an asymptote inclined to the edge. The time solution shows the manner in which the boundary would expand away from the point of intersection of the edge of the sample ($x=0$) with the asymptote.

When $c=0$, Eq. (6) gives a straight line ($\beta=$const) which, of course, is stationary according to the time solution.

The value of c required to give a circle of radius 3/2 (see Fig. 3) is given by $d\beta/d\theta=1=-c/k(3/2)^2$ or $c=-0.44k$. For other negative values of c, Eq. (6) gives rosette-like curves which are not necessarily closed. Since $d\beta/d\theta$ is independent of the direction of traversal of the curve, the rosettes are symmetrical about their maxima and minima. Appropriate pieces of these curves may be used to construct symmetrical grains. For example, Fig. 4 shows a two-sided grain composed of a rosette curve which is symmetrical about the y axis and meets the x axis at the appropriate 60° angle. The value $c/k=-0.36$ that gives the 60° angle of contact to within 1° was found by trial. For comparison, the dotted curve of Fig. 4 shows a circular arc with a 60° contact to the x axis. This arc would be the shape of a two-dimensional, two-sided soap cell.

B. Invariance under Translation

In order to find a curve obeying Eq. (1) that moves into a succession of configurations generated by translating the original curve, we need to rewrite Eq. (1) in Cartesian form. We may derive it directly from a geometrical argument as we did the polar form of Eq. (1). Instead let us convert Eq. (1) by allowing the polar origin to recede to infinity on the negative y axis of the Cartesian coordinate system. We have the limiting relations $\partial r/\partial t = \partial y/\partial t$ and $\beta + (2\pi - \psi) = \pi/2$ which implies $\sin\psi = -\cos\beta$. Therefore expressing K and $\cos\beta$ in terms of y' and y'', where the prime denotes partial differentiation with respect to x, we have from Eq. (1),

$$\frac{\partial y}{\partial t} = k\frac{y''}{(1+y'^2)^{3/2}}(1+y'^2)^{1/2}$$

or

$$\frac{\partial y}{\partial t} = k\frac{y''}{1+y'^2}. \tag{7}$$

Let us note that if a curve has everywhere a small slope, Eq. (7) becomes approximately $\partial y/\partial t = k(\partial^2 y)/(\partial x^2)$. Evidently displacements of small slope in the curve obey the same differential equation as that describing temperature changes due to heat conduction. Such displacements will therefore be propagated like a temperature pulse in a conductor. The same conclusion applies, of course, to the undulations in a curve whose B at various points, though possibly large, differs everywhere by only a small amount from that of a suitable straight line.

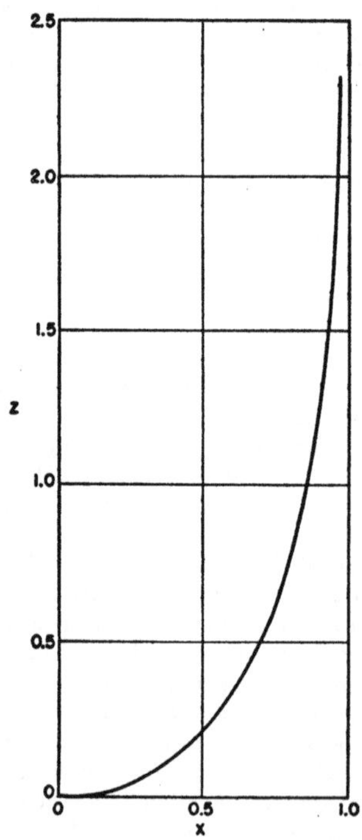

FIG. 5. Translating curve: $z(x) = -(2/\pi)\ln\cos(\pi/2)x$.

To obtain the translating solution of Eq. (7), we substitute $y(x,t) = Vt + z(x)$ where V is the constant speed. We obtain $V/k = z''/(1+z'^2)$. This is a case of Riccati's differential equation[10] and has the solution $z(x) = -(k/V)\ln\cos(V/k)(x+A)+B$. Denoting by b the x coordinate of the asymptote to the curve, we have $b = \pi k/2V$. Re-expressing the solution in terms of b, and choosing $A = B = 0$ to center the curve on the origin, we have $z(x) = -(2b/\pi)\ln\cos(\pi/2b)x$. Figure 5 shows this curve plotted for positive values of x in units of b. Note that the curve moves with a speed $V = \pi k/2b$ which is greater than the speed $V_1 = k/b$ of a quarter circle joined to the asymptote. The curve of Fig. 5 represents the shape of a perfectly mobile boundary of constant $k = M\sigma$ which is moving up the infinite edge to which it was originally parallel.

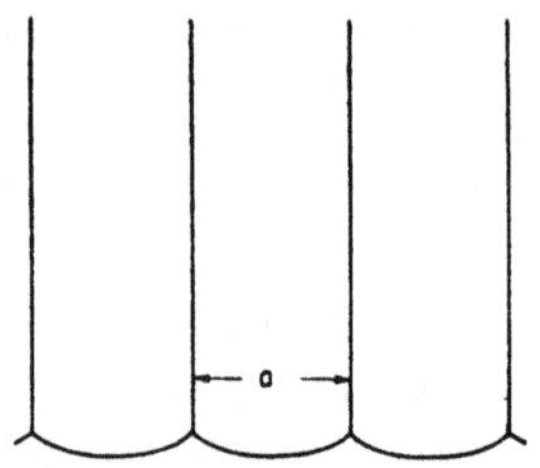

FIG. 6. Curves translating with a speed of $(\pi/3)(k/a)$.

Figure 6 shows a configuration of curves, similar to that sometimes seen in exaggerated grain growth, which translates as a unit. It is constructed by taking a parallel equal-spaced set of lines and capping each adjacent pair with the portion of the curve $z(x)$ for which $(-a/2) \leq x \leq (a/2)$ where a is given by $\tan^{-1}z'(a/2) = \pi/6$. This assures the 120° angles between intersecting curves. To form a stable configuration, the set of lines must either continue indefinitely in both directions along the x axis or they must terminate where the center of some cap meets a fixed vertical wall at right angles. The scalloped boundary then translates upward with a speed S easily shown to be given by $S = k\pi/3a$ where a is the distance between adjacent parallel lines.

C. Invariance under Rotation

For the sake of completeness we consider briefly a curve obeying Eq. (1), which moves into successive configurations obtained by rotating the original curve about the origin. We need a solution to Eq. (1) of the form $r(\theta - \omega t)$ where ω is the angular velocity. Since $\partial/\partial t = -\omega(\partial/\partial\theta)$, we have, by substitution into Eq. (1), $-\omega r(\partial r/\partial\theta) = -k(\partial\beta/\partial\theta)$. If we then divide by $\partial r/\partial\theta$

[10] E. L. Ince, *Ordinary Differential Equations* (Dover Publication, New York, 1926).

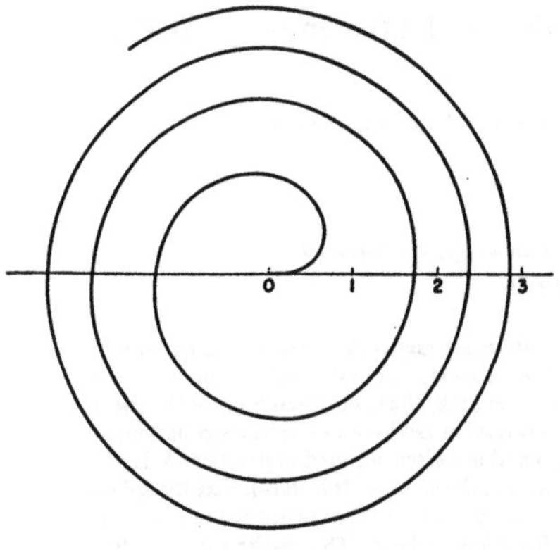

FIG. 7. Rotating curve plotted from $\beta=(\omega/2k)r^2+\omega t$, with $\omega/2k=2.5$ and $t=0$.

and regard β as a function of r and t, we obtain $\partial\beta/\partial r=(\omega/k)r$. Since, for a fixed point on the rotating curve, $\partial\beta/\partial t=\partial\theta/\partial t=\omega$, we obtain for a solution to the equation,

$$\beta=(\omega/2k)r^2+\omega t \tag{8}$$

where the origin of the time is chosen to eliminate the constant. Recalling that $\beta=\tan^{-1}(dy/dx)$, we see that, for a fixed time, Eq. (8) is a differential equation describing a field of line elements in the plane. Choosing the integral curve that passes through the origin, we plot half of it in Fig. 7 (the complete curve is symmetrical about the origin, its point of inflection). Starting from the origin with a slope of $\tan\omega t$, the curve winds around as an unbounded spiral whose turns are separated by approximately $2\pi k/\omega r$ for large r. Such a curve would represent the shape of a spiral dislocation for the idealized case in which it moved in an isotropic medium under the driving force of its curvature only.

IV. STABILITY

It is intuitively plausible that the shapes discussed in the foregoing are stable in the sense that small departures from the ideal shape and curvature will tend to disappear. For example, a small sharply curved bump in an otherwise smooth curve will tend to eliminate itself by moving rapidly toward its center of curvature. If the rule of motion were reversed in sense so that each point of a curve moved away from its center of curvature instead of toward it, a highly unstable condition would result in which small irregularities would become magnified. An analogy to this case would be the behavior of a "rubber" tube which lay in some arbitrary pattern on a flat surface and had a flow of water through it. Suppose the tube were of constant cross section but infinitely extensible with a constant tension. Any loop would be enlarged with a driving force proportional to its curvature by the centrifugal force of water flowing through it.

V. CONCLUSIONS

In order to represent idealized grain boundary motion in two dimensions we have considered the behavior of curves which move so that each point travels toward its center of curvature with a speed that is proportional to the curvature at that point. The geometrical consequences of the rule which have been developed concern the constant rate of change of area enclosed by such curves, and the steady shapes which the curves assume when moving in certain geometrically simple ways (i.e., when magnifying, translating, or rotating). These steady shapes are developed by the idealized boundaries when their initial shape and their constraining conditions are of certain simple types. It is hoped that these results will be helpful in the interpretation of some of the geometrical features of grain growth phenomena.

ACKNOWLEDGMENTS

The author wishes to express gratitude to many members of the Institute for the Study of Metals where part of this work was carried out. Particular thanks are extended to Professor C. S. Smith for many stimulating discussions. Appreciation is also expressed to the members of the Metallurgy Department and the Mathematics Department of the Westinghouse Research Laboratories for helpful discussions.

JOURNAL OF APPLIED PHYSICS VOLUME 34, NUMBER 2 FEBRUARY 1963

Morphological Stability of a Particle Growing by Diffusion or Heat Flow*

W. W. MULLINS
Metallurgy Department, Carnegie Institute of Technology, Pittsburgh, Pennsylvania

AND

R. F. SEKERKA
Physics Department, Harvard University, Cambridge, Massachusetts

(Received 7 May 1962)

The stability of the shape of a spherical particle undergoing diffusion-controlled growth into an initially uniformly supersaturated matrix is studied by supposing an expansion, into spherical harmonics, of an infinitesimal deviation of the particle from sphericity and then calculating the time dependence of the coefficients of the expansion. It is assumed that the pertinent concentration field obeys Laplace's equation, an assumption whose conditions of validity are discussed in detail and are often satisfied in practice. A dispersion law is found for the rate of change of the amplitude of the various harmonics. It is shown that the sphere is stable below and unstable above a certain radius R_c, which is just seven times the critical radius of nucleation theory; analogous conclusions are obtained for the solidification problem. The results for the sphere are used to discuss the stability of nonspherical growth forms.

INTRODUCTION

THE purpose of this paper is to study the stability of the shape of a phase boundary enclosing a particle whose growth during a phase transformation is regulated by the diffusion of material or the flow of heat. The principal approximations used are (1) the neglect of crystallographic factors such as elastic strain energy or anisotropy of interface properties, (2) the description of the thermal or diffusion fields by Laplace's equation (the frequently satisfied conditions for the latter approximation to be valid are discussed explicitly), and (3) the assumption of local equilibrium at each element of interface. The question of stability is studied by introducing a perturbation in the original interface shape and determining whether this perturbation will grow or decay. The method is fundamental in the sense that the velocity of each element of interface is calculated from the basic heat flow or diffusion boundary conditions; no *ad hoc* principles are employed.

In Sec. I, the approximations to be used in Sec. II and III are discussed. In Sec. II, a stability theory is developed for a growing spherical precipitate by supposing an expansion, into spherical harmonics, of an infinitesimal deviation of the particle from sphericity, and then calculating the time dependence of the coefficients of the expansion. The delineation of unstable from stable conditions is easily made on the basis of whether or not any harmonic component shows growth. It is found that the sphere is stable below and unstable above a radius R_c, which is just seven times the critical radius of nucleation theory; analogous results are obtained for a solid sphere growing from a supercooled melt by solidification. In Sec. III, the stability of nonspherical growth forms is briefly discussed by applying and extending the results of Sec. II.

I. DISCUSSION OF APPROXIMATIONS

We consider the diffusion-controlled growth of a nearly spherical precipitate particle from a supersaturated matrix initially of the uniform concentration c_∞. Let the fixed concentration of solute in the precipitate be C and the position dependent concentration of solute in the matrix be c. Denote the equilibrium value of c at a flat interface with the precipitate by c_0, and at a general interface (possibly curved) by c_s.

The diffusion-controlled growth of the precipitate is characterized by the velocity v with which each element of its interface moves according to the equation[1]

$$v=[D/(C-c_s)](\partial c/\partial n), \quad (1)$$

where D is the diffusion coefficient of solute in the matrix and $\partial c/\partial n$ is the normal derivative of the matrix concentration field at the interface. But the concentration field itself is affected by the position and velocity of the interface because of the boundary conditions that must be satisfied there. Therefore, in principle, the time-dependent diffusion problem and the interface motion problem must be solved simultaneously.

To obtain a tractable problem, we restrict our attention to the case in which[2]

$$|(c_\infty-c_s)/(C-c_s)|\lesssim|(c_\infty-c_0)/(C-c_0)|\equiv S_D\ll 1. \quad (2)$$

It is well known that this condition guarantees that the concentration field at a given instant near the growing precipitate, and particularly the gradient on the surface, have nearly the same values as those obtained by solving Laplace's equation while holding fixed the form assumed by the precipitate at the given instant. Indeed, omitting capillarity for the moment, the value $(c_\infty-c_0)/R$ so obtained for the gradient on the surface of a sphere of radius R is precisely the expression found from the rigorous time dependent solution for a growing

---

* This work was supported in part by the U. S. Army Office of Ordnance Research, and in part by a Fulbright and Guggenheim grant to the first author.

[1] Diffusion within the interface is ignored.

[2] The first inequality follows from the condition $|c_0-c_s|\ll|C-c_0|$, which is virtually always fulfilled.

323

sphere[3,4] when this solution is reduced to an approximate form by using condition 2. The physical reason that condition (2) permits this simplification is most easily stated in terms of the sphere problem but also clearly applies to cases of precipitate shapes departing somewhat from sphericity or to cases in which the concentration is not exactly uniform over the surface of the precipitate (e.g. due to capillarity); the quantity Q of solute that results from draining a spherical region of volume $\sim(\frac{4}{3})\pi R^3$ from a concentration c_∞ to that of c_s (certainly a lower limit to the steady state values) must increase the radius R of the precipitate sphere by no more than a small fraction. It follows that

$$Q\simeq(\tfrac{4}{3})\pi R^3(c_\infty-c_s)\simeq 4\pi R^3(dR/R)(C-c_s),$$

where $dR/R\gg 1$; thus $C-c_s\gg c_\infty-c_s$ in accord with condition 2. Rearranging Eq. (2) gives an equivalent condition in terms of the supersaturation, namely, $|(c_\infty-c_0)/c_0|\ll|(C-c_0)/c_0|$.

II. FIRST-ORDER STABILITY THEORY FOR A SPHERICAL PARTICLE UNDERGOING DIFFUSION-CONTROLLED OR THERMALLY-CONTROLLED GROWTH

A. Diffusion Case

Consider a nearly spherical precipitate particle growing in an originally uniformly supersaturated solution (c_∞) under the conditions of Sec. I which permit Laplace's equation to be used to calculate the diffusion field near the precipitate particle. We first investigate the behavior of an infinitesimal distortion of shape caused by a single spherical harmonic[5] Y_{lm}; the behavior of an arbitrary infinitesimal distortion may then be obtained by superposition since all equations are linear. We take the equation of the distorted sphere to be

$$r=\rho(\theta,\varphi)=R+\delta Y_{lm}(\theta,\varphi), \tag{3}$$

where δ is very small, so that powers higher than the first may be neglected (ρ, R and δ depend on the time). If we assume that dilute solution theory holds, the equilibrium concentration $c_s(\theta,\varphi)$ on the surface $\rho(\theta,\varphi)$ is determined by the capillarity condition

$$c_s=c_0+c_0\Gamma_D K, \tag{4}$$

where K is the mean curvature,[6] and $\Gamma_D=(\gamma\Omega)/(RT)$ is a capillary constant in which γ is the interfacial free energy, Ω the increment of precipitate volume per mole of added solute, R the gas constant, and T the absolute temperature; typically $\Gamma_D\approx 10^{-7}$ cm. By utilizing the fact that the curvature K of a surface z (x,y) which deviates only slightly from flatness (i.e., z=const.) may be calculated from the expression $K=-\nabla^2 z$, it is easily shown that the curvature $K(\theta,\varphi)$ at all points of the slightly distorted sphere $r(\theta,\varphi)=R+\delta g(\theta,\varphi)$ is given by the expression

$$K(\theta,\varphi)=(2/R)[1-(\delta g/R)]-(\delta\Lambda g/R^2), \tag{5}$$

where Λ is the angular part of the Laplacian operator expressed in spherical coordinates, i.e.,

$$\Lambda=\frac{1}{\sin\theta}\frac{\partial}{\partial\theta}\left(\sin\theta\frac{\partial}{\partial\theta}\right)+\frac{1}{\sin^2\theta}\frac{\partial^2}{\partial\varphi^2}.$$

Substituting Y_{lm} for g in Eq. (5), utilizing the well known fact that $\Lambda Y_{lm}=-l(l+1)Y_{lm}$, and then substituting the resulting expression for K into Eq. (4), we obtain for the equilibrium concentration $c_s(\theta,\varphi)$ on the distorted sphere the expression

$$c_s(\theta,\varphi)=c_0[1+(2\Gamma_D/R)+(l+2)(l-1)(\Gamma_D\delta Y_{lm}/R^2)]. \tag{6}$$

We seek a solution of Laplace's equation which reduces to Eq. (6) on the surface given by Eq. (3). It is sufficient to consider the expression

$$c(r,\theta,\varphi)=A/r+B\delta Y_{lm}/r^{l+1}+c_\infty,$$

which satisfies Laplace's equation and which reduces to the proper value c_∞ at $r=\infty$. The value of this expression on the surface $\rho(\theta,\varphi)$ [Eq. (3)] to first order in δ is

$$(A/R)(1-\delta Y_{lm}/R)+B\delta Y_{lm}/R^{l+1}+c_\infty.$$

Since this expression must equal that given by Eq. (6), and since the expansion in harmonics is unique, A and B are easily determined by equating coefficients of like harmonics. The resulting solution $c(r,\theta,\varphi)$ of Laplace's equation which obeys all boundary conditions is[7]

$$c(r,\theta,\varphi)=\frac{(c_0-c_\infty)R+2c_0\Gamma_D}{r}+\frac{\{(c_0-c_\infty)R^l+c_0\Gamma_D R^{l-1}l(l+1)\}\delta Y_{lm}}{r^{l+1}}+c_\infty. \tag{7}$$

Since the deviation of the tangent plane of the precipitate particle from that of the original sphere is infinitesimal at all points, it suffices to take the radial derivative to find the velocity $v=(dR/dt)+(d\delta/dt)Y_{lm}$ [time

[3] H. S. Carslaw and J. C. Yaeger *Conduction of Heat in Solids* (Oxford University Press, New York, 1959).

[4] The solution given in reference 3 is stated in terms of the solidification problem. The conversion to the present problem is easily made following the correspondence given in Sec. IIB.

[5] By Y_{lm} we denote a member of the real and complete orthonormal set of eigenfunctions of the angular part of the Laplacian operator.

[6] $K\equiv(1/R_1)+(1/R_2)$, where R_1 and R_2 are the signed principal radii of curvature. The sign of K is such that $K=2/R>0$ on the growing sphere of radius R.

[7] With $\Gamma_D=0$, Eq. (7) represents the electrostatic potential around a nearly spherical conductor whose shape is given by Eq. (3) and whose potential is c_0-c_∞ above that of infinity.

derivative of Eq. (3)]. We obtain

$$v=\frac{dR}{dt}+\frac{d\delta}{dt}Y_{lm}=\frac{D}{C-c_s}\left(\frac{\partial c}{\partial r}\right)_{r\to\rho},$$

$$=\frac{D}{C-c_s}\left\{\frac{c_\infty-c_R}{R}+\left[(l-1)\frac{c_\infty-c_0}{R^2}-\frac{c_0\Gamma_D}{R^3}[l(l+1)^2-4]\right]\delta Y_{lm}\right\},\quad (8)$$

where $c_R=c_0[1+(2\Gamma_D/R)]$ is the concentration on the undistorted sphere. Equating coefficients of Y_{lm}, we obtain[8] for the rate of growth $\dot{\delta}=d\delta/dt$ of the amplitude of the spherical harmonic the expression

$$\dot{\delta}_l=\frac{c_0D(l-1)}{(C-c_R)R^2}\left[\frac{c_\infty-c_0}{c_0}-\frac{\Gamma_D}{R}[(l+1)(l+2)+2]\right]\delta_l,$$

$$=\frac{D(l-1)}{(C-c_R)R}\left[G-\frac{c_0\Gamma_D}{R^2}(l+1)(l+2)\right]\delta_l,\quad (9)$$

where $G=(c_\infty-c_R)/R$ is the normal concentration gradient at the surface of the undistorted sphere; the amplitude has been labeled with the subscript l to indicate the harmonic to which it corresponds.

Equation (9) shows that $\dot{\delta}_l$ is composed of two terms; a positive term proportional to G, which represents a gradient effect favoring growth of the harmonic, and a negative term proportional to Γ_D, which represents a capillary effect favoring decay of the harmonic. The pure gradient effect is realized when γ (and hence Γ_D) vanishes so that the surface assumes a uniform concentration [Eq. (4)]. The isoconcentrates are then bunched together above the protuberances and are rarified above the depressions of the perturbation. The corresponding focusing of diffusion flux away from the depressions onto the protuberances increases the amplitude of the perturbation; we may view the process as an incipience of the so-called point effect of diffusion.

The pure capillary effect is realized when $G=0$. Because the concentration is higher on the protuberances than in the depressions [Eq. (4)], there is a flow of material from the former to the latter which reduces the amplitude of the perturbation; the process is driven by the reduction in surface area (and hence energy) that accompanies the approach to sphericity of the particle.

In general both gradient and capillary effects are present and the question of stability reduces to the study of which effect dominates. Evidently, all harmonics for which the bracket of Eq. (9) is positive must grow; they correspond to values of l which satisfy the inequality

$$(l+1)(l+2)+2<R(c_\infty-c_0)/\Gamma_D c_0.\quad (10)$$

All higher harmonics for which the inequality is reversed must decay. (A given value of l corresponds to an average wavelength or distance between nodes of the harmonic of $\lambda\simeq 2\pi R/l$.)

Again it follows from Eq. (9), that a harmonic of a given l will grow or will decay according, respectively, to whether the radius of the sphere is greater than or is less than the critical value

$$R_c(l)=[(1/2)(l+1)(l+2)+1]R^*,\quad (11)$$

where $R^*\equiv 2\Gamma_D/[(c_\infty-c_0)/c_0]$. The quantity R^* is the critical nucleation radius (corresponding to $c_R=c_\infty$) above which the sphere itself grows and below which it shrinks.

An arbitrary infinitesimal perturbation may be resolved into a harmonic spectrum by standard methods. Since all of the equations we have considered are linear, it is clear that the time derivative of the coefficient of each harmonic component in the spectrum independently obeys Eq. (9). The evolution of the perturbation, as long as it remains sufficiently small, is then determined by integration. As long as any harmonics show growth, the sphere is not truly stable as the corresponding components of an arbitrary perturbation will grow. The condition under which at least the second harmonic ($l=2$) grows[9] is that R exceed the critical value $R_c\equiv R_c(2)$ given from Eq. (11) by

$$R_c=7R^*=14\Gamma_D/[(c_\infty-c_0)/c_0]$$
$$\simeq 1.4\times 10^{-6}\text{ cm}/[(c_\infty-c_0)/c_0],\quad (12)$$

where typical values are used in the last step. We conclude that the sphere becomes unstable at a radius of only seven times that which it must have to grow.[10] Figure 1 shows a plot of R_c as a function of $(c_\infty-c_0)/c_0$; for a ten percent supersaturation, spheres bigger than $\sim 0.14\,\mu$ are unstable.

Although any harmonic obeying inequality (10) will grow, the growth may be very slow unless the radius R differs appreciably from the critical value $R_c(l)$ for which $\dot{\delta}_l=0$. It is easy to show from Eqs. (8) and (9) that

$$\frac{\dot{\delta}_l/\delta_l}{\dot{R}/R}=(l-1)\left[1-\frac{R_c(l)}{R}\right]\left[1-\frac{R^*}{R}\right]^{-1}.\quad (13)$$

Thus if R is several times the critical value $R_c(l)$ for a given harmonic the fractional rate of increase of the amplitude of the harmonic is $l-1$ times that of the radius of the sphere.

Equation (13) shows that the largest value of $(\dot{\delta}_2/\delta_2)/$

[8] The first-order term in δ which results from the product of the expansion of $(C-c_s)^{-1}$ with the term $(c_\infty-c_R)/R$ of Eq. (8) is easily shown to be $\sim(c_\infty-c_R)/(C-c_R)$ times the second (capillary) term in square brackets; it is therefore neglected.

[9] To the first order in δ, the first harmonic ($l=1$) merely translates the sphere a distance δ as reflected by the factor $(l-1)$ in Eq. (9); consequently, we are led to investigate the case $l=2$.

[10] Very similar conclusions have been reached independently by Professor M. Hillert of Institutionen for Metallografi, Stockholm, Sweden (private communication).

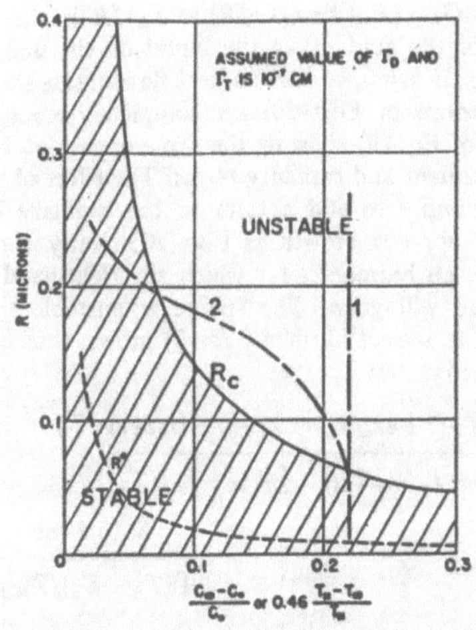

FIG. 1. Solid curve delineates stability of a sphere as a function of its radius and the supersaturation: dashed curve gives critical nucleation radius; broken curves represent two possible growth trajectories of a sphere.

$(\dot{R}/R)$ for the second harmonic ($l=2$), attained when $R \gg R_c$ (capillarity unimportant) is unity. This means that if capillarity is neglected a sphere perturbed by the second harmonic only does not change shape as it grows, a result in accord with the shape-preserving ellipsoidal solutions found by Ham[11] since a sphere perturbed by δY_{20} is, to the first order in δ, an ellipsoid of eccentricity $2\delta/R$. The more severe instabilities characterized by an increase not only of the amplitude δ but also of the ratio of the maximum to minimum polar radii of the harmonic perturbation occur for particle radii exceeding a bound somewhat larger than R_c; the required condition is evidently that the radius of the sphere exceed the minimum value for which Eq. (13) gives $(\dot{\delta}_l/\delta_l)/(\dot{R}/R)>1$, namely, that

$$R>(21/11)R_c(3)=3R_c=21R^*.$$

These conditions are supposed to apply in Fig. 2 which shows a sphere perturbed by the harmonic Y_{30} whose amplitude is progressively increasing.

When some harmonics display growth, Eq. (9) shows that there will be a value l_M of l for which the rate of growth is a maximum. When l_M is large (marked instability), it may be approximated by differentiating the coefficient of δ with respect to l; the result expressed in terms of the average wavelength λ_M of the harmonic is

$$\lambda_M \equiv (2\pi R)/l_M = 2\pi\{3\Gamma_D R/[(c_\infty - c_0)/c_0]\}^{\frac{1}{2}} = \pi(6RR^*)^{\frac{1}{2}}. \quad (14)$$

Since the components corresponding to l_M of an arbitrary local perturbation grow most rapidly, there will be an increasing relative proportion of these components in the developing perturbation which will therefore tend to produce a dimpled condition in the spherical surface on a scale $\sim\lambda_M$ given by Eq. (14).

[11] F. S. Ham, Quart. Appl. Math. 17, 137 (1959).

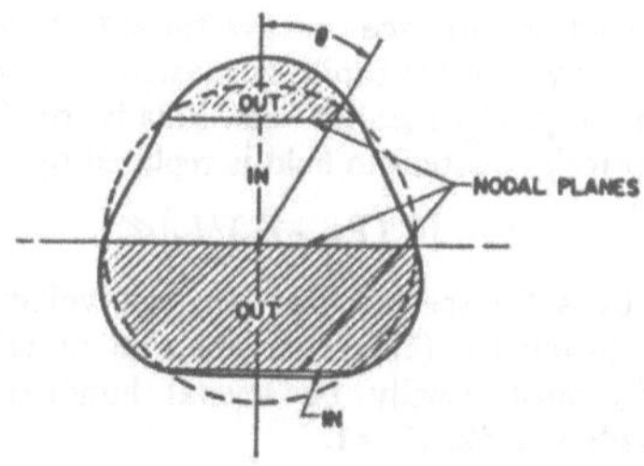

FIG. 2. A sphere perturbed by the harmonic Y_{30}, whose amplitude is progressively increasing. The figure is cylindrically symmetric about the polar axis.

At any particular moment the concentration at infinity c_∞ can not, of course, affect the stability of the interface. It enters Eqs. (8)–(14) only because it determines the local field in the vicinity of the sphere for the particular problem we have considered; for example, c_∞ determines the gradient G through the relation $G=\{c_\infty - c_0[1+(2\Gamma_D/R)]\}/R$. In fact, the latter relation may be used to eliminate c_∞ from Eqs. (8)–(14) as it has been eliminated in the second version of Eq. (9). The significance of the equations in this form is the following: if the field in the vicinity of the sphere (within a distance R or so) is given by $GR(1-R/r)+c_R$, as it is in the particular problem that led to Eqs. (8)–(14), then regardless of the values it assumes beyond, the analysis is momentarily valid and Eqs. (8)–(14) correctly give the corresponding information concerning stability and the values of $\dot{\delta}$.

B. Solidification Case

The problem of the stability of a solid sphere growing in an originally uniformly supercooled melt differs formally from the preceding problem only in that heat flow can occur inside as well as outside the sphere; by contrast, diffusion was restricted to the outside of the growing precipitate. The correspondence between the two problems is well known[8]: thus,[12]

$$v=(-1/L_v)[k_s(\partial T/\partial n)_s + k_L(\partial T/\partial n)_L] \quad (15)$$

replaces Eq. (1), where L_v is the latent heat of freezing per unit volume, k_s and $(\partial T/\partial n)_s$ are the thermal conductivity of the solid and the temperature derivative at the interface along a normal pointing toward the solid, and k_L and $(\partial T/\partial n)_L$ are similar quantities for the liquid. Also

$$T=T_M - T_M\Gamma_T K \quad (16)$$

replaces Eq. (4), where the curvature K is reckoned

[12] We assume the densities of liquid and solid to be the same.

positive for an interface concave toward the solid, and where $\Gamma_T=\gamma/L_v$ is the capillary constant for the present problem; typically $\Gamma_T\simeq10^{-7}$ cm. Finally, condition (2) for Laplace's equation to hold is replaced by

$$S_T=|C_v(T_M-T_\infty)/L_v|\ll1, \tag{17}$$

where C_v is the specific heat per unit volume of the liquid. Inequality (17) often holds in practice since T_M-T_∞ must usually be several hundred degrees centigrade to make $S_T=1$.

We consider again a sphere distorted by a single harmonic described by Eq. (3). Assuming inequality (17) to hold, the Laplacian analysis of heat flow outside the sphere follows exactly the same lines as that for diffusion; the resulting temperature T which obeys Laplace's equation outside the sphere and reduces, to first order in δ, to the correct boundary values on the surface of the distorted sphere is

$$T_1(r,\theta,\varphi)=\frac{(T_M-T_\infty)R-2T_M\Gamma_T}{r}+\frac{\{(T_M-T_\infty)R^l-T_M\Gamma_TR^{l-1}(l+1)\}\delta Y_{lm}}{r^{l+1}}+T_\infty. \tag{18}$$

For the temperature distribution T_2 inside the sphere, it suffices to take the form $A'+B'r^l\delta Y_{lm}$, which satisfies Laplace's equation and which, on the surface ρ becomes, to first order in δ, $A'+B'R^l\delta Y_{lm}$. If, as before, the latter expression is equated to the boundary values that T must assume on the surface [(determined by Eqs. (3), (5), and (16)], the constants A' and B' may be calculated. The resulting temperature distribution within the sphere that satisfies all conditions is

$$T_2(r,\theta,\varphi)=T_M\left(1-\frac{2\Gamma_T}{R}\right)-\frac{T_M(l+2)(l-1)\Gamma_Tr^l}{R^{l+2}}\delta Y_{lm}. \tag{19}$$

By substituting Eqs. (18) and (19) into Eq. (15) in which the normal derivative is replaced by the radial derivative evaluated on $r=\rho$ (Eq. (3)), the velocity v of each element of interface may be calculated. From v, $\dot{\delta}$ is obtained by the same method used in the preceding calculation; the final result, corresponding to Eq. (9) is

$$\dot{\delta}_l=\frac{(l-1)T_Mk_L}{R^2L_v}\left\{\frac{T_M-T_\infty}{T_M}-\frac{\Gamma_T}{R}\times\left[(l+1)(l+2)+2+l(l+2)\frac{k_s}{k_L}\right]\right\}\delta_l$$
$$=\frac{(l-1)k_L}{RL_v}\left\{G_L-\frac{\Gamma_TT_M}{R^2}\times\left[(l+1)(l+2)+l(l+2)\frac{k_s}{k_L}\right]\right\}\delta_l, \tag{20}$$

where $G_L=\{T_M[1-(2\Gamma_T/R)]-T_\infty\}/R$ is the magnitude of the gradient in the liquid on the undistorted sphere. If $k_s=0$, so that no heat flows inside the sphere both forms of Eq. (20) are completely analogous to those of Eq. (9) showing the same opposition between the gradient and capillary terms. The effect of nonzero k_s is simply to add a term to the capillary bracket which opposes growth as Eq. (20) clearly shows. As before, all harmonics for which the right-hand side is positive will grow. The sphere is unstable when at least the second harmonic ($l=2$) grows, that is, when R exceeds

$$R_c=\frac{2\Gamma_T(7+4k_s/k_L)}{S_T\cdot L_v/(C_vT_M)}=\frac{2\Gamma_T(7+4k_s/k_L)}{[(T_M-T_\infty)/T_M]}\simeq\frac{3\cdot10^{-6}\text{ cm}}{[(T_M-T_\infty)/T_M]}, \tag{21}$$

where typical values for a metal are used in the last step and where we must require $S_T\ll1$ for the analysis to be valid. By substituting $0.47(T_M-T_\infty)/T_M$ for $(c_\infty-c_0)/c_0$ in Fig. 1, we make it serve as a plot of the critical value of R in the present case as given by Eq. (21). When the fractional supercooling is one tenth, all spheres larger than $\sim0.30\,\mu$ in radius are unstable. Again, for a really prominent instability, the sphere must be an order of magnitude or so bigger than the value given by Eq. (21). Other points made for the diffusion case are easily extended in a straightforward way to the present case.

III. SMALL-SCALE STABILITY OF NONSPHERICAL GROWTH FORMS: PLANAR APPROXIMATION

For concreteness we consider diffusion-controlled growth; the extension to thermally controlled growth is straightforward paralleling the extension of Sec. II B.

Suppose the surface of a nonspherical form is partitioned into zones (as few as possible) so that within any given zone Q, the gradient G, and the principal radii of curvature R_1 and R_2 are approximately uniform. We consider the stability of the zone Q with respect to undulations of wavelength small compared with the lesser of R_1, R_2, and the length $L(n)$ of Q along the wave normal n (parallel to the surface). In other words, the discussion is limited to waves in Q that may be considered as lying on an infinite plane on which the initial gradient is G; they will be referred to as small-scale waves and the stability which their behavior determines as the small-scale stability of Q. Analysis of stabilities on a larger scale can only be made with respect to a complete unperturbed form specified as a function of time (e.g., a growing sphere).

To obtain the necessary formulas, consider the limiting form taken by the second version of Eq. (9) for small-scale rippling, that is, for large $l=2\pi R/\lambda$, where λ is the scale of the harmonic (e.g., $\lambda=2\pi R/l$ is the equatorial wavelength of the harmonic Y_{ll}); using the notation $\omega=2\pi/\lambda=l/R$ we obtain

$$\dot{\delta}=[D\omega/(C-c_R)][G-c_0\Gamma_D\omega^2]\delta. \quad (22)$$

This shows that the essential behavior of a small-scale undulation depends only on the original gradient G at the unperturbed surface and on the frequency ω of the undulation; the characteristics of the sphere (e.g., R) do not appear explicitly in Eq. (22). Indeed, since small-scale undulations are essentially equivalent to undulations on a plane, Eq. (22) may be derived by applying the methods of Sec. II to calculate $\dot{\delta}$ for the sinusoidal undulation $\delta \sin\omega x$ introduced into an infinite plane on which the field originally had the gradient G (and the value c_R).[13] Evidently then, Eq. (22) may be used to calculate $\dot{\delta}$ for a small-scale undulation on the portion Q of the growth form when the original gradient G at that portion is known. The gradient G may be determined if the local velocity v of the interface is known by using Eq. (1), i.e., $G=v(C-c_s)/D$; it may often be roughly estimated as that present on a sphere of the same mean radius of curvature as Q.

From Eq. (22) the value λ_0 dividing growth from decay is given by

$$\lambda_0=2\pi/\omega_0=2\pi(\Gamma_D c_0/G)^{\frac{1}{2}}=2\pi\{\Gamma_D D c_0/[v(C-c_s)]\}^{\frac{1}{2}}, \quad (23)$$

and the value λ_M corresponding to maximum growth rate is given by

$$\lambda_M=2\pi/\omega_M=\sqrt{3}\lambda_0. \quad (24)$$

If these values are small compared with the local radii of curvature R_1 and R_2 and the length $L(n)$, they are meaningful and small-scale instability occurs (for waves with $\lambda>\lambda_0$ and normal along n); if the values are large compared with either R_1, R_2 or L (for all n), all small-scale waves decay and small-scale stability results.

IV. DISCUSSION

The numerous shape-preserving solutions for the diffusion-controlled or thermally-controlled growth of a particle previously given in the literature[3,11,14–18] all omit capillarity; therefore, strictly speaking, these solutions are unstable within the framework of the assumptions on which they are based for, as we have seen, capillarity is the only stabilizing influence during growth [e.g., set $\Gamma_D=\Gamma_T=0$ in Eqs. (9), (20), (22), etc.] Taking capillarity into account does not help much for although it introduces a certain element of stability it also changes the boundary conditions for the growth problem so that the solutions that omit capillarity remain good approximations only if the particle size is large compared with the critical radius R^* of nucleation theory. But we have seen that this is just the condition that guarantees instability of the particle. In a word, it seems that with capillarity included, the shape-preserving solutions are either bad approximations or are unstable. For example, the ellipsoidal shapes growing in an isotropic medium found by Ham[11] seem unlikely to occur since they are either large compared with R^* and, therefore, unstable with respect to perturbations, or they are just a few times R^* in size and are not valid solutions. Concerning the latter point, for example, we know that an ellipsoid departing slightly from sphericity is not a shape-preserving solution under conditions that make the corresponding sphere stable for Eq. (9) shows that the harmonic components comprising the difference between the two shapes decay causing the elliposid to revert to a sphere. There is perhaps a range of one order of magnitude of particle sizes (e.g., $10R^*\lesssim R\lesssim 100R^*$) within which the solutions are not too bad as approximations and are not unduly unstable. We emphasize that the preceding conclusions are based on the assumption that S_D, $S_T\ll 1$.

An isolated sphere growing in a supersaturated matrix would eventually exceed R_c and become unstable; its growth could be represented in Fig. 1 by a vertical trajectory such as broken-line 1 starting from the R^* curve. The average growth of a number of spheres, on the other hand, would correspond to a trajectory such as that of broken-line 2 which eventually curves to the left as the overlapping diffusion fields of the different spheres reduce the effective supersaturation. Although the trajectory could conceivably always remain in the stable zone, it seems more likely that it would penetrate far into the unstable zone during the early stages of precipitation as shown, since at that time, the particle size is usually expected to become several orders of magnitude greater than R^*. That spheres are not seen in the first stages of precipitation is in accord with these theoretical expectations although there are also undoubtedly strong crystallographic factors determining the shape when the interfaces are still coherent. In the latter stages of precipitation, however, the interfaces usually become noncoherent thereby minimizing the crystallographic influences; at the same time, there is a marked tendency toward spheroidization and coarsening of the precipitate particles. The concomitance of coarsening and spheroidization is in significant accord with our calculations, for coarsening implies the average sphere has a diameter $\sim R^*=R_c/7$, and, therefore, must be stable, that is, small irregularities of these spheres tend to disappear. Under these conditions of noncoherency the assumptions of the present analysis should apply fairly well, so that the spheres should again become unstable and show irregular growth if the supersaturation were suddenly raised (e.g., by decreasing the temperature).

[13] Equation (22) applies only to those waves on an infinite plane whose associated field may be described by Laplace's equation.

[14] B. Riemann and H. Weber, *Die Partiellen Differentialgleichungen der Mathematischen Physik* (F. Vieweg und Sohn, Braunschweig, Germany, 1912), Vol. 2, p. 121.

[15] R. Rieck (1924), quoted in reference 17.

[16] C. Zener, J. Appl. Phys. **20**, 950 (1949).

[17] F. C. Frank, Proc. Roy. Soc. (London) **A201**, 586 (1950).

[18] P. V. Danckwerts, Trans. Faraday Soc. **46**, 701 (1950).

The breakdown of a planar liquid–solid interface into a cell structure during unidirectional solidification of a dilute alloy is a more complicated case of an instability involving simultaneous heat flow and diffusion. A forthcoming analysis of this case using the principles developed here gives a theoretical criterion for breakdown which is in essential agreement with observation; this may be taken as a confirmation of the existence of instabilities of the type considered here.

Little can be said about the ultimate form to which development of the harmonics leads, since the behavior of only first-order deviations has been calculated; when the deviations become large, powers of the amplitude higher than the first can no longer be neglected. We may note in general, however, that in the absence of capillarity we would expect an endless regression of bumps on bumps for the instability conclusion could be applied anew to every small portion of a bump. With capillarity included this is no longer true since a bump with a radius curvature $\sim R_c$ should usually be stable against the formation of smaller bumps on its surface (e.g., the tips of the shapes considered by Hillert[19]).

Finally we have considered only the growth of the spheres and other forms. If we wish to consider the dissolution of these forms by diffusion or melting, we have only to change the sign of the gradient G in the various equations for $\dot{\delta}$ [e.g., Eqs. (9), (20), and (22)] to see that complete stability results with respect to all perturbations considered; the capillarity and gradient terms now both suppress the perturbations.

Note added in proof. A perturbation method similar to that used in this investigation has been used by C. Wagner [J. Electrochem. Soc. **103**, 571 (1956)] to study the stability of the shape of a planar interface undergoing diffusion-controlled migration during an oxidation reaction; his treatment does not include capillarity.

[19] M. Hillert, Särtryck ur Jernkontorets Annaler **141**, 757 (1957).

ENERGY RELATIONS AND THE ENERGY-MOMENTUM TENSOR IN CONTINUUM MECHANICS

J. D. Eshelby
Department of the Theory of Materials
University of Sheffield
Sheffield, England

ABSTRACT

The force on a dislocation or point defect, as understood in solid-state physics, and the crack extension force of fracture mechanics are examples of quantities which measure the rate at which the total energy of a physical system varies as some kind of departure from uniformity within it changes its configuration. One may define similarly a force acting on each element of a mobile interface (a phase boundary or martensitic interface, for example).

Methods for calculating such effective forces are reviewed for both quasi-static and dynamic processes, the latter with particular reference to the motion of crack tips. The elastic energy-momentum tensor proves to be a useful tool in such calculations.

1. INTRODUCTION

In solid-state theory, theoretical metallurgy, fracture mechanics, and elsewhere, there are departures from uniformity in a material on various scales which, for want of a better term, we shall call defects. Examples on a microscopic scale are dislocations and point defects in a crystal lattice or their analogs in a continuum theory. On a larger scale there are regions which differ in some way from the bulk of the material, from which they are separated by an interface, for example inclusions of one phase in another, martensitic plates in ferrite, twins and so on. If the two materials on either side of it are in some sense uniform within themselves we may consider the interface itself to be the defect. On a macroscopic level there are cavities and cracks. All these entities can alter their configuration. Dislocations can glide and climb, point defects diffuse, interfaces can migrate, cavities can change their shape, and cracks can expand. The configuration of the defects can be specified by a number, possibly infinite, of parameters. Following the terminology of analytical mechanics and thermodynamics we can call the rate of decrease of the total energy of the system with respect to a parameter the generalized force acting on that parameter, or, in simple cases, on the defect itself. The idea of the force on a lattice defect is now a familiar one (it goes back to an interesting paper by Burton[1]) and the crack extension force of fracture mechanics is a concept of the same kind.

It is always the total energy which is important, the energy of the system we concentrate our attention on, and which contains the defect, plus the energy of the environment with which it interacts, in our case some mechanical loading device. The distinction between the two parts is, in fact, arbitrary. If we strain a test piece in a tensile testing machine an engineer will know where to draw the line between specimen and machine, but an applied mathematician need not. In thermodynamics the matter is handled by introducing enthalpy and Gibbs free energy, quantities which though nominally referring to the system under observation actually relate to the energy of the system plus the energy of its environment.

In this paper, the term elastic energy means, strictly speaking, internal energy under adiabatic conditions and Helmholtz free energy under isothermal ones. We shall rely on the small difference between the two in solids to give us meaningful results in intermediate cases. Of course the difference can sometimes be important, for example, in thermo-elastic effects which can actually contribute to the force on a defect;[2] we shall not consider such effects here.

It is the object of this paper to give some account of these ideas, and their extension to the dynamic case. We shall make extensive use of the properties of the energy-momentum tensor associated with the elastic field. The writer should perhaps admit that this tensor has become an obsession with him since he first noticed its connection with the force on a defect,[3] and no doubt it appears at some points in the argument where one could get along without it. Still, it serves as a convenient thread to tie together the various topics we shall discuss.

2. ELASTICITY IN LAGRANGIAN COORDINATES

For the most part our discussion will apply to finite deformation with a general stress-strain relation. In this section we briefly set out the necessary elastic theory and then in Sec. 3 show how the energy-momentum tensor appears as a formal concept, preparatory to interpreting it in Sec. 4.

We shall use rectangular Cartesian coordinates X_i to label the initial positions of particles of material in the initial unstrained state (Fig. 1*a*). On deformation the particle at X_i suffers a vector displacement $\mathbf{u}$ and its final position is, say, x_i referred to the same coordinate system, so that

$$u_i(X_m) = x_i(X_m) - X_i \tag{1}$$

If we imagine a replica of the coordinate network X_i to be embedded in the material in the initial state, and to deform with it, it becomes the curvilinear network $\tilde{X}_i$ of Fig. 1*b* after deformation. For some purposes it is convenient to refer things to this embedded (convected) coordinate system, but we shall not do so here.

For the stress it will be convenient to use the nominal (Piola-Kirchhoff or Boussinesq) stress p_{ij} defined so that $p_{ij}\,dS$ is the

Fig. 1

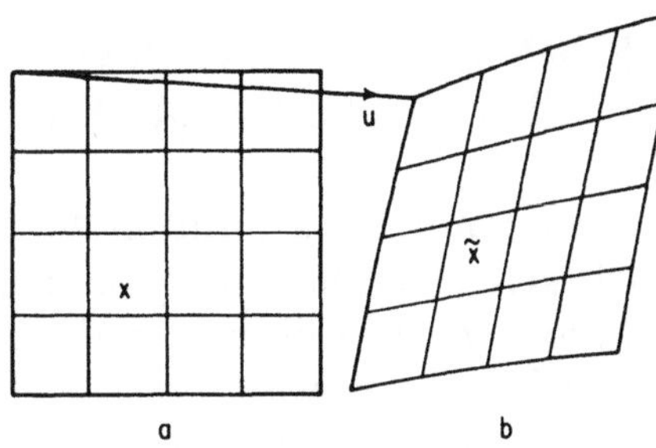

component parallel to the X_i axis of the force on an element of area which, before deformation, was an element of area dS perpendicular to the X_j axis. If W is the strain energy in unit initial volume,

$$p_{ij} = \frac{\partial W(u_{m,n})}{\partial u_{i,j}} \tag{2}$$

In the absence of body forces the equilibrium condition is

$$\frac{\partial p_{ij}}{\partial X_j} = 0 \tag{3}$$

The p_{ij} are not symmetric, but the fact that W depends only on the final shape and size of an element which was originally cubic, and not on its orientation, gives the relation

$$p_{ij} - p_{ji} = p_{jk} u_{i,k} - p_{ik} u_{j,k} \tag{4}$$

If the material is isotropic there is the additional relation

$$p_{ij} - p_{ji} = p_{ki} u_{k,j} - p_{kj} u_{k,i} \tag{5}$$

The displacement plays a double role: it can be regarded as a simple field vector like, say, the electric field E_i but it also represents a displacement of the material in that the material particle originally at X_m is in fact as $x_i = X_i + u_i(X_m)$. However, if, as we shall, we use Lagrangian coordinates X_m rather than the Eulerian coordinates x_m we may for most purposes regard them as the usual rectangular coordinates of theoretical physics and, ignoring their displacement aspect, treat $u_i(X_m)$ as a vector field of the same kind as $E_i(X_m)$ which does not stick out from X_m in any meaningful sense. This brings the elastic field into line with other physical fields and enables us to apply the methods of general field theory.

3. FORMAL DERIVATION OF THE ENERGY-MOMENTUM TENSOR

We are now in a position to describe the mathematical process which generates the energy-momentum tensor whose physical interpretation we shall discuss in the next section. The matter is usually approached by way of the calculus of variations[4,5] but the writer finds the following method clearer and, in any event, it gives all we shall want. To cover later extensions the calculation is somewhat more general than is necessary to deal with static problems in Lagrangian coordinates. The number of components of u_i need not be the same as the number of X_i; later we shall augment X_1, X_2, X_3 by the time variable $t = X_0$. We shall replace the energy density W (or, rather $-W$) by a general Lagrangian density L, and allow it to depend on the u_i as well as the $u_{i,j}$. Although this is not true in Lagrangian coordinates it is in the Eulerian formulation that we shall briefly glance at later.

Suppose, then, that we have a set of quantities $u_i(X_m)$ depending on the independent variables X_m and a function

$$L = L(u_i, u_{i,j}\ X_m) \tag{6}$$

which depends on u_i and its first derivatives $u_{i,j} = \partial u_i/\partial X_j$ and also explicitly on the X_m. We may regard $L(\cdots, \cdots, X_m)$ as a calculating machine which works out the numerical value of L when the appropriate $u_i, u_{i,j}$ are inserted. The explicit dependence on X_m then means that there is a different machine at each point X_m.

If we regard L as the Lagrangian density and require that its integral over a region of X_m space be an extremum we are led to the Euler equations

$$\frac{\partial}{\partial X_j}\frac{\partial L}{\partial u_{i,j}} - \frac{\partial L}{\partial u_i} = 0 \tag{7}$$

We shall use the notation $\partial L/\partial X_i$ to denote the gradient of L, so that $(\partial L/\partial X_i)\,dX_i$ is, to order dX_i, the numerical value of L at $X_i + dX_i$ minus its numerical value at X_i. From it we must distinguish the explicit partial derivative of L with respect to X_i when its other arguments u_i, u_{ij} and the remaining X_m are held constant. We shall denote it by $(\partial L/\partial X_i)_{\text{exp}}$, so that

$$\left(\frac{\partial L}{\partial X_i}\right)_{\text{exp}} = \left.\frac{\partial L(u_i, u_{i,j}, X_m)}{\partial X_i}\right|_{\substack{u_i,\ u_{i,j}\ \text{const.}\\ X_m\ \text{const.},\ m \neq i}} \tag{8}$$

The components of the gradient of L are thus

$$\frac{\partial L}{\partial X_l} = \frac{\partial L}{\partial u_i}u_{i,l} + \frac{\partial L}{\partial u_{i,j}}\frac{\partial u_{i,j}}{\partial X_l} + \left(\frac{\partial L}{\partial X_l}\right)_{\text{exp}} \tag{9}$$

or noting that $\partial u_{i,j}/\partial X_l = \partial u_i/\partial X_i \partial X_l = \partial u_{i,l}/\partial X_j$ and using the rule for differentiating a product

$$\frac{\partial L}{\partial X_l} = \left(\frac{\partial L}{\partial u_i} - \frac{\partial}{\partial X_j}\frac{\partial L}{\partial u_{i,j}}\right)u_{i,l} + \frac{\partial}{\partial X_j}\left(\frac{\partial L}{\partial u_{i,j}}\,u_{i,l}\right) + \left(\frac{\partial L}{\partial X_l}\right)_{\exp} \tag{10}$$

So far Eqs. (7) and (9) are purely mathematical relations. We now make the physical assumption that Eq. (7) actually is the governing equation for the field $u_i(X_m)$. Then the first term on the right of Eq. (10) vanishes and we may rewrite Eq. (10) as

$$\frac{\partial P_{lj}}{\partial X_i} = -\left(\frac{\partial L}{\partial X_l}\right)_{\exp} \tag{11}$$

where

$$P_{lj} = \frac{\partial L}{\partial u_{i,j}}\,u_{i,l} - L\delta_{lj} \tag{12}$$

is the energy-momentum tensor we are seeking.

In most of what follows L will be minus the elastic energy density W. For a uniform material unstressed in the initial state $(\partial W/\partial X_l)$ exp defined as in Eq. (10) is zero. If there is a patch of material where the function W differs from its normal form, but the material is still unstressed in the initial state, we shall say that we have a defect which is a pure inhomogeneity. An example is a small region of abnormally low elastic constants, simulating a lattice vacancy which happens to give rise to no internal stress. Actually we shall be more interested in defects associated with internal stress. One is inclined to think that because internal stress is associated with an incompatible strain there is therefore no displacement function. However, there is one—the displacement associated with Bilby's[6] shape change. One way of generating a state of internal stress is the following.[8,3]

Dice the material into tiny cubes by cuts parallel to the coordinate planes of Fig. 1*a*. Allow each cube to undergo a permanent change of shape and size specified by some suitable finite strain measure, say e_{ij}^T, which is a function of X_m. Apply to each cube surface forces chosen so as to restore its original cubic shape and size. Weld the cubes together again. The coordinate net still has the appearance of Fig. 1*a*, but only because it is held so by a distribution of body force due to the failure of the forces on adjacent cube faces to cancel completely. When these forces are relaxed the network will warp and become, say, the network $\tilde{X}_m$ of Fig. 1*b*. This immediately defines a displacement at X_m, namely the vector joining X_m to the point $\tilde{X}_m$ with the same three coordinate numbers. Naturally there is still a displacement if we deform further by applying external loads. We shall find that dislocations, with their discontinuous displacements, give no trouble.

Wherever $e^T_{ij} = 0$ the energy density is found by inserting $u_{i,j}$ into $W(\ldots, X_m)$. Elsewhere we need to know both $u_{i,j}(X_m)$ and $e_{ij}^T(X_m)$; for example, with the network still held rectangular by the distribution of body force we have a nonzero W at each point, depending on e_{ij}^T even when the $u_{i,j}$ are zero. Whether we absorb the e_{ij}^T – dependence of W into its explicit dependence on X_m, or regard the e_{ij}^T as extra field variables is a matter of choice. We do not need to decide, because all our results will involve integrals taken over surfaces which lie in "good" regions where $e_{ij}^T = 0$. Elsewhere all we need to know is that W exists. In fact we really only need to be assured of the weaker fact that any macroscopic region has a recoverable energy content. The energy does not even necessarily need to be recoverable if we do not call the material's bluff by unloading to see if it is.

4. PHYSICAL INTERPRETATION OF THE STATIC ENERGY-MOMENTUM TENSOR

To get the static energy-momentum tensor associated with an elastic medium we take L in Eq. (12) to be $-W$, where W is the energy density defined in Sec. 2. This gives

$$P_{lj} = W\delta_{lj} - p_{ij}u_{i,l} \tag{13}$$

Wherever W does not explicitly depend on X_m its divergence vanishes,

$$\frac{\partial P_{li}}{\partial X_j} = 0 \tag{14}$$

elsewhere the divergence is equal to $(\partial W/\partial X_l)_{\text{exp}}$.

The elastic energy-momentum tensor has been largely ignored, or at any rate its significance has not been appreciated. For example, Morse and Feshbach[8] set up the complete 4 X 4 array of Sec. 8, give the name "force dyadic" to its spatial part, the dynamic generalization of Eq. (13) (see Eq. (53) below), and then merely wonder if one "can discover a use for the byproduct quantities such as field momentum and force dyadic." However, the energy-momentum tensor is involved in Rice's[9] path-independent integral, which plays a useful role in fracture mechanics.

Sometimes P_{lj} turns up as a convenient auxiliary quantity in ordinary elastic calculations, with no need of interpretation. The following result, which does not seem to have been noticed before, illustrates this.

In the linear theory the elastic energy of a body is given by the surface integral

$$E_{\text{el}} = \frac{1}{2} \int_S p_{ij} u_i \, dS_j \tag{15}$$

but this is not valid in the general nonlinear case. But even then there is a surface integral representation of sorts if the material is homogeneous, namely

$$3E_{\text{el}} = \int_S (X_l P_{lj} + u_l p_{lj}) \, dS_j \tag{16}$$

To prove this we convert to a volume integral by Gauss' theorem, use Eq. (14) (homogeneity) and Eq. (3) (no body force) and an integrand $3W$ is left.

Our interpretation of the physical meaning of P_{lj} will be the following.[3,10,11] Figure 2*a* represents a body subject to surface

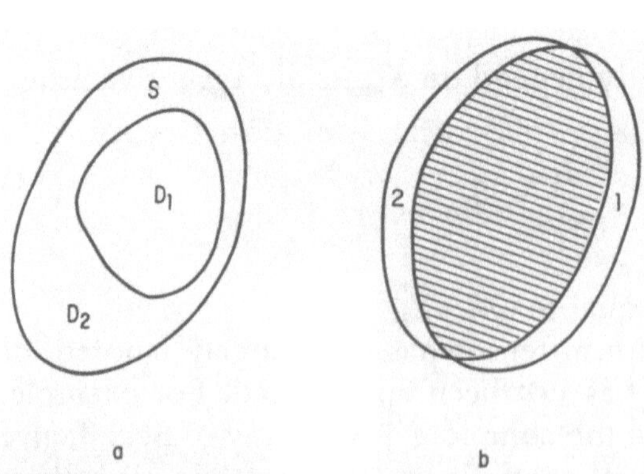

Fig. 2

loading and containing a defect D_1. In addition there may be other defects, symbolized by D_2. We ask what is the change in the total energy of the system (elastic energy plus potential energy of the loading mechanism) if D_1 suffers a small vector displacement $\delta\xi_i$. The answer is that

$$\delta E_{\text{tot}} = -\delta\xi_l F_l \tag{17}$$

where

$$F_l = \int_S P_{lj}\, dS_j \tag{18}$$

and S is any surface surrounding D_1 and isolating it from D_2. F_l is thus the force on the defect in the sense explained in the Introduction. Note that by Eq. (14) S may be deformed without altering F_l provided it remains in "good" material where $(\partial W/\partial X_l)_{\text{exp}} = 0$.

To establish Eq. (7) we first need to know how to find the elastic field in a body containing a defect which has suffered a vector shift $\delta\xi_i$, given the field before the shift. In the linear infinitesimal theory this is straightforward.[3] To find the elastic field of a defect in a finite body one usually begins with a known expression for the displacement field of the defect when situated in an infinite medium, say $u_i^\infty(X_m)$. This will produce a nonzero traction $p_{ij}^\infty n_j$ at the surface of the body. But we require the surface traction to be zero at the surface of the body when it is not externally loaded, so we must apply a surface traction $-p_{ij}^\infty n_j$ which produces, say, a displacement u_i^I, the "image" displacement. In addition, if there is, in fact, external loading which produces a surface traction $p_{ij}^L n_j$, there will be an additional displacement u_i^L so that the net displacement is $u_i(X_m) = u^\infty(X_m) + u^I(X_m) + u^L(X_m)$. It is now clear how we are to find the displacement after the singularity has been displaced by $\delta\xi_i$. We first replace $u_i^\infty(X_m)$ by $u_i^{\infty\prime} = u_i^\infty(X_m - \delta\xi_m)$. If the material is elastically inhomogeneous we must also make the change

$$c_{ijkl}(X_m) \rightarrow c_{kl}(X_m - \delta\xi_m) \tag{19}$$

to ensure that $u_i^{\infty\prime}$ continues to satisfy the equilibrium equations. The relation Eq. (19) is, of course, just the change we should decide to make if the singularity were a pure inhomogeneity. Next we calculate the new u^I and u^L. Obviously the new u^I differs from the old, and, in addition, the new u^L may differ from the old because of the change in Eq. (19). Indeed, in the case of a defect which is a pure inhomogeneity it is only the change in u^L which signals to the loading mechanism that something has happened inside the material.

A little thought shows that the above process is unnecessarily elaborate. We should get the same result if we shifted the entire original displacement field by $\delta\xi_i$, so that

$$u_i(X_m) \rightarrow u_i(X_m - \delta\xi_m) \tag{20}$$

and so forth, coupled with Eq. (19), and then adjusting the surface traction so as to satisfy the original boundary conditions once more. This recipe can be equally well applied to the finite deformation of a solid with any stress-strain relation, and the following will be our prescription for shifting a singularity in the general case.[10,11]

Let $f(X_m)$ denote any quantity associated with the elastic field, in particular the displacement, the stress and the energy density.

Stage (*i*) Make the replacement

$$f(X_m) \rightarrow f(X_m - \delta\xi_m) \tag{21}$$

Stage (*ii*) Adjust the surface tractions until the original boundary conditions are satisfied.

It is worth asking what becomes of Eq. (19) in the general case. We now have to shift the functional form of W: $W(\ldots, X_m) \rightarrow W(\ldots, X_m - \delta\xi_m)$. But since the arguments $u_{i,j}$ which have to be inserted into the energy density function have themselves suffered the change $u_{i,j}(X_m) \rightarrow u_{i,j}(X_m - \delta\xi_m)$ we may just as well write $W[X_m] \rightarrow W[X_m - \delta\xi_m]$ where $W[X_m]$ stands for $W(u_{i,j}(X_m), X_m)$ regarded as a simple function of X_m. This change is already covered by Eq. (21).

In the situation of Fig. 2*a* we want to shift D_1 but not D_2. To do this we simply regard S as the surface of a "body" acted on by a complex loading mechanism made up of the material between S and the surface of the body plus the real external loading mechanism, and apply our previous recipe. Of course the adjustment in stage (*ii*) will change the displacement in the material outside S as well as inside it, but one can show that it does not generate a discontinuity across S.

We now have to find how the elastic energy E_{el} of the body and the potential energy E_{ext} of the loading mechanism change during stage (*i*) and stage (*ii*).

In stage (*i*) E_{el} changes by

$$\delta E_{\text{el}}^{(i)} = \int_V \{W[X_m - \delta\xi_m] - W[X_m]\} dV$$

$$= -\delta\xi_l \int_V \frac{\partial W}{\partial X_l} dV + 0(\delta\xi^2) \tag{22}$$

$$= -\delta\xi_l \int_S W dS_l + 0(\delta\xi^2)$$

by Gauss' theorem. At the beginning of stage (*ii*) the surface traction is $[p_{ij}(X_m) - \delta\xi_l p_{ij,l}(X_m)] n_j + 0(\delta\xi^2)$ and at the end it is $p_{ij}(X_m) n_j$, so that with an error of order $\delta\xi$ it is $p_{ij}(X_m) n_j$ throughout the adjustment. The surface displacement changes from $u_i(X_m) - \delta\xi_l u_{i,l}(X_m) + 0(\delta\xi^2)$ to some final value $u_i^F(X_m)$ which could only be found by detailed calculation. Hence the change in E_{el} during stage (*ii*), equal to the work done on the body during the adjustment, is

$$\delta E_{\text{el}}^{(ii)} = \int_S (u_i^F - u_i + \delta\xi_l u_{i,l}) p_{ij} dS_j + 0(\delta\xi^2) \tag{23}$$

During stages (*i*) and (*ii*) together the increase in the potential energy of the loading mechanism, being equal to minus the work which it does, is

$$\delta E_{\text{ext}} = \int_S (u_i - u_i^F) p_{ij} dS_j + 0(\delta\xi^2) \tag{24}$$

Adding Eq. (22) to Eq. (23) to get the change in total energy we arrive at

$$\delta E_{\text{tot}} = -\delta\xi_l \int_S (W\delta_{lj} - p_{ij} u_{i,l}) dS_j + 0(\delta\xi^2) \tag{25}$$

which is Eq. (17).

The quantity $\delta E_{\text{el}}^{(i)}$ can also be found as follows. In addition to S, the surface of the body, draw the surface S′ derived from S by a shift $-\delta\xi_i$ (Fig. 2*b*). Then $\delta E_{\text{el}}^{(i)}$ is the integral of W over the crescent-shaped volumes 1 and 2, the former being given a positive sign and the latter a negative one. (Of course in the case of 2 we have to suppose that the field is slightly extrapolated beyond S.) This gives the last of the surface integrals contained in Eq. (22) at once, and makes it clear that W in the shaded area of Fig. 2*b*, which appears in the first of them and has cancelled from the last, need not appear in the calculation at all. Thus if we do not know, or do not care to specify, W in the neighborhood of the singularity it does not matter. Even formal infinities in the shaded region cancel out.

Again, in calculating $\delta E_{\text{el}}^{(ii)}$ we supposed that the loading mechanism produced a surface traction which was independent of the slight shift of the point of application consequent on the displacement of the singularity by $\delta\xi_i$, that is, in engineering language we assumed dead loading produced by an ideally soft loading machine.

If this is not so the change in the surface traction which occurs during the change of surface displacement from u_i to u_i^F will be of order $\delta\xi$, but this will only alter δE_{ext} by an amount of order $\delta\xi^2$. Consequently Eq. (17) is also valid for an arbitrarily hard or soft loading mechanism. (We might perhaps feel uncomfortable about the extreme case of infinitely hard loading, i.e., rigidly imposed surface displacements, but then $u_i = u_i^F$ at the surface and Eq. (17) is still valid.)

The fact that Eq. (7) is independent of the hardness or softness of the loading mechanism enables us to treat the situation of Fig. 2*a* without further calculation. Draw an internal surface S isolating D_1 from D_2 and the surface of the body. Then, as already mentioned, we may regard the self-stressed material between S and the surface of the body, together with the loading mechanism properly speaking, as constituting a single loading mechanism of unknown hardness or softness acting on the surface S of the "body" bounded by S, and Eq. (17) may be applied.

Thus, generally, F_l in Eq. (18) gives the force on the singularity (or singularities) inside any surface S, due to other singularities outside S, and to imposed surface tractions produced by any type of loading mechanism.

Although we do not in fact need to know the form of the function W inside S, or even that it exists everywhere, when we *do* know it Eq. (11) shows that we may write

$$F_l = \int_v \left(\frac{\partial W}{\partial X_l}\right)_{\text{exp}} dV \tag{26}$$

with the notation of Eq. (8), so that F_l is the same as the energy of a fictitious body with the energy density $(\partial W(\ldots, X_m)/\partial X_l)_{\text{exp}}$ into which we insert the original displacement gradients $u_{i,l}$. It is unnecessary to know how they change during the operations of stages (*i*) and (*ii*). The following is a similar but more general result.

If we make any small change δW_{exp} in the form of the energy function this will induce a change $\delta u_{i,l}$ in the displacement gradients, and the change in the energy density will be

$$\delta W = \delta W_{\text{exp}} + \frac{\partial W}{\partial u_{i,j}} \delta u_{i,j} = \delta W_{\text{exp}} + p_{ij} \delta u_{i,j} \tag{27}$$

and on integration the change in the elastic energy of the body is, since $p_{ij,j} = 0$,

$$\delta E_{\text{el}} = \int_V \delta W_{\text{exp}} dV + \int_S p_{ij} \delta u_i dS_j \tag{28}$$

The change in the energy of the loading mechanism is

$$\delta E_{\text{ext}} = - \int_S p_{ij} \delta u_i dS_j \tag{29}$$

to the first order, even without dead loading. Hence

$$\delta E_{\text{tot}} = \int_V \delta W_{\text{exp}}(u_{ij}) dV \tag{30}$$

and so, again, we only need to know the change in the functional form of W and the old $u_{i,j}$; the changes $\delta u_{i,j}$ cancel from δE_{tot}, though not from δE_{el} and δE_{ext} taken separately. This result can be of practical use.[12] There are similar results in other parts of theoretical physics, e.g., the Hellman-Feynman theorem in quantum mechanics.

These conditions are related to the concept of complementary energy. If we derive a quantity ϕ from W by the Legendre transformation

$$-\phi = W - \frac{\partial W}{\partial u_{i,j}} u_{i,j} \tag{31}$$

then ϕ is the complementary energy density. In thermodynamic language it is the enthalpy density for adiabatic processes and the Gibbs free energy density for isothermal ones. For the small change contemplated above

$$-\delta\phi = \delta W_{\mathbf{exp}} - \delta p_{ij} u_{i,j} = \delta W_{\mathbf{exp}} - (\delta p_{ij} u_i)_{,j} \tag{32}$$

the terms $\partial u_{i,j} \partial W/\partial u_{i,j}$ and $-p_{ij}\delta u_{i,j}$ cancelling. Hence if

$$\Phi = \int_V \phi dV$$

is the total complementary energy, comparison with Eq. (32) gives

$$-\delta\Phi = \delta E_{\mathrm{tot}} - \int_S \delta p_{ij} u_i dS_j \tag{33}$$

and for dead loading $\delta p_{ij} n_i = \text{const}$, $\delta E_{\mathrm{tot}} = -\delta\Phi$. For other types of loading $-\delta\Phi$ differs from δE_{tot} by a quantity of the first order, in contrast to Eq. (32).

The application of Eq. (18) to dislocations requires a little discussion because the displacement is discontinuous. On the infinitesimal theory the usual dislocations of physical theory have a discontinuity in u_i across a surface but since $u_{i,l}$ does not, Eq. (18) may be applied to them. The same is not true for general Somigliana dislocations,[3] or even for disclinations[13] where the rotation is discontinuous.

For dislocations with a *finite* constant Burgers vector, Eq. (18) will be valid if $u_{i,l}$ is the same at points of the material which were adjacent across the cut in the initial state. To avoid complication we shall only consider the case where the dislocation is created by cutting the material and sliding the faces without separation or interpenetration. If we identify the undeformed material with Frank's[13] reference crystal we must arrange that points originally separated by **b** across the cut are coincident finally. The usual argument of the infinitesimal theory[13,14] may be applied to the finite $u_{i,j}$. It shows that the $u_{i,l}$ (and their gradients $u_{i,lj}$) are the same at points which were opposite one another in the initial state, and this is what we want for Eq. (18) to be applicable, because the integration is with respect to the undeformed state.

The writer has convinced himself that for points *finally* adjacent across the cut $u_{i,j}$ is continuous but $u_{i,jk}$ is not. This fact, if true, should make itself felt in the theory of continuous distributions of dislocations; though there is no trouble with a single infinitesimal dislocation, an isolated bundle of them is, after all, equivalent to a dislocation with a finite Burgers vector.

In the linear theory the displacement may be written $u_i = u_i^\infty + u_i^I + u_i^L + u_i^S$, where, as before, $u_i^\infty + u_i^I$ is the field due to a particular defect divided into its value in an infinite medium and an image term; u_i^L, u_i^S are the displacements produced by external loading and other defects. We can then partition F_l into contributions from external load, other defects, and image effects, say F_l^x, with $x = L, S$ or I. F_l^x is given by the cross-term in Eq. (18) between x-quantities and ∞-quantities. It may be manipulated[10] into the more useful form

$$F_l^x = \int_S (u_i^x p_{ij,l}^\infty - p_{ij}^x u_{i,l}^\infty)\, dS_j \qquad x = L, S, I \tag{34}$$

The quantity which would be denoted by F_l^∞ vanishes if the separation into u_i^∞ and u_i^I has been done properly.

Equation (34) gives the standard results for the forces on defects according to the linear theory. Since most of them can be derived without the help of the energy-momentum tensor this is not very helpful, though once Eq. (18) enabled the writer to decide which of two rival expressions for the interaction between a dislocation and a point defect to attack. The real value of Eq. (18) is that it indicates how far these results are valid. For example, a well-known result in the infinitesimal theory states[13] that the force on a point defect idealized as a center of dilatation is

$$F = -\Delta V \operatorname{grad} p \tag{35}$$

where ΔV is the volume change produced by the introduction of the defect (measured by the expansion of the outer surface of the body) and p is the hydrostatic pressure there would be at the center of the defect if it were not there. Equation (35) can be derived from Eq. (18) by taking S to be a small surface enveloping the defect. In fact the nonlinear theory is grossly inadequate near the defect. What effect does this have on Eq. (35)? If we choose S to be a large enough surface surrounding the defect the quantities which enter P_{lj} will be adequately represented by the linear theory. Since the divergence of P_{lj} is zero on both the nonlinear and linear theories we may contract S to an infinitesimal sphere about the defect and evaluate it using either theory. On the infinitesimal theory the result is Eq. (35) but it holds equally for the nonlinear theory, if p is taken to be not the actual hydrostatic pressure due to the surface loads and other defects (which is meaningless in a nonlinear theory) but rather the hydrostatic pressure which would be calculated from the applied stress and displacement prevailing over the large surface S on the assumption that the linear theory is valid everywhere. Similarly, in other cases results obtained on the linear theory may be used to calculate F_l whenever S in Eq. (18) can in principle be chosen to lie everywhere in regions where the linear theory is adequate.

5. MODIFICATION WHEN THERE ARE BODY FORCES

In deriving Eq. (18) we have supposed that the material inside S is free of body force. However, in ionic crystals a dislocation may carry a charge,[15] and if the crystal is immersed in an electrostatic field there will be a body force acting at its center. There do not seem to be any other sensible cases where a defect is associated with body force, but it is easy to adapt the argument which led to Eq. (18) to cover the general case. The body force inside S is

$$f_i = -\int_S p_{ij}\, dS_j \tag{36}$$

In the shift carried out in stage (i) the electrostatic field (or whatever mechanism is providing the body force) increases its potential energy by $-f_i\delta\xi_l$. In stage (ii) the presence of the body force produces changes in both Eqs. (23) and (24) but they exactly cancel, just as do the terms in $u_i - u_i^F$. So, in all, we have to add a term $+f_i\delta\xi_i$ to the right-hand side of Eq. (17). Hence Eq. (18) must be modified to read

$$F_l = \int_s P^*_{lj}\, dS_i \tag{37}$$

where
$$\begin{aligned} P^*_{lj} &= P_{lj} - p_{lj} = W\delta_{lj} - P_{ij}(\delta_{il} + u_{i,l}) \\ &= W\delta_{lj} - \frac{\partial W}{\partial\left(\dfrac{\partial x_i}{\partial X_j}\right)}\frac{\partial x_i}{\partial X_l} \end{aligned} \tag{38}$$

(see Eq. (1)). The difference in sign between P_{lj} and p_{lj} is merely a matter of convention (see Sec. 8).

It has become traditional to emphasize the difference between the "real" forces acting on the material and the "fictitious" forces acting on defects, but in Eq. (37) the "real" force (Eq. (36)) seems to make a contribution to the "fictitious" force. However, in the case of charged dislocations, the extra term has the required "fictitious" character because when it moves a dislocation does not actually carry marked charges along with it. Rather, atomic rearrangement at its old position eliminates the net charge there, and at its new position atomic rearrangement creates an equal net charge.

One can, of course, replace P_{lj} by P^*_{lj} even when there are no body forces. In an isotropic medium, but not otherwise, P^*_{lj} is, in view of Eq. (5), symmetric, in contrast to P_{lj}.

6. THE FORCE ON AN INTERFACE

So far the energy-momentum tensor has appeared only in an integral over a closed surface. In certain cases it is also possible to give a meaningful interpretation to the force $dF_l = P_{lj}\,dS_j$ on an individual surface element.

Figure 3 represents a specimen in which the material inside a surface S has undergone some kind of transformation. S may be closed or, as indicated by the dotted lines, it may extend to the surface of the specimen. The following are some examples of what we mean by the word "transformation."

In a martensitic transformation region B undergoes a change of natural shape and size. If this is inhibited by the presence of the matrix A, strains are set up inside and outside S. However, adjacent

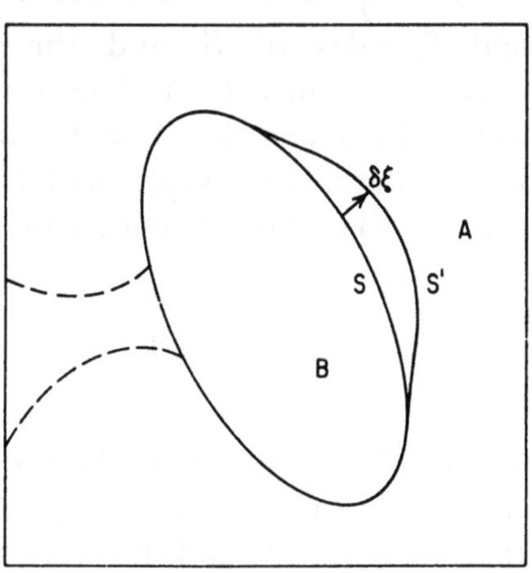

Fig. 3

material particles on opposite sides of S remain adjacent, so that the displacement is continuous. In addition the elastic properties of the material inside B become different from those of A. When a coherent twinned region develops there is also a change of form and elastic properties. Though the material is the same the crystal axes are differently oriented, so that in the linear case, for example, the c_{ijkl} are different in A and B when referred to the same axes. The same is true if S is a grain boundary. In the case of Dauphiné twins[16] in quartz there is a difference of elastic constants in the above sense, but no associated change in form. We may also take S to be the boundary of a cavity, the material inside it having, so to speak, transformed into nothing.

In all these cases the displacement is, or may be taken to be, continuous across S, and we shall assume that this is so in what follows. We also allow an external loading mechanism to act on the outer surface of the specimen.

In the physical cases described, and in others, the boundary may be capable of migrating through the material (growth of martensitic plates, change in the form of a cavity by volume or surface diffusion, and so on). In Fig. 3 migration has made S develop a shallow blister, changing it to S'. The migration may be specified by erecting a small vector $\delta\xi_i$ at each point of S. It is then a sensible question to ask what is the change in the total energy of the system as a result of the migration. The answer is that

$$\delta E_{\text{tot}} = -\int_S \delta\xi_l \,(P_{lj}{}^A - P_{lj}{}^B)\, dS_j \tag{39}$$

where $P_{lj}{}^A$ and $P_{lj}{}^B$ are the energy-momentum tensors on the A and B sides of S, and the normal in $dS_j = n_j dS$ is directed outwards from B to A. The proof is similar to that of Eq. (17). Cut out and remove the A material which lies between S and S' in the initial state, and apply suitable tractions to the raw surface to prevent relaxation. The increase in elastic energy is

$$\delta E_{\text{el}}{}^{(i)A} = -\int \delta\xi_l \, W^A dS_l \tag{40}$$

The displacement on the boundary S' of A is now $u_i{}^A + \delta\xi_l \, u_{i,l}{}^A$ and the traction is $p_{ij}{}^A dS_j + 0(\delta\xi)$. Now alter the displacements to their final values $u_i{}^{FA}$, and the work done on A is

$$\delta E^{(ii)A} = -\int \left(u_i^{FA} - u_i^A - \delta\xi_l u_{i,l}^A\right) p_{ij}^A \, dS_j \tag{41}$$

(Since some of the work expended may have gone towards the potential energy of the part of the loading mechanism associated with A, we cannot label Eq. (41) as purely elastic energy.)

For B we have similarly

$$\delta E_{\text{el}}^{(ii)B} = \int \delta\xi_l W^B dS_l \tag{42}$$

$$\delta E^{(ii)B} = \int \left(u_i^{FB} - u_i^B - \delta\xi_l u_{i,l}^B\right) p_{ij}^B \, dS_j \tag{43}$$

There is no term corresponding to Eq. (24). In a sense A and the part of the actual loading mechanism which acts on its surface provides the loading mechanism for B and conversely. Since u_i is continuous across the interface, we have $u_i^A = u_i^B$, $u_i^{FA} = u_i^{FB}$ for each element dS_j. Also $p_{ij} dS_j$ is continuous across the interface. So, on adding Eq. (40) to Eq. (43) we get Eq. (39).

This expression only gives the change in the elastic energy and the energy of the loading mechanism. If it is to be applied to phase changes we must also include the "chemical" energy, the work required to transform the material which disappeared from A into the material which appeared in B, say

$$\int \delta\xi_l (W_0^B - W_0^A) \, dS_l \tag{44}$$

where $W_0^B - W_0^A$ is the work required to transform a mass of unstressed A into an equal mass of unstressed B, the mass occupying unit volume in the common initial state. More precisely, $W_0^B - W_0^A$ is a change of internal energy in an adiabatic change and of Helmholtz free energy in an isothermal one. The quantity Eq. (44) can be added to Eq. (39) without changing its form by choosing the value of W for $u_{i,j} = 0$, $W(0)$ say, which is so far unspecified, so that $W^A(0) - W^B(0) = W_o^A - W_o^B$. The addition of a constant to W does not, of course, alter Eq. (18), because the extra term is proportional to $\int dS_l$ and so is zero for a closed surface being, by Gauss' theorem, the volume integral of the gradient of unity. It appears that the value ascribed to the $W(0)$ of a particular substance depends on what we propose to transform it into, but in a sense $W(0)$ can be prescribed absolutely since, after all, anything can in principle be transformed into anything else through a suitable intermediary, say the material of a neutron star.

If we use the notation $[Y]$ to denote the discontinuity in Y across S, its value on the side to which the normal points minus its value on the other side, Eq. (39) may be written

$$\delta E_{\text{tot}} = - \int_S \delta\xi_l \left[P_{lj}\right] dS_j$$

$$= - \int_S \delta\xi_l ([W]\,\delta_{lj} - p_{ij}[u_{i,l}]) dS_j \tag{45}$$

It is natural to choose the vector $\delta\xi_l$ which describes the shift of the interface everywhere normal to S, but if we do not it makes no difference. The term $[W]\delta_{lj}$ gives a force parallel to the normal, and so does the term $-p_{ij}[u_{i,l}]$ because as $[u_i] = 0$, by supposition it follows that $[u_{i,l}]\delta\xi'_l = 0$ whenever $\delta\xi'$ lies in S. In fact, it is not hard to see[17] that

$$[u_{i,l}] = \left[\frac{\partial u_i}{\partial n}\right] n_l \tag{46}$$

where $\partial/\partial n$ denotes differentiation along the normal. Thus Eq. (45) is equivalent to the statement that there is an effective normal force

$$F = [W] - \mathbf{T}\cdot\left[\frac{\partial \mathbf{u}}{\partial n}\right] \tag{47}$$

per unit area of interface, where T is the surface traction at the interface.

Equation (45) can be used to find the equilibrium position of phase and twin boundaries in the presence of stresses produced by the transformation itself, or applied externally, or both. Since Eq. (45) must be zero for any small $\delta\xi_l$ the boundary must take up a shape for which Eq. (47) is zero all along it. In the case of a stress-free cavity Eq. (47) becomes the energy density at its surface (a positive quantity). Any increase in its volume leads to a decrease of E_{tot}.[18]

7. RESULTS IN EULERIAN COORDINATES

It is not easy to conduct the argument of Sec. 4 in the Eulerian coordinates x_i of Eq. (1), but we can get the results from the Lagrangian case by introducing a Σ_{lj} defined by $\Sigma_{lj}\,ds_j = P_{lj}\,dS_j$ where ds_j is what dS_j becomes in the final state, so that

$$F_l = \int_S P_{lj}\,dS_j = \int_S \Sigma_{lj}\,ds_j \tag{48}$$

A tedious calculation gives

$$\Sigma_{lj} = w\delta_{lj} - \frac{\partial w}{\partial v_{i,j}}\,v_{i,l} \tag{49}$$

a direct transcription of P_{lj}. Here $v_i(x_j) = x_i - X_i(x_j)$ is the Eulerian displacement as contrasted with the Lagrangian $u_i(X_j) = x_i(X_j) - X_i$, $v_{i,j}$ stands for $\partial v_i/\partial x_j$ and w is the energy density per unit final volume. Since W depends on X_m, w will depend on v_i as well as $v_{i,j}$ but only through the combination $X_m = x_m - v_m$:

$$w = w(v_{i,j}, x_m - v_m) \tag{50}$$

If, similarly, we define the Eulerian stress through $\sigma_{ij}ds_j = p_{ij}\,dS_j$ we get, surprisingly,

$$\sigma_{ij} = \frac{\partial w}{\partial v_{i,j}} + w\delta_{ij} - \frac{\partial w}{\partial v_{r,j}}\,v_{r,i} \tag{51}$$

(Actually Eq. (51) is a disguised form of Hamel's[19] expression for σ_{ij}.) The formalism of Sec. 3, with suitable changes in notation, confirms Eqs. (49) and (51). Equation (49) comes directly from Eq. (12). The equilibrium equation corresponding to Eq. (7) can be written as

$$\frac{\partial}{\partial x_i}\frac{\partial w}{\partial v_{i,j}} - \frac{\partial w}{\partial v_i} = -\left(\frac{\partial w}{\partial x_i}\right)_{\text{exp}} = -\frac{\partial \Sigma_{ij}}{\partial x_j} \tag{52}$$

or $\partial\sigma_{ij}/\partial x_j = 0$ with σ_{ij} as in Eq. (51), by Eq. (50) and Eq. (11). Of course this of itself does not actually prove that σ_{ij} is the Eulerian stress.

One consequence of Eq. (51) is that expressions for radiation pressure,[20] an ordinary force, seem to be formed from Σ_{ij} which we associate with the fictitious force on an inhomogeneity. This is because the first term in Eq. (51) has averaged to zero in a periodic process. The analogy drawn by the writer[10] between the lift on an aerofoil and the supposed Lorentz force on a moving dislocation[13] merely illustrates the same sort of thing and has no direct physical content. (Of course in both these cases there are dynamical terms we have ignored.)

8. THE DYNAMIC ENERGY-MOMENTUM TENSOR

If we supplement the spatial coordinates X_1, X_2, X_3 by $X_0 = t$, the time variable, and take for the Lagrangian density $L = T - W$ where $T = \frac{1}{2} p\dot{u}_i\dot{u}_i$ is the kinetic energy density per unit original volume, the analysis of Sec. 3 gives a 4 X 4 energy-momentum tensor with components

$$P_{ij} = (W - T)\delta_{lj} - p_{ij}u_{i,l} \qquad i, j = 1, 2, 3 \tag{53}$$

$$s_j = P_{0j} = -p_{ij}\dot{u}_i \tag{54}$$

$$g_l = -P_{l0} = -p\dot{u}_i u_{i,l} \tag{55}$$

$$H = P_{00} = T + W \tag{56}$$

If the material is homogeneous and has time-independent properties, Eq. (11) gives the conservation laws

$$\frac{\partial P_{lj}}{\partial X_j} = \frac{\partial g_l}{\partial t} \tag{57}$$

and

$$-\frac{\partial s_j}{\partial X_i} = \frac{\partial H}{\partial t} \tag{58}$$

The space components of P_{lj} only differ from the static case by the kinetic energy term. The vector s_j is the energy flow vector, and $s_j dS_j$ in the flow of energy through dS_j in the direction of the positive normal.[21] H is the energy density.

The vector g_l is variously called the quasi-momentum, pseudomomentum or field momentum. It is also the crystal momentum of lattice and electron theory, in the continuum limit of zero lattice constant, when umklapp processes are irrelevant.

The quasi-momentum density differs from the ordinary momentum density $G_l = \rho \dot{u}_l$. It can be given various formal interpretations. We could use the curvilinear (and now time-dependent) coordinate network $\underset{\sim}{X}_m$ of Fig. 1*b* to describe the motion of a particle moving through the medium, and take the $\underset{\sim}{X}_1$, $\underset{\sim}{X}_2$, $\underset{\sim}{X}_3$ with which it happens to coincide at any instant as its generalized coordinates. Let the particle be a small mesh of side ϵ in Fig. 1*a*. Although its $\underset{\sim}{X}_m$ remain constant, it still has a generalized momentum in the sense of analytical mechanics, say $\epsilon^3 \Pi_l$. Then[10] $g_l = G_l - \Pi_l$. In a linear anisotropic medium the ratio S_l/H gives the group velocity of a disturbance, whereas g_l/H is the slowness, a vector whose direction is that of the phase velocity and whose magnitude is the reciprocal of the magnitude of the phase velocity. (It may sound odd to speak of group velocity for a medium governed by a second-order equation, but as the frequency $\omega/2\pi$ depends on the direction of the wave vector k_l the group velocity $\partial\omega/\partial k_l$ differs from the phase velocity ω/k_l.)

Suppose next that the elastic medium is interacting with some other system by way of a potential $U(u_{i,j}, x_m)$ per unit original volume which depends on the deformation and on the spatial position x_m (Eq. (1)) of the volume element. A good example is provided by electrons moving in the solid and interacting with it via a deformation potential.[13] Then one can show[22,10] that for a closed surface S bounding a volume V

$$\frac{d}{dt}\int_V (G_l - g_l)\,dV = \int_S (p_{lj} - P_{lj} - U\delta_{lj})\,dS_j \tag{59}$$

If the surface integral vanishes the total ordinary momentum and the total quasi-momentum inside S increase at the same rate, and one is as good as the other in drawing up a momentum balance sheet. The surface integral will vanish if the disturbance has not reached S, and U is negligible (or a constant) on S. More important, it vanishes if, as usual in theoretical solid-state calculations, we impose periodic boundary conditions; surface elements separated by a period then given contributions which cancel.

It is perhaps worth remarking that a point charge (electron) interacting with a medium through a deformation potential which only depends on the volume change induces a center of dilatation. Two electrons at rest do not interact, because a center of dilatation produces no hydrostatic pressure, and so the force between them is zero by Eq. (35). When they are moving there is a hydrostatic pressure and hence an interaction which, in most cases, is the one responsible for superconductivity.

The considerations above refer to a medium free of defects. It is difficult to fit moving defects rigorously into the P_{lj}, g_l formalism. However, results obtained otherwise can often be reproduced correctly by assuming that the force on a defect and quasi-momentum are related in the same way as ordinary force and momentum. We return to this point in Sec. 10.

We end this section with some remarks on sign. Here, the sign of g_l has been reversed compared with Refs. 3 and 10, so that now the quasi-momentum of a wave packet points roughly in the direction in which it is going. The sign of P_{lj} was originally chosen so that, in the static case, its surface integral gives the force on a defect in the generally accepted sense, rather than, as would perhaps have been more logical, the equal and opposite force, of unspecified origin, which keeps the defect from moving. (Either sign may be found in the field theory literature.) On the other hand, the surface integral of p_{ij} is equal and opposite to the body force inside the surface. This accounts for the difference in sign between P_{lj} and p_{lj} in Eq. (37) and other places where they occur together. In the dynamic case the surface integral of P_{lj} represents the flow of quasi-momentum into the surface, but the integral of the energy-flow vector s_i gives the outflow of energy; hence the difference of sign in Eqs. (57) and (58).

9. STATIC CRACKS

Apart from their obvious importance in fracture mechanics, cracks, because of the weakness of the singularity at their tips, lend themselves nicely to the illustration and extension of some of the results we have reviewed.

When regarded as a very flat hole, a crack is really a planar defect. Nevertheless, in the two-dimensional case at least, each of the tips of a crack can be regarded as a defect or singularity in its own right. The force on the tip, found by integrating P_{lj} round a loop embracing it, is the same as the crack extension force or energy release rate $\mathcal{G}$ of fracture mechanics. We shall begin by reviewing various formal ways of calculating $\mathcal{G}$ by purely elastic methods, and then, rather crudely, see what happens when one allows for the existence of interatomic forces or plasticity near the tip.

We start with the classical argument[23,24] which gives the energy release rate. For mode I deformation the stress p_{22} just ahead of the crack and the opening displacement just behind it are[25]

$$p_{22}(x) = K(2\pi)^{-1/2}x^{-1/2} \qquad \Delta u_2 = \frac{K}{M}(2\pi)^{-1/2}(-x)^{1/2} \tag{60}$$

where K is the stress intensity factor and M is an elastic modulus which is $\mu/(1-\nu)$ for isotropy. Let the tip advance by $\delta\xi$ and apply forces to the new freshly formed surface so as to close it up again. It is clear from the relation between the old p_{22} and the new Δu_2 shown in Fig. 4*a* that the work extracted from the system when these forces are slowly relaxed is

$$\frac{1}{2}\int_0^{\delta\xi} p_{22}(x)\Delta u_2(x-\delta\xi)dx = \frac{K^2}{2M}\delta\xi \tag{61}$$

The crack extension force or energy release rate $\mathcal{G}$ is, by definition, the energy extracted per unit advance, and so

$$\mathcal{G} = \frac{K^2}{2M} \tag{62}$$

The same method can be applied to modes II and III; for mode II Eq. (62) is still correct with $M = \mu/(1-\nu)$ and for Mode III we must put $M = \mu$.

A crack is equivalent to a pileup of dislocations, in the limit where their Burgers vectors are infinitesimal and they are infinitely numerous.[25] Fig. 4*b* illustrates this for mode II. The first dislocation is locked and the rest are free. For mode I the dislocations are still of edge type, but their Burgers vectors are vertical and for mode III they are screws.

For the discrete pileup of Fig. 4*b* the force on the locked dislocation is, as we have seen, given by the integral

$$F_1 = \int_S P_{1j}dS_j \tag{63}$$

taken over the loop $S = S_1$. But as all the other dislocations are in equilibrium (i.e., there is no force on them) we may extend S to embrace any number of them, provided we do not include the locked

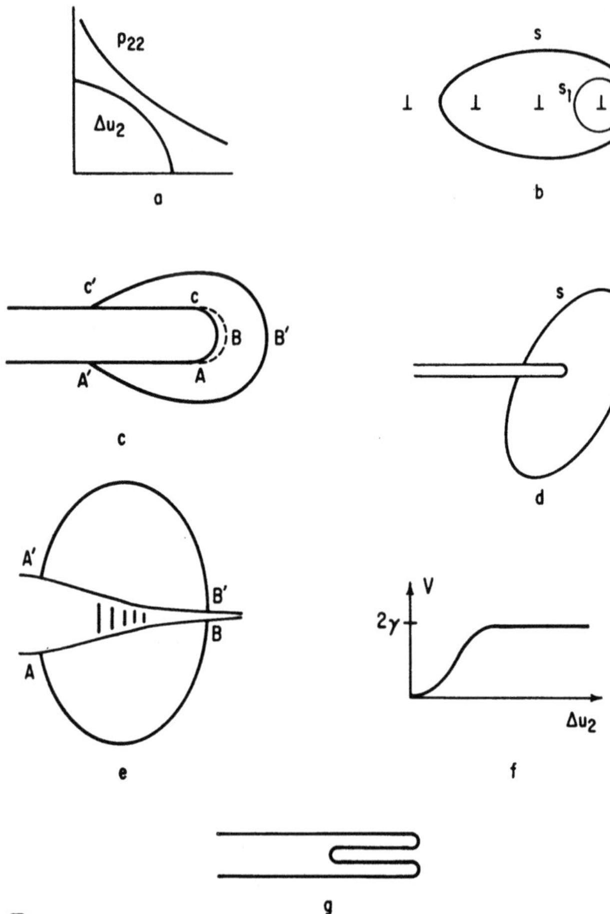

Fig. 4

dislocation at the other end of the crack. We suspect that this result will hold in the continuum limit, so that F_1 is given by Eq. (63) with S any loop embracing the crack tip (Fig. 4*d*). Since F_l and $\mathcal{G}$ both give the energy loss per unit advance we should have $\mathcal{G} = F_1$ One can verify this by trying a few special cases. Actually one can do better.[25] It is possible to write down an expression for the force on the locked dislocation in the discrete case, pass to the limit, and then, using the details of the dislocation-crack analogy show that $F_1 = \mathcal{G}$. Equation (63) is, of course, Rice's[9] path-independent integral expression for $\mathcal{G}$.

One can also get F_1 by regarding the crack as a cavity bounded by flat parallel faces separated by a small but finite distance h and closed by rounded ends (Fig. 4*c*). We can then apply Eq. (45) with $P_{lj}{}^{B}$ zero. If the boundary moves forward by $\delta\xi$ like a trombone slide, $\delta\xi_1 = (\delta\xi, 0, 0)$ is constant over ABC and zero elsewhere so that Eq. (45) gives

$$F_1 = \mathcal{G} = \int_S P_{1j}\, dS_j \tag{64}$$

taken over ABC. The path may be deformed to $A\,A'B'C'C$, and as AA', CC' contribute nothing we are left with an arbitrary loop embracing the tip. We now argue that if the dimensions of S are large compared with h, the elastic field will (as can be verified in simple cases) be almost unaffected by the value of h, and so we can let h tend to zero. The weak point is, of course, that the proof of Eq. (45) involves the removal of material, and the whole point about a hair-line crack is precisely that its advance does not involve the elimination or redistribution of material. The most satisfactory general proof is to repeat word for word for the internal surface S of Fig. 4*d* the argument we applied to the internal surface S of Fig. 2*a*. The fact that S is intersected by the crack leads to no trouble provided the shift made in calculating $\delta E_{el}^{(i)}$ is parallel to the crack. Essentially this argument has been used by Cherepanov[26] and Rice.[27]

It is interesting to see what we get from a crude atomic model, analogous to the Peierls-Nabarro[13] model of a dislocation. In Fig. 4*e* the cracked solid is represented as two elastic blocks pulled open in mode I by externally applied forces and held together by interatomic forces represented by the shading. We suppose that the interatomic forces can be derived from a potential depending on the separation of the blocks, so that, at the faces

$$p_{22} = \frac{\partial V(\Delta u_2)}{\partial(\Delta u_2)} \tag{65}$$

In contrast to the Peierls-Nabarro case, V is not periodic in Δu_1, but has the form shown in Fig. 4*f* with a plateau whose height is 2γ where γ is the surface energy of the material. Let us integrate P_{ij} round the loop S shown in Fig. 4*e*. Because the divergence of P_{ij} is zero the integral can equally well be taken along the upper and lower faces of the crack itself, and we have[9]

$$F_1 = \int_S P_{1j}\, dS_j \tag{66}$$

$$= -\int_A^B p_{22} \frac{\partial}{\partial x_i}(\Delta u_i)\, dX_i \tag{67}$$

$$= \int_B^A p_{22}\, d(\Delta u_i) \tag{68}$$

$$= \int_B^A \frac{\partial V}{\partial \Delta u_i}\, d\Delta u_i \tag{69}$$

$$= V_A - V_B \tag{70}$$

If A lies in the region where the crack is so far open that the interatomic force is practically zero, and B in the region where Δu_i is practically zero, $V_A - V_B$ is 2γ and we have

$$F_1 = \mathcal{G} = 2\gamma \tag{71}$$

or the Griffith condition for brittle cracks.

Equation (68) can also be used in other cases. For example, in the models of Dugdale[28] and of Bilby, Cottrell, and Swinden[29] beyond the crack tip there is a narrow plastic zone in which $p_{22} = \sigma_Y$. If S embraces both tip and plastic zone Eq. (68) gives

$$F_1 = \sigma_Y\, \Delta u_2(\text{tip}) \tag{72}$$

where Δu_2 (tip) is the so-called crack opening displacement. More generally, if the narrow plastic zone exhibits plastic behavior too complicated to be expressed by a constant yield stress, the integral Eq. (68) is still the plastic work required to extend unit length of the zone to rupture.

The argument which gave Eq. (71) can also be applied locally to the three-dimensional case of a penny-shaped crack, or a similar flat crack whose periphery is not necessarily circular. Take X_3 and X_2 to be tangential to the edge and normal to the plane of the crack, and let S be a closed surface which, so to speak, takes a bite out of the crack surface and intercepts a length δl of its edge. Then the process which led from Eq. (66) to Eq. (71) gives

$$\int_S P_{1j}\,dS_j = 2\gamma\delta l \tag{73}$$

which shows that Eq. (62) applies locally all along the edge of a curved crack. This is fortunate since the usual proof is not entirely convincing. For the two-dimensional case we assumed that Δu_2 for the lengthened crack was the same as the original Δu_2 advanced bodily. Although this is wrong one can easily show that the answer is right. But to apply the same argument to a finite length of the edge we must let the crack change its shape by putting out a tongue, and then a certain amount of goodwill is necessary.

10 MOVING CRACKS

Moving cracks sometimes attain speeds not far short of the velocity of transverse elastic waves in the material. Then the static value of $\mathcal{G}$ can no longer be used. In this section we review methods for calculating the dynamic $\mathcal{G}$ first according to pure continuum theory, and then according to the simple Peierls-Nabarro model of the last section. Part of the analysis also applies to defects other than crack tips. Some of the results will then be tried out on a recently published solution for the elastic field near the tip of a crack moving arbitrarily in antiplane strain, and used to give a rather naive treatment of crack branching.

Craggs[30] supposed that the energy release rate per unit time $v\mathcal{G}$ for a crack tip moving with instantaneous velocity v was given by the integral

$$vG = \int_S p_{ij}\,\dot{u}_i\,dS_j \tag{74}$$

taken over a small circle S centered on the tip. Unfortunately, Eq. (74) gives wrong answers in the static limit and, further, its value depends on the shape of S. The right-hand side of Eq. (74) is the integral of minus the normal component of the energy flow vector (Eq. (54)) taken over S, and so gives the rate at which elastic and kinetic energy flows in through S. This energy either leaves via the tip, or goes to increase the energy inside S. So Eq. (74) must be amended to

$$v\mathcal{G} = \int_S p_{ij}\dot{u}_i \, ds_j - \frac{d}{dt}\int_V (W + T)\, dV \tag{75}$$

where V denotes the interior of S. If the displacement has the form

$$u_i = u_i(X_1 - vt, X_2) \tag{76}$$

so that the elastic field is transported rigidly, the energy inside varies only because of the transport of the rigid field into and out of the boundary and

$$\frac{d}{dt}\int (W + T)\, dV = -v \int (W + T)\, dS_1 \tag{77}$$

so that[26,31]

$$v\mathcal{G} = \int_S \left[p_{ij}\dot{u}_i + v(W + T)\delta_{1j}\right] dS_j \tag{78}$$

Alternatively, when Eq. (76) is valid we may replace $\dot{u}_i$ by $-vu_{i,1}$ and cancel v to get

$$\mathcal{G} = \int_S H_{1j}\, dS_j \tag{79}$$

where the tensor

$$H_{lj} = (W + T)\delta_{lj} - p_{ij}u_{i,l} \tag{80}$$

differs from the dynamic P_{lj} only in the sign of the kinetic energy term.

When Eq. (76) does not hold, the displacement can still be written in the form

$$u_i = u_i^0(X_1 - vt, X_2) + u_i'(X_1, X_2, t) \tag{81}$$

in an infinity of ways, but in favorable cases it will be possible to arrange that u_i' is negligible compared with u_i^0 close to the point $X_1 = vt, X_2 = 0$. This is so for all moving crack solutions so far published (compare, for example, with Eq. (93)). If this is so Eqs. (78) and (79) will be correct for small enough S, and we have

$$vF_1 = v\mathcal{G} = \lim_{S\to 0} \int_S \left[p_{ij}\dot{u}_i - v(W + T)\delta_{1j}\right] dS_j \tag{82}$$

or

$$F_1 = \mathcal{G} = \lim_{S\to 0} \int_S H_{1j}\, dS_j \tag{83}$$

We can also arrive at Eq. (83) and hence also Eq. (82), though rather metaphysically, by arguments based on Sec. 8. In the dynamic case

$$\int_S P_{1j}\, dS_j \tag{84}$$

should give the force on the crack tip, plus the rate of change of the quasi-momentum inside S, and so

$$F_1 = \int_S P_{1j}\, dS_j - \frac{d}{dt}\int_V g_1\, dv \tag{85}$$

If Eq. (76) holds the second term is

$$v\int g_1\, dS_1 = \int 2T\, dS_1 \tag{86}$$

because $g_1 = -\rho\dot{u}_i u_{i,1} = \rho\dot{u}_i\dot{u}_i$, and we recover Eq. (83). The rest of the argument goes as before.

If the crack tip is moving with a velocity v_l arbitrarily inclined to the X_1-axis, Eq. (83) becomes

$$F_l = \lim_{S\to 0} \int_S H_{lj}\, dS_i \tag{87}$$

This gives the correct value for the component of force parallel to v_l and, if the displacement has the form

$$u_i = u_i^0(X_m - v_m t) + u'_i(X_m, t) \tag{88}$$

generalizing Eq. (81), the component perpendicular to v_l is zero if u'_i ultimately makes no contribution, and so we need not worry about its significance.

Although in deriving Eq. (81) we have introduced convective terms, the velocity v_l of the singularity does not appear in Eq. (87). This is because v_l is implicitly defined by the field variables themselves. For any nonstatic displacement function we can define a velocity field $v_l(X_m)$ given by the solution of the matrix equation

$$v_l(X_m)u_{i,l} = -\dot{u}_i \tag{89}$$

If u_i has the form of Eq. (88), $v_l(X_m)$ will take the constant value v_l near the singularity, and in the arguments above we have at several points eliminated $v_l = (v, 0, 0)$, or introduced it and cancelled it out by replacing one side of Eq. (89) by the other.

There seems to be nothing really specific to crack tips in the analysis which led to Eq. (87), and this suggests that it gives the force on a defect generally, in three dimensions as well as two. For static problems H_{lj} reduces to P_{lj}, and, as its divergence vanishes, we can ignore the limit sign and recover our original static results. In the dynamic case the divergence vanishes to the extent that the contribution of u'_i (Eq. (88)) is negligible, so that in the limit Eq. (87) should be independent of the shape as well as size of S.

We can test Eq. (87) on some already worked-out examples. But from the present point of view the results are disappointing because, in fact, a crack tip is almost the only sensible singularity for which the force can be worked out cleanly on the basis of continuum mechanics alone, without additional physical considerations. If one evaluates Eq. (87) for a kink, or a mobile point defect (Frenkel caterpillar), with S a sphere of radius r, or a dislocation with S a cylinder of radius r the results diverge, respectively, like r^{-1}, r^{-3} and $\ln r$. This is because each of them has an effective mass, similar to the electromagnetic mass of an electron, which is proportional, respectively to r_o^{-1},[32] r_o^{-3},[3] and $\ln[R(t)/r_o]$[33] where r_o is an appropriate cutoff radius, different in each case, and $R(t)$ depends on the history of the dislocation's motion. We shall see later that crack tips have no effective mass, and that is why for them the calculation goes through nicely. Whenever one has to introduce a cutoff radius in continuum calculations it is a sign that an atomic treatment is really necessary, and so we shall try to extend the model which led to Eq. (71) and apply it not only to cracks but to other defects as well.

Equations (66) and (71) remain valid in the dynamic case so long as S is taken to run along the crack faces, because the extra dynamic term $-T\delta_{ij}$ contributes nothing. However, when we open S out into the loop of Fig. 4*e* quasi-momentum terms appear. For example Eq. (71) becomes

$$\int_S P_{1j}\,dS_j - 2\gamma = \frac{d}{dt}\int_V g_1\,dV \tag{90}$$

where V is the area inside S. This says that the rate of increase of quasi-momentum in V is equal to the flux of it through S (the P_{ij} term) plus the quasi-force -2γ which surface tension exerts. It is negative because it tries to close the crack.

Apart from a few trivial changes in suffixes the analysis leading to Eq. (90) applies also to an edge or screw dislocation in the Peierls-Nabarro model, V becoming the Peierls potential. We should again get Eq. (90) with 2γ equal to the difference in Peierls potential at A and B. But if the loop S completely straddles the core of the dislocation V_A and V_B are the same, both being equal to the minimum of the Peierls potential, and we get simply

$$\int_S P_{1j}\,dS_j = \frac{d}{dt}\int_V g_1\,dV \tag{91}$$

We can treat a Frenkel caterpillar (an idealized crowdion), extended along the X_1 axis, in the same way, imagining that the lattice row of atoms which contains the extra atom interacts with the rest of the crystal through a "Peierls potential." We should again get Eq. (91) now regarded as three dimensional so that S is, say, a sphere centered on the defect and large compared with its width. We could also apply a similar argument to a kink free to glide along the X_1 axis, and once more arrive at Eq. (91).

Equation (91) says that the usual quasi-momentum balance equation is not upset by the presence of the defect. We may say that there is no net force on the defect, which is what is to be expected for a freely moving entity. The fact that the net force on the defect is zero does not, of course, stop us calculating one of the partial forces on it which add up to zero. For example,[34,32] if we supposed that a dislocation or kink merely oscillated about a fixed mean position when an elastic wave was incident on it we should find that Eq. (91) was not satisfied. To prevent it shooting away we should have to apply an additional steady force, say by imposing a constant stress, and this steady force, with reversed sign, is, in a sense, the force which the wave exerts on the defect. For the cases we have considered, Eq. (91) justifies the statement[34] that quasi-momentum and the force on a defect are related like ordinary force and momentum.

To conclude this section we shall apply some of the ideas developed to the simple case of a crack tip moving arbitrarily under antiplane strain. As our results will only apply to the linear theory we shall write

$$X_1 = x, X_2 = y, u_3 = w \tag{92}$$

Initially the medium is stationary and the crack runs along the x axis from $-l$ to 0. At $t = 0$ its right-hand tip begins to move according to the arbitrary law $\xi = \xi(t)$. The displacement near the tip is

$$w = A(\dot{\xi}) B(\xi) W_o \left[\frac{x - \xi}{\beta(\dot{\xi})}, y\right] + 0(R\lambda)^{3/2} \tag{93}$$

where

$$W_o\,(x, y) = \left\{\frac{1}{2}(x^2 + y^2)^{1/2} - \frac{1}{2}x\right\}^{1/2} \tag{94}$$

$$A(\dot{\xi}) = \frac{(1 - \dot{\xi}/c)^{1/4}}{(1 + \dot{\xi}/c)^{1/4}}, \quad \beta(\dot{\xi}) = \left(1 - \frac{\dot{\xi}^2}{c^2}\right)^{1/2} \tag{95}$$

$$B(\xi) = \frac{2}{\pi\mu}\int_o^{\xi} p^o_{zy}\,(x',0)(\xi - x')^{-1/2} dx' \tag{96}$$

R is the distance from the current position of the tip and $p^o_{zy}(x, 0)$ is the stress ahead of the crack before it started to move. Equation (93) ceases to be valid when reflections from the other tip or the surface of the medium overtake the moving tip. Equation (78) gives

$$\mathcal{G}(\xi, \dot{\xi}) = \frac{1}{4}\pi\mu A^2(\dot{\xi}) B^2(\xi) \tag{97}$$

If we equate this to 2γ we have an equation of motion for the crack tip which, solved for $\dot{\xi}$ and integrated, gives its trajectory under the prescribed loading. The acceleration of the tip only enters the term $O(R^{3/2})$ in Eq. (93) and does not affect $\mathcal{G}$. Consequently we may say that the crack tip has no inertia, as stated earlier. Equation (97) can be verified directly in a simple case.[36] After the tip has been moving for a time t the field outside a circle $r = ct$ is unaltered. So if we calculate the total energy in this circle and subtract it from the elastic energy inside the same circle before the motion started the result should be the time-integral of $\dot{\xi}\mathcal{G}$ from 0 to t. For the case of constant $\dot{\xi}$ and constant B this is found to be so. (B is constant if the external loading happens to make $p_{zy}{}^{o}(x, 0)$ strictly proportional to $x^{-1/2}$ and nearly constant in any case at the beginning of the motion.) In a sense this special case establishes Eq. (97) generally, without invoking Eq. (78) if we admit that the term $O(R^{3/2})$ in Eq. (93) does not affect $\mathcal{G}$.

There has been some argument[37] as to whether a crack branches when a critical velocity or, alternatively, a critical $\mathcal{G}$ value is reached. We can use Eq. (97) to give a rather crude discussion. We cannot, of course, really tackle the dynamic problem of a Y-shaped crack (for the static case see a paper by Smith[38]). However, in the case of glass at least, the angle between the branches is often small, and so we consider the limiting case of zero branching angle, Fig. 4g. This only differs from the unbranched case by the presence of an unstressed septum extending back from the tip of the crack. After branching Eq. (97) is still valid but now $\mathcal{G}$ must be equal to 4γ rather than 2γ. Since $B(\xi)$ only depends on the current length, the crack can only provide the extra energy by slowing down so as to increase A^2. In fact we must have

$$A^2(v_2) = 2A^2(v_1) \tag{98}$$

if v_1 and v_2 are the velocities just before and just after branching. Since $A(v)$ is unity for $v = 0$ and decreases monotonically to zero at $v = c$, Eq. (98) has no solution unless v exceeds the value v_{br} which is the root of $A^2(v_{br}) = 2$, namely

$$v_{br} = \frac{3}{5} c = 0.6c \tag{99}$$

If v_1 is only marginally greater than v_{br}, $A(v_2)$ is nearly unity and the branches must move very slowly, but if the crack can restrain itself from branching until v_1 is somewhat above v_{br} the branches can move move off with a reasonable velocity. It seems likely that Eq. (93) is valid generally (i.e., when reflections are not ignored and the applied stresses are, possibly, not constant) provided one replaces $A(v)B(\xi)$ by a function $C(t)$ which cannot be separated into velocity and length factors. If this is granted the values of C_1 and C_2 before and after branching can be related by matching the solutions across the circle of radius $c(t-t_1)$ which separates the two solutions if branching occurs at $t = t_1$. This gives[36] $A_2 C_1 = A_1 C_2$, which leads to the same v_{br} as before. According to our argument, therefore, there is a critical *velocity* for branching. It makes no sense to speak of a critical $\mathcal{G}$, because just before branching $\mathcal{G}$ must have the same value, dictated by the material properties, as it would if branching was not just about to occur.

When the appropriate $A(v)$ becomes available we shall be able to calculate v_{br} for the more realistic case of mode I deformation. In the meanwhile we note that two other critical velocities provided by pure continuum theory are nearly the same for mode I as they are for mode III. The first is the absolute upper limit to the tip velocity, about $0.95c$ (the Rayleigh velocity) for mode I and c for mode III. The other is the Yoffe[39,40] velocity above which the radius across which the stress is a maximum no longer lies dead ahead of the crack; it slightly exceeds $0.7c$ in both cases. If it should turn out that the branching velocity v_{br} is also nearly the same for mode I as it is for mode III, it will also be nearly the same as the value $0.6c$ quoted as a common experimental value for the terminal velocity of a crack when there is no question of branching. (This value is usually related to Roberts and Wells'[41] calculated value $0.6c$, but this is an approximation, and the true terminal velocity in their sense is the Rayleigh velocity.) If in fact v_{br} is nearly the same for modes I and III, and if the agreement with the experimental terminal velocity is not simply a coincidence, we may tentatively give the following interpretation. In glass the initial smooth crack surface (mirror) is succeeded by a so-called mist region. Microscopic examination shows that in it small tongues of crack project from the main crack into the material. This suggests that even when macroscopic branching does not occur when v_{br} is reached, microscopic branching may set in and limit the crack velocity.

11. CONCLUDING REMARKS

We have reviewed some of the energy relations in solid mechanics, and emphasized the usefulness of the energy-momentum tensor. A topic we have not touched on is the use of Rice's path-independent integral in contexts where its interpretation as an effective force is not involved.[9,26,42] There are analogs of the energy-momentum tensor in branches of continuum mechanics other than the theory of elasticity. Some of them might repay investigation.

REFERENCES

1. Burton, C. V.: *Phil. Mag.*, (5) 33:191 (1892).
2. Weiner, J. H.: *J. Appl. Phys.*, 29:1305 (1958).
3. Eshelby, J. D.: *Phil. Trans.*, A244:87 (1951).
4. Hill, E. L.: *Rev. Mod. Phys.*, 23:253 (1951).
5. Corson, E. M.: "Introduction to Tensors, Spinors, and Relativistic Wave-Equations," p. 65, Blackie and Son, London and Glasgow, 1953.
6. Bilby, B. A.: in I. N. Sneddon and R. Hill (eds.), "Prog. Solid Mech.," vol. 1, p. 331, North-Holland Publishing Company, Amsterdam, 1960.
7. Timoshenko, S., and J. N. Goodier: "Theory of Elasticity," p. 425, McGraw-Hill Book Company, New York, 1951.
8. Morse, P. M., and H. Feshbach: "Methods of Theoretical Physics," p. 323, McGraw-Hill Book Company, New York, 1953.
9. Rice, J. R.: *J. Appl. Mech.*, 35:379 (1968).
10. Eshelby, J. D.: in F. Seitz and D. Turnbull (eds.), "Prog. Solid State Physics," vol. 3, p. 79, 1956.
11. ——: in E. M. Rassweiler and W. L. Grube (eds.), "Internal Stresses and Fatigue in Metals," p. 41, North-Holland Publishing Company, Amsterdam, 1959.
12. Ardell, A. J., and R. B. Nicholson: *Acta Met.*, 14:1295 (1966).
13. Nabarro, F. R. N.: "Theory of Crystal Dislocations," Oxford University Press, 1967.
14. Truesdell, C., and R. Toupin: in S. Flügge (ed.), "Encyclopedia of Physics," vol. 3/1, p. 495, Springer-Verlag, Berlin, 1960.
15. Schweinsfeir, R. J., and C. Elbaum: *J. Phys. Chem. Solids*, 28:597 (1967).
16. Thomas, L. A., and W. A. Wooster: *Proc. Roy. Soc.*, A208:43 (1951).
17. Hill, R.: in I. N. Sneddon and R. Hill (eds.), "Prog. Solid Mech.," vol. 2, p. 247, North-Holland Publishing Company, Amsterdam, 1961.
18. Rice, J. R., and D. C. Drucker: *Int. J. Fracture Mech.*, 3:19 (1967).
19. Truesdell, C.: *J. Rat. Mech. Analysis*, 1:178 (1952).
20. Post, E. J.: *Phys. Rev.*, 118:1113 (1960).
21. Love, A. E. H.: "Mathematical Theory of Elasticity," p. 177, Cambridge University Press, 1927.
22. Brenig, W.: *Z. Phys.*, 143:168 (1955).
23. Irwin, G. R.: *J. Appl. Mech.*, 24:361 (1957).
24. Barenblatt, G. I.: in H. L. Dryden and Th. von Kármán (eds.), "Adv. Appl. Mech.," vol. 7, p. 56, 1962.
25. Bilby, B. A., and J. D. Eshelby: in H. Liebowitz (ed.), "Fracture," vol. 1, p. 99, Academic Press, Inc., New York, 1968.
26. Cherepanov, G. P.: *Int. J. Solids Structures*, 4:811 (1968).
27. Rice, J. R.: in H. Liebowitz (ed.), "Fracture," vol. 2, p. 191, Academic Press, Inc., New York, 1968.
28. Dugdale, D. S.: *J. Mech. Phys. Solids*, 8:100 (1960).
29. Bilby, B. A., A. H. Cottrell, and K. H. Swinden: *Proc. Roy. Soc.*, A272:304 (1963).
30. Craggs, J. W.: "Fracture of Solids," p. 51, Interscience, New York, 1963.
31. Atkinson, C., and J. D. Eshelby: *Int. J. Fracture Mech.*, 4:3 (1968).
32. Eshelby, J. D.: *Proc. Roy. Soc.*, A266:222 (1962).
33. ——: *Phys. Rev.*, 90:248 (1953).
34. Nabarro, F. R. N.: *Proc. Roy. Soc.*, A209:278 (1951).
35. Eshelby, J. D.: *J. Mech. Phys. Solids*, 17:177 (1969).

36. ——: in A. S. Argon (ed.), "Physics of Strength and Plasticity," Massachusetts Institute of Technology Press, Cambridge, 1969.
37. Congleton, J., and N. J. Petch: *Phil. Mag.*, 16:749 (1967).
38. Smith, E.: *J. Mech. Phys. Solids*, 16:329 (1968).
39. Yoffe, E. H.: *Phil. Mag.*, 42:739 (1951).
40. McClintock, F. A., and S. P. Sukhatame: *J. Mech. Phys. Solids*, 8:187 (1960).
41. Roberts, D. K., and A. A. Wells: *Engineering*, 178:820 (1954).
42. Hutchinson, J. W.: *J. Mech. Phys. Solids*, 16:13 (1968).

Acta metall. Vol. 33, No. 3, pp. 331–357, 1985

0001-6160/85 $3.00 + 0.00
Pergamon Press Ltd

OVERVIEW No. 41

THE INTERACTIONS OF COMPOSITION AND STRESS IN CRYSTALLINE SOLIDS

F. C. LARCHÉ
Université de Montpellier 2, 34060 Montpellier-Cedex, France

and

J. W. CAHN
Center for Materials Science, National Bureau of Standards, Gaithersburg, MD 20899, U.S.A.

(*Received* 12 *September* 1984)

Abstract—The thermodynamics of stressed crystals that can change phase and composition is examined with particular attention to hypotheses used and approximations made. Bulk and surface conditions are obtained and for each of them practical expressions are given in terms of experimentally measurable quantities. The concept of open-system elastic constants leads to the reformulation of internal elasto-chemical equilibrium problems into purely elastic problems, whose solutions are then used to compute the composition distribution. The atmosphere around a dislocation in a cubic crystal is one of several examples that are completely worked out. The effects of vacancies, and their equilibrium within a solid and near surfaces are critically examined and previous formulas are found to be first order approximations. Consequences of the boundary equations that govern phase changes are studied with several examples. Finally, problems connected with diffusional kinetics and diffusional creep are discussed.

Résumé—Nous étudions, la thermodynamique de cristaux sous contrainte qui peuvent changer de phase et de composition, en prêtant particulièrement attention aux hypothèses utilisées et aux approximations faites. Nous obtenons des conditions dans le matériau massif et en surface et pour chacune d'entr'elles nous donnons des expressions pratiques en fonction de quantités mesurables expérimentalement. Le concept de constantes élastiques d'un système ouvert conduit à reformuler les problèmes d'équilibre élastochimique interne en problèmes purement éleastiques; on utilise alors leurs solutions pour calculer la répartition des compositions. L'atmosphère autour d'une dislocation dans un cristal cubique est l'un des exemples qui sont entièrement résolus. Nous examinons d'une manière les effets des lacunes et leur équilibre critique à l'intérieur d'un solide et au voisinage de la surface et nous montrons que des formules antérieures sont des approximations du premier ordre. Nous étudions les conséquences des équations aux limites qui gouvernent les changements de phases, avec plusieurs exemples. Enfin, nous discutons des problèmes liés à la cinétique de diffusion et au fluage de diffusion.

Zusammenfassung—Das thermodynamische Verhalten von verspannten Kristallen, bei denen sich Phase und Zusammensetzung ändern können, wurde insbesondere im Hinblick auf die benützten Hypothesen und Annahmen untersucht. Volumen- und Oberflächenbedingungen werden ermittelt. Für jede Bedingung werden praktische Ausdrücke mit experimentell meßbaren Größen angegeben. Das Konzept der elastischen Konstanten offener Systeme führt zu einer neuen Formulierung der inneren elastochemischen Gleichgewichtsporbleme mit rein elastischen Problemen, mit deren Lösungen dann die Verteilung der Zusammensetzung berechnet wird. Verschiedene Beispiele werden ausführlich dargestellt, darunter die Wolke um eine Versetzung in einem kubischen Kristall. Der Einfluß der Leerstellen, ihr Gleichgewicht im Innern eines Festkörpers und in der Nähe der Oberfläche werden kritisch untersucht. Früher erhaltene Formeln stellen sich als Näherungen erster Ordnung heraus. Die Folgerungen aus den Gleichungen für die Grenzen, die die Phasenänderungen beschreiben, werden anhand einiger Beispiele behandelt. Zuletzt werden Fragen im Zusammenhang mit Diffusionskinetik und -kriechen diskutiert.

1. INTRODUCTION

The literature of the thermodynamics of solids spans more than a century and has appeared in many fields. It has been marked by long controversies, some even regarding the very existence of equilibrium under conditions of nonhydrostatic stress. The results have been used in applications to global equilibrium problems, and as local equilibrium conditions in non-equilibrium problems of diffusion, creep, electrochemistry and phase changes. The formulations have been gradually generalized to include multi-component anisotropic solids, containing vacancies and other defects, that are nonhydrostatically and nonuniformly stressed. Considerable attention has been given to multi-phase systems and to conditions of equilibrium at interfaces between phases that are in mechanical and thermal contact, that can exchange matter and under conditions of slip or no slip (incoherent and coherent resp.). In view of the importance of the field, a clarification of the controversies seems in order.

Thermodynamics lends itself to many formulations based on different definitions, conventions and notations. When properly done, all these formulations should identify the same measurable quantities and give identical relationships among them. Discrepancies arise when the formulations differ in assumptions that they make about the behaviour of matter. There are also many simplifications that may not be valid or necessary. Invalid assumptions have been made about the laws of thermodynamics and about the conditions for equilibrium. We will examine the main formulations for their assumptions to find their range of validity. Whenever possible we will identify the most general formulation and show how the other formulations follow as special cases, compare predictions, and identify sources of discrepancies. But since general formulations are often more cumbersome to apply, we will examine a set of simple applications to display how one uses the main results in this field.

It may be worthwhile to categorize broadly the main controversies and to illustrate with one simple example how they arise. These are: (1) the question of the existence of equilibrium if diffusion is permitted; (2) the various methods of distinguishing solids from fluids in a formulation. These involve models of solids and constraints on the variations that can occur in solids; (3) the definitions of chemical potential of species inside solids, since in some formulations one cannot arbitrarily add atoms to the interior of a crystal without removing other atoms or destroying vacancies; (4) how one formulates the conditions for equilibrium when the familiar minimum Gibbs-free energy which works only for constant hydrostatic pressure is inapplicable, and when so many different chemical potential conventions have been proposed; and (5) clear distinctions between the accretions that can occur at surfaces and at interior defects such as climbing dislocations, and the addition of atoms to sites inside of crystals.

In addition, there are a variety of simplifications with obvious limitations on the applicability of the results. Among them is one, homogeneity, which has lead to major misconceptions. Many situations will lead to homogeneous systems at equilibrium, but if one requires in tests for equilibrium that all variations keep the system homogeneous, one may constrain the system unnecessarily.

With these controversies in mind, let us examine the simple example of a solid cylinder containing one or more components and a straight axial dislocation. Let us first ignore surface effects and let the cylinder be infinite in all directions. Let there be no restriction on diffusion. If the solid is crystalline, an equilibrium will be reached with the dislocation retained in which the solid is inhomogeneously and nonhydrostatically stressed. If the solid is multicomponent, it will also be compositionally inhomogeneous. The system can reach an equilibrium which of course means that all diffusional flow has ceased, in spite of the shear stresses and the heterogeneity.

If the cylinder had been a highly viscous liquid in which the dislocation had been introduced by a cutting, displacing and welding procedure, the dislocation would disappear on annealing. Equilibrium would not be compatible with shear stress or heterogeneity. It is apparent that crystallinity imposes restrictions on the variations that lead to a different type of equilibrium.

Even in a one component solid, there will be a gradient in the Helmholtz-free energy density at equilibrium. Any definition of a chemical potential, that for a one-component system reduces to the local free energy per atom, cannot subsequently be used by asserting that such chemical potentials must be constant at equilibrium or, if not constant, will lead to diffusional fluxes. Care must be exercised in the definition of chemical potentials in one or multi-component systems to ensure that they are useful.

The constraint which crystallinity imposes in this example is that some of the atoms cannot be moved at will without a counterflux of some other species, including vacancies, to take their place in the crystal structure. At the surface and at the core of dislocations capable of climbing, this constraint does not apply and atoms can be inserted or removed at will.

To illustrate the importance of separate equilibrium conditions at surfaces, let the cylinder in our example have a finite radius and permit surface rearrangement. An equilibrium shape could be reached where transfer of small amounts of any species of atoms from one surface location to another does not change the appropriate free energy. This would be a thermodynamically stationary state in which all fluxes would cease, but it would be metastable or possibly unstable equilibrium because moving the dislocation out of the cylinder would lead to a lowered energy.

2. WHAT IS A SOLID

Formulations of thermodynamics differ considerably in how the essential aspects of solidity are represented mathematically. Many authors purporting to deal specifically with solids, reach conclusions that are the same as for very viscous liquids that may take a long time to reach an equilibrium that does not support shears.

Various models, composed of springs and dashpots, have been proposed to represent the viscoelastic behavior of matter. Whereas the Maxwell model creeps continuously under load, the Meyer–Kelvin–Voigt [1] solid reaches a mechanical equilibrium when the load is entirely carried by the spring. The elements of these solids do not dissolve or diffuse, and Gibbs [2] devised a model of a solid that did both.

Gibbs introduced the idea of a solid component which does not diffuse. Like Meyer–Kelvin–Voigt's solid, it can deform elastically but it always retains its connectivity. In addition Gibbs considered surfaces, where he did permit the solid to grow by accretion or to shrink by vaporization, melting or dissolution into contacting fluids. He also incorporated the concept of a fluid component which can diffuse and distort the solid. He fully developed the thermodynamic properties of such a solid, including its equilibria and revealed a variety of surprising properties. Since the solid component was not involved in any chemical variations except at the surface, there was no need to define a chemical potential in the solid. When the solid was equilibrated with a fluid, the chemical potential of this solid component in the fluid was readily calculated. One important result was that the chemical potential in the saturated fluids in contact with a homogeneously stressed solid depends on the orientation of the surface. There is thus not only no need to define a chemical potential of the solid component, but it does not seem to be definable. The fluid component on the other hand has a defined chemical potential that is constant at equilibrium throughout all phases even if they are heterogeneously stressed. Gibbs' solid is therefore quite active chemically and yet it is different from a fluid. The key was the solid component. Even though this component can dissolve, essential solid properties are obtained.

Gibbs was strongly influenced by the law of definite proportions and required his solid component to be a single element or a stoichiometric compound. If it was a compound, the chemical potential in the saturated fluids is calculated even if the compound dissociates or reacts with the solvent. Modern examples of Gibbs solids are polymer fibers which also can absorb solvent molecules, silicate glasses in which the silicate network is the solid component while modifier ions can diffuse about. A very good example of the kind of equilibrium Gibbs was able to calculate is the bending of a damp wooden beam in which the water redistributes at equilibrium and affects the compliance. Li *et al.* [3] pointed out that mobile interstitials in metals at temperatures where the substitutional atoms did not move was a valid metallurgical example of a Gibbs solid with a fluid component. An example of the equilibria of a dissolving Gibbs solid occurs in stressed electrodes. The equations predict the effect of elastic stress on the electrode potential [4].

Solid state diffusion of every component is counter to the strict definition of Gibbs' solid component. As a result most thermodynamic formulations that permit unrestricted diffusion to take place do not ascribe to the solid any property that differs from a viscous fluid. As the example in the introduction points out, unrestricted diffusion consistent with our knowledge of the solid does permit new kinds of equilibria.

Gibbs' solid component, because it did not diffuse, served as network for defining displacement and hence strain, as well as the local composition of the fluid component. The local energy and entropy density were functions of the local strain and composition. What was needed was a network which continued to define unambiguously the same place in the solid even if all atoms were capable of diffusing. In crystalline structures, the lattice serves this function, and a thermodynamics has been developed. Robin [5] has simply let the lattice itself be the solid component, and has found that "component differences" become the exact analogues of Gibbs fluid components. Instead of modifying Gibbs' concept we have defined a network solid as one in which there is an unambiguous method of locating the same place after diffusion, and where the thermodynamic properties are functions of the strain and local composition defined by this network [6]. Gibbs solid component is one example of such a network; the lattice is another example.

Most of our work has been with simple crystal structures in which there is one type of substitutional site and one type of interstitial site. Atoms of a given species are assumed to be either substitutional or interstitial. The substitutional sites served as a network. Bravais solids where lattice sites are occupied by substitutional atoms are an example. Recently attention has focused on species which could occupy both interstitial or substitutional sites [7], and this has led to the generalization of structures in which many different sites are occupied in a unit cell and where a particular species can occupy several sites. One can even include the case where no species occupies the origin in the unit cell which serves as network marker.

In crystal structures, the network imposes what we have called the network restriction. A site exists, regardless of the species that occupies it, or even if it is empty. Atoms exchange among sites

$$A_I + B_J = A_J + B_I \qquad (2.1)$$

where I and J are different types of sites: sites that are mostly filled are occupied by what are called substitutional atoms, while sites that are mostly vacant are occupied by what are called intersitial atoms.†

Vacancies are capable of diffusing or reacting with atoms on other sites. Letting B be a vacancy, equation (2.1) becomes

$$V_I + A_J = A_I + V \qquad (2.2)$$

†The term interstitial compound is an unfortunate term in which the interstitials are merely small atoms fully occupying a site in the structure [8]. The usual definition of interstitials, that these are atoms occupying sites that are mostly empty, has important consequences in thermodynamic formulations. An empty substitutional site is called a vacancy, while empty interstitial sites are usually ignored, since their concentration or activity in e.g. the law of mass action hardly differs from unity.

where I and J are different sites. If I is an interstitial site, this can also be written

$$A_J = A_I + V_J \tag{2.3}$$

One of the main results of the network restriction is that there is no need to define separate chemical potentials of individual network species. Within the crystal only their differences are ever needed.

The network is unambiguously defined only as long as the structure is not severely distorted. The network can be modified at surfaces and dislocations and these have led to special equilibrium conditions. Of particular interest is the fact that there are differences between solid-fluid interfaces and solid–solid interfaces equilibrium conditions. Two types of solid–solid boundaries have been treated [10]; incoherent interfaces where there are two independent networks with no relationship between them and coherent interfaces where there is an exact correspondence between network sites in the two crystals, and a connectivity across the interface that survives the distortions of a phase change that transfers sites from one crystal to the other. Thus many restrictions in Gibbs' solid have been eliminated. Modern understanding of solid solutions, crystalline defects and diffusion have been incorporated. In addition, solid–solid equilibria, interfaces and phase changes have been considered.

3. DERIVATIONS OF USABLE EQUILIBRIUM CONDITIONS

3.1. Thermodynamic formulation

The basic two laws of thermodynamics are quite general and applicable not only to all equilibrium conditions but also in specifying what cannot happen in nonequilibrium conditions. They often are cumbersome to use, but from them special conditions have been derived (such as constant temperature at equilibrium) that are easier to apply. In addition, there are certain restrictions or constraints that occur commonly that permit even simpler specialized but rigorously applicable procedures to be developed. A good example is the Gibbs free energy. Under the special restriction that temperature, presssure, and the mass of various species be held constant, it can be shown that the laws of thermodyanimcs reduce to the simple condition that the Gibbs free energy monotonically decreases to a minimum. For these common restrictions, it is no longer necessary to start from the basic laws. For equilibrium, one begins with the minimization of Gibbs free energy knowing that this is fully equivalent to the basic laws. The procedure is a general one, subject only to the easily verifiable restrictions on temperature, pressure, and mass. The restrictions are important. When temperature decreases (as in an endothermic reaction held adiabatically), pressure increases or mass is added, the Gibbs free energy can increase and has lost its usefulness as a simple condition for equilibrium.

Whenever we encounter new restrictions or constraints, it is necessary to return to the two basic laws to find new conditions for equilibrium that are general, subject only to the restrictions or constraints. It is important that the restrictions or constraints are verifiable and that they be general enough to include many important situations, but not so general as to lead to cumbersome conditions. The procedures for finding simpler equilibrium conditions subject to new restrictions or constraints are straightforward and if done with mathematical rigor, need only be done once. Applications then follow from these derived conditions. The derivation often identifies the useful free energy. It is dangerous to assert conditions for equilibrium under new restrictions (some type of free energy to be minimized or some potential to be constant) without a derivation that begins with the basic two laws.

There are various derivations in the literature. They differ in the model of "what is a solid" expressed in terms of restrictions on possible variations. They also differ on whether or not they require homogeneity. They differ on whether they begin with the basic two laws, or with some derived law.

It is not difficult to start with the basic laws used by Gibbs; "For the equilibrium of any isolated system, it is necessary and sufficient that in all possible variations in the state of the system which do not alter its entropy, the variation of its energy shall either vanish or be positive" [9, p. 56]. It is quite straightforward to permit the system to be heterogeneous.

Since the general state of a solid is heterogeneous, the energy, entropy and mass of its various components will be integrals over the volume and the minimization procedure is done by standard variational calculus. Such a formulation permits the solid to change its shape by elastic deformation or by a process of network modification which we will call either accretion, dissolution or phase change.

These methods of variational calculus were used by Gibbs every time the system under consideration was not homogeneous; the influence of gravity [9, p. 144], stressed solids [2], surfaces [9, p. 238], multiphase systems [9, p. 64], etc. A variational statement of the first and second laws of thermodynamics for the multicomponent network solid has been carried out [6]. It very neatly produces all the conditions for equilibrium; mechanical, thermal and chemical, in the bulk and at the interfaces. There is usually no need to assume linearity, ideality, or isotropy. The derived equations identify and define important functions and usually can be manipulated to suggest methods of measurement.

The imposed constraints are incorporated into the formulation as Lagrange multipliers and this introduces quantities which must be constant throughout the system at equilibrium. Since sites in a unit cell or a network exist whether occupied by atoms or not, vacancies appear as a conserved species within a network. We formulated three different rules for the transfer of material across an interface [10]. Network sites could be added or subtracted to the solid at solid–fluid and at incoherent solid–solid interfaces. At a coherent solid–solid interface, a single network describes both solids, and during phase changes, sites are transferred but do not change their relative location.

3.2. State variables and notations

The procedure outlined can be followed once the state variables have been identified. With network solids, a strain can be defined. The energy density is assumed to be a function of that strain (either the usual small strain, or the deformation gradient to include the cases of large strains), of the entropy density, and of the density of the various atomic or molecular species.

The choice of the strain or deformation gradient as a state variable that describes the mechanical state of the solid by no means exhausts the possible choices. Continuum mechanicians and others [11–14] have described much more complex solids, where higher gradients of displacement or composition come in the picture. We feel that our choice is sufficient to describe many metallurgical materials. In any case, thermodynamics uses as input data the results of measurements of mechanical and thermal properties, and inadequate specification of state variables would become apparent.

Only small strain theory will be explicitly used here. The relations that are valid without this approximation have been derived [10, 15], and effects that might modify the small strain results will be mentioned and discussed in the course of this article.

The reference state for strain in the solid is quite arbitrary. It can be at zero stress, or under hydrostatic pressure, and at any arbitrary constant composition. It merely serves to identify the same point x' in a solid after composition change and strain. For many elastic energy equations, a convenient reference state is zero stress. There are also useful standard states for thermodynamic quantities. These are often at hydrostatic stress that is not zero and at definite compositions. As a result there are advantages to be flexible about the reference state for strain. We will try to point out in each application which reference state we have used.

When the point x' of a solid is displaced by u, the small strain is defined by†

$$E_{ij} = \tfrac{1}{2}(u_{i,j} + u_{j,i}). \tag{3.1}$$

A change of reference state from x' to $x''(x')$ where $x'' - x' = v$ leads in the small strain approximation to a new strain E''_{ij} given by

$$E''_{ij} = E_{ij} + \tfrac{1}{2}(v_{i,j} + v_{j,i}). \tag{3.2}$$

The density of energy, entropy and component I are respectively denoted by ε, s, ρ_I. Because the elementary volume of solid is affected by its state of strain, densities per unit volume in the deformed state always contain a strain effect. As such they are not very convenient to use. Much better variables are the densities per unit volume in the reference state. These will be noted by primed symbols. The relations between primed and unprimed densities are

$$\varepsilon'/\varepsilon = s'/s = \rho'_I/\rho_I = \rho'_0/\rho_0 \tag{3.3}$$

$$= V_0/V'_0 = 1 + E_{kk} \tag{3.4}$$

where ρ_0 is the molar density of lattice sites, and its inverse V_0 is the molar volume of lattice sites.

All of our chemical densities ρ_I and ρ'_I will be atomic or molar densities (moles/volume). This is especially preferred to mass densities when we consider vacancies as a species. It is useful to introduce dimensionless composition variables

$$c_I = \rho'_I/\rho'_0 = \rho_I/\rho_0.$$

This is the classical mole fraction for single site substitutional alloys. For an interstitial alloy with no vacancies on the substitutional sites, c_I given above is the molal composition. The mole fraction $\tilde{c}_I$ is then

$$\tilde{c}_I = \rho_I/(\rho_0 + \rho_I) = c_I/(1 + c_I)$$

which reduces to c_I at small concentrations.

3.3. Lagrange multipliers

From the entropy constraint comes the standard condition that the temperature is everywhere equal to a Lagrange multiplier, and is therefore constant. It allows us to define a Helmholtz free energy density by a Legendre transform

$$f' = \varepsilon' - \theta s' \tag{3.5}$$

which we subsequently use because it is more convenient in many practical applications.

From the conservation of mass conditions come Lagrange multipliers that differ substantially from standard fluid equilibrium, a direct consequence of the network constraint. As with fluids, conservation of N chemical components lead to N Lagrange multipliers that are constants at equilibrium. Whereas for fluids they can be identified with N chemical potentials, for a system consisting of a network solid containing N substitutional species only $N - 1$ quantities can be identified with physical processes replacing one specie with another on a site. The quantities thus identified with Lagrange multipliers differences we have called diffusion potentials. The notation is M_{IK}, where K is the dependent

†All vectors and tensors are expressed in terms of components with respect to an orthonormal axis system. Small subscripts like i and j are understood to have value 1, 2 or 3. Repeated indices are understood to be summed (Einstein convention) and subscripts preceded by a comma are derivatives, e.g.

$$E_{ii} = E_{11} + E_{22} + E_{33}$$

$$u_{i,j} = \partial u_i/\partial x_j.$$

species. Vacancies are considered a species that can be ignored in some applications. Because of their definition as Lagrange multipliers, the M_{IK}, like the temperature are constants, and take on a precise local meaning everywhere within the system

$$M_{IK} = \text{constant everywhere within the system.}$$
$$= (1/\rho_0')(\partial f'/\partial c_{IK})_{\theta, E_{ij}}. \tag{3.6}$$

Since the c_I are not independent, we have introduced the differential operator

$$(\partial/\partial c_{IK}) = (\partial/\partial c_I)_{c_J \neq I,K} \tag{3.7}$$

for a unit composition increase of species I, an equal decrease in species K, holding the composition of all other substitutional species on that site fixed. For binaries we drop the subscripts and adopt the convention $c = c_1$ and $(\partial/\partial c_{12}) = (\partial/\partial c)$.

From this definition we have

$$M_{IJ} + M_{JK} + M_{KI} = 0 \tag{3.8}$$

$$M_{IJ} = -M_{JI};\ M_{II} = 0. \tag{3.9}$$

In the case of equilibrium with a fluid, M_{IK} is equal to the difference in chemical potential of I and K in the fluid

$$M_{IK} = \mu_I^L - \mu_K^L. \tag{3.10}$$

If the vacancy is chosen as K, we have

$$M_{Iv} = \mu_I^L \tag{3.11}$$

It might seem natural to use the M_{Iv}, and keep the formalism of hydrostatic thermodynamics. This has been done in a number of formulations (7). However, it has practical drawbacks (see Section 5.5), and we have found it preferable to keep the flexibility of choice for the dependent species K.

The Nth Lagrange multiplier which we will call μ_K can not be identified in many problems. It is eliminated from all equilibrium calculations for internal equilibrium of a crystal away from surfaces and dislocations that can climb. It also is eliminated from all equilibrium calculations at coherent boundaries. Only in fluids, at incoherent boundaries and climbable dislocations can we identify μ_K with the chemical potential of the K specie.

The chemical potentials of interstitials are constant and equal to the chemical potentials of the corresponding species in the other phases

$$M_I = \mu_I^L. \tag{3.12}$$

We shall see in section 5 where multisite solids are considered, that there is no need to differentiate between substitutional and interstitial sites. An increase of composition of the interstitial species I, holding the composition of all other interstitial species fixed, results in an equivalent decrease of vacancies on interstitial sites. But unlike vacancies on substitutional sites, vacancies on interstitial sites always have a concentration close to the total number of possible sites and can be dropped from consideration. In order to standardize and simplify the notation, we also call these chemical potentials diffusion potentials, and in order to simplify the notation in the various expressions M_{IK} is understood to represent all diffusion potentials.

The restriction in the number of potentials that are necessary to calculate an equilibrium is a direct consequence of the crystalline nature of the solid and therefore should apply to the same solid under hydrostatic stress. It can be shown (Appendix 1) that in this case, the previous equations together with the boundary conditions to be discussed thereafter, are strictly equivalent to the standard conditions for equilibrium between fluids.

3.4. *Mechanical equilibrium*

The variational calculus gives us [6, 10] the very standard form of the mechanical equilibrium equation. It states that the divergence of the stress tensor is zero

$$T_{ij,j} = 0. \tag{3.13}$$

This equation is also true for the large strain case, but the derivative is with respect to variables x rather than x', a distinction that is not made in the small strain approximation. Large strain forms involving x' have been obtained [15].

3.5. *Interface conditions*

Along each interface, there are conditions for mechanical equilibrium, and a condition for phase change equilibrium. They both depend on the nature of this interface.

3.5.1. Solid–fluid interfaces. For solid–fluid interfaces, the mechanical equations state that the normal is a principal direction of stress. The principal value associated with it is equal in magnitude to the pressure in the liquid and opposite in sign. The pressure is here the classical thermodynamic pressure, which is positive in fluids, and the convention for stress is such that the stress corresponding to a tension is positive.

The phase change equation can be written

$$f - \Sigma \mu_I^L \rho_I = -P \tag{3.14}$$

where μ_I^L are the chemical potentials in the fluid, while the ρ_I and f pertain to the solid. Because of the $(N-1)$ equalities (3.10)

$$f - \sum_{I \neq K} M_{IK}\rho_I - \mu_K^L \rho_0 = -P. \tag{3.15}$$

Because $M_{KK} = 0$ the summation over all species is the same as the summation over all species but K. We can therefore drop the restriction and adopt the notation that Σ without any qualification means summation over all species I.

To simplify notation it is convenient to define the ω function as

$$\omega \equiv f - \Sigma M_{IK}\rho_I - \mu_K \rho_0 \tag{3.16}$$

where μ_K is the Lagrange multiplier associated with the Kth species. At this stage neither ω nor μ_K have physical meaning. Once all the equilibrium equations are written they will have a specific meaning, or are eliminated. In a fluid ω is equal to minus the pressure, and thus because $\mu_K = \mu_K^L$ equation (3.15) could be rewritten

$$\omega^s = \omega^L \tag{3.17}$$

We should emphasize that these equations are between unprimed quantities, that are usually not convenient to use for solids. The conversion follows equations (3.4) and gives

$$\omega'^s = -P(1 + E_{kk}). \tag{3.18}$$

3.5.2. Incoherent interfaces. Along an incoherent solid–solid boundary, the equilibrium equations are

$$T_{ij}^\alpha n_j^\alpha = \omega^\alpha n_i^\alpha \tag{3.19}$$

$$T_{ij}^\beta n_j^\beta = \omega^\beta n_i^\beta \tag{3.20}$$

$$\omega^\alpha = \omega^\beta \tag{3.21}$$

where n_i^α (resp. n_i^β) are the components of the normal to the interface oriented from α to β (resp. β to α). They all contain ω and hence the Lagrange multiplier μ_K.

Equations (3.19) and (3.20) imply that the normal is a principal stress axis and that in this case ω is the value of that principal stress. Multiplication of (3.19) by n_i^α and summation over i gives

$$\omega^\alpha = T_{ij}^\alpha n_i^\alpha n_j^\alpha. \tag{3.22}$$

From (3.20) we can obtain a similar expression for ω^β. Therefore ω^α and ω^β are identified for this problem.

Using the definition of ω we obtain

$$\mu_K = V_0^\beta(f^\beta - \Sigma\, M_{IK}\rho_I^\beta - T_{ij}^\beta n_i^\beta n_j^\beta). \tag{3.23}$$

Substituting this value of μ_K in (3.21) and (3.19) gives the equivalent system of equations

$$\mu_K = V_0^\alpha(f^\alpha - \Sigma\, M_{IK}\rho_I^\alpha - T_{ij}^\alpha n_i^\alpha n_j^\alpha)$$
$$= V_0^\beta(f^\beta - \Sigma\, M_{IK}\rho_I^\beta - T_{ij}^\beta n_i^\beta n_j^\beta) \tag{3.24}$$

$$T_{ij}^\alpha n_j^\alpha = -T_{ij}^\beta n_j^\beta = T_{kl}^\alpha n_k^\alpha n_l^\alpha n_i^\alpha. \tag{3.25}$$

Equations (3.24) and (3.25) contain only known quantities and are the usable ones. Equation (3.23) can be interpreted as a definition for the chemical potential of the K species and it is constant along the interface. Along an incoherent interface we can then calculate a chemical potential for every specie, something which is not possible at any other location within the bulk of the α and β phase. Let us note that each side of equation (3.24) depends on what specie is chosen for K. Because the expression

$$\Sigma\, M_{IK}\left(\frac{\rho_I^\alpha}{\rho_0^\alpha} - \frac{\rho_I^\beta}{\rho_0^\beta}\right) = \Sigma\, M_{IK}(c_I^\alpha - c_I^\beta)$$

is independent of K, the equation itself is independent of this choice. A comparison of (3.23) and (3.15) shows the similarities between solid–fluid and incoherent solid–solid equilibria.

3.5.3. Coherent solid interfaces. In a coherent solid–solid equilibrium, the mechanical boundary conditions

$$T_{ij}^\alpha n_j^\alpha = -T_{ij}^\beta n_j^\beta \tag{3.26}$$

indicate that the tractions (but not necessarily the stress tensor) are continuous across the interface. If the same reference state for strain is chosen for α and β the phase change equation (cf. Appendix 2) reads

$$V_0' f^\alpha - \Sigma\, M_{IK} c_I^\alpha + V_0'(-T_{ij}^\alpha n_i'^\alpha n_j'^\alpha + 2\Omega_{ij}^\alpha T_{jk}^\alpha n_i'^\alpha n_k'^\alpha)$$
$$= V_0' f^\beta - \Sigma\, M_{IK} c_I^\beta + V_0'(-T_{ij}^\beta n_i'^\beta n_j'^\beta$$
$$+ 2\Omega_{ij}^\beta T_{jk}^\beta n_i'^\beta n_k'^\beta) \tag{3.27}$$

where Ω_{ij} is the small rotation tensor

$$\Omega_{ij} = \tfrac{1}{2}(u_{i,j} - u_{j,i}). \tag{3.28}$$

For this type of interface equilibrium, the Lagrange multiplier μ_K has disappeared from the equations. In contrast to the two cases treated before, no definition of individual chemical potential for each specie arises, even at the interface. As we will see none are needed to solve problems. This is a direct consequence of the restrictions in a fully coherent phase change, where no network site is created or destroyed.

4. THE DATA BASE

We have identified a number of important thermodynamic quantities that determine the state of a system, and a number of functions of these state variables that enter into the equations of equilibrium. We now examine how one might determine these quantities from the usual quantities that are measured and available in compilations. They turn out to be identical to those used in ordinary solution thermodynamics and elasticity.

4.1. Geometric variables

The lattice constants are readily determined nonlinear functions of composition, temperature and stress. From the lattice constants in the reference state we can compute ρ_0'. From a comparison of the lattice shape in the actual state and the reference state, we can compute the strain or, if the strain is large, the deformation gradient. Since the actual state and the reference state are usually chosen to be at the same temperature but not necessarily at the same composition, the strain E_{ij} is a sum of a contribution due to composition change with no change in stress, E_{ij}^c, and one due to stress. The general case when neither contribution is isotropic has been treated [15]. The tensor E_{ij}^c is subject to the same crystal symmetry restrictions as the thermal expansion tensor [17]. For the present we will concentrate mostly on the isotropic case. Defining k such that

$$E_{ij}^c = k\delta_{ij} \tag{4.1}$$

and assuming Hooke's law of linear elasticity we can write

$$E_{ij} = \left(k - \frac{\nu}{E} T_{kk}\right)\delta_{ij} + \frac{1+\nu}{E} T_{ij}. \tag{4.2}$$

The dilatation E_{kk} is given by

$$E_{kk} = \frac{1-2\nu}{E} T_{kk} + 3k \tag{4.3}$$

In cubic crystals, E^c_{ij} is also isotropic, so that formula (4.1) is still valid.

The constant ρ_0' appears repeatedly in various formulas because elastic energy naturally appears as energy per unit volume, whereas other energies will be per mole. ρ_0' is the conversion factor that transforms one into the other. Its inverse V_0' is the molar volume of the lattice sites. Combining (3.3), (3.4) and (4.3) we have for isotropic solids

$$V_0/V_0' = \rho_0'/\rho_0 = 1 + \frac{1-2\nu}{E} T_{kk} + 3k. \tag{4.4}$$

The derivative of E^c_{ij} with respect to composition in binary alloys also occurs commonly

$$\eta_{ij} = \mathrm{d}E^c_{ij}/\mathrm{d}c. \tag{4.5}$$

For systems with orthogonal axes

$$\eta_{ii} = (\partial \ln a_i/\partial c)(\text{no summation}) \tag{4.6}$$

where the a_i are the lattice parameters. When E^c_{ij} is isotropic

$$\eta_{ij} = (\mathrm{d}k/\mathrm{d}c)\delta_{ij} = \eta\delta_{ij}. \tag{4.7}$$

In binary isotropic and cubic systems η is also related to the partial molar volumes

$$\eta = (\bar{V}_1 - \bar{V}_2)/3V_0'. \tag{4.8}$$

If η is constant

$$k = (c - c_0)(\bar{V}_1 - \bar{V}_2)/3V_0' \tag{4.9}$$

where c_0 is the composition of the reference state chosen to measure the strain. It is to be emphasized that the anisotropic and nonlinear versions of these equations are readily available (15).

4.2. *Thermochemical quantities*

The two important quantities to be determined are f' and M_{IK}. There are several convenient paths of integration from a hydrostatic state, where these quantities can be determined with standard thermodynamic methods, to the actual stressed state. We begin with the differential of f'

$$\mathrm{d}f' = T_{ij}\mathrm{d}E_{ij} - s'\mathrm{d}\theta + \rho_0'\Sigma M_{IK}\mathrm{d}c_I. \tag{4.10}$$

The function ϕ', defined by a Legendre transform

$$\phi' = f' - T_{ij}E_{ij} \tag{4.11}$$

proves to be useful. Its differential

$$\mathrm{d}\phi' = -E_{ij}\mathrm{d}T_{ij} - s'\mathrm{d}\theta + \rho_0'\Sigma M_{IK}\mathrm{d}c_I \tag{4.12}$$

permits us to deduce the following Maxwell relation

$$-\rho_0'(\partial M_{IK}/\partial T_{ij})_{c_I} = (\partial E_{ij}/\partial c_{IK})_{T_{kl}}. \tag{4.13}$$

Hooke's law at constant composition is

$$T_{ij} = C_{ijkl}(E_{kl} - E^c_{kl}) \tag{4.14}$$

or

$$E_{ij} = E^c_{ij} + S_{ijkl}T_{kl} \tag{4.15}$$

where the C_{ijkl} are moduli of elasticity, and the S_{ijkl} compliances. Both are composition and temperature dependent. From (4.15) we deduce

$$\left(\frac{\partial E_{ij}}{\partial c_{IK}}\right)_{T_{kl}} = \left(\frac{\partial E^c_{ij}}{\partial c_{IK}}\right) + \left(\frac{\partial S_{ijkl}}{\partial c_{IK}}\right) T_{kl}. \tag{4.16}$$

Chemical potentials are assumed known at a hydrostatic pressure P, and composition $c_1, c_2, \ldots$

$$M_{IK}(P, c_1, c_2, \ldots) = \mu_I(P, c_1, c_2, \ldots) - \mu_K(P, c_1, c_2, \ldots). \tag{4.17}$$

It is customary to define standard chemical potentials μ^0_I and activity coefficients such that

$$\mu_I(P, c) = \mu^0_I(P) + R\theta \ln \gamma_I c_I \tag{4.18}$$

where γ_I is chosen for convenience, depending on the problem, that it approaches 1 either for dilute or concentrated solution. Vacancy potentials also are fit to this convention. Since $\mu_v(P, \bar{c}_v) = 0$, where $\bar{c}_v$ is the equilibrium vacancy concentration at P

$$\mu^0_v(P) = -R\theta \ln \gamma_v \bar{c}_v \tag{4.19}$$

where γ_v is the vacancy activity coefficient. If it is constant, the chemical potential of vacancies under pressure P can also be written

$$\mu_v(P, c_v) = R\theta \ln (c_v/\bar{c}_v). \tag{4.20}$$

The expressions for the chemical potentials are introduced into equation (4.13) and the resulting expression integrated along a constant composition path to the stress T_{ij}. For a binary solution

$$M_{12}(T_{ij}, c) = \mu^0_1(P) - \mu^0_2(P) + R\theta \ln \frac{\gamma_1 c}{\gamma_2(1-c)} - V_0'\eta_{ij}T_{ij} - \frac{V_0'}{2}\frac{\mathrm{d}S_{ijkl}}{\mathrm{d}c} T_{ij}T_{kl} + \frac{V_0'}{2}\frac{\mathrm{d}S_{iikk}}{\mathrm{d}c} P^2 - V_0'\eta_{kk}P. \tag{4.21}$$

If the solid is isotropic, this expression becomes

$$M_{12}(T_{ij}, c) = \mu^0_1(P) - \mu^0_2(P) + R\theta \ln \frac{\gamma_1 c}{\gamma_2(1-c)} - V_0'\eta T_{kk} + \frac{V_0'}{2}\frac{\mathrm{d}}{\mathrm{d}c}\left(\frac{\nu}{E}\right)(T_{kk})^2 - \frac{V_0'}{2}\frac{\mathrm{d}}{\mathrm{d}c}\left(\frac{1+\nu}{E}\right) T_{ij}T_{ij} - 3V_0'\eta P + \frac{3V_0'}{2}\frac{\mathrm{d}}{\mathrm{d}c}\left(\frac{1-2\nu}{E}\right) P^2. \tag{4.22}$$

These expressions contain terms both linear and quadratic in stress. They simplify considerably when the elastic coefficients are not composition dependent. Equation (4.22) for instance becomes

$$M_{12}(T_{ij},c)=\mu_1^0(P)-\mu_2^0(P)+R\theta\ln\frac{\gamma_1 c}{\gamma_2(1-c)} - V_0'\eta(T_{kk}+3P). \qquad (4.23)$$

To obtain f' we calculate ϕ' with equation (4.12). It is first integrated along a path of constant composition, from pressure P to stress T_{ij}. Using Hooke's law (4.15), this gives

$$\phi'(T_{ij},c)-\phi'(P,c)=-\tfrac{1}{2}S_{ijkl}T_{ij}T_{kl}-E_{kl}^cT_{kl} +\tfrac{1}{2}S_{ijkk}P^2-E_{kk}^cP \qquad (4.24)$$

and using (4.11)

$$f'(T_{ij},c)-f'(P,c)=\tfrac{1}{2}S_{ijkl}T_{ij}T_{kl}-\tfrac{1}{2}S_{ijkk}P^2. \qquad (4.25)$$

Since under hydrostatic stress, the familiar liquid thermodynamics is valid, the Helmholtz free energy $f'(P,c)$ is known. It may be obtained from the more commonly tabulated molar Gibbs free energy G_m by subtracting PV_0 and dividing by V_0'. This gives

$$f'(P,c)=\rho_0'G_m-P\rho_0'/\rho_0. \qquad (4.26)$$

Since

$$\rho_0'/\rho_0=1+E_{kk}=1+E_{kk}^c-S_{ijkk}P \qquad (4.27)$$

one obtains, after replacement of G_m by its value as a function of composition

$$f'(P,c)=\rho_0'\{c[\mu_1^0(P)+R\theta\ln\gamma_1c] +(1-c)[\mu_2^0(P)+R\theta\ln\gamma_2(1-c)]\} -P(1+E_{kk}^c)+S_{ijkk}P^2. \qquad (4.28)$$

Combination of (4.25) and (4.28) gives the final result

$$f'(T_{ij},c)=\rho_0'\{c[\mu_1^0(P)+R\theta\ln\gamma_1c] +(1-c)[\mu_2^0(P)+R\theta\ln\gamma_2(1-c)]\} -P(1+E_{kk}^c) +\tfrac{1}{2}S_{ijkl}T_{ij}T_{kl}+\tfrac{1}{2}S_{ijkk}P^2. \qquad (4.29)$$

For an isotropic solid, this relation becomes

$$f'(T_{ij},c)=\rho_0'\{c[\mu_1^0(P)+R\theta\ln\gamma_1c] +(1-c)[\mu_2^0(P)+R\theta\ln\gamma_2(1-c)]\} -P(1+3k)-\frac{1}{2}\frac{\nu}{E}(T_{kk})^2 +\frac{1+\nu}{2E}T_{ij}T_{ij}+\frac{3(1-2\nu)}{2E}P^2. \qquad (4.30)$$

Because it always appears in the boundary conditions, the expression for the quantity $V_0'f'-M_{12}c$ is useful. Combining (4.22) and (4.30) we get, in the isotropic case

$$V_0'f'-M_{12}c=\mu_2^0(P)+R\theta\ln\gamma_2(1-c) +V_0'\Big[-P(1+3k) -\frac{1}{2}\frac{\nu}{E}(T_{kk})^2+\frac{1+\nu}{2E}T_{ij}T_{ij} +c\eta(T_{kk}+3P)+\frac{3(1-2\nu)}{2E}P^2 -\frac{1}{2}c\frac{d}{dc}\left(\frac{\nu}{E}\right)(T_{kk})^2 +\frac{1}{2}c\frac{d}{dc}\left(\frac{1+\nu}{E}\right)T_{ij}T_{ij} -\frac{3}{2}c\frac{d}{dc}\left(\frac{1-2\nu}{E}\right)P^2\Big]. \qquad (4.31)$$

When the elastic coefficients are not composition dependent, this becomes

$$V_0'f'-M_{12}c=\mu_2^0(P)+R\theta\ln\gamma_2(1-c) +V_0'[-P(1+3k). -\frac{1}{2}\frac{\nu}{E}(T_{kk})^2+\frac{1+\nu}{2E}T_{ij}T_{ij} +\frac{3(1-2\nu)}{2E}P^2 +c\eta(T_{kk}+3P)]. \qquad (4.32)$$

In a crystal of arbitrary symmetry, this expression is

$$V_0'f'-M_{12}c=\mu_2^0(P)+R\theta\ln\gamma_2(1-c) +V_0'\Big[-P(1+E_{kk}^c) +\frac{1}{2}S_{ijkl}T_{ij}T_{kl}+\frac{1}{2}S_{ijkk}P^2+c\eta_{ij}T_{ij} +\frac{c}{2}\frac{dS_{ijkl}}{dc}T_{ij}T_{kl}+c\eta_{kk}P -\frac{1}{2}c\frac{dS_{ijkk}}{dc}P^2\Big]. \qquad (4.33)$$

Expressions (4.21) to (4.23) apply to substitutional binary solutions. For interstitial binary solutions the integration along a constant composition path from the hydrostatic stress to the stress T_{ij} using (4.13) gives the elastic terms identical to those in (4.21) to (4.23). Because there is no network constraint or interstitial concentration we use (3.12) for M_I and obtain for dilute interstitial solutions

$$M_1(T_{ij},c)=\mu_1^0(P)+R\theta\ln\gamma_1c-V_0'\eta_{ij}T_{ij} -\frac{1}{2}V_0'\frac{dS_{ijkl}}{dc}T_{ij}T_{kl}-V_0'\eta_{kk}P +\frac{V_0'}{2}\frac{dS_{ijkk}}{dc}P^2. \qquad (4.34)$$

Equations for the special cases of isotropy and constant elastic coefficients are like (4.34) except that the elastic terms take the forms they have in (4.22) and

(4.23). We will see in section 5.7 that there is no need to distinguish between interstitial and substitutional solutions. Had we chosen the vacancy on the interstitial site as component 2 we could have obtained (4.34) directly from (4.21) by noting that $\mu_2^0 = 0$ for the vacancy.

5. INTERNAL EQUILIBRIUM

The study of internal equilibrium requires the simultaneous solution of the equations of elasticity and those of chemical equilibrium. The method we have found useful recognizes that the strain is a function of stress and composition. But the composition at equilibrium with a given diffusion potential is determined by the local stress alone. Thus the strain at a given diffusion potential is a function of stress alone. If we obtain this stress-strain function, we can solve these problems as if they were ordinary elastic problems, without any further regard to chemical problems whose effects are now implicitly accounted for.

There are several derivations. The simplest and most easily generalized for large strains and nonlinear effects parallels in its first steps the thermodynamic methods used to derive the relationships between isentropic (adiabatic) and isothermal elasticity. In the first section we review the main results and then apply them to various problems.

5.1. Open-system elastic constants

After a straightforward manipulation of partial derivatives, the following expression, valid for a two-component solid is obtained (Appendix 3)

$$\left(\frac{\partial E_{ij}}{\partial T_{kl}}\right)_{M_{12}} = \left(\frac{\partial E_{ij}}{\partial T_{kl}}\right)_c + V_0' \eta_{ij} \eta_{kl} \bigg/ \left(\frac{\partial M_{12}}{\partial c}\right)_{T_{mn}} \tag{5.1}$$

Making the usual small strain approximations, and an expansion of the strain around $T_{ij} = 0$ produces the constant chemical-potential form of Hooke's law

$$E_{ij} = S^*_{ijkl} T_{kl}. \tag{5.2}$$

The coefficients of the stress have been called open-system compliances, S^*, and are related to the constant composition compliances S by

$$S^*_{ijkl} = S_{ijkl} + V_0' \eta_{ij} \eta_{kl} \bigg/ \left(\frac{\partial M_{12}}{\partial c}\right)_{T_{mn}} \tag{5.3}$$

where $(\partial M_{12}/\partial c)_{T_{mn}}$ is evaluated at $T_{mn} = 0$ and where all the quantities except V_0' are functions of c. The second order terms that are neglected in this expansion have been discussed [15]. Introducing the notation

$$1/\chi = \rho_0' \left(\frac{\partial M_{12}}{\partial c}\right)_{T_{ij}=0} \tag{5.4}$$

i.e.

$$1/\chi = \frac{\rho_0' R\theta}{c}\left(1 + \frac{\partial \ln \gamma_1}{\partial \ln c}\right) \tag{5.5}$$

for interstitial solutions, and

$$1/\chi = \frac{\rho_0' R\theta}{c(1-c)}\left(1 + \frac{\partial \ln \gamma_1}{\partial \ln c}\right)$$

for substitutional binary solutions, the open systems compliances, for isotropic solids are given by

$$\begin{aligned} E^* &= E/(1 + \chi\eta^2 E) \\ \nu^* &= (\nu - \chi\eta^2 E)/(1 + \chi\eta^2 E) \\ (K^{-1})^* &= 3(1 - 2\nu^*)/E^* = K^{-1} + 9\chi\eta^2 \\ G^* &= G \end{aligned} \tag{5.6}$$

where K is the bulk modulus and G the shear modulus.

Far away from spinodals and critical points, the expression (5.3) is not very sensitive to the composition. It is then appropriate to use the values of the open-system constants, at a composition near the average composition of the specimen. The elastic coefficients become constants, and the elastic part of the problem is now independent of the compositional part. For a closed system, the obvious choice is the average composition. For a system that is in contact with a chemical reservoir, the composition at equilibrium under zero stress is usually a good choice. In the case of a very high average stress, the equlibrium composition at some high pressure may be more appropriate. With such replacement of the composition in (5.3) or (5.4) to (5.6), all the solutions of ordinary linear elasticity become directly applicable to elasto-chemical problems.

5.2. Finding the composition field

Finally even though we have eliminated the composition to solve the elastochemical problem, the composition field is easily obtained from the solution. At constant diffusion potential, composition is uniquely determined by the local stress. For a binary for example (4.21) can be solved for the composition

$$\frac{\gamma_1 c}{\gamma_2(1-c)} = \text{constant} \times \exp[\text{elastic terms}/R\theta] \tag{5.7}$$

where

$$\text{constant} = \exp[\{M_{12} - (\mu_1^0 - \mu_2^0)\}/R\theta]. \tag{5.8}$$

A useful linearized version of equation (5.7) is obtained by linearizing the elastic terms of that equation or of (4.21) to (4.23) and differentiating at constant M_{12}, P, and θ. Using (5.5) this gives

$$dc/\chi = -\eta_{ij} dT_{ij} \tag{5.9}$$

or

$$c = c_0 + \chi\eta_{ij} T_{ij} \tag{5.10}$$

where c_0 is a constant of integration and is the composition that an element of unstressed solid would have if it were in equilibrium with the system.

For the isotropic case this becomes

$$c - c_0 = \chi\eta T_{kk}. \tag{5.11}$$

Had we linearized about a hydrostatic pressure P the result would have been

$$c - c(P) = \chi\eta(T_{kk} + 3P). \tag{5.12}$$

There are several ways of evaluating the constants in (5.8) or (5.10), but basically they are all methods of evaluating M_{12} at equilibrium. If the system is in contact with a materials reservoir with specified M_{12} the answer is straightforward. If it is equilibrated with a fluid phase, equation (3.10) applies. If the composition and stress are specified at some point in the system, equation (4.21) can be used. This occurs in some problems where almost all of the solid acts as a reservoir in the sense that most of it is homogeneous in composition and stress, and that transfer of components to small inhomogeneously stressed parts of the system hardly affects the composition of the homogeneous part.

For the typical case of a closed heterogeneous system the overall composition is specified. At equilibrium the diffusion potentials become a constant whose value must be determined as part of the solution. This is a standard procedure in the method of Lagrange multipliers. Equation (5.7) is a one-parameter family of composition profiles. For each assumed value of the parameter M_{12}, we can determine the overall composition by integration. The one that satisfies the specified composition is the solution and this fixes M_{12}.

This procedure is simplified if linearization of (5.7) to give (5.10) is valid, and used to obtain c_0 from which we can obtain M_{12}. We use the conservation of mass for the entire solid of total volume Ω' in the reference state and average composition $\bar{c}$

$$\int_{\Omega'} c\,\mathrm{d}V' = \Omega'\bar{c} \tag{5.13}$$

Substituting (5.10) we obtain

$$c_0 = \bar{c} - \frac{\chi\eta_{ij}}{\Omega'}\int_{\Omega'} T_{ij}\,\mathrm{d}V' \tag{5.14}$$

which can be substituted into (4.21) to (4.23) to obtain M_{12}. Once c_0 is known we have the composition profile of the inhomogeneously stressed system

$$c - \bar{c} = \chi\eta_{ij}\left(T_{ij} - \frac{1}{\Omega'}\int T_{ij}\,\mathrm{d}V'\right) \tag{5.15}$$

or

$$c - \bar{c} = \chi\eta_{ij}(T_{ij} - \bar{T}_{ij})$$

where $\bar{T}_{ij}$ is a component of the volume averaged stress, and χ and η_{ij} are evaluated at c. This is the linearized equation for composition in a closed system.

5.3. Internal equilibrium of vacancies

We consider a single component solid with vacancies as the second component. If, as is often assumed [18], there is no relaxation around a single vacancy at any level of applied stress and the elastic constants do not depend on vacancy concentrations, the diffusion potential M_{v1}, given by equation (4.23), is a function of composition only. Therefore a constant diffusion potential would imply a vacancy composition field that is constant regardless of the stress distribution. Even with these assumptions we will later see (section 6.2) that the local equilibrium vacancy concentration at the interface does depend on stress at the interface.

A more realistic model assumes relaxation. Let the partial molar volume of vacancies differ from the molar volume of the species. If the elastic constants do not depend on vacancy concentration, equation (4.23) yields with $P = 0$

$$M_{v1} = M_{v1}^0 + R\theta\ln\frac{c_v}{1 - c_v} - (\bar{V}_v - \bar{V}_1)T_{kk}/3. \tag{5.16}$$

At equilibrium, this is constant, leading to a vacancy concentration field given by (with $c_v \ll 1$)

$$c_v = \bar{c}_v\exp\left(\frac{\bar{V}_v - \bar{V}}{3R\theta}T_{kk}\right) \tag{5.17}$$

where $\bar{c}_v$ is the equilibrium concentration of vacancies at $P = 0$.

5.4. Dislocation atmospheres

5.4.1. Atmosphere around a dislocation in an isotropic solid. Let us consider a substitutional two-component infinite isotropic solid, with a negligible concentration of vacancies. A straight edge dislocation with a Burgers vector of magnitude b is located in the solid along the z axis. If the sizes of components 1 and 2 are different, there will be a segregation around the dislocation. This problem has been solved, considering one of the atoms as a defect [19]. This means that its concentration has to be relatively small. Indeed in many cases only vacancies or interstitials are considered. These are unnecessary restrictions as we shall see.

Far from the dislocation, the solid is at composition c_0, and stress-free. Therefore we can think of this far-away solid as a chemical reservoir. The solid with the dislocation and its atmosphere has the same diffusion potential as the stress-free solid at c_0. For convenience, we choose the solid at c_0 as the reference for strain. Since we have shown that under small strain approximation, the elastic part of the problem is equivalent to a constant composition problem with the open-system elastic coefficients, equation (5.6), the stress field, with the atmosphere present, is given by

$$T_{rr} = T_{\varphi\varphi} = \frac{-Gb\sin\varphi}{2\pi(1 - \nu^*)r}$$

$$T_{r\varphi} = \frac{Gb\cos\varphi}{2\pi(1 - \nu^*)r}$$

$$T_{zz} = \frac{-Gb\nu^*\sin\varphi}{\pi(1 - \nu^*)r} \tag{5.18}$$

and the composition field is, to a first approximation, using equations (5.11) and (5.18)

$$\Delta c = -\chi\eta \frac{(1+\nu^*)Gb \sin\varphi}{(1-\nu^*)\pi r} \tag{5.19}$$

(these equations correct an algebraic error in Ref [6]). Replacing the open systems constant by their values, we finally obtain

$$T_{rr} = T_{\varphi\varphi} = \frac{-Gb(1+\chi\eta^2 E)\sin\varphi}{2\pi(1-\nu+2\chi\eta^2 E)r}$$

$$T_{r\varphi} = \frac{Gb(1+\chi\eta^2 E)\cos\varphi}{2\pi(1-\nu+2\chi\eta^2 E)r}$$

$$T_{zz} = \frac{-Gb(\nu-\chi\eta^2 E)\sin\varphi}{\pi(1-\nu+2\chi\eta^2 E)r}$$

$$\Delta c = \frac{-\chi\eta(1+\nu)Gb\sin\varphi}{\pi(1-\nu+2\chi\eta^2 E)r} \tag{5.20}$$

where the subscript 0 has been dropped from all the variables since all of them have to be evaluated at composition c_0, including the Burgers vector magnitude. In our case (substitutional solution), χ is given by equation (5.5) and η by (4.7) and (4.8).

We first note that, since χ is positive for a stable solid solution, the stresses are decreased, by a fraction of the order of $\chi\eta^2 E$. This factor tends to zero for highly dilute solutions. But for a concentrated solution, it can be significant. Taking an ideal solution, $c_0 = 0.5$, $\rho_0' = 10^5\,\mathrm{mol\,m^{-3}}$, $R\theta = 10^4\,\mathrm{J\,mol^{-1}}$, $E = 10^{11}\,\mathrm{Nm^{-2}}$, and $\eta = 0.1$ gives a value of 0.25 for $\chi\eta^2 E$. This change in the stress field, which is readily obtained here, has, to our knowledge, not been calculated within the framework of the defects model.

At low concentration, the following approximation holds

$$\chi \simeq c_0 V_0'/R\theta$$

and

$$\chi\eta \simeq \frac{c_0(\bar{V}_1 - \bar{V}_2)}{3R\theta}$$

and we can neglect $2\chi\eta^2 E$ in comparison to $(1-\nu)$ obtaining thereby the classical point-defect solution

$$\Delta c \simeq \frac{-c_0(\bar{V}_1-\bar{V}_2)(1+\nu)Gb\sin\varphi}{3\pi R\theta(1-\nu)r}.$$

But it is to be emphasized that the composition equation (5.7) can be solved exactly by numerical methods. Our result is more general in that it includes in a self-consistent way all the interactions that may be present, specifically in concentrated solutions, between the defects themselves and the defects and the matrix. In particular, it takes into account the nonideality of the solid solutions in a phenomenological way that is model independent. If no measured value is available for the activity coefficient function γ_1, specific statistical mechanical models [20–22] can of course be used and the result directly introduced in the value of χ.

5.4.2. Dislocation atmosphere in a cubic crystal. Analytic expressions are rarely known for the elastic fields caused by point-forces in a medium of arbitrary symmetry [23]. Hence the usual integral methods for calculating atmospheres cannot be used. On the other hand the introduction of open system compliances is not restricted to isotropic solids, and formulas have been developed for the most general elastic solids [15]. Because the elastic field has been found for several cases of dislocations in these nonisotropic single-component crystals, the concept is most valuable.

By a simple substitution of the open-system elastic coefficients, the same elastic calculations are valid for solid solutions equilibrated to constant diffusion potentials. The composition fields are given to first order by equation (5.10) or more exactly from the solution of equation (5.7). We shall treat the case of a [111] screw dislocation in a cubic crystal. The x_3 axis is along the dislocation, the x_2 axis is along [110] and x_1 along [112]. The stress field has been given by Steeds [24]. Because the equations are rather long, we shall derive only the composition field. In cubic crystals, the change in composition with stress is given to first order by

$$\Delta c = \chi\eta T_{kk} \tag{5.21}$$

as for the isotropic case. At constant composition, T_{kk} has the value

$$T_{kk} = \frac{3b\delta\sin 3\phi}{2\sqrt{2}\,\pi r(1-\delta\cos^2 3\phi)(1-\delta)^{1/2} S} \tag{5.22}$$

with

$$\delta = \frac{2S^2(s_{11}-2S/3)}{9(s_{44}+4S/3)\,[(s_{11}-S/3)(s_{11}-2S/3)-(s_{12}+S/3)^2]}$$

and the s_{ij} are the standard two indices compliances, referred to the cube axis. For cubic crystals, the open system compliances are

$$s_{ij}^* = s_{ij} + \chi\eta^2 \qquad i \text{ and } j < 3 \tag{5.23}$$

$$s_{ij}^* = s_{ij} \qquad i \text{ and } j > 3$$

therefore

$$S^* = S$$

and

$$\delta^* = \frac{2S^2(s_{11}^*-2S/3)}{9(s_{44}+4S/3)\,[(s_{11}^*-S/3)(s_{11}^*-2S/3)-(s_{12}^*+S/3)^2]}$$

Combining (5.21), (5.22), and (5.23), we obtain the composition field

$$\Delta c = \frac{3\chi\eta b\delta^*\sin 3\phi}{2\sqrt{2}\,\pi r(1-\delta^*\cos^2 3\phi)(1-\delta^*)^{1/2} S} \tag{5.24}$$

where all the constants that depend on the material have to be taken at c_0, the composition far away from the dislocation. This result, which is obtained by a simple algebraic manipulation has, to our knowledge, never been obtained by other methods.

5.4.3. Dislocation atmospheres: nonlinear effects. At constant diffusion potentials, when the composition changes from the unstressed to the stressed state are small, we have shown that the strain is linearly related to the stress, as in the usual theory of elasticity. But this law has a smaller range of applicability than in the constant composition case. The thermodynamics of solutions introduce nonlinear terms in the stress–strain law. When the strain is expanded as a function of stress, we have identified four second-order effects [15]: (a) nonlinear stress–strain laws at constant composition, due, for instance, to rearrangement of interstitial atoms into sites that become nonequivalent under stress; (b) change of compliances with composition; (c) deviation from Vegard's law; (d) nonlinearity of the solution thermodynamics. The first two effects have been considered within the framework of defects theories. It does not seem that the two others have been treated [25]. Since solution of nonlinear elastic problems have been found [26], they can be used, with the second-order open-system compliances, to find second-order effects on dislocation atmospheres.

5.5. Internal equilibrium of a binary substitutional solid with vacancies

We have seen in Section 4 that, for a binary substitutional solid with vacancies, in equilibrium with a fluid, the following is true

$$M_{1v} = \mu_1^L \tag{5.25}$$

$$M_{2v} = \mu_2^L \tag{5.26}$$

where μ_1^L and μ_2^L are the chemical potentials of species 1 and 2 in the fluid, It seemed therefore rather natural to use these equations, that have the same form as those for fluid equilibrium, rather than the mathematically equivalent

$$M_{12} = \mu_1^L - \mu_2^L \tag{5.27}$$

$$M_{v2} = -\mu_2^L. \tag{5.28}$$

From a theoretical point of view, there is no difference. Although these equations are valid for nonlinear inhomogeneous and anisotropic solids, we give as an example expressions for constant elastic coefficients and isotropy

$$M_{1v} = M_{1v}^0 + R\theta \ln \frac{\gamma_1 c_1}{\gamma_v c_v} - \frac{\bar{V}_1 - \bar{V}_v}{3V_0'} T_{kk} \tag{5.29}$$

$$M_{12} = M_{12}^0 + R\theta \ln \frac{\gamma_1 c_1}{\gamma_2 c_2} - \frac{\bar{V}_1 - \bar{V}_2}{3V_0'} T_{kk}. \tag{5.30}$$

The concentration of vacancies is small compared to c_1 and c_2. Measurement of c_v, γ_v and $\bar{V}_v$ are therefore subject to potentially large errors. These affect equations (5.25), (5.26), and (5.28) but not (5.27). For computational purposes, it is then better to use the second formulation. Besides, if we are only interested in the composition c_1 and c_2, we can neglect the vacancies and use only equation (5.30) for equilibrium calculations. By keeping the flexibility of choice for the dependent substitutional species, we can eliminate species whose concentration has been found to have a negligible effect on the chemical behavior of the solid solutions, including vacancies, even if they are essential to the mechanisms by which chemical equilibrium is attained.

5.6. Multisite solids

Up to this point, we have focused our attention to crystalline solids that are most common in the metallurgical world, where there is only one substitutional site, that is highly occupied, and an interstitial site that is lightly occupied. But in many instances crystals have several nonequivalent sites, occupied by mixed species of atoms or molecules or vacancies. The fraction of empty sites can vary for each type of site from 0 to 1. In the description we can or course eliminate sites that are and remain empty. They don't contribute to the energy or entropy of the system. For all other sites, we can describe their status by the densities of the atoms and the densities of vacancies on each of them. As for the substitutional site with which we have been dealing in the preceding section, there will be a constraint condition: the total density of atoms and vacancies is constant for each site. Using the method described in section 4, it can be shown that at equilibrium, the diffusion potentials are constant, equal on all sites, and equal to the corresponding difference in chemical potentials when equilibrated with a fluid

$$M_{IK}^1 = M_{IK}^2 \ldots = M_{IK}^v = \mu_I^L - \mu_K^L \tag{5.31}$$

where the superscripts label the different sites. There are cases where there is no species K that is present on all sites, or where it is not convenient to use the same K-species for all sites. The formulas can easily be transformed, using equations (3.8) and (3.9)

$$M_{IK} + M_{KJ} = M_{IJ}. \tag{5.32}$$

If a species is not present on one site, it cannot be used as the dependent species on that site, and its diffusion potential equation drops from the set of equations (5.31). The vacancies are to be considered as a species, since an exchange of an i-site vacancy for a j-site vacancy produces no change of state, exactly as the exchange of a K atom on a i-site with a K atom on a j-site.

Equations (5.31) govern the equilibrium partitioning of I atoms on the different sites. If only the total density is of interest, one can interpret equations (5.31) differently. They state that along an equilibrium path, the Helmoltz free energy density is only a function of the total density of the $(N-1)$ independent species.† Calling M_{IK} the common value of the diffusion potential for each site, we have

$$df' = s'd\theta + \Sigma M_{IK} d\rho'_I. \quad (5.33)$$

Equation (5.33) shows that the formulas developed in the preceding section can also be applied, with the total density of each species as composition variables (or the ratio ρ'_I/ρ'_0, ρ'_0 being a chosen total density, like the total density of sites, or the density of sites I, $(I = 1, \ldots, \nu)$ whatever is most useful).

In the equations used in Section 5, the interstitial site was sparsely occupied, and we used equation (4.34) for the diffusion potential of this species. But rigorously its diffusion potential is M_{1v}, where v are the vacancies on interstitial sites

$$M_{1v} = M^0_{1v} - R\theta \ln \frac{\gamma_1 c_1}{\gamma_v c_v} + \text{elastic terms}. \quad (5.34)$$

If there are ν interstitial sites per substitutional site, $\gamma_v c_v$ tends to one as c_v tends to ν. Therefore, in dilute interstitial solutions

$$M_{1v} \simeq \mu^0_1 + R\theta \ln \gamma_1 c_1 + \text{elastic terms} \quad (5.35)$$

which is the expression we have used. In almost all cases, site occupancy is either high or low. Phase transformations occur before intermediate occupancy is reached. But hydrogen in metals is an important case where the occupancy can span all the possible composition field without a phase change. In such cases, the rigorous diffusion potential has to be used. Equations for the internal equilibrium between sites have been given, with the preceding approximation by Li *et al.* [27]. It is clear that there is no need to make the distinction between interstitial and substitutional atoms. A single formalism with multisite occupation is possible, and avoids confusion that can arise if a specie occupies both substitutional and interstitial sites [7]. For most metallurgical examples species do seem to occupy only one site.

We next turn to phase change equilibrium at solid–fluid interfaces. The case of a stoichiometric compound already illustrates the principal features. Let species A completely occupy **a** equivalent sites α per unit cell, species B **b** equivalent sites β, etc. Because there is only one species on each site we cannot define a diffusion potential. In the liquid each species has a well defined chemical potential. The equation for equilibrium is

$$f - (\mathbf{a}\mu^L_A + \mathbf{b}\mu^L_B + \mathbf{c}\mu^L_C \ldots)\rho_0 = -P \quad (5.36)$$

where ρ_0 is the total density of sites in a unit cell. This is a straightforward expression of chemical equilibrium for the dissolution of the compound $A_\mathbf{a}B_\mathbf{b}C_\mathbf{c}\ldots$, which continues to hold under stress. It is Gibbs' equation (393) [9] since he quite clearly considered solids to be compounds (CP) and defined a single chemical potential μ^{CP} for them in the fluid even if they dissociated

$$\mu^{CP} = \mathbf{a}\mu_A + \mathbf{b}\mu_B + \mathbf{c}\mu_C + \ldots \quad (5.37)$$

In defining μ^{CP} there is a rigid adherence to a law of definite proportions dictated by the numbers of equivalent sites fully occupied in the crystal structure.

If we now let the α sites be occupied by several species I, J, K including vacancies we obtain diffusion potentials. Choosing species K as the counterspecies the equilibrium equation is

$$f - \rho_0 \Sigma M_{IK} c^\alpha_I - \rho_0(\mathbf{a}\mu^L_K + \mathbf{b}\mu^L_B \ldots) = -P. \quad (5.38)$$

The term in the parenthesis is the chemical potential for the stoichiometric compound $K_\mathbf{a}B_\mathbf{b}C_\mathbf{c}\ldots$ There are obvious advantages to choosing K to be the major species on site α. If site α is a lightly occupied interstitial site the compound is $B_\mathbf{b}C_\mathbf{c}\ldots$ and μ_K is set to zero.

If several sites are each occupied by more than one species the equations are not changed if a different species is chosen as counter species for each site. If the same species is chosen as counter species of several sites the terms combine. In particular if the same counter species K is used for all sites we obtain

$$f - \rho_0 \sum_I \sum_\alpha M_{IK} c^\alpha_I - (\mathbf{a} + \mathbf{b} + \mathbf{c} + \ldots)\mu_K \rho_0 = -P. \quad (5.39)$$

Summing over all sites we obtain

$$f - \rho_0 \Sigma M_{IK} c_I - (\mathbf{a} + \mathbf{b} + \mathbf{c} + \ldots)\mu_K \rho_0 = -P. \quad (5.40)$$

This is identical with equation (3.15) if we redefine ρ_0 in terms of atom site density instead of unit cell densities.

6. INTERFACE EQUILIBRIA

In this section we illustrate various aspects of equilibria involving three kinds of interfaces that stressed solids can have but ignoring capillary effects. Most of our examples will be uniformly stressed, and have only as many components as are necessary to illustrate the points to be made. When the solid is multicomponent and nonuniformly stressed, the interior equilibria can be solved by the methods of the open-system elastic constants of the previous section. This converts a multicomponent elastic and thermochemical problem into an elastic problem alone, although possibly a nonlinear one.

6.1. Change of solubility with stress

Our first example is a Gibbs solid—a pure substance for instance—in equilibrium at pressure P with a fluid in which it can dissolve along a flat interface. Forces are applied to the solid so that its state of

†When a function $F(x_1, x_2, \ldots, x_n)$ is such that, for all values of the x_i

$$\partial F/\partial x_1 = \partial F/\partial x_2 = \ldots = \partial F/\partial x_n$$

then F is a function only of the sum $(x_1 + x_2 + \ldots x_n)$.

stress is now T_{ij}. To maintain mechanical equilibrium, one of the principal values of T_{ij} is $-P$, and the corresponding principal direction of stress is normal to the fluid–solid interface. What is the change in the chemical potential of the fluid necessary to keep the system in chemical equilibrium? The only equation, besides mechanical equation, is the boundary conditions, equation (3.18) which becomes for a one component linear elastic solid

$$f' - \mu^L \rho'_0 = -P(1 + E_{kk}). \qquad (6.1)$$

Following Gibbs [9, p. 196], we compare this equilibrium with that of the same solid phase equilibrated under hydrostatic stress with the same fluid. Using bars to indicate the values of the thermodynamic quantities in this equilibrium we write

$$\bar{f}' - \bar{\mu}^L \rho'_0 = -P(1 + \bar{E}_{kk}). \qquad (6.2)$$

Subtracting these two equations, we obtain

$$f' - \bar{f}' + P(E_{kk} - \bar{E}_{kk}) = \rho'_0(\mu^L - \bar{\mu}^L) \qquad (6.3)$$

$(f' - \bar{f}')$ is the elastic energy stored in the solid on going from pressure P to stress state T_{ij} and $P(E_{kk} - \bar{E}_{kk})$ is the work done on the solid by the liquid. The l.h.s. of equation (6.3) is thus the work that has to be done to bring a hydrostatically stressed solid to the nonhydrostatic state while surrounded by the liquid. It is necessarily positive, and the fluid in equilibrium with a nonhydrostatically stressed solid is always supersaturated with respect to precipitating a hydrostatically stressed solid by the amount given in (6.3). If we let c_L and $\bar{c}_L$ be the concentration of the solid component in the fluid in equilibrium with respect to the nonhydrostatically and hydrostatically stressed solid, we can use equation (4.30) to obtain

$$\rho'_0 R\theta \ln(\gamma_L c_L / \bar{\gamma}_L \bar{c}_L) = -\frac{\nu}{2E}(T_{kk})^2 + \frac{1+\nu}{2E} T_{ij}T_{ij} + \frac{3(1-2\nu)}{2E}P^2 + \frac{1-2\nu}{E}T_{kk}P. \qquad (6.4)$$

Let t_1, t_2, and $-P$ be the principal values of stress. If the change in solubility is small, and the solution is dilute or ideal, we get

$$\frac{c_L - \bar{c}_L}{\bar{c}_L} = \frac{1}{2\rho'_0 R\theta E}[t_1^2 + t_2^2 - 2\nu t_1 t_2 + 2(1-\nu)(t_1 + t_2 + P)P]. \qquad (6.5)$$

Because $-1 < \nu < 1/2$, the right hand side of equation (6.5) is positive, except of course when $t_1 = t_2 = -P$, where it is zero. The solubility of the solid in the liquid is always increased when a stress is applied to the solid. The solution is supersaturated with respect to a hydrostatically stressed solid at pressure P, a classical result that was derived by Gibbs.

We now turn to the case of a two-component solid in equilibrium with a melt. We have two conditions from equilibrium

$$f' - \mu_1^L \rho'_1 - \mu_2^L \rho'_2 = -P(1 + E_{kk}) \qquad (6.6)$$

$$M_{12} = \mu_1^L - \mu_2^L. \qquad (6.7)$$

We compare again to the equilibrium of the solid with the fluid under pressure P.

$$\bar{f}' - \bar{\mu}_1^L \bar{\rho}'_1 - \bar{\mu}_2^L \rho'_2 = -P(1 + \bar{E}_{kk}) \qquad (6.8)$$

$$\bar{M}_{12} = \bar{\mu}_1^L - \bar{\mu}_2^L \qquad (6.9)$$

Subtraction of (6.8) from (6.6) and (6.9) from (6.7) gives two equations for the change of composition in the fluid and the solid to maintain equilibrium under stress.

Assuming for simplicity (i) $P = 0$, (ii) terminal solutions (i.e. both solid and liquid are dilute solutions), (iii) no change in elastic coefficients with composition, we get

$$R\theta \ln\left(\frac{1-c}{1-\bar{c}}\right) + V'_0\left[-\frac{1}{2}\frac{\nu}{E}(t_1 + t_2)^2 + \frac{1+\nu}{2E} \times (t_1^2 + t_2^2) + c\eta(t_1 + t_2)\right] = R\theta \ln\left(\frac{1-c_L}{1-\bar{c}_L}\right) \qquad (6.10)$$

$$R\theta \ln\frac{c(1-\bar{c})}{\bar{c}(1-c)} - V'_0\eta(t_1 + t_2) = R\theta \ln\frac{c_L(1-\bar{c}_L)}{\bar{c}_L(1-c_L)}. \qquad (6.11)$$

As usual, this system of equations can be solved numerically, or, if the changes are small, we can linearize the equations, and solve with Cramer's rule.

6.2. Vacancies equilibrium in a one component solid

Consider a cylinder of isotropic hydrostatically stressed solid in contact with a fluid in which it cannot dissolve at pressure P, with an equilibrium concentration of vacancies $\bar{c}_v$. A load is applied that produces a stress whose components are T_{zz}, $T_{rr} = T_{\theta\theta}$. We want to calculate the equilibrium concentration of vacancies along the surfaces S_r and S_z. Since the components of the solid don't appear in the fluid, there is no equation like (3.12). But the phase change equation (3.15) applies, and in this case since μ_K is identified with $\mu_v^L = 0$, the equation becomes

$$V'_0 f' - (1 - c_v)M_{1v} = -PV'_0(1 + E_{kk}) \qquad (6.12)$$

where $-P$ is the normal traction. Let us first adopt Herring's simplifying assumptions (a) that there is no volume relaxation around vacancies, (b) that there is no change in elastic constants with vacancy concentration, and (c) that the solid obeys the law of dilute solutions. Using (4.32) we get (i) under pressure $\bar{P}$

$$\mu_v^0(\bar{P}) + R\theta \ln \bar{c}_v = 0 \qquad (6.13)$$

(ii) under stress, along S_z

$$\mu_0^v(\bar{P}) + R\theta \ln c_v^z + V'_0\left[-\bar{P} - \frac{1}{2}\frac{\nu}{E}(2T_{rr} + T_{zz})^2 + \frac{1+\nu}{2E}(2T_{rr}^2 + T_{zz}^2) + \frac{3(1-2\nu)}{2E}\bar{P}^2\right] = V'_0 T_{zz}\left[1 + \frac{1-2\nu}{E}(2T_{rr} + T_{zz})\right] \qquad (6.14)$$

(iii) under stress, along S_r

$$\mu_v^0(P) + R\theta \ln c_v^r + V_0'\left[-P - \frac{1}{2}\frac{\nu}{E}(2T_{rr} + T_{zz})^2 + \frac{1+\nu}{2E}(2T_{rr}^2 + T_{zz}^2) + \frac{3(1-2\nu)}{2E}P^2\right] = V_0'T_{rr}\left[1 + \frac{1-2\nu}{E}(2T_{rr} + T_{zz})\right]. \quad (6.15)$$

It is quite clear that c_v^r and c_v^z are different, unless $T_{rr} = T_{zz}$, i.e. when the system is under hydrostatic stress. Since we have assumed no relaxation around vacancies, $\eta = 0$, and therefore according to equation (4.23), M_{1v} is different on S_z and S_r. As a result, a vacancy flux will appear. This is further discussed in section 8.4.

Making the further assumption that $P = 0$, and neglecting quadratic terms in stress, subtraction of (6.13) from (6.14) and (6.15) gives

$$\ln(c_v^r/\bar{c}_v) = V_0'T_{rr}/R\theta$$
$$\ln(c_v^z/\bar{c}_v) = V_0'T_{zz}/R\theta. \quad (6.16)$$

This is Herring's [18, 28] well known formula: to first order in stress, only the normal pressure affects the equilibrium vacancy concentration at an interface. We will get the same results, whether this interface is a solid–fluid interface or an incoherent solid–solid interface.

The order of magnitude of the quadratic terms can be easily obtained by making $T_{rr} = 0$ so that linear terms disappear in (6.15). We obtain, along S_r

$$\ln(c_v^r/\bar{c}_v) = V_0'T_{zz}^2/2E\,R\theta. \quad (6.17)$$

Within the small strain approximation, this effect is less than 1% of the effect on S_z. But there are cases where it might be significant (cf. section 8.4).

Conditions (a), (b) and (c) can easily be removed through the use of the general formulas developed in section 4. As an example we treat the case where there is a volume relaxation around a vacancy. Using (4.32), assuming $P = 0$, and following the above procedure, we get, to first order in stress

$$\ln(c_r^r/\bar{c}_r) = \frac{V_0'}{R\theta}\left[T_{rr} - \eta c_r^r(2T_{rr} + T_{zz})\right] \quad (6.18)$$

$$\ln(c_r^z/\bar{c}_r) = \frac{V_0'}{R\theta}[T_{zz} - \eta c_r^r(2T_{rr} + T_{zz})]. \quad (6.19)$$

The corrective term, proportional to η, contains the trace of the stress tensor. As such other components than the normal pressure influence the vacancy concentration at a particular interface, if elastic relaxation around vacancies are taken into account.

6.3. *Using open-system elastic constants for multicomponent phase equilibrium*

For the general multicomponent phase-equilibrium under stress, the fact that the M_{IK} are constant gives $(N-1)$ relationships between stress and composition. As shown earlier it is possible to solve these equations for composition as a function of stress and obtain the strain E_{ij}^c that results from composition changes. The result is a stress-strain relation at constant M_{IK}. This relationship was used to solve elastic problems within a single phase as if it were composed of a single component.

These same relationships apply to each individual phase in a multiphase equilibrium, but the phase change boundary conditions of section 3.5 contain a similar coupling between stress and composition. In the present section we shall demonstrate that by using open-system-elastic constants, the compositional part of these equations can also be eliminated. In fact this method allows us to treat multicomponent equilibrium as if each phase were a one-component purely elastic part of the system, and that for such a solid, the ω function is equal to the elastic energy apart from a constant [cf. equation (3.16)]. Finally once the elastic problem has been solved, the composition field is obtained by the methods of section 5.2.

We will use as an example binary isotropic linear solids, although the proof can be made for a multicomponent anisotropic system. We shall further assume constant elastic coefficients, and that, at zero stress and potential M_{12}, the composition is c. Let Δc be the change of composition due to a change of stress. Expanding f' around the unstressed state we find using (3.6) and (5.4)

$$f'(T_{ij}, c + \Delta c) = f'(0,c) + \rho_0' M_{12}\Delta c + (\Delta c)^2/\chi - \frac{\nu}{2E}(T_{kk})^2 + \frac{1+\nu}{2E}T_{ij}T_{ij}. \quad (6.20)$$

Let us consider the function

$$f'^* = f'(0,c) - \frac{\nu^*}{2E^*}(T_{kk})^2 + \frac{1+\nu^*}{2E^*}T_{ij}T_{ij} \quad (6.21)$$

where we have added to the free energy of the solid under zero stress and at potential M_{12}, an elastic energy computed with open-system elastic constants at M_{12}. Replacing these constants by their values (5.6) we obtain

$$f'^* = f'(0,c) + \frac{1+\nu}{2E}T_{ij}T_{ij} - \frac{\nu}{2E}(T_{kk})^2 + \frac{1}{2}\chi\eta^2(T_{kk})^2. \quad (6.22)$$

But the change in composition Δc is given by (5.11) so that (6.22) can be written

$$f'^* = f'(0,c) + \frac{1+\nu}{2E}T_{ij}T_{ij} - \frac{\nu}{2E}(T_{kk})^2 + (\Delta c)^2/2\chi. \quad (6.23)$$

The function $[f' - \rho_0'(c + \Delta c)M_{12}]$ that appears repeatedly in the phase change boundary equations cf. (3.24) and (3.27), is thus obtained as

$$f' - \rho_0'(c + \Delta c)M_{12} = f'^* - \rho_0' c M_{12}. \quad (6.24)$$

Or, if we replace M_{12} and $f'(0,c)$

$$f' - \rho_0'(c + \Delta c)M_{12} = -\frac{v^*}{2E^*}(T_{kk})^2 + \frac{1+v^*}{2E^*}T_{ij}T_{ij} - \rho_0'\mu_2(0,c). \quad (6.25)$$

Thus the various phase change boundary conditions are expressed in terms of an open-system Helmholtz free energy for each phase. This free energy has the same form as a Helmholtz free energy of a one-component phase. Its elastic constants are the open-system elastic constants of section (5.1). The reference state of each phase is the unstressed multicomponent phase with the same value of M. Its composition is c in (6.24) and (6.25), its lattice parameter is used to define strain, and its constant composition elastic constants are to be used in equation (5.3) or (5.6) to calculate the open-system constants.

By examination of (6.25), we can see that the use of these open-system constants allows us to treat, as far as the stress is concerned, any multicomponent system just as if it were a one-component system. Thus elastic solutions developed for one component inclusions, for instance [23], can now be used for similar multicomponent inclusions.

After finding the stress field, the results of section 5.2 can be used to obtain the composition field.

An interesting consequence of the preceding results occurs in a binary system in which both phases have the same conventional elastic constants. In an infinite single component system the Bitter–Crum theorem [16] holds. There is no elastic interaction between particles. The system is degenerate with respect to particle shape and dispersion. In a binary system if the χ or η's differ, the open system elastic constants would differ even if the conventional elastic constants do not. As a result there is now elastic interaction between particles, that is entirely the result of the compliance due to composition changes.

7. PARTIAL EQUILIBRIUM—LOCAL EQUILIBRIUM

When the general conditions for equilibrium are not satisfied, the system will tend to equilibrium. The rates of various processes are usually so different that in the time scale of an experiment we may often assume that some processes have reached equilibrium while others have not occurred at all. In this section we briefly discuss these partial equilibria. When processes are too fast for thermal and chemical relaxation, we obtain the results of classical adiabatic elasticity. The relation between isothermal constant composition elastic coefficients S^{θ}_{ijkl} and adiabatic elastic coefficients S^{s}_{ijkl} is a well known thermodynamic result [17]

$$S^{\theta}_{ijkl} = S^{s}_{ijkl} + \alpha_{ij}\alpha_{kl}\left(\frac{\partial\theta}{\partial s}\right)_{T_{ij}} = S^{s}_{ijkl} + \alpha_{ij}\alpha_{kl}\theta/C^{T} \quad (7.1)$$

α_{ij} is the thermal expansion coefficient, and C^T the heat capacity, both at constant stress.

When thermal and elastic equilibration occur but no diffusion or interface motion, we have classical isothermal elasticity. Comparing equation (5.3) and (7.1) we note that they are quite similar except that temperature instead of compositional derivatives are used. Thus the relationship between adiabatic, isothermal, and open-system elastic constants is one of increasing equilibration first with thermal and then with materials reservoirs.

Diffusion of some species, e.g. interstitials, often is orders of magnitude faster than that of other species. Such a partial equilibrium, called paraequilibrium [29], is often reached in phase transformations of multicomponent alloys. Only hydrostatic cases seem to have been treated. When stresses are important the modification from corresponding binary interstitial alloy problems seems straightforward.

Interface processes, crystal growth or dissolution and grain growth all involve network modification processes that may be quite slow. Grain boundary sliding may not occur. For calculation of such partial equilibria, the corresponding equilibrium equations must be suppressed. Polycrystalline averages of the properties can be used to obtain corresponding averages for stress and composition fields.

The most common partial equilibrium occurs when all processes except diffusion have relaxed to equilibrium. The only suppressed condition is that M_{IK} need be constant, but M_{IK} remains continuous across all interfaces that have reached equilibrium. This partial equilibrium is called local equilibrium at interfaces.

Many experiments are done under conditions where partial equilibrium is maintained while some or all of the remaining variables are observed while they relax to equilibrium. The laws of most of the relaxation processes have been studied. Interface relaxation is complicated and often nonlinear. On the other hand, heat flow in response to thermal gradients is coupled with elasticity and constitutes the subject of thermoelasticity. Diffusion in response to nonuniformity of the M_{IK} is also well understood, regardless of whether the origin of the gradients in M_{IK} are from composition gradients, stress gradients or interface conditions. The next section examines a set of problems involving diffusional equilibration under isothermal conditions with local equilibrium assumed.

8. DIFFUSIONAL KINETICS AND CREEP

Many problems of diffusion involve stress. In diffusional creep the applied stress is the motivating force for the diffusion. Compositional heterogeneity results in a self-stress that affects diffusion in a way that is too often ignored in the diffusion calculation. As we have seen, stress affects the diffusion potential and interface equilibrium conditions. It has an effect both on the rate and direction of the diffusional flux within each grain and on the boundary conditions to the diffusion equations at each interface.

Often only some of the effects of stress have been considered, or approximations have been made that ignored effects of the same order or larger than the effects considered. In this section we will examine the effects of stress on diffusion and creep, inside the grains and at interfaces, and with both applied stresses and the self-stresses that arise from the compositional inhomogeneity.

We begin with a formulation for multicomponent diffusion that is consistent with our thermodynamic formulation and has the proper invariances with respect to arbitrary choices of the species K. We then examine problems of inhomogeneous stress when the network is unaltered. Much of this was the subject of a recent overview [30] in which a hierarchy of increasingly difficult problems were discussed. We next turn our attention to diffusional network alteration phenomena, such as creep and phase change, both under applied stress and self-stress. Because of the importance of vacancies in this problem, interesting phenomena occur even in one-component systems. We reformulate and simplify the general equations to examine a few problems of diffusional creep in a one-component system with vacancies.

8.1. Multicomponent diffusion in isothermal network solids

As shown in [31] the invariant formulation of substitutional multicomponent diffusion flux J_I in an isothermal isotropic or cubic network solid† is given by

$$-J_I = \sum_{J=1}^{N} B_{IJ} \operatorname{grad} M_{JK} \quad I = 1, \ldots N \qquad (8.1)$$

B_{IJ} is a mobility, function of composition and stress at a given temperature. It has been shown that the B_{IJ} are independent of the choice of the species K. There are $(N-1)$ chemical species plus vacancies. There are $(2N-1)$ independent network restrictions on the B_{IJ}

$$\sum_I B_{IJ} = 0 \quad J = 1, \ldots N \qquad (8.2)$$

$$\sum_J B_{IJ} = 0 \quad I = 1, \ldots N. \qquad (8.3)$$

As a result there are $(N-1)^2$ independent coefficients which is the expected number of phenomenological coefficients for the diffusion of $(N-1)$ interacting species without a network constraint. It is also the number expected for $(N-1)$ interstitial species. For a one-component solid with vacancies there is only one term

$$\mathbf{J}_1 = -\mathbf{J}_v = B_{v1} \operatorname{grad} M_{v1}. \qquad (8.4)$$

Similarly for the diffusion of a single interstitial species there is one term

$$-\mathbf{J}_1 = B_1 \operatorname{grad} M_1. \qquad (8.5)$$

For a two-component substitutional solution there are four independent B. With vacancies as the K species the M_{vv} terms disappear and we have

$$-\mathbf{J}_1 = B_{11} \operatorname{grad} M_{1v} + B_{12} \operatorname{grad} M_{2v}$$
$$-\mathbf{J}_2 = B_{21} \operatorname{grad} M_{1v} + B_{22} \operatorname{grad} M_{2v}$$
$$-\mathbf{J}_v = B_{v1} \operatorname{grad} M_{1v} + B_{v2} \operatorname{grad} M_{2v}. \qquad (8.6)$$

with the restrictions that

$$B_{11} + B_{21} + B_{v1} = 0$$
$$B_{12} + B_{22} + B_{v2} = 0.$$

Using species 2 as the K species we have the same coefficients in different combinations with the diffusion potential M

$$-\mathbf{J}_1 = B_{11} \operatorname{grad} M_{12} + B_{1v} \operatorname{grad} M_{v2}$$
$$-\mathbf{J}_2 = B_{21} \operatorname{grad} M_{12} + B_{2v} \operatorname{grad} M_{v2}$$
$$-\mathbf{J}_v = B_{v1} \operatorname{grad} M_{12} + B_{vv} \operatorname{grad} M_{v2}.$$

The knowledge that B remains the same in various formulations should permit flexibility both in gathering of data and in formulating and applications.

Stress affects both B and M in the flux equations. B is affected by the level of stress alone. We expand about a stress state which can be either zero

$$B_{JKij} = B^0_{JKij}(c, \theta) + B^1_{JKijkl}(c, \theta) T_{kl} + \ldots \qquad (8.8)$$

or some other convenient state T^0

$$B_{JKij}(c, \theta, T) = B^0_{JKij}(c, \theta, T^0) + B^1_{JKijkl}(c, \theta, T^0)(T_{kl} - T^0_{kl}) \qquad (8.9)$$

The gradient of M depends on the stress and the stress gradient. From the Maxwell equation (4.13) the coefficient of the stress gradient is the strain produced by a unit composition change

$$\left(\frac{\partial M_{JK}}{\partial T_{ij}}\right) = -V'_0 \left(\frac{\partial E_{ij}}{\partial c_{JK}}\right)_{T_{kl}} \qquad (8.10)$$

which is precisely defined and readily estimated from lattice parameter-composition data. For cubic or isotropic cases

$$\partial M_{JK} / \partial T_{ij} = -V'_0 \eta_{JK} \delta_{ij} \qquad (8.11)$$

and

$$\nabla M_{JK} = R\theta[(\nabla c_J / c_J) - (\nabla c_K / c_K)] - V'_0 \eta_{JK} \nabla(trT). \qquad (8.12)$$

Strictly this should be at the actual stress, but in most cases data for unstressed crystals should be adequate, and leads to a linear formulation. Combining (8.1) with (8.12) and retaining only terms linear in T we obtain for cubic or isotropic cases

$$-J_I = -A_I(\nabla trT) + \rho'_0 \sum_{J \neq K} D_{IJ(K)} \nabla c_J. \qquad (8.13)$$

where the factor ρ'_0 needs to be introduced since the c are defined to be dimensionless rather than molar

†The reference geometry for diffusion is usually the unstressed state. With the notation we have used, the fluxes should be noted with a prime. Since there is no confusion possible, we shall drop it here.

densities and

$$A_I = V_0' \sum_J B_{IJ} \eta_{JK}$$

$$D_{IJ(K)} = V_0' R\theta B_{IJ} \left(\frac{1}{c_J} + \frac{1}{c_K} \right). \qquad (8.14)$$

Because diffusion fluxes and gradients are independent of the choice of K, A_I and the B_{IJ} can be shown also to be independent of that choice, but to be consistent the $D_{IJ(K)}$ must depend on the choice in the way shown in (8.14). To avoid large uncertainties in the $D_{IJ(K)}$ it is again clearly advantageous to choose K to be the major species, rather than vacancies.

8.2. Diffusion without network changes

Conservation of matter is expressed by the equation

$$\rho_0' \frac{\partial c_I}{\partial t} + \operatorname{div} \mathbf{J}_I = 0. \qquad (8.15)$$

Compositional heterogeneity produces a long-range stress field and changing compositions change this field. Since stress and stress gradients affect B and M, the stress and diffusion equations have to be solved simultaneously. It has been common to ignore this mutual interaction and study either the stress resulting from diffusion or the effect of stress on diffusion alone. When the ignored effects are small, this is valid, but for most cases it is not.

A straightforward technique for solving the stress and diffusion equations has been developed [30]. As in section 5 the relationship between elastic stress and an arbitrary composition field often remains solvable and can be used to eliminate stress from the diffusion equation. Plastic stress accommodation would render this technique invalid.

A hierarchy of increasingly complicated problems was examined for cases of diffusion in binary alloys in which there was no applied stress. All stress was due to compositional heterogeniety alone.

The mutual interaction in most cases is a major factor. In the case of spinodal decomposition, it can change the sign of the diffusional flux and is responsible for the metastability between the chemical and coherent spinodal [32]. The stress effect is so long ranged that compositional heterogeneity can affect diffusion elsewhere. Fick's law which states that the flux depends only on local gradients is often not valid. Because this stress effect is proportional to the local concentration it can be neglected in dilute solutions.

Interface boundary conditions for diffusion in interstitial solutions have been examined for cases in which the network is chemically inactive. The boundary condition is a simple continuity of M at a fixed location in the reference state. It depends on the level of stress at the boundary. For local equilibrium equation (5.7) is applicable.

8.3. Diffusion with self-stress and phase-change at the boundary

In our previous work [30] on the effect of self-stress on diffusion the network was conserved at the boundary. There are many metallurgical problems, such as diffusion controled phase growth, where the network is not conserved, but where equilibrium prevails at the interface. This equilibrium is governed by equation (5.7) and a phase-change equation that depends on the nature of the boundary.

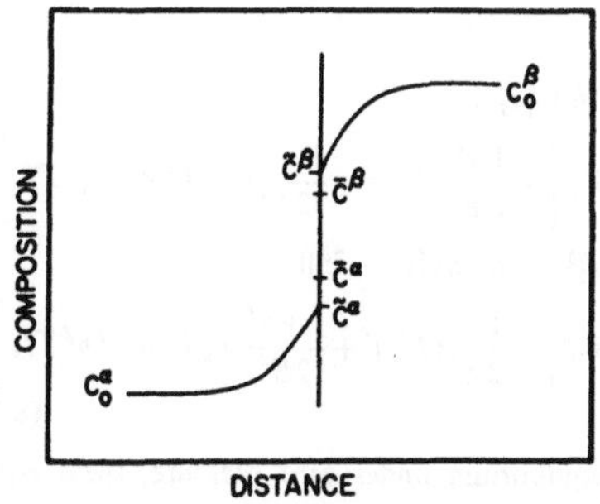

Fig. 1. Compositions in a self-stressed diffusion couple with an incoherent interface. The compositions far away from the interface are c_0^α and c_0^β. The self-stress generated by the composition gradient has shifted the equilibrium composition at the boundary to $\tilde{c}^\alpha$, $\tilde{c}^\beta$ from their unstressed phase diagram values of $\bar{c}^\alpha$, $\bar{c}^\beta$.

Self-stress is what we call the stress that is the result of sample heterogeneity. Generally its value at a point is a function of the composition distribution everywhere. For special geometries its value becomes a simple expression involving principally the local composition, and the effects of self-stress on the thermodynamic variables can be expressed in terms of the local composition only reducing self-stress problems to composition problems.

One such geometry is the semi-infinite solid with concentration fields that are functions only of the distance from the surface. We will consider the case of a semi-infinite couple, with diffusion in α and β, and an incoherent boundary. Under pressure P, the equilibrium compositions are $\bar{c}^\alpha$ and $\bar{c}^\beta$. When diffusion takes place, the compositions are c_0^α and c_0^β far away from the boundary, and $\tilde{c}^\alpha$ and $\tilde{c}^\beta$ at the boundary (Fig. 1). We shall further assume, for simplicity, that the pressure P is zero, and that the diffusing sample is under zero external pressure. This implies that the tractions are zero at the α–β boundary. We also assume no change of elastic constant with composition for either phase. Under these hypotheses, the mechanical equilibrium at the interface, equation (3.25) is always fulfilled. Equations (5.7) and (3.24) become, using (4.22) and (4.25)

$$\mu_1^{0\alpha} - \mu_2^{0\alpha} + R\theta \ln \frac{\tilde{\gamma}_1^\alpha \tilde{c}^\alpha}{\tilde{\gamma}_2^\alpha (1 - \tilde{c}^\alpha)} - V_0'^\alpha \eta^\alpha T_{kk}^\alpha$$

$$= \mu_1^{0\beta} - \mu_2^{0\beta} + R\theta \ln \frac{\tilde{\gamma}_1^\beta \tilde{c}^\beta}{\tilde{\gamma}_2^\beta (1 - \tilde{c}^\beta)} - V_0'^\beta \eta^\beta T_{kk}^\beta \qquad (8.16)$$

and

$$\mu_2^{0\alpha} + R\theta \ln[\bar{\gamma}_2^\alpha(1-\bar{c}^\alpha)] + V_0'^\alpha\left[-\frac{1}{2}\frac{\nu^\alpha}{E^\alpha}(T_{kk}^\alpha)^2 + \frac{1+\nu^\alpha}{2E^\alpha}T_{ij}^\alpha T_{ij}^\alpha + \bar{c}^\alpha\eta^\alpha T_{kk}^\alpha\right]$$
$$= \mu_2^{0\beta} + R\theta \ln[\bar{\gamma}_2^\beta(1-\bar{c}^\beta)] + V_0'^\beta\left[-\frac{1}{2}\frac{\nu^\beta}{E^\beta}(T_{kk}^\beta)^2 + \frac{1+\nu^\beta}{2E^\beta}T_{ij}^\beta T_{ij}^\beta + \bar{c}^\beta\eta^\beta T_{kk}^\beta\right]. \quad (8.17)$$

At equilibrium under zero pressure, these equations become

$$\mu_1^{0\alpha} - \mu_2^{0\alpha} + R\theta \ln\frac{\bar{\gamma}_1^\alpha\bar{c}^\beta}{\bar{\gamma}_2^\alpha(1-\bar{c}^\alpha)} = \mu_1^{0\beta} - \mu_2^{0\beta} + R\theta \ln\frac{\bar{\gamma}_1^\beta\bar{c}^\beta}{\bar{\gamma}_2^\beta(1-\bar{c}^\beta)} \quad (8.18)$$

$$\mu_2^{0\alpha} + R\theta \ln[\bar{\gamma}_2^\alpha(1-\bar{c}^\alpha)] = \mu_2^{0\beta} + R\theta \ln[\bar{\gamma}_2^\beta(1-\bar{c}^\beta)]. \quad (8.19)$$

We first have to find the stress field. In a half-space specimen, we have found [30] that its trace depends only on the local composition

$$T_{kk}^\alpha = -2Y^\alpha\eta^\alpha(\bar{c}^\alpha - c_0^\alpha) \quad (8.20)$$

$$T_{kk}^\beta = -2Y^\beta\eta^\beta(\bar{c}^\beta - c_0^\beta) \quad (8.21)$$

Where $Y = E/(1-\nu)$. Introducing these values in (8.16) and (8.17), and after subtraction of (8.18) from (8.16) and (8.19) from (8.17), we obtain the system of equations to solve for $\bar{c}^\alpha$ and $\bar{c}^\beta$. As we have seen before, it can be solved numerically or, if $(\bar{c}^\alpha - c_0^\alpha)$ and $(\bar{c}^\beta - c_0^\beta)$ are small, it can be linearized, and the resulting system of equations solved by Cramer's rule.

Under the assumption that there is no normal stress across the α–β interface, a common tangent construction is possible (see Appendix 4 for the demonstration). To the Helmholtz free energy per mole we have to add the elastic energy per mole, which is just a function of the local composition. Its value is

$$f_{el} = \frac{V_0'E}{1-\nu}\eta^2(\bar{c}^\alpha - c_0^\alpha)^2 \quad (8.22)$$

where V_0' is the molar volume at composition c_0. The construction is shown on Fig. 2. This type of construction has been used by Hillert [33] for the case of massive transformation in which case it is proper to assume that the phase which is forming is homogeneous and by Purdy *et al.* [34] for diffusion induced grain boundary migration.

8.4. *Effect of vacancies: general formulation*

When vacancies, in addition to providing a mechanism for diffusion, also interact with the stress, and provide a means of creating or destroying network at an interface, new phenomena appear, in particular diffusional creep. In this section, we consider only one-component systems, where these effects are not obscured by the phenomena previously described in this chapter. We first formulate the creep as a boundary value problem and then turn our attention to specific creep problems.

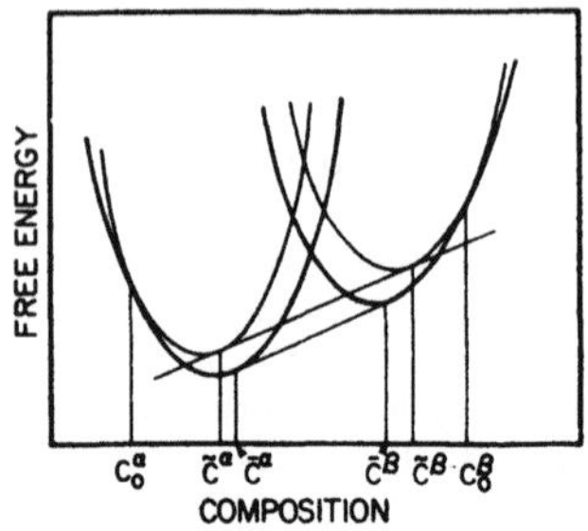

Fig. 2. Common tangent construction that gives the composition of Fig. 1. The unstressed free energies (heavy lines) are shifted by an amount equal to the elastic energy $V_0'E\eta^2(c-c_0)^2/(1-\nu)$ to give the light curves. The common tangent construction gives c^α and c^β.

The partial differential equation

The flux of vacancies J is given by

$$-J_i = \rho_0' B_{ij}(M_{vl})_{,j} \quad (8.23)$$

where B_{ij} is a tensor function of the temperature θ, c_v (the concentration of vacancies) and the stress. An expansion around $T = 0$ gives

$$B_{ij} = B_{ij}^0(c_v,\theta) + B_{ijkl}^1(c_v,\theta)T_{kl} + \ldots \quad (8.24)$$

The coefficient of order 0 is given by

$$B_{ij}^0 = D_{ij}c_v(1-c_v)/R\theta\rho_0' \quad (8.25)$$

where D_{ij} is the self-diffusion matrix. Usually it is not very much dependent on the vacancy concentration.

The tensors B_{ij}^0 and B_{ij}^1 being properties of a crystalline material follow the rules of crystalline symmetries. For isotropic materials

$$B_{ij}^0 = B^0\delta_{ij} \quad (8.26)$$

and

$$B^0 = c_v(1-c_v)D/R\theta\rho_0'. \quad (8.27)$$

The tensor B_{ijkl}^1 has the same form as an elastic tensor for an isotropic material

$$B_{ijkl}^1 T_{kl} = \beta T_{kk}\delta_{ij} + \gamma T_{ij} \quad (8.28)$$

where β and γ are two constants. This equation reveals that if the tensor B_{ij} is stress dependent, it introduces a stress-coupled anisotropy in an otherwise isotropic diffusion coefficient.

Neglecting second order effects in stress in M_{vl}, that is, assuming that the elastic coefficients do not depend on vacancy concentration, the gradient of the vacancies diffusion potential can be written

$$(M_{vl})_{,j} = \frac{R\theta}{c_v(1-c_v)}\left[1 + \frac{\partial\ln\gamma_v}{\partial\ln c_v}\right](c_v)_{,j} - V_0'\eta_{kl}T_{kl,j}. \quad (8.29)$$

If dilute solution laws apply, this equation simplifies into

$$(M_{vi})_{,j} = \frac{R\theta}{c_v}(c_v)_{,j} - V'_0\eta_{kl}T_{kl,j} \quad (8.30)$$

which, for isotropic material becomes

$$\nabla M_{v1} = (R\theta/c_v)\nabla c_v - V'_0\eta\nabla(trT). \quad (8.31)$$

The conservation equation is expressed as usual

$$\rho'_0\frac{\partial c_v}{\partial t} + J_{i,i} = s\rho'_0. \quad (8.32)$$

The source and sink terms, which is the number of vacancies created per unit volume, come, for instance, from the vacancy source at a moving dislocation. The complete diffusion equation for vacancies is obtained by combining (8.23) with (8.32)

$$\rho'_0\frac{\partial c_v}{\partial t} = \rho'_0 s + [B_{ij}(M_{v1})_{,j}]_{,i}. \quad (8.33)$$

In an isotropic solution, one gets

$$\frac{\partial c_v}{\partial t} = s + D\nabla^2 c_v - \frac{DV'_0\eta}{R\theta}\nabla c_v\cdot\nabla T_{kk} - \frac{c_v DV'_0\eta}{R\theta}\nabla^2 T_{kk}. \quad (8.34)$$

where we have neglected the stress dependence of B_{ij}.

When the relaxation of the lattice around a vacancy can be neglected, the last two terms of the r.h.s. disappear, and one obtains the simple equation

$$\frac{\partial c_v}{\partial t} = s + D\nabla^2 c_v. \quad (8.35)$$

Initial conditions

The initial conditions consist in a given vacancy concentration field. For steady state, these conditions are not needed. They are unimportant at long times, as long as a steady state can be reached.

Boundary conditions

The boundary conditions depends of course on the problem that is treated. The most useful seems to be given by an equilibrium condition along all surfaces of the solid. Written for an isotropic solid, constant elastic coefficients, a reference pressure $P = 0$ (with an equilibrium vacancy concentration c_v), dilute solution behavior, and a reference composition $c_v = 0$ for strain, this reads (equations 3.18 and 4.31)

$$\mu_v^0(0) + R\theta\ln c_v = -PV'_0\left(1 + \frac{1-2\nu}{E}T_{kk} + 3c_v\eta_v\right) - V'_0\left[-\frac{1}{2}\frac{\nu}{E}(T_{kk})^2 + \frac{1+\nu}{2E}T_{ij}T_{ij} - (1-c_v)\eta_v T_{kk}\right] \quad (8.36)$$

or

$$R\theta\ln(c_v/\bar{c}_v) = -PV'_0\left(1 + \frac{1-2\nu}{E}T_{kk} + 3c_v\eta_v\right) - V'_0\left[-\frac{1}{2}\frac{\nu}{E}(T_{kk})^2 + \frac{1+\nu}{2E}T_{ij}T_{ij} - (1-c_v)\eta_v T_{kk}\right]. \quad (8.37)$$

Since $c_v \ll 1$, these equations can be simplified into

$$\mu_v^0(0) + R\theta\ln c_v = -PV'_0\left(1 + \frac{1-2\nu}{E}T_{kk}\right) - V'_0\left[-\frac{1}{2}\frac{\nu}{E}(T_{kk})^2 + \frac{1+\nu}{2E}T_{ij}T_{ij} - \eta_v T_{kk}\right]. \quad (8.38)$$

Because it is the dominant term linear in stress, the r.h.s. is usually $-PV'_0$. Only this term was taken into account in Herring's theory of diffusional creep. We shall see in the next section cases where the quadratic terms are important for new effects.

Network modification along the surfaces due to the vacancy flux is simply given by

$$n'_i\left(\frac{\partial n'_i}{\partial t} + V_0 J_i\right) = 0 \quad (8.39)$$

where the x'_i are the coordinates of a point of the interface. This equation tells us that the shape of the specimen changes as diffusion takes place, due to the vacancy creation and annihilation at the surfaces.

Stress equilibrium

Up to now we have been concerned with the diffusion equation. Stress equilibrium in this quasistatic model obeys the partial differential equation (3.13)

$$T_{ij,j} = 0 \quad (8.40)$$

with proper boundary conditions. In most problems they will be given in terms of tractions along the surface. It is important to note that, because of the network modifications there, they are specified on a changing (and usually unknown) surface.

To specify the problem fully in term of stress, we need the Beltrami–Mitchell equations [11, 30]. For isotropic materials, the expression is

$$(1+\nu)T_{ij,kk} + T_{kk,ij} + E\eta\left[\frac{1+\nu}{1-\nu}\delta_{ij}(c_v)_{,kk} + (c_v)_{,ij}\right] = 0 \quad (8.41)$$

8.5. Some creep problems

8.5.1. Herring's classical problems: diffusional viscosity of a polycrystalline solid. Let us first show that with Herring's assumptions and approximations [18] the equations presented in section 8.4 become identical to his starting equations. Only steady state is considered. There is no volume change associated with a vacancy (i.e. the average volume of a vacancy is equal to the atomic volume). This implies $\eta = 0$; therefore the interactions between stress and composition appear only in the boundary condition pertaining to network modification. Furthermore all terms nonlinear in stress are neglected, and the reference pressure is zero. The solution of atoms and vacancies is ideal (i.e. there is no interactions with vacancies and their concentration is very small). Finally, there is no source term within a grain.

With these approximations, the diffusion equation (8.33) becomes

$$\nabla^2 M_{r1} = 0. \qquad (8.42)$$

The expression for the diffusion potential is

$$M_{1r} = \mu_1^0(0) - \mu_r^0(0) + R\theta \ln[(1 - c_v)/c_v] \qquad (8.43)$$

and the boundary condition (8.29) becomes

$$\mu_r^0(0) + R\theta \ln c_r = -PV_0'. \qquad (8.44)$$

Subtracting (8.44) from (8.43), and neglecting $\ln(1 - c_r)$, one gets

$$M_{1r} = \mu_1^0 + PV_0' \qquad (8.45)$$

This is the boundary condition used by Herring [his equation (2)] for the partial differential equation (8.42) since our P equals his $-P_{zz}$. The stress equilibrium equation is the same, and he implicitly used condition (8.37) to get the rate of displacement of the interface [e.g. to go from (3) to (4) in his paper]. Thus within the assumptions explicitly spelled out at the beginning of this section, we recover Herring's equations and boundary conditions.

His solutions combined a mean field (the average of the stress tensor within a grain is equal to the applied stress) and a perturbation analysis (the shape of the grain does not change as diffusion proceeds).

The formulation of the problem with fewer assumptions is possible using the equations of the previous section which contains important additional terms in the diffusion equation (8.33) and boundary conditions (8.29). We next explore a few problems chosen to illustrate the physical consequences of these additional terms.

8.5.2. *Quadratic effects.* Usually the linear term of the r.h.s. of (8.36) is the dominant one, but, whenever the specimen surfaces are all immersed in a fluid of constant pressure, this term is constant and at steady state does not contribute any gradients. Under these conditions the higher order terms are the only ones present. We consider two examples in which we approximate conditions for which P is constant over the surfaces of interest.

The first treated by Roitburd [35] is a pore in a specimen under uniaxial stress in which he examined the shape change by vacancy fluxes that redistributed material around the pore. Other vacancy sinks and sources were assumed so far away that fluxes between them and pores could be neglected. Because P in the pore is constant, the effects depend entirely on the quadratic terms. The result of the calculation is that a spherical pore will distort to an oblate spheroid with the minor axis along the stress axis. Because this conclusion arises from quadratic terms the same result is obtained regardless of whether the specimen is under tension or compression.

A closely related problem is a long single crystal rod of nonuniform cross section under a uniaxial load applied at the ends. If the characteristic length of the nonuniformities is short compared to the specimen length, we may examine the redistribution of material along the lateral surfaces by vacancy flux and ignore the fluxes between these surfaces and the specimen ends. Along the surface P is again constant. If we assume $\eta_v = 0$ and that the elastic constant are independent of c_v (8.36) becomes

$$\mu_r^0(0) + R\theta \ln c_r =$$

$$-V_0'\left[-\frac{1}{2}\frac{\nu}{E}(T_{kk})^2 + \frac{1+\nu}{2E}T_{ij}T_{ij}\right]. \qquad (8.46)$$

The r.h.s. is minus elastic energy of the solid. Let us note that the rod is unstable to necking. A small indentation (or any change in cross section) will produce a higher stress at its root (or at the minimum cross section). Vacancy flux will remove material from the root (or at minimum cross section) and deposit it nearby at a place of lowered elastic energy. The rod is unstable to necking by diffusion creep regardless of whether it is under tension or compression. This is the same result as Roitburd's pore, which can be considered an internal notch.

This counterintuitive result is consistent with thermodynamics. Consider the work done by the loading system, applied force times distance moved. The compliance of a rod with nonuniform cross section increases if the rod necks down, and thus the load system does work on the specimen. Conversely if the rod were to become more uniform under load, its compliance would decrease and it would have to do work on the load system. This would be in violation of thermodynamic principles.

Another interesting result of equation (8.46) is the case of a uniform rod, in which we again can ignore the ends as vacancy sources or sinks. The equation states that for $\eta_r = 0$ and elastic constants independent of c_r the equilibrium vacancy concentration is a maximum at zero stress, and is lowered equally by tensile and compressive stresses. This result is again understood if we realize that the cross-section will be reduced if vacancies leave the system, increasing the specimen's compliance. The result will be modified if we assume that the elastic constants are a function of c_r and if we let η_r differ from zero, but for small changes it will not affect the sign.

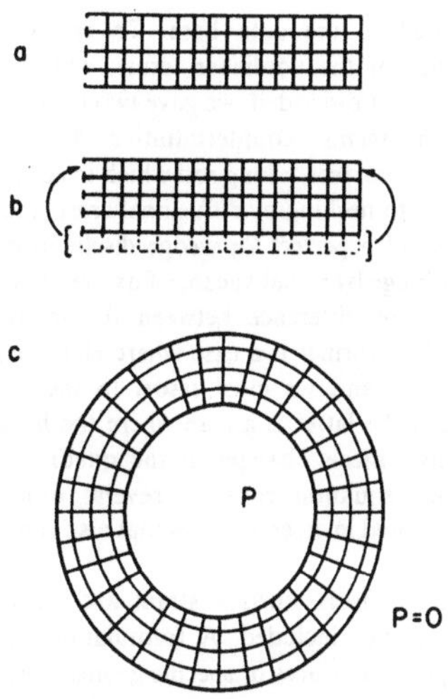

Fig. 3. Radial vacancy fluxes that remove layers from the inner surfaces and deposit them on the outer surface of a 2π wedge disclination do not enlarge the disclination and therefore no work is done by any pressure difference. To see this, consider the cross section (c) of 2π wedge disclination made by elastically bending the perfect crystal (a) into a circular cylindrical shell and joining the ends. The 2π wedge disclination after radial diffusion is unchanged because it can be made from (b) which is identical to (a) except for translation of bottom layers to top. It will therefore reach the same equilibrium geometry in the presence of the pressure differences.

8.5.3. Balancing quadratic and linear effects. The 2π wedge disclination. Linear effects do not automatically dominate quadratic effects. An interesting example where both are present and cancel identically is a hollow tube composed of a 2π wedge disclination in which there is a pressure difference between the inside and outside of the tube.

To form the 2π wedge disclination we take a rectangular sheet of a perfect single crystal, bend it into a tube and weld the seam to insure perfect matching of lattice planes (Fig. 3).

At this stage there are tangential compressive stresses at the inner surface and tensile stresses at the outer surfaces. M_{1v} at the two surfaces is the same because the stresses at the two surfaces have the same magnitude. Because of this the system reaches a vacancy equilibrium in this heterogeneously stressed system, in which vacancy gradients and stress gradients combine to give a constant M_{1v} throughout.

Now apply a pressure difference between the inside and outside and permit vacancy flow. It is readily shown that in spite of the pressure difference the value of M_{1v} at the inner surface equals that at the outer surface. In the presence of the higher pressure at the inside there is a change in elastic free energy density, a reduction at the inner surface and an increase at the outer surface, and vice versa if the sign of the pressure difference is changed. The elastic energy is quadratic in the stress, but the change in stress due to the imposed pressure difference is linear in ΔP. The result is that the linear terms in P in M_{1v} cancel identically the changes in the quadratic terms in the tangential stresses. The linear and quadratic terms balance identically to give the same M_{1v} at the two surfaces. Again an equilibrium is reached in which M_{1v} is constant throughout and vacancy concentration gradients compensate for stress gradients.

This surprising result that the 2π wedge disclination will not creep by vacancy flow even when there is a pressure difference can also be understood by considering the consequence of the transfer of an entire plane of atoms from the inside to the outside. If we start with either of the flat single crystal plates and create the disclination we see that the tube is the same whether the atom layer is transferred or not (Fig. 3).

9. SUMMARY AND CONCLUSIONS

We have reviewed and applied the thermodynamics that has been developed for multicomponent multiphase stressed crystalline solids. We have found equilibria in which the solids were neither homogeneous in stress nor composition. We have considered equilibria for three types of multiphase contact, solid–fluid, incoherent and coherent solid–solid. We have also examined simple nonequilibrium cases where potential gradients determine diffusion. Diffusional creep in particular was used to illustrate the importance of a full thermodynamic treatment.

Crystalline solids differ fundamentally from liquids in that they possess long range three-dimensional translational order. This implies that we can define a lattice and site occupancy. The number density and type of sites is known, and a local change in composition can only be made by redistributing atoms and vacancies among these sites. This fundamental restriction in the interior of a crystalline solid introduces important differences in the thermodynamics of solids compared to that of liquids. Because these restrictions apply at coherent boundaries but not at other boundaries, we find different equilibrium conditions at the various boundaries.

The equations that result from the thermodynamics constitute a set of coupled partial differential, algebraic equations and boundary conditions for stress and composition. For the kinetics, the diffusion equations are added. Although full nonlinear and large strain formulations exist, we have concentrated on examples where the essential features were displayed with small-strain approximations and linearized thermodynamics.

The thermodynamics has resulted in identifying and precisely definining the important phenomenological quantities needed for predictive calculation. The definitions in particular are important and much of the controversy in the literature is judged to be the result of inadequate definitions of quantities.

Furthermore the necessary data needed for evaluating the equations turn out to be computable from classically measured quantities, such as free energies of hydrostatically stressed solid solutions, elastic coefficients and lattice parameters.

One important method to solve the equilibrium equations uses the notion of open-system elasticity. This method eliminates the composition variable from the system of equations, and leaves a purely elastic problem to be solved. Central to the method are the open-system elastic constants, and in this paper we show that the same technique applies to multiphase solid equilibria. With this technique a large number of elastochemical problems are now solved, because they become identical to solved problems of chemically homogeneous elastic solids. Once the stress field is known, only algebraic equations have to be solved to obtain the composition in the solid. As an example of the use of this concept, we have solved the dislocation atmosphere (stress field and composition field) in an isotropic and a cubic solid, automatically taking into account, in a self-consistent way the thermodynamics of the solid solutions. Another example is the inclusion problem, although we have not found in the literature the shapes that satisfy the phase equilibrium boundary condition other than sphere, circular rod and plate.

The question of the need for definining separate chemical potentials for each chemical species inside the solid has been a subject of controversy ever since Gibbs. We hope that we have shown that problems of equilibria can be solved without defining or using them. Gibbs' famous example of a homogeneously stressed solid which gave three different chemical potential when equilibrated with three fluids each at a pressure equal to minus a principal stress should alert everyone to the danger of attempting a definition. Of course our $M_{I\nu}$ could be construed to be a chemical potential of the Ith specie, but we prefer for clarity to retain the vacancy as the counter specie.

Questions of species that occupy more than one site needed to be addressed. As our section 5.6 shows, the classical notion of chemical reactions among species on different sites very nicely resolves any confusion. Treating interstitials as atoms occupying sites that are mostly empty resulted in a unified treatment and clearly demonstrated the principle that make it possible to develop a treatment in which interstitial seem to require a different treatment.

We have supplemented an earlier overview on the effect of self-stress on diffusion by adding boundary conditions that permit phases to grow or shrink at the interface.

Diffusional creep is an important field in which the linearized and simplified treatment of Herring was an important first step. However Herring's definitions were not precise and this has led to much later confusion. We have presented a detailed derivation of a fuller treatment in which each term is fully defined and related to the data base. To emphasize the importance of the nonlinear terms, which Herring alluded to, but discarded, we gave two examples each of which seems counterintuitive but thermodynamically correct, a long rod which in compression is unstable to necking by diffusional creep and a tube composed of a perfect 2π wedge disclination which does not bulge by radial vacancy flux even when there is a pressure difference between the interior and exterior. The former is a case where Herring's linear term is zero and we must resort to the quadratic terms, and the latter is a case where the linear term identically cancels changes in the quadratic terms. The fuller equation contains several other terms usually ignored in creep theories that also can become important.

Capillary effects (surface strain and surface free energy) are not included. A formulation exists for some type of interfaces or specific geometries [36, 37]. Theories of equilibrium of stressed solids with capillarity effects for the three type of interfaces considered here are being developed [38].

Although the elastic energy is usually small compared to the free energy change resulting from a composition change, there are domains where the interactions of composition and stresses are likely to be important. Self-stresses resulting from the presence of defects or heterogeneity of the material can have sizable consequences. The depression of the consolute critical point and the spinodal is a well known example. In systems without critical points coherent equilibrium is also strongly affected. Coherent phase diagram features have recently been found [39, 40] that differ markedly from incoherent phase diagrams. The equations that could be used to calculate these phase diagrams have been obtained in sections 3 and 4.

Interesting consequences originate from the long range nature of the elastic forces. For instance this introduces nonlocal effects in the diffusion equation. Under hydrostatic pressure, a multi-phase incoherent dispersion at equilibrium is degenerate with respect to the shape of the phases, i.e. the equilibrium is independent of the shape of the precipitates. Under a more general state of stress (coherent precipitates, for instance), this simple result is no longer valid. The equilibrium equations have to be solved on an unknown boundary and the equilibrium shape is to be determined as part of the solution (a so-called free boundary-problem). With the use of the open-system elastic constants such problems can be expressed as a purely elastic problem. The phase equilibrium boundary conditions is the one that makes the problem different from classical elastic inclusion problems for which a shape is imposed. The solutions of the elastic equations of general shape will not be consistent with the phase equilibrium boundary condition. The catalog of the shapes that produce an elastic field that in turn satisfies this condition has not yet been found. The introduction of capillarity would modify this condition. Work has been done on the subject [41].

Acknowledgements—We are grateful to W. C. Johnson and R. F. Sekerka for helpful discussions and criticism. We are especially grateful to M. Hillert for questioning the need to treat interstitials differently from other species. Out of our discussion with him the ideas of section 5.6 evolved. J. Hirth kindly called our attention to misprints in [6] and [37] which have been corrected in this article.

REFERENCES

1. C. Truesdell and R. A. Toupin, The *Classical Field Theories, in Encyclopedia of Physics* (edited by S. Flugge), Vol. III/I. Springer, Berlin (1960).
2. J. W. Gibbs, *Scientific Papers*, Vol. 1, pp. 184-218. Longman, London (1906).
3. J. C. M. Li, R. A. Oriani and L. S. Darken, *Z. Phys. Chem. Neue Folge* **49**, 271 (1966).
4. L. Yang, G. T. Horne and G. M. Pound, *Proc. Symp. on Physical Metallurgy of Stress Corrosion Cracking*, Pittsburgh, p. 29. Interscience, New York (1959).
5. P.-Y. F. Robin, *Am. Mineralogist* **59**, 1286 (1974).
6. F. C. Larché and J. W. Cahn, *Acta metall.* **21**, 1051 (1973).
7. W. W. Mullins, in *Proc. Int. Conf. on Solid–Solid Phase Transformations*, Pittsburgh, p. 49. Met. Soc. A.I.M.E., (1982).
8. L. H. Bennett, A. J. McAllister and R. E. Watson, *Physics Today* **30**, 34 (1977).
9. J. W. Gibbs, *Scientific Papers*, Vol. 1. Longman, London (1906).
10. F. C. Larché and J. W. Cahn, *Acta metall.* **26**, 1579 (1978).
11. L. E. Malvern, *Introduction to the Mechanics of a Continuous Medium*. Prentice-Hall, NJ (1969).
12. J. D. Van der Waals, (translated by J. S. Rowlinson), *J. Stat. Phys.* **20**, 197 (1979).
13. J. W. Cahn and J. E. Hilliard, *J. chem. Phys.* **28**, 258 (1958).
14. E. W. Hart, *Phys. Rev.* **113**, 412 (1958).
15. F. C. Larché and J. W. Cahn, *Acta metall.* **26**, 53 (1978).
16. F. Bitter, *Phys. Rev.* **37**, 1527 (1931). M. M. Crum, as cited by F. R. N. Nabarro, *Proc. R. Soc.* **A175**, 519 (1940).
17. J. F. Nye, *Physical Properties of Crystals*. Clarendon Press, Oxford (1957).
18. C. Herring, *J. appl. Phys.* **21**, 437 (1950).
19. A. H. Cottrell, *Report of a Conference on Strength of Solids*. The Physical Society, London (1948).
20. N. Louat, *Proc. Phys. Soc.* **B69**, 459 (1956).
21. D. N. Beshers, *Acta metall.* **6**, 521 (1958).
22. R. A. Johnson, *Phys. Rev.* **B24**, 7383 (1981).
23. J. D. Eshelby, *Adv. Solid State Phys.* **3**, 79 (1956).
24. J. W. Steeds, *Introduction to Anisotropic Elasticity Theory of Dislocations*. Clarendon Press, Oxford (1973).
25. R. W. Balluffi and A. V. Granato, in *Dislocations in Solids*, (edited by F. R. N. Nabarro), Vol. 4, p. 2. North-Holland, Amsterdam (1979).
26. B. K. D. Gairola, in *Dislocations in Solids* (edited by F. R. N. Nabarro), Vol. 1, p. 223. North-Holland, Amsterdam (1979).
27. J. C. M. Li, F. V. Nolfi and C. A. Johnson, *Acta metall.* **19**, 749 (1971).
28. C. Herring, in *The Physics of Powder Metallurgy* (edited by W. E. Kingston). McGraw-Hill, New York (1951).
29. M. Hillert, in *Alloy Phase Diagrams* (edited by L. W. Bennett, T. B. Massalski, and B. C. Giesen). North-Holland, Amsterdam (1983).
30. F. C. Larché and J. W. Cahn, *Acta metall.* **30**, 1835 (1982).
31. J. W. Cahn and F. C. Larché, *Scripta metall.* **17**, 927 (1983).
32. J. W. Cahn, *Acta metall.* **9**, 795 (1961).
33. M. Hillert, *Metall. Trans.* **15A**, 411 (1984).
34. K. Tashiro and G. R. Purdy, *Scripta metall.* **17**, 455 (1983).
35. A. L. Roitburd, *Soviet. Phys. Solid St.* **23**, 622 (1981).
36. J. W. Cahn, *Acta metall.* **28**, 1333 (1980).
37. J. W. Cahn and F. C. Larché, *Acta metall.* **30**, 51 (1982).
38. J. I. Alexander and W. C. Johnson. To be published.
39. R. O. Williams, *Metall. Trans. A* **11**, 247 (1980).
40. J. W. Cahn and F. C. Larché, *Acta metall.* **32**, 1915 (1984).
41. W. C. Johnson and J. W. Cahn, *Acta metall.* **32**, 1925 (1984).

APPENDIX 1

Solid–liquid equilibrium under hydrostatic stress

We consider the case of a substitutional binary solid. In equilibrium with a fluid under hydrostatic stress (for instance if it is entirely surrounded by the fluid), the mechanical equilibrium equations (3.13) and (3.14) implies that the stress is equal to

$$T_{ij} = -P\delta_{ij} \tag{A1.1}$$

where P is the pressure in the fluid. The stress being uniform, the constancy of the diffusion potential implies that the composition is uniform. Therefore the solid is uniform. The boundary condition

$$f - \mu_1^L\rho_1 - \mu_2^L\rho_2 = -P \tag{A1.2}$$

can be combined with the equation for the diffusion potential

$$M_{12} = \mu_1^L - \mu_2^L \tag{A1.3}$$

to give

$$\mu_1^L = (f + P + \rho_2 M_{12})V_0 \qquad \text{(A1.4)}$$

$$\mu_2^L = (f + P - \rho_1 M_{12})V_0. \qquad \text{(A1.4)}$$

Because the *solid is uniform*, these expressions are valid everywhere. The quantities on the right hand side of (A1.4) and (A1.5) depend only on the value of the state variables. Let us call them μ_1^s and μ_2^s

$$\mu_1^s \equiv (f + P + \rho_2 M_{12})V_0 \tag{A1.6}$$

$$\mu_2^s \equiv (f + P - \rho_1 M_{12})V_0. \tag{A1.7}$$

Elimination of M_{12} between these two equations give

$$f = -P + \rho_1\mu_1^s + \rho_2\mu_2^s$$

and, because of the uniformity, we can multiply by V_0 to get the total Helmholtz free energy

$$F = -PV_0 + N_1\mu_1^s + N_2\mu_2^s$$

where N_1 and N_2 are the total number of moles of components 1 and 2 respectively. The differential of f' is

$$df' = T_{ij}dE_{ij} + M_{12}d\rho_1'$$

M_{12} is replaced by its value obtained from (A1.6) and (A1.7). Using the definition of ρ_1', and after multiplication by V_0', one obtains

$$dF = -PdV_0 + \mu_1^s dN_1 + \mu_2^s dN_2$$

Therefore

$$\mu_1^s \equiv \left(\frac{\partial F}{\partial N_1}\right)_{V,N_2}$$

$$\mu_2^s \equiv \left(\frac{\partial F}{\partial N_2}\right)_{V,N_1}$$

We have recovered all the classical formula for fluid–fluid equilibrium. Despite network constraints, a solid under hydrostatic stress behaves as if it were a fluid.

APPENDIX 2

The boundary conditions for coherent phase change: Small strain approximation

The full large strain boundary condition for coherent phase change is [15]

$$\omega'^{\alpha} - \omega'^{\beta} - \mathbf{n}'^{\alpha} \cdot (\mathbf{F}^{\alpha T} - \mathbf{F}^{\beta T}) \cdot (\partial f'^{\alpha}/\partial \mathbf{F}) \cdot \mathbf{n}'^{\alpha} = 0 \quad \text{(A2.1)}$$

where the same reference state is chosen for both phases. The superscript T stands for transpose and $\mathbf{F}$ is the deformation gradient. $(\partial f'/\partial \mathbf{F})$ is the first Piola–Kirchhoff tensor $\mathbf{T}_R$. It is related to the Cauchy stress tensor $\mathbf{T}$ by

$$\mathbf{T}_R = J\mathbf{T} \cdot (\mathbf{F}^{-1})^T \quad \text{(A2.2)}$$

where J is the determinant of $\mathbf{F}$. In the small strain approximation, the displacement tensor is given, to first order in the derivatives $u_{i,j}$, by [11]

$$\mathbf{F} = I + E + \mathbf{\Omega} + O(u_{i,j}^2) \quad \text{(A2.3)}$$

where E is the small strain tensor, [equation (3.1)], Ω the small rotation tensor, and I the unit tensor. To the same approximation, its inverse is given by

$$\mathbf{F}^{-1} = \mathbf{I} - \mathbf{E} - \mathbf{\Omega} + O(u_{i,j}^2) \quad \text{(A2.4)}$$

Using these equations we get

$$\mathbf{n}'^{\alpha} \cdot (\mathbf{F}^{\alpha T} - \mathbf{F}^{\beta T}) \cdot \mathbf{T}_R^{\alpha} \cdot \mathbf{n}'^{\alpha} \approx \mathbf{n}'^{\alpha} \cdot (\mathbf{E}^{\alpha} - \mathbf{E}^{\beta} - \Omega^{\alpha} + \Omega^{\beta}) \cdot \mathbf{T} \cdot \mathbf{n}'^{\alpha} \quad \text{(A2.5)}$$

Dropping terms of order $u_{i,j}^2$, and since, for an arbitrary 3×3 tensor

$$\mathbf{n}' \cdot \mathbf{A} \cdot \mathbf{n}' = \mathbf{n}' \cdot \mathbf{A}^{\mathrm{T}} \cdot \mathbf{n}'$$

we finally obtain

$$\mathbf{n}' \cdot (\mathbf{F}^{\mathrm{T}} \cdot \mathbf{T}_R) \cdot \mathbf{n}' = \mathbf{n}' \cdot \mathbf{T} \cdot \mathbf{n}' - 2\mathbf{n}' \cdot \mathbf{\Omega} \cdot \mathbf{T} \cdot \mathbf{n}'. \quad \text{(A2.6)}$$

Since the same reference state has been chosen for α and β, the following equalities hold

$$\begin{aligned} \rho_I'^{\alpha} &= \rho_0' c_I^{\alpha} & \rho_I'^{\beta} &= \rho_0' c_I^{\beta} \\ J^{\alpha} &= \rho_0'/\rho_0^{\alpha} & J^{\beta} &= \rho_0'/\rho_0^{\beta}. \end{aligned} \quad \text{(A2.7)}$$

Using (A2.1) and (A2.5) we finally obtain

$$V_0'(f'^{\alpha} - f'^{\beta}) - \sum M_{IK}(c_I^{\alpha} - c_I^{\beta}) - V_0'(E_{ij}^{\alpha} - \Omega_{ij}^{\alpha} - E_{ij}^{\beta} - \Omega_{ij}^{\beta})\, T_{jk} n_i'^{\alpha} n_k'^{\alpha} = 0 \quad \text{(A2.6)}$$

The various terms are seen to be energies per mole of lattice sites. It is then easy to make a change of reference volume (like the stress free state for each phase). To the level of approximation used in linear elasticity, this won't affect the various terms of this equation.

M. Gurton and P. Voorhees (Proc. Roy. Soc. Lond. A **440** 323–343 (1993)) have shown that the last term of equation (A2.6) can simply be written $V_0'(E_{ij}^{\alpha} - E_{ij}^{\beta})\, T_{ij}$.

APPENDIX 3

Derivation of the open-system elastic stiffness and compliance tensor

All the calculations are done at constant temperature, so that all the partial derivatives are understood to be at constant temperature. We first treat the case of a binary solid, then generalize to a multicomponent solid.

A3.1. Binary solid

To simplify the notation we take ρ' to be ρ_1'. The differential of the stress can be written

$$\mathrm{d}T_{ij} = \left(\frac{\partial T_{ij}}{\partial E_{kl}}\right)_{\rho'} \mathrm{d}E_{kl} + \left(\frac{\partial T_{ij}}{\partial \rho'}\right)_{E_{kl}} \mathrm{d}\rho' \quad \text{(A3.1)}$$

or

$$\mathrm{d}T_{ij} = \left(\frac{\partial T_{ij}}{\partial E_{kl}}\right)_{M_{12}} \mathrm{d}E_{kl} + \left(\frac{\partial T_{ij}}{\partial M_{12}}\right)_{E_{kl}} \mathrm{d}M_{12}. \quad \text{(A3.2)}$$

The differential of the diffusion potential is

$$\mathrm{d}M_{12} = \left(\frac{\partial M_{12}}{\partial \rho'}\right)_{E_{ij}} \mathrm{d}\rho' + \left(\frac{\partial M_{12}}{\partial E_{ij}}\right)_{\rho'} \mathrm{d}E_{ij}. \quad \text{(A3.3)}$$

Replacing $\mathrm{d}\rho'$ from (A3.3) into (A3.1) yield

$$\mathrm{d}T_{ij} = \left[\left(\frac{\partial T_{ij}}{\partial E_{kl}}\right)_{\rho'} - \left(\frac{\partial T_{ij}}{\partial \rho'}\right)_{E_{kl}} \left(\frac{\partial M_{12}}{\partial E_{kl}}\right)_{\rho'} \Big/ \left(\frac{\partial M_{12}}{\partial \rho'}\right)_{E_{kl}}\right] \mathrm{d}E_{kl} + \left[\left(\frac{\partial T_{ij}}{\partial \rho'}\right)_{E_{kl}} \Big/ \left(\frac{\partial M_{12}}{\partial \rho'}\right)_{E_{kl}}\right] \mathrm{d}M_{12} \quad \text{(A3.4)}$$

and the coefficient of the term $\mathrm{d}E_{kl}$ is the $(ijkl)$ component of the open-system stiffness tensor.

Using the stress–strain relationship (4.14) and the Maxwell relation

$$\left(\frac{\partial T_{ij}}{\partial \rho'}\right)_{E_{kl}} = \left(\frac{\partial M_{12}}{\partial E_{ij}}\right)_{\rho'} \quad \text{(A3.5)}$$

one gets

$$\left(\frac{\partial M_{12}}{\partial E_{ij}}\right)_{\rho'} = -C_{ijkl}\eta_{kl} + \frac{\mathrm{d}C_{ijkl}}{\mathrm{d}\rho'}(E_{kl} - E_{kl}^{c}). \quad \text{(A3.6)}$$

The value of M_{12} as a function of E_{ij} rather than T_{ij} is obtained from (4.14) by using

$$S_{ijkl} C_{klmn} = \delta_{im}\delta_{jn}. \quad \text{(A3.7)}$$

Neglecting strain dependent terms, we finally get

$$C_{ijkl}^{*} = C_{ijkl} - \chi C_{ijmn} C_{klpq} \eta_{mn} \eta_{pq} \quad \text{(A3.8)}$$

Because of the linearity, we have

$$S_{ijkl}^{*} C_{klmn}^{*} = \delta_{im}\delta_{jn} \quad \text{(A3.9)}$$

where S_{ijkl}^{*} are the open-system compliances. Combining (A3.8) and (A3.9) gives

$$S_{ijkl}^{*} = S_{ijkl} + \chi \eta_{ij} \eta_{kl} \quad \text{(A3.10)}$$

where η_{ij} are defined by (4.4).

A3.2. Multicomponent solids

We follow the same derivation as above. The differential of the stess tensor is

$$\mathrm{d}T_{ij} = \left(\frac{\partial T_{ij}}{\partial E_{kl}}\right)_{\rho'} \mathrm{d}E_{kl} + \sum_{I \neq K} \left(\frac{\partial T_{ij}}{\partial \rho_{IK}'}\right) \mathrm{d}\rho_{IK}' \quad \text{(A3.11)}$$

The differential of the potentials are

$$\mathrm{d}M_{IK} = \sum_{J \neq K} \left(\frac{\partial M_{IK}}{\partial \rho_{JK}'}\right) \mathrm{d}\rho_{JK}' + \left(\frac{\partial M_{IK}}{\partial E_{ij}}\right)_{\rho'} \mathrm{d}E_{ij} \quad \text{(A3.12)}$$

$\mathrm{d}\rho_J'$ can be obtained from this sytem of linear equation by Kramer's rule

$$\mathrm{d}\rho_J' = \left\{\sum_{I \neq K} \left[\mathrm{d}M_{IK} - \left(\frac{\partial M_{IK}}{\partial E_{ij}}\right)_{\rho'} \mathrm{d}E_{ij}\right] A_{IJ}\right\} \Big/ \chi \quad \text{(A3.13)}$$

where χ is the determinant

$$\chi = \left|\frac{\partial M_{IK}}{\partial \rho_{JK}'}\right| = \rho_0' \left|\frac{\partial M_{IK}}{\partial c_{JK}}\right|$$

and A_{IJ} is the minor of the $(\partial M_{IK}/\partial \rho_{JK}')$ term of D. Replacing $\mathrm{d}\rho_I'$ by its value in (A3.11) and using the Maxwell relation

$$\left(\frac{\partial T_{ij}}{\partial \rho_{IK}'}\right)_{E_{kl}} = \left(\frac{\partial M_{IK}}{\partial E_{ij}}\right)_{\rho'} \quad \text{(A3.14)}$$

we get

$$\left(\frac{\partial T_{ij}}{\partial E_{kl}}\right)_{M_{IK}} = \left(\frac{\partial T_{ij}}{\partial E_{kl}}\right)_{\rho'} - \chi \sum_{I,J \neq K} \left(\frac{\partial T_{ij}}{\partial \rho_{IK}'}\right)_{E_{pq}} \left(\frac{\partial T_{kl}}{\partial \rho_{JK}'}\right)_{E_{pq}} A_{IJ} \quad \text{(A3.15)}$$

Using (A3.9), Hooke's law, and neglecting strain dependent

terms we finally get

$$S^{*}_{ijkl} = S_{ijkl} + \chi \sum_{I,J \neq K} \left(\frac{\partial E^{c}_{ij}}{\partial c_{IK}}\right)\left(\frac{\partial E^{c}_{lK}}{\partial c_{JK}}\right) A_{IJ} \qquad \text{(A3.16)}$$

where χ is the determinant

$$\chi = \rho'_0 \left| \frac{\partial M_{IK}}{\partial c_J} \right|$$

and A_{IJ} the minor of the (IJ) term of χ/ρ'_0.

APPENDIX 4

A common tangent construction

Let ξ^k be three unit vectors normal to each other, such that ξ^3 is the normal to the interface, with components ξ^k_i. The vectors λ^k are defined by

$$\lambda^k_i = E_{ij}\xi^k_j \qquad \text{(A4.1)}$$

Since the determinant $|\xi^k_j|$ has the value 1 the system of equations (A4.1) constitute a valid linear change of variable. Using the chain rule, we obtain, considering the ξ^k as fixed

$$\left(\frac{\partial f'}{\partial E_{ij}}\right)_{E_{kl\neq ij}} = T_{ij} = \left(\frac{\partial f'}{\partial \lambda^k_i}\right)_{\lambda^m_n} \xi^k_j \qquad \text{(A4.2)}$$

After multiplication by ξ^k_j and summation on j one gets

$$T_{ij}\xi^k_j = \left(\frac{\partial f'}{\partial \lambda^k_i}\right)_{\lambda^m_n}. \qquad \text{(A4.3)}$$

Let us define the free energy ζ' by

$$\zeta' = f' - \lambda^3_i\left(\frac{\partial f'}{\partial \lambda^3_i}\right) \qquad \text{(A4.4)}$$

$$= f' - T_{ij}n_jE_{ik}n_k \qquad \text{(A4.5)}$$

and it is easy to show that

$$M_{IK} = \left(\frac{\partial f'}{\partial \rho'_{IK}}\right)_{E_{ij}} = \left(\frac{\partial \zeta'}{\partial \rho'_{IK}}\right)_{\lambda^1_i,\lambda^2_i,T_{ij}n_j} \qquad \text{(A4.6)}$$

The conditions for equilibrium at an incoherent interface [equation (3.24)] can be written

$$\hat{f}^{\alpha} - \sum c^{\alpha}_I\left(\frac{\partial \hat{f}^{\alpha}}{\partial c_I}\right) - T^{\alpha}_{ij}n^{\alpha}_i n^{\alpha}_j V'^{\alpha}_0$$

$$= \hat{f}^{\beta} - \sum c^{\beta}_I\left(\frac{\partial \hat{f}^{\beta}}{\partial c_I}\right) - T^{\beta}_{ij}n^{\beta}_i n^{\beta}_j V'^{\beta}_0 \qquad \text{(A4.7)}$$

where quantities such as $\hat{f}$ are just $f'V'_0$, i.e. quantities for one mole of lattice sites.

If the normal pressure is zero, so that $T_{ij}n_j = 0$ it becomes equivalent to

$$\zeta^{\alpha} - \sum c^{\alpha}_I\left(\frac{\partial \zeta^{\alpha}}{\partial c_I}\right)_{\lambda^1_i\lambda^2_i} = \zeta^{\beta} - \sum c^{\beta}_I\left(\frac{\partial \zeta^{\beta}}{\partial c_I}\right)_{\lambda^1_i\lambda^2_i} \qquad \text{(A4.8)}$$

which together with

$$M^{\alpha}_{IK} = M^{\beta}_{IK}$$

which can then be written

$$\left(\frac{\partial \zeta^{\alpha}}{\partial c_I}\right)_{\lambda^1_i\lambda^2_i} = \left(\frac{\partial \zeta^{\beta}}{\partial c_I}\right)_{\lambda^1_i\lambda^2_i} \qquad \text{(A4.9)}$$

show that c_{IK} can be obtained by a tangent construction to ζ, which, in this case is just equal to $\hat{f}$.

II

Papers on Continuum Mechanics

Offprint from "Archive for Rational Mechanics and Analysis",
Volume 104, Number 3, 1988, pp. 195–221

Multiphase thermomechanics with interfacial structure 1. Heat conduction and the capillary balance law

MORTON E. GURTIN

1. Introduction

In [1986g, 1988g] I began the development of a nonequilibrium thermomechanics of two-phase continua, a development based on *dynamical* statements of the thermomechanical laws in conjunction with GIBBS's notion of a sharp phase-interface endowed with energy and entropy. I have since come to realize that there is an additional balance law appropriate to the interface $\mathscr{s}$. This law, which represents **balance of capillary forces**, has the form[1]

$$\int_{\partial c} \mathbf{C}\boldsymbol{v} + \int_{c} \boldsymbol{\pi} = 0, \tag{1.1}$$

with c an arbitrary subsurface of $\mathscr{s}$ and $\boldsymbol{v}$ the outward unit normal to the boundary curve ∂c of c. Here $\mathbf{C}(\boldsymbol{x}, t)$, the **capillary stress**, is a linear transformation of tangent vectors into (not necessarily tangent) vectors, while $\boldsymbol{\pi}(\boldsymbol{x}, t)$, the **interaction**, is a vector field; $\mathbf{C}(\boldsymbol{x}, t)$ represents microforces exerted across ∂c in response to the creation of new surface; $\boldsymbol{\pi}(\boldsymbol{x}, t)$ characterizes the interaction between the interface and the bulk material. I view (1.1) as a *balance law which is supplementary to the usual laws for forces and moments.*

Balance of capillary forces has the local form

$$\operatorname{div}_{\mathscr{s}} \mathbf{C} + \boldsymbol{\pi} = \mathbf{0}, \tag{1.2}$$

with $\operatorname{div}_{\mathscr{s}}$ the surface divergence on $\mathscr{s}$. The *normal* component of (1.2) arises in previous theories, emerging as an Euler-Lagrange equation corresponding to the requirement that a global Gibbs function be stationary.[2] This is not surprising: balance laws often follow as Euler-Lagrange equations, an example being balance

[1] In the absence of external supplies.

[2] HERRING [1951], CAHN & HOFFMAN [1972, 1974]; the notion of capillary forces is implied by these authors. Another special case of (1.1), the requirement that $\mathbf{C}\boldsymbol{v}$ be continuous across a corner, was derived variationally by HERRING [1951] and used by HERRING and others to discuss the formation of facets.

of forces in elastostatics. Such variational derivations underline the consistency between theories and, what is more important, often point the way toward a correct statement of the relevant law. On the other hand, *such derivations tend to obscure the fundamental nature of balance laws as basic axioms in any general dynamical framework which includes dissipation.*

In [1986g, 1988g] I derived the normal component of (1.2) as a consequence of the second law. While this approach might seem advantageous, the more general framework necessitates a more complicated constitutive theory: in the present study, capillary balance furnishes a relation between the interfacial temperature and the curvature, orientation, and normal velocity of the interface; this relation is a constitutive *postulate* in [1986g, 1988g].

It is the purpose of this paper to develop a fairly complete thermomechanics based on capillary balance as an independent axiom. To avoid inessential complications that might obscure an understanding of this law, attention is limited to nondeformable bodies in the absence of diffusion. The addition of diffusion is elementary; the extension to deformable bodies will be the subject of [1988s].

The theory is based on three physical laws: balance of capillary forces, balance of energy, growth of entropy. A fundamental assumption underlying balance of energy is that interfacial forces supply power to the interface through the velocity $V\boldsymbol{m}$, where $\boldsymbol{m}$ is a unit-normal field for the interface and V is the corresponding *normal velocity*; in particular, the power expended on the interface by the capillary stress has the form

$$\int_{\partial c} (V\boldsymbol{m}) \cdot \mathbf{C}\boldsymbol{\nu}, \qquad (1.3)^3$$

indicating a sharp departure from classical ideas. Power is generally the product of a force acting on a material point (particle) and the velocity of the point. Here material points do not move, but power is expended, and this power is reckoned by the motion of the phase boundary.[4]

A fairly general constitutive theory is considered for the interface. The free energy f and entropy s are allowed to depend on the temperature θ, and—to have a theory of sufficient generality to model crystal growth—also on the orientation $\boldsymbol{m}$ and the normal velocity V; in addition, constitutive equations are given for the symmetric and normal components,[5] $\mathbf{C}_{\text{sym}}$ and c, of the capillary stress, and for the normal component π of the interaction:

$$\begin{gathered} f = \hat{f}(\theta, \boldsymbol{m}, V), \qquad s = \hat{s}(\theta, \boldsymbol{m}, V), \\ \mathbf{C}_{\text{sym}} = \hat{\mathbf{C}}_{\text{sym}}(\theta, \boldsymbol{m}, V), \qquad c = \hat{c}(\theta, \boldsymbol{m}, V), \\ \pi = \hat{\pi}(\theta, \boldsymbol{m}, V). \end{gathered} \qquad (1.4)$$

[3] *Cf.* Remark 4.4.

[4] Conceptually, it is useful to identify the interface with a collection of particles transported with velocity $V\boldsymbol{m}$. This view is emphasized by ANGENENT & GURTIN [1988a].

[5] At each point the capillary stress $\mathbf{C}$ maps tangent vectors into vectors in $\mathbb{R}^3$; we write $\mathbf{C} = \mathbf{T} + \boldsymbol{m} \otimes c$, where $\mathbf{T}$ maps tangent vectors into tangent vectors; $\mathbf{C}_{\text{sym}}$ is then the symmetric part of $\mathbf{T}$.

Use of the second law[6] then leads to the following list of constitutive restrictions:

(i) the free energy, the entropy, and the normal and symmetric components of the capillary stress are independent of the normal velocity V and

$$\hat{s}(\theta, \boldsymbol{m}) = -\partial_\theta \hat{f}(\sigma, \boldsymbol{m}), \qquad \hat{c}(\theta, \boldsymbol{m}) = -\partial_{\boldsymbol{m}} \hat{f}(\theta, \boldsymbol{m}),$$
$$\hat{\mathbf{C}}_{\mathrm{sym}}(\theta, \boldsymbol{m}) \text{ is a surface tension of amount } \hat{f}(\theta, \boldsymbol{m}); \tag{1.5}$$

[7]

(ii) the interaction $\hat{\pi}(\theta, V, \boldsymbol{m})$ equals the jump in bulk free energy across the interface minus a drag force of the form $\beta(\theta, \boldsymbol{m}, V)\, V,\ \beta \geqq 0$.

In classical theories of melting—in which the interface is devoid of structure—changes of phase occur at the transition temperature θ_M, which is the temperature at which the bulk free energies of the two phases coincide. Within our theory a flat and stationary interface has temperature θ_M, but a curved and moving interface need not; in fact, the relation

$$u = f(\boldsymbol{m})\, H + \partial_{\boldsymbol{m}} \partial_{\boldsymbol{m}} f(\boldsymbol{m}) \cdot \mathbf{L} - \beta(\boldsymbol{m})\, V, \tag{1.6}$$

for the (dimensionless) *temperature difference*

$$u = \frac{\theta - \theta_M}{\theta_M}$$

as a function of V, the curvature tensor $\mathbf{L}$, and the corresponding mean curvature H, follows from capillary balance as an *approximation appropriate to a weak interface*; that is, to an interface for which the interfacial densities are small and the dependence on V weak. Here $f(\boldsymbol{m}) = \hat{f}(\theta_M, \boldsymbol{m})$, $\beta(\boldsymbol{m}) = \beta(\theta_M, \boldsymbol{m}, 0)$, and we have chosen a scaling in which the latent heat l satisfies $l = 1$.

The usual heat equation in bulk combined with (1.6) and a similar approximation for balance of energy lead to the partial differential equations and free-boundary conditions:

$$C_i u^{\cdot} = -\operatorname{div} q, \quad q = -K_i \nabla u \quad \text{in bulk,}$$
$$\left.\begin{aligned} u &= \mathbf{B}(\boldsymbol{m}) \cdot \mathbf{L} - \beta(\boldsymbol{m})\, V \\ V &= [q] \cdot \boldsymbol{m} \end{aligned}\right\} \quad \text{on the interface,} \tag{1.7}$$

Here C_i and K_i, respectively, denote the (appropriately scaled) bulk specific heat and bulk conductivity (tensor) for phase i ($i = 1, 2$); q is the heat flux; $[q]$ is the jump in q across the interface; $\mathbf{B}(\boldsymbol{m}) \cdot \mathbf{L}$ represents the first two terms on the right side of (1.6).

Global growth conditions are found for the system (1.7) in a bounded domain under various boundary conditions. In particular, for a bounded solid $S(t)$ in a

[6] In the manner of COLEMAN & NOLL [1963]. This procedure is applied in [1988a], but the absence of capillary balance severely complicates the analysis. See also MURDOCH [1976], who obtains restrictions for a materially stationary interface in a deforming continua.

[7] $\partial_z g$ denotes the derivative (usually partial) of g with respect to z. In particular, $\partial_{\boldsymbol{m}} g$ is the surface gradient with respect to $\boldsymbol{m}$ on the surface of the unit ball in $\mathbb{R}^3$.

liquid melt, enclosed by a container B whose boundary ∂B is held at the spatially constant temperature $U = U(t)$,

$$F(\mathcal{s})^{\cdot} + U \operatorname{vol}(S)^{\cdot} + (C/2)\left\{\int_B (u - U)^2\right\}^{\cdot} \leqq 0, \tag{1.8}$$

where

$$F(\mathcal{s}) = \int_{\mathcal{s}} f(m)$$

is the total interfacial free-energy at the transition temperature.

We also consider perfect conductors, which are materials with infinite thermal conductivity. We give a plausibility argument, based on our general theory, leading to the conclusion that, for a boundary held at the temperature $U = U(t)$, the motion of the interface is governed by the relation[8]

$$\beta(m)\, V - \mathbf{B}(m) \cdot \mathbf{L} = -U(t) \tag{1.9}$$

in conjunction with various kinematical equations relating V and $\mathbf{L}$ to the motion of the interface. For an interface consistent with (1.9),

$$F(\mathcal{s})^{\cdot} + U \operatorname{vol}(S)^{\cdot} \leqq 0. \tag{1.10}$$

2. Primitive quantities

We consider a body consisting of two phases separated, at each time t, by an **interface** $\mathcal{s}(t)$, and write $B_i(t)$ for the subregion of the body occupied by phase i. We assume that the body occupies all of $\mathbb{R}^3$; that the $B_i(t)$ are closed regions with $\mathbb{R}^3$ as their union and $\mathcal{s}(t)$ as their intersection; and that $\mathcal{s}(t)$ is a smoothly propagating surface.[9] We orient $\mathcal{s}(t)$ by a choice of unit normal field $m(x, t)$, called the **orientation** of $\mathcal{s}(t)$, chosen so that

$m(x, t)$ coincides with the outward unit normal to $\partial B_1(t)$.

We write $V(x, t)$ for the **normal velocity** of $\mathcal{s}(t)$ in the direction $m(x, t)$, $\mathbf{L}(x, t)$ for the **curvature tensor** on $\mathcal{s}(t)$, and $H = \operatorname{tr} \mathbf{L}$ for (twice) the **mean curvature**.

The thermodynamics of the body is described by three types of fields: bulk fields that describe the bulk behavior of the individual phases; superficial fields that describe the behavior of the interface; external supplies that describe the

---

[8] Generalizing "flow by mean curvature" in which $V = H$ (*cf.* BRAKKE [1978], ALLEN & CAHN [1979], GAGE & HAMILTON [1986], GRAYSON [1987]). Consequences of (1.9), for the motion of an interfacial curve in $\mathbb{R}^2$, will be discussed by ANGENENT & GURTIN [1988a].

[9] Concerning surfaces, we use the notation and many of the results of GURTIN & MURDOCH [1974], MURDOCH [1976, 1978], and GURTIN [1986g, 1988g]; these are discussed in the Appendix. To agree with standard terminology, we take $\mathbf{L} = -\nabla_{\mathcal{s}} m$, rather than $\mathbf{L} = \nabla_{\mathcal{s}} m$ as was done in [1986g, 1988g].

interaction between the body and the external world. In particular, we have the following primitive quantities:[10]

bulk fields

$\varepsilon(x, t)$, **bulk internal energy** (volume),
$\eta(x, t)$, **bulk entropy** (volume),
$\theta(x, t)$, **absolute temperature**,
$q(x, t)$, **heat flux** (area),

interfacial fields

$e(x, t)$, **interfacial internal energy** (area),
$s(x, t)$, **interfacial entropy** (area),
$\mathbf{C}(x, t)$, **capillary stress** (length),
$\pi(x, t)$, **interaction** (area),

external supplies

$q(x, t)$, **bulk heat supply** (volume),
$r(x, t)$, **interfacial heat supply** (area),
$b(x, t)$, **capillary supply** (area).

Here ε, η, θ, and q are bulk scalar fields; π is a bulk vector field; e, s, and r are superficial scalar fields; $\mathbf{C}$ is a superficial tensor field; π and b are superficial vector fields. (Superficial and bulk fields are defined in the Appendix.)

We assume that the

temperature is continuous across the interface; (2.1)

generally, we will not specify regularity hypotheses other than to note that *the remaining bulk fields are allowed to suffer jump discontinuities across the interface.*

3. Basic laws

3.1. Balance of capillary forces

Let $c(t)$ be a sufficiently regular subsurface of $\mathscr{I}(t)$, and let $\nu(x, t)$, a vector field tangential to $\mathscr{I}(t)$, be the outward unit normal to the boundary curve $\partial c(t)$. The integrals

$$\int_{\partial c} \mathbf{C}\nu, \quad \int_{c} \pi, \quad \int_{c} b \tag{3.1}$$

represent forces involved with the creation of new surface: the first gives the force exerted across ∂c by the interface; the second and third give forces exerted on c by the bulk material and by the external world.

We write $\mathbf{C}_{\text{sym}}$, $\mathbf{C}_{\text{skw}}$, and c for the **symmetric**, **skew**, and **normal** components of the capillary stress $\mathbf{C}$ (*cf.* the paragraph containing (A8)). The interfacial force $\mathbf{C}\nu$ in (3.1) is then the sum of a tangential force $(\mathbf{C}_{\text{sym}} + \mathbf{C}_{\text{skw}})\,\nu$ and a normal force $(c \cdot \nu)\,m$.

[10] (volume) is shorthand for "per unit volume", and so forth.

We postulate that each subsurface $c(t)$ be consistent with **balance of capillary forces**

$$\int_{\partial c} \mathbf{C}\nu + \int_{c} b + \int_{c} \pi = \mathbf{0}. \tag{3.2}$$

This law has an equivalent local form, which is easily derived using the surface divergence theorem (A16):

$$\operatorname{div}_{\sigma} \mathbf{C} + b + \pi = \mathbf{0} \tag{3.3}$$

with $\operatorname{div}_{\sigma}$ the surface divergence on σ (*cf.* (A13)).

We regard the normal velocity V as *intrinsic* in the sense that superficial forces supply power to the interface through the velocity Vm. The next result characterizes this power.

Theorem of expended power.

$$\int_{\partial c} (Vm) \cdot \mathbf{C}\nu + \int_{c} Vm \cdot (b + \pi) = - \int_{c} (V\mathbf{C}_{\text{sym}} \cdot \mathbf{L} + c \cdot m^{\circ}). \tag{3.4}$$

Proof. By (A7), (A8), (A11), (A12), (A14), and (B1),

$$\begin{aligned} \operatorname{div}_{\sigma} (V\mathbf{C}^T m) &= Vm \cdot \operatorname{div}_{\sigma} \mathbf{C} + \mathbf{C} \cdot \nabla_{\sigma}(Vm), \\ \mathbf{C} \cdot \nabla_{\sigma}(Vm) &= -V\mathbf{C}_{\text{sym}} \cdot \mathbf{L} - c \cdot m^{\circ}; \end{aligned} \tag{3.5}$$

(3.3) and (3.5) imply (3.4). □

The left side of (3.4) gives the total power expended on c, while the right side catalogs the manner in which this power is used: $-V\mathbf{C}_{\text{sym}} \cdot \mathbf{L}$ represents power expended in creating new surface; $-c \cdot m^{\circ}$ represents power expended in changing the orientation of the interface. Note that the skew part of $\mathbf{C}$ does not expend power. Note also that for $\mathbf{C}_{\text{sym}}$ a surface tension σ, $V\mathbf{C}_{\text{sym}} \cdot \mathbf{L} = V\sigma H$

If we introduce the normal components

$$\pi = \pi \cdot m, \quad b = b \cdot m \tag{3.6}$$

of the interaction and capillary supply, then, by $(\text{A14})_{2\text{-}4}$ and (3.6), the normal component of (3.3) has the form

$$\mathbf{C}_{\text{sym}} \cdot \mathbf{L} + \operatorname{div}_{\sigma} c + b + \pi = 0; \tag{3.7}$$

this relation is central to what follows.

Remark 3.1. The balance law (3.2) should be viewed as a conservation law over and above the usual balance laws for forces and moments. When the current theory is extended to include deformations of the body, balance of forces across the interface yields the equation

$$\operatorname{div}_{\sigma} \mathbf{T} + f = -[T]\, m \tag{3.8}$$

for the bulk and interfacial Cauchy stress tensors T and $\mathbf{T}$ and the interfacial body force f (*cf.* GURTIN & MURDOCH [1974], ALEXANDER & JOHNSON [1985, 1986], LEO [1987], and FONSECA [1988]); in this extended theory (3.2) and (3.8) are separate balance laws (GURTIN & STRUTHERS [1988s]), and for that reason one should not attempt to identify $\mathbf{C}$, $\boldsymbol{b}$, and $\boldsymbol{\pi}$ with $\mathbf{T}$, f, and $[T]\,\boldsymbol{m}$, as tempting as this might seem. Introducing conservation laws which bear formal resemblance to the usual balance laws for forces and moments, but are, in fact, supplementary to such laws, has been fruitful in other theories, especially those which model internal structure; an example is ERICKSEN's [1961] theory of liquid crystals (*cf.* TRUESDELL & NOLL [1965], Section 127). □

Remark 3.2. Let $\boldsymbol{r}(\boldsymbol{x}) = \boldsymbol{x} - \boldsymbol{x}_0$. Then given any sufficiently regular subsurface c of σ,

$$M(c) = \int_{\partial c} \boldsymbol{r} \wedge \mathbf{C}\boldsymbol{v} + \int_{c} \boldsymbol{r} \wedge \boldsymbol{\pi} + \int_{c} \boldsymbol{r} \wedge \boldsymbol{b}$$

is the **total moment** (about $\boldsymbol{x}_0$) exerted on c by the capillary stress, the interaction, and the capillary supply. By (A17) and (3.3),

$$M(c) = - \int_{\partial c} (\boldsymbol{m} \wedge \boldsymbol{c} + 2\mathbf{C}_{\text{skw}}\mathbf{P}).$$

Thus *the existing forces, by themselves, do not satisfy balance of moments*: $-\boldsymbol{m} \wedge \boldsymbol{c} - 2\,\mathbf{C}_{\text{skw}}\,\mathbf{P}$ represents a distributed couple, per unit length of σ, that must be ***balanced by surface couples in the interaction between the interface and the bulk material.*** Such couples are regarded as indeterminate in the present theory. □

3.2. Balance of energy. Growth of entropy

Consider an arbitrary fixed subbody Ω with $\boldsymbol{n}$ the outward unit normal on $\partial\Omega$, and let c be the portion of σ that lies in Ω:

$$c(t) = \Omega \cap \sigma(t).$$

The internal energy and internal entropy of Ω are given by

$$\int_{\Omega} \varepsilon + \int_{c} e, \qquad \int_{\Omega} \eta + \int_{c} s.$$

Ω can lose energy and entropy because of the possible motion of the interface relative to $\partial\Omega$; these flows are represented by the quantities outflow(e, Ω) and outflow(s, Ω) defined by (B2) (*cf.* [1986g], p. 218). The integrals

$$- \int_{\partial\Omega} \boldsymbol{q} \cdot \boldsymbol{n}, \qquad - \int_{\partial\Omega} (\boldsymbol{q}/\theta) \cdot \boldsymbol{n}, \qquad \int_{\Omega} q, \qquad \int_{\Omega} q/\theta,$$

respectively, measure bulk heat and entropy flow into Ω by conduction, and heat and entropy supplied directly to Ω by the external world. We also allow the external world to supply heat and entropy to the interface; these supplies are represented by the terms

$$\int_{c} r, \qquad \int_{c} r/\theta.$$

Finally, the power expended on Ω is given by the left side of (3.4), but without the term involving $\boldsymbol{\pi}$ (since $\boldsymbol{\pi}$ represents interactions *within* Ω).

In view of this discussion, the *first two laws* for Ω have the form:

balance of energy

$$\left\{\int_{\Omega}\varepsilon+\int_{c}e\right\}^{\cdot}+\text{outflow}(e,\Omega)=-\int_{\partial\Omega}q\cdot n+\int_{\Omega}q+\int_{\partial c}(Vm)\cdot\mathbf{C}\nu$$
$$+\int_{c}Vm\cdot b+\int_{c}r, \qquad (3.9)^{11}$$

growth of entropy

$$\left\{\int_{\Omega}\eta+\int_{c}s\right\}^{\cdot}+\text{outflow}(s,\Omega)\geqq-\int_{\partial\Omega}(q/\theta)\cdot n+\int_{\Omega}q/\theta+\int_{c}r/\theta. \qquad (3.10)^{12}$$

In view of the general balance theorem (Appendix B), we are led to the classical local **bulk relations**

$$\begin{aligned}\varepsilon^{\cdot}&=-\operatorname{div}q+q,\\ \eta^{\cdot}&\geqq-\operatorname{div}(q/\theta)+q/\theta,\end{aligned} \qquad (3.11)$$

in conjunction with the **interface conditions**

$$\begin{aligned}[\varepsilon]\,V+e^{\circ}-eHV&=-[q]\cdot m+r+\operatorname{div}_{\sigma}(Vc)+Vb,\\ -[\eta]\,V+s^{\circ}-sHV&\geqq(-[q]\cdot m+r)/\theta.\end{aligned} \qquad (3.12)$$

Crucial to the derivation of (3.12) is the assumption that θ be continuous across the interface.

Remark 3.3. We have taken the normal velocity of the interface as the kinematic variable that characterizes the manner in which capillary forces expend power; tangential motion does not induce power. As is consistent with a "constraint" of this form, we leave as *indeterminate* the tangential component of the interaction π and therefore concern ourselves only with π, b, and the *normal component* (3.7) of balance of forces (3.3). Moreover, the skew component $\mathbf{C}_{\text{skw}}$ of the capillary force enters neither (3.7) nor the interface conditions (3.12), and hence will not appear in any of the subsequent results. However, while irrelevant to our further discussions, $\mathbf{C}_{\text{skw}}$ does appear when discussing interfaces with corners. □

Remark 3.4. Heat flow within the interface is easily accounted for by the addition of an energy flow

$$-\int_{\partial c}h\cdot\nu$$

[11] *Cf.* [1988g]. Similar versions of the first law, but *without* the capillary stress, are contained in the work of MOECKEL [1975], FERNANDEZ-DIAZ & WILLIAMS [1979], and GURTIN [1986g].
[12] *Cf.* [1986g, 1988g].

to the right side of (3.9), and by the addition of an entropy flow

$$-\int_{\partial c} (h/\theta)\cdot \nu$$

to the right side of (3.10), where $h(x, t)$, a tangential vector field, is the interfacial heat flux. □

3.3. *Free-energy inequalities*

The subsequent analysis is simplified if we introduce the **bulk** and **interfacial free energies**

$$\psi = \varepsilon - \theta\eta, \quad f = e - \theta s. \tag{3.13}$$

The local relations (3.11) may then be combined to give the bulk **free-energy inequality**

$$\Gamma := \psi^{\cdot} + \eta\theta^{\cdot} + \theta^{-1}\, q\cdot g \leqq 0, \tag{3.14}$$

where

$$g = \nabla\theta \tag{3.15}$$

is the **temperature gradient.** Further, in view of (3.7) and (B1), (3.12) imply the **interfacial free-energy inequality**:

$$\gamma := f^{\circ} + s\theta^{\circ} + c\cdot m^{\circ} + V\{-fH + \mathbf{C}_{\mathrm{sym}}\cdot \mathbf{L} - [\psi] + \pi\} \leqq 0. \tag{3.16}$$

Remark 3.5. The global axioms (3.2) and (3.9) for force and energy balance are together equivalent to the corresponding local relations (3.3), $(3.11)_1$, and $(3.12)_1$. Further, granted force and energy balance, the global axiom (3.10) expressing growth of entropy is equivalent to the free-energy inequalities (3.14) and (3.16).

Remark 3.6. The difference between the left and right sides of (3.10) is

$$-\int_{\Omega} (\Gamma/\theta) - \int_{c} (\gamma/\theta);$$

thus $-\Gamma/\theta$ is the *bulk entropy-production* per unit volume, $-\gamma/\theta$ is the *interfacial entropy-production* per unit area. □

4. Constitutive equations. Thermodynamic restrictions

4.1. *Bulk and interfacial constitutive equations*

We consider, for the two phases ($i = 1, 2$), **bulk constitutive equations** of the form

$$\psi = \psi_i(\theta, g), \quad \eta = \eta_i(\theta, g), \quad q = q_i(\theta, g) \tag{4.1}$$

with $g = \nabla\theta$ the temperature gradient. These are supplemented by **interfacial constitutive equations**:

$$\begin{gathered} f = \hat{f}(\theta, m, V), \quad s = \hat{s}(\theta, m, V), \\ \mathbf{C}_{\text{sym}} = \hat{\mathbf{C}}_{\text{sym}}(\theta, m, V), \quad c = \hat{c}(\theta, m, V), \\ \pi = \hat{\pi}(\theta, m, V). \end{gathered} \tag{4.2}$$

In view of Remark 3.3, we leave as *indeterminate* the tangential component of the interaction π, and, since the skew component $\mathbf{C}_{\text{skw}}$ of the capillary stress is irrelevant to all subsequent discussions, we do not discuss its corresponding constitutive behavior.

Note that, by (3.13), the constitutive equations (4.1) and (4.2) induce auxiliary relations for the internal energies:

$$\varepsilon = \varepsilon_i(\theta, g), \quad e = \hat{e}(\theta, m, V). \tag{4.3}$$

4.2. *Thermodynamic restrictions*

Given any time interval T, any temperature field $\theta(x, t)$, $(x, t) \in \mathbb{R}^3 \times T$, and any motion of the interface $\mathscr{I}(t)$, $t \in T$, the constitutive equations (4.1) and (4.2) may be used to compute a corresponding **process** $(\psi, \eta, q, f, s, \mathbf{C}_{\text{sym}}, c, \pi)$. The local capillary balance (3.7) and the local energy balances $(3.11)_2$ and $(3.12)_2$ then determine the capillary supply b and the heat supplies q and r needed to support the process.[13] Granted this, Remark 3.5 implies that the global law of entropy growth (3.10) will be satisfied if and only if the free-energy inequalities (3.14) and (3.16) are satisfied.

Definition. The **constitutive equations are compatible with thermodynamics** if given any temperature field and any motion of the interface, the corresponding process satisfies (3.14) and (3.16).

The inequality (3.14), when required to hold for all temperature fields, is equivalent to the requirement that $\psi_i(\theta, g)$ and $\eta_i(\theta, g)$ be independent of g and satisfy

$$\eta_i(\theta) = -\partial_\theta \psi_i(\theta), \quad q_i(\theta, g) \cdot g \leq 0 \tag{4.4}$$

(COLEMAN & NOLL [1963], COLEMAN & MIZEL [1963]). The next theorem, our main result, gives corresponding restrictions for the interfacial constitutive equations.

[13] One might object to the premise of the availibility of arbitrary supplies, especially the capillary supply b. Allowing a supply for each balance law is an assumption now standard in continuum mechanics. Assumptions of this form are generally tacit throughout physics. Indeed, statical equations are often derived from the requirement that a functional, for example a global free energy, be stationary. The corresponding analyses generally require arbitrary variation of the underlying state, with the assumption left tacit that suitable supplies are available to support such variations.

Compatibility theorem. *The constitutive equations are compatible with thermodynamics if and only if, in addition to* (4.4), *the following restrictions are satisfied*: (i) $\hat{f}(\theta, \boldsymbol{m}, V)$, $\hat{\mathbf{C}}_{\mathrm{sym}}(\theta, \boldsymbol{m}, V)$, $\hat{c}(\theta, \boldsymbol{m}, V)$ *and* $\hat{s}(\theta, \boldsymbol{m}, V)$ *are independent of the normal velocity* V *and satisfy*

$$\begin{gathered}\hat{s}(\theta, \boldsymbol{m}) = -\partial_\theta \hat{f}(\theta, \boldsymbol{m}), \qquad \hat{c}(\theta, \boldsymbol{m}) = -\partial_{\boldsymbol{m}} \hat{f}(\theta, \boldsymbol{m}), \\ \hat{\mathbf{C}}_{\mathrm{sym}}(\theta, \boldsymbol{m}) = \hat{f}(\theta, \boldsymbol{m})\, \mathbf{I}(\boldsymbol{m}),\end{gathered} \tag{4.5}$$

so that the free energy determines entropy and the normal and symmetric components of the capillary stress, and the symmetric component is a surface tension of amount $\hat{f}(\theta, \boldsymbol{m})$;

(ii) *the interaction* $\hat{\pi}(\theta, V, \boldsymbol{m})$ *has the form*

$$\hat{\pi}(\theta, \boldsymbol{m}, V) = \psi_2(\theta) - \psi_1(\theta) - \hat{\beta}(\theta, \boldsymbol{m}, V)\, V, \tag{4.6}$$

where the **kinetic coefficient** $\hat{\beta}(\theta, \boldsymbol{m}, V)$ *is consistent with*

$$\hat{\beta}(\theta, \boldsymbol{m}, V) \geqq 0. \tag{4.7}$$

Proof. We must show that (i) and (ii) are equivalent to the requirement that the interfacial inequality (3.16) hold in all processes. In view of the constitutive equations (4.1) and (4.2), (3.16) is equivalent to the inequality

$$\begin{gathered}(\partial_\theta \hat{f} + \hat{s}) \cdot \theta^\circ + (\partial_{\boldsymbol{m}} \hat{f} + \hat{c}) \cdot \boldsymbol{m}^\circ + (\partial_V \hat{f}) \cdot V^\circ \\ + V\{-\hat{f} H + \hat{\mathbf{C}}_{\mathrm{sym}} \cdot \mathbf{L} + (\psi_1 - \psi_2) + \hat{\pi}\} \leqq 0,\end{gathered} \tag{4.8}$$

where, for convenience, we have omitted all arguments.

Assume that (4.8) holds for all temperature fields and motions of the interface. In view of the Variation Lemma ([1988g]), the rates θ°, $\boldsymbol{m}^\circ$, and V° in (4.8) may be specified independently of the other quantities; this leads to all of the assertions of (i) except those concerning $\hat{\mathbf{C}}_{\mathrm{sym}}$, and also to the inequality

$$V\{-\hat{f}(\theta, \boldsymbol{m})\, H + \hat{\mathbf{C}}_{\mathrm{sym}}(\theta, \boldsymbol{m}, V) \cdot \mathbf{L} + (\psi_1 - \psi_2)(\theta) + \hat{\pi}(\theta, \boldsymbol{m}, V)\} \leqq 0.$$

Since the dependence on $\mathbf{L}$ is linear, $\hat{\mathbf{C}}_{\mathrm{sym}}$ must be consistent with (i), and the remaining inequality implies (4.6). Conversely, the assertions (i) and (ii) trivially yield (4.8) in all processes. □

Remark 4.1. By definition, π and V are components with respect to the same direction, so that, for V positive, π may be regarded as a force *in the direction of motion* exerted on the interface by the bulk material. Equation (4.5) gives this force as the sum of two terms. The first term is a force $[\psi]$ which is positive if the phase into which the interface is moving has higher free energy (and is thus less stable) than the other phase. The second term $-\hat{\beta} V$ is, by (4.7), negative, and represents a drag force, a force on the interface which opposes its motion. □

Remark 4.2. The paper [1988g] begins with an arbitrary energy flux $\boldsymbol{j}\cdot\boldsymbol{v}$ in place of $(V\boldsymbol{m})\cdot\mathbf{C}\boldsymbol{v} = V\boldsymbol{c}\cdot\boldsymbol{v}$ in (3.9) and shows, as a consequence of the second law, that $\boldsymbol{j}$ necessarily has the form $V\boldsymbol{c}$. However, the more general framework necessitates a more complicated constitutive theory: in the present study, capillary balance furnishes a relation between interfacial temperture, curvature, orientation, and normal velocity; such a relation is a constitutive *postulate* in [1986g, 1988g]. Moreover, the structure of the constitutive theory in [1988g] leads to the (somewhat strange) requirement that the interfacial entropy be independent of orientation. □

To avoid repeated hypotheses, we now make the following:

Assumption. *We assume, for the remainder of the paper, that the constitutive equations are compatible with thermodynamics, and that the capillary supply and the bulk and interfacial heat supplies vanish*:

$$b = 0, \quad q = 0, \quad r = 0. \tag{4.9}$$

Note that (3.13) and (4.5) yield the **Gibbs relations**

$$f^\circ = -s\theta^\circ - \boldsymbol{c}\cdot\boldsymbol{m}^\circ, \quad e^\circ = \theta s^\circ - \boldsymbol{c}\cdot\boldsymbol{m}^\circ, \tag{4.10}$$

while (3.14), (3.16), (4.4), $(4.5)_3$, (4.6), (4.10), and Remark 3.6 imply that

$$\theta\Gamma = \boldsymbol{q}\cdot\boldsymbol{g}, \quad \gamma = -\beta V^2. \tag{4.11}$$

5. The general free-boundary problem

5.1. Bulk equations. Interface conditions

The equations derived thus far combine to form an important free-boundary problem for the temperature. The differential equation, to be satisfied in bulk, is balance of energy $(3.11)_2$. If we let

$$C_i(\theta) = \partial_\theta\varepsilon_i(\theta) \tag{5.1}$$

denote the **bulk specific heats**, and assume that the heat flux is given by Fourier's law

$$\boldsymbol{q}_i(\theta, \nabla\theta) = -\boldsymbol{K}_i(\theta)\,\nabla\theta \tag{5.2}$$

with $\boldsymbol{K}_i(\theta)\in \operatorname{lin}(\mathbb{R}^3,\mathbb{R}^3)$ the **conductivity tensor** for phase i, then balance of energy has the form

$$C_i(\theta)\,\theta^\cdot = \operatorname{div}\{\boldsymbol{K}_i(\theta)\,\nabla\theta\}. \tag{5.3}$$

Equally important are the conditions expressing force and energy balance for the interface. The latter is given by $(3.12)_1$. By $(4.5)_3$, the equation (3.7) expressing normal force-balance has the form

$$\pi = -fH - \operatorname{div}_{\scriptscriptstyle\mathcal{J}} c, \tag{5.4}$$

or equivalently, using (4.6) and writing $\beta = \hat{\beta}(\theta, \boldsymbol{m}, V)$,

$$[\psi] = \beta V - fH - \operatorname{div}_{\mathscr{s}} \boldsymbol{c}. \qquad (5.5)^{14}$$

The basic equations which govern the evolution of the interface are the bulk equations (5.3), the interface conditions $(3.12)_1$ and (5.5), and the appropriate interfacial constitutive equations:

bulk equations

$$C_i(\theta)\,\theta^{\cdot} = \operatorname{div}\{\boldsymbol{K}_i(\theta)\,\nabla\theta\}, \qquad (5.6)$$

interface conditions

$$\begin{gathered}[\psi_i(\theta)] = \beta V - fH - \operatorname{div}_{\mathscr{s}} \boldsymbol{c},\\ [\varepsilon_i(\theta)]\,V = -[\boldsymbol{K}_i(\theta)\,\nabla\theta]\cdot\boldsymbol{m} + e^{\circ} - eHV - \operatorname{div}_{\mathscr{s}}(V\boldsymbol{c}), \qquad (5.7)\\ f = \hat{f}(\theta, \boldsymbol{m}), \quad e = \hat{e}(\theta, \boldsymbol{m}), \quad \beta = \hat{\beta}(\theta, \boldsymbol{m}, V), \quad \boldsymbol{c} = -\partial_{\boldsymbol{m}}\hat{f}(\theta, \boldsymbol{m}).\end{gathered}$$

Here we have used the obvious notation for the jump in a bulk constitutive function; for example,

$$[\psi_i(\theta)] = \psi_2(\theta) - \psi_1(\theta), \quad [\boldsymbol{K}_i(\theta)\,\nabla\theta] = \boldsymbol{K}_2(\theta)\,\nabla\theta - \boldsymbol{K}_1(\theta)\,\nabla\theta.$$

5.2. Initial conditions. Boundary conditions

Appropriate **initial conditions** are:

$$\begin{gathered}\theta(\boldsymbol{x}, 0) = \theta_0(\boldsymbol{x}) \quad \text{for all } \boldsymbol{x} \in \mathbf{R}^3,\\ B_i(0) = B_{0i}\end{gathered} \qquad (5.8)$$

with $\theta_0(\boldsymbol{x})$ the prescribed initial temperature and B_{0i} the prescribed initial phase regions.

Since the body (the region of space occupied by the two phases) is all of $\mathbf{R}^3$, conditions at infinity are required. Such conditions are standard if the interface is finite.

Thus far we have limited our discussion to unbounded bodies. If the **body** B is a *bounded region* (fixed in time), then boundary conditions are required. When the interface $\mathscr{s}(t)$ touches the boundary, conditions expressing balance of capillary forces are needed at the juncture of the interface and the boundary; these require a detailed description of the *boundary interface* between the individual phases and

[14] Within a *statical* theory ($V = 0$) HERRING [1951] and CAHN & HOFFMAN [1974] derive an equation of this form as a necessary condition for the free energy to be a minimum. With $\boldsymbol{c}$ and V zero, (5.5) is usually referred to as the Gibbs-Thomson relation (*cf.*, *e.g.*, MULLINS & SEKERKA [1964], eq. (3b)). In [1986g, 1988g] the relation (5.5) (with $\beta = 0$) follows as a consequence of the second law, but the derivation requires a constitutive equation (for the interfacial temperature) which in a sense replaces the law of capillary balance.

∂B, a description beyond the scope of this paper. Here we shall restrict our attention to situations in which the interface does not touch the boundary; in the same spirit, when discussing boundary conditions away from $\mathscr{A}(t)$, we will ignore the effects of a boundary interface. Appropriate boundary conditions are then a prescription of

$$\left.\begin{array}{l}\theta(\boldsymbol{x}, t) \text{ on a portion of } \partial B \text{ and} \\ \boldsymbol{q}(\boldsymbol{x}, t) \cdot \boldsymbol{n}(\boldsymbol{x}) \text{ on the remainder, with} \\ \boldsymbol{n} \text{ the outward unit normal to } \partial B .\end{array}\right\} \tag{5.9}$$

The free-boundary problem described by (5.6)–(5.9) is extremely difficult, chiefly because of the nonlinearities inherent in the free-boundary conditions (5.7). For that reason we shall develop, in the next section, an approximate theory for weak interfaces.

6. Weak interfaces

6.1. Behavior near the transition temperature

We assume that there is a *unique* temperature θ_M, called the **transition temperature**, at which the bulk free energies coincide:

$$\psi_1(\theta_M) = \psi_2(\theta_M). \tag{6.1}$$

Remark 6.1. In the absence of interfacial structure (*i.e.*, for f, c, and β identically zero) $(5.7)_1$ yields $[\psi] = 0$, so that $\theta = \theta_M$. This is a free-boundary condition of the classical (Stefan) theory of melting. As we shall see, within the current framework the interfacial temperature will generally *not* equal the transition temperature. □

The difference

$$l = \varepsilon_2(\theta_M) - \varepsilon_1(\theta_M) \tag{6.2}$$

in energy between phases at the transition temperature is the **latent heat**, which we assume to be nonzero:

$$l \neq 0. \tag{6.3}$$

By (3.13), $(4.4)_1$, and (6.2),

$$\partial_\theta\{\psi_2(\theta) - \psi_1(\theta)\}|_{\theta=\theta_M} = -l/\theta_M. \tag{6.4}$$

We are interested in behavior near the transition temperature and therefore introduce the (dimensionless) **temperature difference**

$$u = \frac{\theta - \theta_M}{\theta_M}. \tag{6.5}$$

Then, by (6.1), (6.2), and (6.4), for u small,

$$\begin{aligned}\psi_2(\theta) - \psi_1(\theta) &= -lu + O(u^2),\\ \varepsilon_2(\theta) - \varepsilon_1(\theta) &= l + O(u).\end{aligned} \tag{6.6}$$

6.2. *Approximate conditions for weak interfaces*

We now derive approximate interface conditions appropriate to a **weak interface**; that is, an interface whose free energy, internal energy, and kinetic coefficient are small, and whose kinetic coefficient depends only weakly on V:

$$\begin{aligned}\hat{f}(\theta, \boldsymbol{m}) = \delta f^*(\theta, \boldsymbol{m}), \quad \hat{e}(\theta, \boldsymbol{m}) = \delta e^*(\theta, \boldsymbol{m}),\\ \hat{\beta}(\theta, \boldsymbol{m}, V) = \delta\beta^*(\theta, \boldsymbol{m}, \delta V).\end{aligned} \tag{6.7}$$

Here $\delta > 0$ is small and the starred quantities are $O(1)$ in magnitude. By $(4.5)_2$,

$$\hat{c}(\theta, \boldsymbol{m}) = \delta c^*(\theta, \boldsymbol{m}).$$

For convenience, let

$$\begin{aligned}f_0(\boldsymbol{m}) = \hat{f}(\theta_M, \boldsymbol{m}), \quad \boldsymbol{c}_0(\boldsymbol{m}) = \hat{\boldsymbol{c}}(\theta_M, \boldsymbol{m}),\\ \beta_0(\boldsymbol{m}) = \hat{\beta}(\theta_M, \boldsymbol{m}, 0).\end{aligned} \tag{6.8}$$

Then, if we argue formally, it is clear from $(6.6)_1$ and $(5.7)_1$ that $u = O(\delta)$, so that, by (6.6) and (6.7), the interface conditions (5.7) have the asymptotic forms

$$\begin{aligned}lu &= -\beta_0(\boldsymbol{m})\, V + f_0(\boldsymbol{m})\, H + \operatorname{div}_s \boldsymbol{c}_0(\boldsymbol{m}) + O(\delta^2),\\ lV &= -[\boldsymbol{K}_i(\theta)\, \nabla\theta] \cdot \boldsymbol{m} + O(\delta).\end{aligned}$$

Neglecting higher-order terms and, for convenience, dropping the subscript zero, we are led to the **approximate interface conditions**:

$$\begin{aligned}lu &= -\beta(\boldsymbol{m})\, V + f(\boldsymbol{m})\, H + \operatorname{div}_s \boldsymbol{c}(\boldsymbol{m}),\\ lV &= -[\boldsymbol{K}_i(\theta)\, \nabla\theta] \cdot \boldsymbol{m}.\end{aligned} \tag{6.9}$$ [15]

By $(4.5)_2$,

$$\boldsymbol{c}(\boldsymbol{m}) = -\partial_{\boldsymbol{m}} f(\boldsymbol{m}),$$

and we may use (A11) to write $(6.9)_1$ in the form

$$lu = -\beta(\boldsymbol{m})\, V + f(\boldsymbol{m})\, H + \partial_{\boldsymbol{m}}\partial_{\boldsymbol{m}} f(\boldsymbol{m}) \cdot \mathbf{L},$$

showing that the interfacial temperature generally depends on the entire curvature tensor, rather than simply on the mean curvature. If we let $\mathbf{B}(\boldsymbol{m})$ denote the linear

[15] *Cf.* [1988 g], eq. (6.5). The boundary condition $(6.9)_2$ is a classical Stefan condition. Free-boundary conditions of the form $lu = -\beta(\boldsymbol{m})\,V$ were introduced by FRANK [1958] and used by CHERNOV [1963a, b]; $lu = fH$ was introduced by MULLINS & SEKERKA [1963, 1964]; $lu = -\beta V + fH$ was used by VORONKOV [1964]. See also SEIDENSTICKER [1966], TARSHIS & TILLER [1966], and the review articles by SEKERKA [1968, 1973, 1984], CHERNOV [1971, 1974], DELVES [1974], and LANGER [1980].

transformation from $m^\perp$ into $\mathbf{R}^3$ defined by

$$\mathsf{B}(m)\,w = f(m)\,w + [\partial_m \partial_m f(m)]\,w \quad \text{for } w \in m^\perp, \tag{6.10}$$

then $(6.9)_1$ has the succinct form

$$lu = -\beta(m)\,V + \mathsf{B}(m) \cdot \mathsf{L}. \tag{6.11}$$

7. Free-boundary problems for weak interfaces

7.1. The quasi-linear and quasistatic problems

We now consider free-boundary problems based on the approximate interface conditions $(6.9)_2$ and (6.11) in conjunction with (5.6) *linearized* about the transition temperature θ_M:

$$C_i u^{\cdot} = \operatorname{div}(K_i \nabla u), \qquad C_i = C_i(\theta_M), \qquad K_i = K_i(\theta_M).$$

We label phases so that *phase* 2 *has the higher internal energy at the transition temperature.* Then

$$l > 0, \tag{7.1}$$

and to avoid an unnecessary constant, we rescale by defining $C_i^* = C_i/(\theta_M l)$, $K_i^* = K_i/(\theta_M l)$, $f^* = f/l$, and $\beta^* = \beta/l$. Then, dropping the star superscript, we are led to the **quasilinear system**:

$$\begin{aligned} &C_i u^{\cdot} = \operatorname{div}(K_i \nabla u) \quad \text{in bulk}, \\ &\left.\begin{aligned} u &= -\beta(m)\,V + \mathsf{B}(m) \cdot \mathsf{L} \\ V &= -[K_i \nabla u] \cdot m \end{aligned}\right\} \quad \text{on the interface}, \end{aligned} \tag{7.2}$$

with $\mathsf{B}(m)$ given by (6.10). Note that, by $(4.4)_2$ and (4.7),

$$\beta(m) \geqq 0, \qquad K_i \text{ is positive semi-definite}.$$

The *quasilinear problem* is stated by (7.2) supplemented by the initial conditions (5.8) and the boundary conditions (5.9) (with θ replaced by u and with $q = -K_i \nabla u$).

Generally; one expects the interface to move slowly in comparison to the time scale for heat conduction. With this in mind, we consider the **quasistatic system** which neglects the terms $C_i u^{\cdot}$ in the bulk equations:

$$\begin{aligned} &\operatorname{div}(K_i \nabla u) = 0 \quad \text{in bulk}, \\ &\left.\begin{aligned} u &= -\beta(m)\,V + \mathsf{B}(m) \cdot \mathsf{L} \\ V &= -[K_i \nabla u] \cdot m \end{aligned}\right\} \quad \text{on the interface}. \end{aligned} \tag{7.3}$$

The *quasi-static problem* is stated by (7.3) suplemented by $(5.8)_2$ and (5.9). (The condition $(5.8)_1$, involving the prescription of $u(x, 0)$, is dropped.) If the body is infinite, the boundary conditions (5.9) are replaced by conditions at infinity.

In discussing the above problems, it is assumed that the interface does not touch ∂B; in particular, the initial data must be consistent with this assumption.

7.2. Growth theorems

We now establish Lyapunov functions for solutions of the quasi-linear and quasi-static systems. We restrict our attention to a *bounded* body and to the following two types of boundary conditions:

(i) **isolated boundary**:

$$n \cdot K_l \nabla u = 0 \quad \text{on } \partial B \quad \text{for all time;} \tag{7.4}$$

(ii) **thermally uniform boundary**:

$$u = U \quad \text{on } \partial B \quad \text{for all time.} \tag{7.5}$$

In (ii), $U = U(t)$, a function of time alone, is the prescribed **boundary temperature.**

By (6.8),

$$F(\mathscr{S}) = \int_{\mathscr{S}} f(m) \tag{7.6}$$

is the **total interfacial free energy** at the transition temperature, and, by (4.11) and Remark 3.6,

$$\mathscr{D}(u) = \sum_{l=1,2} \int_{B_l} \nabla u \cdot K_l \nabla u + \int_{\mathscr{S}} \beta(m) V^2 \geqq 0 \tag{7.7}$$

is, within the approximation of a weak interface, proportional to the **total production of entropy.**

Growth theorem.[16] *Let u be a solution of the quasi-linear system with* $C_1 = C_2 = C$.

(i) *If the boundary is isolated,*

$$\left\{\operatorname{vol}(B_1) - C \int_B u\right\}^{\cdot} = 0,$$

$$\left\{F(\mathscr{S}) + \tfrac{1}{2} C \int_B u^2\right\}^{\cdot} = -\mathscr{D}(u) \leqq 0. \tag{7.8}$$

(ii) *If the boundary is thermally uniform,*

$$F(\mathscr{S})^{\cdot} + U \operatorname{vol}(B_1)^{\cdot} + \tfrac{1}{2} C \left\{\int_B (u - U)^2\right\}^{\cdot} = -\mathscr{D}(u) \leqq 0. \tag{7.9}$$

Let u be a solution of the quasistatic system.

(iii) *If the boundary is isolated,*

$$\operatorname{vol}(B_1)^{\cdot} = 0, \quad F(\mathscr{S})^{\cdot} = -\mathscr{D}(u) \leqq 0. \tag{7.10}$$

(iv) *If the boundary is thermally uniform,*

$$F(\mathscr{S})^{\cdot} + U \operatorname{vol}(B_1)^{\cdot} = -\mathscr{D}(u) \leqq 0. \tag{7.11}$$

[16] *Cf.* [1986g], Sections 10, 11; [1988g], eqs. (7.9), (7.10). We write vol (D) for the volume of regions D in $\mathbb{R}^3$.

Proof. The proof is based on three identities. The first, a direct consequence of the divergence theorem, asserts that

$$\sum_{i=1,2} \int_{B_i} \operatorname{div} h = \int_{\partial B} h \cdot n - \int_{\mathscr{s}} [h] \cdot m \tag{7.12}$$

for any bulk vector field h. The other two identities are:

$$\int_{\mathscr{s}} V = \operatorname{vol}(B_1)^{\cdot}, \qquad \int_{\mathscr{s}} uV = -F(\mathscr{s})^{\cdot} - \int_{\mathscr{s}} \beta(m) V^2. \tag{7.13}$$ [17]

The first of (7.13) follows from $(B3)_1$. We now sketch a proof of $(7.13)_2$; this proof uses only interface condition

$$u = -\beta(m) V + f(m) H + \partial_m \partial_m f(m) \cdot \mathsf{L} \tag{7.14}$$

and the fact that $\mathscr{s}$ is a *closed* surface. In view of $(B3)_2$ with $\Omega = B$, (7.6), (A11), (B1), and the surface divergence theorem (A15) with $\mathscr{c} = \mathscr{s}$,

$$\begin{aligned} F(\mathscr{s})^{\cdot} &= \int_{\mathscr{s}} f(m)^{\circ} - \int_{\mathscr{s}} f(m) HV, \\ &= \int_{\mathscr{s}} \partial_m f(m) \cdot m^{\circ} - \int_{\mathscr{s}} f(m) HV, \\ &= - \int_{\mathscr{s}} \{\partial_m \partial_m f(m) \cdot \mathsf{L} + f(m) H\} V; \end{aligned}$$

and $(7.13)_2$ follows from (7.14).

Let u be a solution of the quasi-linear system with $C_1 = C_2$. Since u is continuous across the interface

$$\left\{\int_B u^p\right\}^{\cdot} = \int_B (u^p)^{\cdot}, \tag{7.15}$$

$p = 1, 2$. Let q be the bulk field defined by $q = K_i \nabla u$ in B_i. Assume that the boundary is isolated in the sense of (7.4). Then $(7.8)_1$ follows from (7.12) with $h = q$, $(7.2)_{1,3}$, $(7.13)_1$, and (7.15); while $(7.8)_2$ is a consequence of (7.12) with $h = uq$, $(7.2)_{1,2}$, (7.7), $(7.13)_2$, and (7.15). On the other hand, assume that the boundary is thermally uniform. Then, since $U(t)$ is independent of x,

$$\int_{\partial B} uq \cdot n = U \int_{\partial B} q \cdot n; \tag{7.16}$$

(7.9) follows from (7.12) with $h = q$ and with $h = uq$, (7.2), (7.7), (7.13), (7.15), and (7.16).

Finally, (7.10) and (7.11) follow from (7.8) and (7.9) with $C = 0$. □

Remark 7.1. In view of the agreement (7.1), phase 2 has higher internal energy at the transition temperature; thus for a solid-liquid system $B_1(t)$ would be the region occupied by the solid phase. If the boundary is supercooled, then $U < 0$, and the fact that $U \operatorname{vol}(B_1)$ is negative in (7.9) and (7.11) at least indicates the tendency of the solid phase to grow. □

---

[17] [1988g], eq. (6.6).

For an isotropic material the quasi-static system reduces to the Mullins-Sekerka system:[18]

$$\Delta u = 0 \quad \text{in bulk},$$

$$u = -\beta V + fH, \qquad V = -[k_i \nabla u] \cdot m \quad \text{on the interface},$$

with β, f, and k_i scalar constants and Δ the laplacian. In this case (7.10) becomes

$$\mathrm{vol}\,(B_1)^{\cdot} = 0, \quad f\,\mathrm{area}\,(\mathscr{s})^{\cdot} = -\mathscr{D}(u) \leqq 0,$$

while (7.11) takes the form

$$f\,\mathrm{area}\,(\mathscr{s})^{\cdot} + U\,\mathrm{vol}\,(B_1)^{\cdot} = -\mathscr{D}(u) \leqq 0.$$

An analogous simplification holds for the quasilinear problem.

8. Perfect conductors

Consider the quasilinear system (7.2) for a bounded region with boundary held at the spatially constant temperature $U(t)$ (cf. (7.5)). We now discuss the asymptotic form this system takes when *the conductivity of each phase is large*. Precisely, we consider (7.2) and (7.5) with

$$\boldsymbol{K}_i \quad \text{replaced by } \delta^{-1}\,\boldsymbol{K}_i \tag{8.1}$$

under the assumption that δ is small. In a formal perturbation for u in powers of δ, it is clear that the lowest-order term, also written u, should be consistent with

$$\begin{gathered}\operatorname{div}(\boldsymbol{K}_i \nabla u) = 0 \text{ in bulk}, \quad [\boldsymbol{K}_i \nabla u] \cdot \boldsymbol{m} = 0 \text{ on the interface},\\ u = U \quad \text{on } \partial B\end{gathered} \tag{8.2}$$

as well as the interface condition $u = -\beta(\boldsymbol{m})\,V + \mathbf{B}(\boldsymbol{m}) \cdot \mathbf{L}$. Under reasonable assumptions, the problem (8.2) has the unique solution $u(\boldsymbol{x}, t) \equiv U(t)$; the only equation then left to solve is the free-boundary condition for u:

$$\beta(\boldsymbol{m})\,V - \mathbf{B}(\boldsymbol{m}) \cdot \mathbf{L} = -U(t) \quad \text{on the interface}. \tag{8.3}$$

This equation, together with kinematical conditions for $\mathscr{s}(t)$, forms a boundary-value problem for the evolution of the interface.

Let $F(\mathscr{s})$ be defined by (7.6) and $\mathscr{D}(\mathscr{s})$ by

$$\mathscr{D}(\mathscr{s}) = \int_{\mathscr{s}} \beta(\boldsymbol{m})\,V^2.$$

Then $(7.13)_2$ with $u = U$ and $(7.13)_1$ yield the

Growth theorem for perfect conductors. *For an interface consistent with* (8.3),

$$F(\mathscr{s})^{\cdot} + U\,\mathrm{vol}\,(B_1)^{\cdot} = -\mathscr{D}(\mathscr{s}) \leqq 0. \tag{8.4}$$

[18] [1963, 1964], although Mullins & Sekerka take $\beta = 0$.

Remark 8.1. If, instead of a thermally-uniform boundary, we consider an isolated boundary, then the condition $u = U$ in (8.2) is replaced by $(K_i \nabla u) \cdot n = 0$ on ∂B; (8.2) thus modified still has the solution $u(x, t) \equiv U(t)$, but now $U(t)$ is *indeterminate*. On the other hand, for δ in (8.1) small but nonzero, (7.8) is satisfied (granted $C_1 = C_2 = C$); we therefore expect that an *isolated perfect conductor* is described by the interface equations

$$\beta(m)\, V - \mathbf{B}(m) \cdot \mathbf{L} = -U(t), \qquad \mathrm{vol}\,(B_1)^{\cdot} = C_0 U^{\cdot}\,, \tag{8.5}$$

where $C_0 = C\,\mathrm{vol}\,(B)$. Within the approximations underlying the linear heat equation, $(8.5)_2$ is the requirement that the internal energy be constant. Moreover, (8.4) remains valid; thus, by use of (8.5),

$$\{F(\mathscr{o}) + \tfrac{1}{2}\, C_0 U^2\}^{\cdot} = -\mathscr{D}(\mathscr{o}) \leqq 0. \tag{8.6}$$

Remark 8.2. For materials which have both large conductivity and *small specific heat*, one might consider the previous analysis with (8.1) supplemented by replacement of C_i by δC_i. In this instance, the arguments leading to (8.3) and (8.4) remain unchanged, but (8.5) is replaced by the system

$$\beta(m)\, V - \mathbf{B}(m) \cdot \mathbf{L} = -U(t), \qquad \mathrm{vol}\,(B_1)^{\cdot} = 0, \tag{8.7}$$

while (8.6) reduces to

$$F(\mathscr{o})^{\cdot} = -\mathscr{D}(\mathscr{o}) \leqq 0. \tag{8.8}$$

Acknowledgment. I greatly acknowledge valuable discussions with S. AGNENENT, N. MILIC, and A. STRUTHERS. This work was supported by the Army Research Office and by the National Science Foundation.

Appendix on surfaces

A. Surfaces

We use the notation and many of the results of GURTIN & MURDOCH [1975], MURDOCH [1976, 1978], and GURTIN [1986g, 1988g]. Given inner product spaces $\mathscr{V}$ and $\mathscr{W}$, $\mathrm{lin}\,(\mathscr{V}, \mathscr{W})$ is the space of linear transformations from $\mathscr{V}$ into $\mathscr{W}$; $\mathrm{lin}\,(\mathscr{V}, \mathscr{W})$ is equipped with inner product $A \cdot B = \mathrm{tr}\,(AB^T)$. Here tr denotes the **trace**, B^T is the **transpose** of B, and we write $u \cdot v$ for the **inner product** of u and v, regardless of the space in question. Further, $A \in \mathrm{lin}\,(\mathscr{V}, \mathscr{V})$ is **symmetric** if $A = A^T$, **skew** if $A = -A^T$; more generally, each $A \in \mathrm{lin}\,(\mathscr{V}, \mathscr{V})$ admits a unique additive decomposition into **symmetric** and **skew parts**

$$\tfrac{1}{2}\,(A + A^T) \quad \text{and} \quad \tfrac{1}{2}\,(A - A^T). \tag{A1}$$

The **tensor product** of $v \in \mathscr{V}$ and $w \in \mathscr{W}$ is the transformation $v \otimes w \in \mathrm{lin}\,(\mathscr{W}, \mathscr{V})$ defined by $(v \otimes w)\, z = (w \cdot z)\, v$ for all $z \in \mathscr{W}$.

Let m be a *unit vector*. $\mathbf{I}(m) \in \mathrm{lin}\,(m^{\perp}, \mathbb{R}^3)$ is the **inclusion** of $m^{\perp}$ into $\mathbb{R}^3$: $\mathbf{I}(m)$ maps $a \in m^{\perp}$ into a considered as a vector in $\mathbb{R}^3$. $\mathbf{P}(m)$ is the **perpendicu-**

lar projection from $\mathbb{R}^3$ onto the plane $m^\perp$: for each $a \in \mathbb{R}^3$, $\mathbf{P}(m)\, a \in m^\perp$ is defined by

$$\mathbf{P}(m)\, a = a - (a \cdot m)\, m. \tag{A2}$$

We consider $\mathbf{P}(m)$ as an element of lin $(\mathbb{R}^3, m^\perp)$. Thus the codomain[19] of $\mathbf{P}(m)$ is $m^\perp$ and not $\mathbb{R}^3$; with this agreement,

$$\mathbf{I}(m)^T = \mathbf{P}(m). \tag{A3}$$

Let $\mathscr{S}$ denote a smooth, oriented **surface** in $\mathbb{R}^3$ with unit normal field $m(x)$, the **orientation** of $\mathscr{S}$. Then $m(x)^\perp$ is the **tangent plane** to $\mathscr{S}$ at $x \in \mathscr{S}$. We use the shorthand

$$\mathbf{I}(x) = \mathbf{I}(m(x)), \quad \mathbf{P}(x) = \mathbf{P}(m(x)), \tag{A4}$$

so that $\mathbf{P}(x)$ is the projection onto the tangent plane at x, while $\mathbf{I}(x)$ is the inclusion of the tangent plane into $\mathbb{R}^3$.

We will consistently use the following terminology:

superficial scalar or vector field: a scalar or vector field on $\mathscr{S}$;

tangential vector field: a superficial vector field whose values are tangential to $\mathscr{S}$;

superficial tensor field: a field $\mathbf{C}$ on $\mathscr{S}$ with values $\mathbf{C}(x) \in \text{lin}\,(m(x)^\perp, \mathbb{R}^3)$;

tangential tensor field: a superficial tensor field $\mathbf{C}$ whose values satisfy $\mathbf{C}(x)\, a \in m(x)^\perp$ for each $a \in m(x)^\perp$.

For $\mathbf{C}$ a superficial tensor field:

$$\mathbf{C} \textit{ tangential} \Leftrightarrow \mathbf{C} = \mathbf{IPC} \Leftrightarrow \mathbf{C}^T m = 0. \tag{A5}$$

The first implication in (A5) is immediate. To derive the second, note that $m \cdot \mathbf{C}a = a \cdot \mathbf{C}^T m$ for $a \in m^\perp$, and $m \cdot \mathbf{C}a = 0$ for all such a if and only if $\mathbf{C}$ is tangential.

Let $\mathbf{T}$ be a tangential tensor field. Although $\mathbf{T}(x)$ maps tangent vectors into tangent vectors, we consider the codomain of $\mathbf{T}(x)$ to be $\mathbb{R}^3$. Postmultiplying by $\mathbf{P}(x)$ transforms $\mathbf{T}(x)$ to an element of lin $(m(x)^\perp, m(x)^\perp)$, premultiplying by $\mathbf{P}(x)$ extends $\mathbf{T}(x)$ to an element of lin $(\mathbb{R}^3, \mathbb{R}^3)$, and neither of these adjustments changes its essential character:

$$\mathbf{T}(x)\, a = \mathbf{P}(x)\, \mathbf{T}(x)\, a = \mathbf{T}(x)\, \mathbf{P}(x)\, a \textit{ for every } a \in m(x)^\perp, \textit{ but}$$

$$\mathbf{T}(x) \in \text{lin}\,(m(x)^\perp, \mathbb{R}^3), \quad \mathbf{P}(x)\mathbf{T}(x) \in \text{lin}\,(m(x)^\perp, m(x)^\perp), \quad \mathbf{T}(x)\mathbf{P}(x) \in \text{lin}\,(\mathbb{R}^3, \mathbb{R}^3).$$

Moreover,

$$\begin{gathered}\mathbf{PT} \textit{ symmetric} \Leftrightarrow \mathbf{TP} \textit{ symmetric}, \\ \mathbf{PT} \textit{ skew} \Leftrightarrow \mathbf{TP} \textit{ skew}.\end{gathered} \tag{A6}$$

Indeed, let $\mathbf{PT}$ be symmetric. Then $\mathbf{PT} = \mathbf{T}^T\mathbf{I}$. Also, by (A5), $\mathbf{T} = \mathbf{IPT}$ and $\mathbf{T}^T = \mathbf{T}^T\mathbf{IP}$. Thus $\mathbf{TP} = \mathbf{IPTP} = \mathbf{IT}^T\mathbf{IP} = \mathbf{IT}^T$, and $\mathbf{TP}$ is symmetric. The remaining assertions of (A6) are proved analogously.

[19] We very carefully identify the domain $\mathscr{V}$ and codomain $\mathscr{W}$ of transformations in lin $(\mathscr{V}, \mathscr{W})$; identification of the codomain is crucial, since the domain of the transpose is the codomain of the original map.

Guided by (A6), we refer to a tangential tensor field $\mathbf{T}$ as **symmetric** or **skew** according as $\mathbf{P}(x)\,\mathbf{T}(x)$ (or equivalently $\mathbf{T}(x)\,\mathbf{P}(x)$) is symmetric or skew at each $x \in \mathcal{S}$. With this terminology, *the inclusion* $\mathbf{I}$ *is a symmetric tangential tensor field.* A further consequence of this definition is that

$$\mathbf{T} \cdot \mathbf{F} = 0 \quad \textit{for } \mathbf{T} \textit{ symmetric and } \mathbf{F} \textit{ skew}. \tag{A7}$$

Similarly, we define the **trace**, $\operatorname{tr}\mathbf{T}$, of a tangential field $\mathbf{T}$ by $\operatorname{tr}\mathbf{T} = \operatorname{tr}(\mathbf{PT}) = \operatorname{tr}(\mathbf{TP})$.

Let $\mathbf{T}$ be a *tangential* tensor field. Then $\mathbf{T}$ admits the unique decomposition

$$\mathbf{T} = \mathbf{T}_{\text{sym}} + \mathbf{T}_{\text{skw}},$$

where $\mathbf{T}_{\text{sym}}$ and $\mathbf{T}_{\text{skw}}$, respectively, are *symmetric and skew* tangential tensor fields called the symmetric and skew parts of $\mathbf{T}$. In fact,

$$\mathbf{T}_{\text{sym}} = \tfrac{1}{2}\,\mathbf{I}(\mathbf{PT} + \mathbf{T}^T\mathbf{I}), \quad \mathbf{T}_{\text{skw}} = \tfrac{1}{2}\,\mathbf{I}(\mathbf{PT} - \mathbf{T}^T\mathbf{I});$$

i.e., e.g., the symmetric part of $\mathbf{T}$ is the symmetric part of $\mathbf{P}(x)\,\mathbf{T}(x) \in \operatorname{lin}(m(x)^\perp, m(x)^\perp)$ postmultiplied by $\mathbf{I}(x)$ to convert to $\operatorname{lin}(m(x)^\perp, \mathbb{R}^3)$.

Each superficial tensor $\mathbf{C}$ admits the unique decomposition

$$\mathbf{C} = \mathbf{C}_{\text{sym}} + \mathbf{C}_{\text{skw}} + m \otimes c, \tag{A8}$$

where $\mathbf{C}_{\text{sym}}$ is a *symmetric tangential* tensor field, $\mathbf{C}_{\text{skw}}$ is a *skew tangential* tensor field, and c is a *tangential vector field.* Indeed,

$$c = \mathbf{C}^T m, \tag{A9}$$

while $\mathbf{C}_{\text{sym}}$ and $\mathbf{C}_{\text{skw}}$ are the symmetric and skew parts of the *tangential* tensor field $\mathbf{C} - m \otimes \mathbf{C}^T m$. We will refer to $\mathbf{C}_{\text{sym}}$, $\mathbf{C}_{\text{skw}}$, and c, respectively, as the **symmetric, skew,** and **normal components** of $\mathbf{C}$. If for some scalar field σ,

$$\mathbf{C} = \mathbf{C}_{\text{sym}} = \sigma\mathbf{I}, \tag{A10}$$

then $\mathbf{C}$ is a **surface tension** σ.

We write $\nabla_{\mathcal{S}}$ for the **surface gradient.**[20] For ϕ a superficial scalar field, $\nabla_{\mathcal{S}}\phi$ is a tangential vector field; for v a superficial vector field, $\nabla_{\mathcal{S}} v$ is a superficial tensor field. The trace of $\mathbf{P}\,\nabla_{\mathcal{S}} v$ is the **surface divergence** of v:

$$\operatorname{div}_{\mathcal{S}} v = \operatorname{tr} \mathbf{P}(\nabla_{\mathcal{S}} v).$$

The superficial tensor field

$$\mathbf{L} = -\nabla_{\mathcal{S}} m \tag{A11}$$

is the **curvature tensor.** A classical result is that

$$\mathbf{L} \textit{ is tangential and symmetric}. \tag{A12}$$

We write

$$H = \operatorname{tr} \mathbf{L}$$

for (twice) the **mean curvature.**

[20] For $z = z(t)$ a curve on $\mathcal{S}$, $\phi(z)^{\cdot} = \nabla_{\mathcal{S}}\phi(z) \cdot z^{\cdot}$, $v(z)^{\cdot} = [\nabla_{\mathcal{S}} v(z)]\, z^{\cdot}$; for v tangential, $\mathbf{P}\,\nabla_{\mathcal{S}} v$ is the covariant derivative of v.

Let $\mathbf{C}$ be a superficial tensor field. Then $\operatorname{div}_\sigma \mathbf{C}$ is the unique vector field on σ with the property

$$\boldsymbol{a} \cdot \operatorname{div}_\sigma \mathbf{C} = \operatorname{div}_\sigma (\mathbf{C}^T \boldsymbol{a}) \tag{A13}$$

for all vectors $\boldsymbol{a}$.

The surface gradient and surface divergence obey the usual laws for the differentiation of scalar products and inner products (*cf.* GURTIN & MURDOCH [1975], eq. (2.17)). Less standard are the **identities**:

$$\begin{aligned} \operatorname{div}_\sigma (\mathbf{C}^T \boldsymbol{v}) &= \boldsymbol{v} \cdot \operatorname{div}_\sigma \mathbf{C} + \mathbf{C} \cdot \nabla_\sigma \boldsymbol{v}, \\ \operatorname{div}_\sigma \mathbf{C} &= \operatorname{div}_\sigma (\mathbf{C}_{\text{sym}} + \mathbf{C}_{\text{skw}}) + (\operatorname{div}_\sigma c)\, \boldsymbol{m} - \mathbf{L}c, \\ \boldsymbol{m} \cdot \operatorname{div}_\sigma \mathbf{C}_{\text{skw}} &= 0, \\ \boldsymbol{m} \cdot \operatorname{div}_\sigma \mathbf{C}_{\text{sym}} &= \mathbf{C}_{\text{sym}} \cdot \mathbf{L}. \end{aligned} \tag{A14}$$

Here $\boldsymbol{v}$ is a superficial vector field, while $\mathbf{C}$ is a superficial tensor field with $\mathbf{C}_{\text{sym}}$, $\mathbf{C}_{\text{skw}}$, and c the corresponding symmetric, skew, and normal components.

The identities $(\text{A14})_{1,2}$ are easily derived using (A13) and (A8), while $(\text{A14})_{3,4}$ are consequences of $(\text{A14})_1$ with $\boldsymbol{v} = \boldsymbol{m}$ in conjunction with (A5), (A7), and (A11).

If c is a sufficiently regular subsurface of σ whose boundary curve ∂c is sufficiently smooth, and $\boldsymbol{v}$ is a *tangential* vector field, then the **surface divergence theorem** asserts that

$$\int_{\partial c} \boldsymbol{v} \cdot \boldsymbol{\nu} = \int_{c} \operatorname{div}_\sigma \boldsymbol{v}, \tag{A15}$$

where $\boldsymbol{\nu}$, a vector field tangential to σ, is the outward unit normal to ∂c. For $\mathbf{C}$ a superficial tensor field and $\boldsymbol{a}$ a constant vector, $\mathbf{C}^T\boldsymbol{a}$ is tangential, and (A13) and (A15) yield

$$\int_{\partial c} \mathbf{C}\boldsymbol{\nu} = \int_{c} \operatorname{div}_\sigma \mathbf{C}. \tag{A16}$$

Given vectors $\boldsymbol{a}, \boldsymbol{b} \in \mathbb{R}^3$, we define

$$\boldsymbol{a} \wedge \boldsymbol{b} = \boldsymbol{a} \otimes \boldsymbol{b} - \boldsymbol{b} \otimes \boldsymbol{a}.$$

Further, we write

$$\boldsymbol{r}(\boldsymbol{x}) = \boldsymbol{x} - \boldsymbol{x}_0$$

for the position vector from a fixed point $\boldsymbol{x}_0 \in \mathbb{R}^3$. We then have the following identity (*cf.* GURTIN & MURDOCH [1975], p. 305), valid for $\mathbf{C}$ a sufficiently smooth superficial tensor field:

$$\int_{\partial c} \boldsymbol{r} \wedge \mathbf{C}\boldsymbol{\nu} = \int_{c} (\boldsymbol{r} \wedge \operatorname{div}_\sigma \mathbf{C} + \mathbf{I}\mathbf{C}^T - \mathbf{C}\mathbf{P}),$$

or equivalently, by (A8) and the definitions given in the paragraph containing (A7),

$$\int_{\partial c} \boldsymbol{r} \wedge \mathbf{C}\boldsymbol{\nu} = \int_{c} \{\boldsymbol{r} \wedge \operatorname{div}_\sigma \mathbf{C} - \boldsymbol{m} \wedge c - 2\, \mathbf{C}_{\text{skw}}\mathbf{P}\}, \tag{A17}$$

with the tangential field c here viewed as having values in $\mathbb{R}^3$.

B. Smoothly propagating surfaces

Suppose now that $\mathscr{s}(t)$ is a smoothly propagating surface ([1988g], Appendix D), so that, for each time t, $\mathscr{s}(t)$ divides $\mathbb{R}^3$ into closed regions $B_1(t)$ and $B_2(t)$ with $\mathbb{R}^3$ as their union and $\mathscr{s}(t)$ as their intersection. We orient $\mathscr{s}(t)$ by choosing, as unit normal field $\boldsymbol{m}(\boldsymbol{x}, t)$, the outward unit normal to $\partial B_1(t)$. We write $V(\boldsymbol{x}, t)$ for the **normal velocity** of $\mathscr{s}(t)$ in the direction $\boldsymbol{m}(\boldsymbol{x}, t)$.

We use the following notation regarding time derivatives: for $\phi = \phi(t)$, $\phi^{\cdot} = d\phi/dt$; for ϕ a bulk field, $\phi^{\cdot}(\boldsymbol{x}, t) = \partial_t\phi(\boldsymbol{x}, t)$; for ϕ a superficial scalar or vector field, ϕ° is the **normal time-derivative**, the time derivative following the interface (*cf.* [1986g], eq. (4.4); [1988g], eq. (D5)). Then (*cf.* [1988g], eq. (D15))

$$\boldsymbol{m}^{\circ} = -\nabla_{\mathscr{s}} V. \tag{B1}$$

Superficial fields and **tangential fields** are as specified in Appendix A, but here they are defined for all $\boldsymbol{x} \in \mathscr{s}(t)$ and all t; in the same spirit, **bulk fields** are fields on $\mathbb{R}^3$ for all time. The assertion that a relation or inequality is satisfied "**in bulk**" signifies that it holds "in the interiors of $B_1(t)$ and $B_2(t)$ for all t"; similarly, the quantifier "**on the interface**" is shorthand for "on $\mathscr{s}(t)$ for all t".

For $\boldsymbol{v}$ a bulk vector field or a bulk tensor field, $\operatorname{div}\boldsymbol{v}$ is the corresponding divergence. For ϕ a bulk field we write $[\phi]$ for the **jump** in ϕ across the interface (the limit from phase 2 minus that from phase 1).

Let Ω be a (sufficiently regular) closed region of space with outward unit normal $\boldsymbol{n}$, and let

$$\mathscr{c}(t) = \Omega \cap \mathscr{s}(t), \qquad \Omega_i(t) = \Omega \cap B_i(t).$$

For g a superficial scalar field,

$$\operatorname{outflow}(g, \Omega) = \int_{\partial \mathscr{c}} gVp/(1-p^2)^{1/2}, \qquad p = \boldsymbol{m}\cdot\boldsymbol{n}; \tag{B2}$$

this integral represents the rate at which g is carried out of Ω across $\partial\Omega$ due to the motion of the interface. We then have the identities:

$$\left\{\int_{\Omega} g\right\}^{\cdot} = \int_{\Omega_1} g^{\cdot} + \int_{\Omega_2} g^{\cdot} - \int_{\mathscr{c}} [g]\, V, \tag{B3}$$ [21]

$$\left\{\int_{\mathscr{c}} g\right\}^{\cdot} + \operatorname{outflow}(g, \Omega) = \int_{\mathscr{c}} (g^{\circ} - gHV).$$

The next result, which we state without proof,[22] allows the reduction of global balance laws to differential equations and jump conditions. In the statement of this theorem Ω and $\mathscr{c}$ are as specified above, while $\boldsymbol{\nu}$, a vector field tangential to $\mathscr{s}$, is the outward unit normal to $\partial\mathscr{c}$.

[21] $(B3)_1$ is standard; I am not aware of a rigorous proof of $(B3)_2$ (*cf.* SCRIVEN [1960], MOECKEL [1975]), although a proof is given by ANGENENT & GURTIN [1988g] for $\mathscr{s}$ a curve in $\mathbb{R}^2$.

[22] Essential ingredients of the proof are the identities (B3); *cf.* the proof of (6.4) and (6.5) of [1986g] and (2.5) of [1988g].

General balance theorem. *Let α and β be bulk scalar fields. $\boldsymbol{u}$ a bulk vector field, $\boldsymbol{v}$ a tangential vector field, g and b superficial scalar fields, all sufficiently smooth. Then:*

(i) *the balance law*

$$\left\{\int_{\Omega} \alpha + \int_{c} g\right\}^{\cdot} + \text{outflow}(g, \Omega) = \int_{\partial\Omega} \boldsymbol{u} \cdot \boldsymbol{n} + \int_{\Omega} \beta + \int_{\partial c} \boldsymbol{v} \cdot \boldsymbol{\nu} + \int_{c} b \tag{B4}$$

holds for all Ω if and only if

$$\begin{aligned} \alpha^{\cdot} &= \operatorname{div} \boldsymbol{u} + \beta \quad \textit{in bulk}, \\ -[\alpha]\, V + g^{\circ} - gHV &= [\boldsymbol{u}] \cdot \boldsymbol{m} + \operatorname{div}_s \boldsymbol{v} + b \quad \textit{on the interface}. \end{aligned} \tag{B5}$$

(ii) (B4) *holds with "=" replaced by "$\geqq$" if and only if* (B5) *holds with the same replacement.*

References

[1951] Herring, C., Surface tension as a motivation for sintering, *The Physics of Powder Metallurgy* (ed. W. E. Kingston) McGraw-Hill, New York.

[1958] Frank, F. C., On the kinematic theory of crystal growth and dissolution processes, *Growth and Perfection of Crystals* (eds. R. H. Doremus, B. W. Roberts, D. Turnbull) John Wiley, New York.

[1960] Scriven, L. E., Dynamics of a fluid interface, Chem. Eng. Sci. **12**, 98–108.

[1961] Ericksen, J. L., Conservation laws for liquid crystals, Trans. Soc. Rheol. **5**, 23–34.

[1963] Chernov, A. A., Crystal growth forms and their kinetic stability [in Russian], Kristallografiya **8**, 87–93. English Transl. Sov. Phys. Crystall. **8**, 63–67 (1963).

[1963] Chernov, A. A., Application of the method of characteristics to the theory of the growth forms of crystals [in Russian], Kristallografiya **8**, 499–505. English Transl. Sov. Phys. Crystall. **8**, 401–405 (1964).

[1963] Coleman, B. D., & V. J. Mizel, Thermodynamics and departures from Fourier's law of heat conduction, Arch. Rational Mech. Anal. **13**, 245–261.

[1963] Coleman, B. D., & W. Noll, The thermodynamics of elastic materials with heat conduction and viscosity, Arch. Rational Mech. Anal. **13**, 167–178.

[1963] Mullins, W. W., & R. F. Sekerka, Morphological stability of a particle growing by diffusion or heat flow, J. Appl. Phys. **34**, 323–329.

[1964] Mullins, W. W., & R. F. Sekerka, Stability of a planar interface during solidification of a dilute binary alloy, J. Appl. Phys. **35**, 444–451.

[1964] Voronkov, V. V., Conditions for formation of mosaic structure on a crystallization front [in Russian], Fizika Tverdogo Tela **6**, 2984–2988. English Transl. Sov. Phys. Solid State **6**, 2378–2381 (1965).

[1965] Truesdell, C., & W. Noll, The non-linear field theories of mechanics, *Handbuch der Physik*, vol. III/3 (ed. S. Flügge), Springer-Verlag, Berlin.

[1966] Seidensticker, R. G., Stability considerations in temperature gradient zone melting, *Crystal Growth* (ed. H. S. Peiser), Pergamon, Oxford (1967).

[1966] Tarshis, L. A., & W. A. Tiller, The effect of interface-attachment kinetics on the morphological stability of a planar interface during solidification, *Crystal Growth* (ed. H. S. Peiser), Pergamon, Oxford (1967).

[1968] SEKERKA, R. F., Morphological stability, J. Crystal Growth **3**, **4**, 71–81.
[1971] CHERNOV, A. A., Theory of the stability of face forms of crystals [in Russian], Kristallografiya **16**, 842–863. English Transl. Sov. Phys. Crystall. **16**, 734–753 (1972).
[1972] CHERNOV, A. A., Theory of the stability of face forms of crystals, Sov. Phys. Crystallog. **16**, 734–753.
[1972] HOFFMAN, D. W., & J. W. CAHN, A vector thermodynamics for anisotropic surfaces—1. Fundamentals and applications to plane surface junctions, Surface Sci. **31**, 368–388.
[1973] SEKERKA, R. F., Morphological stability, *Crystal Growth: an Introduction*, North-Holland, Amsterdam.
[1974] CAHN, J. W., & D. W. HOFFMAN, A vector thermodynamics for anisotropic surfaces—2. curved and faceted surfaces, Act. Metall. **22**, 1205–1214.
[1974] CHERNOV, A. A., Stability of faceted shapes, J. Crystal Growth **24/25**, 11–31.
[1974] DELVES, R. T., Theory of interface instability, *Crystal Growth* (ed. B. R. PAMPLIN), Pergamon, Oxford.
[1975] GURTIN, M. E., & A. I. MURDOCH, A continuum theory of elastic material surfaces, Arch. Rational Mech. Anal. **57**, 291–323.
[1975] MOECKEL, G. P., Thermodynamics of an interface, Arch. Rational Mech. Anal. **57**, 255–280.
[1976] MURDOCH, A. I., A thermodynamic theory of elastic material interfaces, Q. J. Mech. Appl. Math. **29**, 245–275.
[1978] BRAKKE, K. A., *The Motion of a Surface by its Mean Curvature*, Princeton University Press.
[1978] MURDOCH, A. I., Direct notation for surfaces with application to the thermodynamics of elastic material surfaces of second grade, Res. Rept. ES 78–134, Dept. Eng. Sci., U. Cincinnati.
[1979] ALLEN, S. M., & J. W. CAHN, A macroscopic theory for antiphase boundary motion and its application to antiphase domain coarsening, Act. Metall. **27**, 1085–1095.
[1979] FERNANDEZ-DIAZ, J., & W. O. WILLIAMS, A generalized Stefan condition, Zeit. Angew. Math. Phys. **30**, 749–755.
[1980] LANGER, J. S., Instabilities and pattern formation in crystal growth, Rev. Mod. Phys. **52**, 1–27.
[1984] SEKERKA, R. F., Morphological instabilities during phase transformations, *Phase Transformations and Material Instabilities in Solids* (ed. M. E. GURTIN), Academic Press, New York.
[1985] ALEXANDER, J. I. D., & W. C. JOHNSON, Thermomechanical equilibrium in solid-fluid systems with curved interfaces, J. Appl. Phys. **58**, 816–824.
[1986] GAGE, M., & R. S. HAMILTON, The heat equation shrinking convex plane curves, J. Diff. Geom. **23**, 69–96.
[1986g] GURTIN, M. E., On the two-phase Stefan problem with interfacial energy and entropy, Arch. Rational Mech. Anal. **96**, 199–241.
[1986] JOHNSON, W. C., & J. I. D. ALEXANDER, Interfacial conditions for thermomechanical equilibrium in two-phase crystals, J. Appl. Phys. **58**, 816–824.
[1987] GRAYSON, M. A., The heat equation shrinks embedded plane curves to round points, J. Diff. Geom. **26**, 285–314.
[1987] LEO, P. H., The effect of elastic fields on the morphological stability of a precipitate grown from solid solution, Ph. D. Thesis, Dept. Metall. Eng. Mat. Sci., Carnegie-Mellon U., Pittsburgh, PA.

[1988g] GURTIN, M. E., Toward a nonequilibrium thermodynamics of two phase materials, Arch. Rational Mech. Anal., **100**, 275–312.
[1988a] ANGENENT, S., & M. E. GURTIN, Multiphase thermomechanics with interfacial structure. 2. Evolution of an isothermal interface. Forthcoming.
[1988s] GURTIN, M. E., & A. STRUTHERS. Forthcoming.
[1988] FONSECA, I., Interfacial energy and the Maxwell rule, Arch. Rational Mech. Anal. Forthcoming.

Department of Mathematics
Carnegie Mellon University
Pittsburgh, PA

(Received April 28, 1988)

Acta metall. Vol. 37, No. 12, pp. 3119–3138, 1989
0001-6160/89 $3.00 + 0.00

OVERVIEW NO. 86

THE EFFECT OF SURFACE STRESS ON CRYSTAL–MELT AND CRYSTAL–CRYSTAL EQUILIBRIUM

P. H. LEO[1]† and R. F. SEKERKA[2]

[1]Department of Aerospace Engineering and Mechanics, University of Minnesota, Minneapolis, MN 55455 and [2]Mellon College of Science, Carnegie-Mellon University, Pittsburgh, PA 15213, U.S.A.

(*Received* 21 *September* 1988; *in revised form* 14 *November* 1988)

Abstract—The effect of surface stress on the equilibrium conditions at crystal–melt, coherent crystal–crystal and greased crystal–crystal interfaces is investigated by using a variational method to test for equilibrium. In all three cases, the interface between the phases is modelled as a Gibbsian dividing surface, and the excess internal energy associated with the interface is allowed to depend on both the deformation of the interface and the crystallographic normal to the interface. The position of an interface can vary due to both deformation at the interface and transformation between the two phases at the interface (accretion), and so we define a special variation that accounts for both. Thus, surface stress appears explicitly in both the force and energy balances at crystal–melt and coherent crystal–crystal interfaces. In particular, an interfacial strain energy term appears in the energy balance at these interfaces; this term gives the energy of deforming the interface against the force associated with the surface stress, and is a new result from this analysis. Anisotropy also appears in this energy balance through a term that can be expressed by using Cahn and Hoffman's ξ-vector. Finally, it is shown that a greased crystal–crystal system differs from crystal–melt and coherent crystal–crystal systems in that two independent deformations and crystallographic normals can be defined at a greased interface. However, by partitioning the excess energy associated with a greased interface between these deformations and normals, one can reduce the equilibrium conditions at a greased interface to those that obtain if the two crystals would interact only through a thin fluid layer at the interface.

Résumé—On étudie l'effet de la contrainte superficielle sur les conditions d'équilibre aux interfaces cristal–métal fondu, cristal cristal–cohérent et cristal cristal–lubrifie en utilisant une méthode variationnelle pour tester l'équilibre. Dans les trois cas, on modélise l'interface entre les phases comme une surface de séparation de Gibbs, et l'on fait dépendre l'énergie interne en excès associée à l'interface à la fois de la déformation de l'interface et de la normale cristallographique à l'interface. La position d'une interface peut varier par suite de la déformation à l'interface et de la tranformation entre les deux phases à l'interface (accroissement d'une phase), et l'on définit ainsi une variation particulière qui tient compte des deux. La contrainte superficielle apparaît donce de façon explicite dans les équilibres de la force et de l'énergie aux interfaces cristal–métal fondu et cristal cristal–cohérent. Plus précisement, un terme d'énergie de déformation interfaciale apparaît dans le bilan d'énergie sur ces interfaces; ce terme donne l'énergie de déformation de l'interface contre la force associée à la contrainte superficielle, ce qui est un nouveau résultat fourni par notre analyse. L'anisotropie apparaît également dans ce bilan d'énergie à travers un terme qui peut être exprimé à l'aide du vecteur de cahnet Hoffman. On montre enfin qu'un système cristal cristal–lubrifié diffère des deux autres systèmes étudiés, dans la mesure où l'on peut définir deux déformations et deux normales cristallographiques indépendantes pour une interface lubrifiée. Cependant, si l'on répartit l'énergie en excès associée à une interface lubrifiée entre ces déformations et ces normales, on peut réduire les conditions d'équilibre sur une surface lubrifiée à celles que l'on obtiendrait si les deux cristaux n'interagissaient qu'à travers une mince couche fluide à l'interface.

Zusammenfassung—Der Einfluß der Oberflächenspannung auf die Gleichgewichtsbedingungen an Grenzflächen zwischen Kristall und Schmelze, an kohärenten Grenzflächen zwischen zwei Kristallen und an geschmierten Grenzflächen zwischen zwei Kristallen wird mittels einer Variationsmethode zur Prüfung des Gleichgewichts untersucht. In allen drei Fällen wird die Grenzfläche als eine Gibbssche Trennfläche behandelt; die mit der Grenzfläche zusammenhängende zusätzliche innere Energie wird als abhängend von der Verformung der Grenzfläche und von der kristallografischen Normalen auf die Grenzfläche angesehen. Die Lage der Grenzfläche kann sich ändern als Folge der Verformung an der Grenzfläche und der Umwandlung zwischen den beiden Phasen an der Grenzfläche (Wachstum); hierzu definieren wir eine entsprechende spezielle Variation. Die Oberflächenspannung erscheint also explizit sowohl in der Kräfte- wie auch der Energiebilanz an den Kristall–Schmelze- und Kohärenten Kristall-Kristall-Grenzflächen. Insbesondere erscheint ein Energie-Term der Grenzflächenverzerrung an diesen Grenzflächen. Dieser Term beschreibt die Energie, die Grenzfläche gegen die mit der Oberflächenspannung zusammenhängende Kraft zu verformen und ist ein neues Ergebnis dieser Analyse. Die Anisotropie erscheint in dieser

†Formerly, Department of Metallurgical Engineering and Materials Science, Carnegie-Mellon University, Pittsburgh, PA 15213, U.S.A.

Energiebilanz auch über einen Term, der mittels des ξ-Vektors von Cahn und Hoffman ausgedrückt werden kann. Schließlich wird gezeigt, daß ein geschmiertes Kristall–Kristall-System sich von den beiden anderen Systemen darin unterscheidet, daß zwei unabhängige Verformungen und kristallografische Normalen an dieser Grenzfläche definiert werden können. Allerdings können die Gleichgewichts bedingungen an einer geschmierten Grenzfläche auf die reduziert werden, die man erhält, wen zwei Kristalle an der Grenzfläche nur über eine dünne Flüssigkeitschicht wechselwirken, indem man die die mit der geschmierten Grenzfläche zusammenhängende zusätzliche Energie aufteilt.

1. INTRODUCTION

The thermodynamic description of a crystalline solid differs from that of a fluid in many respects. For example, Larche and Cahn [1, 2] and Mullins and Sekerka [3] have shown how the internal energy and chemical potentials of a crystal depend, in general, on the stresses and strains that can arise in crystalline solids. This dependence will be reflected in the conditions that describe both single-phase and multiphase equilibrium. Further, if one considers the equilibrium between a crystal and its melt or between two crystals, then one must take into account the continuity conditions that describe the interphase interface [2].

The excess properties associated with the dividing surface between a crystal and its melt or between two crystals can also distinguish crystalline phases from fluid phases. Gibbs [4] showed that the surface of a solid is fundamentally different from that of a fluid because one can distinguish between straining a solid surface and creating new surface at constant strain, whereas these two processes are indistinguishable at fluid surfaces. Gibbs' work was followed by the work of Shuttleworth [5], Herring [6] and Nicholson [7], among others, who associate a "surface stress" with the change in energy associated with deforming a solid surface, and a "surface free energy" with the change in energy of creating new surface at constant strain. More recently, Cahn [8], Mullins [9, 10], Cahn and Larche [11] and Alexander and Johnson [12, 13] have begun to include the surface free energy and surface stress in describing two-phase equilibrium.

The intent of this paper is to describe how surface stress and surface energy enter the general conditions that describe the equilibrium between a crystal and its melt and between two crystals. In particular, we model the interface between the two phases as a Gibbsian dividing surface such that the excess energy of the interface depends upon both the deformation at the interface and the crystallographic orientation of the interface. The dependence of the excess energy on the interfacial deformation leads to a surface stress, while the dependence of the excess energy on the crystallographic orientation leads to a vector force related to Cahn and Hoffman's ξ-vector [14, 15]. Also, following Larche and Cahn [2], we model a crystal–crystal interface as being either coherent or greased, where at a coherent interface we enforce a strict one-to-one matching of points across the interface, while we insist that no gaps develop at a greased interface, and that the adjacent phases at the interface are free to slide with respect to one another.

Once we construct this model system, we use a variational method to test for equilibrium. Our thermodynamic treatment of the interface hinges on understanding how the variational movement of the interface is affected by both deformations at the interface and a phase transformation between the two phases, which we call accretion, at constant strain. Thus, by defining a special interfacial variation that accounts for both processes, we derive equilibrium conditions that show how surface anisotropies, surface stress and surface free energy enter the force and energy balances at the interface.

Previous treatments of this subject have not fully accounted for both accretion and deformation at crystal–melt and crystal–crystal interfaces. Gurtin and Murdoch [16] have shown how surface stress affects the mechanical equilibrium of a solid when no accretion is allowed. Cahn [8] and Mullins [9, 10] have separately considered how surface stress and surface free energy affect equilibrium for the special case of an isotropic crystalline sphere immersed in a fluid. Alexander and Johnson [12, 13] have derived general equilibrium conditions at crystal–melt and both coherent and greased crystal–crystal interfaces; however, they do not properly treat accretion and deformation at the interface, so their results are incomplete.

The interfacial variation that we define enables us to deduce the complete set of equilibrium conditions at arbitrarily shaped crystal–melt, coherent crystal–crystal and greased crystal–crystal interfaces. As would be expected, our results reduce to the results of Cahn and Mullins for the case of a crystalline sphere immersed in a fluid. Moreover, our interfacial variation can be used to find particularly simple forms for the continuity conditions at crystal–melt, coherent and greased interfaces, and so provides a link with Larche and Cahn's treatment of coherent and greased crystal–crystal equilibrium in the absence of surface excess properties [2].

We consider first the equilibrium between a crystal and its melt, and between two crystals across a coherent interface. In the crystal–melt case, we assume that the interface behaves as a two-dimensional solid, so we "attach" the interface to the solid phase. In the coherent crystal–crystal case, we find that the two phases share a common reference state, so there is no special attachment of the interface to either phase. In both of these cases, the fact that we can use a single reference state to measure the displacements in the entire system allows us to perform the equilibrium calculation in either the reference state or the actual state; it turns out to be convenient to use the reference state for this calculation, although we have obtained the same results by working in the actual state [17]. The other model of the interface that we consider is the greased crystal–crystal interface, which we envision as being diametrically opposite in nature to the coherent interface. The points on either side of a greased interface are free to slide with respect to each other, so there is no common reference state for the two phases, but rather a different reference state for each phase. Since we do not want to attach the interface to one phase or the other, the excess energy associated with the interface must account for the properties of both of these "dual" reference states, and we perform the equilibrium calculation in the actual state.

We adopt the following working hypothesis to position the matematical dividing surface that separates the system into two phases: suppose that we have made a physically sensible choice of the dividing surface and partitioned the energy (entropy, mass, etc.) of the system accordingly. We assume that any small changes in the position of the dividing surface lead to negligibly small corrections to this partitioning, so that, from a practical standpoint, the equilibrium conditions that result are independent of the position of the dividing surface. While it is clear that a more complete theory of crystal–melt and crystal–crystal equilibrium should deal with the question of where to place the dividing surface, we believe that our approach offers the possibility of gaining insight into the roles of surface stress and surface free energy at solid surfaces and interfaces without introducing the complexities involved in positioning the dividing surface.

2. CRYSTAL–MELT AND COHERENT CRYSTAL–CRYSTAL EQUILIBRIUM

In this section, we consider first the general question of how to take a variation at an interface that moves due to both deformation and accretion, and we show how this variation can be used to describe the continuity conditions at both crystal–melt and coherent crystal–crystal interfaces. Next, we consider what variables are necessary to determine the energy of a crystal phase and a fluid phase, as well as the excess energy associated with an interface, and we set up the variational calculation to test for equilibrium. Then, by using the interfacial variation that we define, we derive the general equations that describe equilibrium between a crystal and its melt and between two crystals across a coherent interface.

†We use primes throughout the paper to denote quantities evaluated in the reference state of the system.

2.1. The variation at an accreting interface

If a thermodynamic system that is doing no work against external forces is in equilibrium at constant total entropy and mass, then all variations in the total energy of the system from its equilbrium state must be nonnegative [4]. In order to calculate this total energy variation, one integrates the pointwise variations of the bulk energy densities over the volumes of the bulk phases, and then integrates the pointwise variations of the excess energy density over the interface separating the phases. For bulk phases, these pointwise changes can be accounted for in two distinct ways. One can take a Langrangian variation, denoted by δ, in which one fixes on a small volume element at a reference, or material, point and follows it from the actual (equilibrium) state to some varied state. The other possible variation is an Eulerian variation, denoted by δ^E, in which one considers a fixed volume element in the system, and calculates the energy change in that volume element between the varied and actual states.

However, we must introduce new ideas in order to correctly measure the changes that can occur at a surface or interface. These ideas can be illustrated by using Fig. 1. Figure 1 shows some actual state of a solid [α in Fig. 1(c)], the reference state of which is pictured in Fig. 1(a). We insist, for reasons we discuss later, that this reference state be homogeneous. The actual state is then imagined to undergo small variations that bring it to its varied state [Fig. 1(d)]. The reference state for the varied state is pictured in Fig. 1(b), and is identical to the reference state in Fig. 1(a) except that the position of the interface has changed due to accretion. This change in the position of the surface between Fig. 1(a) and (b) is measured along some "accretion vector" $\delta \mathbf{a}'^{\alpha}$, which, following Larche and Cahn [2], we take to be

$$\delta \mathbf{a}'^{\alpha} = \delta y^{\alpha} \mathbf{n}' \tag{1}$$

at every point on the reference surface in Fig. 1(a), where $\mathbf{n}'$ is the normal to that surface and is taken to point outward from α,† and δy^{α} is the magnitude of the accretion, and is positive if α grows. This change in the position of the interface due to accretion must be considered when one takes a variation at the interface. That is, the interface in the varied state differs from the interface in the actual state not only because it occupies a new spatial position, but also because, due to accretion, it encompasses a new set of reference points. Therefore, we must define a new, interfacial variation that is appropriate for solid surfaces that accrete. One possible variation, which we label $\tilde{\delta}$, can be defined by using Fig. 1 to write

$$\tilde{\delta}\phi^{xs} = \phi^{xs}(C^v) - \phi^{xs}(B) \tag{2}$$

for some function ϕ^{xs} on the interface. That is, $\tilde{\delta}\phi^{xs}$ is the change in the value of ϕ^{xs} between points on the interface in the varied state [Fig. 1(d)] and points on the interface in the actual state [Fig. 1(c)], when these

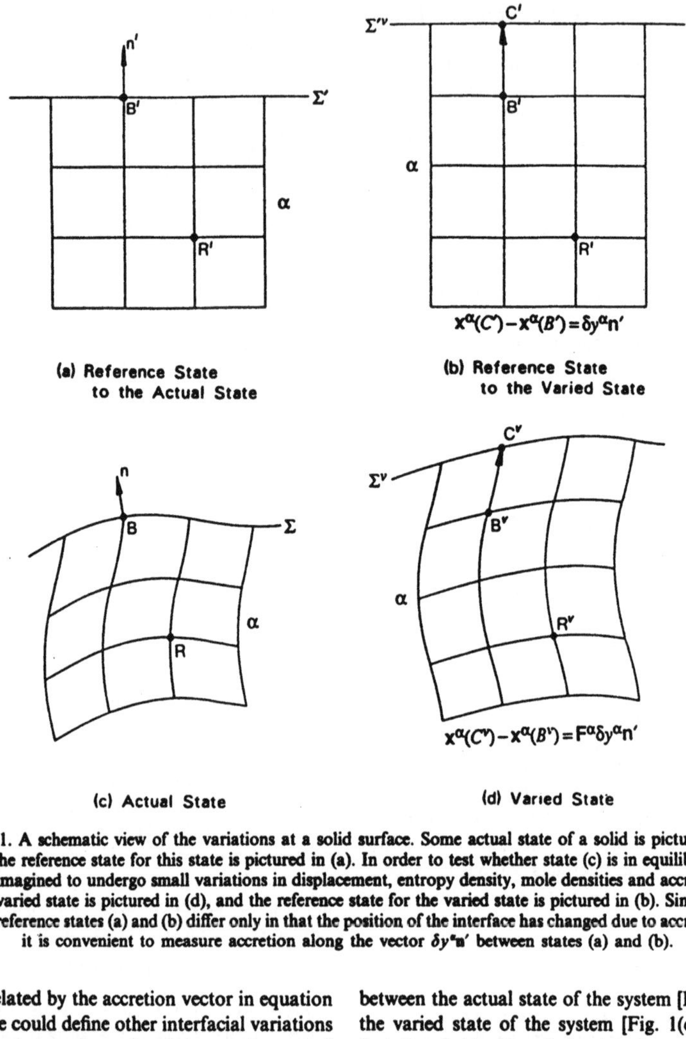

Fig. 1. A schematic view of the variations at a solid surface. Some actual state of a solid is pictured in (c); the reference state for this state is pictured in (a). In order to test whether state (c) is in equilibrium, it is imagined to undergo small variations in displacement, entropy density, mole densities and accretion; this varied state is pictured in (d), and the reference state for the varied state is pictured in (b). Since the two reference states (a) and (b) differ only in that the position of the interface has changed due to accretion, it is convenient to measure accretion along the vector $\delta y^\alpha \mathbf{n}'$ between states (a) and (b).

points are related by the accretion vector in equation (1). While we could define other interfacial variations by a different mapping of points on the actual interface to those on the varied interface, the δ variation defined in equation (2) turns out to be especially convenient for this calculation.

The definition of the interfacial variation in equation (2) is a major new idea in our development of the thermodynamics of solid surfaces and interfaces, and its use in the variational calculations leads directly to our understanding of the effect of surface stress on crystal–melt and crystal–crystal equilibrium. Before we present the details of the variational calculation, we show that the interfacial variation defined by equation (2) can be used to describe continuity at crystal–melt and coherent crystal–crystal interfaces.

2.2. *Crystal–melt interface*

Consider the crystalline solid α in Fig. 1 to be in contact with its fluid melt. We want to describe the change in the position of the crystal–melt interface between the actual state of the system [Fig. 1(c)] and the varied state of the system [Fig. 1(d)]. Consider first the fluid side of the interface. We take an Eulerian view of the fluid, so we measure the total normal movement of the fluid surface as

$$\delta y^F \mathbf{n}^F = -\delta y^F \mathbf{n}, \tag{3}$$

where $\mathbf{n}^F$ is the unit normal, from fluid to solid, of the interface in the actual state, so $\mathbf{n} = -\mathbf{n}^F$ is the corresponding unit normal from solid to fluid, and δy^F is the magnitude of the total normal movement of the fluid surface and is positive if the fluid grows at the expense of the solid. In order to describe the movement of the solid side of the interface, we use the framework that has just been developed. The interfacial variation of the position $\mathbf{x}^\alpha$ of the solid surface is given by equation (2) as

$$\delta \mathbf{x}^\alpha = \mathbf{x}^\alpha(C^v) - \mathbf{x}^\alpha(B) = \mathbf{x}^\alpha(C^v) - \mathbf{x}^\alpha(B^v) + \mathbf{x}^\alpha(B^v) - \mathbf{x}^\alpha(B). \tag{4}$$

This expression can be evaluated by using Fig. 1(b) to note that $\mathbf{x}^\alpha(C') - \mathbf{x}^\alpha(B') = \delta y^\alpha \mathbf{n}'$ maps onto the vector

$$\begin{aligned}\mathbf{x}^\alpha(C^v) - \mathbf{x}^\alpha(B^v) &= \mathbf{F}^\alpha\cdot[\mathbf{x}^\alpha(C') - \mathbf{x}^\alpha(B')] \\ &= \mathbf{F}\cdot\mathbf{n}'\delta y^\alpha \qquad (5)\end{aligned}$$

in the varied state [Fig. 1(d)], where the bulk deformation gradient tensor

$$\mathbf{F} = \mathbf{x}\overleftarrow{\nabla}'_{\mathbf{X}} \qquad (6)$$

is the gradient of spatial points $\mathbf{x}$ over reference points $\mathbf{X}$. (We have denoted the gradient operator in equation (6) with a prime in order to clearly distinguish it from the gradient operator ∇_x taken over spatial coordinates $\mathbf{x}$.) Throughout the paper, we use a single dot to indicate a contraction on one index, so $\mathbf{a}\cdot\mathbf{b} = a_i b_i$ (sum on i) is a scalar, $(\mathbf{A}\cdot\mathbf{b})_i = A_{ik} b_k$ (sum on k) are the components of a vector, and $(\mathbf{A}\cdot\mathbf{B})_{ij} = A_{ik} B_{kj}$ (sum on k) are the components of a second rank tensor. The absence of a dot indicates a dyadic product, so $\mathbf{ab}$ is a second rank tensor with components $(\mathbf{ab})_{ij} = a_i b_j$. Further, we distinguish between gradients operating from the right and from the left by using notation similar to Malvern's [18], in this notation scheme, the ij component of the tensor product $\mathbf{a}\overleftarrow{\nabla}$ of a vector $\mathbf{a}$ and the gradient operator ∇ is $\partial a_i/\partial x_j$ and the ij component of $\overrightarrow{\nabla}\mathbf{a}$ is $\partial a_j/\partial x_i$.

The evaluation of $\tilde{\delta}\mathbf{x}^\alpha$ is now completed by identifying $\mathbf{x}^\alpha(B^v) - \mathbf{x}^\alpha(B) = \delta\mathbf{u}^\alpha$ as the Lagrangian variation of the displacement vector $\mathbf{u} = \mathbf{x} - \mathbf{X}$; we find that

$$\tilde{\delta}\mathbf{x}^\alpha = \delta\mathbf{u}^\alpha + \mathbf{F}^\alpha\cdot\mathbf{n}'\delta y^\alpha \qquad (7)$$

which is identical to the variational movement $\delta\mathbf{s}^\alpha$ of a solid surface defined by Larche and Cahn [2]. However, while Larche and Cahn identify this variation only for the movement of a solid surface, we will use the $\tilde{\delta}$ variation for all quantities associated with a solid surface or interface. The condition that the solid and the fluid stay in contact at the interface through the variation is simply

$$\mathbf{n}\cdot\tilde{\delta}\mathbf{x}^\alpha = \mathbf{n}\cdot(\delta\mathbf{u}^\alpha + \mathbf{F}^\alpha\cdot\mathbf{n}'\delta y^\alpha) = -\delta y^{\mathrm{F}}. \qquad (8)$$

2.3. Coherent crystal–crystal interface

Consider a system that consists of two crystalline solids (α and β) in contact across a coherent interface (Fig. 2). We again want to describe the movement of the interface between the actual state [Fig. 2(c)] and some varied state [Fig. 2(d)]. Following Larche and Cahn [2] and Robin [19], we note that since the α and β lattices match across a coherent interface, one can define a special homogeneous reference state in which the α and β lattices are indistinguishable [Fig. 2(a)]. For instance, one reference state that could be used

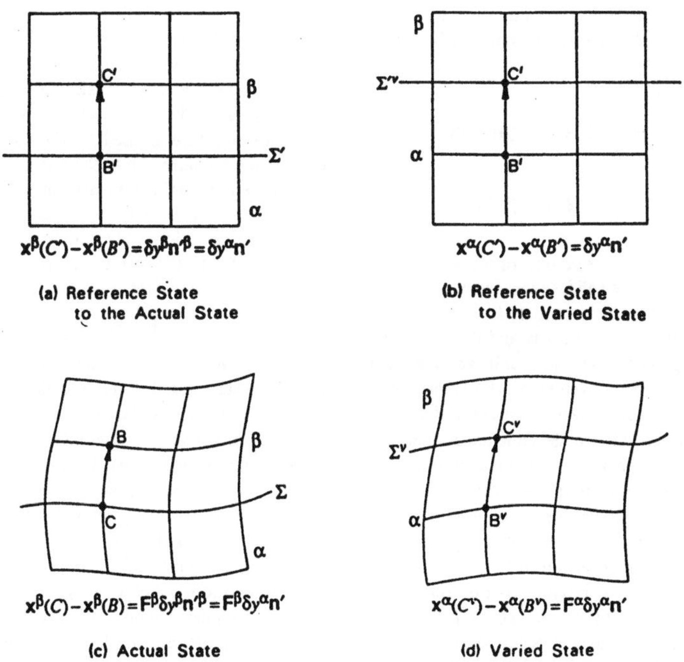

Fig. 2. A schematic view of the variations at a coherent crystal–crystal interface. Since the lattices of the two phases match across a coherent interface, one can choose a reference state for a coherent system in which the two phases are indistinguishable. Thus, the reference state (a) that corresponds to the actual state (c) and the reference state (b) that corresponds to the varied state (d) are identical except that the position of the interface has changed due to accretion.

is the Larche–Cahn example in which β is stress free and the α lattice is stressed so that it matches the β lattice [2]. Since the two phases are indistinguishable in this common reference state, the accretive movement of the interface measured in this reference state must be the same in both phases [Figs 2(a) and (b)], so

$$\delta y^{\alpha} = -\delta y^{\beta} \tag{9}$$

where δy^{α} is the magnitude of the accretion to α and is positive if α grows, and δy^{β} is the magnitude of the accretion to β and is positive if β grows.

Another constraint can be found for the coherent interface from the condition that the lattice is continuous at the interface. From Fig. 2 and the definition of $\mathfrak{F}$, one can show that

$$\mathfrak{F}\mathbf{x}^{\alpha} = \mathbf{x}^{\alpha}(C^{\prime\prime}) - \mathbf{x}^{\alpha}(B) = \delta\mathbf{u}^{\alpha} + \mathbf{F}^{\alpha}\cdot\mathbf{n}'\delta y^{\alpha} \tag{10}$$

and

$$\mathfrak{F}\mathbf{x}^{\beta} = \mathbf{x}^{\beta}(C^{\prime\prime}) - \mathbf{x}^{\beta}(B) = \delta\mathbf{u}^{\beta} - \mathbf{F}^{\beta}\cdot\mathbf{n}'\delta y^{\beta} \tag{11}$$

by using the same logic that led to equation (7). The negative sign in equation (11) arises from the convention that $\mathbf{n}'$ points from α to β. Since the two phases match at the interface

$$\mathfrak{F}\mathbf{x}^{\alpha} = \mathfrak{F}\mathbf{x}^{\beta}. \tag{12}$$

Equation (12) is the boundary condition required to maintain coherence at a deforming and accreting crystal-crystal interface; this condition is identical to Larche and Cahn's coherency condition [2].

2.4. Energy densities

We take the energy E^{S} of a crystalline solid to be given by [2, 3]

$$E^{S} = \int_{V'} e'(\mathbf{F}, s', \rho_i')\, d^3X \tag{13}$$

where V' is the reference volume of the solid, e' is the energy density of the solid, taken per unit volume of this reference state, and s' and ρ_i' $(i = 1, \ldots, K)$ are, respectively, the entropy density and the mole density of species i, both taken per unit volume of the reference state. For convenience, we assume that the set of ρ_i' includes only massive species and can all be independently varied through some unspecified defect mechanism; while this convention is a point of difference between the Mullins and Sekerka [3] and Larche and Cahn [1, 2, 20] developments of the thermodynamics of solids, it is not the focus of this work. We note finally that the inclusion of $\mathbf{F}$ in the variable set for the energy density implies that the solid behaves elastically [21] and that the reference state from which displacements are measured is homogeneous [3].

We take an Eulerian view of the fluid melt, so we write the total energy E^{F} of the fluid as [2, 3]

$$E^{F} = \int_{V^F} e^{F}(s^{F}, \rho_i^{F})\, d^3x \tag{14}$$

where V^{F} is the volume of the fluid in the actual state, and e^{F}, s^{F} and ρ_i^{F} $(i = 1, \ldots, K)$ are the energy, entropy and molar densities of species i of the fluid, all per unit volume of the actual state.

We take the interface to be a two-dimensional continuum to which we assign thermodynamic potential functions. We take the reference state of the interface (i.e. the state from which we measure the displacement of the interface) to be contiguous with the reference state of the solid in the crystal–melt case and with the common reference state in the coherent crystal–crystal case, so if the bulk reference state is homogeneous, then so is the interfacial reference state. We take the total interfacial energy to be

$$E^{xs} = \int_{\Sigma'} e'^{xs}(\hat{\mathbf{F}}, \mathbf{n}', s'^{xs}, \Gamma_i')\, d^2X \tag{15}$$

where the integration is over the reference interface Σ' [i.e. the interface pictured in Figs 1(a) and 2(a)]. In equation (15), e'^{xs}, s'^{xs} and $\Gamma_i'(i = 1, \ldots, K)$ are the excess energy density, entropy density and molar density of species i per unit area of the reference state, respectively. We assume that the Γ_i' are independent.

The terms in the variable set for e'^{xs} that distinguishes a solid surface from a fluid surface are the surface deformation gradient tensor $\hat{\mathbf{F}}$ and the normal vector $\mathbf{n}'$ to the reference interface Σ'. The reference normal $\mathbf{n}'$ explicitly allows for interfacial anisotropies that may arise due to the crystallographic orientation at the surface. We note that the normal $\mathbf{n}'$ to the reference interface is used in the variable set (15) rather than the normal $\mathbf{n}$ to the interface in the actual state since $\mathbf{n}'$ contains information on the crystalline orientation of the interface independent of any subsequent deformations. These deformations are then accounted for by the surface deformation gradient $\hat{\mathbf{F}}$, which is defined as [16]

$$\hat{\mathbf{F}} = \mathbf{x}(\Sigma)\hat{\nabla}_{\Sigma'} \tag{16}$$

where $x(\Sigma)$ is the position vector to the interface Σ in the actual state [i.e. the interface pictured in Figs 1(c) and 2(c), and $\nabla_{\Sigma'}$ is the surface gradient operator [22, 16], taken over the reference interface Σ'.† The tensor $\hat{\mathbf{F}}$ maps vectors that lie in the reference interface Σ' onto the corresponding vectors in the actual interface Σ, so is formally defined as a tensor operator in two dimensions [16]. However, as discussed in the Appendix, we consider $\hat{\mathbf{F}}$ to be a tensor operator in three dimensions in order to eliminate some mathematical complexity while retaining the essential ideas and mathematics involved in defining tensor operations on surfaces and interfaces. Finally, we note for future comparison with the greased crystal–crystal system that since there is continuity of lattice at a coherent interface, $\hat{\mathbf{F}}^{\alpha} = \hat{\mathbf{F}}^{\beta} \equiv \hat{\mathbf{F}}$ at a coherent interface.

†We note that there may be anisotropies implicit in the response of e'^{xs} to variations in $\hat{\mathbf{F}}$. That is, the interface may behave as an elastically anisotropic membrane.

2.5. The variational calculation

The condition for equilibrium when no external work is done on the system is that, at constant entropy and mass, all virtual variations Δ from the equilibrium state increase the total energy of the system, so that

$$\Delta E^{\text{tot}} = \Delta E^{\alpha} + \Delta E^{\beta} + \Delta E^{\text{xs}} \geq 0 \tag{17}$$

in equilibrium. We now calculate these variations, assuming them to be small.

The energy variation of a crystalline solid α with an energy given by equation (13) has been calculated in previous works [2, 12, 13], and is

$$\Delta E^{\alpha} = \int_{V^{\alpha}} \left[-(\mathbf{T}_R^{\alpha} \cdot \overleftarrow{\nabla}_X) \cdot \delta \mathbf{u}^{\alpha} + \frac{\partial e'^{\alpha}}{\partial s'^{\alpha}} \delta s'^{\alpha} + \frac{\partial e'^{\alpha}}{\partial \rho_i'^{\alpha}} \delta \rho_i'^{\alpha} \right] \mathrm{d}^3 X + \int_{\Sigma'} [(\mathbf{T}_R^{\alpha} \cdot \mathbf{n}') \cdot \delta \mathbf{u}^{\alpha} + e'^{\alpha} \delta y^{\alpha}] \, \mathrm{d}^2 X \tag{18}$$

where we identify

$$\frac{\partial e'}{\partial \mathbf{F}} = \mathbf{T}_R \tag{19}$$

as the first Piola–Kirchoff stress tensor in the bulk [18]. Since the integration in equation (18) is over the reference state of the system, the variations are Lagrangian. When β is also a solid, the expression for ΔE^{β} takes the same form as equation (18), except that we replace 'α' with 'β' and, since $\mathbf{n}'$ points from α to β, we replace $\mathbf{n}'$ with $-\mathbf{n}'$.

The energy variation of a fluid phase with an energy given by equation (14) is

$$\Delta E^{\mathrm{F}} = \int_{V^{\mathrm{F}}} \left[\frac{\partial e^{\mathrm{F}}}{\partial s^{\mathrm{F}}} \delta^{\mathrm{E}} s^{\mathrm{F}} + \frac{\partial e^{\mathrm{F}}}{\partial \rho_i^{\mathrm{F}}} \delta^{\mathrm{E}} \rho_i^{\mathrm{F}} \right] \mathrm{d}^3 x + \int_{\Sigma} e^{\mathrm{F}} \delta y^{\mathrm{F}} \mathrm{d}^2 x \tag{20}$$

where the volume integral results from a Taylor expansion of the energy density in equation (14), and the surface integral accounts for the volume change due to the total movement of the interface. The variations $\delta^{\mathrm{E}} s^{\mathrm{F}}$ and $\delta^{\mathrm{E}} \rho_i^{\mathrm{F}}$ are Eulerian variations since the integration is over the actual state of the fluid.

The variation of the excess interfacial energy in equation (15) is given by

$$\Delta E^{\text{xs}} = \int_{\Sigma'} \left[\frac{\partial e'^{\text{xs}}}{\partial \hat{\mathbf{F}}} : \mathcal{S}\hat{\mathbf{F}} + \mathbf{q}' \cdot \mathcal{S}\mathbf{n}' + \frac{\partial e'^{\text{xs}}}{\partial s'^{\text{xs}}} \delta s'^{\text{xs}} + \frac{\partial e'^{\text{xs}}}{\partial \Gamma_i'} \delta \Gamma_i' \right] \mathrm{d}^2 X + \int_{\Sigma'} \kappa' e'^{\text{xs}} \delta y^{\alpha} \mathrm{d}^2 X \tag{21}$$

where we define

$$\mathbf{q}' = \frac{\partial e'^{\text{xs}}}{\partial \mathbf{n}'} - \mathbf{n}' \left(\frac{\partial e'^{\text{xs}}}{\partial \mathbf{n}'} \cdot \mathbf{n}' \right) \tag{22}$$

in order to take in account the fact that since $\mathbf{n}'$ is a unit vector, only two components of $\mathbf{n}'$, and hence of $\partial e'^{\text{xs}}/\partial \mathbf{n}'$, can be independently varied. The first integral in equation (21) arises from a Taylor expansion, following the $\mathcal{S}$ variation of the interface, of the excess energy density in equation (15), and the second integral accounts for the change in area of the reference interface due to accretion. The symbol ':' in equation (21) indicates a tensor scalar product, taken such that $\mathbf{A}:\mathbf{B} = A_{ij}B_{ij}$ (sum on i and j), and κ' is the curvature of the interface in the reference state, taken so that if the α phase is a sphere of radius R', then $\kappa' = 2/R'$. By analogy to equation (19), and following Alexander and Johnson [12, 13], we write

$$\frac{\partial e'^{\text{xs}}}{\partial \hat{\mathbf{F}}} = \hat{\mathbf{T}}_R, \tag{23}$$

where $\hat{\mathbf{T}}_R$ is the first Piola–Kirchoff surface stress tensor as defined by Gurtin and Murdoch [16], and has the same tensor properties as the surface deformation gradient $\hat{\mathbf{F}}$.

Equation (21) reveals how the $\mathcal{S}$ variation enters the calculation of ΔE^{xs}. We find that in order to properly calculate the roles of surface anisotropy and surface stress at the interface, we must carefully calculate the variations $\mathcal{S}\mathbf{n}'$ and $\mathcal{S}\hat{\mathbf{F}}$. Consider first the variation $\mathcal{S}$. By using equation (2) and the notation in Fig. 1, we see that

$$\mathcal{S}\mathbf{n}' = \mathbf{n}'(C^v) - \mathbf{n}'(B) = \mathbf{n}'(C') - \mathbf{n}'(B'). \tag{24}$$

That is, $\mathcal{S}\mathbf{n}'$ represents the difference in the normal vectors between points in the reference state after accretion [Fig. 1(c)] and before accretion [Fig. 1(a)], when these points are connected by the accretion vector $\delta y^{\alpha}\mathbf{n}'$. This variation is given by [21, 12]

$$\mathcal{S}\mathbf{n}' = -\overleftarrow{\nabla}_{\Sigma'} \delta y^{\alpha} \tag{25}$$

so we see that both $\mathcal{S}\mathbf{n}'$ and the vector $\mathbf{q}'$ defined in equation (22) are tangent to the reference interface Σ'.

Equation (25) shows that the $\mathcal{S}$ variation of the reference normal $\mathbf{n}'$ depends only on the accretive variation δy^{α}. In contrast, we now show that the variation $\mathcal{S}\hat{\mathbf{F}}$ depends on both the accretive variation δy^{α} and the mechanical variation $\delta \mathbf{u}$. By using equations (2) and (16) and the notation in Fig. 1, $\mathcal{S}\hat{\mathbf{F}}$ can be written as

$$\mathcal{S}\hat{\mathbf{F}} = \hat{\mathbf{F}}(C^v) - \hat{\mathbf{F}}(B) = \mathbf{x}(C^v)\overleftarrow{\nabla}_{\Sigma'^v} - \mathbf{x}(B)\overleftarrow{\nabla}_{\Sigma'} \tag{26}$$

where Σ' and Σ'^v indicate the positions of the interface in the reference state before [Fig. 1(a)] and after [Fig. 1(b)] accretion, respectively, and $\mathbf{x}(B)$ and $\mathbf{x}(C^v)$ are the position vectors to the actual interface Σ [Fig. 1(c)] and the varied interface Σ^v [Fig. 1(d)], respectively. If we define the operation $\mathcal{S}(\overleftarrow{\nabla}_{\Sigma'}) = \overleftarrow{\nabla}_{\Sigma'^v} - \overleftarrow{\nabla}_{\Sigma'}$ that accounts for the accretive movement of the reference interface, then

$$\mathcal{S}\hat{\mathbf{F}} = (\mathcal{S}\mathbf{x})\overleftarrow{\nabla}_{\Sigma'} + \mathbf{x}(B)\mathcal{S}(\overleftarrow{\nabla}_{\Sigma'}) \tag{27}$$

to first order in the variation. The idea that we can

account for accretion by varying the surface gradient operator has no analog in the bulk, and results directly from the definition of the interfacial variation.

We can now evaluate $\delta\hat{\mathbf{F}}$. From equation (7)

$$\delta\mathbf{x} = \delta\mathbf{x}^{\alpha} = \delta\mathbf{u}^{\alpha} + \mathbf{F}^{\alpha}\cdot\mathbf{n}'\delta y^{\alpha} \tag{28}$$

at a crystal–melt interface, and from equations (10), (11) and (12)

$$\delta\mathbf{x} = \delta\mathbf{x}^{\alpha} = \delta\mathbf{u}^{\alpha} + \mathbf{F}^{\alpha}\cdot\mathbf{n}'\delta y^{\alpha} = \delta\mathbf{x}^{\beta} = \delta\mathbf{u}^{\beta} - \mathbf{F}^{\beta}\cdot\mathbf{n}'\delta y^{\beta} \tag{29}$$

at a coherent crystal-crystal interface. The variation $\delta(\hat{\nabla}_{\Sigma'})$ is found as follows: the position of the reference interface prior to accretion [Σ' in Fig. 1(a)] can be written as $\mathbf{r}'(u, v)$, where u and v are parameters choosen to describe the surface [22]. Since the reference interface after accretion [Σ'^{v} in Fig. 1(b)] is related to Σ' through the accretion vector in equation (1), its position is given by $\mathbf{r}'^{v}(u, v) = \mathbf{r}'(u, v) + \delta y^{\alpha}(u, v)\mathbf{n}'(u, v)$. By using the definition of the surface gradient operator given in equation (A7) of the Appendix, one can calculate $\delta(\hat{\nabla}_{\Sigma'})$; this result may be combined with equations (28) and (29) to yield

$$\delta\hat{\mathbf{F}} = (\delta\mathbf{u}^{\alpha} + \mathbf{F}^{\alpha}\cdot\mathbf{n}'\delta y^{\alpha})\hat{\nabla}_{\Sigma'} + \hat{\mathbf{F}}\cdot(\hat{\nabla}_{\Sigma'}\delta y^{\alpha})\mathbf{n}' + \hat{\mathbf{F}}\cdot\mathbf{L}'\delta y^{\alpha} \tag{30}$$

for both the crystal–melt and coherent crystal–crystal cases, where $\hat{\mathbf{F}} = \hat{\mathbf{F}}(B)$ is the surface deformation gradient in the actual state of the system and

$$\mathbf{L}' = -(\mathbf{1} - \mathbf{n}'\mathbf{n}')\mathbf{n}'\hat{\nabla}_{\Sigma'} \tag{31}$$

is the Weingarten map to the reference interface Σ' [16], and is defined such that trace $\mathbf{L}'$ gives the negative of the mean curvature of the reference interface, and det$\mathbf{L}'$ gives the negative of the Gaussian curvature of the reference interface. We also reiterate that in our notation, $(\hat{\nabla}_{\Sigma'}\delta y^{\alpha})\mathbf{n}'$ is the dyad composed of the two vectors $\hat{\nabla}_{\Sigma'}\delta y^{\alpha}$ and $\mathbf{n}'$.

We now substitute equations (25) and (30) into equation (21) to find the following general form for ΔE^{xs}

$$\Delta E^{xs} = \int_{\Sigma'}\Big\{\hat{\mathbf{T}}_R:[(\delta\mathbf{u}^{\alpha} + \mathbf{F}^{\alpha}\cdot\mathbf{n}'\delta y^{\alpha})\hat{\nabla}_{\Sigma'} + \hat{\mathbf{F}}\cdot(\hat{\nabla}_{\Sigma'}\delta y^{\alpha})\mathbf{n}' + \hat{\mathbf{F}}\cdot\mathbf{L}'\delta y^{\alpha}] - \mathbf{q}'\cdot\hat{\nabla}_{\Sigma'}\delta y^{\alpha} + \frac{\partial e'^{xs}}{\partial s'^{xs}}\delta s'^{xs} + \frac{\partial e'^{xs}}{\partial \Gamma'_i}\delta\Gamma'_i + \kappa' e'^{xs}\delta y^{\alpha}\Big\}\,\mathrm{d}^2X. \tag{32}$$

Equation (32) can be simplified as follows: first, one can show that the term $\hat{\mathbf{T}}_R:[\hat{\mathbf{F}}\cdot(\hat{\nabla}_{\Sigma'}\delta y^{\alpha})\mathbf{n}']$ vanishes by using the definition of the tensor scalar product [21] and equation (A1) of the Appendix. Secondly, one can show by using equation (2.17b) in Gurtin–Murdoch [16] and the surface divergence theorem [16] that if Σ' is a closed surface, then

$$\int_{\Sigma'}\hat{\mathbf{T}}_R:[(\delta\mathbf{u}^{\alpha} + \mathbf{F}^{\alpha}\cdot\mathbf{n}'\delta y^{\alpha})\hat{\nabla}_{\Sigma'}]\,\mathrm{d}^2X$$

$$= -\int_{\Sigma'}(\hat{\mathbf{T}}_R\cdot\hat{\nabla}_{\Sigma'})\cdot(\delta\mathbf{u}^{\alpha} + \mathbf{F}^{\alpha}\cdot\mathbf{n}'\delta y^{\alpha})\,\mathrm{d}^2X \tag{33}$$

and

$$\int_{\Sigma'} -\mathbf{q}'\cdot\hat{\nabla}_{\Sigma'}\delta y^{\alpha}\,\mathrm{d}^2X = \int_{\Sigma'}(\mathbf{q}'\cdot\hat{\nabla}_{\Sigma'})\delta y^{\alpha}\,\mathrm{d}^2X. \tag{34}$$

The remaining term can be simplified by again using the definition of the tensor scalar product, the fact that trace $(\mathbf{AB})$ = trace $(\mathbf{BA})$ and identity (2.18) in Gurtin and Murdoch [16]; we find that

$$\hat{\mathbf{T}}_R:(\hat{\mathbf{F}}\cdot\mathbf{L}') = [(\hat{\mathbf{F}}^{\mathrm{T}}\cdot\hat{\mathbf{T}}_R)\cdot\hat{\nabla}_{\Sigma'}]\cdot\mathbf{n}', \tag{35}$$

where a superscript 'T' indicates the transpose of a tensor. In total, then, equation (32) reduces to

$$\Delta E^{xs} = \int_{\Sigma'}\Big\{-(\hat{\mathbf{T}}_R\cdot\hat{\nabla}_{\Sigma'})\cdot(\delta\mathbf{u}^{\alpha} + \mathbf{F}^{\alpha}\cdot\mathbf{n}'\delta y^{\alpha}) + [(\hat{\mathbf{F}}^{\mathrm{T}}\cdot\hat{\mathbf{T}}_R)\cdot\hat{\nabla}_{\Sigma'}]\cdot\delta y^{\alpha}\mathbf{n}' + (\mathbf{q}'\cdot\hat{\nabla}_{\Sigma'})\delta y^{\alpha} + \frac{\partial e'^{xs}}{\partial s'^{xs}}\delta s'^{xs} + \frac{\partial e'^{xs}}{\partial \Gamma'_i}\delta\Gamma'_i + \kappa' e'^{xs}\delta y^{\alpha}\Big\}\,\mathrm{d}^2X. \tag{36}$$

Finally, we must ensure that both the crystal–melt and the coherent crystal–crystal systems maintain constant total entropy and mass throughout the variation. These constraints take the general form

$$\Delta\Phi = \int_{V'^{\alpha}}\delta\phi'^{\alpha}\,\mathrm{d}^3X + \int_{\Sigma'}\phi'^{\alpha}\delta y^{\alpha}\,\mathrm{d}^2X + \int_{V^{F}}\delta^{E}\phi^{F}\,\mathrm{d}^3x + \int_{\Sigma}\phi^{F}\delta y^{F}\,\mathrm{d}^2x + \int_{\Sigma'}\delta\phi'^{xs}\,\mathrm{d}^2X + \int_{\Sigma'}\kappa'\phi'^{xs}\delta y^{\alpha}\,\mathrm{d}^2X = 0 \tag{37}$$

for the crystal melt case, and

$$\Delta\Phi = \int_{V'^{\alpha}}\delta\phi'^{\alpha}\,\mathrm{d}^3X + \int_{\Sigma'}\phi'^{\alpha}\delta y^{\alpha}\,\mathrm{d}^2X + \int_{V'^{\beta}}\delta\phi'^{\beta}\,\mathrm{d}^3X + \int_{\Sigma'}\phi'^{\beta}\delta y^{\beta}\,\mathrm{d}^2X + \int_{\Sigma'}\delta\phi'^{xs}\,\mathrm{d}^2X + \int_{\Sigma'}\kappa'\phi'^{xs}\delta y^{\alpha}\,\mathrm{d}^2X = 0 \tag{38}$$

for the coherent crystal–crystal case, where $\Phi = S$, $\phi' = s'$ and $\phi^{F} = s^{F}$ for the entropy constraint, and $\Phi = N_i$ $(i = 1, \ldots, K)$, $\phi' = \rho'_i$, $\phi^{F} = \rho^{F}_i$ and $\phi'^{xs} = \Gamma'_i$ for the mass constraint. The first four terms in equations (37) and (38) account for the change in Φ in the two bulk phases, and the last two terms account for the change in Φ at the interface.

2.6. *Results*

Consider a closed system that consists of either a crystalline solid in contact with its melt, or two crystalline solids in contact across a coherent interface. If we insist that the system does no work against external forces (e.g. by either extending the outer boundary of the system to infinity or holding it rigidly), than the method of Lagrange multipliers requires that in equilibrium

$$\Delta E^{\rm tot} - \theta \Delta S^{\rm tot} - \mu_i \Delta N_i^{\rm tot} \geqslant 0 \tag{39}$$

where θ is the Lagrange multiplier associated with the entropy constraint, and the μ_i ($i = 1, \ldots, K$) are the Lagrange multipliers associated with the K mass constraints. Applying equation (39) to the two systems described above yields the following results:

2.6.1. Crystal–melt. For the crystal–melt case, we substitute equations (18), (20), (36) and (37) into equation (39) to find

$$\begin{aligned}
&\Delta E^{\rm tot} - \theta \Delta S^{\rm tot} - \mu_i \Delta N_i^{\rm tot} \\
&= \int_{V^\alpha} \bigg[-(\mathbf{T}_R^\alpha \cdot \bar{\nabla}'_X) \cdot \delta \mathbf{u}^\alpha + \left(\frac{\partial e'^\alpha}{\partial s'} - \theta \right) \delta s'^\alpha \\
&\qquad + \left(\frac{\partial e'^\alpha}{\partial \rho'_i} - \mu_i \right) \delta \rho_i'^\alpha \bigg] \mathrm{d}^3 X \\
&+ \int_{V^{\rm F}} \left[\left(\frac{\partial e^{\rm F}}{\partial s^{\rm F}} - \theta \right) \delta^{\rm E} s^{\rm F} + \left(\frac{\partial e^{\rm F}}{\partial \rho_i^{\rm F}} - \mu_i \right) \delta^{\rm E} \rho_i^{\rm F} \right] \mathrm{d}^3 x \\
&+ \int_{\Sigma'} \left[\left(\frac{\partial e'^{\rm xs}}{\partial s'^{\rm xs}} - \theta \right) \delta s'^{\rm xs} + \left(\frac{\partial e'^{\rm xs}}{\partial \Gamma'_i} - \mu_i \right) \delta \Gamma'_i \right] \mathrm{d}^2 X \\
&+ \int_{\Sigma'} [\omega'^\alpha \delta y^\alpha + (\mathbf{T}_{\rm R}^\alpha \cdot \mathbf{n}') \cdot \delta \mathbf{u}^\alpha - (\hat{\mathbf{T}}_{\rm R} \cdot \bar{\nabla}_{\Sigma'}) \\
&\qquad \times (\delta \mathbf{u}^\alpha + \mathbf{F}^\alpha \cdot \mathbf{n}' \delta y^\alpha) \\
&+ \{ (\hat{\mathbf{F}}^{\rm T} \cdot \hat{\mathbf{T}}_{\rm R}) \cdot \bar{\nabla}_{\Sigma'} \} \cdot \delta y^\alpha \mathbf{n}' + (\mathbf{q}' \cdot \bar{\nabla}_{\Sigma'}) \delta y^\alpha \\
&+ \kappa' \sigma' \delta y^\alpha] \mathrm{d}^2 X + \int_\Sigma \omega^{\rm F} \delta y^{\rm F} \mathrm{d}^2 x \geqslant 0
\end{aligned} \tag{40}$$

where

$$\omega' = e' - \theta s' - \mu_i \rho'_i \tag{41}$$

is the grand potential density, measured per unit volume of the reference state, of a bulk phase, and

$$\sigma' = e'^{\rm xs} - \theta s'^{\rm xs} - \mu_i \Gamma'_i \tag{42}$$

is the grand potential density, or surface free energy density,† associated with the interface, and measured per unit area of the reference state.

The variations in the first three integrals in equation (40) are independent, and lead to the equilibrium conditions

$$\mathbf{T}_{\rm R}^\alpha \cdot \bar{\nabla}'_{\rm X} = 0 \quad \text{in } \alpha \tag{43}$$

$$\frac{\partial e'^\alpha}{\partial s'} = \frac{\partial e^{\rm F}}{\partial s^{\rm F}} = \frac{\partial e'^{\rm xs}}{\partial s'^{\rm xs}} = \theta \quad \text{uniform} \tag{44}$$

and

$$\frac{\partial e'^\alpha}{\partial \rho'_i} = \frac{\partial e^{\rm F}}{\partial \rho_i^{\rm F}} = \frac{\partial e'^{\rm xs}}{\partial \Gamma'_i} = \mu_i \quad \text{uniform.} \tag{45}$$

†We have consistently used the term "surface free energy" to denote this function. Gibbs [4] and later Herring [6] have also called this function the surface tension, but we avoid this terminology so as not to confuse σ with the surface stress.

These conditions are the usual conditions of mechanical equilibrium and uniformity of temperature θ and chemical potential μ_i. The mechanical condition (43) is equivalent to [18, 21]

$$\mathbf{T} \cdot \bar{\nabla}_{\rm x} = 0 \tag{46}$$

where $\nabla_{\rm x}$ is the gradient operator over spatial coordinates and

$$\mathbf{T} = J^{-1} \mathbf{T}_{\rm R} \cdot \mathbf{F}^{\rm T} \tag{47}$$

is the Cauchy stress in the bulk [18, 21] where $J = \det \mathbf{F}$ is the Jacobian of the bulk deformation gradient $\mathbf{F}$.

The equilibrium conditions at the crystal–melt interface are found from the last two integrals in equation (40). We first use the boundary condition (8) at the crystal–melt interface to eliminate the variation $\delta y^{\rm F}$, so the remaining variations $\delta \mathbf{u}^\alpha$ and δy^α are independent. Nanson's formula [18] is then used to convert the integral over Σ to an integral over Σ'. We find that the equilibrium conditions that apply on the reference interface Σ' are

$$\mathbf{T}_{\rm R}^\alpha \cdot \mathbf{n}' - J \omega^{\rm F} \mathbf{n}' \cdot (\mathbf{F}^\alpha)^{-1} - (\hat{\mathbf{T}}_{\rm R} \cdot \bar{\nabla}_{\Sigma'}) = 0 \tag{48}$$

which arises from the independent variation $\delta \mathbf{u}^\alpha$, and

$$\begin{aligned}
&\omega'^\alpha - J \omega^{\rm F} + \kappa' \sigma' + (\mathbf{q}' \cdot \bar{\nabla}_{\Sigma'}) \\
&\qquad + [(\hat{\mathbf{F}}^{\rm T} \cdot \hat{\mathbf{T}}_{\rm R}) \cdot \bar{\nabla}_{\Sigma'} - \mathbf{F}^{\alpha \rm T} \cdot (\hat{\mathbf{T}}_{\rm R} \cdot \bar{\nabla}_{\Sigma'})] \cdot \mathbf{n}' = 0
\end{aligned} \tag{49}$$

which arises from the independent variation δy^α. Equation (48) expresses the equilibrium balance of forces at the interface. Equation (49) is an energy balance at the interface, and so sets the grand potential density ω'^α in the solid as a function of the fluid pressure $p = -\omega^{\rm F}$, a capillary term $\kappa' \sigma'$, an anisotropy term $\mathbf{q}' \cdot \bar{\nabla}_{\Sigma'}$ and two terms that contain the surface stress. We will show in the discussion that the terms $\kappa' \sigma'$ and $\mathbf{q}' \cdot \bar{\nabla}_{\Sigma'}$ can be combined into a single term by defining a vector force analogous to Cahn and Hoffman's ξ-vector [14, 15] [see equation (114)]. Also, an alternate form for the energy balance (49) can be found by solving equation (48) for $\hat{\mathbf{T}}_{\rm R} \cdot \bar{\nabla}_{\Sigma'}$ and substituting the result into equation (49); we find

$$\begin{aligned}
&\omega'^\alpha - (\mathbf{T}_{\rm R}^\alpha \cdot \mathbf{n}') \cdot (\mathbf{F}^\alpha \cdot \mathbf{n}') + \kappa' \sigma' + (\mathbf{q}' \cdot \bar{\nabla}_{\Sigma'}) \\
&\qquad + [(\hat{\mathbf{F}}^{\rm T} \cdot \hat{\mathbf{T}}_R) \cdot \bar{\nabla}_{\Sigma'}] \cdot \mathbf{n}' = 0
\end{aligned} \tag{50}$$

which shows ω'^α as a function of the bulk stress, capillarity, anisotropy and a single term involving the surface stress. We note that the fact that the surface stress appears explicitly in the energy balance at the crystal–melt interface arises from our definition of the variation at an interface, and is one of the new results from this analysis.

It is also possible to express the crystal–melt equilibrium conditions on the actual interface Σ. This is accomplished by using Nanson's formula to transform the second Σ' integral in equation (40) to an integral over Σ. Then, by using the chain rule to relate the gradient $\nabla_{\Sigma'}$ over the actual surface to the gradient ∇_Σ over the reference surface [see equation (77)], equation (47) and the corresponding equation relating the Cauchy surface stress $\hat{\mathbf{T}}$ to the first Piola–Kirchoff surface stress $\hat{\mathbf{T}}_{R'}$ [16]

$$\hat{\mathbf{T}} = \hat{J}^{-1}\hat{\mathbf{T}}_R \cdot \hat{\mathbf{F}}^T \quad (51)$$

where $\hat{J} = \det \hat{\mathbf{F}}$ is the Jacobian of the surface deformation gradient $\hat{\mathbf{F}}$, we find that on Σ

$$\mathbf{T}^{\alpha}\cdot\mathbf{n} - \omega^{F}\mathbf{n} - (\hat{\mathbf{T}}\cdot\hat{\nabla}_{\Sigma}) = 0 \quad (52)$$

and

$$\omega^{\alpha} - \omega^{F} + \left(\frac{\hat{J}}{J}\right)\kappa'\sigma + \left(\frac{\hat{J}}{J}\right)[(\hat{\mathbf{F}}\cdot\mathbf{q})\cdot\hat{\nabla}_{\Sigma}] + \left(\frac{\hat{J}}{J}\right)^2 [(\hat{\mathbf{F}}^T\cdot\hat{\mathbf{T}})\cdot\hat{\nabla}_{\Sigma} - \mathbf{F}^{\alpha T}\cdot(\hat{\mathbf{T}}\cdot\hat{\nabla}_{\Sigma})]\cdot(\mathbf{n}\cdot\mathbf{F}^{\alpha}) = 0 \quad (53)$$

where ω^{α} is taken per unit volume in the actual state, both σ and e^{xs} are taken per unit area of the interface in the actual state, and we define

$$\mathbf{q} = \frac{\partial e^{xs}}{\partial \mathbf{n}'} - \mathbf{n}'\left(\frac{\partial e^{xs}}{\partial \mathbf{n}'}\cdot\mathbf{n}'\right). \quad (54)$$

The force balance in equation (52) corresponds to that given by equation (48), while the energy balance in equation (53) corresponds to that given by equation (49). We can also combine equations (52) and (53) to write an energy balance on Σ similar in form to equation (50); however, since this reveals no new information, we do not include it here.

2.6.2. Coherent crystal–crystal. For the coherent crystal–crystal case, we substitute equations (18) (applied to both the α and β phases), (36) and (38) into equation (39) to get

$$\begin{aligned}
\Delta E^{tot} - \theta\Delta S^{tot} - \mu_i \Delta N_i^{tot} &= \int_{V'^{\alpha}}\left[-(\mathbf{T}_R^{\alpha}\cdot\nabla'_X)\cdot\delta\mathbf{u}^{\alpha} + \left(\frac{\partial e'^{\alpha}}{\partial s'} - \theta\right)\delta s'^{\alpha} + \left(\frac{\partial e'^{\alpha}}{\partial \rho'_i} - \mu_i\right)\delta\rho_i'^{\alpha}\right]d^3X \\
&+ \int_{V'^{\beta}}\left[-(\mathbf{T}_R^{\beta}\cdot\nabla'_X)\cdot\delta\mathbf{u}^{\beta} + \left(\frac{\partial e'^{\beta}}{\partial s'} - \theta\right)\delta s'^{\beta} + \left(\frac{\partial e'^{\beta}}{\partial \rho'_i} - \mu_i\right)\delta\rho_i'^{\beta}\right]d^3X \\
&+ \int_{\Sigma'}\left[\left(\frac{\partial e'^{xs}}{\partial s'^{xs}} - \theta\right)\delta s'^{xs} + \left(\frac{\partial e'^{xs}}{\partial \Gamma'_i} - \mu_i\right)\delta\Gamma'_i\right]d^2X \\
&+ \int_{\Sigma'}[\omega'^{\alpha}\delta y^{\alpha} + (\mathbf{T}_R^{\alpha}\cdot\mathbf{n}')\cdot\delta\mathbf{u}^{\alpha} + \omega'^{\beta}\delta y^{\beta} - (\mathbf{T}_R^{\beta}\cdot\mathbf{n}')\cdot\delta\mathbf{u}^{\beta} + \kappa'\sigma'\delta y^{\alpha} \\
&- (\hat{\mathbf{T}}_R\cdot\hat{\nabla}_{\Sigma'})\cdot(\delta\mathbf{u}^{\alpha} + \mathbf{F}^{\alpha}\cdot\mathbf{n}'\delta y^{\alpha}) + \{(\hat{\mathbf{F}}^T\cdot\hat{\mathbf{T}}_R)\cdot\hat{\nabla}_{\Sigma'}\}]\cdot\delta y^{\alpha}\mathbf{n}' \\
&+ (\mathbf{q}'\cdot\hat{\nabla}_{\Sigma'})]\, d^2X \geqslant 0. \quad (55)
\end{aligned}$$

The variations in the first three integrals are independent, and lead to the equilibrium conditions

$$\mathbf{T}_R^{\alpha}\cdot\nabla'_X = 0 \quad \text{in } \alpha \quad (56)$$

$$\mathbf{T}_R^{\beta}\cdot\nabla'_X = 0 \quad \text{in } \beta \quad (57)$$

$$\frac{\partial e'^{\alpha}}{\partial s'} = \frac{\partial e'^{\beta}}{\partial s'} = \frac{\partial e'^{xs}}{\partial s'^{xs}} = \theta \quad \text{uniform} \quad (58)$$

and

$$\frac{\partial e'^{\alpha}}{\partial \rho'_i} = \frac{\partial e'^{\beta}}{\partial \rho'_i} = \frac{\partial e'^{xs}}{\partial \Gamma'_i} = \mu_i \quad \text{uniform} \quad (59)$$

where the mechanical equilibrium conditions in equations (56) and (57) can also be expressed in the form of equation (46) in both the α and β phases.

The equilibrium conditions at the interface are found by considering the last integral in equation (55) and using the boundary conditions (9) and (12) to eliminate δy^{β} and $\delta\mathbf{u}^{\beta}$, so that the remaining variations are independent. This process yields the equilibrium conditions on Σ'

$$\mathbf{T}_R^{\alpha}\cdot\mathbf{n}' - \mathbf{T}_R^{\beta}\cdot\mathbf{n}' - (\hat{\mathbf{T}}_R\cdot\hat{\nabla}_{\Sigma'}) = 0 \quad (60)$$

and

$$\begin{aligned}
&\omega'^{\alpha} - \omega'^{\beta} + \kappa'\sigma' + (\mathbf{q}'\cdot\hat{\nabla}_{\Sigma'}) + (\mathbf{T}_R^{\beta}\cdot\mathbf{n}') \times ([\mathbf{F}^{\beta} - \mathbf{F}^{\alpha}]\cdot\mathbf{n}') \\
&+ [(\hat{\mathbf{F}}^T\cdot\hat{\mathbf{T}}_R)\cdot\hat{\nabla}_{\Sigma'} - \mathbf{F}^{\alpha T}\cdot(\hat{\mathbf{T}}_R\cdot\hat{\nabla}_{\Sigma'})]\cdot\mathbf{n}' = 0. \quad (61)
\end{aligned}$$

Equation (60) is the usual balance of forces on Σ', and results from the independent variation $\delta\mathbf{u}^{\alpha}$. One can, by following the development of equation (52), use Nanson's formula and equations (47) and (51) to show that this balance takes the form $\mathbf{T}^{\alpha}\cdot\mathbf{n} - \mathbf{T}^{\beta}\cdot\mathbf{n} - (\hat{\mathbf{T}}\cdot\hat{\nabla}_{\Sigma}) = 0$ when reckoned on the actual interface Σ. The equilibrium condition (61) is the energy balance at a coherent crystal–crystal interface, and arises from the independent variation δy^{α}. This equation can also be expressed on Σ, but the result turns out to be unrevealing and unnecessarily complicated, since the special reference state in which the α and β lattices match was chosen to minimize the complexity of this calculation. The term $(\mathbf{T}_R^{\beta}\cdot\mathbf{n}')\cdot([\mathbf{F}^{\beta} - \mathbf{F}^{\alpha}]\cdot\mathbf{n}')$ is similar to the term that arises in the equilibrium of bulk phases across a coherent interfaces [2, 19]. However, if we solve equation (60) for $\hat{\mathbf{T}}_R\cdot\hat{\nabla}_{\Sigma'}$ and substitute the result into equation (61), we find that we can write the energy balance at a coherent interface as

$$\begin{aligned}
&\omega'^{\alpha} - \omega'^{\beta} + \kappa'\sigma' + (\mathbf{q}'\cdot\hat{\nabla}_{\Sigma'}) + (\mathbf{T}_R^{\beta}\cdot\mathbf{n}')\cdot(\mathbf{F}^{\beta}\cdot\mathbf{n}') \\
&- (\mathbf{T}_R^{\alpha}\cdot\mathbf{n}')\cdot(\mathbf{F}^{\alpha}\cdot\mathbf{n}') + [(\hat{\mathbf{F}}^T\cdot\hat{\mathbf{T}}_R)\cdot\hat{\nabla}_{\Sigma'}]\cdot\mathbf{n}' = 0. \quad (62)
\end{aligned}$$

Equation (62) shows that the grand potential densities ω'^{α} and ω'^{β} are coupled at a coherent interface through the capillarity $\kappa'\sigma'$, the anisotropy term $\mathbf{q}'\cdot\hat{\nabla}_{\Sigma'}$, the bulk stresses $\mathbf{T}_R^{\alpha}$ and $\mathbf{T}_R^{\beta}$ and a single surface stress term, which is identical to the surface stress term in the energy balance (50) at a crystal–melt interface. We also note that equation (62) is invariant with respect to interchanging the α and β labels, since, by our convention, $\mathbf{n}'$ would go to $-\mathbf{n}'$ and κ' would go to $-\kappa'$ upon such a switch. This invariance is missing in the Alexander-Johnson formulation of this equation [13].

3. GREASED CRYSTAL–CRYSTAL EQUILIBRIUM

The calculation of the equilibrium conditions at a greased crystal–crystal interface is more complicated than the corresponding calculation at a coherent interface. Since phases are free to slide with respect to each other at a greased interface, there is no common reference state for the two phases, as there is in the coherent case. Therefore, we must allow for "dual" reference states for the greased crystal–crystal system. This is illustrated in Fig. 3. The actual state of some system that consists of two solids in contact across a greased interface is pictured in Fig. 3(c). In this picture, the material points labelled A and B occupy the same spatial point $\mathbf{x}$ at the interface. However, because of the possibility of sliding at the interface, the images of A and B in the reference state [Fig. 3(a)], A' and B' respectively, stay on the interface, but need no longer occupy the same spatial position. That is, every point on the interface in the actual state can have an image in two distinct reference states.

This concept of dual reference states has major implications in modelling the greased interface. Since there is no longer a single reference state for the system, we perform the equilibrium calculation in the actual state. Also, rather than attaching the interface to one phase or the other, we allow the excess energy associated with the interface to depend on the deformations of both the dual reference states as well as the normal vectors to the interface in both reference states. Further, the absence of a single reference state for the two phases implies that the accretions δy^{α} and δy^{β} of the two phases are independent. Thus, it is possible that a gap can form between the two phases in the reference state for the varied state [Fig. 3(b)] (though this is not pictured in Fig. 3).

However, we require that there can be no gap

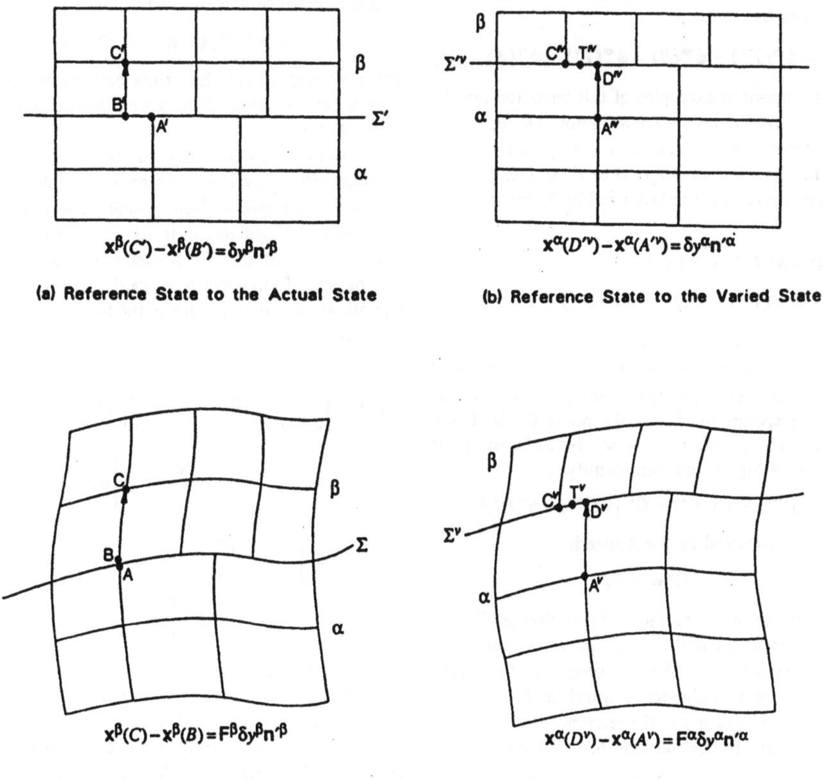

Fig. 3. A schematic view of the variations at a greased crystal–crystal interface. Since the lattices of the two phases α and β do not match across a greased interface, one must choose a different reference state for each phase. Thus, the accretion vectors $\delta y^{\alpha}\mathbf{n}'^{\alpha}$ and $\delta y^{\beta}\mathbf{n}'^{\beta}$ need not be the same, though we have drawn them so that there is no gap between the phases in state (b).

between the phases as the actual state moves to the varied state. This condition that the two phases stay contiguous at the interface through the variation can be expressed as [2]

$$\tilde{\delta}\mathbf{x}^{\alpha}\cdot\mathbf{n} = \tilde{\delta}\mathbf{x}^{\beta}\cdot\mathbf{n} \equiv \delta N^{\mathrm{T}} \tag{63}$$

at every point $\mathbf{x}$ on the interface in the actual state [Fig. 3(c)], where $\tilde{\delta}\mathbf{x}^{\alpha}$ and $\tilde{\delta}\mathbf{x}^{\beta}$ are defined in equations (10) and (11), and the unit normal $\mathbf{n}$ at $\mathbf{x}$ points from α to β. The quantity δN^{T} is the total normal movement of the interface between the actual state and the varied state. Consistent with our model of the greased interface, equation (63) places no constraint on the tangential movements at the interface, and so gives only one constraint among the variations $\tilde{\delta}\mathbf{u}^{\alpha}$, $\tilde{\delta}\mathbf{u}^{\beta}$, δy^{α} and δy^{β} at the interface.

Since we will perform the variational calculation in the actual state, we define an interfacial variation $\tilde{\Delta}$ for this problem between points connected by the vector $\delta N^{\mathrm{T}}\mathbf{n}$. Specifically, consider a point T^{v} in the varied state [Fig. 3(d)] such that $\mathbf{x}(T^{\mathrm{v}}) - \mathbf{x}(B) = \mathbf{x}(T^{\mathrm{v}}) - \mathbf{x}(A) = \delta N^{\mathrm{T}}\mathbf{n}$, where $\mathbf{x}(B) = \mathbf{x}(A)$ is some spatial point in the actual state. The interfacial variation of a function ϕ^{xs} defined on the interface is now given by

$$\tilde{\Delta}\phi^{\mathrm{xs}} = \phi^{\mathrm{xs}}(T^{\mathrm{v}}) - \phi^{\mathrm{xs}}(B) = \phi^{\mathrm{xs}}(T^{\mathrm{v}}) - \phi^{\mathrm{xs}}(A). \tag{64}$$

Two important examples of this variation are the variations $\tilde{\Delta}\mathbf{X}^{\alpha}$ and $\tilde{\Delta}\mathbf{X}^{\beta}$. The variation $\tilde{\Delta}\mathbf{X}^{\alpha}$ represents the difference in the position of the points in the α reference state that correspond to T^{v} and A. By using equation (64) and the notation in Fig. 3, this variation is

$$\begin{aligned}\tilde{\Delta}\mathbf{X}^{\alpha} &= \mathbf{x}^{\alpha}(T'^{\mathrm{v}}) - \mathbf{x}^{\alpha}(A') \\ &= \mathbf{x}^{\alpha}(T'^{\mathrm{v}}) - \mathbf{x}^{\alpha}(D'^{\mathrm{v}}) + \mathbf{x}^{\alpha}(D'^{\mathrm{v}}) - \mathbf{x}^{\alpha}(A')\end{aligned} \tag{65}$$

where T'^{v} is the material point in the reference state in Fig. 3(b) that corresponds to T^{v}, A' is the material point in the reference state for the α phase in Fig. 3(a) that corresponds to A, and the point D'^{v} is chosen so that $\mathbf{x}^{\alpha}(D'^{\mathrm{v}}) - \mathbf{x}^{\alpha}(A') = \delta y^{\alpha}\mathbf{n}'^{\alpha}$. Now, by using the notation in Fig. 3, one can identify

$$\mathbf{x}^{\alpha}(T'^{\mathrm{v}}) - \mathbf{x}^{\alpha}(D'^{\mathrm{v}}) = \hat{\mathbf{G}}^{\alpha}\cdot[\mathbf{x}^{\alpha}(T^{\mathrm{v}}) - \mathbf{x}^{\alpha}(D^{\mathrm{v}})] \tag{66}$$

where, as discussed in the Appendix

$$\hat{\mathbf{G}}^{\alpha} = \mathbf{X}^{\alpha}\tilde{\nabla}_{\Sigma} \tag{67}$$

is the inverse of the surface deformation gradient $\hat{\mathbf{F}}$ in the α phase, so is the tensor operator, defined in three dimensions, that has the property of mapping vectors tangent to the actual interface Σ onto their corresponding vectors in the reference state for the α phase. We find from equations (63), (65) and (66) that

$$\begin{aligned}\tilde{\Delta}\mathbf{X}^{\alpha} &= \delta y^{\alpha}\mathbf{n}'^{\alpha} + \hat{\mathbf{G}}^{\alpha}\cdot\{\mathbf{x}^{\alpha}(T^{\mathrm{v}}) - \mathbf{x}^{\alpha}(A) \\ &\qquad - [\mathbf{x}^{\alpha}(D^{\mathrm{v}}) - \mathbf{x}^{\alpha}(A)]\} \\ &= \delta y^{\alpha}\mathbf{n}'^{\alpha} - \hat{\mathbf{G}}^{\alpha}\cdot[\tilde{\delta}\mathbf{x}^{\alpha} - \mathbf{n}(\tilde{\delta}\mathbf{x}^{\alpha}\cdot\mathbf{n})] \\ &= \delta y^{\alpha}\mathbf{n}'^{\alpha} - \hat{\mathbf{G}}^{\alpha}\cdot\tilde{\delta}\mathbf{x}^{\alpha}\end{aligned} \tag{68}$$

where the last equality results directly from equation (A3) of the Appendix.

A similar analysis on the β side of the interface yields

$$\begin{aligned}\tilde{\Delta}\mathbf{X}^{\beta} &= \delta y^{\beta}\mathbf{n}'^{\beta} - \hat{\mathbf{G}}^{\beta}\cdot[\tilde{\delta}\mathbf{x}^{\beta} - \mathbf{n}(\tilde{\delta}\mathbf{x}^{\beta}\cdot\mathbf{n})] \\ &= \delta y^{\beta}\mathbf{n}'^{\beta} - \hat{\mathbf{G}}^{\beta}\cdot\tilde{\delta}\mathbf{x}^{\beta}\end{aligned} \tag{69}$$

where $\hat{\mathbf{G}}^{\beta} = \mathbf{X}^{\beta}\tilde{\nabla}_{\Sigma}$ maps a vector in the actual interface Σ to its corresponding vector in the β reference state, and we again use equation (A3) of the Appendix. Also, we emphasize that because of the dual reference states, $\hat{\mathbf{G}}^{\alpha}$ and $\hat{\mathbf{G}}^{\beta}$ are different at a greased interface.

3.1. *The variational calculation*

We are now in position to calculate the equilibrium conditions for a greased crystal-crystal system. Since we work in the actual state, we write the bulk energy densities as $e^{\alpha} = e(\mathbf{G}^{\alpha}, s^{\alpha}, \rho_i^{\alpha})$ and $e^{\beta} = e(\mathbf{G}^{\beta}, s^{\beta}, \rho_i^{\beta})$, where the deformations $\mathbf{G} \equiv \mathbf{F}^{-1} = \mathbf{X}\tilde{\nabla}_x$ take the actual state back to the appropriate reference state, and the quantities s and ρ_i are the entropy and mole densities, respectively, per unit actual volume. The excess energy density per unit actual area associated with the interface is taken to be

$$e^{\mathrm{xs}} = e^{\mathrm{xs}}(\hat{\mathbf{G}}^{\alpha}, \hat{\mathbf{G}}^{\beta}, \mathbf{n}'^{\alpha}, \mathbf{n}'^{\beta}, s^{\mathrm{xs}}, \Gamma_i) \tag{70}$$

where s^{xs} and Γ_i are the excess entropy and mole density per unit area of the actual interface, respectively. By taking the variable set for e^{xs} to include $\hat{\mathbf{G}}^{\alpha}$, $\hat{\mathbf{G}}^{\beta}$, $\mathbf{n}'^{\alpha}$ and $\mathbf{n}'^{\beta}$, we ensure that our model of a greased interface will be invariant with respect to interchanging the α and β labels. Even though we will perform the equilibrium calculation in the actual state of the system, we must use the reference normals in the variable set (70) in order to properly account for the crystallographic orientations of the two phases at the interface.

The variation of E^{α} is found to be

$$\begin{aligned}\Delta E^{\alpha} = \int_{V^{\alpha}}\Bigg[\left(\frac{\partial e^{\alpha}}{\partial \mathbf{G}^{\alpha}}\cdot\tilde{\nabla}_x\right)\cdot\delta^{\mathrm{E}}\mathbf{u}^{\alpha} + \frac{\partial e^{\alpha}}{\partial s}\,\delta^{\mathrm{E}}s^{\alpha} \\ + \frac{\partial e^{\alpha}}{\partial \rho_i}\,\delta^{\mathrm{E}}\rho_i^{\alpha}\Bigg]\mathrm{d}^3x \\ + \int_{\Sigma}\Bigg[-\left(\frac{\partial e^{\alpha}}{\partial \mathbf{G}^{\alpha}}\cdot\mathbf{n}\right)\cdot\delta^{\mathrm{E}}\mathbf{u}^{\alpha} \\ + e^{\alpha}\delta N^{\mathrm{T}}\Bigg]\mathrm{d}^2x\end{aligned} \tag{71}$$

so has the same form as equation (18), except that the variations are now Eulerian since the integration is over the actual state, and we cannot identify the derivative $\partial e/\partial \mathbf{G}$ as a stress tensor. Equation (71) can also be used for ΔE^{β} by replacing 'α' with 'β', and, since $\mathbf{n}$ points from α to β and δN^{T} is measured from α to β, by replacing $\mathbf{n}$ with $-\mathbf{n}$ and δN^{T} with $-\delta N^{\mathrm{T}}$. The Eulerian variation $\delta^{\mathrm{E}}\mathbf{u}$ that appears in equation (71) is related to the Lagrangian variation $\delta\mathbf{u}$ by $\mathbf{G}\cdot\delta\mathbf{u} = \delta^{\mathrm{E}}\mathbf{u}$, which can be derived by writing $\mathbf{u} =$

$\mathbf{x}(\mathbf{X}, t) - \mathbf{X} = \mathbf{x} - \mathbf{X}(\mathbf{x}, t)$, identifying $\delta\mathbf{u} \equiv (\partial\mathbf{u}/\partial t)_{\mathbf{X}}\delta t$ and $\delta^{E}\mathbf{u} \equiv (\partial\mathbf{u}/\partial t)_{\mathbf{x}}\delta t$, and calculating $\partial/\partial t$ of the function $\mathbf{u}[\mathbf{x}(\mathbf{X}, t), t]$ at constant $\mathbf{X}$.

The variation of the excess energy associated with the interface is

$$\Delta E^{xs} = \int_{\Sigma}\left[\frac{\partial e^{xs}}{\partial\hat{\mathbf{G}}^{\alpha}}:\tilde{\Delta}\hat{\mathbf{G}}^{\alpha} + \frac{\partial e^{xs}}{\partial\hat{\mathbf{G}}^{\beta}}:\tilde{\Delta}\hat{\mathbf{G}}^{\beta} + \frac{\partial e^{xs}}{\partial s^{xs}}\tilde{\Delta}s^{xs} + \mathbf{q}^{\alpha}\cdot\tilde{\Delta}\mathbf{n}'^{\alpha} + \mathbf{q}^{\beta}\cdot\tilde{\Delta}\mathbf{n}'^{\beta} + \frac{\partial e^{xs}}{\partial\Gamma_i}\tilde{\Delta}\Gamma_i\right]\mathrm{d}^2x + \int_{\Sigma}\kappa e^{xs}\,\delta N^{\mathrm{T}}\,\mathrm{d}^2x \tag{72}$$

where, following equation (54), we define

$$\mathbf{q}^{\alpha} = \frac{\partial e^{xs}}{\partial\mathbf{n}'^{\alpha}} - \mathbf{n}'^{\alpha}\left(\frac{\partial e^{xs}}{\partial\mathbf{n}'^{\alpha}}\cdot\mathbf{n}'^{\alpha}\right) \tag{73}$$

and

$$\mathbf{q}^{\beta} = \frac{\partial e^{xs}}{\partial\mathbf{n}'^{\beta}} - \mathbf{n}'^{\beta}\left(\frac{\partial e^{xs}}{\partial\mathbf{n}'^{\beta}}\cdot\mathbf{n}'^{\beta}\right). \tag{74}$$

We must now evaluate the variations $\tilde{\Delta}\hat{\mathbf{G}}^{\alpha}$, $\tilde{\Delta}\hat{\mathbf{G}}^{\alpha}$, $\tilde{\Delta}\mathbf{n}'^{\alpha}$ and $\tilde{\Delta}\mathbf{n}'^{\beta}$. Consider first the variation $\tilde{\Delta}\mathbf{n}'^{\alpha}$. From equations (64), (65) and Fig. 3, we see that the variation $\tilde{\Delta}\mathbf{n}'^{\alpha}$ asks for the difference in the normal vectors to Σ'' [Fig. 3(b)] and Σ' [Fig. 3(a)] at points in the α phase connected by the vector $\tilde{\Delta}\mathbf{X}^{\alpha}$. This variation can be shown to be [22]

$$\begin{aligned}\tilde{\Delta}\mathbf{n}'^{\alpha} &= \mathbf{n}'^{\alpha}\cdot(\tilde{\Delta}\mathbf{X}^{\alpha}\vec{\nabla}_{\Sigma'})\\ &= -\vec{\nabla}_{\Sigma'}\delta y^{\alpha} + \mathbf{n}'^{\alpha}\cdot(\hat{\mathbf{G}}^{\alpha}\cdot[\delta\mathbf{x}^{\alpha} - \mathbf{n}(\delta\mathbf{x}^{\alpha}\cdot\mathbf{n})])\vec{\nabla}_{\Sigma'}\\ &= -\vec{\nabla}_{\Sigma'}\delta y^{\alpha} + \mathbf{n}'^{\alpha}\cdot(\hat{\mathbf{G}}^{\alpha}\cdot\delta\mathbf{x}^{\alpha})\vec{\nabla}_{\Sigma'}\end{aligned} \tag{75}$$

where the last equality follows from equation (A3) of the Appendix.

Similarly, one can show that

$$\begin{aligned}\tilde{\Delta}\mathbf{n}'^{\beta} &= \mathbf{n}'^{\beta}\cdot(\tilde{\Delta}\mathbf{X}^{\beta}\vec{\nabla}_{\Sigma'})\\ &= -\vec{\nabla}_{\Sigma'}\delta y^{\beta} + \mathbf{n}'^{\beta}\cdot(\hat{\mathbf{G}}^{\beta}\cdot[\delta\mathbf{x}^{\beta} - \mathbf{n}(\delta\mathbf{x}^{\beta}\cdot\mathbf{n})])\vec{\nabla}_{\Sigma'}\\ &= -\vec{\nabla}_{\Sigma'}\delta y^{\beta} + \mathbf{n}'^{\beta}\cdot(\hat{\mathbf{G}}^{\beta}\cdot\delta\mathbf{x}^{\beta})\vec{\nabla}_{\Sigma'}.\end{aligned} \tag{76}$$

The gradients that appear in equations (75) and (76) are taken over the reference interface to the α and β phases, respectively, since we are interested in how the appropriate $\mathbf{n}'$ varies with position on the reference interface. We can relate the gradient $\nabla_{\Sigma'}$ in either α or β to the gradient ∇_{Σ} over the interface in the actual state by using the chain rule to write [12]

$$(\mathbf{a}\vec{\nabla}_{\Sigma'})\cdot\hat{\mathbf{G}}^{*} = \mathbf{a}\vec{\nabla}_{\Sigma} \tag{77}$$

for any vector $\mathbf{a}$ defined on the surface, where the superscript '$*$' indicates either α or β, depending on which phase Σ' corresponds to. We will apply equation (77) later in the analysis.

The quantities $\tilde{\Delta}\hat{\mathbf{G}}^{\alpha}$ and $\tilde{\Delta}\hat{\mathbf{G}}^{\beta}$ are evaluated as follows: The logic that led to equation (27) allows us to write $\tilde{\Delta}\hat{\mathbf{G}} = (\tilde{\Delta}\mathbf{X})\vec{\nabla}_{\Sigma} + \mathbf{X}\tilde{\Delta}(\vec{\nabla}_{\Sigma})$ to first order in the variation, where the term $\tilde{\Delta}(\vec{\nabla}_{\Sigma}) = \vec{\nabla}_{\Sigma^{v}} - \vec{\nabla}_{\Sigma}$ accounts for the change in the surface gradient operator as the interface moves from the actual state to the varied state. Since the position of the varied interface $\mathbf{r}^{v}$ can be written as $\mathbf{r}^{v} = \mathbf{r} + \delta N^{\mathrm{T}}\mathbf{n}$, where $\mathbf{r}$ is the position of the actual interface, we find that the same process that led to equation (30) leads to the results

$$\tilde{\Delta}\hat{\mathbf{G}}^{\alpha} = (\delta y^{\alpha}\mathbf{n}'^{\alpha} - \hat{\mathbf{G}}^{\alpha}\cdot\delta\mathbf{x}^{\alpha})\vec{\nabla}_{\Sigma} + \hat{\mathbf{G}}^{\alpha}\cdot(\vec{\nabla}_{\Sigma}\delta N^{\mathrm{T}})\mathbf{n} + \hat{\mathbf{G}}^{\alpha}\cdot\mathbf{L}\delta N^{\mathrm{T}} \tag{78}$$

and

$$\tilde{\Delta}\hat{\mathbf{G}}^{\beta} = (\delta y^{\beta}\mathbf{n}'^{\beta} - \hat{\mathbf{G}}^{\beta}\cdot\delta\mathbf{x}^{\beta})\vec{\nabla}_{\Sigma} + \hat{\mathbf{G}}^{\beta}\cdot(\vec{\nabla}_{\Sigma}\delta N^{\mathrm{T}})\mathbf{n} + \hat{\mathbf{G}}^{\beta}\cdot\mathbf{L}\delta N^{\mathrm{T}} \tag{79}$$

where $\mathbf{L} = -(\mathbf{1} - \mathbf{nn})\mathbf{n}\vec{\nabla}_{\Sigma}$ is the Weingarten map of the actual interface, and we have used equations (68) and (69).

The variation of the excess energy is now given by

$$\begin{aligned}\Delta E^{xs} = \int_{\Sigma}\Big\{&\frac{\delta e^{xs}}{\partial\hat{\mathbf{G}}^{\alpha}}:[(\delta y^{\alpha}\mathbf{n}'^{\alpha} - \hat{\mathbf{G}}^{\alpha}\cdot\delta\mathbf{x}^{\alpha})\vec{\nabla}_{\Sigma} + \hat{\mathbf{G}}^{\alpha}\cdot(\vec{\nabla}_{\Sigma}\delta N^{\mathrm{T}})\mathbf{n} + \hat{\mathbf{G}}^{\alpha}\cdot\mathbf{L}\delta N^{\mathrm{T}}]\\ &+\frac{\partial e^{xs}}{\partial\hat{\mathbf{G}}^{\beta}}:[(\delta y^{\beta}\mathbf{n}'^{\beta} - \hat{\mathbf{G}}^{\beta}\cdot\delta\mathbf{x}^{\beta})\vec{\nabla}_{\Sigma} + \hat{\mathbf{G}}^{\beta}\cdot(\vec{\nabla}_{\Sigma}\delta N^{\mathrm{T}})\mathbf{n} + \hat{\mathbf{G}}^{\beta}\cdot\mathbf{L}\delta N^{\mathrm{T}}]\\ &-\mathbf{q}^{\alpha}\cdot\vec{\nabla}_{\Sigma'}\delta y^{\alpha} + \mathbf{q}^{\alpha}\cdot[\mathbf{n}'^{\alpha}\cdot(\hat{\mathbf{G}}^{\alpha}\cdot\delta\mathbf{x}^{\alpha})\vec{\nabla}_{\Sigma'}]\\ &-\mathbf{q}^{\beta}\cdot\vec{\nabla}_{\Sigma'}\delta y^{\beta} + \mathbf{q}^{\beta}\cdot[\mathbf{n}'^{\beta}\cdot(\hat{\mathbf{G}}^{\beta}\cdot\delta\mathbf{x}^{\beta})\vec{\nabla}_{\Sigma'}]\\ &+\frac{\partial e^{xs}}{\partial s^{xs}}\tilde{\Delta}s^{xs} + \frac{\partial e^{xs}}{\partial\Gamma_i}\tilde{\Delta}\Gamma_i + \kappa e^{xs}\delta N^{\mathrm{T}}\Big\}\mathrm{d}^2x.\end{aligned} \tag{80}$$

Equation (80) can be reduced as follows: consider first the terms that arise from the dependence of e^{xs} on the normals. By using equation (2.17a) in Gurtin–Murdoch [16], equation (A4) in the Appendix and the chain rule, equation (77), we can write

$$\begin{aligned}\mathbf{n}'^{*}\cdot(\hat{\mathbf{G}}^{*}\cdot\delta\mathbf{x}^{*})\vec{\nabla}_{\Sigma'} &= -(\mathbf{n}'^{*}\vec{\nabla}_{\Sigma'})\cdot\hat{\mathbf{G}}^{*}\cdot\delta\mathbf{x}^{*}\\ &= -(\mathbf{n}'^{*}\vec{\nabla}_{\Sigma})\cdot\delta\mathbf{x}^{*}\end{aligned} \tag{81}$$

where the superscript '$*$' again refers to either α or β, and we note that $\mathbf{n}'\vec{\nabla}_{\Sigma'} = (\mathbf{n}'\vec{\nabla}_{\Sigma'})^{\mathrm{T}}$ [16]. Also, it is important to realize that because of the property of $\hat{\mathbf{G}}$ given by equation (A3) of the Appendix, no information about the normal component of the variation $\delta\mathbf{x}$ can be inferred from equation (81).

We can also use the chain rule and the fact that $\hat{\mathbf{G}}^{-1} = \hat{\mathbf{F}}$ to write

$$\int_{\Sigma} -\mathbf{q}^{*}\cdot\vec{\nabla}_{\Sigma'}\delta y^{*} = \int_{\Sigma} -\mathbf{q}^{*}\cdot(\vec{\nabla}_{\Sigma}\delta y^{*})\cdot\hat{\mathbf{F}}^{*} = \int_{\Sigma} -(\hat{\mathbf{F}}^{*}\cdot\mathbf{q}^{*})\cdot(\vec{\nabla}_{\Sigma}\delta y^{*}). \tag{82}$$

Equation (82) can be further reduced by using equation (2.17b) in Gurtin–Murdoch [16], the surface divergence theorem and the assumption that the interface is closed; we find

$$\int_{\Sigma} -(\hat{\mathbf{F}}^{*}\cdot\mathbf{q}^{*})\cdot(\vec{\nabla}_{\Sigma}\delta y^{*}) = \int_{\Sigma}[(\hat{\mathbf{F}}\cdot\mathbf{q}^{*})\cdot\vec{\nabla}_{\Sigma}]\delta y^{*}. \tag{83}$$

We now consider the terms that arise from the dependence of e^{xs} on $\hat{\mathbf{G}}^\alpha$ and $\hat{\mathbf{G}}^\beta$. We find that the terms $(\partial e^{xs}/\partial \hat{\mathbf{G}}^*):[\hat{\mathbf{G}}^*\cdot(\vec{\nabla}_\Sigma \delta N^T)\mathbf{n}]$ vanish by using the definition of the tensor scalar product and equation (A3) of the Appendix. Further, the same process that led to equation (35) leads to the result

$$\frac{\partial e^{xs}}{\partial \hat{\mathbf{G}}^*}:\hat{\mathbf{G}}^*\cdot\mathbf{L}\delta N^T = \left[\left(\hat{\mathbf{G}}^{*T}\cdot\frac{\partial e^{xs}}{\partial \hat{\mathbf{G}}^*}\right)\cdot\vec{\nabla}_\Sigma\right]\cdot\delta N^T\mathbf{n}. \quad (84)$$

The remaining terms are reduced by using equation (2.17b) in Gurtin and Murdoch [16], the surface divergence theorem and the assumption that the interface is closed; this follows the development of equation (33).

The final expression for ΔE^{xs} is given by

$$\Delta E^{xs} = \int_\Sigma \Bigg[-\left(\frac{\partial e^{xs}}{\partial \hat{\mathbf{G}}^\alpha}\cdot\vec{\nabla}_\Sigma\right)\cdot(\delta y^\alpha \mathbf{n}'^\alpha - \hat{\mathbf{G}}^\alpha\cdot\vec{\delta}\mathbf{x}^\alpha)$$
$$-\left(\frac{\partial e^{xs}}{\partial \hat{\mathbf{G}}^\beta}\cdot\vec{\nabla}_\Sigma\right)\cdot(\delta y^\beta \mathbf{n}'^\beta - \hat{\mathbf{G}}^\beta\cdot\vec{\delta}\mathbf{x}^\beta)$$
$$+\left[\left(\hat{\mathbf{G}}^{\alpha T}\cdot\frac{\partial e^{xs}}{\partial \hat{\mathbf{G}}^\alpha}+\hat{\mathbf{G}}^{\beta T}\cdot\frac{\partial e^{xs}}{\partial \hat{\mathbf{G}}^\beta}\right)\cdot\vec{\nabla}_\Sigma\right]\cdot\delta N^T\mathbf{n}$$
$$+[(\hat{\mathbf{F}}^\alpha\cdot\mathbf{q}^\alpha)\cdot\vec{\nabla}_\Sigma]\delta y^\alpha - \mathbf{q}^\alpha\cdot(\mathbf{n}'^\alpha\vec{\nabla}_\Sigma)\cdot\vec{\delta}\mathbf{x}^\alpha$$
$$+[(\hat{\mathbf{F}}^\beta\cdot\mathbf{q}^\beta)\cdot\vec{\nabla}_\Sigma]\delta y^\beta - \mathbf{q}^\beta\cdot(\mathbf{n}'^\beta\vec{\nabla}_\Sigma)\cdot\vec{\delta}\mathbf{x}^\beta$$
$$+\frac{\partial e^{xs}}{\partial s^{xs}}\tilde{\Delta}s^{xs}+\frac{\partial e^{xs}}{\partial \Gamma_i}\tilde{\Delta}\Gamma_i+\kappa e^{xs}\delta N^T\Bigg]\mathrm{d}^2x. \quad (85)$$

Finally, the constraints on the system of constant entropy and mass can be written in the form [following equation (38)]

$$\Delta\Phi = \int_{V^\alpha}\delta^E\phi^\alpha \mathrm{d}^3x + \int_\Sigma \phi^\alpha\delta N^T\mathrm{d}^2x$$
$$+\int_{V^\beta}\delta^E\phi^\beta \mathrm{d}^3x - \int_\Sigma \phi^\beta\delta N^T\mathrm{d}^2x$$
$$+\int_\Sigma \tilde{\Delta}\phi^{xs}\mathrm{d}^2x + \int_\Sigma \kappa\phi^{xs}\delta N^T\mathrm{d}^2x = 0, \quad (86)$$

where $\Phi = S$ and $\phi = s$ for the entropy constraint, and $\Phi = N_i$ $(i = 1,\ldots,K)$ and $\phi = \rho_i$ $(\phi^{xs} = \Gamma_i)$ for the mass constraints.

3.2. Results

By substituting equations (71) (applied to both α and β), (85), (86) and (63) into equation (39), we find that in equilibrium

$$\Delta E^{tot} - \theta\Delta S^{tot} - \mu_i\Delta N_i^{tot}$$
$$= \int_{V^\alpha}\Bigg[\left(\frac{\partial e^\alpha}{\partial \mathbf{G}}\cdot\vec{\nabla}_x\right)\cdot(\mathbf{G}^\alpha\cdot\delta\mathbf{u}^\alpha)+\left(\frac{\partial e^\alpha}{\partial s}-\theta\right)\delta^E s^\alpha$$
$$+\left(\frac{\partial e^\alpha}{\partial \rho_i}-\mu_i\right)\delta^E\rho_i^\alpha\Bigg]\mathrm{d}^3x$$
$$+\int_{V^\beta}\Bigg[\left(\frac{\partial e^\beta}{\partial \mathbf{G}}\cdot\vec{\nabla}_x\right)\cdot(\mathbf{G}^\beta\cdot\delta\mathbf{u}^\beta)$$
$$+\left(\frac{\partial e^\beta}{\partial s}-\theta\right)\delta^E s^\beta+\left(\frac{\partial e^\beta}{\partial \rho_i}-\mu_i\right)\delta^E\rho_i^\beta\Bigg]\mathrm{d}^3x$$
$$+\int_\Sigma\Bigg[\left(\frac{\partial e^{xs}}{\partial s^{xs}}-\theta\right)\tilde{\Delta}s^{xs}+\left(\frac{\partial e^{xs}}{\partial \Gamma_i}-\mu_i\right)\tilde{\Delta}\Gamma_i\Bigg]\mathrm{d}^2x$$
$$+\int_\Sigma\Bigg\{\omega^\alpha\delta N^T-\omega^\beta\delta N^T+\kappa\sigma\delta N^T$$
$$-\left(\frac{\partial e^\alpha}{\partial \mathbf{G}}\cdot\mathbf{n}\right)\cdot(\mathbf{G}^\alpha\cdot\delta\mathbf{u}^\alpha)+\left(\frac{\partial e^\beta}{\partial \mathbf{G}}\cdot\mathbf{n}\right)\cdot(\mathbf{G}^\beta\cdot\delta\mathbf{u}^\beta)$$
$$-\left(\frac{\partial e^{xs}}{\partial \hat{\mathbf{G}}^\alpha}\cdot\vec{\nabla}_\Sigma\right)\cdot(\delta y^\alpha\mathbf{n}'^\alpha-\hat{\mathbf{G}}^\alpha\cdot\vec{\delta}\mathbf{x}^\alpha)$$
$$-\left(\frac{\partial e^{xs}}{\partial \hat{\mathbf{G}}^\beta}\cdot\vec{\nabla}_\Sigma\right)\cdot(\delta y^\beta\mathbf{n}'^\beta-\hat{\mathbf{G}}^\beta\cdot\vec{\delta}\mathbf{x}^\beta)$$
$$+\left[\left(\hat{\mathbf{G}}^{\alpha T}\cdot\frac{\partial e^{xs}}{\partial \hat{\mathbf{G}}^\alpha}+\hat{\mathbf{G}}^{\beta T}\cdot\frac{\partial e^{xs}}{\partial \hat{\mathbf{G}}^\beta}\right)\cdot\vec{\nabla}_\Sigma\right]\cdot\delta N^T\mathbf{n}$$
$$+[(\hat{\mathbf{F}}^\alpha\cdot\mathbf{q}^\alpha)\cdot\vec{\nabla}_\Sigma]\delta y^\alpha-\mathbf{q}^\alpha\cdot(\mathbf{n}'^\alpha\vec{\nabla}_\Sigma)\cdot\vec{\delta}\mathbf{x}^\alpha$$
$$+[(\hat{\mathbf{F}}^\beta\cdot\mathbf{q}^\beta)\cdot\vec{\nabla}_\Sigma]\delta y^\beta$$
$$-\mathbf{q}^\beta\cdot(\mathbf{n}'^\beta\vec{\nabla}_\Sigma)\cdot\vec{\delta}\mathbf{x}^\beta\Bigg\}\mathrm{d}^2x \geqslant 0. \quad (87)$$

The variations in the first three integrals are independent, and give the equilibrium conditions

$$\frac{\partial e^\alpha}{\partial \mathbf{G}}\cdot\vec{\nabla}_x = 0 \quad \text{in } \alpha \quad (88)$$

$$\frac{\partial e^\beta}{\partial \mathbf{G}}\cdot\vec{\nabla}_x = 0 \quad \text{in } \beta \quad (89)$$

$$\frac{\partial e^\alpha}{\partial s}=\frac{\partial e^\beta}{\partial s}=\frac{\partial e^{xs}}{\partial s^{xs}}=\theta \quad \text{uniform} \quad (90)$$

and

$$\frac{\partial e^\alpha}{\partial \rho_i}=\frac{\partial e^\beta}{\partial \rho_i}=\frac{\partial e^{xs}}{\partial \Gamma_i}=\mu_i \quad \text{uniform.} \quad (91)$$

Equations (88) and (89) can be put in a more common form by writing the Cauchy stress as

$$\mathbf{T} = \omega\mathbf{1} - \mathbf{G}^T\left(\frac{\partial e}{\partial \mathbf{G}}\right)_{s,\rho_i} = \omega\mathbf{1} - \mathbf{G}^T\left(\frac{\partial \omega}{\partial \mathbf{G}}\right)_{\theta,\mu_i} \quad (92)$$

where $\mathbf{1}$ is the identity tensor in three dimensions. The first equality in equation (92) can be derived from equations (47) and (19) by using the product rule for differentiation and the chain rule applied to the function $e[\mathbf{G}(\mathbf{F})] = e(\mathbf{F}^{-1})$, while the second equality follows by using equation (41) to construct the function $\omega(\mathbf{G}, \theta, \mu_i)$ from $e(\mathbf{G}, s, \rho_i)$. There are two terms in equation (92) because the energy density e depends on the deformation through both the total energy in some volume and the volume measure itself. Equation (92) can be used along with the chain rule applied to the function $\omega[\mathbf{G}(\mathbf{x})]$ to show that equations (88) and (89) are equivalent to equation (46) in each of α and β.

The equilibrium conditions on the interface Σ are found by considering the last integral in equation (87), and by using the boundary condition (63) to eliminate one of the variations. In particular, because of the properties of $\hat{\mathbf{G}}$ given by equations (A3) and (A4) in the Appendix, it is convenient to use equation (63) to eliminate the variation $\delta\mathbf{u}^\beta\cdot\mathbf{n}$, so the remaining variations—$\delta\mathbf{u}^\alpha\cdot\mathbf{n}$, the two components of $\delta\mathbf{u}^\alpha$ tangent to Σ, the two components of $\delta\mathbf{u}^\beta$ tangent to Σ, δy^α and δy^β—are independent. These independent variations lead to equilibrium conditions on Σ that are quite complicated, primarily due to the dependence of e^{xs} (hence σ) on both $\hat{\mathbf{G}}^\alpha$ and $\hat{\mathbf{G}}^\beta$. However, we can find a more revealing set of equations by adopting a model in which the surface excess energy e^{xs} is partitioned in the form.

$$e^{xs}(\hat{\mathbf{G}}^\alpha, \hat{\mathbf{G}}^\beta, \mathbf{n}'^\alpha, \mathbf{n}'^\beta, \theta, \mu_i) = e^{xs,\alpha}(\hat{\mathbf{G}}^\alpha, \mathbf{n}'^\alpha, s^{xs}, \Gamma_i) + e^{xs,\beta}(\hat{\mathbf{G}}^\beta, \mathbf{n}'^\beta, s^{xs}, \Gamma_i) \quad (93)$$

which effectively decouples the two sides of the interface. We note that while this partitioning is arbitrary, we believe it to be consistent with the idea that a greased interface promotes minimal contact between the two phases it separates. The partitioning of the surface energy in equation (93) leads to the two surface stresses [following equation (92)]

$$\hat{\mathbf{T}}^\alpha = \sigma^\alpha\hat{\mathbf{1}} - \hat{\mathbf{G}}^{\alpha T}\cdot\left(\frac{\partial e^{xs,\alpha}}{\partial \hat{\mathbf{G}}^\alpha}\right)_{\mathbf{n}'^\alpha, s, \rho_i} = \sigma^\alpha\hat{\mathbf{1}} - \hat{\mathbf{G}}^{\alpha T}\cdot\left(\frac{\partial \sigma^\alpha}{\partial \hat{\mathbf{G}}^\alpha}\right)_{\mathbf{n}'^\alpha, \theta, \mu_i} \quad (94)$$

and

$$\hat{\mathbf{T}}^\beta = \sigma^\beta\hat{\mathbf{1}} - \hat{\mathbf{G}}^{\beta T}\cdot\left(\frac{\partial e^{xs,\beta}}{\partial \hat{\mathbf{G}}^\beta}\right)_{\mathbf{n}'^\beta, s, \rho_i} = \sigma^\beta\hat{\mathbf{1}} - \hat{\mathbf{G}}^{\beta T}\cdot\left(\frac{\partial \sigma^\beta}{\partial \hat{\mathbf{G}}^\beta}\right)_{\mathbf{n}'^\beta, \theta, \mu_i} \quad (95)$$

where $\hat{\mathbf{1}}$ is the identity tensor on the interface Σ, as defined in equations (A5) and (A6) of the Appendix, and the two surface free energies σ^α and σ^β arise directly from the partitioning of the surface excess energy.

Now we substitute equation (93) into the last integral of equation (87) and use equation (63) to eliminate the variation $\delta\mathbf{u}^\beta\cdot\mathbf{n}$, so that all the remaining variations are independent. Then, by using equations (92), (94), (95) and the chain rule,† we find the following equilibrium conditions on Σ

$$\mathbf{n}\cdot(\mathbf{T}^\alpha\cdot\mathbf{n}) - \mathbf{n}\cdot(\mathbf{T}^\beta\cdot\mathbf{n}) - \mathbf{n}\cdot[(\hat{\mathbf{T}}^\alpha + \hat{\mathbf{T}}^\beta)\cdot\hat{\nabla}_\Sigma] = 0 \quad (96)$$

$$\mathbf{T}^\alpha\cdot\mathbf{n} - \mathbf{n}(\mathbf{n}\cdot\mathbf{T}^\alpha\cdot\mathbf{n}) = \hat{\mathbf{T}}^\alpha\cdot\hat{\nabla}_\Sigma - \mathbf{n}[\mathbf{n}\cdot(\hat{\mathbf{T}}^\alpha\cdot\hat{\nabla}_\Sigma)] \quad (97)$$

$$\mathbf{T}^\beta\cdot\mathbf{n} - \mathbf{n}(\mathbf{n}\cdot\mathbf{T}^\beta\cdot\mathbf{n}) = -\hat{\mathbf{T}}^\beta\cdot\hat{\nabla}_\Sigma + \mathbf{n}[\mathbf{n}\cdot(\hat{\mathbf{T}}^\beta\cdot\hat{\nabla}_\Sigma)] \quad (98)$$

†Specifically, the chain rule is applied to the function e^{xs*} to eliminate the terms $\mathbf{q}^*\cdot\mathbf{n}'^*\hat{\nabla}_\Sigma$ in equation (87).

$$\omega^\alpha - \mathbf{n}\cdot(\mathbf{T}^\alpha\cdot\mathbf{n} - \hat{\mathbf{T}}^\alpha\cdot\hat{\nabla}_\Sigma) + \left(\frac{\hat{J}^\alpha}{J^\alpha}\right)\kappa'^\alpha\sigma^\alpha + \left(\frac{\hat{J}^\alpha}{J^\alpha}\right)[(\hat{\mathbf{F}}^\alpha\cdot\mathbf{q}^\alpha)\cdot\hat{\nabla}_\Sigma] + \left(\frac{\hat{J}^\alpha}{J^\alpha}\right)^2\cdot[(\hat{\mathbf{F}}^{\alpha T}\cdot\hat{\mathbf{T}}^\alpha)\cdot\hat{\nabla}_\Sigma - \mathbf{F}^{\alpha T}\cdot(\hat{\mathbf{T}}^\alpha\cdot\hat{\nabla}_\Sigma)]\cdot(\mathbf{n}\cdot\mathbf{F}^\alpha) = 0 \quad (99)$$

and

$$\omega^\beta - \mathbf{n}\cdot(\mathbf{T}^\beta\cdot\mathbf{n} + \hat{\mathbf{T}}^\beta\cdot\hat{\nabla}_\Sigma) + \left(\frac{\hat{J}^\beta}{J^\beta}\right)\kappa'^\beta\sigma^\beta + \left(\frac{\hat{J}^\beta}{J^\beta}\right)[(\hat{\mathbf{F}}^\beta\cdot\mathbf{q}^\beta)\cdot\hat{\nabla}_\Sigma] + \left(\frac{\hat{J}^\beta}{J^\beta}\right)^2\cdot[(\hat{\mathbf{F}}^{\beta T}\cdot\hat{\mathbf{T}}^\beta)\cdot\hat{\nabla}_\Sigma - \mathbf{F}^{\beta T}\cdot(\hat{\mathbf{T}}^\beta\cdot\hat{\nabla}_\Sigma)]\cdot(\mathbf{n}\cdot\mathbf{F}^\beta) = 0 \quad (100)$$

where κ'^α is the curvature of the reference interface as measured from α and evaluated at the point on the reference interface in α that corresponds to some spatial point $\mathbf{x}$, and κ'^β is the reference curvature as measured from β and evaluated at the the point on the reference interface in β that corresponds to the same $\mathbf{x}$.

Equation (96) expresses the continuity of the normal component of the tractions at the interface, and arises from the independent variation $\delta\mathbf{u}^\alpha\cdot\mathbf{n}$. Equations (97) and (98) express the condition that there are no net forces tangent to the interface in α and β, respectively, and arise from the independent variations of the components of $\delta\mathbf{u}^\alpha$ and $\delta\mathbf{u}^\beta$ that lie tangent to Σ. (In contrast, since the tangential components of the variations of the displacement are not independent at a coherent interface, we found that a coherent interface can support tangential tractions.) Equations (96) (one condition), (97) (two conditions) and (98) (two conditions) specify five independent conditions for mechanical equilibrium at the interface. We note that equations (96)–(98) can be rearranged into the vector equation that describes the balance of the tractions at the interface, $\mathbf{T}^\alpha\cdot\mathbf{n} - \mathbf{T}^\beta\cdot\mathbf{n} = (\hat{\mathbf{T}}^\alpha + \hat{\mathbf{T}}^\beta)\cdot\hat{\nabla}_\Sigma$ (three conditions), plus two conditions, given by either equations (97) or (98), on the tangential components of the tractions at the interface. Also, we see that in the limit of zero excess quantities at the interface, equation (96) simply describes the continuity of pressure at the interface and equations (97) and (98) require that tangential tractions vanish at the interface.

Equations (99) and (100) set the values of the grand potential densities ω^α and ω^β on either side of a greased interface, and arise from the independent variations δy^α and δy^β. If we define a pressure

$$p = -\mathbf{n}\cdot(\mathbf{T}^\alpha\cdot\mathbf{n} - \hat{\mathbf{T}}^\alpha\cdot\hat{\nabla}_\Sigma) = -\mathbf{n}\cdot(\mathbf{T}^\beta\cdot\mathbf{n} + \hat{\mathbf{T}}^\beta\cdot\hat{\nabla}_\Sigma) \quad (101)$$

where the equality follows from equation (96), then equations (99) and (100) can both be written in the form

$$\omega^* + p + \left(\frac{\hat{J}^*}{J^*}\right)\kappa'^*\sigma^* + \left(\frac{\hat{J}^*}{J^*}\right)[(\hat{\mathbf{F}}^*\cdot\mathbf{q}^*)\cdot\hat{\nabla}_\Sigma]$$

$$+\left(\frac{\hat{J}^*}{J^*}\right)^2[(\hat{\mathbf{F}}^{*T}\cdot\hat{\mathbf{T}}^*)\cdot\hat{\nabla}_\Sigma - \mathbf{F}^{*T}\cdot(\hat{\mathbf{T}}^*\cdot\hat{\nabla}_\Sigma)]\cdot(\mathbf{n}\cdot\mathbf{F}^*)=0 \quad (102)$$

where, as usual, '$*$' can be either 'α' or 'β'. If we compare this form of the energy balance at a greased interface with the energy balance (53) at a crystal–melt interface, then we see that each side of the greased interface behaves as though it were in contact with a fluid at the pressure $p=-\omega^F$ given by equation (101). By following this analogy further, we note that both equations (99) and (100) could be rearranged so that the grand potential is given in terms of the bulk stress, capillary, an anisotropy term and a single surface stress term; this would parallel the derivation of equation (50).

4. DISCUSSION

Before we begin the discussion, we summarize the results of our analysis. This equilibrium conditions that apply on the reference interface between a crystal and its melt take the form of a force balance

$$\mathbf{T}_R^\alpha\cdot\mathbf{n}' - J\omega^F\mathbf{n}'\cdot(\mathbf{F}^\alpha)^{-1} - (\hat{\mathbf{T}}_R\cdot\hat{\nabla}_{\Sigma'})=0 \quad (103)$$

and an energy balance, which can be written either as

$$\omega'^\alpha - J\omega^F + \kappa'\sigma' + (\mathbf{q}'\cdot\hat{\nabla}_{\Sigma'}) + G^{\alpha}=0 \quad (104)$$

where $\mathbf{q}'$ is given by equation (22), and we define

$$G^{\alpha}=[(\hat{\mathbf{F}}^T\cdot\hat{\mathbf{T}}_R)\cdot\hat{\nabla}_{\Sigma'} - \mathbf{F}^{\alpha T}\cdot(\hat{\mathbf{T}}_R\cdot\hat{\nabla}_{\Sigma'})]\cdot\mathbf{n}',$$

or as

$$\omega'^\alpha - (\mathbf{T}_R^\alpha\cdot\mathbf{n}')\cdot(\mathbf{F}^\alpha\cdot\mathbf{n}') + \kappa'\sigma' + (\mathbf{q}'\cdot\hat{\nabla}_{\Sigma'}) + [(\hat{\mathbf{F}}^T\cdot\hat{\mathbf{T}}_R)\cdot\hat{\nabla}_{\Sigma'}]\cdot\mathbf{n}'=0. \quad (105)$$

At a coherent crystal–crystal interface, the equilibrium conditions expressed on the reference interface consist of a force balance

$$\mathbf{T}_R^\alpha\cdot\mathbf{n}' - \mathbf{T}_R^\beta\cdot\mathbf{n}' - (\hat{\mathbf{T}}_R\cdot\hat{\nabla}_{\Sigma'})=0 \quad (106)$$

and an energy balance, which can be written either as

$$\omega'^\alpha - \omega'^\beta + \kappa'\sigma' + (\mathbf{q}'\cdot\hat{\nabla}_{\Sigma'}) + (\mathbf{T}_R^\beta\cdot\mathbf{n}')\cdot([\mathbf{F}^\beta - \mathbf{F}^\alpha]\cdot\mathbf{n}') + G^{\alpha}=0 \quad (107)$$

or as

$$\omega'^\alpha - \omega'^\beta + \kappa'\sigma' + (\mathbf{q}'\cdot\hat{\nabla}_{\Sigma'}) + (\mathbf{T}_R^\beta\cdot\mathbf{n}')\cdot(\mathbf{F}^\beta\cdot\mathbf{n}') - (\mathbf{T}_R^\alpha\cdot\mathbf{n}')\cdot(\mathbf{F}^\alpha\cdot\mathbf{n}') + [(\hat{\mathbf{F}}^T\cdot\hat{\mathbf{T}}_R)\cdot\hat{\nabla}_{\Sigma'}]\cdot\mathbf{n}'=0. \quad (108)$$

Finally, at a greased crystal–crystal interface, we find that the equilibrium conditions that apply on the actual interface Σ are given by a balance of the normal components of the force on the interface

$$\mathbf{n}\cdot(\mathbf{T}^\alpha\cdot\mathbf{n}) - \mathbf{n}\cdot(\mathbf{T}^\beta\cdot\mathbf{n}) - \mathbf{n}\cdot[(\hat{\mathbf{T}}^\alpha+\hat{\mathbf{T}}^\beta)\cdot\hat{\nabla}_\Sigma]=0 \quad (109)$$

two conditions on the forces tangent to the interface

$$\mathbf{T}^\alpha\cdot\mathbf{n} - \mathbf{n}(\mathbf{n}\cdot\mathbf{T}^\alpha\cdot\mathbf{n}) = \hat{\mathbf{T}}^\alpha\cdot\hat{\nabla}_\Sigma - \mathbf{n}[\mathbf{n}\cdot(\hat{\mathbf{T}}^\alpha\cdot\hat{\nabla}_\Sigma)] \quad (110)$$

and

$$\mathbf{T}^\beta\cdot\mathbf{n} - \mathbf{n}(\mathbf{n}\cdot\mathbf{T}^\beta\cdot\mathbf{n}) = -\hat{\mathbf{T}}^\beta\cdot\hat{\nabla}_\Sigma + \mathbf{n}[\mathbf{n}\cdot(\hat{\mathbf{T}}^\beta\cdot\hat{\nabla}_\Sigma)] \quad (111)$$

and two conditions that set the grand potential density on either side of the interface

$$\omega^\alpha - \mathbf{n}\cdot(\mathbf{T}^\alpha\cdot\mathbf{n} - \hat{\mathbf{T}}^\alpha\cdot\hat{\nabla}_\Sigma) + \left(\frac{\hat{J}^\alpha}{J^\alpha}\right)\kappa'^\alpha\sigma^\alpha + \left(\frac{\hat{J}^\alpha}{J^\alpha}\right)[(\hat{\mathbf{F}}^\alpha\cdot\mathbf{q}^\alpha)\cdot\hat{\nabla}_\Sigma] + \left(\frac{\hat{J}^\alpha}{J^\alpha}\right)^2[(\hat{\mathbf{F}}^{\alpha T}\cdot\hat{\mathbf{T}}^\alpha)\cdot\hat{\nabla}_\Sigma - \mathbf{F}^{\alpha T}\cdot(\hat{\mathbf{T}}^\alpha\cdot\hat{\nabla}_\Sigma)]\cdot(\mathbf{n}\cdot\mathbf{F}^\alpha)=0 \quad (112)$$

and

$$\omega^\beta - \mathbf{n}\cdot(\mathbf{T}^\beta\cdot\mathbf{n} + \hat{\mathbf{T}}^\beta\cdot\hat{\nabla}_\Sigma) + \left(\frac{\hat{J}^\beta}{J^\beta}\right)\kappa'^\beta\sigma^\beta + \left(\frac{\hat{J}^\beta}{J^\beta}\right)[(\hat{\mathbf{F}}^\beta\cdot\mathbf{q}^\beta)\cdot\hat{\nabla}_\Sigma] + \left(\frac{\hat{J}^\beta}{J^\beta}\right)^2[(\hat{\mathbf{F}}^{\beta T}\cdot\hat{\mathbf{T}}^\beta)\cdot\hat{\nabla}_\Sigma - \mathbf{F}^{\beta T}\cdot(\hat{\mathbf{T}}^\beta\cdot\hat{\nabla}_\Sigma)]\cdot(\mathbf{n}\cdot\mathbf{F}^\beta)=0 \quad (113)$$

where $\mathbf{q}^\alpha$ and $\mathbf{q}^\beta$ are given by equations (73) and (74), respectively.

The basic idea implicit in our calculation of the equilibrium conditions at crystal–melt, coherent crystal–crystal and greased crystal–crystal interfaces is that, because of accretion, the variation one considers at these interfaces is fundamentally different from any bulk variation. The variation $\mathring{\delta}$ that we define at every point on a crystal–melt and coherent crystal–crystal interfaces takes accretion into account by following the interface as it accretes in the reference state, while the variation $\tilde{\Delta}$ that we define at every point on a greased crystal–crystal interface follows the total normal movement of the interface from the actual state to the varied state. By using these pointwise variations to calculate the total variation of the excess energy associated with the interface, we find the equilibrium conditions in equations (103)–(113), which show that both the surface free energy and the surface stress contribute to the energy balance at crystal–melt, coherent crystal–crystal and greased crystal–crystal interfaces.

The calculations for the crystal–melt and coherent crystal–crystal cases have been performed in the reference state of the system. This is possible because in both cases, we only need to consider one reference state—the reference state of the solid in the crystal–melt case, and the reference state that is common to both α and β in the coherent crystal–crystal case. However, the possibility of sliding at a greased interface leads to dual reference states in the greased system, which requires that we perform the equilibrium calculation for the greased crystal–crystal system in its actual state. While the calculation in the actual state is more complicated than the calculation in the reference state, we note that the equilibrium calculations for both the crystal–melt and coherent crystal–crystal systems can be performed in the actual state of the system with no change in the final results [17].

Another difference between the greased crystal–crystal system and either the crystal–melt or coherent crystal–crystal system is the following: because of the dual reference states needed to describe a greased interface, one can define two surface deformation gradients and two reference normals at that interface. In contrast, there is only one surface deformation gradient and reference normal at a crystal–melt or coherent crystal–crystal interface. Since we do not want to attach the greased interface to one phase or the other, we allow the excess energy density associated with a greased interface to depend on both surface deformations and reference normals. The resulting equilibrium conditions are quite complicated, and do not lend themselves to simple interpretation [17], so we have chosen to simplify the results by partitioning the surface free energy according to equation (93). In this limit, we find that the greased crystal–crystal system is equivalent to a system in which the two crystalline phases are each in equilibrium with a thin layer of fluid at the interface, and so are coupled only through the fluid pressure. Accordingly, we associate $\hat{\mathbf{T}}^{\alpha}$ and σ^{α} with the surface stress and surface free energy between α and the fluid layer, and $\hat{\mathbf{T}}^{\beta}$ and σ^{β} with the surface stress and surface free energy between β and the fluid layer.

By taking this interpretation of the greased crystal–crystal system, we can now focus the remainder of the discussion on the equilibrium of crystal–melt and coherent crystal–crystal systems. Consider first the interfacial balance of forces, given by equation (103) at a crystal–melt interface and by equation (106) at a coherent interface. These equations, which have been derived previously by Gurtin and Murdoch [16], Cahn [8], Cahn and Larche [11] and Alexander and Johnson [12, 13], show that contribution of the interface to the force balance is through the surface divergence of the surface stress.

Of more interest in the present analysis are the energy balances in either equations (104) or (105) at a crystal–melt interface, and in either equations (107) or (108) at a coherent interface. These equations ensure that the system is in equilibrium with respect to a transformation from one phase to another by balancing the bulk energies, as given by the grand potential densities, with energy terms that arise from capillarity, interfacial anisotropies, bulk stresses and the surface stress.

We first consider the capillary term $\kappa'\sigma'$ and the anisotropy term $\mathbf{q}'\cdot\hat{\nabla}_{\Gamma'}$. These terms account for the energy of newly accreted material, where, roughly, $\kappa'\sigma'$ gives the energy per unit volume of the reference state associated with the amount of new material and $\mathbf{q}'\cdot\hat{\nabla}_{\Gamma'}$ gives the energy per unit volume of the reference state associated with the crystallographic orientation of the new material.† Following Alexander and Johnson [12], we can define

$$\xi' = \sigma'\mathbf{n}' + \mathbf{q}' \tag{114}$$

which is analogous to Cahn and Hoffman's ξ-vector [14, 15]. In fact, since at constant surface deformation $\hat{\mathbf{F}}$ we can formally write the differential expressions

$$\mathrm{d}\sigma' = \left(\frac{\partial \sigma'}{\partial \mathbf{n}'}\right)_{\theta,\mu_i}\cdot \mathrm{d}\mathbf{n}' - s'^{xs}\,\mathrm{d}\theta - \Gamma_i'\,\mathrm{d}\mu_i \tag{115}$$

and

$$\mathrm{d}e'^{xs} = \left(\frac{\partial e'^{xs}}{\partial \mathbf{n}'}\right)_{s'^{xs},\Gamma_i}\cdot \mathrm{d}\mathbf{n}' + \theta\,\mathrm{d}s'^{xs} + \mu_i\,\mathrm{d}\Gamma_i' \tag{116}$$

we see from equation (42) that

$$\begin{aligned}\mathbf{q}' &\equiv \left\{\left(\frac{\partial e'^{xs}}{\partial \mathbf{n}'}\right)_{s'^{xs},\Gamma_i} - \mathbf{n}'\left[\left(\frac{\partial e'^{xs}}{\partial \mathbf{n}'}\right)_{s'^{xs},\Gamma_i}\cdot\mathbf{n}'\right]\right\}\\ &= \left\{\left(\frac{\partial \sigma'}{\partial \mathbf{n}'}\right)_{\theta,\mu_i} - \mathbf{n}'\left[\left(\frac{\partial \sigma'}{\partial \mathbf{n}'}\right)_{\theta,\mu_i}\cdot\mathbf{n}'\right]\right\}.\end{aligned} \tag{117}$$

Thus, at constant $\hat{\mathbf{F}}$, θ and μ_i, we can define a ξ-vector in the reference state of our system that corresponds exactly to the Cahn–Hoffmann definition. By taking this analogy further, we see that σ' is the work, measured per unit area in the reference state, of expanding or contacting the interface during accretion, while $\mathbf{q}'$ gives the force, measured per unit length in the reference state, that resists rotations of the interface during accretion.

Since we use the reference state to measure accretion independently of deformation, we find that both the capillary term $\kappa'\sigma'$ and the anisotropy term $\mathbf{q}'\cdot\hat{\nabla}_{\Gamma'}$ are associated with the interface in the reference state of the system. That is, the amount of newly accreted material is proportional to the area of the reference interface, and hence to κ'. Also, by taking the excess energy of the interface to depend on the normal to the interface in the reference state, we separate the effects of crysallographic anisotropies at the interface from those of any subsequent deformations. The newly accreted material is then deformed into the actual state, so the energy of the accreted material in its actual state will depend not only on the surface free energy and the anisotropy of the surface free energy, but also on the bulk stresses and the surface stresses. Of course, the particular choice of the reference state does not affect the final results. However, we note that we chose a homogeneous reference state in order to write the energy densities in equations (13) and (15) as functions of the appropriate deformation gradient. This assumption is especially important at the interface, since the interfacial variation involves a change in the reference state itself. If we had chosen an inhomogeneous reference state, then in order to fully isolate surface stress effects in the analysis, we would need to know not only the deformation from the reference interface to the actual interface, but also the

†One can use equations (81) and (82) to show that this energy is given by $(\hat{\mathbf{F}}\cdot\mathbf{q})\cdot\hat{\nabla}_{\Sigma}$ when measured per unit volume of the actual state.

deformation between the reference interface and the interface in some homogeneous state.

Having accounted for the energy of newly accreted material by measuring accretion in the reference state, we now consider how deforming this accreted material into its equilibrium configuration affects the energy balances at crystal–melt and coherent crystal–crystal interfaces. In this regard, we have included the alternate expressions (104) and (105) at the crystal–melt interface and (107) and (108) at the coherent interface in order to better understand the roles of the bulk stress and the surface stress in the deformation of the newly accreted material.

The forms of the energy balance given in equation (104) at the crystal–melt interface and equation (107) at the coherent interface are most easily compared to previous work on this subject. We find that in both the crystal–melt system and the coherent system, the term $G^{\alpha\alpha}$ arises directly from the effect of accretion on the variation at the interface, and in particular on the variation $\delta\hat{\mathbf{F}}$ in equation (30). If $G^{\alpha\alpha}$ vanishes, then our equations match those derived by Alexander and Johnson at crystal–melt [12] and coherent interfaces [13], since Alexander and Johnson include surface excess properties but do not consider the effect of accretion on the interfacial variations. We note in particular that when surface stress (hence $G^{\alpha\alpha}$) vanishes, the term $(\mathbf{T}_R^{\beta}\cdot\mathbf{n}')\cdot([\mathbf{F}^{\beta}-\mathbf{F}^{\alpha}]\cdot\mathbf{n}')$ that appears in the energy balance (107) at a coherent interface has been identified by Robin [19] and Larche and Cahn [2] as the bulk elastic energy per unit reference volume required to maintain coherence during accretion.

While it is convenient to define $G^{\alpha\alpha}$ as the "new" surface stress effect in the energy balances (104) and (107), a much more appealing physical understanding of the role of surface stress in the equilibrium energy balance at crystal–melt and coherent crystal–crystal interfaces can be found by examining equations (105) and (108). If we first consider the energy balance (105) at the crystal–melt interface, we find that the grand potential density ω'^{α} of the solid is balanced by the terms $\kappa'\sigma'$ and $\mathbf{q}'\cdot\tilde{\nabla}_{\Gamma'}$, a term $(\mathbf{T}_R^{\alpha}\cdot\mathbf{n}')\cdot(\mathbf{F}^{\alpha}\cdot\mathbf{n}')$ arising from the bulk stress, and the single term $[(\hat{\mathbf{F}}^{T}\cdot\hat{\mathbf{T}}_R)\cdot\tilde{\nabla}_{\Gamma'}]\cdot\mathbf{n}'$ arising from the surface stress. As we have seen, the terms $\kappa'\sigma'$ and $\mathbf{q}'\cdot\tilde{\nabla}_{\Gamma'}$ give the energy of creating new, undeformed material at the interface, so the remaining terms give the energy of stretching this material into its equilibrium configuration. Then, since $(\mathbf{T}_R^{\alpha}\cdot\mathbf{n}')\cdot(\mathbf{F}^{\alpha}\cdot\mathbf{n}')$ is the elastic energy of displacing the interface an amount $\mathbf{F}^{\alpha}\cdot\mathbf{n}'$ against the force $\mathbf{T}_R^{\alpha}\cdot\mathbf{n}'$ arising from the bulk stress, we see that $[(\hat{\mathbf{F}}^{T}\cdot\hat{\mathbf{T}}_R)\cdot\tilde{\nabla}_{\Gamma'}]\cdot\mathbf{n}'$ must give the additional elastic energy of displacing the interface against the excess force associated with the surface stress. That is, the term $[(\hat{\mathbf{F}}^{T}\cdot\hat{\mathbf{T}}_R)\cdot\tilde{\nabla}_{\Gamma'}]\cdot\mathbf{n}'$ can be identified as the "interfacial elastic energy" resulting from the surface stress.

The energy balance (108) at a coherent crystal–crystal interface has a similar interpretation. However, since the two phases at a coherent interface are coupled through the shared reference state and the concomitant matching of accretion vectors, equation (9), we find that the energy balance at the interface sets only the difference between the grand potential densities. This difference in the grand potential densities is balanced by the terms $\kappa'\sigma'$ and $\mathbf{q}'\cdot\tilde{\nabla}_{\Gamma'}$, the difference in the elastic energies arising from the bulk stresses, $(\mathbf{T}_R^{\beta}\cdot\mathbf{n}')\cdot(\mathbf{F}^{\beta}\cdot\mathbf{n}')-(\mathbf{T}_R^{\alpha}\cdot\mathbf{n}')\cdot(\mathbf{F}^{\alpha}\cdot\mathbf{n}')$, and the elastic energy $[(\hat{\mathbf{F}}^{T}\cdot\hat{\mathbf{T}}_R)\cdot\tilde{\nabla}_{\Gamma'}]\cdot\mathbf{n}'$ owing to the surface stress. Thus, we identify $(\mathbf{T}_R^{\beta}\cdot\mathbf{n}')\cdot(\mathbf{F}^{\beta}\cdot\mathbf{n}')-(\mathbf{T}_R^{\alpha}\cdot\mathbf{n}')\cdot(\mathbf{F}^{\alpha}\cdot\mathbf{n}')$ as the bulk elastic energy of deforming newly accreted material at the interface into a coherent state, and $[(\hat{\mathbf{F}}^{T}\cdot\hat{\mathbf{T}}_R)\cdot\tilde{\nabla}_{\Gamma'}]\cdot\mathbf{n}'$ as the additional elastic energy of performing this deformation against the excess force associated with the surface stress.

Finally, we consider the equilibrium of an isotropic solid sphere immersed in a fluid, in order to examine how the equilibrium conditions derived here simplify in this special case. Cahn [8] and Mullins [9, 10] have studied this problem, and both find the equilibrium conditions that apply on the interface in the actual state

$$p^{S}-p^{F}-\frac{2f}{R}=0 \tag{118}$$

and

$$\omega^{S}-\omega^{F}+\frac{2\sigma}{R}=0, \tag{119}$$

where p^{S} is the hydrostatic pressures in the solid (so $\mathbf{T}^{\alpha}=-p^{S}\mathbf{1}$), p^{F} is the pressure in the fluid, R is the actual radius of the solid sphere (so $2/R$ is the actual curvature of the sphere), σ is the isotropic surface free energy and f is the scalar function both Cahn and Mullins choose for the surface stress.

In order to reduce the relevant conditions that apply on the crystal–melt interface in the actual state, equations (52) and (53), we first choose the reference state for the solid to be a sphere of radius R_0. Then, since all the deformations are hydrostatic, we have $\mathbf{F}^{\alpha}=(1+\psi)\mathbf{1}$ and $\hat{\mathbf{F}}=(1+\psi)\hat{\mathbf{1}}$, where $\psi=(R-R_0)/R_0$. Further, we take the surface stress to be hydrostatic, so $\hat{\mathbf{T}}=f\hat{\mathbf{1}}$. By using the fact that $\omega^{F}=-p^{F}$ along with identities (2.17d) and (2.20) from Gurtin and Murdoch [16] one can show that the force balance in equation (52) reduces exactly to equation (118), plus the condition that $\tilde{\nabla}_{\Sigma}f=0$, so f must be constant over the surface of the solid sphere.

Consider next the energy balance (53) at the interface. Since both f and ψ are constants, one finds that the two surface stress terms in equation (53) cancel, so there is no explicit surface stress contribution to this energy balance. Then, since $J^{\alpha}=\det\mathbf{F}^{\alpha}=(1+\psi)^3$ and $\hat{J}=\det\hat{\mathbf{F}}=(1+\psi)^2$, one can show from the definition of ψ that equation (53) reduces to equation (119). Therefore, despite the apparent complexity of our interfacial equilibrium conditions (52) and (53), they reduce identically to the Cahn and Mullins results for a solid sphere immersed in a fluid. We conclude that the symmetry and homogeneity associated with the particular problem of a solid sphere in a fluid offsets the need to carefully define the interfacial variation; however, this is not the case when more general situations are investigated.

5. SUMMARY

We have extended the thermodynamics of solids to two phase systems that include surface excess quantities. This formulation is predicated on defining variations at crystal–melt, coherent crystal–crystal and greased crystal–crystal interfaces that include accretive effects, and so give a proper accounting of the roles of surface stress and surface free energy in the equilibrium conditions at these interfaces. This leads to equilibrium conditions between two crystalline solids and between a crystal and its melt in which the surface stress influences both the mechanical force balance at the two-phase interface and the energy balance at the interface. We have also shown that dual reference states are needed to describe a greased interface, and that these dual reference states lead to a model of a greased interface in which the two phases are each in equilibrium with a thin fluid layer at the interface, and hence with each other.

Our purpose in deriving these equilibrium conditions is to better understand how coherency, surface stress and surface free energy can affect the kinetics of solid-state phase transformations. Specifically, we will employ these equilibrium conditions in nonequilibrium situations via the postulate of local equilibrium in order to find the local equilibrium concentrations at coherent and greased crystal–crystal interfaces. These concentrations can then be used as boundary conditions in models that describe the kinetics of solid-state precipitation processes. This approach will be the basis of future work in which we consider the effect of elastic fields on the morphological stability of a precipitate grown from solid solution.

Note added in proof—It has recently come to our attention that the partial derivative with respect to $\mathbf{n}'$ that we use in equations (22), (54), (73), (74) and (117) must be carefully defined to account for the fact that the domain of $\hat{\mathbf{F}}$ also depends on $\mathbf{n}'$. Gurtin and Struthers [23] have solved this problem by using a rotation tensor to expand the domain of $\hat{\mathbf{F}}$ and defining a derivative with respect to $\mathbf{n}'$ that follows the moving interface.

Acknowledgements—We gratefully acknowledge many interesting and useful discussions with M. E. Gurtin, W. C. Johnson and W. W. Mullins. This work was conducted in partial fulfillment of the Ph.D. requirements of P. H. Leo in the Department of Metallurgical Engineering and Materials Science at Carnegie-Mellon University. This work was supported by the National Science Foundation under Grant No. DMR-8409397/8645461.

REFERENCES

1. F. Larche and J. W. Cahn, *Acta metall.* **21**, 1051 (1973).
2. F. C. Larche and J. W. Cahn, *Acta metall.* **26**, 1579 (1978).
3. W. W. Mullins and R. F. Sekerka, *J. chem. Phys.* **82**, 5192 (1985).
4. J. W. Gibbs, *Scientific Papers, vol. 1: Thermodynamics*, Dover, New York (1961).
5. R. Shuttleworth, *Proc. Phys. Soc.* **A63**, 444 (1950).
6. C. Herring, in *The Physics of Power Metallurgy* (edited by W. E. Kingston). McGraw-Hill, New York (1951).
7. M. M. Nicholson, *Proc. R. Soc.* **A228**, 490 (1955).
8. J. W. Cahn, *Acta metall.* **28**, 1333 (1980).
9. W. W. Mullins, in *Proc. Int. Conf. on Solid–Solid Phase Trans.* (edited by H. I. Aaronson, D. E. Laughlin, R. F. Sekerka and C. M. Wayman), TMS-AIME, Warrendale, Pa (1981).
10. W. W. Mullins, *J. chem. Phys.* **81**, 1436 (1984).
11. J. W. Cahn and F. Larche, *Acta metall.* **30**, 51 (1982).
12. W. C. Johnson and J. I. D. Alexander, *J. appl. Phys.* **58**, 816 (1985).
13. W. C. Johnson and J. I. D. Alexander, *J. appl. Phys.* **59**, 2735 (1985).
14. D. W. Hoffman and J. W. Cahn, *Surf. Sci.* **31**, 368 (1972).
15. J. W. Cahn and D. W. Hoffman, *Acta metall.* **22**, 1205 (1974).
16. M. E. Gurtin and A. I. Murdoch, *Arch. Rat. Mech. Anal.* **30**, 225 (1975).
17. P. H. Leo, Ph.D. dissertation, Carnegie-Mellon Univ. (1987).
18. L. E. Malvern, *Introduction to the Mechanics of a Continuous Medium*. Prentice-Hall, Englewood Cliffs, N.J. (1969).
19. P. Y. Robin, *Am. Miner.* **59**, 1286 (1974).
20. F. C. Larche and J. W. Cahn, *Acta metall.* **33**, 331 (1985).
21. M. E. Gurtin, *An Introduction to Continuum Mechanics*. Academic Press, New York (1981).
22. C. E. Weatherburn, *Differential Geometry of Three Dimensions*. Cambridge Univ. Press (1955).
23. M. E. Gurtin and A. Struthers, private communication.

APPENDIX

Surface Tensors

In this appendix, we briefly discuss the notation and ideas that we use to define surface (or interfacial) tensors, and in particular the surface tensors $\hat{\mathbf{F}}$, $\hat{\mathbf{G}}$ and $\hat{\mathbf{I}}$ used in the analysis. The mathematical description of these tensor functions has been given by Gurtin and Murdoch (GM) [16], and we refer the interested reader to this work and to Weatherburn's book on differential geometry [22] for a more complete discussion of this material.

We begin by describing what we mean by a surface tensor. Consider a surface Σ, which can represent either the free surface of some solid or some two-phase interface. The position vector to this surface is given by $\mathbf{r}(u, v)$, where u and v are arbitrary parameters we choose to describe the surface. Then, at each point on Σ, we can construct a local coordinate system that consists of two unit vectors $\mathbf{r}_u = \partial\mathbf{r}/\partial u$ and $\mathbf{r}_v = \partial\mathbf{r}/\partial v$ tangent to the surface, and the unit normal $\mathbf{n}$ to the surface. The plane described by the two vectors $\mathbf{r}_u$ and $\mathbf{r}_v$ at each point on Σ defines the tangent space T_x to the surface Σ at that point. A tangent vector is any vector defined in this tangent space. We define a surface tensor to be any linear operator, with the appropriate tensor transformation properties, that maps the tangent vectors in T_x onto some new set of vectors; thus, a surface tensor is a function of position on Σ.

As an example of a surface tensor, consider the surface deformation gradient $\hat{\mathbf{F}}$. Physically, as defined by equation (16), $\hat{\mathbf{F}}$ maps vectors tangent to the reference interface Σ' onto the corresponding vectors tangent to the actual interface Σ, so by our definition is a surface tensor defined on Σ'. Mathematically, $\hat{\mathbf{F}}$ is a linear operator whose domain is the tangent space T'_x to the reference interface Σ' and whose range (or codomain) is the tangent space T_x to the actual interface Σ.

Considering $\hat{\mathbf{F}}$ to be a linear operator from the tangent space T'_x to the tangent space T_x is mathematically elegant in that it immediately reveals that $\hat{\mathbf{F}}$ is a mapping from the reference surface to the actual surface. However, it is a very inconvenient definition in performing the variational calculation because it requires that one define new tensors that shift between two- and three-dimensional spaces in order to account for whether vectors are defined in three dimensions or in some two-dimensional tangent space. We have therefore chosen to extend both the domain and range of $\hat{\mathbf{F}}$ to three dimensions. This approach results in no loss of mathematical or physical accuracy, and leads to the following identities:

Since $\hat{\mathbf{F}}$ operates only on vectors tangent to Σ', we extend its domain to three dimensions in such a way that

$$\hat{\mathbf{F}}\cdot\mathbf{n}'=0 \tag{A1}$$

so that for any vector $\mathbf{a}$, $\hat{\mathbf{F}}\cdot\mathbf{a}=\hat{\mathbf{F}}\cdot[\mathbf{a}-\mathbf{n}'(\mathbf{a}\cdot\mathbf{n}')]$ is the vector in the actual state that corresponds to the vector $\mathbf{a}-\mathbf{n}'(\mathbf{a}\cdot\mathbf{n}')$, which is tangent to the reference surface Σ'. Further, since we know that physically, $\hat{\mathbf{F}}$ maps from T'_x to T_x, it is clear that when we extend the range of $\hat{\mathbf{F}}$ to three dimensions

$$(\hat{\mathbf{F}}\cdot\mathbf{a})\cdot\mathbf{n}=0 \tag{A2}$$

so $\hat{\mathbf{F}}\cdot\mathbf{a}$ is tangent to the actual surface Σ.

These same ideas can be used to describe the tensors $\hat{\mathbf{G}}$ and $\hat{\mathbf{I}}$. Consider the tensor $\hat{\mathbf{G}}$ defined by equation (67). The tensor $\hat{\mathbf{G}}$ gives the mapping from vectors in the tangent space T_x of the actual surface to their counterparts in the tangent space T'_x to the reference surface, and so is the inverse of $\hat{\mathbf{F}}$. Thus, we extend the domain of $\hat{\mathbf{G}}$ to three dimensions in such a way that

$$\hat{\mathbf{G}}\cdot\mathbf{n}=0 \tag{A3}$$

so $\hat{\mathbf{G}}\cdot\mathbf{a}=\hat{\mathbf{G}}\cdot[\mathbf{a}-\mathbf{n}(\mathbf{a}\cdot\mathbf{n})]$ is the vector in the reference state that corresponds to the vector $\mathbf{a}-\mathbf{n}(\mathbf{a}\cdot\mathbf{n})$, which is tangent to the actual surface. Σ. Since this mapped vector lies tangent to the reference surface Σ', extending the range of $\hat{\mathbf{G}}$ to three dimensions leads to the result

$$(\hat{\mathbf{G}}\cdot\mathbf{a})\cdot\mathbf{n}'=0 \tag{A4}$$

Finally, consider the identity tensor $\hat{\mathbf{I}}$ defined on the actual surface Σ [see equations (94) and (95)]. This tensor has the property that for any vector $\mathbf{t}$ in the tangent space T_x, $\hat{\mathbf{I}}\cdot\mathbf{t}=\mathbf{t}$. Then, we can extend both the domain and range of $\hat{\mathbf{I}}$ to three dimensions without affecting the operation of $\hat{\mathbf{I}}$ by taking

$$\hat{\mathbf{I}}\cdot\mathbf{n}=0 \tag{A5}$$

such that

$$\hat{\mathbf{I}}\cdot\mathbf{a}=\hat{\mathbf{I}}\cdot[\mathbf{a}-\mathbf{n}(\mathbf{a}\cdot\mathbf{n})]=\mathbf{a}-\mathbf{n}(\mathbf{a}\cdot\mathbf{n}). \tag{A6}$$

It is useful to note in conclusion that the properties discussed above are given only to reconcile our approach to the tensors $\hat{\mathbf{F}}$, $\hat{\mathbf{G}}$ and $\hat{\mathbf{I}}$ with the more formal GM approach. That is, if we actually calculate $\hat{\mathbf{F}}$, $\hat{\mathbf{G}}$ or $\hat{\mathbf{I}}$ for some situation, then they would naturally satisfy equations (A1)–(A6). For example, if we consider the coordinate system $(\mathbf{r}_u, \mathbf{r}_v, \mathbf{n})$ discussed earlier and define the reciprocal vectors $\mathbf{r}_u^+$ and $\mathbf{r}_v^+$ to $\mathbf{r}_u$ and $\mathbf{r}_v$, respectively [22], then we can write $\hat{\mathbf{I}}$ as the dyad $\mathbf{r}_u\mathbf{r}_u^+ + \mathbf{r}_v\mathbf{r}_v^+$, which clearly satisfies equation (A5) and (A6). Similarly, if we use Weatherburn's definition of the surface gradient operator [22],

$$\nabla_\Sigma=\mathbf{r}_u^+\frac{\partial}{\partial u}+\mathbf{r}_v^+\frac{\partial}{\partial v} \tag{A7}$$

and explicitly calculate $\hat{\mathbf{F}}$ from equation (16), then the resulting dyad would naturally satisfy equations (A1) and (A2). Thus, the properties given by equations (A1)–(A6) are intrinsic to the physical tensors $\hat{\mathbf{F}}$, $\hat{\mathbf{G}}$ and $\hat{\mathbf{I}}$, and only serve to match our physical understanding of these tensors with their more abstract mathematical definitions.

Offprint from "Archive for Rational Mechanics and Analysis", Volume 108, Number 4, 1989, pp. 323–391

Multiphase Thermomechanics with Interfacial Structure 2. Evolution of an Isothermal Interface

SIGURD ANGENENT & MORTON E. GURTIN

1. Introduction

Paper 1 [1988g][1] of this series began an investigation whose goal is a thermomechanics of two-phase continua based on Gibbs's notion of a sharp phase-interface endowed with thermomechanical structure. In that paper a balance law, balance of capillary forces, was introduced and then applied in conjunction with suitable statements of the first two laws of thermodynamics; the chief results are thermodynamic restrictions on constitutive equations, exact and approximate free-boundary conditions at the interface, and a hierarchy of free-boundary problems. The simplest versions of these problems (the Mullins-Sekerka problems) are essentially the classical Stefan problem with the free-boundary condition $u = 0$ for the temperature replaced by the condition $u = hK$, where K is the curvature of the free-boundary and $h > 0$ is a material constant. This dependence on curvature renders the problem difficult, and apart from numerical studies involving linearization stability, there are almost no supporting theoretical results.

For perfect conductors the theory seems far more tractable;[2] there the temperature is constant, and the free-boundary problem reduces to a single set of evolution equations for the interface.

In this paper we develop further the theory of perfect conductors, but to avoid the severe geometric complications associated with the motion of surfaces in $\mathbf{R}^3$, we restrict our attention to interfaces that evolve as curves in $\mathbf{R}^2$. For any such interface, we write $T(\theta)$ for the unit tangent, $N(\theta)$ for the unit normal, and θ for the angle from a fixed coordinate axis to $N(\theta)$.

We begin with a fairly thorough description of the basic laws, which are

[1] See also [1986g, 1988gg].

[2] The theory of perfect conductors might be applicable to small interfaces, where bulk effects are small, or to interfaces of arbitrary size in superconductors such as solid helium in which heat flow is insignificant (*cf.* MARIS & ANDREEV [1987]). A mechanical theory of this type might also model the motion of grain boundaries (*cf.* ALLEN & CAHN [1979]).

balance of capillary forces and a mechanical version of the second law, and we derive corresponding thermodynamic restrictions on constitutive equations.[3] In particular, we show that the capillary force $C(\theta)$ must be related to the *interfacial energy*[4] $f(\theta)$ through the relation

$$C(\theta) = f(\theta)\, T(\theta) + f'(\theta)\, N(\theta). \tag{1.1}$$

Balance of capillary forces in conjunction with the thermodynamically reduced constitutive equations lead to an evolution equation which relates the normal velocity V to the curvature K; this relation has the form[5]

$$\beta(\theta)\, V = [f(\theta) + f''(\theta)]\, K - F, \tag{1.2}$$

with F the (constant) energy-difference between bulk phases, and $\beta(\theta) > 0$ a kinetic coefficient which measures the drag opposing interfacial motion. The relation (1.2), when combined with purely kinematical conditions for an evolving curve and applied on a convex section of the interface, results in a single partial differential equation for the velocity $V = V(\theta, t)$:

$$\Phi(\theta)\, V_t = [V + \Psi(\theta)]^2\, [V_{\theta\theta} + V], \tag{1.3}$$

where

$$\Phi(\theta) = [f(\theta) + f''(\theta)]/\beta(\theta), \qquad \Psi(\theta) = F/\beta(\theta).$$

For $\Phi(\theta) > 0$ this equation is parabolic[6] and yields a theory which seems quite similar in structure to its isotropic counterpart based on $V = K - F$. There is, however, no compelling physical reason to exclude energies $f(\theta)$ for which $f(\theta) + f''(\theta) < 0$ over certain ranges of the angle θ;[7] for such ranges the equation (1.3) is backward-parabolic and corresponding evolution problems are generally not well posed. We show that a necessary condition for the *statical* stability of the interface is that $f(\theta) + f''(\theta) \geqq 0$, and for that reason use the terms *strictly stable*, *stable* or *unstable* according as $f(\theta) + f''(\theta) > 0$, $f(\theta) + f''(\theta) \geqq 0$, or $f(\theta) + f''(\theta) < 0$.

We begin our analysis of (1.3) by restricting attention to interfacial energies that are *strictly stable*. We deduce steady solutions of (1.3) for which the interface is convex and infinite, in the shape of a bump. The bump recedes in one solution and advances in the other; for the receding bump the kinetic coefficient can be arbitrary, but the advancing bump requires a nonconvex polar diagram for $\beta(\theta)$.

---

[3] The underlying proofs are more transparent in $\mathbb{R}^2$ than $\mathbb{R}^3$, and for that reason we rederive many results which could simply be taken from [1988g].

[4] We use the term energy in a generic sense; as to what thermodynamic potential the energy actually represents depends on what thermodynamic theory this purely mechanical theory is meant to "approximate".

[5] There is a large and growing literature concerning the evolution equation $V = K$ (*cf.*, *e.g.*, BRAKKE [1978], GAGE [1986], GAGE & HAMILTON [1986], GRAYSON [1987], and ANGENENT [1988]).

[6] Aside from the trivial degeneracy ($V = -\Psi(\theta)$) which occurs at inflection points.

[7] Material scientists often consider such models (*cf.*, *e.g.*, GJOSTEIN [1963], CAHN & HOFFMAN [1974]).

We next analyze the global behavior of a *smooth* interface as measured by its perimeter $L(t)$ and enclosed area $A(t)$. Our main result, based on the assumption of a stable interfacial energy, is most easily stated in terms of a bounded solid in an infinite liquid bath.

$$\left.\begin{array}{l}\text{If the bath is } \textit{not} \text{ supercooled, then } A(t)\to 0; \text{ if the bath is}\\ \text{supercooled, then initially small interfaces have } L(t)\to 0,\\ \text{initially large interfaces have } A(t)\to\infty.\end{array}\right\}\quad(1.4)$$

We show that, for the case in which $A(t)\to\infty$, the isoperimetric ratio $L(t)^2/4\pi A(t)$ remains bounded as $t\to\infty$. We show further that if (for a nonconvex interface) one defines a finger as a section of the interface between inflection points, then the total number of fingers as well as the total curvature of each finger cannot increase with time. These results presume the existence of a smooth, simple (non self-intersecting) interface. In this regard, it is clear that in certain circumstances the interface can pierce itself as it evolves.

We next consider energies $f(\theta)$ which are unstable for certain values of θ. Here we find it convenient to introduce a global definition of stability which is based on ideas of WULFF [1901], HERRING [1951b], and FRANK [1963]. We define global stability in terms of the convexity of the Frank diagram, which is the polar diagram of the reciprocal function $f(\theta)^{-1}$; we refer to the convex sections of this diagram as the globally-stable (GS) sections, to the remaining sections as the globally-unstable sections. These definitions are consistent: $f(\theta)$ is stable on GS sections; $f(\theta)$ is unstable somewhere within each globally-unstable section.

One way of treating unstable energies is to allow the interface to be *nonsmooth* with corners which correspond to jumps in θ across the globally-unstable sections. Balance of capillary forces for corresponding "weak solutions" of the evolution equations leads to the requirement that $C(\theta)$ be continuous across each such corner; interestingly, this requirement is automatically met.

In contrast to standard results for a strictly stable energy, the presence of corners leads to the possibility of *facets* (flat sections); in fact, to the presence of wrinklings, where a wrinkling is a series of facets with normals that oscillate between two fixed values. We show that such wrinklings are dynamically stable: the lengths of the individual facets do not increase with time.

The use of corners leads to *free-boundary* problems for the evolution of the interface, as the positions of the corners are not generally known *a priori*. We discuss these problems in some detail.

Material scientists often consider interfacial energies that are continuous but have derivatives which suffer jump discontinuities.[8] We study such interfaces; as before, we use corners to remove the globally instable sections. We show that, in agreement with statical results,[9] discontinuities in $f'(\theta)$ lead to facets in the evolving interface. We show further that the result (1.4) remains valid for nonsmooth, nonstable energies.

Following TAYLOR's [1978] statical treatment of crystal shapes, we consider a particular class of nonsmooth energies, called *crystalline*, for which the GS

[8] *Cf., e.g.*, HERRING [1951ab], CAHN & HOFFMAN [1974].

[9] *Cf., e.g.*, TAYLOR [1978].

sections are isolated points (that is, for which the Frank diagram touches the boundary of its convex hull only at discrete points). An interesting property of crystalline energies is that their evolution is governed by a system of *ordinary differential equations* involving only nearest-neighbor interactions. We solve these equations for a rectangular crystal; the corresponding solution shows that, in situations for which the crystal shrinks (*cf.* (1.4)), the corresponding *isoperimetric ratio* generally *tends to infinity*. This is in sharp contrast to an isotropic interface, which shrinks to a round point.[10]

I. The thermomechanics of evolving curves

2. Kinematics

This chapter discusses the kinematics of smooth curves which evolve smoothly in time, and forms the basis of our theory of the motion of phase interfaces in $\mathbb{R}^2$.

2.1. Curves

A **curve** is a smooth map $p \mapsto \boldsymbol{r}(p)$ from an interval of $\mathbb{R}$ into $\mathbb{R}^2$ such that:

(i) $\boldsymbol{r}_p$ never vanishes;
(ii) the domain of $\boldsymbol{r}$ is either all of $\mathbb{R}$ or a bounded interval $[P, Q]$;
(iii) if the domain is $\mathbb{R}$, either $\boldsymbol{r}$ is periodic[11] or

$$|\boldsymbol{r}(p)| \to \infty \quad \text{as } |p| \to \infty.$$

The set Range ($\boldsymbol{r}$) is then called the **trace** of $\boldsymbol{r}$.

We will classify curves $\boldsymbol{r}$ as follows: $\boldsymbol{r}$ is **bounded** or **unbounded** according as its trace is bounded or unbounded; $\boldsymbol{r}$ is **closed** if the domain is $\mathbb{R}$ and $\boldsymbol{r}$ is *periodic*, $\boldsymbol{r}$ has **endpoints** if its domain is a bounded interval. A nonclosed curve is **simple** if it is one-to-one; a closed curve is **simple** if given any $p, q \in \mathbb{R}$, $\boldsymbol{r}(p) = \boldsymbol{r}(q)$ only when $p - q$ is a multiple of the minimal period of $\boldsymbol{r}$.

Let $\boldsymbol{r}$ be a curve. An **arc-length map** for $\boldsymbol{r}$ is a smooth mapping $s(p)$ from the domain of $\boldsymbol{r}$ into $\mathbb{R}$ such that

$$s_p = |\boldsymbol{r}_p|. \tag{2.1}$$

We assume henceforth that an arc-length map is prescribed. Since the **arc length** $s = s(p)$ is an invertible function of p, any function $\phi(p)$ may be considered a function $\phi(s)$, and *vice versa*.

The vector

$$T(s) = \boldsymbol{r}_s(s) \tag{2.2}$$

[10] GAGE [1984], GAGE & HAMILTON [1986], GRAYSON [1987].

[11] A function ϕ on $\mathbb{R}$ is **periodic** if there is a $\lambda > 0$ such that $\phi(p) = \phi(p + \lambda)$ for all $p \in \mathbb{R}$; λ is then a **period** of ϕ and the infimum of all periods is the **minimal period** of ϕ. (The minimal period of a curve $\boldsymbol{r}$ is strictly positive since $|\boldsymbol{r}_p| \neq 0$.)

defines a (unit) **tangent** to the curve in the direction of increasing p. We define a corresponding (unit) **normal** $N(s)$ through the requirement that $\{T, N\}$ be a positively-oriented orthonormal basis of $\mathbf{R}^2$, and we define the **angle** $\theta(s)$, as a smooth function of s, through[12]

$$N = (\cos\theta, \sin\theta), \qquad T = (\sin\theta, -\cos\theta). \tag{2.3}$$

We will refer to the range of the function $s \mapsto \theta(s)$ as the **angle range** (Figure 2A). Note that N and T may be considered as functions of θ, in which case

$$N_\theta = -T, \qquad T_\theta = N. \tag{2.4}$$

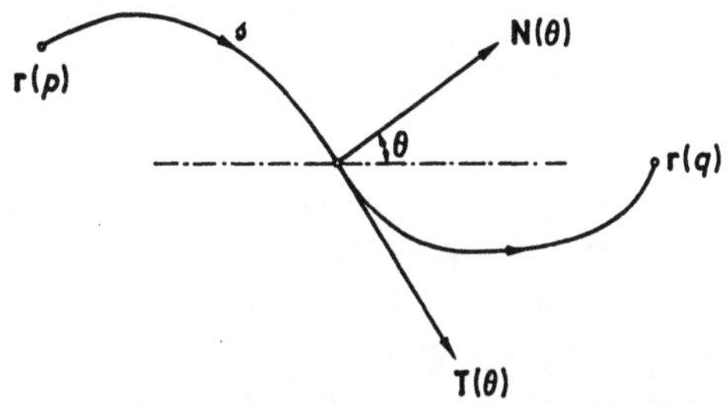

Fig. 2A. Sign conventions for curves

The function

$$K(s) = \theta_s(s) \tag{2.5}$$

is the **curvature**; by (2.4), $K(s)$ obeys the Frenet formulas:

$$N_s = -KT, \qquad T_s = KN. \tag{2.6}$$

Let r be a curve with trace $\mathscr{c}$ and normal N. Then r is a **boundary curve if** r is simple and either closed or unbounded. By the Jordan-curve theorem, $\mathscr{c}$ then divides $\mathbf{R}^2$ into two regions,[13] and one of these regions, Ω, say, will have N as *outward* normal; we will refer to Ω as the **reference region.**

A curve is **convex** if K never vanishes; in view of (2.5), the mapping $s \mapsto \theta(s)$ is then invertible and we may use θ in place of s or p as independent variable. In particular, we may parametrize the curve itself by θ, giving a function $r(\theta)$; granted this, (2.2) and (2.5) yield

$$r_\theta = K^{-1}T. \tag{2.7}$$

Note that, because of our sign convention for curvature, $K < 0$ for a boundary curve whose reference region is bounded and strictly convex.

Useful in the study of convex curves is the **support function**

$$h = r \cdot N; \tag{2.8}$$

[12] This defines the function $\theta(s)$ up to a multiple of 2π.

[13] We use the term **region** as a synonym for connected open set.

by (2.4) and (2.7),

$$r = \not{h}N - \not{h}_\theta T, \quad \not{h}_{\theta\theta} + \not{h} = -K^{-1}. \tag{2.9}$$

We now give three useful lemmas concerning convex curves.

Lemma 1. *Two convex curves with the same angle range and the same curvature are equal modulo a translation.*

Proof. By $(2.9)_2$, the difference between the support functions of the two curves $r_1(\theta)$ and $r_2(\theta)$ must have the form $a \cdot N(\theta)$, and this with $(2.9)_1$ implies that $r_1(\theta) - r_2(\theta) = a$. □

Lemma 2. *Consider a curve whose curvature is not identically zero, and suppose that the curvature has the form $K(\theta(s))$ with $K(\theta)$ a smooth function on the angle range. Then the curve is convex.*

Proof. Let Υ denote the angle range, and let Γ be a connected component of the set $\{\theta \in \Upsilon : K(\theta) \neq 0\}$. We must show that $\Gamma = \Upsilon$. Assume that $\Gamma \neq \Upsilon$. Then there is a boundary point θ_0 of Γ in $\mathbb{R}$ with $\theta_0 \in \Upsilon$ and $K(\theta_0) = 0$. Since $K(\theta)$ is smooth up to θ_0, $|K(\theta)| \leqq C|\theta - \theta_0|$ near θ_0. But on Γ, $s = s(\theta)$ and $s_\theta(\theta) = K(\theta)^{-1}$, so that $|s(\theta)| \to \infty$ as $\theta \to \theta_0$, a contradiction. □

Lemma 3. *Let $K(\theta)$ be a smooth, 2π-periodic function on $\mathbb{R}$. Then the restriction of $K(\theta)$ to an open interval $\mathcal{O}$ is the curvature of a convex boundary curve if and only if either* (a) *or* (b) *is satisfied:*

(a) *$\mathcal{O} = \mathbb{R}$, $K(\theta)$ is nonvanishing, and*

$$\int_0^{2\pi} K(\theta)^{-1} e^{i\theta}\, d\theta = 0; \tag{2.10}$$

(b) *$\mathcal{O}$ is a bounded interval (θ_1, θ_2) with $\theta_1 - \theta_2 \leqq \pi$; $K(\theta)$ is nonvanishing on (θ_1, θ_2); $K(\theta_1) = K(\theta_2) = 0$.*

In case (a) *the boundary curve is closed; in case* (b) *the boundary curve is unbounded. In either case a boundary curve with angle range $\mathcal{O}$ and curvature $K(\theta)$ is generated by* $(2.9)_1$ *with $\not{h}$ any solution of* $(2.9)_2$.

Proof. Note first that if $g(\theta)$ is a smooth 2π-periodic function on $\mathbb{R}$, then

$$\int_0^{2\pi} [g(\theta) + g''(\theta)]\, e^{i\theta}\, d\theta = 0, \tag{2.11}$$

an assertion which follows immediately upon integrating the term $g''(\theta)\, e^{i\theta}$ twice by parts.

Suppose that the restriction of $K(\theta)$ to an open interval $\mathcal{O}$ is the curvature of a convex boundary curve r. Then r is simple and either closed or unbounded. Assume that r is closed. Then $\mathcal{O} = \mathbb{R}$ and (2.10) is a consequence of (2.11) and $(2.9)_2$. Thus (a) is satisfied.

Assume that r is unbounded. Then $|r(p)| \to \infty$ as $p \to \pm\infty$. so that

$$|s(p)| \to \infty \quad \text{as } p \to \pm\infty. \tag{2.12}$$

Further, since r is simple, $\mathcal{O}$ is a bounded interval (θ_1, θ_2) with $\theta_1 - \theta_2 \leqq \pi$. Trivially, $K(\theta)$ is nonzero on (θ_1, θ_2). Further, $s_\theta(\theta) = K(\theta)^{-1}$ and, by (2.12), $|s(\theta)| \to \infty$ as $\theta \to \theta_1, \theta_2$; thus, since $K(\theta)$ is smooth up to θ_1 and θ_2, $K(\theta_1) = K(\theta_2) = 0$. Therefore (b) is satisfied.

Conversely, suppose that $\mathcal{O}$ and $K(\theta)$ satisfy either (a) or (b). Let $\not{h}(\theta)$ be any solution of $(2.9)_2$ on $\mathcal{O}$ and define $r(\theta)$ by $(2.9)_1$.

Case (a). By (2.10), $r(\theta)$ is 2π-periodic and defines a closed curve parametrized by θ. Further, K is the curvature of r, so that r is convex. Moreover, θ is the angle and $r(0) = r(2\pi)$, so that r is simple. Thus r is a convex boundary curve.

Case (b). Let $s(\theta)$ be any solution of $s_\theta(\theta) = K(\theta)^{-1}$ on $\mathcal{O}$. Since $K(\theta)$ is smooth up to θ_1 and θ_2, $|K(\theta)| \leqq C|\theta - \theta_0|$ for all θ near θ_1 and θ_2. Thus, since $K(\theta_1) = K(\theta_2) = 0$, $|s(\theta)| \to \infty$ as $\theta \to \theta_1, \theta_2$ and, by $(2.9)_1$, $|r(\theta)| \to \infty$ as $\theta \to \theta_1, \theta_2$. The mapping $\theta \mapsto s(\theta)$ is therefore a smooth bijection of (θ_1, θ_2) onto $\mathbf{R}$. If we write $\theta = \theta(s)$, $|r(s)| \to \infty$ as $|s| \to \infty$; hence $r(s)$ is a convex, unbounded curve, parametrized by arc length, with $\theta(s)$ the angle function and $K(\theta)$ the curvature. Since $\theta_1 - \theta_2 \leqq \pi$, r is simple; hence r is a convex boundary curve.

The final assertions of the lemma are clear from the preceding analysis. □

2.2. Evolving curves

Roughly speaking, an evolving curve is a smooth family of curves $p \mapsto r(p, t)$, where t, the **time**, ranges in a half-open interval $[0, T)$, called the **duration** of r. We will use evolving curves to model the motion of phase interfaces in $\mathbf{R}^2$. For a given motion r, the underlying physics must be independent of the choice of **parameter** p, and hence can involve r only through *intrinsic* quantities such as curvature and normal velocity, which are independent of parametrization. On the other hand, this invariance allows us to use any parametrization we wish. In fact, we shall restrict our attention to parametrizations with $r_t(p, t) \perp r_p(p, t)$; such parametrizations greatly simplify the analysis, chiefly because the *velocity* $r_t(p, t)$ is equal to the normal time-derivative $r^\circ(p, t)$, an intrinsic quantity.

Precisely, an **evolving curve**[14] is a smooth mapping $(p, t) \mapsto r(p, t)$ with the following properties:

(i) the domain of r is either $\mathbf{R} \times [0, T)$ or a set of the form

$$\{(p, t): p \in [P(t), Q(t)],\ t \in [0, T)\}, \tag{2.13}$$

where $P, Q: [0, T) \to \mathbf{R}$ $(P < Q)$ are smooth functions;

(ii) $r(\cdot, t)$ is a *curve* for each $t \in [0, T)$;

(iii) $r_t(p, t) \perp r_p(p, t)$ for all (p, t) **(orthogonality)**.

[14] We use the term "normally evolving curve" in Appendix B.

We will refer to $r(P(t), t)$ and $r(Q(t), t)$ as the **initial** and **terminal points** (or collectively, as the **endpoints**) of r, and to the interval $[P(t), Q(t)]$ (or $\mathbb{R}$) as the **parameter interval** at time t.

Let r be an evolving curve: r is bounded, unbounded, closed, simple, convex, or has endpoints, according as $r(\cdot, t)$ has that property for each $t \in [0, T)$; a *restriction* r_0 of r is an **evolving subcurve** of r if modulo a translation of time, r_0 is a *bounded* evolving curve; r is an **evolving facet** if its trace $\mathscr{S}(t)$ is a segment of a straight line at each t.

Let an evolving curve r be given. An **arc-length map** for r is a smooth mapping $s(p, t)$ such that $s(\cdot, t)$ is an arc-length map for the curve $r(\cdot, t)$ at each t. It is not difficult to construct an arc-length map for r, and any two such maps differ by a smooth function of time. We assume henceforth that an arc-length map is prescribed. Since $s = s(p, t)$ is an invertible function of p, any function $\phi(p, t)$ may be considered a function $\phi(s, t)$, and *vice-versa*. We will refer to $\phi(s, t)$ as the **arc-length description** of ϕ.

We write[15] ϕ° for the **normal time-derivative** of ϕ, the time derivative holding p fixed. In particular, we define the **normal velocity** $V(s, t)$ through the identity

$$r^\circ = VN, \tag{2.14}$$

the **arc velocity** $v(s, t)$ through

$$v = s^\circ. \tag{2.15}$$

Given a function $\phi(s, t)$,

$$\phi^\circ = \phi_t + v\phi_s, \tag{2.16}$$

with ϕ_t the time derivative holding s fixed; thus

$$(\phi^\circ)_s = (\phi_t + v\phi_s)_s = \phi_{st} + v\phi_{ss} + v_s\phi_s = (\phi_s)^\circ + v_s\phi_s. \tag{2.17}$$

Transport identities.

$$v_s = -KV, \quad \theta^\circ = V_s, \quad K^\circ = V_{ss} + K^2V. \tag{2.18}$$

Proof. By (2.1) and (2.2), $T = |r_p|^{-1} r_p$. Thus $J = s_p = |r_p|$ satisfies

$$J^\circ = |r_p|^{-1} r_p \cdot (r_p)^\circ = T \cdot (VN)_s J = T \cdot (N_s) VJ,$$

and, in view of (2.6), $J^\circ = -JKV$. On the other hand, (2.15) yields $J^\circ = v_p = v_sJ$ and $(2.18)_1$ follows.

Let e be a fixed unit vector. Then, by (2.2), (2.3), and (2.17),

$$(r_s)^\circ \cdot e = (T \cdot e)^\circ = (N \cdot e)\,\theta^\circ,$$

$$(r_s)^\circ \cdot e = (r^\circ)_s \cdot e - v_sT \cdot e,$$

while (2.6), (2.14), and $(2.18)_1$ imply

$$(r^\circ)_s = V_sN + v_sT.$$

---

[15] *Cf.* Appendix B. We write $\phi^{\cdot}(t)$ for the derivative of a function $\phi(t)$ of time alone.

The last three relations yield $(2.18)_2$, since e is arbitrary (*cf.* (2.16)). Finally, $(2.18)_3$ follows from (2.17) with $\phi = \theta$, (2.5), and $(2.18)_{1,2}$. □

Note that, trivially, for an evolving facet (*cf.* (2.5) and $(2.18)_1$),

$$K = \theta_s = 0, \quad v_s = 0. \tag{2.19}$$

An endpoint $R(t) = r(P(t), t)$ is a **normal trajectory** if $R^{\cdot}(t) \cdot T(P(t), t) = 0$. Since $R^{\cdot}(t) = r_p(P(t), t)\, P^{\cdot}(t) + r^{\circ}(P(t), t)$, $r(P(t), t)$ is a normal trajectory if and only if P is independent of t.

Proposition. *Let r have endpoints, and let $S(t)$ denote the arc length and $\Theta(t)$ the angle of an endpoint $R(t)$. Then*

$$\begin{aligned} R^{\cdot}(t) &= V(S(t), t)\, N(S(t), t) + [S^{\cdot}(t) - v(S(t), t)]\, T(S(t), t), \\ \Theta^{\cdot}(t) &= V_s(S(t), t) + [S^{\cdot}(t) - v(S(t), t)]\, K(S(t), t), \end{aligned} \tag{2.20}$$

and, if the endpoint $R(t)$ is a normal trajectory,

$$S^{\cdot}(t) = v(S(t), t), \quad \Theta^{\cdot}(t) = V_s(S(t), t). \tag{2.21}$$

Proof. Since $R(t) = r(S(t), t), \Theta(t) = \theta(S(t), t)$, the identities (2.20) follow from (2.16), (2.14), (2.5), and $(2.18)_2$. If $R(t)$ is a normal trajectory, then $R(t) = r(P, t)$ with P constant; thus, since $S(t) = s(P, t)$, $(2.21)_1$ follows from (2.15), and, in view of $(2.20)_2$, this yields $(2.21)_2$. □

For a *convex* evolving curve the mapping $s \mapsto \theta(s, t)$ is invertible and we may use θ and t in place of s and t as independent variables. Then

$$K^{\circ} = K_t + K_\theta \theta^{\circ} \tag{2.22}$$

(with K_t the derivative of K with respect to t holding θ fixed).

Proposition. *For a convex evolving curve with curvature, normal velocity, and arc velocity expressed as functions of (θ, t),*

$$K_t = K^2(V_{\theta\theta} + V), \quad v_\theta = -V. \tag{2.23}$$

In addition, if r has endpoints, and if the angle θ at an endpoint $R(t)$ has the constant value θ_0, then

$$R^{\cdot}(t) = V(\theta_0, t)\, N(\theta_0) - V_\theta(\theta_0, t)\, T(\theta_0). \tag{2.24}$$

Proof. Clearly,

$$V_s = V_\theta K, \quad V_{ss} = V_{\theta\theta} K^2 + V_\theta K_\theta K,$$

and these relations, (2.22), and $(2.18)_{2,3}$ yield $(2.23)_1$. On the other hand, $v_s = v_\theta K$ and $(2.23)_2$ follows from $(2.18)_1$. Finally, (2.20) with $\Theta^{\cdot} = 0$ and $V_s = V_\theta K$ imply (2.24). □

The next definition will be useful in discussing evolving curves that represent interfaces between phases. An **interfacial motion** is an evolving curve r with $r(\cdot, t)$ a *boundary curve* at each t. The trace $\mathscr{A}(t)$ of r then divides $\mathbb{R}^2$ into two regions. The region $\Omega(t)$ with $N(s, t)$ as *outward* normal is called the **reference region**; without loss in generality, *$\Omega(t)$ is taken to be the bounded region interior to $\mathscr{A}(t)$ when r is closed* (Figure 2B).

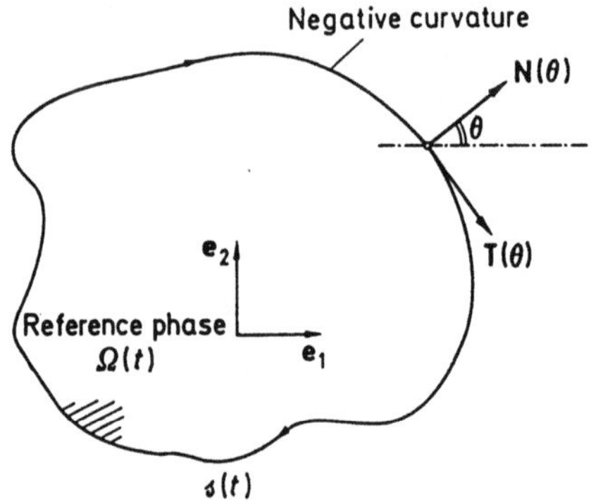

Fig. 2B. Sign conventions for interfacial motions. $\Omega(t)$, the region occupied by the reference phase, has N as outward unit normal and is the region interior to the trace $\mathscr{A}(t)$

By a **steady motion** we mean an *interfacial motion* r with the following property: there is a vector $U \neq 0$ such that, for some choice of arc-length map,

$$r(s, t) = r_0(s) + tU; \tag{2.25}$$

U is then the **steady velocity**, the curve r_0 the **portrait**.

Proposition. *Given a boundary curve r_0 and a vector $U \neq 0$, there is a unique*[16] *steady motion r with r_0 as portrait and U as steady velocity.*

Proof. Assume that $r_0(s)$ is parametrized by arc length and let $(s, t) \mapsto k(s, t)$ $(\mathbb{R} \times [0, \infty) \to \mathbb{R}^2)$ be the time-dependent curve defined by (2.25). Then $v(s, t)$ as defined in $(B3)_2$ of Appendix B, is independent of t and given by

$$v(s) = -U \cdot T(s), \tag{2.26}$$

with $T(s)$ the tangent for $r_0(s)$. Thus k obeys hypothesis (iii) of Theorem 1 (Appendix B) and there is a reparametrization ϕ of k such that $r = k \circ \phi$ is an evolving curve. Writing $\phi(p, t) = (\phi(p, t), t)$, and differentiating $r(p, t) = k(\phi(p, t), t)$ with respect to p yields the conclusion that $\phi(p, t)$ is an **arc-length map** for r. Thus $k(s, t)$ is an arc-length description of $r(p, t)$, which is the desired conclusion. If g is a second steady motion with k as an arc-length description, then, trivially, g must be a reparametrization of r. □

[16] The term "*unique*", when used relative to evolving curves, will always signify "*unique up to a reparametrization*".

As is clear from the proof of Theorem 1 (Appendix B), a parameter change $\phi(p, t) = (s(p, t), t)$ that converts the arc-length description (2.25) to a description $r(p, t)$ as an evolving curve has $s(p, t)$ a solution of the initial-value problem

$$s^\circ(p, t) = v(s(p, t)), \qquad s(p, 0) = p; \tag{2.27}$$

in this case $v(s)$, given by (2.26), is the arc velocity and the parameter p is the initial arc-length.

Proposition. *For a steady motion the normal velocity $V(s)$, the curvature $K(s)$, the normal $N(s)$, the tangent $T(s)$, and the angle $\theta(s)$ are independent of time and,*

$$V(s) = \mathbf{U} \cdot \mathbf{N}(s). \tag{2.28}$$

Proof. By (2.25), $T(s)$ and (hence) $N(s)$ and $\theta(s)$ are independent of t. Further, since $VN = r^\circ = r_t + vr_s$ and $T = r_s$, $V(s)$ is independent of t and given by (2.28). Finally, since $K = T_s \cdot N$ (*cf.* (2.6)), $K(s)$ is also independent of t. □

For a convex, steady motion, $V(\theta)$ and $K(\theta)$ are independent of time, and

$$V(\theta) = \mathbf{U} \cdot \mathbf{N}(\theta). \tag{2.29}$$

By a **steadily evolving bump** we mean a convex, steady motion such that

$$K(\theta)\, \mathbf{U} \cdot \mathbf{N}(\theta) \quad \text{never vanishes.} \tag{2.30}$$

A steadily evolving bump is **advancing** or **receding** according as

$$K(\theta)\, \mathbf{U} \cdot \mathbf{N}(\theta) < 0 \quad \text{or} \quad K(\theta)\, \mathbf{U} \cdot \mathbf{N}(\theta) > 0 \tag{2.31}$$

(Figure 2C). Note that steadily evolving bumps are necessarily *unbounded.*

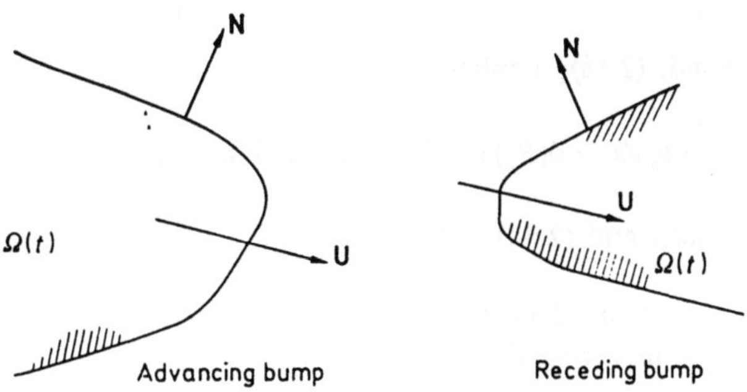

Fig. 2C. Steadily evolving bumps. The steady velocity is given by $\mathbf{U}$

Finally, we note that a trivial example of a steady motion is a **steadily evolving facet**; such motions are completely determined by $\mathbf{U}$ and the corresponding angle θ (≡ constant).

2.3. *Integral identities*

Let r be an evolving curve with arc length lying in an interval $[S_1(t), S_2(t)]$. We use the notation

$$\int_{\sigma(t)} \phi\, ds = \int_{S_1(t)}^{S_2(t)} \phi(s,t)\, ds, \qquad \int_{\partial\sigma(t)} \phi = \phi(S_2(t),t) - \phi(S_1(t),t);$$

trivially,

$$\int_{\partial\sigma(t)} \phi = \int_{\sigma(t)} \phi_s\, ds.$$

The intrinsic measure of length on the curve is arc length. In general, the endpoints of r will not be normal trajectories; hence $\sigma(t)$ will lose arc length across its boundary at a rate given by $\{v(S_2(t),t) - S_2^{\cdot}(t)\} - \{v(S_1(t),t) - S_1^{\cdot}(t)\}$. Thus, for $\phi(s,t)$ a smooth function

$$\begin{aligned} \text{outflow}\,(\phi, \partial\sigma)\,(t) := {} & \phi(S_2(t),t)\,\{v(S_2(t),t) - S_2^{\cdot}(t)\} \\ & - \phi(S_1(t),t)\,\{v(S_1(t),t) - S_1^{\cdot}(t)\} \end{aligned} \tag{2.32}$$

represents the rate at which ϕ is *carried out* of $\sigma(t)$ across $\partial\sigma(t)$ with this loss in arc length; note that, by (2.21),

$$\left.\begin{array}{l} \text{outflow}\,(\phi, \partial\sigma)\,(t) = 0 \text{ when the} \\ \text{endpoints are normal trajectories.} \end{array}\right\} \tag{2.33}$$

Transport theorem for integrals.

$$\left(\frac{d}{dt}\right) \int_{\sigma(t)} \phi\, ds + \text{outflow}\,(\phi, \partial\sigma)\,(t) = \int_{\sigma(t)} (\phi^{\circ} - \phi\, KV)\, ds. \tag{2.34}$$

Proof. By (2.16) (suppressing the argument t where convenient),

$$\left(\frac{d}{dt}\right) \int_{S_1(t)}^{S_2(t)} \phi(s,t)\, ds = S_2^{\cdot}\, \phi(S_2) - S_1^{\cdot}\, \phi(S_1) + \int_{S_1}^{S_2} (\phi^{\circ} - v\phi_s)\, ds.$$

On the other hand, $(2.18)_1$ implies

$$\int_{S_1}^{S_2} v\phi_s\, ds = \phi(S_2)\, v(S_2) - \phi(S_1)\, v(S_1) + \int_{S_1}^{S_2} \phi KV\, ds.$$

The last two results and (2.32) yield (2.34). □

If we take $\phi = 1$ in (2.34) and appeal to (2.20) and (2.32), we arrive at an important identity involving the **length**

$$L(t) := \text{length}\,(\sigma(t)).$$

Transport theorem for length.

$$L^{\cdot}(t) = R_2^{\cdot}(t) \cdot T_2(t) - R_1^{\cdot}(t) \cdot T_1(t) - \int_{\sigma(t)} KV\, ds \tag{2.35}$$

if $\boldsymbol{r}$ *has initial and terminal points* $\boldsymbol{R}_1(t)$ *and* $\boldsymbol{R}_2(t)$ *with corresponding tangents* $\boldsymbol{T}_1(t)$ *and* $\boldsymbol{T}_2(t)$;

$$L^{\cdot}(t) = -\int_{\mathscr{o}(t)} KV\,ds \tag{2.36}$$

if $\boldsymbol{r}$ *is closed, or if its endpoints are normal trajectories.*

2.4. Piecewise-smooth evolving curves

We now extend some of the previous definitions and results to curves that are continuous but not smooth. For convenience, we write **"PS"** as an abbreviation for **"piecewise smooth"**.

Let $\boldsymbol{r} = \{\boldsymbol{r}_1, \boldsymbol{r}_2, \ldots\}$ denote a finite or countably infinite list of evolving curves $\boldsymbol{r}_i$, called **arcs of** $\boldsymbol{r}$, of equal duration $[0, T)$, with $[P_i(t), Q_i(t)]$ the parameter interval for $\boldsymbol{r}_i$ at time t. Then $\boldsymbol{r}$ is a **PS evolving curve** if:

(i) at each **juncture** i (*i.e.*, each i with $\boldsymbol{r}_i$ and $\boldsymbol{r}_{i+1}$ arcs of $\boldsymbol{r}$),

$$\boldsymbol{r}_i(Q_i(t), t) = \boldsymbol{r}_{i+1}(P_{i+1}(t), t) =: \boldsymbol{R}_i(t); \tag{2.37}$$

(ii) there is an integer $N > 0$ such that either

(a) $\boldsymbol{r}$ consists of N arcs, in which case $\boldsymbol{r}_1$ and $\boldsymbol{r}_N$ are **terminal arcs**, and $\boldsymbol{r}$ has **endpoints** $\boldsymbol{r}_{\text{init}}(t) = \boldsymbol{r}_1(P_1(t), t)$ and $\boldsymbol{r}_{\text{term}}(t) = \boldsymbol{r}_N(Q_N(t), t)$; or

(b) $\boldsymbol{r}$ consists of an infinite number of arcs, but $\boldsymbol{r}_i = \boldsymbol{r}_{i+N}$ for all i; in this case $\boldsymbol{r}$ is **closed**, and the smallest such integer N is the **essential number of arcs** of $\boldsymbol{r}$.

In addition: $\boldsymbol{r}_i$ is an **internal arc** if both i and $i-1$ are junctures; $\boldsymbol{R}_i(t)$ and $\boldsymbol{R}_i^{\cdot}(t)$ are the **position** and **total velocity of the juncture** i; $\mathscr{o}(t)$, the **trace** of $\boldsymbol{r}$, is the union of the individual traces

$$\mathscr{o}_i(t) = \{\boldsymbol{r}_i(p, t) \colon p \in [P_i(t), Q_i(t)]\}.$$

An **evolving subcurve** $\boldsymbol{r}_0$ of $\boldsymbol{r}$ is defined in the obvious manner, as is the phrase **"$\boldsymbol{r}$ is simple"**.

Let $\boldsymbol{r}$ be a PS evolving curve. An **arc-length map** for $\boldsymbol{r}$ is a list $\{s_1, s_2, \ldots\}$ with $s_i(p, t)$ an arc-length map for $\boldsymbol{r}_i$ and

$$s_i(Q_i(t), t) = s_{i+1}(P_{i+1}(t), t) =: S_i(t) \tag{2.38}$$

at each juncture i. It is not difficult to construct an arc-length map for $\boldsymbol{r}$; granted one is prescribed, we can define arc length s at time t by $s = s_i(p, t)$ for any i and $p \in [P_i(t), Q_i(t)]$. This allows us to consider the tangent, normal, orientation, curvature, normal velocity, and arc velocity as functions $T(s, t)$, $N(s, t)$, $\theta(s, t)$, $K(s, t)$, $V(s, t)$, and $v(s, t)$ of arc length and time. These functions will generally suffer jump discontinuities across $s = S_i(t)$; with this in mind, given any function $\phi(s, t)$, we write

$$\phi_i^{\pm}(t) := \phi(S_i(t) \pm 0, t). \tag{2.39}$$

We associate with each juncture i three functions of time:

$$\begin{aligned} \varkappa_i &= T_i^+ \cdot N_i^- = -T_i^- \cdot N_i^+, \\ k_i &= 2\varkappa_i/(1 + T^+ \cdot T^-), \\ v_i &= (V_i^+ + V_i^-)/2; \end{aligned} \tag{2.40}$$

k_i is the **transition curvature** and v_i the **average velocity** of the juncture. Note that

$$k_i = 2 \sin \vartheta_i/(1 + \cos \vartheta_i) = 2 \tan (\vartheta_i/2), \quad \vartheta_i = \theta_i^+ - \theta_i^-, \tag{2.41}$$

so that k_i, as a function of ϑ_i, is strictly increasing on $[0, \pi)$, is asymptotic to ϑ_i for ϑ_i small, and tends to ∞ as $\vartheta_i \to \pi$. (In the literature, it is more common to refer to ϑ_i as the curvature of the corner.) Note also that for r_i and r_{i+1} convex, the PS evolving curve $\{r_i, r_{i+1}\}$ is convex (in the usual sense for continuous curves) if and only if k_i and the curvatures of r_i and r_{i+1} have the same sign.

Corner conditions. *At each juncture i*

$$\begin{aligned} V_i^+ N_i^+ + (S_i^\cdot - v_i^+) T_i^+ = V_i^- N_i^- + (S_i^\cdot - v_i^-) T_i^- &= R_i^\cdot, \\ R_i^\cdot \cdot (T_i^+ - T_i^-) &= k_i v_i. \end{aligned} \tag{2.42}$$

Proof. The identities $(2.42)_1$ are a direct consequence of $(2.20)_1$ and (2.37). The inner product of

$$V_i^+ N_i^+ + (R_i^\cdot \cdot T_i^+) T_i^+ = V_i^- N_i^- + (R_i^\cdot \cdot T_i^-) T_i^-$$

with T_i^+ and T_i^- gives two equations, whose difference yields $(2.42)_2$. □

The next proposition gives an evolution equation for the **length**

$$L(t) = \text{length}\,(\mathscr{o}(t))$$

of r. As this result shows (and as is clear from $(2.42)_2$), $k_i v_i$ measures the rate at which the juncture i generates arc length.

Transport of length.

(i) *Let r have N arcs. Further, let $R_{\text{init}}(t)$ and $R_{\text{term}}(t)$ denote the endpoints of r with $T_{\text{init}}(t)$ and $T_{\text{term}}(t)$ the corresponding tangents. Then*

$$L^\cdot(t) = R_{\text{term}}^\cdot(t) \cdot T_{\text{term}}(t) - R_{\text{init}}^\cdot(t) \cdot T_{\text{init}}(t) - \int_{\mathscr{o}(t)} KV\,ds - \sum_{i=1}^{N-1} k_i v_i. \tag{2.43}$$

(ii) *Let r be closed with N the essential number of arcs. Then*

$$L^\cdot(t) = - \int_{\mathscr{o}(t)} KV\,ds - \sum_{i=1}^{N} k_i v_i. \tag{2.44}$$

Proof. We apply (2.35) to each arc of r, add the resulting equations, and appeal to $(2.42)_2$ and (2.42). □

A **PS interfacial motion** is a *closed*[17] PS evolving curve $\boldsymbol{r}$ with $\boldsymbol{r}(\cdot, t)$ a *boundary curve* at each t. As before, the **reference region** $\Omega(t)$, with outward normal $\boldsymbol{N}$, is the bounded region interior to the trace $\partial\Omega(t)$ of $\boldsymbol{r}$. A standard result is the following equation for the evolution of $A(t) = \text{area}\,(\Omega(t))$:

$$A^{\cdot}(t) = \int_{\partial\Omega(t)} V\,ds. \tag{2.45}$$

3. Basic laws[18]

We consider a body which occupies all of $\mathbb{R}^2$ and consists of two phases separated, at each time t, by an interface. We assume that the interface, as a function of t, is an *interfacial motion* $\boldsymbol{r}$. Let $\mathscr{I}$ denote the trace of $\boldsymbol{r}$. The curve $\mathscr{I}(t)$ represents the **interface** at time t; by definition, $\mathscr{I}(t)$ divides $\mathbb{R}^2$ into two sets, the regions occupied by the two phases. The reference region, $\Omega(t)$, has $\boldsymbol{N}$ as its outward unit normal; we will refer to the phase occupying $\Omega(t)$ as the **reference phase.**

3.1. Balance of forces[19]

Consider an interfacial motion $\boldsymbol{r}$. The micromechanics of the interface is described by two functions of s and t: $\boldsymbol{C}(s, t)$, the force exerted along the interface at s; and $\boldsymbol{b}(s, t)$, the force exerted on $\mathscr{I}(t)$ per unit length. $\boldsymbol{C}(s, t)$ is the **capillary force**; if we write

$$\boldsymbol{C} = \sigma\boldsymbol{T} + \xi\boldsymbol{N}, \tag{3.1}$$

then $\sigma(s, t)$ is the **surface tension,** $\xi(s, t)$ is the **surface shear.**

Balance of internal forces is, to us, the requirement that, if $\mathscr{c}$ is the trace of an arbitrary evolving subcurve $\boldsymbol{r}_0$ of $\boldsymbol{r}$, then

$$\int_{\partial\mathscr{c}(t)} \boldsymbol{C} + \int_{\mathscr{c}(t)} \boldsymbol{b}\,ds = \boldsymbol{0} \qquad (3.2)^{20}$$

for the duration of $\boldsymbol{r}_0$. This law has the local form

$$\boldsymbol{C}_s + \boldsymbol{b} = \boldsymbol{0}, \tag{3.3}$$

or equivalently, by (3.1) and (2.6),

$$\xi_s + \sigma K + b = 0, \qquad \sigma_s - \xi K + b_{\tan} = 0, \tag{3.4}$$

[17] With the exception of Section 9.6, the underlying PS curves will be bounded.

[18] The underlying physics is discussed at greater length in [1988g] (see also [1986g, 1988gg]).

[19] [1988g].

[20] (3.2) should be viewed as a conservation law over and above the usual (gross) balance laws for forces and moments (*cf.* the discussion of [1988g]).

where

$$b = N \cdot b, \quad b_{\tan} = T \cdot b$$

are the normal and tangential components of $\boldsymbol{b}$.

Motion *tangential* to the interface depends on the choice of parameterization and is hence irrelevant to the underlying physics; the intrinsic evolution of the interface is *normal* to itself, through the velocity $\boldsymbol{r}^\circ$. As is consistent with a "constraint" of this type, we leave $b_{\tan}$ as *indeterminate* and consider only the normal component $(3.4)_1$ of the force balance.

We assume that the normal force b is the sum of two terms:

$$b = b_{\text{ext}} + \lambda;$$

b_{ext}, the **external force**, represents the normal force exerted on the interface by the external world;[21] λ, the **interactive force**, gives the normal force exerted on the interface by the bulk material. With this decomposition, the **normal force-balance** takes the form

$$\xi_s + \sigma K + \lambda + b_{\text{ext}} = 0. \tag{3.5}$$

3.2. Energy. The second law

We associate with each interfacial motion an **interfacial energy** $f(s, t)$ per unit length. In addition, the individual phases possess bulk energies; in accord with our tacit assumption of isothermal conditions, we assume that the energy of each phase is constant, and we write F for the **energy of the reference phase** minus that of the other phase.

Let R denote a fixed bounded region of space, and let

$$\Omega_R(t) = \Omega(t) \cap R, \qquad \mathscr{s}_R(t) = \mathscr{s}(t) \cap R, \tag{3.6}$$

so that $\Omega_R(t)$ is the portion of $\Omega(t)$ in R, while $\mathscr{s}_R(t)$ (assumed nonempty) is the portion of $\mathscr{s}(t)$ in R. We assume that the boundary of R is sufficiently smooth that $\mathscr{s}_R$ is the trace of an evolving subcurve $\boldsymbol{r}_R$ of $\boldsymbol{r}$, at least on a sufficiently small time interval. Then (modulo an inconsequential constant) the total energy of R is given by

$$F \operatorname{area}(\Omega_R(t)) + \int_{\mathscr{s}_R(t)} f\,ds.$$

Interfacial energy is carried out of R whenever the normal trajectories of the interface cross ∂R; in view of (2.32), this *outflow* is given by the quantity outflow $(f, \partial\mathscr{s}_R)(t)$.

The terms

$$\int_{\partial\mathscr{s}_R} C \cdot r^\circ = \int_{\partial\mathscr{s}_R} \xi V, \qquad \int_{\mathscr{s}_R} b_{\text{ext}} V\,ds$$

[21] This force is essential to the thermodynamical development of Section 4 (*cf.* [1988g], Footnote 13). In later sections we will restrict attention to interfacial motions with $b_{\text{ext}} = 0$.

represent power supplied to R by the portion of the interface outside of R and by the external world. (For convenience, we write $\mathscr{I}_R$ for $\mathscr{I}_R(t)$.) The surface tension σ and the interaction λ do not supply power: σ because it acts in a direction orthogonal to the velocity $\boldsymbol{r}^\circ$; λ because it represents interactions *within* R.

The second law for R is the assertion that the rate at which the energy increases plus the energy outflow cannot be greater than the power supplied to R:

$$\left(\frac{d}{dt}\right)\left\{\dot{F}\,\text{area}\,(\Omega_R) + \int_{\mathscr{I}_R} f\,ds\right\} + \text{outflow}\,(f, \partial\mathscr{I}_R) \leqq \int_{\partial\mathscr{I}_R} \xi V + \int_{\mathscr{I}_R} b_{\text{ext}}\, V\,ds \tag{3.7}$$

during the duration of $\boldsymbol{r}_R$. This **global energy-inequality** is assumed to hold for every such region R.

The transport theorem (2.34), the identity (2.45), and the fact that R is arbitrary imply that

$$FV + f^\circ - fKV - (\xi V)_s - b_{\text{ext}}\, V \leqq 0; \tag{3.8}$$

hence $(2.18)_2$ and (3.5) yield the **local energy-inequality**

$$f^\circ - \xi\theta^\circ + (\sigma - f)\,KV + (\lambda + F)\,V \leqq 0. \tag{3.9}$$

Remark. One could also consider, as an additional postulate, an energy inequality for the *interface itself* of the form

$$\left(\frac{d}{dt}\right)\int_{c} f\,ds + \text{outflow}\,(f, \partial c) \leqq \int_{\partial c} \xi V + \int_{c} bV\,ds \tag{3.10}$$

with c the trace of an arbitrary evolving subcurve. (Note that we use $b = b_{\text{ext}} + \lambda$ to account for the power supplied to the interface by the bulk material.) As we shall see, (3.10) follows as a *consequence* of our constitutive assumptions. Note that, by $(3.4)_1$, (3.10) has the local form

$$f^\circ - \xi\theta^\circ + (\sigma - f)\,KV \leqq 0. \tag{3.11}$$

4. Constitutive equations. Consequences of thermodynamics. Stability

4.1. Constitutive equations. The compatibility theorem

As **constitutive equations** we allow the energy, surface tension, capillary shear, and interactive force to depend on the orientation of the interface through a dependence on θ, and on the kinetics of the interface through a dependence on V:

$$\begin{aligned} f &= \hat{f}(\theta, V), & \xi &= \hat{\xi}(\theta, V), \\ \sigma &= \hat{\sigma}(\theta, V), & \lambda &= \hat{\lambda}(\theta, V). \end{aligned} \tag{4.1}$$ [22]

[22] Each of these functions is assumed to be 2π-periodic with respect to θ.

The first three relations characterize the interface, the last models the interaction between the interface and the bulk material.

Given an interfacial motion $\boldsymbol{r}$, the constitutive equations may be used to compute a corresponding **process** $(f, \xi, \sigma, \lambda)$. The normal force-balance (3.5) then determines the external force b_{ext} needed to support the process. Granted this, the second law (3.7) will hold in every region R if and only if the local-energy inequality (3.9) is satisfied. This should motivate the following definition: the **constitutive equations are compatible with thermodynamics** if given any interfacial motion, the corresponding process satisfies (3.9).

Compatibility theorem.[23] *The constitutive equations are compatible with thermodynamics if and only if:*

(i) *the energy, surface tension, and surface shear are independent of V and satisfy*

$$\hat{\sigma}(\theta) = \hat{f}(\theta), \quad \hat{\xi}(\theta) = \hat{f}_\theta(\theta); \tag{4.2}$$

(ii) *the interactive force has the form*

$$\begin{gathered} \hat{\lambda}(\theta, V) = -F - \beta(\theta, V)\, V, \\ \beta(\theta, V) \geqq 0. \end{gathered} \tag{4.3}$$

Proof. The following simple result will be useful:

Let $\phi(x)$ be smooth and satisfy $\phi(x)\, x \geqq 0$ for all $x \in \mathbf{R}$. Then there is a smooth function $\mu(x) \geqq 0$ such that $\phi(x) = \mu(x)\, x$. (4.4)

The proof is simple: $\phi(x)\, x$ has a minimum at $x = 0$; thus $\phi(0) = 0$, so that $\mu(x) = x^{-1}\phi(x) \geqq 0$ is well defined and smooth at $x = 0$.

To prove the theorem, we note that, in view of the constitutive equations, (3.9) is equivalent to the inequality

$$\begin{aligned} [\hat{f}_\theta(\theta, V) - \hat{\xi}(\theta, V)]\, \theta^\circ + \hat{f}_V(\theta, V)\, V^\circ + [\hat{\sigma}(\theta, V) - \hat{f}(\theta, V)]\, KV& \\ + [\hat{\lambda}(\theta, V) + F]\, V \leqq 0.& \end{aligned} \tag{4.5}$$

Assume that (4.5) holds for all motions of the interface, It is a simple matter to construct an interfacial motion for which, at some point and time, the fields θ, V, K, θ°, and V° have arbitrary values (*cf.* the Variation Lemma of GURTIN [1988]); this implies (i) and the inequality

$$\{F + \hat{\lambda}(\theta, V)\}\, V \leqq 0. \tag{4.6}$$

Assertion (ii) follows from (4.6) and (4.4) with $x = V$ and $\phi(x) = -F - \hat{\lambda}(\theta, x)$.

Conversely, the assertions (i) and (ii) trivially yield (4.5) in all processes. □

[23] This is a special case of a more general theorem [1988g] in which the bulk material is allowed to conduct heat.

We will refer to $\beta(\theta, V)$ as the **kinetic coefficient**.

By definition, λ and V are components with respect to the same direction, so that, for V positive, λ may be regarded as a force *in the direction of motion* exerted on the interface by the bulk material. Equation (4.3) gives this force as the sum of two terms. The first term is a force $-F$ which is positive if the phase into which the interface is moving has higher energy (and is thus less stable) than the other phase. The second term $-\beta V$ is, by $(4.3)_1$, negative, and represents a *drag force* opposing interfacial motion. Note that, for small values of the velocity V, (4.3) has the approximation

$$\begin{aligned} \lambda &= -F - \beta_0(\theta)\, V, \\ \beta_0(\theta) &= \beta(\theta, 0) \geqq 0. \end{aligned} \tag{4.7}$$

Some important consequences of the compatibility theorem are expressed in the following remarks.

Remark 1. The relations (4.2) imply (3.11) with "$\leqq$" replaced by "$=$", and this yields the interfacial energy-inequality (3.10) as an *equality*. Thus *the interface does not dissipate energy*; energy is dissipated at most in the interaction between the interface and the bulk material. The right side of (3.7) minus the left side gives this dissipation, which a simple calculation shows to be

$$\int_{\sigma_R} \beta(\theta, V)\, V^2\, ds. \tag{4.8}$$

Thus $\beta(\theta, V)\, V^2$ represents the *energy dissipated by the interaction* per unit length.

Remark 2. By (2.4) and (4.2), the *capillary force* (3.1) *may be regarded as a function of orientation*:

$$C = C(\theta) = \hat{f}(\theta)\, T(\theta) + \hat{f}'(\theta)\, N(\theta). \tag{4.9}$$

The relations (4.2) also imply that $\sigma_s = \xi K$ in every process. Therefore, by $(3.4)_2$, *tangential forces are balanced with* $b_{\text{tan}} = 0$, which obviates the need for constraint forces. Thus, granted $b_{\text{tan}} = b_{\text{ext}} = 0$, we may use (4.3) to write the capillary balance law (3.2) in the equivalent integral form

$$\int_{\partial c(t)} C(\theta) = \int_{c(t)} (F + \beta(\theta, V)\, V)\, N\, ds \tag{4.10}$$

for every c, or, by (3.5), (4.2), and (4.3), in the *equivalent* local form

$$\beta(\theta, V)\, V = [f(\theta) + f''(\theta)]\, K - F. \tag{4.11}$$

A final consequence of the compatibility theorem is the following result, which is essentially a statement of the second law for evolving curves.

Corollary. *Consider a curve* r *that evolves according to the normal force-balance* (3.5) *and the thermodynamic relations* (4.2) *and* (4.3). *Let* ȷ *denote the trace of* r, *and let* $S_1(t) < S_2(t)$ *denote the arc lengths corresponding to the end-*

points of $\mathscr{s}(t)$. *Then*

$$\left(\frac{d}{dt}\right) \int_{\mathscr{s}(t)} f\,ds + F \int_{\mathscr{s}(t)} V\,ds = - \int_{\mathscr{s}(t)} \beta(\theta, V)\, V^2\,ds + \int_{\mathscr{s}(t)} b_{\mathrm{ext}}\, V\,ds + W_2(t) - W_1(t),$$

$$W_i(t) = C(S_i(t), t) \cdot \left(\frac{d}{dt}\right) r(S_i(t), t). \tag{4.12}$$

Proof. By (4.2) and $(2.18)_2$, $f^\circ - fKV = \xi V_s - \sigma KV$. If we substitute this relation into (2.34) with $\phi = f$, integrate the term ξV_s by parts, and appeal to (2.32), (3.5), $(2.20)_1$, and (3.1), we arrive at (4.12). □

4.2. *General assumptions. Admissibility for evolving curves*

Since there is no danger of confusion, we will use the shorthand:

$$f(\theta) = \hat{f}(\theta).$$

Further, to avoid repeated hypotheses, we will assume, for the remainder of the paper, that the following hypotheses are satisfied:

Assumptions.

(i) *The constitutive equations are compatible with thermodynamics.*
(ii) *The interfacial energy and kinetic coefficient satisfy*:

$$f(\theta) > 0, \quad \beta(\theta, V) > 0, \qquad \beta(\theta, V) \text{ is independent of } V. \tag{4.13}$$

(iii) *We henceforth restrict attention to evolving curves that correspond to vanishing external forces*:

$$b_{\mathrm{ext}} = 0. \tag{4.14}$$

We will refer to an evolving curve as **admissible** if it is consistent with (4.11). By Remark 2 of Section 4.1, admissibility for an evolving curve is equivalent to the requirement that the curve be consistent with balance of capillary forces.

4.3. *Stability of the interfacial energy*

The following calculation leads to a condition on f which ensures that straight line segments locally minimize inerfacial energy. Let ℓ denote an oriented, straight line segment with initial point r_0 and terminal point r_1, and let r denote a (not necessarily admissible) evolving curve whose initial and terminal points are *fixed* at r_0 and r_1, respectively, and whose trace satisfies $\mathscr{s}(0) = \ell$.

Let $\mathcal{F}(t)$ denote the energy of $\mathscr{o}(t)$:

$$\mathcal{F}(t) = \int_{\mathscr{o}(t)} f(\theta)\, ds.$$

Then

$$\mathcal{F}^{\cdot}(0) = 0, \quad \mathcal{F}^{\cdot\cdot}(0) = [f(\theta_0) + f''(\theta_0)] \int_{\ell} (V_s)^2\, ds, \tag{4.15}$$

with θ_0 the angle of ℓ. We will establish (4.15) at the end of the section. Thus[24] *a necessary and sufficient condition that* $\mathcal{F}(t)$ *have a strict local minimum at* $t = 0$ *is that*

$$f(\theta_0) + f''(\theta_0) > 0.$$

This proposition should motivate the following definition. The interfacial energy f is **strictly stable at** θ, **stable at** θ, or **unstable at** θ according as

$$f(\theta) + f''(\theta) > 0, \quad f(\theta) + f''(\theta) \geqq 0, \quad f(\theta) + f''(\theta) < 0; \tag{4.16}$$

f is: **strictly stable** if it is strictly stable for all $\theta \in \mathbb{R}$; **stable** if it is stable for all $\theta \in \mathbb{R}$. As we shall see in Section 6, the partial differential equation describing the evolution of the interface will be *parabolic* where the interfacial energy is strictly stable and *backward parabolic* where f is unstable. Since $f(\theta) > 0$, the interfacial energy cannot be unstable for all θ.

By (2.4) and (4.9),

$$C'(\theta) = [f(\theta) + f''(\theta)]\, N(\theta), \tag{4.17}$$

Thus, if the interfacial energy is strictly stable, then given any angle α, $N(\alpha) \cdot C(\theta)$ increases strictly with θ for $\alpha - \frac{\pi}{2} \leqq \theta \leqq \alpha + \frac{\pi}{2}$ and decreases strictly with θ for $\alpha + \frac{\pi}{2} \leqq \theta \leqq \alpha + \frac{3\pi}{2}$.

We now prove (4.15). By (2.33) and (2.34)

$$\mathcal{F}^{\cdot}(t) = \int_{\mathscr{o}(t)} [f'(\theta)\, \theta^{\circ} - f(\theta)\, KV]\, ds,$$

and, since $V = 0$ at the endpoints, while $K = 0$ and $\theta = \theta_0$ at $t = 0$, we may use $(2.18)_2$ to conclude that $(4.15)_1$ is satisfied. Similarly,

$$\mathcal{F}^{\cdot\cdot}(t) = \int_{\mathscr{o}(t)} [f'(\theta)\, \theta^{\circ\circ} + f''(\theta)\, (\theta^{\circ})^2 - f(\theta)\, K^{\circ}\, V]\, ds \text{ at } t = 0.$$

But $\theta^{\circ} = V_s$ and, by (2.17) and (2.18), $\theta^{\circ\circ} = KVV_s + (V^{\circ})_s$, $K^{\circ}V = V_{ss}V + K^2V^2$; thus

$$\mathcal{F}^{\cdot\cdot}(t) = \int_{\mathscr{o}(t)} [f''(\theta_0)\, (V_s)^2 - f(\theta_0)\, V_{ss}V]\, ds \quad \text{at } t = 0,$$

and integrating the last term by parts we arrive at $(4.15)_2$.

---

[24] *Cf.* Herring [1951 b], Frank [1963], Gjostein [1963], Gruber (Gjostein [1963]), Taylor [1978], Fonseca [1988].

II. Smooth interfacial motions

5. Evolution equations for the interface

We now discuss the equations that describe *admissible* evolving curves; that is, evolving curves whose evolution is governed by the constitutive equations and the capillary balance law.

5.1. Isotropic interface

For an isotropic interface f and β are constants. Without loss in generality, we set $f = \beta = 1$; (4.11) then reduces to[25]

$$V = K - F. \tag{5.1}$$

A complete set of partial differential equations for an admissible evolving curve consists of (5.1) supplemented by the kinematical conditions (2.18) (*cf.* (2.16)) satisfied by all evolving curves:

$$\begin{aligned} V &= K - F, & \theta_t + v\theta_s &= V_s, \\ K_t + vK_s &= V_{ss} + K^2 V, & v_s &= -KV \end{aligned} \tag{5.2}$$

(where the subscript t denotes the time derivative holding s fixed). The domains of the underlying fields in the arc-length description are not known *a priori*, since s varies in the interval $[0, L(t)]$ with $L(t) = \text{length}(\mathscr{A}(t))$. However, (2.36) relates $L^{\cdot}(t)$ to KV, and we can introduce the rescaled variable $s^* = s/L(t)$.[26]

When the curve is *convex*, the system (5.2) takes a particularly simple form; indeed, (5.1) and (2.23) yield

$$K_t = K^2[K_{\theta\theta} + K - F], \tag{5.3}$$

(with K_t the time derivative holding θ fixed).

5.2. Anisotropic interface

5.2.1. Basic equations. For convenience, we define

$$\Phi = \beta^{-1}(f + f''), \quad \Psi = \beta^{-1} F; \tag{5.4}$$

then (4.11) becomes

$$V = \Phi(\theta)\, K - \Psi(\theta). \tag{5.5}$$ [27]

[25] ALLEN & CAHN [1979] and RUBINSTEIN, STERNBERG, & KELLER [1987] deduce the equation $V = K$ as a formal approximation to the Landau-Ginzburg equation. Evolution according to $V = K$ is discussed by many authors; *cf.* BRAKKE [1978], SETHIAN [1985], ABRESCH & LANGER [1986], GAGE [1984, 1986], GAGE & HAMILTON [1986], GRAYSON [1987], HUISKEN [1987], OSHER & SETHIAN [1987], and the references therein.

[26] ABRESCH & LANGER [1986].

[27] The special case $V = -\Psi(0)$ was introduced by FRANK [1958].

A complete system of equations for an admissible curve consists of (5.5) in conjunction with $(5.2)_{2-4}$. When the curve is *convex* this system reduces to

$$K_t = K^2[\Phi K - \Psi]_{\theta\theta} + K^2[\Phi K - \Psi] \tag{5.6}$$

(with K_t the derivative holding θ fixed). This equation is also valid for nonconvex motions, at least locally where $K \neq 0$.

Remark. The term of highest order on the right side of (5.6) is $K^2 \Phi K_{\theta\theta}$; thus (5.6) is *parabolic* for $\Phi(\theta) > 0$, backward parabolic for $\Phi(\theta) < 0$. (Note that (5.6) degenerates at $K = 0$.) By (4.13), (4.16), and (5.4), *parabolicity is equivalent to the strict stability of the interfacial energy, while backward parabolicity is equivalent to instability* (cf. (4.16)). There is no compelling physical reason to suppose that the interfacial energy is strictly stable; in fact, material scientists often consider energies which are unstable[28] for particular ranges of the orientation θ. Since $f(\theta) > 0$ and periodic, at worst we can have an equation which is backward parabolic for some but not all values of θ.

Note that, by (5.5), the general equation (5.6), when expressed in terms of the normal velocity $V(\theta, t)$, has the form

$$\Phi(\theta)\, V_t = [V + \Psi(\theta)]^2\, [V_{\theta\theta} + V]. \tag{5.7}$$

5.2.2. Equations when the curve is the graph of a function. Locally, an evolving curve may be represented as the graph of a function $y = h(x, t)$, provided the x and y axes are chosen appropriately. Consider the choice indicated in Figure 5A (with orientation such that arc length increases with increasing x) and let

$$p = h_x. \tag{5.8}$$

Then

$$p \tan\theta = -1, \qquad h_t = (\sin\theta)^{-1}\, V, \qquad K = h_{xx}(1 + p^2)^{-\frac{3}{2}}, \tag{5.9}$$

and the evolution equation (5.5) takes the form

$$\begin{gathered} h_t = Q(p)\, h_{xx} - B(p), \\ Q(p) = \Phi(\theta) \sin^2\theta, \qquad B(p) = \Psi(\theta)/\sin\theta, \end{gathered} \tag{5.10}$$

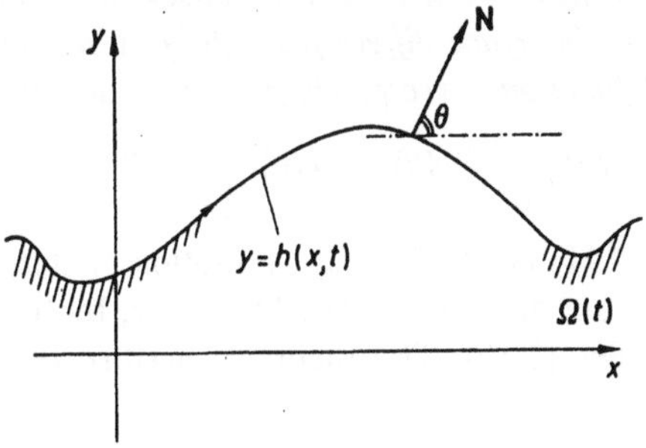

Fig. 5A. Sign conventions when the curve is a graph $y = h(x, t)$

[28] *Cf., e.g.*, GJOSTEIN [1963], CAHN & HOFFMAN [1974].

or, upon differentiating with respect to x,

$$p_t = [Q(p)\, p_x - B(p)]_x, \tag{5.11}$$

which is in conservation form.

6. Stationary interfaces. Steady interfacial motions

We assume throughout this section that the interface is strictly stable. Then

$$f(\theta) + f''(\theta) > 0$$

for all θ, and we can write

$$w(\theta) = [f(\theta) + f''(\theta)]^{-1}. \tag{6.1}$$

Note that, by (2.11),

$$\int_0^{2\pi} w(\theta)^{-1}\, e^{i\theta}\, d\theta = 0. \tag{6.2}$$

This section is restricted to evolving curves and interfacial motions that are admissible; to avoid repetition, *we shall omit the term "admissible" in most of the ensuing discussion.*

6.1. Stationary interfaces

By a **stationary interface** we mean an interfacial motion which is independent of time. A trivial consequence of (5.5) is that for $F = 0$ the unbounded time-independent facets form the complete collection of stationary interfaces. The next theorem establishes the existence of stationary interfaces for $F \neq 0$.

Wulff's theorem.[29] *Assume that* $F \neq 0$. *Then*

$$r(\theta) = F^{-1}\, [f'(\theta)\, T(\theta) - f(\theta)\, N(\theta)] \tag{6.3}$$

defines a stationary interface which is closed, convex, and parametrized by angle, and any other stationary interface differs from (6.3) *by at most a translation. The curvature and support function corresponding to* (6.3) *are given by*

$$K(\theta) = Fw(\theta), \qquad \not{h}(\theta) = -f(\theta)/F. \tag{6.4}$$

Proof. In view of (4.11) (with $V = 0$), a motion is stationary if and only if it is convex with curvature given by $(6.4)_1$. Moreover, by (6.2), the function $K(\theta)$ defined by $(6.4)_1$ satisfies (2.10). The theorem therfore follows from Lemma 3 (Section 2.1). □

[29] WULFF [1901]. See also DINGHAS [1944], TAYLOR [1978]. The bounded region Ω with boundary defined by (6.3) is actually a Wulff region in the sense of (7.12): Ω has least *interfacial* energy among all regions Γ with area (Γ) = area (Ω).

Let Q denote the orthogonal transformation that rotates vectors clockwise by $\frac{\pi}{2}$. An interesting consequence of (4.9) is that the capillary force $C(\theta)$ corresponding to (6.3) is given by

$$C(\theta) = FQ\boldsymbol{r}(\theta).$$

6.2. Steadily evolving facets

By (2.29), (5.5), and (6.1), steady interfacial motions $\boldsymbol{r}$ evolve according to

$$K(\theta) = [F + \beta(\theta)\,\boldsymbol{U}\cdot\boldsymbol{N}(\theta)]\,w(\theta), \tag{6.5}$$

where $\boldsymbol{U}\,(\neq \boldsymbol{0})$ is the steady velocity. It is convenient to introduce the vector potential

$$\boldsymbol{\beta}(\theta) = \beta(\theta)\,\boldsymbol{N}(\theta), \tag{6.6}$$

whose locus forms the *polar diagram* Polar $(\boldsymbol{\beta})$ of $\boldsymbol{\beta}$, and to write (6.5) in the form

$$K(\theta) = G(\theta)\,w(\theta), \qquad G(\theta) = F + \boldsymbol{U}\cdot\boldsymbol{\beta}(\theta). \tag{6.7}$$

Appealing to the proposition containing (2.29), we see that a steady motion is a steadily evolving facet if and only if the corresponding angle θ is identically constant and a solution of

$$G(\theta) = 0. \tag{6.8}$$

The equation (6.8) has a simple geometric solution. To state this solution concisely, we introduce the following terminology. For $\boldsymbol{d} \neq \boldsymbol{0}$, let $\ell(\boldsymbol{d})$ denote the straight line

$$\ell(\boldsymbol{d}) = \{\boldsymbol{x}\colon\ \boldsymbol{d}\cdot\boldsymbol{x} = |\boldsymbol{d}|^2\}; \tag{6.9}$$

(6.9) defines a one-to-one correspondence between nonzero vectors and lines that do not pass through the origin; we will refer to $\boldsymbol{d}$ as the **support vector** for $\ell = \ell(\boldsymbol{d})$. In the same spirit, for $\boldsymbol{d} \neq \boldsymbol{0}$, we write $\ell = \ell(0\boldsymbol{d})$ for the line through the origin perpendicular to $\boldsymbol{d}$ and refer to ℓ as the line with **support vector** $0\boldsymbol{d}$. Then

$$\left.\begin{array}{l} G(\theta_0) = 0 \ \text{ if and only if the line with support vector} \\ (-F/|\boldsymbol{U}|^2)\,\boldsymbol{U} \ \text{ intersects the polar diagram Polar}\,(\boldsymbol{\beta}) \text{ at } \boldsymbol{\beta}(\theta_0); \end{array}\right\} \tag{6.10}$$

hence we have the following result (Figure 6A).

Theorem. *Given any vector* $\boldsymbol{U} \neq \boldsymbol{0}$, *there is a steadily evolving facet with steady velocity* $\boldsymbol{U}$ *and angle* θ_0 *if and only if the line with support vector* $(-F/|\boldsymbol{U}|^2)\,\boldsymbol{U}$ *intersects* Polar $(\boldsymbol{\beta})$ *at* $\boldsymbol{\beta}(\theta_0)$.

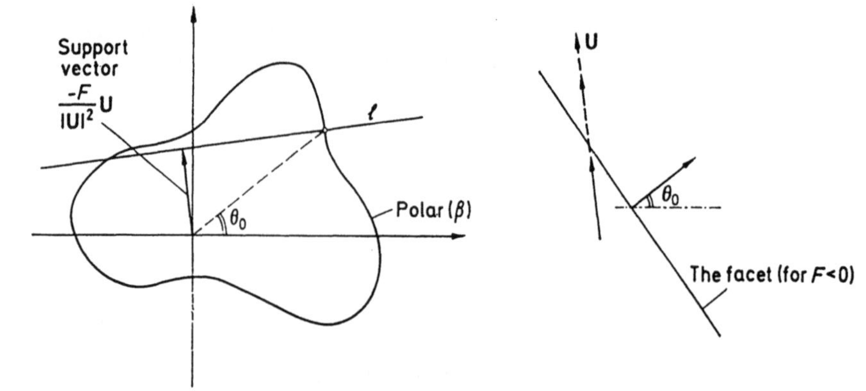

Fig. 6A. When the line ℓ with support vector $\dfrac{-F}{|\boldsymbol{U}|^2}\,\boldsymbol{U}$ intersects Polar (β) at θ_0, then there is a steadily evolving facet with steady velocity $\boldsymbol{U}$ and direction θ_0.

6.3. Steady motions that are not flat

Let F and $\boldsymbol{U} \neq \boldsymbol{0}$ be given. Let $\boldsymbol{r}$ be a steady motion corresponding to F with $\boldsymbol{U}$ as steady velocity, and assume that $\boldsymbol{r}$ is **not flat** in the sense that its curvature is not identically zero. Then Lemma 2 (Section 2.1) implies that $\boldsymbol{r}$ is *convex*. By definition, $\boldsymbol{r}$ is a boundary curve at each t, and hence is simple and either closed or unbounded. Assume that $\boldsymbol{r}$ is closed. Let $U = |\boldsymbol{U}|$, $\boldsymbol{e} = \boldsymbol{U}/U$. Then (2.10) and (6.7) yield

$$\int_0^{2\pi} [G(\theta)\, w(\theta)]^{-1}\, \boldsymbol{e} \cdot \boldsymbol{N}(\theta)\, d\theta = 0. \tag{6.11}$$

Let $M(\theta, \boldsymbol{e}, U)$ denote the left side of (6.11). By (6.2) and $(6.7)_2$, $M(\theta, \boldsymbol{e}, 0) = 0$. Further, differentiating $M(\theta, \boldsymbol{e}, U)$ with respect to U yields the conclusion that $M(\theta, \boldsymbol{e}, U)$ is strictly monotone in U. Thus (6.11) is possible only if $U = 0$, which violates the definition of a steady motion; hence $\boldsymbol{r}$ cannot be closed. Thus $\boldsymbol{r}$ is *unbounded.*

Therefore, appealing to Lemma 3 of Section 2.1, the angle range of $\boldsymbol{r}$ is a bounded interval (θ_1, θ_2), and this interval and the accompanying curvature $K(\theta)$ must be a solution of the following problem:

$$\left.\begin{array}{l}\text{find an angle interval } (\theta_1, \theta_2),\ \theta_2 - \theta_1 \leqq \pi, \text{ such that } K(\theta), \text{ defined}\\ \text{by (6.5), does not vanish on } (\theta_1, \theta_2) \text{ and has } K(\theta_1) = K(\theta_2) = 0.\end{array}\right\} \tag{6.12}$$

Conversely, if (θ_1, θ_2) is consistent with (6.12), then Lemma 3 of Section 2.1 implies that $K(\theta)$ restricted to (θ_1, θ_2) is the curvature of a convex, bounded curve $\boldsymbol{r}_0$. By virtue of the proposition following (2.25), $\boldsymbol{r}_0$ is the portrait and $\boldsymbol{U}$ the steady velocity of a (unique) steady motion $\boldsymbol{r}$; trivially, $\boldsymbol{r}$ has $K(\theta)$ as its curvature.

Thus we are reduced to solving (6.12); since $w(\theta) > 0$, (θ_1, θ_2) is a solution of (6.12) if and only if

$$\theta_1, \theta_2, 0 < \theta_1 - \theta_2 \leqq \pi, \text{ are } \textit{consecutive} \text{ zeros of } G(\theta). \tag{6.13}$$

To facilitate the discussion of such zeros, let us agree to call a line ℓ a **chord for** Polar (β) **between** θ_1 **and** θ_2 if $0 < \theta_1 - \theta_2 \leqq \pi$ and ℓ intersects Polar (β) at $\beta(\theta_1)$ and $\beta(\theta_2)$, but not at any other point $\beta(\theta)$ with $\theta \in (\theta_1, \theta_2)$. In view of (6.10), (6.13) is then equivalent to the requirement that

the line ℓ with support vector $(-F/|U|^2)\,U$
be a chord for Polar (β) between θ_1 and θ_2.

Given a line ℓ with ℓ a chord for Polar (β) between θ_1 and θ_2, we write

$$\ell|_{(\theta_1,\theta_2)} := \{x: x = \beta(\theta_1) + a[\beta(\theta_2) - \beta(\theta_1)],\ a \in (0, 1)\}$$

for the segment of ℓ between $\beta(\theta_1)$ and $\beta(\theta_2)$.

Continuing as before, let F and $U \neq 0$ be given, and let r be a steady motion corresponding to F and U. Let (θ_1, θ_2) be the angle range for r, so that the line ℓ with support vector $(-F/|U|^2)\,U$ is a chord for Polar (β) between θ_1 and θ_2. Choose $\theta \in (\theta_1, \theta_2)$. Then there is a unique point $x \in \ell|_{(\theta_1,\theta_2)}$ such that $x = \alpha N(\theta)$ for some $\alpha \geqq 0$; in addition,

$$\left.\begin{array}{l}\beta(\theta) > \alpha \text{ or } \beta(\theta) < \alpha \text{ according as } \ell|_{(\theta_1,\theta_2)} \\ \text{is interior or exterior to Polar } (\beta).\end{array}\right\} \tag{6.14}$$

Since $x \in \ell$, $x \cdot U = -F$; thus (6.6) and (6.7) yield

$$G(\theta) = [\beta(\theta) - \alpha]\, U \cdot N(\theta).$$

Further $K = wG$ with $w > 0$ and $K(\theta) \neq 0$; hence

$$K(\theta)\, U \cdot N(\theta) = C[\beta(\theta) - \alpha], \qquad C > 0,$$

and we conclude, with the aid of (2.30), (2.31), and (6.14), that r is a *steadily evolving bump*, which recedes or advances according as $\ell|_{(\theta_1,\theta_2)}$ is interior or exterior to Polar (β).

The results established above are summarized in the next theorem, in which F and $U \neq 0$ are assumed prescribed.

Theorem on steady motions. *Let ℓ denote the line with support vector $(-F/|U|^2)\,U$. Let r be a nonflat steady motion which corresponds to F and has U as steady velocity. Then r is a steadily evolving bump. If (θ_1, θ_2) is the angle range of r, then ℓ is a chord for* Polar (β) *between θ_1 and θ_2, and r is receding or advancing according as $\ell|_{(\theta_1,\theta_2)}$ is interior or exterior to* Polar (β).

Conversely if ℓ is a chord for Polar (β) *between θ_1 and θ_2, then there is a unique steady motion r corresponding to F with (θ_1, θ_2) as angle range and U as steady velocity.*

Corollary. *There are no advancing bumps if* Polar (β) *is convex, and none when* $F = 0$.

Remark. For an isotropic material ($f = \beta = 1$) and $F = 0$, (6.12) reduces to $K(\theta) = U \cdot N(\theta)$; if $U = U(1, 0)$, $U > 0$, there are two steady motions, one with angle range $(-\frac{\pi}{2}, \frac{\pi}{2})$, the other with angle range $(\frac{\pi}{2}, \frac{3\pi}{2})$, and both motions are receding bumps.[30]

Remark. It is generally believed that dendritic growth requires diffusion in the bulk material. It is interesting that a steadily advancing bump is possible even without diffusion. In the present theory such growth is a consequence of anisotropy in the kinetic coefficient and results when certain orientations suffer drag forces sufficiently lower than neighboring orientations.

7. Global behavior for an interface with stable energy

In this section we analyze the global behavior of the interface under the assumption of a *strictly stable* interfacial energy; in particular, we consider the general anisotropic equation

$$\beta(\theta)\, V = [f(\theta) + f''(\theta)]\, K - F \tag{7.1}$$

with

$$f(\theta) + f''(\theta) > 0, \quad \beta(\theta) > 0 \tag{7.2}$$

for all $\theta \in \mathbb{R}$ (*cf.* (4.13), (4.16)).

7.1. Existence of interfacial motions from a prescribed initial curve

The existence of an interface evolving from a given initial configuration is ensured by the

Existence theorem. *Let f and β be C^∞. Let Ω_0 be a given initial domain, which we assume to be bounded with boundary a Lipschitz-continuous simple closed curve. Then there is a unique, maximal family of domains $\Omega(t)$ $(0 \leqq t < T_{max})$ such that:*

(i) *$\partial\Omega(t)$ is a C^2 simple closed curve continuous for $0 < t < T_{max}$;*
(ii) *the evolution of $\partial\Omega(t)$ is governed by* (7.1);
(iii) $\Omega(0) = \Omega_0$.

In fact, this solution is C^∞ for $0 < t < T_{max}$.

A proof of this theorem is given by ANGENENT [1988], who shows that, for $T_{max} < \infty$, one of the following must be true:

(E1) $\sup_{s \in \mathbb{R}} |K(s, t)| \to \infty$ as $t \to T_{max}$;
(E2) K and its derivatives remain bounded as $t \to T_{max}$, so that $\partial\Omega(t)$ converges to a C^∞ curve Γ; however Γ is not simple.

The condition (E1) will occur whenever the interface shrinks to a point, or whenever the interface develops a kink; (E2) indicates the formation of self-intersections or self-tangencies (Figure 7A).

[30] This solution is known; it is referred to as the "grim reaper" by geometers (*cf.* GRAYSON [1987], p. 298).

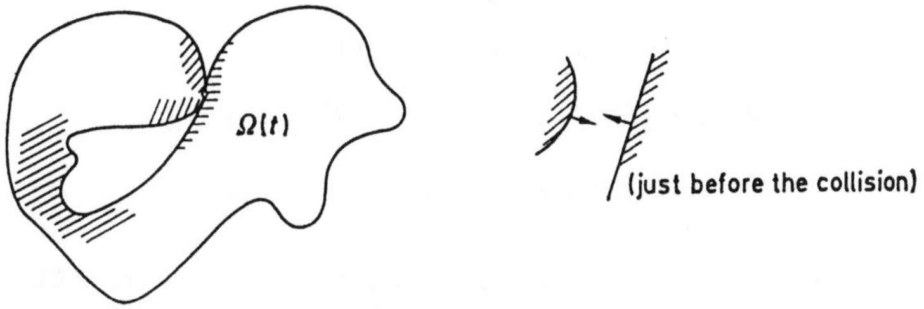

Fig. 7A. A simply connected region may evolve to a multiply connected region

When Ω_0 is smooth, $\partial\Omega(t)$ admits a parametrization $(p, t) \mapsto \boldsymbol{r}(p, t)$ as an admissible interfacial motion of duration $[0, T_{max})$. In the next section we will study the behavior of such motions as measured by their **perimeter** $L(t)$ and **enclosed area** $A(t)$,

$$L(t) = \text{length}\,(\partial\Omega(t)), \qquad A(t) = \text{area}\,(\Omega(t)). \tag{7.3}$$

We will, however, restrict our attention to motions, termed regularly maximal, whose singularity at $t = T_{max}$ (for $T_{max} < \infty$) is not too pathological.

Precisely, an admissible interfacial motion with duration $[0, T)$ is **regularly maximal** if either $T = \infty$ or

$$T < \infty \quad \text{and } A(t) \to 0 \quad \text{as} \quad t \to T. \tag{7.4}$$

Regularly maximal motions cannot be extended beyond $t = T$, but for T finite exhibit fairly regular behavior as $t \to T$: they either explode or disappear. This class of motions does not include motions that develop self-tangencies, self-intersections, or kinks at $t = T$.

7.2. *Growth and decay of the interface*

Let $\boldsymbol{r}$ denote an admissible interfacial motion. For convenience, we write

$$\mathscr{F}(t) := \int_{\partial\Omega(t)} f(\theta)\, ds \tag{7.5}$$

for the **total interfacial energy.** The next result, essentially the second law, is an immediate consequence of (3.7) and the discussion of the paragraph containing (4.8).

Growth theorem[31]

$$\mathscr{F}^{\cdot}(t) + FA^{\cdot}(t) = -\int_{\partial\Omega(t)} \beta(\theta)\, V^2\, ds \leqq 0. \tag{7.6}$$

[31] GURTIN [1988g].

It is convenient to rewrite (7.1) in the form

$$V = \Phi(\theta)\,K - \Psi(\theta),$$
$$\Phi(\theta) = [f(\theta) + f''(\theta)]/\beta(\theta), \quad \Psi(\theta) = F/\beta(\theta). \tag{7.7}$$

Also, for any 2π-periodic function $g(\theta)$ on $\mathbb{R}$, we write

$$g_{av} = (2\pi)^{-1}\int_0^{2\pi} g(\theta)\,d\theta, \quad g_{max} = \sup_{\theta\in\mathbb{R}} g(\theta), \quad g_{min} = \inf_{\theta\in\mathbb{R}} g(\theta). \tag{7.8}$$

If we use (2.5) to change variable in $(7.8)_1$ from θ to s, using the fact that as s increases from 0 to $L(t)$, σ goes from 0 to -2π, we arrive at

$$\int_{\partial\Omega(t)} g(\theta)\,K\,ds = -2\pi\,g_{av}. \tag{7.9}$$

Note that, by (7.2),

$$\Phi_{av} > 0, \quad \Phi_{min} > 0.$$

Given any bounded region Γ, we will refer to the number

$$\text{isoper}\,(\Gamma) := \text{length}\,(\partial\Gamma)^2/4\pi\,\text{area}\,(\Gamma) \tag{7.10}$$

as the **isoperimetric ratio** for Γ; then

$$\text{isoper}\,(\Gamma) \geqq 1 \quad \text{(isoperimetric inequality)} \tag{7.11}$$

with equality holding if and only if $\partial\Gamma$ is a circle.[32]

The following generalization of the isoperimetric ratio is useful. Let $e(\theta)$ be a continuous, piecewise smooth, strictly positive, 2π-periodic function on $\mathbb{R}$. Then the **Wulff ratio for** $e(\theta)$ is the number

$$W(e) = (4\pi)^{-1}\inf\left\{\int_{\partial\Gamma} e(\theta)\,ds\right\}^2\Big/\text{area}\,(\Gamma) \tag{7.12}$$

with the infimum taken over all bounded regions Γ with $\partial\Gamma$ piecewise smooth. This infimum is actually attained:[33] minima Γ are called **Wulff regions for** $e(\theta)$, and are convex regions, unique modulo translation and scaling. When $e = 1$, the minimum is a circle and $W(e) = 1$. More generally, taking Γ to be a circle yields $W(e) \leqq (e_{av})^2$, so that

$$0 < (e_{min})^2 \leqq W(e) \leqq (e_{av})^2. \tag{7.13}$$

The inequality $F > 0$ occurs when the reference phase has higher bulk energy than the other phase; in this instance (7.6) indicates a tendency for the less stable reference phase to shrink. On the other hand, $F < 0$ when the reference phase has lower bulk energy; here $FA(t)$ is negative and of the wrong sign for a

[32] The condition isoper $(\Omega(t)) \to \infty$ might indicate the formation of a dendritic structure. In this connection GURTIN [1986g] (*cf.* eq. (7.6)) discusses $A(t) \to 0$, $L(t) \to L_0 > 0$.

[33] *Cf.* TAYLOR [1978], who uses the term "Wulff crystal" rather than "Wulff region".

Lyapunov function, indicating a tendency for the more stable reference phase to grow, at least in situations for which area dominates length. The next theorem shows that this is indeed the case. In fact, we show that for $F \geqq 0$ the reference phase shrinks to zero; for $F < 0$ the reference phase shrinks to zero when initially small, but grows unboundedly when initially large.

Theorem on the growth of the reference phase. *Consider a regularly maximal admissible interfacial motion with duration* $[0, T)$.

(i) *If* $F \geqq 0$, *then* $T < \infty$ *and* $A(t) \to 0$ *as* $t \to T$.

(ii) *If* $F < 0$, *then*:

(a) *if* $L(0)$ *is sufficiently small then* $T < \infty$ *and* $A(t) \to 0$ *as* $t \to T$;

(b) *if* $A(0)$ *is sufficiently large, then* $T = \infty$ *and* $A(t) \to \infty$ *as* $t \to \infty$; *even so, the isoperimetric ratio* isoper $(\Omega(t))$ *remains bounded*:

$$\limsup_{t \to \infty} \text{isoper}\,(\Omega(t)) \leqq [(\beta^{-1})_{\text{av}}]^2 / W(\beta^{-1}); \tag{7.14}$$

thus, for $\beta =$ constant, isoper $(\Omega(t)) \to 1$.

Proof. We begin with the identities:

$$\begin{aligned} A^{\cdot}(t) &= -2\pi\Phi_{\text{av}} - F \int_{\partial\Omega(t)} \beta(\theta)^{-1}\, ds, \\ L^{\cdot}(t) &= -2\pi\Psi_{\text{av}} - \int_{\partial\Omega(t)} \Phi(\theta)\, K^2\, ds. \end{aligned} \tag{7.15}$$

To derive $(7.15)_1$ we integrate $(7.7)_1$ over $\partial\Omega(t)$ and use (2.45) and (7.9); to derive $(7.15)_2$ we multiply $(7.7)_1$ by K, integrate over $\partial\Omega(t)$, and use (2.36) and (7.9).

If we can show that $L(t) \to 0$ in finite time provided the solution persists that long, then we can conclude from (7.4) that $T < \infty$ and $A(t) \to 0$ as $t \to T$; we cannot conclude (from this alone) that $L(t) \to 0$ as $t \to T$.

Assume that $F \geqq 0$. Then, by $(7.15)_2$ and the remark made in the previous paragraph, (i) follows.

Assume that $F < 0$. By (7.5), $f_{\min} L \leqq \mathscr{F} \leqq f_{\max} L$; thus (7.6) and $(7.15)_1$ yield

$$\mathscr{F}^{\cdot} \leqq -FA^{\cdot} \leqq -2\pi\,|F|\,\Phi_{\text{av}} + (F^2/f_{\min}\beta_{\min})\,\mathscr{F},$$

which implies (iia).

Next, $(7.15)_2$ yields

$$L(t) \leqq L(0) + 2\pi\,|F|\,(\beta^{-1})_{\text{av}}\, t. \tag{7.16}$$

On the other hand, (7.12) yields

$$\int_{\partial\Omega(t)} \beta(\theta)^{-1}\, ds \geqq 2[\pi W(\beta^{-1})\, A(t)]^{\frac{1}{2}};$$

therefore, by $(7.15)_1$,

$$\begin{aligned} A^{\cdot}(t) &\geqq D A(t)^{\frac{1}{2}} - C, \\ C = 2\pi\Phi_{\text{av}} > 0, &\quad D = 2\,|F|\,[\pi W(\beta^{-1})]^{\frac{1}{2}} > 0. \end{aligned} \tag{7.17}$$

By (7.4), (7.16), and (7.17), if $A(0)$ is sufficiently large, then $T = \infty$ and $A(t) \to \infty$ as $t \to \infty$. In fact, $(7.17)_1$ is easily integrated to give

$$A(t)^{\frac{1}{2}} + \varkappa \ln [A(t)^{\frac{1}{2}} - \varkappa] \geqq A(0)^{\frac{1}{2}} + \varkappa \ln [A(0)^{\frac{1}{2}} - \varkappa] + Dt/2, \tag{7.18}$$

where $\varkappa = C/D$. Since $A(t) \to \infty$ as $t \to \infty$, (7.10), (7.16), and (7.18) imply (7.14). □

Conjecture. Consider case (iib) of the last theorem, in which $F < 0$ and $A(0)$ is sufficiently large that the interface grows without bound. We conjecture that, as $t \to \infty$,

$$\partial\Omega(t) \textit{ is asymptotic to a Wulff region for } \beta(\theta)^{-1}. \tag{7.19}$$

Our argument in support of (7.19) is as follows. As the interface grows the curvature term in (5.5) should ultimately be negligible. Granted this, $V \simeq -F\beta(\theta)^{-1}$ and the total energy dissipated

$$\int_{\partial\Omega(t)} \beta(\theta)\, V^2\, ds$$

(*cf.* (4.8)) should be asymptotic to

$$\mathscr{D} = F^2 \int_{\partial\Omega(t)} \beta(\theta)^{-1}\, ds.$$

On the other hand, as $\Omega(t)$ grows, bulk energy should dominate interfacial energy; hence the total energy should be asymptotic to

$$\mathscr{E} = F \operatorname{area} (\Omega(t)).$$

It seems reasonable to expect that the interface should ultimately minimize both $\mathscr{D}$ and $\mathscr{E}$. Thus, since $F < 0$, and since $\mathscr{E}$ scales as $\mathscr{D}^2$, one might expect the ultimate shape of the interface to minimize $\mathscr{D}^2/|\mathscr{E}|$, which is exactly what a Wulff region for $\beta(\theta)^{-1}$ does.

Remarks.

(1) Assume that (7.19) is valid. Then for β constant $\Omega(t)$ is asymptotic to a circle as $t \to \infty$. More generally, it follows from properties of Wulff regions that $\partial\Omega(t)$ will have a *smooth* asymptotic shape if and only if $\beta(\theta)$ has a *strictly convex* polar diagram, Polar (β); if not, the asymptotic shape will have corners in which the angle jumps across the Maxwell lines (Appendix A) of Polar (β).

(2) When Polar (e^{-1}) is strictly convex, Wulff regions for $e(\theta)$ have the *common isoperimetric ratio*

$$\operatorname{isoper}_{\text{Wulff}}(e) := (e_{\text{av}})^2 / W(e). \tag{7.20}$$

Thus, for Polar (β) strictly convex, (7.14) yields

$$\limsup_{t \to \infty} \operatorname{isoper} (\Omega(t)) \leqq \operatorname{isoper}_{\text{Wulff}} (\beta^{-1}).$$

(3) Assume that $F<0$. One can conclude from the proof of the last theorem that for the *interface initially a circle*,

$$\begin{gathered} A(t)\to 0 \quad \text{if } L(0)<\delta_0\ell, \qquad A(t)\to\infty \quad \text{if } L(0)>\ell, \\ \ell=2\pi\beta_{\max}\Phi_{\mathrm{av}}/|F|, \qquad \delta_0=\beta_{\min}f_{\min}/\beta_{\max}f_{\mathrm{av}}. \end{gathered} \tag{7.21}$$

ℓ represents a *critical circumference* for a circular interface, while $\delta_0\in(0,1]$ is a measure of the underlying *anisotropy*; for initial circumferences between $\delta_0\ell$ and ℓ, (7.21) furnishes no information. For an *isotropic material*, $\delta_0=1$, and the reference phase grows or shrinks according as the initial circumference is greater than or less than $\ell=2\pi f/|F|$.

7.3. Evolution of curvature. Fingers

Let $\boldsymbol{r}$ denote a bounded, admissible interfacial motion. The next theorem shows that the total curvature between inflection points cannot increase.

Theorem. *Let $\boldsymbol{c}$ be the trace of an evolving subcurve of $\boldsymbol{r}$. Assume that the curvature does not change sign on $\boldsymbol{c}$ and vanishes at the end points of $\boldsymbol{c}$. Then*[34]

$$\left(\frac{d}{dt}\right)\int\limits_{\boldsymbol{c}(t)}|K|\,ds\leqq 0. \tag{7.22}$$

In fact, if $\Theta(t)$ denotes the interval of angles $\Theta(s,t)$ for arc lengths s comprising $\boldsymbol{c}(t)$, then $\Theta(t)$ nests as t increases.[35]

Proof. Let $[S_1(t),S_2(t)]$ denote the arc-length interval corresponding to $\boldsymbol{c}(t)$. We will give a proof for $K(s,t)\geqq 0$ on $[S_1(t),S_2(t)]$. (The proof for $K\leqq 0$ is analogous.) For any function $\phi(s,t)$, let $\phi_i(t)=\phi(S_i(t),t)$. By hypothesis,

$$K_i(t)=0, \tag{7.23}$$

so that

$$(K_s)_1\geqq 0, \quad (K_s)_2\leqq 0. \tag{7.24}$$

By (5.5), $V=U-\dot{\Psi}$ with $U=\Phi(\theta)K$. and, since $\Phi>0$, (7.23) and (7.24) yield the conclusion that (7.24) holds with K replaced by U. On the other hand, since $\Psi=\Psi(\theta)$, $\Psi(\theta)_s=\Psi'(\theta)K=0$ at $S_1(t)$ and $S_2(t)$. Thus (7.24) holds with $(K_s)_i$ replaced by $(V_s)_i$, and, in view of (7.23) and $(2.20)_2$, with $(K_s)_i$ replaced by $(\theta_i)^{\cdot}$. The desired conclusions follow from these assertions and (since $K=|K|$) from the identity

$$\left(\frac{d}{dt}\right)\int\limits_{\boldsymbol{c}(t)}K\,ds=\dot{\theta}_2(t)-\dot{\theta}_1(t). \quad\square \tag{7.25}$$

---

[34] *Cf.* BRAKKE [1978], Prop. 2, p. 230 and ALBRESCH & LANGER [1986] for the case $V=K$.

[35] *Cf.* GRAYSON [1987], Lemma 1.9(iii), for the case $V=K$.

For a convex interface,

$$\left(\frac{d}{dt}\right) \int_{\mathscr{o}(t)} |K|\, ds = 0$$

(*cf.* (7.8)). But one can prove more. Intuitively, dividing the motion into subcurves on which K does not change sign, and then appealing to (7.22) on each subcurve, makes the following result plausible. We will give a careful proof of this theorem as well as of the remaining results of this section in [1989ag].

Theorem.[36]

$$\left(\frac{d}{dt}\right) \int_{\mathscr{o}(t)} |K|\, ds \leqq 0. \tag{7.26}$$

The next result shows that an initially convex interface will remain convex for all time. To state the theorem concisely, we use the term **inflection point** for an interfacial point[37] at which the curvature changes sign.

Theorem. *The number of inflection points cannot increase with time.*

Remark. Roughly speaking, a *finger* may be defined as a section of the interface between inflection points. The last theorem and (7.22) then have the following corollary:

$$\left.\begin{array}{l}\textit{the total number of fingers as well as the total cur-}\\ \textit{vature of each finger cannot increase with time.}\end{array}\right\} \tag{7.27}$$

The final result, essentially a consequence of (5.10) and the comparison theorem[38] for parabolic equations, shows that nested interfaces remain nested.

Theorem. *Let $\Omega(t)$ and $\Omega^*(t)$, $0 \leqq t < T$, be reference regions for two admissible interfacial motions (corresponding to the same, β, f, and F). Assume that $\Omega(0) \subset \Omega^*(0)$. Then $\Omega(t) \subset \Omega^*(t)$ for $0 \leqq t < T$.*

III. Interfacial motions with corners

8. Corners. The Frank diagram

When the interfacial energy is not stable, an admissible interfacial motion must, at each time, exhibit orientations for which the evolution equations are *backward parabolic*. Two ways of overcoming this difficulty are: (i) to regularize

[36] *Cf.* ALBRESCH & LANGER [1986] for the case $V = K$.

[37] This definition makes sense: from the parabolicity of (5.10), straight line segments in the interfacial curve disappear immediately.

[38] *Cf.* PROTTER & WEINBERGER [1967].

the evolution equations; (ii) to allow corners which correspond to jumps in θ across the unstable portions of $f(\theta)$. Here we restrict our attention to (ii).[39]

8.1. Corners

In this section we will make extensive use of the relation (4.9) expressing the capillary force as a function of the angle θ:

$$C = C(\theta) = f(\theta)\,T(\theta) + f'(\theta)\,N(\theta). \tag{8.1}$$

Corners are defined by a jump discontinuity in the dependence of θ on s. An immediate consequence of balance of forces (3.2) is the continuity of $C(\theta(s, t))$ with respect to s. Thus at a corner defined by a jump in orientation from θ^- to θ^+ we must have $C(\theta^-) = C(\theta^+)$.[40] This discussion should motivate the following definition. Let θ^-, θ^+ be *distinct* angles with $C(\theta^-) = C(\theta^+)$. Then the *ordered pair* $\{\theta^-, \theta^+\}$ is: a **corner** if $|\theta^+ - \theta^-| < \pi$; a **cusp** if $\theta^+ - \theta^- = \pi$. One should visualize $\{\theta^-, \theta^+\}$ as representing a jump in angle *from* θ^- *to* θ^+ *as arc length increases.* If $\{\theta^-, \theta^+\}$ is a corner, then $\{\theta^+, \theta^-\}$ is a corner.

Cusp and corner theorem. *Cusps are not possible. Corners are not possible when the interfacial energy is strictly stable. In fact, if $\{\theta^-, \theta^+\}$ is a corner (labelled so that $0 < \theta^+ - \theta^- < \pi$), then either f is unstable somewhere in (θ^-, θ^+) or $f(\theta) + f''(\theta) \equiv 0$ on (θ^-, θ^+).*

Proof. Let $\{\theta; \theta + \pi\}$ be a cusp. Since $N(\theta) = -N(\theta + \pi)$ and $T(\theta) = -T(\theta + \pi)$, if we take the inner product of $T(\theta)$ with $C(\theta) = C(\theta + \pi)$ we conclude, with the aid of (8.1), that $f(\theta) = -f(\theta + \pi)$, which contradicts the assumption $f > 0$.

Next, since $N(\theta) = (\cos\theta, \sin\theta)$, we may use (4.17) to conclude that $C(\theta^-) = C(\theta^+)$ if and only if

$$\int_{\theta^-}^{\theta^+} [f(\theta) + f''(\theta)] \cos(\theta + \alpha)\, d\theta = 0$$

for all $\alpha \in \mathbf{R}$, and this implies the remaining assertions of the theorem. □

8.2. The Frank diagram

In this section we will use the notation and terminology of Appendix A. Here, because the interfacial energy is smooth, the polar diagrams we will encounter will not have sharp spots; for that reason we will give a direct proof of certain assertions, even though these assertions actually follow from the more general results of Appendix A.

---

[39] *Cf.* § 9.7.

[40] HERRING [1951a], eq. (19).

The **Frank diagram**[41] is the polar diagram of $f(\theta)^{-1}$, and is hence the locus of the **Frank potential**

$$\sigma(\theta) = f(\theta)^{-1} N(\theta). \tag{8.2}$$

The Frank potential and the capillary force (8.1) have an interesting relation. First of all, by (2.4),

$$C(\theta) = -f(\theta)^2 \sigma'(\theta), \tag{8.3}$$

so that the capillary force is *tangent* to the Frank diagram and points in the direction of decreasing θ. Further, (A1) with $g(\theta) = f(\theta)^{-1}$ and (T2) of Appendix A yield the following theorem (*cf.* Figure 8A).

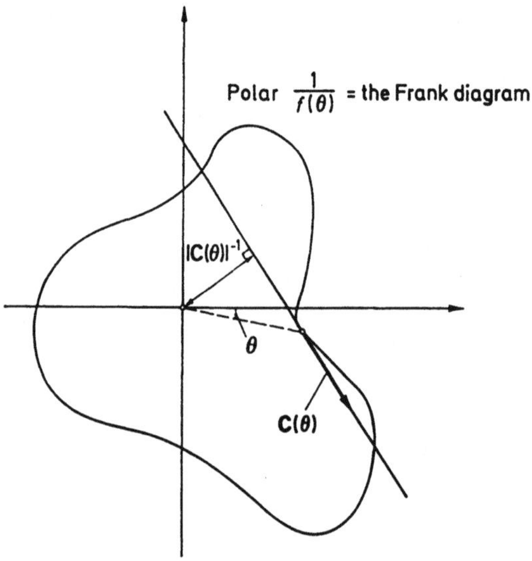

Fig. 8A. The Frank diagram = Polar (f^{-1}). The capillary force $C(\theta)$ is the negative of the supporting tangent of the Frank diagram

Theorem. *The capillary force is the negative of the supporting tangent of the Frank diagram. Thus*

$$C(\theta) = -\sigma^*(\theta), \tag{8.4}$$

and $|C(\theta)|^{-1}$ *is the support function of the Frank diagram.*

In view of this result, $\{\theta^-, \theta^+\}$ is a corner if and only if θ_1 and θ_2 have a common supporting tangent (relative to the Frank diagram).

The next theorem[42] establishes the existence of corners for unstable interfacial energies and shows how this energy may be decomposed into stable sections separated by corners.

[41] FRANK [1963].

[42] The ideas underlying this theorem are due to WULFF [1901], HERRING [1951b], FRANK [1963], GJOSTEIN [1963], and GRUBER (*cf.* GJOSTEIN [1963]).

Convexity-stability theorem.

(i) *The Frank diagram is convex if and only if f is stable.*
(ii) *More generally, f is stable on the globally-convex sections of the Frank diagram. If* (θ^-, θ^+) *is an open interval separating two adjacent globally-convex sections, then* $\{\theta^-, \theta^+\}$ *is a corner and f is unstable somewhere in* (θ^-, θ^+).

Proof. Let $k_F(\theta)$ denote the curvature of the Frank diagram, so that $k_F(\theta)$ is given by (A5) with $g(\theta) = f(\theta)^{-1}$. Then

$$k_F(\theta) \geqq 0 \quad \text{if and only if} \quad f(\theta) + f''(\theta) \geqq 0, \tag{8.5}$$

and (i) and the first assertion of (ii) follows.

Suppose that (θ^-, θ^+) is an open interval separating two adjacent, globally convex sections. Then (θ_1, θ_2) is an angle interval for a Maxwell line, and the final assertion of (ii) follows from (ii) of the Maxwell theorem (Appendix A). □

This theorem should motivate the following terminology in which we use "GS" as shorthand for the term **"globally-stable"**. The globally convex sections of the Frank diagram will be referred to as **GS sections of the energy**; angles θ that belong to GS sections will be referred to as **GS angles**; an open interval (θ^-, θ^+) that separates two *adjacent* GS sections will be referred to as a **globally unstable section**; the corresponding corners $\{\theta^-, \theta^+\}$ and $\{\theta^+, \theta^-\}$ will be referred to as **GS corners** and the angles θ^- and θ^+ as **GS corner angles.**

We will refer to an energy f as **regular** if:

(R1) the GS sections of f are finite in number;
(R2) each GS angle θ is strictly stable;
(R3) the Maxwell lines (Appendix A) of the Frank diagram are mutually disjoint.

(*Cf.* Figure 8B.) A consequence of the proposition containing (8.3) and the properties of the convex hull is the following alternative characterization of regular energies.

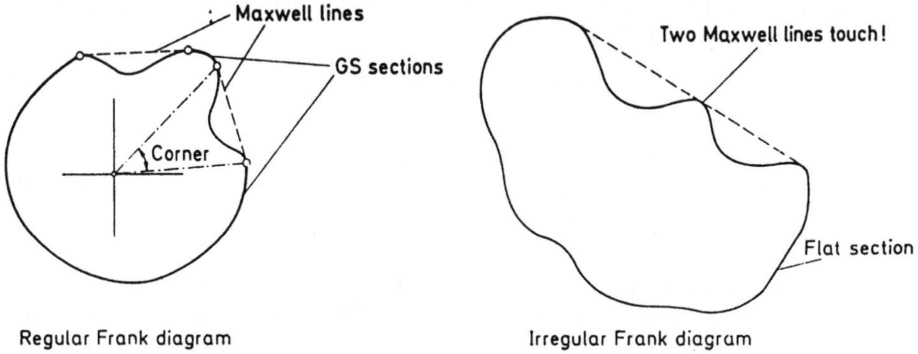

Fig. 8B. Examples of Frank diagrams

Proposition. *Regular interfacial energies have the following properties:*
(R4) *GS sections are not singletons.*
(R5) *Let* $\{\theta^-, \theta^+\}$ *be a corner with* θ^- *and* θ^+ *GS angles. Then* $\{\theta^-, \theta^+\}$ *is a GS corner, and (aside from* $\{\theta^+, \theta^-\}$*) there is no other corner involving* θ^- *or* θ^+.
In fact, granted (R1) *and* (R2), (R3) *is equivalent to either* (R4) *or* (R5).

9. Unstable interfacial energies. Motions with corners

Consider now an energy f such that

$$f \text{ is not stable but regular.} \tag{9.1}$$

It seems reasonable to consider "motions" in which the interface at each time is a piecewise smooth closed curve whose regular arcs and "corners", respectively, correspond to GS sections and GS corners of f. For such an "interfacial motion" the evolution equations are *parabolic*, since the nonparabolic portions are removed by corners, but the positions of the corners as functions of time are not known *a priori* and hence constitute *free boundaries*. The next section begins our discussion of such motions.

9.1. Corner conditions for piecewise smooth evolving curves that are admissible

Let $\boldsymbol{r} = \{\boldsymbol{r}_1, \boldsymbol{r}_2, \ldots\}$ be a PS (piecewise smooth) evolving curve. We will use the notation and terminology of Section 2.4; thus $\boldsymbol{R}_i(t)$ and $\dot{\boldsymbol{R}}_i(t)$ are the position and total velocity of the juncture i; s is the arc length; $\boldsymbol{T}(s, t)$, $\boldsymbol{N}(s, t)$, $\theta(s, t)$, $K(s, t)$, $V(s, t)$, and $v(s, t)$ are the tangent, normal, angle, curvature, normal velocity, and arc velocity.

Our interest is in PS evolving curves that are consistent with balance of capillary forces and have arcs which correspond to GS sections of the energy. Such curves are the subject of the next definition.

Let $\boldsymbol{r}$ be a PS evolving curve. Then $\boldsymbol{r}$ is **admissible** if:[43]
(A1) $\theta(s, t)$ is always GS;
(A2) the capillary force $C(\theta)$ defined by (8.1) is consistent with capillary balance in the form (*cf.* (4.10))

$$\int_{\partial c(t)} C(\theta) = \int_{c(t)} (F + \beta(\theta)\, V)\, N\, ds; \tag{9.2}$$

whenever c is the trace of an evolving subcurves of $\boldsymbol{r}$.
Given an admissible PS evolving curve, smoothness dictates that on each evolving arc the angle $\theta(s, t)$ belong to a single GS section. Further, we may, without

[43] There is a slight ambiguity in our use of the word "admissible": in Section 4.2, which concerned *smooth* evolving curves, admissibility meant consistency with balance of capillary forces and the underlying constitutive equations. Here and in what follows we require, in addition, that admissible evolving curves have angles $\theta(s, t)$ that are GS.

loss in generality, assume that on adjacent arcs $\theta(s, t)$ belongs to different GS sections; were this not the case $\theta(s, t)$ would be continuous across the juncture of the arcs, and the arcs may be combined to form a single (smooth) evolving curve.

Alternative characterization of admissibility. Let $\mathbf{r}$ *be a PS evolving curve consistent with* (A1). *Then* $\mathbf{r}$ *is admissible if and only if*:
(A3) *each of the arcs* $\mathbf{r}_i$ *evolves according to*

$$V = \Phi(\theta)\,K - \Psi(\theta),$$
$$\Phi(\theta) = [f(\theta) + f''(\theta)]/\beta(\theta), \qquad \Psi(\theta) = F/\beta(\theta); \tag{9.3}$$

(A4) *each of the pairs* $\{\theta(S_i(t) - 0, t), \theta(S_i(t) + 0, t)\}$ *(i a juncture) is independent of time and is a GS corner of the interfacial energy.*

Proof. Note that, by Remark (ii) of Section 4.1,

$$\text{(A2) is equivalent to (A3) and the continuity in } s \text{ of } C(\theta(s, t)). \tag{9.4}$$

Thus, in view of the definition of a corner, (A3) and (A4) yield admissibility. Conversely, suppose $\mathbf{r}$ is admissible. Then, by (9.4), (A3) holds and each $\{\theta(S_i(t) - 0, t), \theta(S_i(t) + 0, t)\}$ is a corner. Thus, in view of (9.1) and the last proposition of Section 8.2, (A4) follows. □

In view of (A4), the corner-angles

$$\theta_i^{\pm} := \theta(S_i(t) \pm 0, t) \tag{9.5}$$

are *constants*, as are the quantities (*cf.* (2.40))

$$\varkappa_i = \mathbf{T}(\theta_i^+) \cdot \mathbf{N}(\theta_i^-) = -\mathbf{T}(\theta_i^-) \cdot \mathbf{N}(\theta_i^+). \tag{9.6}$$

It is often convenient to use (θ, t) as independent variables on a given convex arc, irrespective of the sign the curvature takes on the other arcs.

Corner conditions for admissible PS evolving curves. *At each juncture i*:
(CC1) *the capillary force* $\mathbf{C}(s, t)$ *is continuous across* $s = S_i(t)$;
(CC2) $K(S_i(t) \pm 0, t)\,\varkappa_i \geqq 0, \quad K(S_i(t) - 0, t)\,K(S_i(t) + 0, t) \geqq 0$;
(CC3) $V_s(S_i(t) \pm 0, t) = [v(S_i(t) \pm 0, t) - S_i^{\cdot}(t)]\,K(S_i(t) \pm 0, t)$;
(CC4) *if the evolving arcs* $\mathbf{r}_i$ *and* $\mathbf{r}_{i+1}$ *are convex, then* $V(\theta, t)\,\mathbf{N}(\theta) - V_\theta(\theta, t) \times \mathbf{T}(\theta, t)$ *is continuous across* $\{\theta_i^-, \theta_i^+\}$ *and its value at* $\theta = \theta_i^{\pm}$ *is the total velocity* $\mathbf{R}_i^{\cdot}(t)$.

Proof. (9.4) yields (CC1), (CC3) follows from $(2.20)_2$ and (A4), and (CC4) follows from $(2.42)_1$, (2.24), and (2.18). We have only to prove (CC2). Since $\{\theta_i^-, \theta_i^+\}$ is a GS corner of f, one of the intervals (θ_i^-, θ_i^+) (θ_i^+, θ_i^-) is a globally unstable section of f. Assume that (θ_i^-, θ_i^+) is such a section. Then, by property (ii) of admissible PS evolving curves, $\theta \leqq \theta_i^-$ on $\mathscr{s}_i$, while $\theta \geqq \theta_i^+$ on $\mathscr{s}_{i+1}$, so that $K(S_i(t) \pm 0, t) \geqq 0$. On the other hand, since $0 < \theta_i^+ - \theta_i^- < \pi$, (9.6) implies that $\varkappa_i \geqq 0$, so that (CC2) is satisfied. A similar argument applies when (θ_i^+, θ_i^-) is a globally unstable section of f. □

Remark. It should be emphasized that the corner inequalities (CC2) are based on the hypothesis that the underlying arcs correspond to adjacent GS sections of the interfacial energy. These inequalities imply that *corners preserve local convexity*. In particular (*cf.* (2.40)),

$$\left.\begin{array}{l}\textit{if two adjacent arcs are convex, then their curvatures as well as the}\\ \textit{transition curvature of the corner between them are of the same sign.}\end{array}\right\} \quad (9.7)$$

9.2. Facetings. Evolving curves with wrinkles

Let $\boldsymbol{r} = \{\boldsymbol{r}_1, \boldsymbol{r}_2, \ldots, \boldsymbol{r}_N\}$ be an admissible PS evolving curve. Then $\boldsymbol{r}$ is a **faceting** if each of its arcs is a *facet*, and if no two adjacent arcs combine to form a single facet. On each facet, $\theta(s, t)$ is independent of s; if $\theta(s, t) \equiv$ constant on each facet, then $\boldsymbol{r}$ has **fixed orientations.** An example of a faceting with fixed orientations is a **wrinkling** (Figure 9A); here there are *fixed* angles θ_{odd} and θ_{even} such that, for all t,

$$\begin{aligned} \theta(s, t) &= \theta_{\text{odd}} \quad \text{on } \mathscr{s}_i(t) \text{ for } i \text{ odd},\\ \theta(s, t) &= \theta_{\text{even}} \quad \text{on } \mathscr{s}_i(t) \text{ for } i \text{ even}; \end{aligned} \qquad (9.8)$$

in this case $\boldsymbol{r}$ is a **wrinkling between angles** θ_{odd} **and** θ_{even}. Note that, by $(9.3)_1$, for a wrinkling the normal velocity V of each facet is constant (in space and time) with

$$\begin{aligned} V &= -\Psi(\theta_{\text{odd}}) \quad \text{on } \mathscr{s}_i \text{ for } i \text{ odd},\\ V &= -\Psi(\theta_{\text{even}}) \quad \text{on } \mathscr{s}_i \text{ for } i \text{ even}. \end{aligned} \qquad (9.9)$$

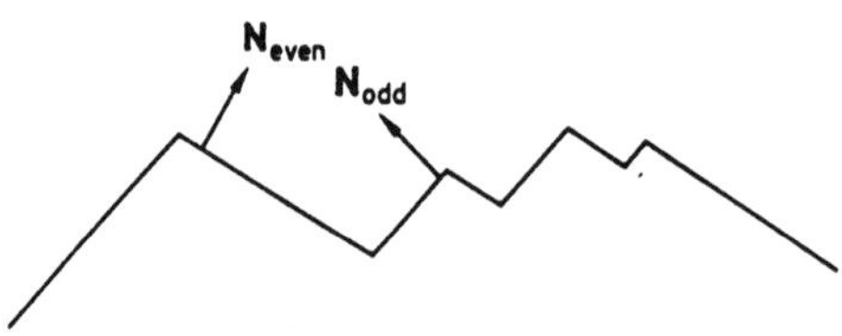

Fig. 9A. Wrinkling between the angles θ_{odd} and θ_{even}. Here $N_{\text{odd}} = N(\theta_{\text{odd}})$ and $N_{\text{even}} = N(\theta_{\text{even}})$

By definition, on a facet, $\theta(t) = \theta(s, t)$ is independent of s. Therefore its normal velocity is given by $V = -\Psi(\theta(t))$ and $V_s = 0$. Thus, by $(2.18)_2$, $\theta^{\cdot} = 0$; since $\theta_t = \theta^{\cdot} - v\theta_s$, $\theta_t = 0$. Thus the angle θ is identically constant on a facet. Consider now a faceting. Each internal facet meets two corners, $\{\alpha^-, \alpha^+\}$ and $\{\theta^-, \theta^+\}$, say. Since the orientation of the facet is constant, $\alpha^+ = \theta^-$. Our assumption that the energy be regular then implies that $\alpha^- = \theta^+$ (*cf.* (R5)). Thus and by (A1) we have the following

Proposition.[44] *The only possible facetings are wrinklings between the fixed angles of a GS corner.*

Let r_A and r_B be admissible evolving curves, and let $r = \{r_1, r_2, \ldots, r_N\}$ be a faceting. Then r **connects** r_A and r_B if $\{r_A, r_1, r_2, \ldots, r_N, r_B\}$ is an admissible PS evolving curve. The next theorem, the main result of this section, shows that wrinkles decay with time.

Wrinkle-decay theorem. *Let r be a wrinkling that connects admissible evolving curves r_A and r_B. Then the total length of the wrinkling decreases with time. In fact, the lengths of the initial and terminal facets decrease with time, while the lengths of the internal facets remain constant.*

Our proof is based on two subsidiary results.

Proposition. *The length of each internal facet of a wrinkling is independent of time. Thus (since the orientation of each facet is fixed) the system of internal facets behaves as a rigid body undergoing translational motion.*

Proof. Let r_i be an internal facet (Figure 9B). The two adjacent facets $r_{i\pm1}$ have the same normal velocity $V_{i\pm1} = -\Psi(\theta_{i\pm1})$, so that the distance d between them does not change with time. The middle facet r_i will move relative to its two neighbors, but it will always meet them at the fixed angle $|\theta_i - \theta_{i-1}| = \alpha$. Therefore its length $L(t) = d/\sin\alpha$ is also independent of time. □

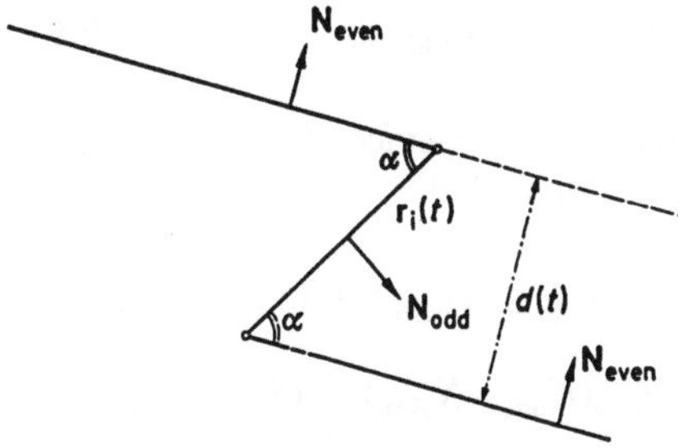

Fig. 9B. The length of an internal facet $r_i(t)$ does not change with time

Let $r = \{r_1, r_2, \ldots, r_N\}$ be a wrinkling that connects admissible evolving curves r_A and r_B. It is convenient to use the following notation: $\dot{R}_A$, θ_A, T_A, N_A, K_A, and V_A, respectively, denote the total velocity, orientation, tangent, normal, curvature, and normal velocity of the **terminal** point of r_A; an analogous notation applies to the corresponding quantities associated with the **initial** point of r_B; $\varkappa_A$ denotes the corner curvature between r_A and the initial facet; $\varkappa_B$ denotes the corner curvature between the terminal facet and r_B; L_A and L_B, respectively,

[44] Here the smoothness of $f(\theta)$ is crucial. In Section 10.3 we will show that certain nonsmooth energies exhibit more general types of facetings.

denote the lengths of the initial and terminal facets; given any function $g(\theta)$, we write $g_{\text{odd}} = g(\theta_{\text{odd}})$, $g_{\text{even}} = g(\theta_{\text{even}})$, with θ_{odd} and θ_{even} as in (9.8).

Properties of wrinklings that connect evolving curves. *Let* $\boldsymbol{r} = \{\boldsymbol{r}_1, \boldsymbol{r}_2, \ldots, \boldsymbol{r}_N\}$ *be a wrinkling that connects admissible evolving curves* $\boldsymbol{r}_A$ *and* $\boldsymbol{r}_B$. *Then for N even:*

$$\left.\begin{aligned} \theta_A &= \theta_{\text{even}}, \quad & \boldsymbol{R}_A^{\boldsymbol{\cdot}} \cdot \boldsymbol{N}_{\text{odd}} &= -\Psi_{\text{odd}}, \\ \theta_B &= \theta_{\text{odd}}, \quad & \boldsymbol{R}_B^{\boldsymbol{\cdot}} \cdot \boldsymbol{N}_{\text{even}} &= -\Psi_{\text{even}}; \end{aligned}\right\} \tag{9.10}$$

for N odd:

$$\theta_A = \theta_B = \theta_{\text{even}}, \quad \boldsymbol{R}_A^{\boldsymbol{\cdot}} \cdot \boldsymbol{N}_{\text{odd}} = \boldsymbol{R}_B^{\boldsymbol{\cdot}} \cdot \boldsymbol{N}_{\text{odd}} = -\Psi_{\text{odd}}; \tag{9.11}$$

in either case:

$$\left.\begin{aligned} \Phi(\theta_A)\, K_A = -\varkappa_A L_A^{\boldsymbol{\cdot}}, \quad \Phi(\theta_B)\, K_B = -\varkappa_B L_B^{\boldsymbol{\cdot}} \quad & \text{if } N \geqq 2, \\ \Phi(\theta_A)\,[K_A - K_B] = -\varkappa_A L_A^{\boldsymbol{\cdot}} \quad & \text{if } N = 1. \end{aligned}\right\} \tag{9.12}$$

Proof. We will establish only those results which concern A. The conclusions concerning B are verified analogously. Assume that $N \geqq 2$. Then the corner condition $(2.42)_1$, applied at $s = S_A(t)$ to the corner between $\boldsymbol{r}_A$ and $\boldsymbol{r}_1$ and at $s = S_1(t)$ to the corner between $\boldsymbol{r}_1$ and $\boldsymbol{r}_2$, yields the relations

$$\boldsymbol{R}_A^{\boldsymbol{\cdot}} = -\Psi_{\text{odd}} \boldsymbol{N}_{\text{odd}} + (S_A^{\boldsymbol{\cdot}} - v_1)\, \boldsymbol{T}_{\text{odd}},$$

$$-\Psi_{\text{odd}} \boldsymbol{N}_{\text{odd}} + (S_1^{\boldsymbol{\cdot}} - v_1)\, \boldsymbol{T}_{\text{odd}} = -\Psi_{\text{even}} \boldsymbol{N}_{\text{even}} + (S_1^{\boldsymbol{\cdot}} - v_2)\, \boldsymbol{T}_{\text{even}}.$$

Thus $\boldsymbol{R}_A^{\boldsymbol{\cdot}} \cdot \boldsymbol{N}_{\text{odd}} = -\Psi_{\text{odd}}$. Further, since $V_A = \boldsymbol{R}_A^{\boldsymbol{\cdot}} \cdot \boldsymbol{N}_{\text{even}}$ and $\varkappa_A = \boldsymbol{T}_{\text{odd}} \cdot \boldsymbol{N}_{\text{even}}$ (*cf.* (9.6)), the above relations yield

$$V_A = -\Psi_{\text{odd}}(\boldsymbol{N}_{\text{odd}} \cdot \boldsymbol{N}_{\text{even}}) + (S_A^{\boldsymbol{\cdot}} - v_1)\, \varkappa_A,$$

$$-\Psi_{\text{odd}}(\boldsymbol{N}_{\text{odd}} \cdot \boldsymbol{N}_{\text{even}}) + (S_1^{\boldsymbol{\cdot}} - v_1)\, \varkappa_A = -\Psi_{\text{even}}.$$

Thus $V_A = -\Psi_{\text{even}} + (S_A - S_1)^{\boldsymbol{\cdot}}\, \varkappa_A$. But $L_A = S_1 - S_A$ and, by (9.3), $V_A + \Psi_{\text{even}} = \Phi(\theta_A)\, K_A$; thus $\Phi(\theta_A)\, K_A = -\varkappa_A L_A^{\boldsymbol{\cdot}}$.

We have established all of the results for A except $(9.12)_2$. Thus let $N = 1$. Applying $(2.42)_1$ to the corners at $s = S_A(t)$ and $s = S_B(t)$, taking the inner product of the resulting relations with $\boldsymbol{N}_{\text{even}}$, and then subtracting the two relations yields $V_A - V_B = (S_A - S_B)^{\boldsymbol{\cdot}}\, \varkappa_A$. But $\theta_A = \theta_B$ and $L_A = L_B = S_B - S_A$; in view of (9.3), $(9.12)_2$ follows. □

Proof of the wrinkle-decay theorem. For $N \geqq 2$ the proof follows from (CC2), $(9.12)_1$, and the proposition following the wrinkle-decay theorem. For $N = 1$, (9.6) yields $\varkappa_B = -\varkappa_A$, so that $(K_A - K_B)/\varkappa_A = (K_A/\varkappa_A) + (K_B/\varkappa_B)$; (CC2) and $(9.12)_2$ then yield $L_A^{\boldsymbol{\cdot}} \leqq 0$. □

9.3. Curves that are convex except for wrinkles

A **PS admissible evolving curve is convex** if each of its arcs is convex. The remark (9.7) renders this definition meaningful. In particular, at each t the corresponding PS curve $\mathscr{A}(t)$ is convex in the usual sense for continuous curves. For convenience, we write "**CPS**" as an abbreviation for "*convex, piecewise smooth*".

An admissible PS evolving curve $\mathbf{r}$ is **convex except for wrinkles** if $\mathbf{r}$ consists of CPS evolving curves, called **convex sections**, connected by wrinklings. Each convex section is the union of convex *arcs* (which are smooth). A convex section $\mathbf{r}_C$ is **internal** if $\mathbf{r}$ is closed, or if the initial and terminal arcs of $\mathbf{r}_C$ are internal arcs of $\mathbf{r}$; in this case the initial and terminal points of $\mathbf{r}_C$ connect to wrinklings. It is generally most convenient to take (θ, t) as independent variables on each convex section and to express the underlying system of evolution equations in terms of the normal velocity $V(\theta, t)$. For an internal convex section, (5.7), (CC4), (9.10), and (9.12) yield an interesting system of equations for $V(\theta, t)$.

Evolution equations for convex sections. *Let* $\mathbf{r}$ *be a PS evolving curve that is closed, admissible, and convex except for wrinkles. Consider a convex section of* $\mathbf{r}$, *let* θ_A *and* θ_B, *respectively, denote the angles corresponding to the initial and terminal points of the section, and let* $\{\gamma_A, \theta_A\}$ *and* $\{\theta_B, \gamma_B\}$ *denote the corners at these terminal points. Then the evolution of* $V(\theta, t)$ *for this section is governed by the following conditions*:

(E1) *on each convex arc*:

$$\Phi(\theta)\, V_t = [V + \Psi(\theta)]^2\, [V_{\theta\theta} + V];$$

(E2) $VN - V_\theta T$ *is continuous across corners separating convex arcs*;

(E3) *at the initial and terminal points*:

$$[V(\theta_A, t)\, N(\theta_A) - V_\theta(\theta_A, t)\, T(\theta_A)] \cdot N(\gamma_A) = -\Psi(\gamma_A),$$

$$[V(\theta_B, t)\, N(\theta_B) - V_\theta(\theta_B, t)\, T(\theta_B)] \cdot N(\gamma_B) = -\Psi(\gamma_B).$$

Remark. It seems reasonable to expect that (E1)–(E3), with compatible initial data, yield a well-posed problem for $V(\theta, t)$ on each convex section. Thus wrinklings between convex sections essentially decouple these sections from each other, at least until the wrinkles decay. If $V(\theta, t)$ and (hence) $K(\theta, t)$ are known on each section, then the evolution of the wrinkles (from prescribed initial positions) is easily determined using (9.12) and properties (i) and (ii) of wrinklings. Granted this is done, the initial and terminal positions of the convex sections are known as functions of time, and this data and a knowledge of the corresponding curvatures yields the complete evolutionary behavior of the convex sections, and hence of the complete evolving curve.

9.4. Equations near a corner when the curve is a graph

Consider the situation shown in Figure 9C, in which an evolving PS curve is represented, in a neighborhood of a corner $\{\theta^-, \theta^+\}$, as the graph of a function

$y = h(x, t)$. Here x ranges in an interval (x_0, x_1); $t \in [0, T)$; $x = \zeta(t)$ is the position at time t of the corner; and the curve is oriented so that arc length increases with increasing x. Then $h(x, t)$ is continuous and piecewise smooth, with a jump discontinuity in $p = h_x$ at the **free boundary** $x = \zeta(t)$.

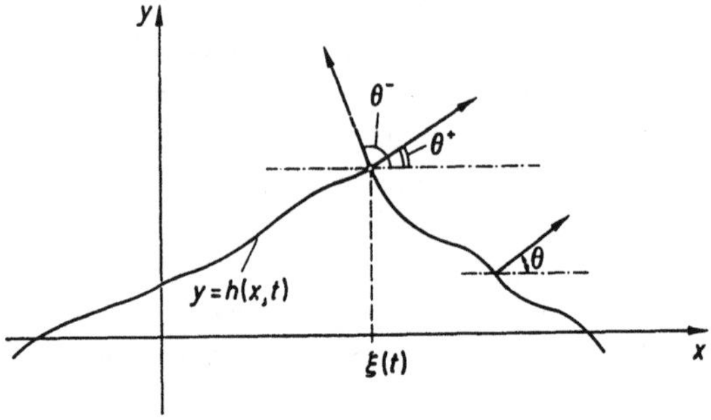

Fig. 9C. A corner when the evolving curve is a graph $y = h(x, t)$; $x = \xi(t)$ marks the corner

The function p satisfies (5.11) away from the free boundary and is consistent with two free-boundary conditions. The first of these, a direct consequence of (5.9), is given by

$$p(\zeta(t) \pm 0, t) = P^{\pm} \tag{9.13}$$

with

$$P^{\pm} = -\cot \theta^{\pm}.$$

The second condition is more complicated. For any function $\phi(x, t)$, let $\phi^{\pm}(t) = \phi(\zeta(t) \pm 0, t)$. By $(2.20)_1$,

$$V^{\pm} = \boldsymbol{R}^{\boldsymbol{\cdot}} \cdot \boldsymbol{N}^{\pm}; \tag{9.14}$$

thus, since $\boldsymbol{N} = (\cos\theta, \sin\theta)$, if we eliminate the y-component of $\boldsymbol{R}^{\boldsymbol{\cdot}}$ between the two equations (9.14), and use the fact that $\zeta^{\boldsymbol{\cdot}}$ is the x-component of $\boldsymbol{R}^{\boldsymbol{\cdot}}$, we arrive at an expression for $\zeta^{\boldsymbol{\cdot}}$ as a function of $V^{\pm}$ and $\theta^{\pm}$; the expressions $V = \Phi K - \Psi$ and $K = p_x(1 + p^2)^{-3/2}$ then lead to the free-boundary condition:

$$\zeta^{\boldsymbol{\cdot}} = A^{+}(p_x)^{+} - A^{-}(p_x)^{-} - C \tag{9.15}$$

with $A^{\pm}$ and C the constants defined by

$$A^{\pm} = \Phi(\theta^{\pm})\, D^{\pm}/W^{\pm}, \qquad C = \Psi(\theta^{+})\, D^{+} - \Psi(\theta^{-})\, D^{-},$$

$$W^{\pm} = (1 + (P^{\pm})^2)^{\frac{3}{2}}, \quad D^{+} = \sin\theta^{-}/a, \quad D^{-} = \sin\theta^{+}/a,$$

$$a = \sin(\theta^{-} - \theta^{+}).$$

The basic system of equations then consists in (5.11) away from $x = \zeta(t)$ supplemented by (9.13) and (9.15) at $x = \zeta(t)$. A change independent variable renders this system more transparent. Thus let

$$u(x, t) = \begin{cases} A^{-}(p(x, t) - P^{-}) & \text{for } x < \zeta(t) \\ A^{+}(p(x, t) - P^{+}) & \text{for } x > \zeta(t), \end{cases}$$

so that $u(x, t)$ is continuous across $x = \zeta(t)$. Further, let Q and B be as specified in (5.10), and define

$$Q^{\pm}(u) = Q((u/A^{\pm}) + P^{\pm}), \quad B^{\pm}(u) = A^{\pm}B((u/A^{\pm}) + P^{\pm}).$$

Then the system under consideration reduces to the *partial differential equations*

$$\left.\begin{aligned} u_t &= [Q^{-}(u)\,u_x - B^{-}(u)]_x \quad \text{for } x < \zeta(t), \\ u_t &= [Q^{+}(u)\,u_x - B^{+}(u)]_x \quad \text{for } x > \zeta(t), \end{aligned}\right\} \tag{9.16}$$

in conjunction with the *free-boundary conditions*

$$\begin{gathered} u(\zeta(t) \pm 0, t) = 0, \\ u_x(\zeta(t) + 0, t) - u_x(\zeta(t) - 0, t) = \zeta^{\cdot}(t) + C. \end{gathered} \tag{9.17}$$

Apart from the constant C, which may be transferred from (9.17) to (9.16) by the coordinate change $x^* = x + Ct$, (9.17) *are exactly the free-boundary conditions of the classical Stefan problem.*

9.5. Stationary interfaces and steady interfacial motions with corners

Stationary interfaces and steady interfacial motions are defined as for smooth interfaces,[45] except that we now add the requirement that all angles be GS angles and all corners GS corners.

An argument analogous to that given in Section 5.1 then implies that $r(\theta)$ defined on the set of GS angles θ by

$$r(\theta) = F^{-1}[f(\theta)\,N(\theta) - f'(\theta)\,T(\theta)]$$

yields a closed, convex stationary interface that is PS. (The remark given in the last paragraph of Section 6.1 and the fact that θ jumps only at GS corners implies that $r(\theta)$ is continuous across such jump.)

We also have the possibility of steady motions with corners. Let $F \neq 0$ be given. Let $\{\theta_1, \theta_2\}$, $0 < \theta_2 - \theta_1 < \pi$, be a GS corner, let ℓ be a line with ℓ a chord for Polar (β) between θ_1 and θ_2 (if β is strictly convex, then exactly one such line exists), and compute U by the requirement that $-(F/|U|^2)\,U$ be the support vector for ℓ (*cf.* Section 6.2). Then any (stationary) infinite wrinkling in which θ jumps back and forth between θ_1 and θ_2 is a portrait of a steady interfacial motion with steady velocity U. One might refer to this motion as a *steady wrinkling* (Figure 9D).

Other solutions are possible. For example, when the Frank diagram and the polar diagram of β are of the form shown in Figure 9E, there is a steadily receding bump (with a corner) as shown. Similarly, one can construct advancing bumps with corners for nonconvex Frank and β-diagrams of certain prescribed shapes.

[45] To begin with, our definition of a PS evolving curve is restricted to bounded curves, but the extension to unbounded curves is straightforward.

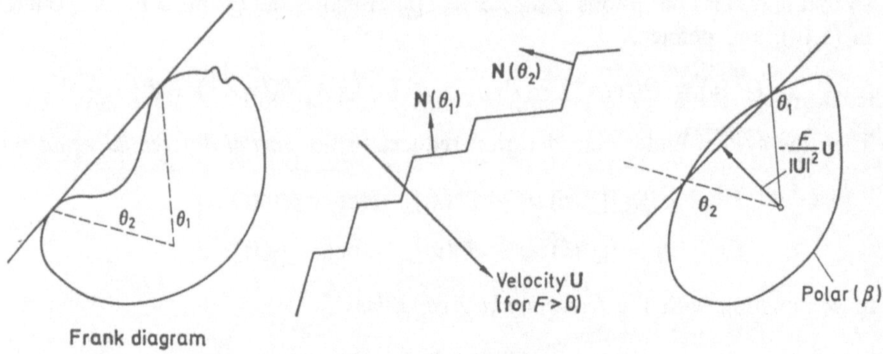

Fig. 9 D. Construction of a steady wrinkling

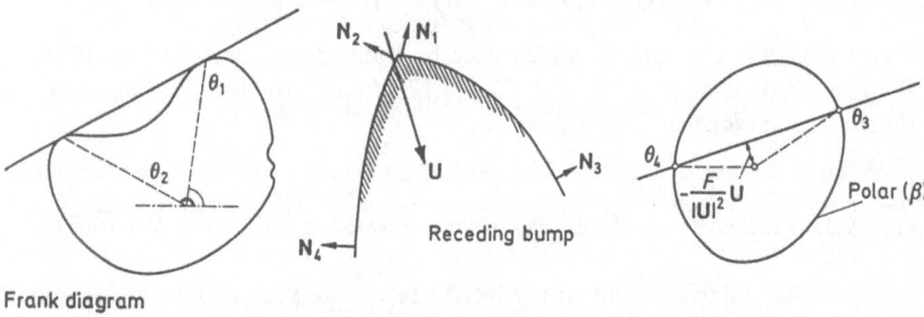

Fig. 9E. Construction of a receding bump

9.6. Existence

Let Ω_0 be a bounded, simply connected domain in $\mathbf{R}^2$. Then $\partial\Omega_0$ is a GS **boundary** if it is piecewise $C^{2+\alpha}$ for some $\alpha \in (0, 1)$, if its outward normal always points in a GS direction, and if all of its corners are GS corners (*cf.* Section 8.2).

Existence theorem.[46] *Let f and β be C^∞. Let Ω_0 be a given initial domain, assumed to be admissible. Then there is a unique, maximal family of domains $\Omega(t)$ $(0 \leqq t < T_{\max})$ such that:*

(i) *$\partial\Omega(t)$ is an admissible PS evolving curve;*
(ii) *$\Omega(0) = \Omega_0$.*
In fact, this evolving curve is piecewise C^∞ for $0 < t < T_{\max}$.

Further, for $T_{\max} < \infty$, as $t \to T_{\max}$ either (E1) or (E2) (of Section 7.1) must hold, or

(E3) an arc of $\partial\Omega(t)$ shrinks to zero.

The condition (E3) is possible with the facets of a wrinkling and with arcs whose initial and terminal points correspond to the same corner angle.

[46] ANGENENT & GURTIN [1989].

9.7. *A note on regularized equations*

Another method of treating situations in which the evolution equations are backward parabolic is to develop a suitable regularization of these equations. Such a regularization will be discussed elsewhere; under certain simplifying assumptions (among them that $\beta = 1$) this regularization reduces to the following fourth-order parabolic system for a convex section:

$$K_t = K^2(V_{\theta\theta} + V),$$

$$V = \Phi(\theta)\, K - \varepsilon K[(K^2)_{\theta\theta} + K^2] - F,$$

with $\varepsilon > 0$ a small constant.

IV. Nonsmooth interfacial energies

10. Interfacial energies with sharp spots

Material scientists often consider interfacial energies that are continuous but have derivatives which suffer jump discontinuities.[47] We now discuss energies of this type.

10.1. *Sharp spots. The capillary set. Stability*

By an **interfacial energy with sharp spots**[48] we mean a 2π-periodic, strictly positive function $f(\theta)$ on $\mathbf{R}$ that is smooth except for sharp spots (Appendix A). The **sharp spots** are then the angles across which $f'(\theta)$ jumps, the remaining angles are **smooth spots**. Tacit in this definition is the requirement that the number of sharp spots not be zero.

Let θ_0 be a sharp spot. By (8.1), the capillary force[49] $\hat{C}(\theta)$ is discontinuous across θ_0: the tangential component $\sigma = f(\theta)$ is continuous, but the normal component $\xi = f'(\theta)$ is not:

$$\hat{C}(\theta_0 + 0) - \hat{C}(\theta_0 - 0) = [f'(\theta_0 + 0) - f'(\theta_0 - 0)]\, N(\theta_0).$$

If we think of the energy at a sharp spot as the limit of a sequence of smooth, locally-convex energies, then it seems reasonable to allow the capillary shear ξ at θ_0 to have values between the two extremes $f'(\theta_0 \pm 0)$. With this in mind, we

[47] *Cf., e.g.*, HERRING [1951ab], CAHN & HOFFMAN [1974].

[48] We use this terminology to avoid confusion: a sharp spot marks a loss in smoothness for the *interfacial energy*; corners denote jumps in orientation that are consistent with capillary balance, and hence denote possible discontinuous tangencies for an *evolving curve*.

[49] It is convenient to write $\hat{C}(\theta)$, rather than $C(\theta)$, for the capillary force defined by (8.1) away from sharp spots, and to reserve $C(s, t)$ for the capillary force on an evolving curve.

define the **capillary** set $\{\hat{C}(\theta_0)\}$ at θ_0 to be the vector fan between $\hat{C}(\theta_0 - 0)$ and $\hat{C}(\theta_0 + 0)$; $\{\hat{C}(\theta_0)\}$ is thus the set of all vectors of the form

$$C = f(\theta_0)\, T(\theta_0) + \xi N(\theta_0) \tag{10.1}$$

with ξ in the closed interval bounded by the numbers $f'(\theta_0 - 0)$ and $f'(\theta_0 + 0)$. It is convenient to also use this terminology at smooth spots, in which case the capillary set is the *singleton* $\{\hat{C}(\theta)\}$. The proposition following (8.4) then generalizes:

Proposition. *The capillary set is the negative of the supporting tangent-fan of the Frank diagram*:

$$\{\hat{C}(\theta)\} = -\{\sigma^*(\theta)\}. \tag{10.2}$$

There is a natural extension of the notions of **corners** and **cusps**: we simply replace the condition $\hat{C}(\theta^-) = \hat{C}(\theta^+)$ by

$$\{\hat{C}(\theta^-)\} \cap \{\hat{C}(\theta^+)\} \neq \emptyset;$$

the theorem on common tangents (Apperdix A) then implies that θ^- and θ^+ have a common supporting tangent:

Corner-force theorem. *Cusps are not possible. If* $\{\theta^-, \theta^+\}$ *with* $0 < \theta^+ - \theta^- < \pi$ *is a corner, then* $\{\hat{C}(\theta^-)\} \cap \{\hat{C}(\theta^+)\}$ *consists in exactly one vector* $\boldsymbol{a}$, *and* $\boldsymbol{a}$ *points in the same direction as* $\sigma(\theta^-) - \sigma(\theta^+)$. *We write* $\boldsymbol{a} = \hat{C}(\theta^-, \theta^+)$ *and refer to* $\hat{C}(\theta^-, \theta^+) = \hat{C}(\theta^+, \theta^-)$ *as the* **corner force** *corresponding to* $\{\theta^-, \theta^+\}$.

We can also extend the terminology conerning stability to sharp spots. Indeed, the paragraph containing (4.16) with $f''(\theta)$ considered as a distribution yields the following definitions: f is **strictly-stable** or **unstable at a sharp spot** θ_0 according as

$$f'(\theta_0^+) > f'(\theta_0^-) \quad \text{or} \quad f'(\theta_0^+) < f'(\theta_0^-). \tag{10.3}$$

With these definitions *the convexity-stability theorem of Section 8 holds without change.*

GS sections and **GS corners** of the interfacial energy are defined as for smooth f. **GS angles** (whether sharp spots or smooth spots) are, as before, angles that belong to GS sections; GS sharp spots are then, necessarily, *strictly stable.*

The *convexification* of the Frank diagram is the polar diagram of a function $\Sigma(\theta)$ and is hence the locus of a vector potential

$$\boldsymbol{\Sigma}(\theta) = \Sigma(\theta)\, N(\theta),$$

the **convexified Frank potential**. On GS sections $\boldsymbol{\Sigma}(\theta)$ coincides with $\sigma(\theta)$; between such sections $\boldsymbol{\Sigma}(\theta)$ coincides with the Maxwell lines of the Frank diagram. The function

$$\Sigma^*(\theta) = \Sigma(\theta)^{-2}\, \boldsymbol{\Sigma}'(\theta) = -\Sigma(\theta)^{-1}\, T(\theta) + \Sigma(\theta)^{-2}\, \Sigma'(\theta)\, N(\theta) \tag{10.4}$$

is the supporting tangent of the the convexified Frank diagram. By (i) of the Maxwell theorem (Appendix A), we have the following analog of the "thermodynamic relation" (8.4) (*cf.* Figure 10A).

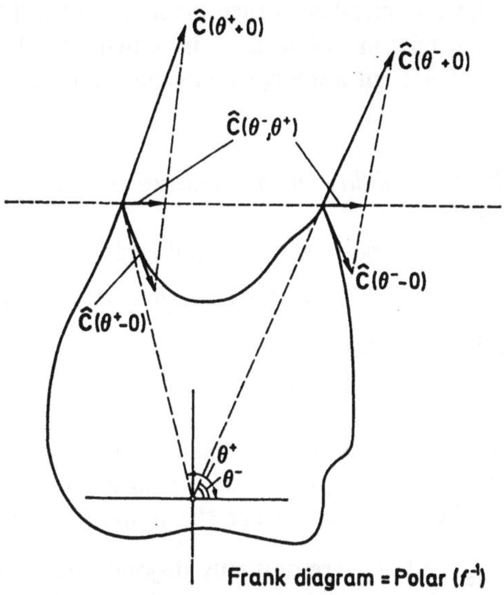

Fig. 10A. The corner force $\hat{C}(\theta^{-}, \theta^{+})$

Proposition. *Let* $\{\theta^{-}, \theta^{+}\}$ *with* $0 < \theta^{+} - \theta^{-} < \pi$ *be a GS corner. Then, for all* $\theta \in (\theta^{-}, \theta^{+})$,

$$\hat{C}(\theta^{-}, \theta^{+}) = -\Sigma^{*}(\theta). \tag{10.5}$$

Remark. Let $\{\theta_0, \theta_1\}$ be a GS corner with θ_0 a sharp spot. Let $\mathcal{N}$ be a fixed neighborhood of θ_0 containing no sharp spots other than θ_0. Consider $f(\theta)$ as the limit, as $\varepsilon \to 0$, of energies $f_\varepsilon(\theta)$ which are C^1 on $\mathcal{N}$, which are *strictly stable* on $\ell_\varepsilon := (\theta_0 - \varepsilon, \theta_0 + \varepsilon)$, and which satisfy $f(\theta) = f_\varepsilon(\theta)$ outside ℓ_ε. Let $C_\varepsilon(\theta)$ denote the capillary force for $f_\varepsilon(\theta)$, and consider the set

$$\mathscr{C}_\varepsilon = \{C_\varepsilon(\theta) : \theta \in \ell_\varepsilon \text{ is a GS angle of } f_\varepsilon(\theta)\}.$$

The limit $\mathscr{C}$ of the sets $\mathscr{C}_\varepsilon$ (that is, the intersection of the sets $\mathscr{C}_\varepsilon$ over all sufficiently small ε) in some sense represents the globally stable set of capillary forces at θ_0. It is not difficult to show that $\mathscr{C}$ is independent of the choice of $f_\varepsilon(\theta)$; in fact, $\mathscr{C}$ is the negative of the supporting tangent-fan $\{\Sigma^{*}(\theta_0)\}$ of the *convexified* Frank diagram. Since $\{\Sigma^{*}(\theta_0)\} \subset \{\sigma^{*}(\theta_0)\}$, $\mathscr{C}$ is a subset of the capillary set $\{\hat{C}(\theta_0)\}$. For these reasons, we refer to $\{-\Sigma^{*}(\theta_0)\}$ as the **GS capillary set** at θ_0.

Regularity for an interfacial energy with sharp spots is defined exactly as in Section 8.2.

Remark. A major difference between smooth energies and energies with sharp spots is that for the latter, regularity does not rule out singleton GS sections, which are possible at sharp spots. Further, granted regularity, if $\{\theta^-, \theta^+\}$ $(0 < \theta^+ - \theta^- < \pi)$ is a corner with θ^- and θ^+ GS angles, then $\{\theta^-, \theta^+\}$ is a GS corner. Moreover, there is another corner involving θ^- (other than $\{\theta^+, \theta^-\}$) if and only if $\{\theta^-\}$ is a singleton GS section, in which case the corner is of the form $\{\alpha, \theta^-\}$ $(0 < \alpha < \pi)$; an analogous assertion applies to θ^+.

10.2. Admissibility in the presence of sharp spots

We consider now an energy f with the following properties:

f is an interfacial energy with sharp spots; f is regular.

The following notation is useful:

$\mathscr{H}$:= the *convexified* Frank diagram of f,
$\mathscr{I}$:= the set of GS angles of f,
$\mathscr{I}_{sm}$:= the *interior* of the set of GS angles which are smooth spots[50] of $\mathscr{H}$,
$\mathscr{I}_{smco}$:= the set of all smooth spots of $\mathscr{H}$ which are GS corner angles,
$\mathscr{I}_{sh}$:= the set of GS angles which are sharp spots of $\mathscr{H}$.

Note that $\mathscr{I}_{sm}$, $\mathscr{I}_{smco}$, and $\mathscr{I}_{sh}$ are mutually disjoint and have $\mathscr{I}$ as their union. We use the term **smooth GS part** to designate the closure of a connected component of $\mathscr{I}_{sm}$.

Let $\boldsymbol{r}$ be a PS evolving curve. Fix t and the angle θ_0, and let $\mathscr{R}$ be the set of all s such that $\theta(s, t) = \theta_0$. We will refer to the closures of the connected components of $\mathscr{R}$ as the **θ_0-sections** at time t.

We use the following terminology for arcs $\boldsymbol{r}_i$ of $\boldsymbol{r}$: (i) $\boldsymbol{r}_i$ is **curved** if, at each t, $\theta_s(s, t) = 0$ at most at a discrete set of points of $\sigma_i(t)$; (ii) if $\boldsymbol{r}_i$ is a facet with $\theta_i(t)$ the corresponding angle, then $\boldsymbol{r}_i$ is **maximal** if $\sigma_i(t)$ is a $\theta_i(t)$-section at each t.

At sharp spots θ_0 there is no uniquely defined capillary force; instead there is a vector fan $\{\hat{C}(\theta_0)\}$ of possible forces. It thus seems reasonable to allow the capillary force $C(s, t)$ associated with a given evolving curve to vary within $\{\hat{C}(\theta_0)\}$ whenever $\theta(s, t) = \theta_0$. This should motivate the following notion of admissibility.

A PS evolving curve $\boldsymbol{r}$ is **admissible** if:[51]

(AP1) $\theta(s, t)$ is always GS;

(AP2) there is an **associated capillary force** $C(s, t)$ such that $C(s, t) \in \{\hat{C}(\theta(s, t))\}$ for all (s, t) and

$$\int_{\partial c(t)} C = \int_{c(t)} (F + \beta(\theta)\, V)\, N\, ds \tag{10.6}$$

whenever c is the trace of an evolving subcurve of $\boldsymbol{r}$;

[50] By the smooth and sharp spots of $\mathscr{H}$ we mean the smooth and sharp spots of $\Sigma(\theta)$. Sharp spots of $\mathscr{H}$ are GS sharp spots of $f(\theta)$, but the converse is not always true.

[51] We will also need one technical assumption: for $\theta_0 \in \mathscr{I}_{sh}$, the set of all s such that $\theta(s, t) = \theta_0$ has a finite number of connected components.

(AP3) each arc of r is either a maximal facet or a curved arc, and adjacent curved arcs correspond to different smooth GS parts.

The associated capillary force is, in fact, unique, a result we shall prove by showing that $C(s, t)$ is essentially characterized by the *convexified Frank diagram.*

Let r be a *closed* PS evolving curve. Fix t and the angle θ_0 with $\theta_0 \in \mathscr{I}_{sh}$, and consider a given θ_0-section $\mathscr{S} = [H, S]$. Then $\mathscr{S}$ is **trivial** or **nontrivial** according as $H = S$ or $S > H$. Further:

(T1) $\theta(s, t)$ **increases across** $\mathscr{S}$ if, for all sufficiently small $\varepsilon > 0$,

$$\theta(H - \varepsilon, t) < \theta_0, \quad \theta(S + \varepsilon, t) > \theta_0;$$

the statement "$\theta(s, t)$ **decreases across** $\mathscr{S}$" has an analogous meaning; in either case we refer to $\mathscr{S}$ as **transitional**;

(T2) $\mathscr{S}$ is a **local maximum** for $\theta(s, t)$ if, for all sufficiently small $\varepsilon > 0$,

$$\theta(H - \varepsilon, t_0) \leqq \theta_0, \quad \theta(S + \varepsilon, t_0) \leqq \theta_0;$$

the statement "$\mathscr{S}$ a **local minimum** for $\theta(s, t)$" has an analogous meaning; in either case we refer to $\mathscr{S}$ as **nontransitional.**

Finally, a function $g(s)$ **goes from** a **to** b if $g(H + 0) = a$ and $g(S - 0) = b$.

The degree of smoothness assumed for PS evolving curves ensures that each $\mathscr{S}$ is either transitional or nontransitional. Indeed, if $\mathscr{S} = [H, S]$, then there is an $\varepsilon > 0$ such that θ is different from θ_0 everywhere in $(H - \varepsilon, H) \cup (S, S + \varepsilon)$; the continuity of θ then yields the following four possibilities: $\pm(\theta - \theta_0) > 0$ on $(H - \varepsilon, H)$, $\pm(\theta - \theta_0) > 0$ on $(S, S + \varepsilon)$.

Characterization of the capillary force. *Let r be a closed,*[52] *admissible PS evolving curve. Then there is exactly one capillary force $C(s, t)$ associated with r, and $C(s, t)$ has the following properties*:

(i) *$C(s, t)$ belongs to the GS capillary set $\{-\Sigma^*(\theta(s, t))\}$, so that $C(s, t) = -\Sigma^*(\theta(s, t))$ whenever $\theta(s, t)$ is a smooth spot of $\mathscr{H}$.*

(ii) *Let θ_0 a sharp spot of $\mathscr{H}$, and let $\mathscr{S} = [H, S]$ be a θ_0-section at some fixed time $t \in (0, T)$.*

(F1) *If $\mathscr{S}$ is transitional, then $\mathscr{S}$ is nontrivial, and $C(s, t)$ varies linearly with s on $\mathscr{S}$, going from $-\Sigma^*(\theta_0 - 0)$ to $-\Sigma^*(\theta_0 + 0)$ or from $-\Sigma^*(\theta_0 + 0)$ to $-\Sigma^*(\theta_0 - 0)$ according as $\theta(s, t)$ increases or decreases across $\mathscr{S}$.*

(F2) *If $\mathscr{S}$ is nontransitional, then $C(s, t)$ is constant on $\mathscr{S}$ with value $-\Sigma^*(\theta_0 - 0)$ or $-\Sigma^*(\theta_0 + 0)$ according to as $\mathscr{S}$ is a local maximum or a local minimum for $\theta(s, t)$.*

Proof. By (AP1) and (10.2), $C(s, t) = -\Sigma^*(\theta(s, t))$ whenever $\theta(s, t)$ is a smooth spot. Further, this result, (F1), and (F2), imply that $C(s, t)$ is uniquely determined

[52] For r not closed $C(s, t)$ is determined uniquely (and (F1), (F2) hold) on the internal arcs of r, and also on the terminal arcs provided they are not sharp-spot facets (facets whose angles are sharp spots). Boundary conditions are needed to determine $C(s, t)$ uniquely on terminal sharp-spot facets.

with $C(s, t) \in \{-\Sigma^*(\theta(s, t))\}$ for all (s, t). Thus we have only to establish (F1) and (F2). We will prove only (F1), and only when $\theta(s, t)$ increases across $\mathcal{S}$; the remaining assertions are proved analogously. Since the time t is fixed, we shall suppress it as an argument. Suppose that

$$\theta_0 + \delta \text{ is GS for all sufficiently small } \delta > 0. \tag{10.7}$$

Then $\theta(S + 0) = \theta_0$ and

$$C(S - 0) = C(S + 0) = \hat{C}(\theta_0 + 0) = -\sigma^*(\theta_0 + 0) = -\Sigma^*(\theta_0 + 0). \tag{10.8}$$

On the other hand, if (10.7) is not satisfied, then there is a corner $\{\theta_0, \theta_1\}$ such that $\theta(S + 0) = \theta_1$, and, by appeal to (10.5) and the corner-force theorem,

$$C(S - 0) = C(S + 0) = \hat{C}(\theta_0, \theta_1) = -\Sigma^*(\theta_0 + 0). \tag{10.9}$$

Similar results hold at H. The results (10.8) and (10.9) and their counterparts for H have the following consequences: (i) $\mathcal{S}$ is nontrivial, for otherwise $\Sigma^*(\theta_0 - 0) = \Sigma^*(\theta_0 + 0)$, which is not possible when θ_0 is a sharp spot of the convexified Frank diagram; (ii) $C(s, t)$ goes from $-\Sigma^*(\theta_0 - 0)$ to $-\Sigma^*(\theta_0 + 0)$.

Finally, since $\theta(s) \equiv \theta_0$ on $\mathcal{S}$, capillary balance (10.6) must there have the local form

$$C_s = [F + \beta(\theta_0)\, V]\, N(\theta_0) \tag{10.10}$$

with $V = V(t)$ independent of s. Thus $C(s)$ varies *linearly* on $\mathcal{S}$. □

Let

$$\begin{aligned} \Delta(\theta) &:= [\Sigma^*(\theta - 0) - \Sigma^*(\theta + 0)] \cdot N(\theta) \\ &= \Sigma(\theta)^{-2}\, [\Sigma'(\theta - 0) - \Sigma'(\theta + 0)] \geqq 0 \end{aligned} \tag{10.11}$$

(*cf.* (A11) of Appendix A). Because of (F1), evolving facets on which $\theta \equiv \theta_0$, with θ_0 a sharp spot, are to be expected. By (F1) and (10.10), the length $L(t)$ and the normal velocity $V(t)$ of a "transitional" facet are related by

$$\pm \Delta(\theta_0) = L(t)\, [F + \beta(\theta_0)\, V(t)] \tag{10.12}$$

with the plus or minus sign chosen according as $\theta(s, t)$ increases or decreases across the corresponding θ_0-section. For a "nontransitional" facet, (10.12) remains valid, but with $\Delta(\theta) = 0$.

Remark 1. Let $S_i(t)$ be the arc length at a juncture i of a closed, admissible PS evolving curve. Then $\theta_i^- := \theta(S_i(t) - 0, t)$ and $\theta_i^+ := \theta(S_i(t) + 0, t)$ are independent of time, and either:

(i) $\{\theta_i^-, \theta_i^+\}$ is a GS corner, in which case the juncture is **nontrivial**; or
(ii) $\theta_i^- = \theta_i^+ \in \mathcal{I}_{sh}$ in which case the juncture is **trivial**.

Remark 2. There are exactly three possibilities for an arc r_i of an admissible PS evolving curve.

(1) $\mathbf{r}_i$ is a curved arc. Since each boundary angle of a smooth GS part must belong to $\mathscr{S}_{smco} \cup \mathscr{S}_{sh}$, it is a clear from (F1) that (on $\mathbf{r}_i$) $\theta(s, t)$ must belong to a *single* smooth GS part $\mathscr{S}_i$. In this case $\mathbf{r}_i$ evolves according to

$$\begin{gathered} V = \Phi(\theta)\,K - \beta(\theta)^{-1}\,F, \\ \Phi(\theta) = \beta(\theta)^{-1}\,[f(\theta) + f''(\theta)] > 0, \end{gathered} \tag{10.13}$$

with $f''(\theta)$ computed by restricting $f(\theta)$ to $\mathscr{S}_i$. Further, the initial and terminal orientations, θ_{init} and θ_{term}, are constants belonging to $\mathscr{S}_{smco} \cup \mathscr{S}_{sh}$ (*cf.* the paragraph containing (9.4)). We assign a **transition number** to $\mathbf{r}_i$ as follows: the transition number is $+1$, -1, or 0 according to $\theta_{term} > \theta_{init}$, $\theta_{term} < \theta_{init}$, or $\theta_{term} = \theta_{init}$.

(2) $\mathbf{r}_i$ is a facet with orientation $\theta_i \in \mathscr{S}_{smco}$. Then $\mathbf{r}_i$ evolves according to

$$\beta(\theta_i)\,V = -F.$$

(3) $\mathbf{r}_i$ is a facet with orientation $\theta_i \in \mathscr{S}_{sh}$. Then either $\theta(s, t)$ increases across $\sigma_i(t)$ for all t, or decreases across $\sigma_i(t)$ for all t, or $\sigma_i(t)$ is nontransitional for all t; we define the **transition number** χ_i for $\mathbf{r}_i$ to be $+1, -1$, or 0, respectively, for these three possibilities. Then $\mathbf{r}_i$ evolves according to (*cf.* (10.12))

$$\chi_i\,\Delta(\theta_i) = L_i(t)\,[F + \beta(\theta_i)\,V(t)] \tag{10.14}$$

with $L_i(t)$ the length of $\mathbf{r}_i$.

10.3. Crystalline energies

Let $f(\theta)$ be an interfacial energy with sharp spots. Then $f(\theta)$ is **crystalline**[53] if its *convexified Frank diagram is a polygon*, and if the vertices of this polygon form the *complete* set of GS angles (*cf.* Figure 10B). Such energies are clearly *regular*.

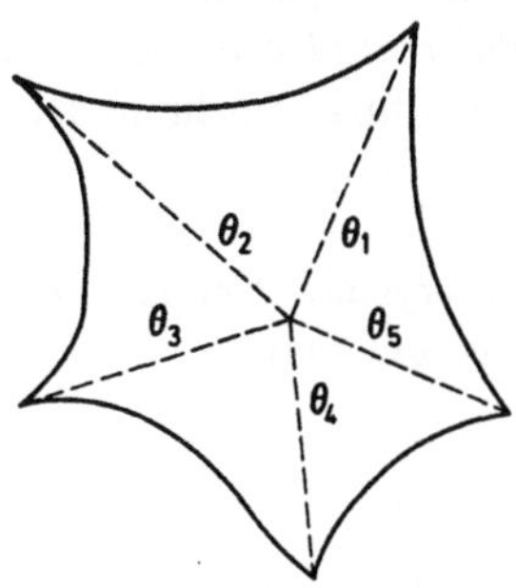

Fig. 10B. The Frank diagram of a crystalline energy f. The GS angles are the angles corresponding to the five sharp spots

[53] *Cf.* TAYLOR [1978], who studies the stable equilibria of interfaces corresponding to crystalline energies.

Let $f(\theta)$ be crystalline, and consider an admissible PS evolving curve $\boldsymbol{r}$. Each arc of $\boldsymbol{r}$ must have $\theta(s, t)$ equal to one of the GS angles, and hence must have $\theta(s, t) \equiv$ constant. Thus we have the following[54]

Proposition. *Facetings with fixed orientation are the only admissible PS evolving curves for a crystalline energy.*

By an **evolving crystal** we mean a faceting $\boldsymbol{r}$ that is simple and *closed.* In view of the proposition, these are the only admissible interfacial motions consistent with a crystalline energy. As before, we let the reference region $\Omega(t)$ denote the bounded region *interior* to the corresponding curve at each time, so that $\Omega(t)$ has $\boldsymbol{N}$ as its outward unit normal. We will refer to $\boldsymbol{r}$ as **essentially convex** if $\Omega(t)$ is a convex region at each t. (Note that $\boldsymbol{r}$ cannot be convex as an evolving curve, since $K = 0$ on each facet.)

Let $\boldsymbol{r} = \{\boldsymbol{r}_1, \boldsymbol{r}_2, \ldots, \boldsymbol{r}_N\}$ be an evolving crystal, and let θ_i, V_i, and L_i denote the orientation, normal velocity, and length of $\boldsymbol{r}_i$. Further, for any function $g(\theta)$ let $g_i = g(\theta_i)$, and define

$$\varrho_{i,j} := [\boldsymbol{T}_j \cdot \boldsymbol{N}_i]^{-1}, \qquad \alpha_{i,j} := (\boldsymbol{T}_i \cdot \boldsymbol{T}_j)\, \varrho_{i,j}. \tag{10.15}$$

Remark. From (10.11) and the definition of a crystalline energy, it is clear that each of the (fixed) angles $\theta_i \in \mathcal{I}_{\mathrm{sh}}$; thus

$$\Delta_i := \Delta(\theta_i) > 0. \tag{10.16}$$

Moreover, our agreement that $\boldsymbol{N}$ be outward implies that, for an essentially convex crystal, θ decreases across each facet and

$$\varrho_{i,i+1} < 0, \qquad \alpha_{i,i+1} < 0. \tag{10.17}$$

Curves governed by smooth energies have, as evolution equations, a fairly complex system of *partial differential equations*; the next theorem shows that, in contrast, crystals evolve according to a finite set of *ordinary differential equations.*

Evolution equations for a crystal. *Crystals corresponding to a crystalline energy evolve according to the equations*

$$\begin{aligned} \dot{L}_i &= [\alpha_{i,i+1} + \alpha_{i-1,i}]\, V_i - \varrho_{i-1,i} V_{i-1} - \varrho_{i,i+1} V_{i+1}, \\ V_i &= (\beta_i)^{-1}\, [-F + \chi_i(\Delta_i / L_i)] \end{aligned} \tag{10.18}$$

at each juncture i.

---

[54] Facetings are as defined in Section 9.2, except that the underlying notion of admissibility is as defined in Section 10.2.

Proof. The second of (10.18) is (10.12). To derive the second, let v_i denote the arc velocity and $[S_i, S_{i+1}]$ the arc-length interval for $\boldsymbol{r}_i$. Then

$$\begin{aligned} V_{i-1}N_{i-1} + [S_i^{\boldsymbol{\cdot}} - v_{i-1}]\,\boldsymbol{T}_{i-1} &= V_iN_i + [S_i^{\boldsymbol{\cdot}} - v_i]\,\boldsymbol{T}_i, \\ V_iN_i + [S_{i+1}^{\boldsymbol{\cdot}} - v_i]\,\boldsymbol{T}_i &= V_{i+1}N_{i+1} + [S_{i+1}^{\boldsymbol{\cdot}} - v_{i+1}]\,\boldsymbol{T}_{i+1}. \end{aligned} \tag{10.19}$$

If we take the inner product of these equations with $\boldsymbol{T}_i$ and then add the resulting equations, we find that

$$\begin{aligned} L_i^{\boldsymbol{\cdot}} = &-(\varrho_{i-1,i})^{-1}\,V_{i-1} - (\varrho_{i,i+1})^{-1}\,V_{i+1} \\ &+ [S_{i+1}^{\boldsymbol{\cdot}} - v_{i+1}]\,(\boldsymbol{T}_i \cdot \boldsymbol{T}_{i+1}) - [S_i^{\boldsymbol{\cdot}} - v_{i-1}]\,(\boldsymbol{T}_i \cdot \boldsymbol{T}_{i-1}). \end{aligned} \tag{10.20}$$

Next, we take the inner product of (10.19) with $\boldsymbol{N}_i$; the result is

$$\begin{aligned} (\varrho_{i-1,i})^{-1}\,[S_i^{\boldsymbol{\cdot}} - v_{i-1}] &= -V_i + V_{i-1}(\boldsymbol{T}_i \cdot \boldsymbol{T}_{i-1}), \\ (\varrho_{i-1,i})^{-1}\,[S_{i+1}^{\boldsymbol{\cdot}} - v_{i+1}] &= V_i - V_{i+1}(\boldsymbol{T}_i \cdot \boldsymbol{T}_{i+1}). \end{aligned} \tag{10.21}$$

The relations (10.20) and (10.21) combine to give $(10.18)_1$. □

A geometric derivation of (10.18) follows upon noting that, for $\vartheta_i = \theta_i - \theta_{i-1}$ (Figure 10C),

$$L_i^{\boldsymbol{\cdot}} = -[\cot\vartheta_i + \cot\vartheta_{i+1}]\,V_i + (\sin\vartheta_i)^{-1}\,V_{i-1} + (\sin\vartheta_{i+1})^{-1}\,V_{i+1},$$

which is easily obtained, at least formally, by giving the time a "small" increment.

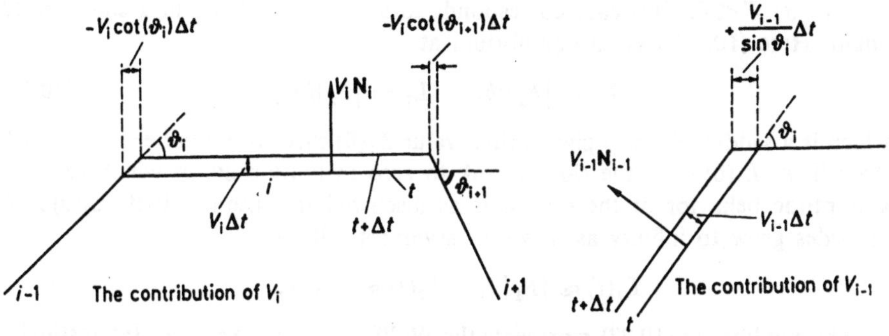

Fig. 10C. Derivation of $L_i^{\boldsymbol{\cdot}}(t)$, expressed as a linear combination of V_{i-1}, V_i and V_{i+1}

10.3.1. Evolution of a rectangular crystal. A simple example that yields useful information occurs when the convexified Frank diagram is quadrilateral with vertices at $\theta = 0, \pi/2, \pi, 3\pi/2$. A corresponding evolving crystal, if essentially convex, is rectangular at each t with sides having these angles as orientations. Therefore $L_1 = L_3$, $L_2 = L_4$, and by (10.15), $\varrho_{i,i+1} = -1$, $\alpha_{i,i+1} = 0$. Thus, defining

$$\begin{aligned} F_1 &= F(\beta_2 + \beta_4)/\beta_2\beta_4, & F_2 &= F(\beta_1 + \beta_3)/\beta_1\beta_3, \\ \delta_1 &= (\varDelta_4\beta_2 + \varDelta_2\beta_4)/\beta_2\beta_4, & \delta_2 &= (\varDelta_3\beta_1 + \varDelta_1\beta_3)/\beta_1\beta_3, \end{aligned} \tag{10.22}$$

the evolution equations (10.18) reduce to

$$\begin{aligned} L_1^{\cdot} &= -F_1 - \delta_1/L_2, \\ L_2^{\cdot} &= -F_2 - \delta_2/L_1 \end{aligned} \tag{10.23}$$

with (*cf.* (10.16))

$$\operatorname{sgn} F_i = \operatorname{sgn} F, \quad \delta_i > 0. \tag{10.24}$$

Case 1: $F = 0$. Solutions of (10.23) approach zero in *finite* time T. With the definition

$$\delta = \delta_1/\delta_2,$$

we see that there is a constant $C > 0$ such that

$$L_1(t) = CL_2(t)^{\delta}. \tag{10.25}$$

Thus for $\delta = 1$ the *isoperimetric ratio* $\varrho(t) = [\text{length}\,(\partial\Omega)]^2/4\pi\,\text{area}\,(\Omega)]$ is constant, but for $\delta \neq 1$

$$\varrho(t) \to \infty \quad \text{as } t \to T. \tag{10.26}$$

For $\delta > 1$, $L_1(t)$ approaches zero faster than $L_2(t)$, so that the crystal shrinks to a point, but is ultimately in the shape of a "needle oriented by θ_2 and θ_4". This is in sharp contrast to an interfacial motion for an isotropic interface (*cf.* Section 5.1); there the interface shrinks to a round point.[55]

Case 2: $F > 0$. Solutions still approach zero in finite time T. The result (10.25) holds asymptotically, and the discussion of Case 1 for $\delta \neq 1$ is appropriate.

Case 3: $F < 0$. This case corresponds to a crystal evolving in a supercooled liquid. Here (10.23) has an equilibrium at

$$L_2 = |F_1|/\delta_1, \quad L_1 = |F_2|/\delta_2, \tag{10.27}$$

which is a saddle. For any given initial value $L_1(0)$ there is a number $l > 0$ such that: (i) if $L_2(0) < l$ the sides shrink to zero in finite time, in which case the asymptotic behavior of the crystal is as discussed in Case 1; (ii) if $L_2(0) > l$, the sides grow to infinity as $t \to \infty$, asymptotically as

$$L_1(t) \approx |F_1|\,t, \quad L_2(t) \approx |F_2|\,t.$$

The equilibrium (10.27) represents the Wulff shape of the crystal. Interestingly, none of the asymptotic shapes of the crystal are of this form.

11. General global behavior

Section 7 established results for the growth and shrinking of an interface whose energy is smooth and stable. We now generalize these results. We assume that

f is a (not necessarily stable) regular
interfacial energy with sharp spots.

---

[55] GAGE [1984], GAGE & HAMILTON [1986], GRAYSON [1987].

Let r be a **PS interfacial motion** that is **admissible** in the sense of (AP1), (AP2), and (AP3),[56] and let $L(t)$, $A(t)$, and $\mathcal{F}(t)$ denote the perimeter, area, and total interfacial energy as defined in (7.3) and (7.5).

Theorem.

$$\mathcal{F}^{\cdot}(t) + FA^{\cdot}(t) = -\int_{\partial\Omega(t)} \beta(\theta)\, V^2\, ds \leqq 0. \tag{11.1}$$

Proof. Our first step will be to show that (4.12) (with $b_{ext} = 0$) holds for each arc r_i of r. It is clear from its proof that (4.12) holds for r_i in cases (1) and (2) of Remark 2 (Section 10.2). We now show that (4.12) is also satisfied in case (3). For this case N and T are constant, $K \equiv 0$, and V is independent of arc length. Let $R_1(t)$ and $R_2(t)$ denote the initial and terminal points of r_i, with $C_1(t)$ and $C_2(t)$ the corresponding capillary forces. Since $f(\theta_i) \equiv$ constant, (2.35) yields

$$\left(\frac{d}{dt}\right) \int_{\sigma_i(t)} f(\theta_i)\, ds = f(\theta_i)\, [R_2^{\cdot} - R_1^{\cdot}] \cdot T,$$

while (10.6) implies

$$F \int_{\sigma_i(t)} V\, ds = -\int_{\sigma_i(t)} \beta(\theta_i)\, V^2\, ds + V[C_2 - C_1] \cdot N.$$

Further, by (10.1), $f(\theta_i) = C_1 \cdot T = C_2 \cdot T$; this relation, $(2.20)_1$, and the last two identities imply (4.12).

Thus (4.12) holds on each arc of r. If we apply (4.12) to each such arc, add the resulting equations, and use (2.45), (2.37), and the fact that, by (10.6), $C(s, t)$ is continuous in s, we arrive at (11.1). □

Let $e(\theta)$ be a continuous, piecewise smooth, strictly positive, 2π-periodic function on $\mathbb{R}$. We need a generalization of the Wulff ratio (7.12), a generalization in which the exterior normals of the underlying regions Γ are restricted to point in GS directions.[57] This is easily accomplished by replacing $e(\theta)$ by $+\infty$ whenever θ is not GS. Thus, if

$$e^*(\theta) = \begin{cases} e(\theta) & \text{for } \theta \text{ a GS angle} \\ +\infty & \text{otherwise,} \end{cases} \tag{11.2}$$

the **GS Wulff ratio for** $e(\theta)$ is the number

$$W_{GS}(e) := (4\pi)^{-1} \inf \left\{ \int_{\partial\Gamma} e^*(\theta)\, ds \right\}^2 \Big/ \text{area}\,(\Gamma) \tag{11.3}$$

with the infimum taken over all bounded regions Γ with $\partial\Gamma$ piecewise smooth. Corresponding minima Γ are called **GS Wulff regions for** $e(\theta)$.

[56] *Cf.* the paragraphs containing (2.45) and (10.6).

[57] GS is with respect to the interfacial energy $f(\theta)$, so that $W_{GS}(e)$ depends also on $f(\theta)$.

Let $\{\Theta_m^-, \Theta_m^+\}$, $m = 1, 2, \ldots, M$, denote the GS corners in the order encountered as the Frank diagram is traversed in the *clockwise* direction, so that $\Theta_1^- > \Theta_1^+ \geqq \Theta_2^- > \Theta_2^+ \geqq \Theta_3^-$, and so forth;[58] for each such corner, let[59]

$$k_m := 2\tan(\vartheta_m/2) < 0, \quad \vartheta_m = \Theta_m^+ - \Theta_m^-. \tag{11.4}$$

We define the **GS average** g_{GSav} of a function $g(\theta)$ by

$$2\pi g_{\mathrm{GSav}} := \int_{\theta \in \mathcal{S}} g(\theta)\, d\theta + \tfrac{1}{2} \sum_{m=1}^{M} |k_m|\, [g(\Theta_m^+) + g(\Theta_m^-)]. \tag{11.5}$$

Theorem on the growth of the reference phase. *Consider a regularly maximal*[60] *admissible PS interfacial motion with duration* $[0, T)$.

(i) *If* $F \geqq 0$, *then* $T < \infty$ *and* $A(t) \to 0$ *as* $t \to T$.
(ii) *If* $F < 0$, *then*:
  (a) *if* $L(0)$ *is sufficiently small, then* $T < \infty$ *and* $A(t) \to 0$ *as* $t \to T$;
  (b) *if* $A(0)$ *is sufficiently large, then* $T = \infty$ *and* $A(t) \to \infty$ *as* $t \to \infty$; *Even so*, isoper $(\Omega(t))$ *remains bounded*:

$$\limsup_{t \to \infty} \operatorname{isoper}(\Omega(t)) \leqq [(\beta^{-1})_{\mathrm{GSav}}]^2 / W_{\mathrm{GS}}(\beta^{-1}). \tag{11.6}$$

Proof. Our first step will be to show that

$$\begin{aligned} A^{\cdot}(t) &= -C - F \int_{\partial\Omega(t)} \beta(\theta)^{-1}\, ds, \\ \mathcal{F}^{\cdot} &\leqq FC + [F^2/(f_{\min}\beta_{\min})]\, \mathcal{F}, \\ L^{\cdot}(t) &\leqq -2\pi F (\beta^{-1})_{\mathrm{GSav}}, \end{aligned} \tag{11.7}$$

where $C > 0$ is a constant.

To prove $(11.7)_1$, let N denote the essential number of arcs of $\boldsymbol{r}$, let

$$\ell_{\mathrm{arc}} := \{i : \boldsymbol{r}_i \text{ is a curved arc}, \ 1 \leqq i \leqq N\},$$

$$\ell_{\mathrm{sh}} := \{i : \boldsymbol{r}_i \text{ is a facet with } \theta_i \in \mathcal{S}_{\mathrm{sh}}, \ 1 \leqq i \leqq N\},$$

and, for each $i \in \ell_{\mathrm{arc}}$, let $\mathcal{S}_i$ denote the smooth GS part corresponding to $\boldsymbol{r}_i$. Since $\boldsymbol{N}$ is the outward normal to $\partial\Omega$, the interface is negatively (clockwise) oriented as a closed curve. Moreover, because of the regularity of f, as s increases, $\theta(s, t)$ varies continuously through smooth GS parts, crosses sharp spots to *adjacent* smooth GS parts, or jumps across GS corners to *adjacent* sharp spots or smooth GS parts. Thus:

[58] We use this numbering scheme since for a closed, convex curve the angle decreases with decreasing arc length.

[59] The k_m are M numbers associated with a given interfacial energy; these should be differentiated from the transition curvatures (2.41), which correspond to the actual junctures in an interfacial motion (*cf.* (11.9)).

[60] In the sense of the sentence containing (7.4). Here it must be kept in mind that we have, as yet, no existence theorem appropriate to an energy with sharp spots, although we do have an existence theorem for a smooth but unstable energy (*cf.* Section 9.6).

(i) The sum of the transition numbers for the totality of $\mathscr{I}_i$, $i \in \ell_{\text{arc}}$, corresponding to a given smooth GS part is -1, and

$$\mathscr{I}_{\text{sm}} = \text{interior}\left(\bigcup_{i \in \ell_{\text{arc}}} \mathscr{I}_i\right).$$

(ii) For each $\theta_0 \in \mathscr{I}_{\text{sh}}$ there is at least one $i \in \ell_{\text{sh}}$ such that $\theta_i = \theta_0$, and the sum of the transition numbers χ_i of the $\boldsymbol{r}_i$ with $i \in \ell_{\text{sh}}$ and $\theta_i = \theta_0$ is -1.

Because of these conclusions, and since $K = \theta_s$, for any continuous, 2π-periodic function $g(\theta)$,

$$\begin{aligned} \sum_{i \in \ell_{\text{arc}}} \int_{\sigma_i} g(\theta)\, K\, ds &= - \int_{\theta \in \mathscr{I}_{\text{sm}}} g(\theta)\, d\theta, \\ \sum_{i \in \ell_{\text{sh}}} \chi_i g(\theta_i)\, \Delta(\theta_i) &= - \sum_{\theta \in \mathscr{I}_{\text{sh}}} g(\theta)\, \Delta(\theta). \end{aligned} \tag{11.8}$$

Thus, integrating $V(s, t)$ over $\partial\Omega(t)$, we conclude, with the aid of (2.45) and the remark containing (10.14), that $(11.7)_1$ holds with

$$C = \int_{\theta \in \mathscr{I}_{\text{sm}}} \Phi(\theta)\, d\theta + \sum_{\theta \in \mathscr{I}_{\text{sh}}} \beta(\theta)^{-1} \Delta(\theta) > 0.$$

Next, by (7.5), $f_{\min} L \leqq \mathscr{F}$; (11.1) and $(11.7)_1$ therefore yield $(11.7)_2$. To verify $(11.7)_3$, let

$$\mathscr{J} = \text{the set of nontrivial junctures } j,\ 1 \leqq j \leqq N.$$

Each $j \in \mathscr{J}$ will have transition curvature k_j defined by (2.41). These transition curvatures are related to the numbers ℓ_m defined by (11.4): for each j there is a corresponding $m = m(j)$ such that

$$\begin{aligned} k_j &= \ell_m \quad \text{and} \quad \{\theta_j^-, \theta_j^+\} = \{\Theta_m^-, \Theta_m^+\}, \quad \text{or} \\ k_j &= -\ell_m \quad \text{and} \quad \{\theta_j^-, \theta_j^+\} = \{\Theta_m^+, \Theta_m^-\}. \end{aligned} \tag{11.9}$$

Further, if we let $J(m)$ denote the set of all j with $|k_j| = |\ell_m|$,

$$\sum_{\in J(m)} k_j = \ell_m. \tag{11.10}$$

If j is a juncture for a facet whose angle is a sharp spot of the Frank diagram, and if χ denotes the transition number of the facet, then

$$\chi k_j \geqq 0. \tag{11.11}$$

The verification of $(11.7)_3$ is based on (2.44), which we write in the form

$$L^{\boldsymbol{\cdot}}(t) = - \int_{\partial\Omega(t)} KV\, ds - \sum_{j \in \mathscr{J}} k_j v_j \tag{11.12}$$

with v_j the average velocity $(2.40)_3$ of the juncture j. By Remark 2 of Section 10.2 and $(11.8)_1$,

$$\int_{\partial\Omega(t)} KV\, ds = \sum_{i \in \ell_{\text{arc}}} \int_{\sigma_i} [\Phi(\theta)\, K^2 - F\beta(\theta)^{-1} K]\, ds \geqq F \int_{\theta \in \mathscr{I}_{\text{sm}}} \beta(\theta)^{-1}\, d\theta. \tag{11.13}$$

Let $j \in \mathcal{J}$ be the juncture of an arc r_i of r. Then, on the arc at that juncture,

$$k_j V \geqq -k_j \beta(\theta)^{-1} F. \tag{11.14}$$

Indeed, (11.13) holds trivially if r_i is a facet with angle $\theta \in \mathcal{S}_{\text{smco}}$; for r_i a facet with angle $\theta \in \mathcal{S}_{\text{sh}}$, (11.14) is a consequence of (10.15), (10.11), and (11.11); if r_i is curved, then (11.14) follows from (10.13) and (CC2) of Section (9.1) (which also holds in the present circumstances). By $(2.40)_3$, (11.10), and (11.14),

$$\sum_{j \in \mathcal{J}} k_j v_j \geqq -\tfrac{1}{2} F \sum_{m=1}^{M} \ell_m \left[\beta(\Theta_m^+)^{-1} + \beta(\Theta_m^-)^{-1}\right]. \tag{11.15}$$

Since $\ell_m < 0$, (11.5), (11.12), (11.13), and (11.15) imply $(11.7)_3$.

Finally, an argument the same as that given in the paragraph containing (7.16) shows that (11.7) yields all of the desired conclusions. □

Conjecture. Consider case (iib) of the last theorem. We conjecture that, as $t \to \infty$,

$\partial\Omega(t)$ is asymptotic to a GS Wulff region for $\beta(\theta)^{-1}$.

Remark. Suppose that the energy is *crystalline* with $\theta_1 > \theta_2 > \ldots > \theta_M$ the complete set of GS angles. Then $\{\theta_m, \theta_{m+1}\}$, $m = 1, 2, \ldots, M$ $(\theta_{M+1} = \theta_1)$ are the GS corners,

$$\ell_m = 2 \sin \vartheta_m / (1 + \cos \vartheta_m), \qquad \vartheta_m = \theta_{m+1} - \theta_m,$$

and it is not difficult to verify, using the Wulff construction,[61] that

$$W_{\text{GS}}(1) = (2\pi)^{-1} \sum_{m=1}^{M} |\ell_m| = (1)_{\text{GSav}}.$$

Further, as is clear from (11.3), $W_{\text{GS}}(1)$ represents the minimum value of the isoperimetric ratio isoper (Γ) over all polygons Γ whose outward unit normals are limited to the GS directions $\theta_1, \theta_2, \ldots, \theta_M$. Thus, when $\beta =$ constant, (11.9) yields, for case (iib) of the last theorem,

$$\text{isoper}\,(\Omega(t)) \to (2\pi)^{-1} \sum_{m=1}^{M} |\ell_m|$$

as $t \to \infty$.

Acknowledgment. We greatly acknowledge valuable discussions with A. STRUTHERS. This work was supported by the Army Research Office and by the National Science Foundation.

Appendix A. Polar diagrams

Let $g(\theta)$ be a 2π-periodic, *strictly positive* function on $\mathbb{R}$. Then g is **smooth except for sharp spots** if:

[61] *Cf.* TAYLOR [1978].

(i) g is continuous;
(ii) g' and g'' are continuous except possibly for a finite number of jump discontinuities.

The angles at which g' suffers jump discontinuities will be referred to as **sharp spots**; all other angles will be referred to as **smooth spots.**

Let $N(\theta)$ and $T(\theta)$ be defined by (2.3). The **polar diagram** Polar (g) of g is the *simple closed curve* in $\mathbf{R}^2$ defined by the vector function

$$\boldsymbol{g}(\theta) := g(\theta)\, N(\theta).$$

We orient Polar (g) by θ, so that Polar (g) is positively (counter-clockwise) oriented as a closed curve. Since $N' = -T$,

$$\boldsymbol{g}'(\theta) = -g(\theta)\, T(\theta) + g'(\theta)\, N(\theta)$$

whenever θ is a smooth spot; thus

$$n(\theta) := |\boldsymbol{g}'(\theta)|^{-1} [g(\theta)\, N(\theta) + g'(\theta)\, T(\theta)]$$

defines an outward unit normal to Polar (g). A tangent vector on Polar (g) more useful than $\boldsymbol{g}'(\theta)$ is the **supporting tangent**

$$\begin{aligned} g^*(\theta) &:= g(\theta)^{-2}\, \boldsymbol{g}'(\theta) \\ &= -g(\theta)^{-1}\, T(\theta) + g(\theta)^{-2}\, g'(\theta)\, N(\theta), \end{aligned} \tag{A1}$$

which is well defined for θ a smooth spot. This notion has an extension which is defined for all angles. By the **vector fan** between vectors a and b we mean the set of all vectors c such that

$$c = a + \alpha(b - a), \quad 0 \leqq \alpha \geqq 1.$$

For each θ, the **supporting tangent-fan** $\{g'(\theta)\}$ is the vector fan defined by $g^*(\theta - 0)$ and $g^*(\theta + 0)$, and is hence the *set* of vectors of the form

$$-g(\theta)^{-1}\, T(\theta) + \zeta N(\theta), \tag{A2}$$

where ζ varies in the closed interval of $\mathbf{R}$ bounded by $g(\theta)^{-2}\, g'(\theta - 0)$ and $g(\theta)^{-2}\, g'(\theta + 0)$. This definition has some simple consequences for θ a sharp spot: let γ and Γ, respectively, designate the smaller and larger values of $g^*(\theta \pm 0) \cdot N(\theta)$, and choose $a, b \in \{g^*(\theta)\}$; then

$$\begin{aligned} g^*(\theta + 0) - g^*(\theta - 0) &= g(\theta)^{-2}\, [g'(\theta + 0) - g'(\theta - 0)]\, N(\theta), \\ \gamma \leqq a \cdot N(\theta) \leqq \Gamma, &\quad a - b \text{ is parallel to } N(\theta). \end{aligned} \tag{A3}$$

Somewhat less trivial are the following:

Properties of the supporting tangent.

(T1) *Let* $a \in \{g'(\theta)\}$, *and let* ℓ *be the line through* $\boldsymbol{g}(\theta)$ *in the direction* a. *Then the perpendicular distance from* ℓ *to the origin is* $|a|^{-1}$.

(T2) $|g^*(\theta)|^{-1}$ *is the support function of* Polar (g):

$$\boldsymbol{g}(\theta) \cdot n(\theta) = |g^*(\theta)|^{-1} \quad \textit{for } \theta \textit{ a smooth spot.}$$

(T3) *$g^*(\theta)$ is constant on a connected open subset ℓ of* Polar (g) *if and only if ℓ is a straight line.*

Proof. Let $a \in \{g^*(\theta)\}$, so that $a = -g(\theta)^{-1}\, T(\theta) + \zeta N(\theta)$. Omitting the argument θ, the *unit* vector $u = |a|^{-1}\, [g^{-1}\, N + \zeta T]$ is orthogonal to a, so that $d := |g \cdot u|$ is the perpendicular distance from ℓ to the origin. But $d = |a|^{-1}$. Thus (T1) is valid. (T2) is an obvious consequence of (T1), and (T3) follows from (T2). □

Theorem on common tangents. *Let* $\{g^*(\theta_1)\} \cap \{g^*(\theta_2)\} \neq \emptyset$. *Then*
(CT1) $|\theta_2 - \theta_1| \neq \pi$;
(CT2) $\{g^*(\theta_1)\} \cap \{g^*(\theta_2)\}$ *is a single vector* a;
(CT3) *if* $\theta_2 - \theta_1 < \pi$, a *points in the same direction as* $g(\theta_2) - g(\theta_1)$.
In this case we will refer to a as the **common supporting tangent** *for θ_1 and θ_2.*

Proof. Assume that $\theta_2 - \theta_1 \leqq \pi$. Let $a \in \{g^*(\theta_1)\} \cap \{g^*(\theta_2)\}$, and let ℓ_i denote the line through $g(\theta_i)$ in the direction a. It suffices to show that $\ell_1 = \ell_2$. Suppose not. Then ℓ_1 and ℓ_2 are parallel, and, by (T1), the origin is equidistant from them. Thus $T(\theta_1) \cdot a$ and $T(\theta_2) \cdot a$ are of opposite sign. But by (A2), $T(\theta_i) \cdot a = g(\theta_i)^{-1} > 0$, a contradiction. □

Corollary. *Let $P(g)$ be convex. Let $\theta_1, \theta_2 \in \mathbf{R}$ with $0 < \theta_2 - \theta_1 < \pi$ and $\{g^*(\theta_1)\} \cap \{g^*(\theta_2)\} \neq \emptyset$. Then $g(\theta)$, $\theta_1 \leqq \theta \leqq \theta_2$, is a straight line.*

The *curvature* $k_g(\theta)$ of Polar (g) (positive for Polar (g) strictly convex) is given by

$$k_g = [g^2 - gg'' + 2(g')^2]/[(g')^2 + g^2]^{\frac{3}{2}} \tag{A5}$$

whenever g' and g'' exist. By (A1),

$$\begin{aligned}(d/d\theta)\, g^*(\theta) &= -A(\theta)\, k_g(\theta)\, N(\theta),\\ A &= g^{-3}[(g')^2 + g^2]^{-\frac{3}{2}} > 0.\end{aligned} \tag{A6}$$

Theorem. *Let θ_1 and θ_2 have a common supporting tangent. Then one of the following three conditions must hold:*
(C1) *$P(g)$ is a straight line between θ_1 and θ_2;*
(C2) *$g'(\theta + 0) > g'(\theta - 0)$ at some sharp spot $\theta \in [\theta_1, \theta_2]$;*
(C3) *$k_g(\theta) < 0$ at some smooth spot $\theta \in [\theta_1, \theta_2]$.*

Proof. Let a be the common supporting tangent, and let b be the unit vector with

$$b \cdot a = 0 \quad \text{and} \quad b \cdot N(\theta) > 0 \quad \text{for all } \theta \in [\theta_1, \theta_2]. \tag{A7}$$

Suppose that neither (C2) nor (C3) is satisfied: for $\theta \in [\theta_1, \theta_2]$,

$$g'(\theta + 0) \leqq g'(\theta - 0) \quad \text{for } \theta \text{ sharp}, \quad k_g(\theta) \geqq 0 \text{ for } \theta \text{ smooth}. \tag{A8}$$

Let Θ denote the set of sharp spots in (θ_1, θ_2). If we integrate $(d/d\theta)\, g^*(\theta)$ from $\theta_1 + 0$ to $\theta_2 - 0$ using (A6) and (A3)$_1$, and then take the inner product of the resulting relation with $\boldsymbol{b}$, we find that

$$[g^*(\theta_2 - 0) - g^*(\theta_1 + 0)] \cdot \boldsymbol{b}$$
$$= \sum_{\theta \in \Theta} c(\theta)\, [g'(\theta + 0) - g'(\theta - 0)] - \int_{\theta_1}^{\theta_2} C(\theta)\, k_g(\theta)\, d\theta, \qquad \text{(A9)}$$
$$c(\theta), C(\theta) > 0 \quad \text{on } [\theta_1, \theta_2];$$

thus, by (A8) and (A7),

$$[g^*(\theta_2 - 0) - \boldsymbol{a}] \cdot \boldsymbol{b} - [g^*(\theta_1 + 0) - \boldsymbol{a}] \cdot \boldsymbol{b} \leqq 0. \qquad \text{(A10)}$$

Next, in view of (A3)$_{2,3}$ and (A8)$_1$,

$$g^*(\theta_2 - 0) - \boldsymbol{a} = \alpha_2 N(\theta_2), \quad \alpha_2 \geqq 0,$$
$$g^*(\theta_1 + 0) - \boldsymbol{a} = \alpha_1 N(\theta_1), \quad \alpha_1 \leqq 0,$$

and we may use (A7) to conclude that (A10) holds with "$\leqq$" replaced by "$=$"; hence (A9) vanishes, and this yields (A8) with inequalities replaced by equalities. Thus, by (A3)$_1$ and (A6), $g^*(\theta)$ is constant on (θ_1, θ_2); in view of (T3), this implies (C1). □

The convex hull[62] of Polar (g) is a polar diagram Polar (G) of a function $G(\theta)$. We will refer to Polar (G) as the **convexification** of $g(\theta)$. Let

$$\boldsymbol{G}(\theta) := G(\theta)\, N(\theta).$$

The set Polar $(g) \cap$ Polar (G) on which the polar diagram coincides with its convex hull is important. This subset of $\mathbf{R}^2$ is conveniently identified with a set of angles, namely

$$C(g) := \{\theta \in R \colon g(\theta) = G(\theta)\}.$$

We will refer to the *connected components* of $C(g)$ as the **globally convex sections** of Polar (g). The portion of Polar (g) that is *disjoint* from Polar (G) is the union of open line-semgents; the closures of these line segments will be referred to as the **Maxwell lines** of Polar (g). Let m be a Maxwell line with end points $g(\theta_1)$ and $g(\theta_2)$, $\theta_1 < \theta_2$; we will refer to (θ_1, θ_2) as the **angle interval for** m and to θ_1 and θ_2 as **Maxwell angles**.

[62] More precisely, boundary of the convex hull of Polar (g). In our discussion of curves the term "convex" means "strictly convex". Here our terminology is ambiguous, as the boundary of the convex hull will generally not be strictly convex. Thus the globally convex sections of Polar (g) are allowed to have subsets with vanishing curvature.

Maxwell theorem. *Let θ belong to a globally convex section of* Polar (g). *Then:*

$$g(\theta) = G(\theta), \quad \{G^*(\theta)\} \subset \{g^*(\theta)\},$$
$$g'(\theta + 0) \leqq G'(\theta + 0) \leqq G'(\theta - 0) \leqq g'(\theta - 0), \tag{A11}$$
$$N(\theta) \cdot [G^*(\theta + 0) - G^*(\theta - 0)] \leqq 0$$

with inequality in $(\text{A11})_4$ *if θ is a sharp spot of g.*

Let (θ_1, θ_2) be an angle interval for a Maxwell line. Then:

(i) *θ_1 and θ_2 have a common supporting tangent G_0 with G_0 the constant value of $G^*(\theta)$ on (θ_1, θ_2);*

(ii) *either* (C2) *or* (C3) *holds on* (θ_1, θ_2).

Proof. $(\text{A11})_1$ is obvious; $(\text{A11})_3$ follows directly from the properties of the convex hull; $(\text{A11})_3$ implies $(\text{A11})_2$ and, by virtue of (A1), also $(\text{A11})_4$. Further $G^*(\theta) \equiv G_0$ on (θ_1, θ_2) follows from (T3) applied to G rather than g, while $(\text{A11})_2$ yields $G_0 \in \{g^*(\theta_1)\} \cap \{g^*(\theta_2)\}$; the desired conclusion in (i) then follows from the theorem on common tangents. Finally, between θ_1 and θ_2, $P(G)$ is a straight line disjoint from $P(g)$; thus (C1) is not possible, and so either (C2) or (C3) is satisfied. □

Appendix B. Invariance under reparametrization

A suitable discussion of invariance requires a class of time-dependent curves broader than the class of evolving curves. Let T, $0 < T \leqq \infty$, be fixed. A **time-dependent interval** is a set of the form $\mathbf{R} \times [0, T)$ or a set of the form

$$\{(p, t): p \in [P(t), Q(t)],\ t \in [0, T)\} \tag{B1}$$

with $P, Q: [0, T) \to \mathbf{R}$ $(P < Q)$ smooth functions. A **time-dependent curve** is a smooth mapping $(p, t) \mapsto \boldsymbol{r}(p, t)$ such that:

(i) the domain of $\boldsymbol{r}$ is a time-dependent interval;

(ii) $\boldsymbol{r}(\cdot, t)$ is a *curve* for each $t \in [0, T)$.

If Domain $(\boldsymbol{r})$ has the form (B1), then $\boldsymbol{r}$ has **endpoints** $\boldsymbol{r}(P(t), t)$ and $\boldsymbol{r}(Q(t), t)$.

We write $\mathscr{C}$ for the set of time-dependent curves. Let $\boldsymbol{r} \in \mathscr{C}$. Then $\boldsymbol{r}$ **evolves normally** if

$$\boldsymbol{r}_t(p, t) \cdot \boldsymbol{r}_p(p, t) = 0$$

for all $(p, t) \in$ Domain $(\boldsymbol{r})$; thus the term "**evolving curve**" as used in the main body of the paper is here synonymous with **normally evolving curve**.

Let $\boldsymbol{r} \in \mathscr{C}$. We consider the tangent $T(p, t) := \boldsymbol{r}_p(p, t)/|\boldsymbol{r}_p(p, t)|$ and normal $N(p, t)$ to $\boldsymbol{r}$ as functions of $(p, t) \in$ Domain $(\boldsymbol{r})$, and similarly for the **normal velocity**

$$V(p, t) := \boldsymbol{r}_t(p, t) \cdot N(p, t). \tag{B2}$$

Further we define

$$J(p, t) := |\boldsymbol{r}_p(p, t)|,$$
$$v(p, t) := -J(p, t)^{-1}\, \boldsymbol{r}_t(p, t) \cdot T(p, t). \tag{B3}$$

Let $Z(t)$ be given with $(Z(t), t) \in \text{Domain}(\boldsymbol{r})$ for some interval of t. Then the curve $t \mapsto \boldsymbol{r}(Z(t), t)$ is a **normal trajectory** provided

$$T(Z(t), t) \cdot (d/dt)\, \boldsymbol{r}(Z(t), t) = 0.$$

But by (B3),

$$T(Z(t), t) \cdot (d/dt)\, \boldsymbol{r}(Z(t), t) = J(Z(t), t)\, [dZ(t)/dt - v(Z(t), t)], \tag{B4}$$

so that $v(p, t)$ gives the rate at which the parameter p changes with time following a normal trajectory. This discussion should motivate the following definition.

Choose $(p_0, t_0) \in \text{Domain}(\boldsymbol{r})$. The function $t \mapsto Z(t)$, *maximally* defined as the solution of the problem

$$dZ(t)/dt = v(Z(t), t), \qquad Z(t_0) = p_0, \tag{B5}$$

is the **normal parameter-trajectory** through (p_0, t_0).

Let $\boldsymbol{r} \in \mathscr{C}$, let Φ be a smooth function on Domain $(\boldsymbol{r})$, and choose $(p, t) \in$ Domain $(\boldsymbol{r})$. Then the **normal time-derivative** $\Phi^\circ(p, t)$ of Φ at (p, t) is defined as follows:

$$\Phi^\circ(p, t) = (d/d\tau)\, \Phi(Z(\tau), \tau)\big|_{\tau = t}, \tag{B6}$$

with $Z(\tau)$ the normal parameter-trajectory through (p, t). Clearly, $\boldsymbol{r}^\circ \cdot N = \boldsymbol{r}_t \cdot N$, while (B4) and (B5) yield $\boldsymbol{r}^\circ \cdot T = 0$; hence (B2) implies

$$\boldsymbol{r}^\circ(p, t) = V(p, t)\, N(p, t). \tag{B7}$$

The next proposition is easily verified.

Proposition 1. *Let* $\boldsymbol{r} \in \mathscr{C}$. *Then the following are equivalent:*
(i) $\boldsymbol{r}$ *evolves normally*;
(ii) *the normal parameter-trajectories are of the form* $Z = \text{constant}$;
(iii) $\boldsymbol{r}_t \equiv \boldsymbol{r}^\circ$;
(iv) $v \equiv 0$.

Moreover, if $\boldsymbol{r}$ *evolves normally, and if* Φ *is a smooth function on* Domain $(\boldsymbol{r})$, *then*

$$\Phi^\circ = \Phi_t. \tag{B8}$$

Let $\boldsymbol{r} \in \mathscr{C}$. By a **parameter change** for $\boldsymbol{r}$ we mean a smooth bijection $\boldsymbol{\phi}$, from a time-dependent interval onto Domain $(\boldsymbol{r})$, of the form

$$(p, t) \mapsto \boldsymbol{\phi}(p, t) = (\phi(p, t), t), \qquad \phi_p > 0;$$

if $\boldsymbol{r}$ is closed we require, in addition, that there exist a smooth function $\omega > 0$ on $[0, T)$ such that

$$\phi(p + \omega(t), t) = \phi(p, t) + \lambda(t) \tag{B9}$$

for all $(p, t) \in \mathbb{R} \times [0, \infty)$, where $\lambda(t)$ is the minimal period of $\boldsymbol{r}(\cdot, t)$. Given a parameter change $\boldsymbol{\phi}$ for $\boldsymbol{r}$, the function $\boldsymbol{r} \circ \boldsymbol{\phi}$ on Domain$(\boldsymbol{\phi})$ defined by

$$(\boldsymbol{r} \circ \boldsymbol{\phi})(p, t) = \boldsymbol{r}(\boldsymbol{\phi}(p, t)) = \boldsymbol{r}(\phi(p, t), t) \tag{B10}$$

is also a member of $\mathscr{C}$ (and is closed if $\boldsymbol{r}$ is closed); we refer to $\boldsymbol{r} \circ \boldsymbol{\phi}$ as a **reparametrization** of $\boldsymbol{r}$. This definition, (B4) with $Z(t) = \phi(p, t)$, and the equivalence of (i) and (iv) in Proposition 1 yield

Proposition 2. *Let* $\boldsymbol{r} \in \mathscr{C}$. *Then* $\boldsymbol{r} \circ \boldsymbol{\phi}$ *evolves normally if and only if* $\phi_t(p, t) = v(\phi(p, t), t)$ *for all* $(p, t) \in \text{Domain}(\boldsymbol{\phi})$, *so that each of the functions* $t \mapsto \phi(p, t)$ *is a normal parameter-trajectory. Thus, when* $\boldsymbol{r}$ *evolves normally,* $\boldsymbol{r} \circ \boldsymbol{\phi}$ *evolves normally if and only if* $\phi_t = 0$.

The next result is central; it shows that within a large class of time-dependent curves there is no essential loss of generality in limiting attention to curves that evolve normally.

Theorem 1. *Let* $\boldsymbol{r} \in \mathscr{C}$ *satisfy one of the following three conditions*:
(i) $\boldsymbol{r}$ *is closed*;
(ii) $\boldsymbol{r}$ *has endpoints, and the endpoints are normal trajectories*;
(iii) $\boldsymbol{r}$ *is unbounded, and there are smooth functions* $a: \mathbb{R} \to \mathbb{R}$ *and* $b: [0, T) \to \mathbb{R}$ *such that for all* $(p, t) \in \text{Domain}(\boldsymbol{r})$,

$$|v(p, t)| \leqq a(p)\, b(t). \tag{B11}$$

Then there is a parameter change $\boldsymbol{\phi}$ *for* $\boldsymbol{r}$ *such that* $\boldsymbol{r} \circ \boldsymbol{\phi}$ *is a normally evolving curve.*

Proof. For each p in the initial interval $[P(0), Q(0)]$ or $\mathbb{R}$, let $t \mapsto Z(p, t)$ denote the normal parameter-trajectory through $(p, 0)$. Consider (ii). In this case $Z(P(0), t) = P(t)$ and $Z(Q(0), t) = Q(t)$ for $0 \leqq t < T$; thus, since $Z(p, t)$ is, for p fixed, a maximal solution of (B5), and since $P(0) \leqq p \leqq Q(0)$, $Z(p, t)$ is also defined for $0 \leqq t < T$. In fact, the definition of Z as the solution of (B5) renders the mapping $\boldsymbol{\phi}$ defined by $\boldsymbol{\phi}(p, t) = (Z(p, t), t)$ a smooth bijection of $[P(0), Q(0)] \times [0, T)$ onto Domain ($\boldsymbol{r}$). Further, differentiating $Z_t(p, t) = v(Z(o, t)$, with respect to p, one easily concludes that $Z_p(p, t) > 0$, since it has this property at $p = 0$. Thus $\boldsymbol{\phi}$ is a parameter change for $\boldsymbol{r}$. The last proposition then implies that $\boldsymbol{r} \circ \boldsymbol{\phi}$ is a normally evolving curve.

Consider (iii). Let $\boldsymbol{\phi}_0$ denote the parameter change for $\boldsymbol{r}$ defined by $\boldsymbol{\phi}_0(p, t) = (\phi_0(p), t)$ with $\phi_0(p)$ any solution of $d\phi_0(p)/dp = a(p)$. Then the reparametrization $\boldsymbol{r}_0 \circ \boldsymbol{\phi}_0$ obeys (B11) with $a(p) = 1$. Thus it suffices to consider unbounded curves $\boldsymbol{r} \in \mathscr{C}$ that obey an estimate of the form

$$|v(p, t)| \leqq b(t) \tag{B12}$$

with b continuous on $[0, T)$. This estimate and the definition of Z as the solution of (B5), imply that for each $p \in \mathbb{R}$, $Z(p, t)$ is defined for $0 \leqq t < T$. In fact, arguing as above, the mapping $\boldsymbol{\phi}$ is a parameter change for $\boldsymbol{r}$, and $\boldsymbol{r} \circ \boldsymbol{\phi}$ is a normally evolving curve.

Consider (i). Since $\boldsymbol{r}$ is periodic with period $\lambda(t)$ a smooth function of t, the estimate (B12) again is satisfied. Thus the mapping $\boldsymbol{\phi}$ defined by $\boldsymbol{\phi}(p, t) = (Z(p, t), t)$

is a smooth bijection of $\mathbf{R}\times[0,T)$ onto Domain $(\mathbf{r})$, and $Z_p(p,t)>0$. Since $\mathbf{r}(z,t)=\mathbf{r}(z+\lambda(t),t)$. (B3) yields $v(z+\lambda(t),t)=v(z,t)+\lambda_t(t)$, and this, in turn, leads to the conclusion that $z(p,t):=Z(p,t)+\lambda(t)$ satisfies $z_t(p,t)=v(z(p,t),t)$, $z(p,0)=p+\lambda(0)$. Thus $z(p,t)=Z(p+\lambda(0),t)$, so that $Z(p,t)+\lambda(t)=Z(p+\lambda(0),t)$ for all (p,t), and we have compliance with the condition (B9). Thus $\boldsymbol{\phi}$ is a parameter change for $\mathbf{r}$, and $\mathbf{r}\circ\boldsymbol{\phi}$ is a normally evolving curve. □

We define the **arc-length derivative** $\Phi_s(p,t)$, the **curvature** $K(p,t)$, and the **angle derivative** $\Phi_\theta(p,t)$ through:

$$\begin{aligned}\Phi_s(p,t)&:=\Phi_p(p,t)\,J(p,t)^{-1},\\ K(p,t)&:=\mathbf{N}(p,t)\cdot\mathbf{T}_s(p,t),\\ \Phi_\theta(p,t)&:=K(p,t)^{-1}\,\Phi_s(p,t)\end{aligned}\qquad\text{(B13)}$$

(the last definition being appropriate to (p,t) with $K(p,t)\neq 0$).

To discuss invariance under reparametrization, we now write $\mathbf{T}_\mathbf{r}(p,t)$, $\mathbf{N}_\mathbf{r}(p,t)$, $V_\mathbf{r}(p,t)$, and $K_\mathbf{r}(p,t)$ to make explicit the dependence of these quantities on the time-dependent curve $\mathbf{r}\in\mathscr{C}$ in question. This allows us to consider, for example, the normal velocity as a mapping V that assigns to each $\mathbf{r}\in\mathscr{C}$ a function $(p,t)\mapsto V_\mathbf{r}(p,t)$ on Domain $(\mathbf{r})$.

More generally, a **curve descriptor** is a mapping Ψ that assigns to each $\mathbf{r}\in\mathscr{C}$ a function $(p,t)\mapsto\Psi_\mathbf{r}(p,t)$ on Domain $(\mathbf{r})$. Given a curve descriptor Ψ, we may consider its normal time-derivative, its arc-length derivative, and its angle derivative as curve descriptors; *i.e.*, *e.g.*, $(\Psi_s)_\mathbf{r}:=(\Psi_\mathbf{r})_s$.

A curve descriptor Ψ is intrinsic if it is invariant under reparametrization; that is, if, at each t, its value at p on a reparametrized curve $\mathbf{r}\circ\boldsymbol{\phi}$ is the same as its value at $\phi(p,t)$ on the original curve $\mathbf{r}$. Precisely, Ψ is **intrinsic** if, given any $\mathbf{r}\in\mathscr{C}$ and any parameter change $\boldsymbol{\phi}$ for $\mathbf{r}$,

$$\Psi_{(\mathbf{r}\circ\boldsymbol{\phi})}=(\Psi_\mathbf{r})\circ\boldsymbol{\phi}.$$

The following result is well known.

Invariance theorem. *The following curve descriptors are intrinsic: tangent, normal, normal velocity, and curvature. If a curve descriptor is intrinsic, then so also are its normal time-derivative, its arc-length derivative, and its angle derivative.*

Proof. Let $\boldsymbol{\phi}$ be a parameter change for $\mathbf{r}\in\mathscr{C}$, let $\mathbf{g}=\mathbf{r}\circ\boldsymbol{\phi}$, and write $\mathbf{g}(p,t)=\mathbf{r}(q,t)$, $q=\phi(p,t)$. Since $\mathbf{g}_p=\mathbf{r}_q\phi_p$, $\mathbf{T}_\mathbf{g}(p,t)=\mathbf{g}_p(p,t)/|\mathbf{g}_p(p,t)|=\mathbf{r}_q(q,t)/|\mathbf{r}_q(q,t)|=\mathbf{T}_\mathbf{r}(q,t)$, so that $\mathbf{T}$ and (hence) $\mathbf{N}$ are invariant. Let Ψ be an intrinsic curve-descriptor. Then the same argument applied to $(\text{B13})_1$ yields Ψ_s intrinsic. Thus, by $(\text{B13})_{2,3}$, K and Ψ_θ are invariant. Next, let $Z(t)$ be the normal parameter-trajectory for $\mathbf{g}$ through (p_0,t_0). A simple calculation shows that

$$\left(\frac{d}{dt}\right)\mathbf{r}(\phi(Z(t),t),t)=\left(\frac{d}{dt}\right)\mathbf{g}(Z(t),t)=0,$$

so that $z(t) := \phi(Z(t), t)$ is the normal parameter-trajectory for r through $\phi(p_0, t_0)$. On the other hand, since Ψ is intrinsic,

$$\left(\frac{d}{d\tau}\right) \Psi_g(Z(\tau), \tau) = \left(\frac{d}{d\tau}\right) \Psi_r(z(\tau), \tau). \tag{B14}$$

At $\tau = t_0$, the left side of (B14) is $(\Psi^\circ)_g\,(p_0, t_0)$, the right side is $(\Psi^\circ)_r\,(\phi(p_0, t_0))$; thus the normal time-derivative of an intrinsic curve-descriptor is intrinsic. In particular, $r \mapsto r^\circ$ is intrinsic; hence, by (B7), V is intrinsic. □

References

[1901] WULFF, G., Zur Frage der Geschwindigkeit des Wachsthums und der Auflösung der Krystallflächen, Zeit. Krystall. Min. **34**, 449–530.

[1944] DINGHAS, A., Über einen geometrischen Satz von Wulff für die Gleichgewichtsform von Krystallen, Zeit. Krystall. **105**, 304–314.

[1951 a] HERRING, C., Surface tension as a motivation for sintering, *The Physics of Powder Metallurgy* (ed. W. E. KINGSTON) McGraw-Hill, New York.

[1951 b] HERRING, C., Some theorems on the free energies of crystal surfaces, Phys. Rev. **82**, 87–93.

[1958] FRANK, F. C., On the kinematic theory of crystal growth and dissolution processes, *Growth and Perfection of Crystals* (eds. R. H. DOREMUS, B. W. ROBERTS & D. TURNBULL) John Wiley, New York.

[1963] FRANK, F. C., The geometrical thermodynamics of surfaces, Metal Surfaces: Structure, Energetics, and Kinetics, Am. Soc. Metals, Metals Park, Ohio.

[1963] GJOSTEIN, N. A., Adsorption and surface energy (II): thermal faceting from minimization of surface energy, Act. Metall. **11**, 969–977.

[1967] PROTTER, M., & H. WEINBERGER, *Maximum Principles in Differential Equations*, Prentice Hall, New York.

[1974] CAHN, J. W., & D. W. HOFFMAN, A vector thermodynamics for anisotropic surfaces — 2. curved and faceted surfaces, Act. Metall. **22**, 1205–1214.

[1978] BRAKKE, K. A., *The Motion of a Surface by its Mean Curvature*, Princeton University Press.

[1978] TAYLOR, J. E., Crystalline variational problems, Bull. Am. Math. Soc. **84**, 568–588.

[1979] ALLEN, S. M., & J. W. CAHN, A macroscopic theory for antiphase boundary motion and its application to antiphase domain coarsening, Act. Metall. **27**, 1085–1098.

[1984] GAGE, M., Curve shortening makes convex curves circular, Invent. Math. **76**, **76**, 357–364.

[1985] SETHIAN, J. A., Curvature and the evolution of fronts, Comm. Math. Phys. **101**, 487–499.

[1986] ABRESCH, U., & J. LANGER, The normalized curve shortening flow and homothetic solutions. J. Diff. Geom. **23**, 175–196.

[1986] GAGE, M., & R. S. HAMILTON, The heat equation shrinking convex plane curves, J. Diff. Geom. **23**, 69–95.

[1986] GAGE, M., On an area preserving evolution equation for plane curves, Contemporary Math. **51**, 51–62.

[1986g] GURTIN, M. E., On the two-phase Stefan problem with interfacial energy and entropy, Arch. Rational Mech. Anal. **96**, 199–241.

[1987] GRAYSON, M. A., The heat equation shrinks embedded plane curves to round points, J. Diff. Geom. **26**, 285–314.

[1987] HUISKEN, G., Deforming hypersurfaces of the sphere by their mean curvature, Math. Z. **198**, 138–146.

[1987] MARIS, H. J., & A. A. ANDREEV, The surface of crystalline helium 4, Phys. Today, February, 25–30.

[1987] OSHER, S., & J. A. SETHIAN, Fronts propagating with curvature dependent speed: algorithms based on Hamilton-Jacobi formulations, CAM Report 87–12, Dept. Math., U. California, Los Angeles.

[1987] RUBINSTEIN, J., STERNBERG, P., & J. B. KELLER, Fast reaction, slow diffusion, and curve shortening, Forthcoming.

[1988] ANGENENT, S., Forthcoming.

[1988g] GURTIN, M. E., Multiphase thermomechanics with interfacial structure. 1. Heat conduction and the capillary balance law. Arch. Rational Mech. Anal. **104**, 195–221.

[1988gg] GURTIN, M. E., Multiphase thermomechanics with interfacial structure. Toward a nonequilibrium thermomechanics of two phase materials, Arch. Rational Mech. Anal. **100**, 275–312.

[1988] FONSECA, I., Interfacial energy and the Maxwell rule, Res. Rept. 88–18, Dept. Math., Carnegie Mellon, Pittsburgh.

[1989] ANGENENT, S., & M. E. GURTIN, Forthcoming.

Department of Mathematics
University of Wisconsin
Madison

and

Department of Mathematics
Carnegie Mellon University
Pittsburgh

(Received January 13, 1989)

J. Mech. Phys. Solids Vol. 38, No. 3, pp. 345–360, 1990.

0022–5096/90 $3.00+0.00

ON THE DRIVING TRACTION ACTING ON A SURFACE OF STRAIN DISCONTINUITY IN A CONTINUUM

ROHAN ABEYARATNE‡ and JAMES K. KNOWLES§

‡ Department of Mechanical Engineering, Massachusetts Institute of Technology, Cambridge, MA 02139, U.S.A. and § Division of Engineering and Applied Science, California Institute of Technology, Pasadena, CA 91125, U.S.A.

(*Received* 17 *January* 1989; *in revised form* 1 *June* 1989)

This paper is dedicated to the memory of Eli Sternberg

ABSTRACT

THE notion of the *driving traction* on a surface of strain discontinuity in a continuum undergoing a general thermomechanical process is defined and discussed. In addition, the associated constitutive notion of a *kinetic relation*, in which the normal velocity of propagation of the surface of discontinuity may be a given function of the driving traction and temperature, is introduced for the special case of a thermoelastic material.

1. INTRODUCTION

VARIOUS aspects of the theory of finite elastostatics for materials characterized by non-elliptic elastic potentials have been studied in a number of recent papers; see e.g. ABEYARATNE and KNOWLES (1987a, b, 1988a, b), BALL (1977), BALL and JAMES (1987), ERICKSEN (1975), FOSDICK and MACSITHIGH (1983), GURTIN (1983), JAMES (1981, 1986), KNOWLES and STERNBERG (1978) and SILLING (1988). One feature of such materials is that, under suitable conditions, they can sustain *equilibrium* deformations in which the displacement gradient and stress tensors suffer jump discontinuities across certain surfaces in the body, while the displacement and traction remain continuous; such singular surfaces have been called equilibrium shocks. One area in which the associated theory finds application is that of the continuum-mechanical modeling of a solid in equilibrium with more than one "phase" present. In this setting, an equilibrium shock corresponds to the boundary between two distinct phases of the material (BALL and JAMES, 1987; JAMES, 1981, 1986; SILLING, 1988).

A quasi-static motion involving equilibrium shocks at each instant may be dissipative (KNOWLES, 1979): in any portion of the body that is traversed by a moving shock, the rate of work of the external forces differs from the rate of storage of strain energy by the rate of work done in moving the surface of discontinuity. This latter rate of work can be expressed as the integral over the shock surface of the product of a scalar "driving traction" f with the component of shock velocity normal to the shock itself. At each instant during the motion, the value of f at each point on the singular surface can be calculated from the limiting values of the deformation gradient on the two sides of the surface. If the driving traction f vanishes at all points of the shock for each equilibrium state visited during a quasi-static motion, the motion is dissipation-free. An equilibrium state for which $f = 0$ is said to satisfy the Maxwell condition.

In the presence of equilibrium shocks, the mechanical balance laws lead to the usual field equations at points in the body away from the singular surfaces and require the traction to be continuous across these surfaces. However, on solving even the simplest of boundary-value problems for non-elliptic elastic materials, one encounters a massive failure of uniqueness of solution (ABEYARATNE, 1980; ABEYARATNE and KNOWLES, 1987a). One way to single out a preferred equilibrium field from among the infinitely many available ones is to require that the field be stable in the sense that the associated potential energy be an absolute minimum. If an equilibrium field containing a shock is to be stable in this sense, it is known that the Maxwell condition $f = 0$ is necessary (ABEYARATNE, 1983; ERICKSEN, 1975; GURTIN, 1983; JAMES, 1981).

A somewhat more general point of view holds that the lack of uniqueness in the conventional equilibrium problems for non-elliptic elastic materials arises from a constitutive deficiency associated with particles on the shock surface. Such a view was adopted by ABEYARATNE and KNOWLES (1988a, b) in the one-dimensional context of bar theory. By exploiting an analogy between the problem considered and internal-variable theories of inelastic solids (see e.g. RICE, 1971, 1975), they are led to postulate a supplementary constitutive requirement in the form of a "kinetic relation" between f and the velocity of the shock during a quasi-static motion. This leads to a determinate macroscopic response (or force–elongation relation) for the bar in quasi-static motions. In general, this response exhibits rate- and history-dependence. Two limiting cases of the kinetic relation describe rate-independent behavior: one corresponds to the Maxwell condition and hence to dissipation-free, reversible macroscopic response. The other leads to a force–elongation relation similar to that associated with rate-independent plasticity.

The investigations described above are for the most part carried out within the framework of the purely mechanical theory of non-elliptic elastic materials. They are limited to the study of equilibrium states or one-parameter families of such states (i.e. quasi-static motions), and they are often confined to one-dimensional settings as well. The purpose of the present study is to consider the corresponding issues in a more general, three-dimensional setting in which the material need not be elastic, and both thermal and inertial effects are taken into account. In Section 2, we recall the basic thermodynamic and mechanical laws and the associated field equations, inequalities and jump conditions for a continuum undergoing a thermomechanical process in which displacement and temperature are assumed to be continuous, but their gradients are permitted to jump across a moving surface. We derive a useful representation of the entropy production rate in Section 3, and we use it to introduce the notion of a driving traction acting on a singular surface in an arbitrary continuum during a thermomechanical process of the assumed type. We specialize the earlier results to slow isothermal processes in Section 4. No constitutive assumptions are invoked until Section 5, where we specialize the foregoing results to the case of a continuum composed of a thermoelastic material. Section 6 is devoted to a discussion of the notion of a constitutive "kinetic relation."

2. Balance Laws, Field Equations and Jump Conditions

Consider a body B that occupies a region R in a reference configuration. A motion of the body on a time interval $[t_0, t_1]$ is characterized by a one-parameter family of invertible mappings $\hat{\mathbf{y}}(\cdot, t) : \mathrm{R} \to \mathrm{R}_*$, with

$$\mathbf{y} = \hat{\mathbf{y}}(\mathbf{x}, t) = \mathbf{x} + \mathbf{u}(\mathbf{x}, t) \quad \text{for } \mathbf{x} \in \mathrm{R},\ t \in [t_0, t_1]. \tag{2.1}$$

We assume that the deformation $\hat{\mathbf{y}}$, or equivalently the displacement $\mathbf{u}$, is continuous with piecewise continuous first and second derivatives on $\mathrm{R} \times [t_0, t_1]$. Let $\mathbf{F}(\mathbf{x}, t) = \operatorname{Grad} \hat{\mathbf{y}}(\mathbf{x}, t)$ and $\mathbf{v}(\mathbf{x}, t) = \partial \hat{\mathbf{y}}(\mathbf{x}, t)/\partial t$ stand respectively for the deformation gradient tensor and the particle velocity at points $(\mathbf{x}, t)$ in space-time where they exist.

Let $\rho(\mathbf{x})$ denote the mass density of B at the point $\mathbf{x}$ in the reference configuration, $\mathbf{b}(\mathbf{x}, t)$ the body force per unit mass, and $\boldsymbol{\sigma}(\mathbf{x}, t)$ the nominal stress tensor. At each t, we require $\rho(\cdot)$ and $\mathbf{b}(\cdot, t)$ to be continuous on R, while $\boldsymbol{\sigma}(\cdot, t)$ is to be piecewise continuous with a piecewise continuous gradient on R. The balance laws for linear and angular momentum require

$$\int_{\partial \mathrm{D}} \boldsymbol{\sigma} \mathbf{n} \, \mathrm{d}A + \int_{\mathrm{D}} \rho \mathbf{b} \, \mathrm{d}V = \mathrm{d}/\mathrm{d}t \int_{\mathrm{D}} \rho \mathbf{v} \, \mathrm{d}V, \tag{2.2}$$

$$\int_{\partial \mathrm{D}} \hat{\mathbf{y}} \times \boldsymbol{\sigma} \mathbf{n} \, \mathrm{d}A + \int_{\mathrm{D}} \hat{\mathbf{y}} \times \rho \mathbf{b} \, \mathrm{d}V = \mathrm{d}/\mathrm{d}t \int_{\mathrm{D}} \hat{\mathbf{y}} \times \rho \mathbf{v} \, \mathrm{d}V, \tag{2.3}$$

respectively, at each $t \in [t_0, t_1]$ and for all regular subregions $\mathrm{D} \subset \mathrm{R}$.

Next, let $\mathbf{q}(\mathbf{x}, t)$ denote the nominal heat flux vector, $r(\mathbf{x}, t)$ the heat supply per unit mass and $\varepsilon(\mathbf{x}, t)$ the internal energy per unit mass. At each t, we suppose that $r(\cdot, t)$ is continuous on R and that $\mathbf{q}(\cdot, t)$ is piecewise continuous with a piecewise continuous gradient on R. The internal energy $\varepsilon(\cdot, \cdot)$ is required to be piecewise continuous with piecewise continuous first derivatives on $\mathrm{R} \times [t_0, t_1]$. The first law of thermodynamics requires that at each instant t,

$$\int_{\partial \mathrm{D}} \boldsymbol{\sigma} \mathbf{n} \cdot \mathbf{v} \, \mathrm{d}A + \int_{\mathrm{D}} \rho \mathbf{b} \cdot \mathbf{v} \, \mathrm{d}V + \int_{\partial \mathrm{D}} \mathbf{q} \cdot \mathbf{n} \, \mathrm{d}A + \int_{\mathrm{D}} \rho r \, \mathrm{d}V$$
$$= \mathrm{d}/\mathrm{d}t \int_{\mathrm{D}} \rho \varepsilon \, \mathrm{d}V + \mathrm{d}/\mathrm{d}t \int_{\mathrm{D}} (1/2) \rho \mathbf{v} \cdot \mathbf{v} \, \mathrm{d}V, \tag{2.4}$$

for every regular $\mathrm{D} \subset \mathrm{R}$. Finally, let $\theta(\mathbf{x}, t)$ denote the absolute temperature and $\eta(\mathbf{x}, t)$ the entropy per unit mass. At each t, we assume that $\theta(\cdot, t)$ is continuous with a piecewise continuous gradient on R, while $\eta(\cdot, \cdot)$ is assumed to be piecewise continuous with piecewise continuous first derivatives on $\mathrm{R} \times [t_0, t_1]$. The *rate of entropy production* in a regular region $\mathrm{D} \subset \mathrm{R}$ is defined to be

$$\Gamma(t; \mathrm{D}) \equiv \mathrm{d}/\mathrm{d}t \int_{\mathrm{D}} \rho \eta \, \mathrm{d}V - \int_{\partial \mathrm{D}} \mathbf{q} \cdot \mathbf{n}/\theta \, \mathrm{d}A - \int_{\mathrm{D}} \rho r/\theta \, \mathrm{d}V. \tag{2.5}$$

The Clausius–Duhem version of the second law of thermodynamics requires

$$\Gamma(t;\mathrm{D}) \geqslant 0 \quad \text{for } \mathrm{D} \subset \mathrm{R},\ t \in [t_0, t_1]. \tag{2.6}$$

At a fixed instant t, localization of the balance laws (2.2)–(2.4) and the inequality (2.6) at a point $\mathbf{x}$ at which $\mathbf{F}$, $\dot{\mathbf{F}}$, $\mathbf{v}$, $\mathbf{q}$, $\boldsymbol{\sigma}$, ε and η are all continuous yields the following familiar local results:

$$\left.\begin{aligned} \operatorname{Div}\boldsymbol{\sigma}+\rho\mathbf{b} &= \rho\dot{\mathbf{v}},\\ \boldsymbol{\sigma}\mathbf{F}^{\mathrm{T}} &= \mathbf{F}\boldsymbol{\sigma}^{\mathrm{T}},\\ \boldsymbol{\sigma}\cdot\dot{\mathbf{F}}+\operatorname{Div}\mathbf{q}+\rho r &= \rho\dot{\varepsilon},\\ \operatorname{Div}(\mathbf{q}/\theta)+\rho r/\theta &\leqslant \rho\dot{\eta} \end{aligned}\right\}. \tag{2.7}$$

On the other hand, suppose that $\mathrm{S}(t)$ is a regular surface in R at time t across which some or all of the thermomechanical quantities listed above suffer jump discontinuities. Localization of (2.2)–(2.6) at a point $\mathbf{x}$ on $\mathrm{S}(t)$ yields the following jump conditions:

$$\left.\begin{aligned} &[[\boldsymbol{\sigma}\mathbf{n}]]+\rho[[\mathbf{v}]]V_n = \mathbf{0},\\ &[[\boldsymbol{\sigma}\mathbf{n}\cdot\mathbf{v}]]+[[\rho(\varepsilon+\mathbf{v}\cdot\mathbf{v}/2)]]V_n+[[\mathbf{q}\cdot\mathbf{n}]] = 0,\\ &[[\rho\eta]]V_n+[[\mathbf{q}\cdot\mathbf{n}/\theta]] \leqslant 0, \end{aligned}\right\} \quad \text{on } \mathrm{S}(t), \tag{2.8}$$

where

$$V_n = \mathbf{V}\cdot\mathbf{n}, \tag{2.9}$$

and $\mathbf{V} = \mathbf{V}(\mathbf{x}, t)$ is the velocity of the point $\mathbf{x}$ on the moving surface $\mathrm{S}(t)$. The unit normal $\mathbf{n}$ on the singular surface $\mathrm{S}(t)$ is chosen such that $V_n \geqslant 0$; if $V_n > 0$, the positive side of $\mathrm{S}(t)$ is the side into which $\mathbf{V}$ (and therefore $\mathbf{n}$) points. If $g(\mathbf{x}, t)$ denotes a generic field quantity that jumps across $\mathrm{S}(t)$, we write $[[g(\mathbf{x}, t)]] = \overset{+}{g}(\mathbf{x}, t)-\overset{-}{g}(\mathbf{x}, t)$, where $\overset{+}{g}(\mathbf{x}, t)$ and $\overset{-}{g}(\mathbf{x}, t)$ stand for the limiting values of g at the point $\mathbf{x}$ on $\mathrm{S}(t)$ from the positive and negative sides, respectively.

In addition to the jump conditions listed in (2.8), one also has the kinematic results

$$[[\mathbf{F}]]\boldsymbol{\ell} = \mathbf{0}, \quad [[\mathbf{v}]] = -[[\mathbf{F}]]\mathbf{V} = -V_n[[\mathbf{F}]]\mathbf{n} \quad \text{on } \mathrm{S}(t), \tag{2.10}$$

where $\boldsymbol{\ell}$ is any vector tangent to the singular surface $\mathrm{S}(t)$. The jump conditions in (2.10) are immediate consequences of the smoothness requirements imposed on the deformation (2.1).

Conversely, the field equations (2.7) together with the jump conditions (2.8), (2.10), imply the global balance laws (2.2), (2.3), (2.4), (2.6). All of the above results may be found in TRUESDELL and NOLL (1965).

3. DRIVING TRACTION

In the present section, we first introduce the notion of the driving traction on a surface of discontinuity associated with a thermomechanical process in an arbitrary continuum. The process is assumed to possess the smoothness specified in the preceding section; in particular, both displacement and temperature are required to be continuous across the moving singular surface S(t). We then make use of the driving traction to derive an alternate form for the entropy and energy jump conditions $(2.8)_3$ and $(2.8)_2$, respectively.

The concept of driving traction emerges naturally from an alternate representation for the rate of entropy production defined in (2.5). To obtain this representation, we begin by putting the energy jump condition $(2.8)_2$ into a form more useful for our purposes. In the identity

$$[[\boldsymbol{\sigma}\mathbf{n}\cdot\mathbf{v}]] = (1/2)(\overset{+}{\mathbf{v}}+\overset{-}{\mathbf{v}})\cdot[[\boldsymbol{\sigma}\mathbf{n}]]+(1/2)(\overset{+}{\boldsymbol{\sigma}}\mathbf{n}+\overset{-}{\boldsymbol{\sigma}}\mathbf{n})\cdot[[\mathbf{v}]], \tag{3.1}$$

we replace $[[\boldsymbol{\sigma}\mathbf{n}]]$ and $[[\mathbf{v}]]$ on the right by substituting from $(2.8)_1$ and $(2.10)_2$, respectively. This gives

$$[[\boldsymbol{\sigma}\mathbf{n}\cdot\mathbf{v}]] = -(1/2)\rho[[\mathbf{v}\cdot\mathbf{v}]]V_n-(1/2)V_n(\overset{+}{\boldsymbol{\sigma}}\mathbf{n}+\overset{-}{\boldsymbol{\sigma}}\mathbf{n})\cdot[[\mathbf{F}]]\mathbf{n}. \tag{3.2}$$

By making use of $(2.10)_1$, one can show that

$$(\overset{+}{\boldsymbol{\sigma}}\mathbf{n}+\overset{-}{\boldsymbol{\sigma}}\mathbf{n})\cdot[[\mathbf{F}]]\mathbf{n} = (\overset{+}{\boldsymbol{\sigma}}+\overset{-}{\boldsymbol{\sigma}})\cdot[[\mathbf{F}]], \tag{3.3}$$

where the dot product on the right is that associated with a pair of tensors: $\mathbf{A}\cdot\mathbf{B} = \mathrm{Trace}\,(\mathbf{A}\mathbf{B}^{\mathrm{T}})$. Combining (3.2) and (3.3) yields

$$[[\boldsymbol{\sigma}\mathbf{n}\cdot\mathbf{v}]] = -(1/2)\rho V_n[[\mathbf{v}\cdot\mathbf{v}]]-(1/2)V_n(\overset{+}{\boldsymbol{\sigma}}+\overset{-}{\boldsymbol{\sigma}})\cdot[[\mathbf{F}]] \quad \text{on } \mathrm{S}(t). \tag{3.4}$$

Using (3.4) to replace $[[\boldsymbol{\sigma}\mathbf{n}\cdot\mathbf{v}]]$ in $(2.8)_2$ then supplies the alternate version of the energy jump condition:

$$\rho V_n[[\varepsilon]] = (1/2)V_n(\overset{+}{\boldsymbol{\sigma}}+\overset{-}{\boldsymbol{\sigma}})\cdot[[\mathbf{F}]]-[[\mathbf{q}\cdot\mathbf{n}]] \quad \text{on } \mathrm{S}(t). \tag{3.5}$$

We turn now to (2.5) and assume that, at each instant t, the region $\mathrm{D}\subset\mathrm{R}$ is divided into two parts $\mathrm{D}^+(t)$ and $\mathrm{D}^-(t)$ by the moving surface of discontinuity S(t). Allowing for the jump in specific entropy across S(t), we may write (2.5) as

$$\Gamma(t;\mathrm{D}) = \int_{\mathrm{D}}\rho\dot{\eta}\,\mathrm{d}V-\int_{\mathrm{S}(t)\cap\mathrm{D}}[[\rho\eta]]V_n\,\mathrm{d}A-\int_{\partial\mathrm{D}}\mathbf{q}\cdot\mathbf{n}/\theta\,\mathrm{d}A-\int_{\mathrm{D}}\rho r/\theta\,\mathrm{d}V, \tag{3.6}$$

which is equivalent to

$$\Gamma(t;\mathrm{D}) = \int_{\mathrm{D}}\rho\dot{\eta}\,\mathrm{d}V-\int_{\mathrm{S}(t)\cap\mathrm{D}}[[\rho\eta]]V_n\,\mathrm{d}A-\int_{\partial\overset{+}{\mathrm{D}}\cup\partial\overset{-}{\mathrm{D}}}\mathbf{q}\cdot\mathbf{n}/\theta\,\mathrm{d}A$$

$$-\int_{\mathrm{S}(t)\cap\mathrm{D}}[[\mathbf{q}\cdot\mathbf{n}/\theta]]\,\mathrm{d}A-\int_{\mathrm{D}}\rho r/\theta\,\mathrm{d}V. \tag{3.7}$$

Applying the divergence theorem to the third integral on the right in (3.7) and then utilizing $(2.7)_3$ to eliminate $\mathrm{Div}\,\mathbf{q}$ yields

$$\Gamma(t;\mathrm{D}) = \int_{\mathrm{D}} \left\{ \frac{\rho\dot{\eta}\theta + \boldsymbol{\sigma}\cdot\dot{\mathbf{F}} - \rho\dot{\varepsilon}}{\theta} + \frac{\mathbf{q}\cdot\operatorname{Grad}\theta}{\theta^2} \right\} \mathrm{d}V - \int_{\mathrm{S}(t)\cap\mathrm{D}} \left\{ \frac{[[\mathbf{q}\cdot\mathbf{n}]] + [[\rho\eta\theta]]V_n}{\theta} \right\} \mathrm{d}A, \tag{3.8}$$

where we have made use of the continuity of the temperature θ across $\mathrm{S}(t)$. Finally, version (3.5) of the energy jump condition may be used to eliminate the term $[[\mathbf{q}\cdot\mathbf{n}]]$ in (3.8), yielding the desired representation for the rate of entropy production:

$$\Gamma(t;\mathrm{D}) = \Gamma_{\mathrm{loc}}(t;\mathrm{D}) + \Gamma_{\mathrm{con}}(t;\mathrm{D}) + \Gamma_{\mathrm{s}}(t;\mathrm{D}), \tag{3.9}$$

where

$$\Gamma_{\mathrm{loc}}(t;\mathrm{D}) = \int_{\mathrm{D}} \left\{ \frac{\rho\dot{\eta}\theta + \boldsymbol{\sigma}\cdot\dot{\mathbf{F}} - \rho\dot{\varepsilon}}{\theta} \right\} \mathrm{d}V, \tag{3.10}$$

$$\Gamma_{\mathrm{con}}(t;\mathrm{D}) = \int_{\mathrm{D}} \frac{\mathbf{q}\cdot\operatorname{Grad}\theta}{\theta^2} \mathrm{d}V, \tag{3.11}$$

$$\Gamma_{\mathrm{s}}(t;\mathrm{D}) = \int_{\mathrm{S}(t)\cap\mathrm{D}} \left\{ \frac{[[\rho\varepsilon - \rho\eta\theta - (\overset{+}{\boldsymbol{\sigma}} + \overset{-}{\boldsymbol{\sigma}})\cdot\mathbf{F}/2]]V_n}{\theta} \right\} \mathrm{d}A. \tag{3.12}$$

In (3.9), the total rate of entropy production $\Gamma(t;\mathrm{D})$ at the instant t for the subregion $\mathrm{D} \subset \mathrm{R}$ is decomposed into three parts: Γ_{loc} arises from local dissipation in the material away from the singular surface; Γ_{con} is the entropy production rate due to heat conduction; finally, Γ_{s} represents the contribution to the entropy production rate arising from the moving singular surface $\mathrm{S}(t)$. A similar decomposition in the absence of a surface of discontinuity is given by TRUESDELL and NOLL (1965, §79).

The *internal dissipation* $\delta(\mathbf{x}, t)$ is given by

$$\delta \equiv \theta\dot{\eta} + (1/\rho)\boldsymbol{\sigma}\cdot\dot{\mathbf{F}} - \dot{\varepsilon}; \tag{3.13}$$

see chapter 2 of TRUESDELL (1969). The local entropy production rate $\Gamma_{\mathrm{loc}}(t;\mathrm{D})$ of (3.10) may be written in terms of δ as follows:

$$\Gamma_{\mathrm{loc}}(t;\mathrm{D}) = \int_{\mathrm{D}} \rho(\mathbf{x})\delta(\mathbf{x}, t)/\theta(\mathbf{x}, t)\,\mathrm{d}V. \tag{3.14}$$

Next, it is convenient to introduce the Helmholtz free energy per unit mass $\psi(\mathbf{x}, t)$ defined by

$$\psi(\mathbf{x}, t) = \varepsilon(\mathbf{x}, t) - \theta(\mathbf{x}, t)\eta(\mathbf{x}, t), \quad \mathbf{x} \in R, \quad t \in [t_0, t_1]. \tag{3.15}$$

We now define the (scalar) *driving traction* $f(\mathbf{x}, t)$ on the singular surface $\mathrm{S}(t)$ by

$$f(\mathbf{x}, t) = \rho(\mathbf{x})[[\psi(\mathbf{x}, t)]] - (1/2)\{\overset{+}{\boldsymbol{\sigma}}(\mathbf{x}, t) + \overset{-}{\boldsymbol{\sigma}}(\mathbf{x}, t)\}\cdot[[\mathbf{F}(\mathbf{x}, t)]], \quad \mathbf{x} \in \mathrm{S}(t), \quad t \in [t_0, t_1]. \tag{3.16}$$

By (3.15), (3.16) and (3.12), we may rewrite the contribution $\Gamma_{\mathrm{s}}(t;\mathrm{D})$ to the rate of entropy production due to the moving singular surface in terms of the driving traction $f(\mathbf{x}, t)$, the temperature $\theta(\mathbf{x}, t)$ on $S(t)$ and the normal velocity $V_n(\mathbf{x}, t)$ of $S(t)$:

$$\Gamma_s(t; D) = \int_{S(t) \cap D} f(\mathbf{x}, t) V_n(\mathbf{x}, t)/\theta(\mathbf{x}, t)\, dA. \tag{3.17}$$

We note from (3.16) that, if the thermomechanical process under consideration is smooth in the sense that the free energy ψ and the deformation gradient $\mathbf{F}$ are continuous everywhere at all times, then the driving traction f on any surface vanishes.

Employing the representation for the entropy production rate furnished by (3.9), (3.11), (3.14) and (3.17) in the Clausius–Duhem inequality (2.6) and localizing the result at a point $\mathbf{x}$ away from the singular surface $S(t)$ yields the inequality

$$\rho\delta + (1/\theta)\mathbf{q} \cdot \mathrm{Grad}\,\theta \geqslant 0 \quad \text{on } R - S(t),\ t \in [t_0, t_1]. \tag{3.18}$$

This result, which can also be derived directly from $(2.7)_3$, $(2.7)_4$ and (3.13) and which in fact may be used in place of $(2.7)_4$, may be found in TRUESDELL (1969, p. 34, eq. (2.47)). If the localization of the Clausius–Duhem inequality is instead carried out at a point $\mathbf{x}$ *on* the singular surface, the result is the condition

$$f V_n \geqslant 0 \quad \text{on } S(t),\ t \in [t_0, t_1], \tag{3.19}$$

which may also be derived directly from $(2.8)_3$ with the help of (3.5), (3.15) and (3.16). For the special case of isothermal, quasi-static processes in thermoelastic materials, the counterpart of (3.19) was obtained by KNOWLES (1979).

Conversely, if the local results (3.18) and (3.19) hold, then it follows from (3.10)–(3.12) that

$$\Gamma_{loc}(t; D) + \Gamma_{con}(t; D) \geqslant 0, \quad \Gamma_s(t; D) \geqslant 0, \tag{3.20}$$

and hence from (3.9) that the Clausius–Duhem inequality holds. Also, it is clear that the entropy production rate Γ vanishes for every sub-region D if and only if equality holds in both (3.18) and (3.19). If the inequality in (3.19) is strict on $D \cap S(t)$, we may conclude from (3.9) and (3.17) that the driving traction associated with the moving singular surface makes a positive contribution to the rate of entropy production $\Gamma(t; D)$ and thus represents a dissipative effect.

The alternative version (3.5) of the original energy jump condition $(2.8)_2$ may be rewritten in still another form with the help of (3.15) and (3.16):

$$\rho\theta[[\eta]] V_n = -f V_n - [[\mathbf{q} \cdot \mathbf{n}]] \quad \text{on } S(t),\ t \in [t_0, t_1]. \tag{3.21}$$

The original set of jump conditions (2.8) may be replaced by $(2.8)_1$, (3.21) and (3.19).

Finally, we derive an alternative formula for the driving traction f defined in (3.16). Rearranging (3.3) yields

$$(1/2)(\overset{+}{\sigma} + \overset{-}{\sigma}) \cdot [[\mathbf{F}]] = \mathbf{n} \cdot [[\mathbf{F}^T \sigma]]\mathbf{n} - (1/2)[[\sigma]]\mathbf{n} \cdot (\overset{+}{\mathbf{F}} + \overset{-}{\mathbf{F}})\mathbf{n}, \tag{3.22}$$

while $(2.8)_1$ and $(2.10)_2$ may be used to show that

$$[[\sigma]]\mathbf{n}\cdot(\overset{+}{\mathbf{F}}+\overset{-}{\mathbf{F}})\mathbf{n} = \rho V_n^2\mathbf{n}\cdot[[\mathbf{F}^T\mathbf{F}]]\mathbf{n}. \tag{3.23}$$

Combining (3.22) and (3.23) with (3.16) provides the representation

$$f = \mathbf{n}\cdot[[\rho\psi\mathbf{1}-\mathbf{F}^T\sigma+(1/2)\rho V_n^2\mathbf{F}^T\mathbf{F}]]\mathbf{n}; \tag{3.24}$$

here **1** is the identity tensor.

Note that no constitutive assumptions have yet been made. The foregoing analysis, however, would *not* apply to the classical adiabatic theory of shock waves in gas dynamics (see e.g. COURANT and FRIEDRICHS, 1948), since we assume that the temperature is continuous across the singular surface S(t).

4. Slow Isothermal Processes in a Continuum

In this section, we specialize the foregoing results to the case of slow isothermal processes, by which we mean processes for which the inertia terms in the global balances of momentum (2.2) and (2.3) are replaced by zero, the contribution of the kinetic energy is omitted in the global statement (2.4) of the first law, and the temperature is constant:

$$\theta(\mathbf{x},t) = \theta_0 = \text{constant} \quad \text{for } \mathbf{x}\in \mathrm{R},\ t\in[t_0,t_1]. \tag{4.1}$$

The field equations that follow from these modified balance laws are given by (2.7) with the right-hand side of $(2.7)_1$ replaced by zero and θ replaced by θ_0 in $(2.7)_4$; the corresponding jump conditions are given by (2.8) with $\rho[[\mathbf{v}]]V_n$ and $\rho[[\mathbf{v}\cdot\mathbf{v}]]V_n$ replaced by zero in $(2.8)_1$ and $(2.8)_2$ respectively, and θ replaced by θ_0 in $(2.8)_3$.

When (4.1) holds, Eq. (2.4) (with kinetic energy omitted) can be used to eliminate the heat flux and heat supply terms from the original representation (2.5) for the entropy production rate. After doing so and making use of the definition (3.15) of the free energy ψ, one finds that

$$\Gamma(t;\mathrm{D}) = (1/\theta_0)\left(\int_{\partial\mathrm{D}}\sigma\mathbf{n}\cdot\mathbf{v}\,dA+\int_{\mathrm{D}}\rho\mathbf{b}\cdot\mathbf{v}\,dV-d/dt\int_{\mathrm{D}}\rho\psi\,dV\right). \tag{4.2}$$

Since the contents of the parentheses in (4.2) represent the excess of the rate of mechanical work over the rate of increase of the free energy associated with D, one infers from (4.2) that, for slow isothermal processes, the entropy production rate coincides with the rate of mechanical dissipation per unit temperature.

An alternative expression for the entropy production rate in a slow isothermal process may be obtained by carrying out a calculation entirely analogous to that of Section 3 for general processes. Such a calculation shows that the contribution $\Gamma_{con}(t;\mathrm{D})$ to $\Gamma(t;\mathrm{D})$ vanishes because of (4.1), and that

$$\Gamma(t;\mathrm{D}) = (1/\theta_0)\left(\int_{\mathrm{D}}\rho\delta\,dV+\int_{\mathrm{S}(t)\cap\mathrm{D}}fV_n\,dA\right), \tag{4.3}$$

where δ continues to be given by (3.13), and f is given by

$$f = [[\rho\psi]] - \overset{\pm}{\sigma} \cdot [[\mathbf{F}]] = \mathbf{n} \cdot [[\rho\psi\mathbf{1} - \sigma^{\mathrm{T}}\mathbf{F}]]\mathbf{n}. \tag{4.4}$$

(Note that $(4.4)_1$ can be obtained formally from (3.16) by using (3.3) and $(2.8)_1$ (with $\rho[[\mathbf{v}]]V_n$ replaced by zero), while $(4.4)_2$ may be formally obtained from (3.24) by replacing V_n by zero.) Comparison of (4.2) with (4.3) yields the following identity for slow isothermal processes:

$$\int_{\partial D} \sigma\mathbf{n} \cdot \mathbf{v}\, dA + \int_{D} \rho\mathbf{b} \cdot \mathbf{v}\, dV + \int_{S(t) \cap D} (-f) V_n\, dA = \int_{D} \rho\delta\, dV + d/dt \int_{D} \rho\psi\, dV$$

$$\text{for } D \subset R,\ t \in [t_0, t_1]. \tag{4.5}$$

This mechanical work-energy balance motivates our choice of the name *driving traction* for f: Equation (4.4) suggests that $-f(\mathbf{x}, t)\mathbf{n}$ be interpreted as a traction exerted by the surface S(t) on the body at the point $\mathbf{x}$ at time t. Equivalently, $f(\mathbf{x}, t)$ may be regarded as a normal traction applied to S(t) by the body.

5. Thermoelastic Materials

The preceding discussion has made no use of special constitutive relations for the continuum under consideration. We now assume that, for the material at hand, there is a characterizing internal energy potential $\hat{\varepsilon}(\mathbf{F}, \eta)$ such that

$$\left.\begin{aligned} \varepsilon(\mathbf{x}, t) &= \hat{\varepsilon}(\mathbf{F}(\mathbf{x}, t), \eta(\mathbf{x}, t)), \\ \sigma(\mathbf{x}, t) &= \rho\hat{\varepsilon}_{\mathbf{F}}(\mathbf{F}(\mathbf{x}, t), \eta(\mathbf{x}, t)), \\ \theta(\mathbf{x}, t) &= \hat{\varepsilon}_{\eta}(\mathbf{F}(\mathbf{x}, t), \eta(\mathbf{x}, t)); \end{aligned}\right\} \tag{5.1}$$

moreover, it is assumed that $\hat{\varepsilon}_\eta(\mathbf{F}, \cdot)$ is invertible for every tensor $\mathbf{F}$ with positive determinant. We call such a material *thermoelastic*. To avoid cumbersome formulae, we have assumed that ρ and $\varepsilon(\cdot, \cdot)$ are independent of $\mathbf{x}$, so that the body is homogeneous in the reference configuration.

For a thermoelastic material, (5.1) imply that

$$\rho\dot{\varepsilon} = \sigma \cdot \dot{\mathbf{F}} + \rho\theta\dot{\eta} \quad \text{on } R - S(t), \tag{5.2}$$

so that the internal dissipation δ of (3.13) vanishes. It then follows from (3.14) that

$$\Gamma_{\mathrm{loc}}(t; D) = 0 \tag{5.3}$$

for all t and all $D \subset R$, whence (3.9) becomes

$$\Gamma(t; D) = \Gamma_{\mathrm{con}}(t; D) + \Gamma_s(t; D), \tag{5.4}$$

where Γ_{con} and Γ_s are given by (3.11) and (3.17), respectively. Thus in *every* process in a thermoelastic material, entropy production is due only to heat conduction and the motion of the singular surface S(t).

By $(5.1)_3$ and the assumed invertibility of $\hat{\varepsilon}_\eta(\mathbf{F}, \cdot)$, we may write $\eta = \hat{\eta}(\mathbf{F}, \theta)$ and hence introduce the Helmholtz free energy potential $\hat{\psi}$ for a thermoelastic material through

$$\hat{\psi}(\mathbf{F},\theta) = \hat{\varepsilon}(\mathbf{F},\hat{\eta}(\mathbf{F},\theta)) - \hat{\eta}(\mathbf{F},\theta)\theta. \tag{5.5}$$

Then by (3.15), $\psi(\mathbf{x},t) = \hat{\psi}(\mathbf{F}(\mathbf{x},t),\theta(\mathbf{x},t))$, and (5.1), (5.5) show that, for any process,

$$\sigma(\mathbf{x},t) = \rho\hat{\psi}_{\mathbf{F}}(\mathbf{F}(\mathbf{x},t),\theta(\mathbf{x},t)), \quad \eta(\mathbf{x},t) = -\hat{\psi}_{\theta}(\mathbf{F}(\mathbf{x},t),\theta(\mathbf{x},t)). \tag{5.6}$$

For a thermoelastic material, the driving traction $f(\mathbf{x},t)$ introduced for an arbitrary continuum in (3.16) can be represented in terms of ρ and the function $\hat{\psi}$ characteristic of the material as follows:

$$f = \rho[[\hat{\psi}(\mathbf{F},\theta)]] - (1/2)\rho\{\hat{\psi}_{\mathbf{F}}(\overset{+}{\mathbf{F}},\theta) + \hat{\psi}_{\mathbf{F}}(\bar{\mathbf{F}},\theta)\}\cdot[[\mathbf{F}]], \tag{5.7}$$

where $\overset{+}{\mathbf{F}} = \overset{+}{\mathbf{F}}(\cdot,t)$ and $\bar{\mathbf{F}} = \bar{\mathbf{F}}(\cdot,t)$ represent the limiting values of the deformation gradient tensor on the positive and negative sides of $S(t)$, respectively.

Consider now a slow isothermal process in a thermoelastic material. For every $\mathbf{F}$ with positive determinant, let

$$W(\mathbf{F}) = \rho\hat{\psi}(\mathbf{F},\theta_0) \tag{5.8}$$

so that $(5.6)_1$ may be written in the form

$$\sigma(\mathbf{x},t) = W_{\mathbf{F}}(\mathbf{F}(\mathbf{x},t)). \tag{5.9}$$

With inertia omitted from the global statements (2.2) and (2.3) of the mechanical balance laws, the associated purely mechanical field equations and jump conditions corresponding to $(2.7)_{1,2}$, $(2.8)_1$ and $(2.10)_1$ are

$$\left.\begin{aligned}\operatorname{Div}\sigma + \rho\mathbf{b} &= \mathbf{0},\\ \sigma\mathbf{F}^T &= \mathbf{F}\sigma^T,\end{aligned}\right\} \quad \text{on } R - S(t), \tag{5.10}$$

$$\left.\begin{aligned}[[\sigma\mathbf{n}]] &= \mathbf{0},\\ [[\mathbf{F}]]\ell &= \mathbf{0},\end{aligned}\right\} \quad \text{on } S(t). \tag{5.11}$$

Equations (5.9)–(5.11) are the usual equations of elasto*statics*, with W identified as the elastic potential characteristic of the material; the time t appears merely as a history parameter. Consequently, all mechanical quantities in a slow isothermal process in a thermoelastic material coincide with those of a one-parameter family of equilibrium states; we speak of the corresponding one-parameter family of deformations $\mathbf{x} \to \mathbf{x} + \mathbf{u}(\mathbf{x},t)$ as a quasi-static motion. For materials whose elastic potential W is such that the system (5.9), $(5.10)_1$ does not remain elliptic at all deformations, weak solutions may arise in which $\mathbf{F}$ jumps across the slowly moving singular surface $S(t)$; see KNOWLES and STERNBERG (1978), GURTIN (1983) and ROSAKIS (1988). The jump conditions appropriate for such materials are given by (5.11).

By (4.3) and (5.3), entropy production for slow isothermal processes in thermoelastic materials arises solely from the moving surface of discontinuity $S(t)$; the entropy production rate is given in terms of the driving traction by

$$\Gamma(t;\mathrm{D}) = \Gamma_s(t;\mathrm{D}) = (1/\theta_0)\int_{\mathrm{S}(t)\cap\mathrm{D}} f V_n \,\mathrm{d}A. \tag{5.12}$$

The formula $(4.4)_1$ for the driving traction f may be written in the present circumstances with the help of (5.8) and (5.9) as

$$f = [[W(\mathbf{F})]] - W_{\mathbf{F}}(\overset{\pm}{\mathbf{F}})\cdot[[\mathbf{F}]]. \tag{5.13}$$

The inequality (3.19) continues to hold:

$$f V_n \geqslant 0 \quad \text{on } \mathrm{S}(t). \tag{5.14}$$

The representation (5.12) for the driving traction f is equivalent to one derived by YATOMI and NISHIMURA (1983). An alternative formula for f in the present case which follows from $(4.4)_2$ and (5.8) is

$$f = \mathbf{n}\cdot[[\mathbf{P}(\mathbf{F})]]\mathbf{n}, \tag{5.15}$$

where $\mathbf{n}$ is the unit normal to the surface of discontinuity $\mathrm{S}(t)$, and $\mathbf{P}(\mathbf{F})$ is the *energy-momentum tensor* introduced by ESHELBY (1956, 1970, 1975):

$$\mathbf{P}(\mathbf{F}) = W(\mathbf{F})\mathbf{1} - \mathbf{F}^{\mathrm{T}} W_{\mathbf{F}}(\mathbf{F}). \tag{5.16}$$

In the form (5.15), (5.16) appropriate to isothermal slow processes in a thermoelastic material, the entity f first appeared as the "force on a defect" in the work of ESHELBY (1956). This representation for f is also equivalent to that for the force on the interface between two phases derived by ESHELBY (1970) and discussed by RICE (1975). The formulae (5.15), (5.16) may also be found in KNOWLES (1979).

6. Kinetic Relations

In the purely mechanical setting of the one-dimensional theory of bars of non-elliptic elastic material in tension (ABEYARATNE and KNOWLES, 1988a, b), we have considered a bar lying along the x-axis in an equilibrium state in which there is a single strain discontinuity located at an arbitrary station $x = s$. The force F acting on the bar is assumed to be given, and the stress response of the material is taken to be one for which the stress at first rises with increasing strain, then declines, and finally rises again. For each F in a certain range, there is a one-parameter family, parameter s, of such equilibrium states of physical interest. For each such state, the overall elongation e of the bar depends on both F and s, as do other macroscopic quantities such as the total strain energy E and the total potential energy $U = E - Fe$ in the bar: $e = e(F,s)$, $E = E(F,s)$, $U = U(F,s)$. It turns out that, in this equilibrium theory, $\partial U(F,s)/\partial F = -e$, and $\partial U(F,s)/\partial s = Af$, where A is the cross-sectional area of the bar, and f is the one-dimensional counterpart of the static driving traction given in (5.12). This suggests that the location s of the strain discontinuity plays the role of an "internal variable" whose "conjugate force" is proportional to the driving traction f. In the references cited above, quasi-static motions are considered in which at each instant the bar is in one of the equilibrium states just described. *Admissible* quasi-static motions are those that satisfy the one-dimensional counterpart $f\dot{s} \geqslant 0$ of (5.13). Even with admissibility imposed as a requirement, specifying the force history $F(\tau)$, $0 \leqslant \tau \leqslant t$, in a quasi-static motion fails to determine the current value of elongation $e(t)$, since the shock location (or internal variable) $s(t)$ must be specified as well. This suggests a constitutive deficiency. In internal variable theories designed to model microstructural effects on inelastic macrostructural behavior, a so-called "kinetic relation" giving the time rate of change of the internal, or microstructural, variable as a function of the associated "thermodynamic force" is commonly added as a constitutive postulate; see e.g. Rice (1970, 1971, 1975). In their one-dimensional setting, Abeyaratne and Knowles (1988a, b) adopt this point of view and hence require that the shock location $s(t)$ and the associated driving traction $f(t)$ be related by a kinetic law of the form $\dot{s}(t) = V(f(t))$, where the kinetic response function V is determined by the material. Such a kinetic relation then couples the time-evolution of the location of the surface of strain discontinuity to the local strains on either side of the jump.

The equilibrium problem of the twisting of an infinite medium containing a circular hole considered by Abeyaratne and Knowles (1987a, b) involves cylindrical geometry in plane strain and is essentially one-dimensional because of axial symmetry. The surface of discontinuity is now circular, and its radius plays the role of the location s of the strain jump in the tensile bar problem. The formalism appropriate to internal variables emerges in this problem as well, and an additional constitutive postulate in the form of a kinetic relation could have been proposed on similar grounds. It would take the form

$$V_n(\mathbf{x}, t) = V(f(\mathbf{x}, t)), \quad \mathbf{x} \in \mathrm{S}(t), \tag{6.1}$$

where $\mathbf{x}$ is the position vector to a typical point on the cylindrical surface of discontinuity $\mathrm{S}(t)$, $f(\mathbf{x}, t)$ is given at each t by (5.13) but specialized to plane strain for incompressible materials, and $V_n(\mathbf{x}, t)$ is the component of velocity of the singular surface normal to itself, i.e. in the radial direction. Again, $V(f)$ is a function determined by the material. To be consistent with the admissibility requirement (5.14), V must be such that

$$V(f) f \geqslant 0 \tag{6.2}$$

for all possible values of f.

The consequences of imposing a kinetic relation in the one-dimensional setting of the bar are discussed in detail in Abeyaratne and Knowles (1988a, b) and will not be repeated here. A similar discussion could be provided for the twist problem. While in both cases the *motivation* for the imposition of the kinetic relation as a requirement is taken from internal variable theories of inelastic behavior, the *possibility* of imposing it comes about because of the lack of uniqueness of weak solutions to the relevant equilibrium problem. Indeed, for a given force F in the bar problem, for example, solutions containing a *single* "equilibrium shock" at $x = s$ fail to be unique to the extent that their totality comprises a one-parameter family (parameter s). This is exactly the extent of lack-of-uniqueness needed to make room for the kinetic law as an additional requirement. Thus the equilibrium problem seems to be under-determined to precisely the degree necessary to accommodate a kinetic relation applicable to *quasi-static* motions in the bar. Entirely similar remarks apply to the axially symmetric twist problem.

The examples discussed above suggest and support the proposition that supplementary constitutive information (such as a kinetic relation) is needed at a surface of strain discontinuity, but they do so only in a one-dimensional, purely mechanical setting, and with limitation to quasi-static motions. The remarks that follow pertain to the more general proposal of a constitutive relationship between the driving traction $f(\mathbf{x}, t)$ and the normal velocity $V_n(\mathbf{x}, t)$ of the material singular surface $\mathrm{S}(t)$ in the present setting of three-dimensional, thermomechanical processes in thermoelastic materials with inertial effects taken into account.

Such a proposal may be examined from the perspective of irreversible thermodynamics. In the representation for the entropy production rate Γ provided by (5.4), (3.11) and (3.17), one could view the ratios $(\mathrm{Grad}\,\theta)/\theta^2$ and f/θ as thermodynamic "affinities" and the terms $\mathbf{q}$ and V_n as the corresponding "fluxes"; see chapter 14 of CALLEN (1985), chapter 14 of KESTIN (1968), and Lecture 7 of TRUESDELL (1969) for discussions of these notions. In the theory of irreversible processes, it is customary to postulate a constitutive relationship in which the present value of each flux is a function of the present value of the affinities and perhaps of their past histories as well. In our setting, one might—as a simplest case—postulate a relationship between present values of each flux and its corresponding affinity, as in the theory of "purely resistive" thermodynamical systems (CALLEN, 1985, chapter 14). This would give

$$\mathbf{q}(\mathbf{x}, t) = \mathbf{Q}^*(\mathrm{Grad}\,\theta(\mathbf{x}, t)/\theta^2(\mathbf{x}, t), \theta(\mathbf{x}, t)), \quad \mathbf{x} \in \mathrm{R} - \mathrm{S}(t), \tag{6.3}$$

and

$$V_n(\mathbf{x}, t) = V^*(f(\mathbf{x}, t)/\theta(\mathbf{x}, t), \theta(\mathbf{x}, t)), \quad \mathbf{x} \in \mathrm{S}(t), \tag{6.4}$$

where $\mathbf{Q}^*$ and V^* are functions determined by the material; the constitutive statements (6.3) and (6.4) represent a heat conduction law and a kinetic relation, respectively. The consequence (5.14) of the Clausius–Duhem inequality would then require that

$$V^*(f/\theta, \theta) f/\theta \geqslant 0 \tag{6.5}$$

for all possible values of f/θ. We observe that, if the kinetic response function V^* is continuous, then (6.5) requires that

$$V^*(0, \theta_0) = 0. \tag{6.6}$$

Quasi-static motions arising in the purely mechanical examples discussed in the initial paragraphs of this section may be regarded as embedded in slow isothermal processes taking place in a thermoelastic material. From this point of view, (6.4) would reduce to (6.1) with $V(f) = V^*(f/\theta_0, \theta_0)$, and (6.5) would reduce to (6.2).

If the body is at uniform temperature and in a state of mechanical equilibrium involving an equilibrium shock S at all points of which the Maxwell condition $f = 0$ holds, we say the body is in a *Maxwell state*. If V^* is smooth and departures from a Maxwell state are slight, so that f/θ is small, one might replace (6.4) by its linearization with respect to f/θ. (Linearized kinetic relations are often used in irreversible thermodynamics to describe processes that are "close to thermodynamic equilibrium".) In view of (6.6), this would yield

$$V_n(\mathbf{x}, t) = v(\theta) f(\mathbf{x}, t), \quad \mathbf{x} \in S(t), \tag{6.7}$$

where $v(\theta)$ is a function of temperature; by (6.5), v is necessarily positive. A linear relation of the type (6.7) between the driving traction and the velocity of the singular surface was shown by ABEYARATNE and KNOWLES (1988b) to lead to a conventional type of viscoelastic macroscopic response in the one-dimensional theory of quasi-static motions in tensile bars composed of a particular non-elliptic elastic material.

Note that the singular surface $S(t)$ lies in the region occupied by the body in the reference configuration; thus the driving traction f might more precisely be called the *nominal* driving traction. The kinetic relation (6.4) is thus in a form appropriate to a "Lagrangian", or "material", description. One can readily show that it is objective.

In a slow isothermal process, the moving singular surface $S(t)$ may be thought of as a phase boundary, and the process may be viewed as one in which particles of the body are being transformed from one phase to another. The normal velocity V_n is clearly a measure of the rate at which this takes place, and the relation (6.4) may thus be regarded as controlling the kinetics of the phase transformation. A special form for the function V^* in (6.4) might, for example, be derived from models of phase transitions based on thermal activation theory; see section 3.1 of FINE (1964) or section 1.9 of PORTER and EASTERLING (1981) for related discussions.

Since the kinetic relation (6.4) is to be constitutive, it must apply not only during slow isothermal processes, but also when inertia effects are included. For such dynamical processes, it is to be expected that moving surfaces of discontinuity $S(t)$ can occur which are *not* phase boundaries—e.g. counterparts of ordinary shock waves—and for which a supplementary relation such as (6.4) is thus expected to be inappropriate on physical grounds. From the mathematical point of view, it is the lack of uniqueness of solutions of equilibrium boundary-value problems for non-elliptic elastic materials that allows the prescription of a supplementary kinetic relation for quasi-static motions. Whether there is a precisely analogous lack of uniqueness for the boundary-initial-value problems arising in the *dynamics* of non-elliptic elastic materials, and how in general to distinguish phase boundaries from ordinary shock waves, then become questions of some importance. In the mathematical study of systems of conservation laws in one space dimension (see e.g. DAFERMOS, 1983, 1984; LAX, 1973), it is known that the solution to the initial value problem subject to an "entropy inequality" such as $(2.8)_3$ (or equivalently (3.19)) is unique, provided that the underlying stress–strain curve rises monotonically and is either strictly convex or strictly concave. Material models put forward in connection with the study of phase transformations generally do not satisfy these conditions; for such models, the entropy inequality is not strong enough to secure uniqueness. The fact that the entropy inequality must be supplemented at phase boundaries in order to assure uniqueness in the initial-value problem for such materials reflects the need for an additional requirement of a *constitutive* nature, as distinguished from a fundamental physical principle. A kinetic relation such as (6.4), in which f is now given by the representation (5.7) pertinent to *dynamical* processes and the material function V^* is subject to the requirement (6.5) imposed by the entropy inequality, might provide an appropriate supplementary constitutive requirement. Other approaches to this issue for materials whose stress–strain curves are neither monotonic, concave nor convex have been pursued by a number of authors: see e.g. HATORI (1986), JAMES (1980), OLEINIK (1959), SHEARER (1986) and SLEMROD (1983).

ACKNOWLEDGEMENTS

The authors are grateful to Professor J. R. RICE for a number of helpful criticisms of an earlier draft of this manuscript. In particular, he pointed out to us the alternative representation (3.24) for the driving traction. The work reported here was supported in part by the U.S. Office of Naval Research.

REFERENCES

ABEYARATNE, R. 1980 *J. Elasticity* **10,** 255.
ABEYARATNE, R. 1983 *J. Elasticity* **13,** 175.
ABEYARATNE, R. and KNOWLES, J. K. 1987a *J. Elasticity* **18,** 227.
ABEYARATNE, R. and KNOWLES, J. K. 1987b *J. Mech. Phys. Solids* **35,** 343.
ABEYARATNE, R. and KNOWLES, J. K. 1988a *Int. J. Solids Struct.* **24,** 1021.
ABEYARATNE, R. and KNOWLES, J. K. 1988b *ASME J. appl. Mech.* **10,** 491.
BALL, J. M. 1977 *Archs ration. Mech. Analysis* **63,** 337.
BALL, J. M. and JAMES, R. D. 1987 *Archs ration. Mech. Analysis* **100,** 13.
CALLEN H. B. 1985 *Thermodynamics and an Introduction to Thermostatistics*, Second Edn. Wiley, New York.
COURANT, R. and FRIEDRICHS, K. O. 1948 *Supersonic Flow and Shock Waves*. Interscience, New York.
DAFERMOS, C. M. 1983 In *Systems of nonlinear partial differential equations* (edited by J. M. BALL), pp. 25–70. D. Reidel, Dordrecht.
DAFERMOS, C. M. 1984 In *Rational Thermodynamics* (edited by C. TRUESDELL), pp. 211–218. Springer, New York.
ERICKSEN, J. L. 1975 *J. Elasticity* **5,** 191.
ESHELBY, J. D. 1956 In *Solid State Physics*, Vol. 3 (edited by F. SEITZ and D. TURNBULL), pp. 79–144. Academic Press, New York.
ESHELBY, J. D. 1970 In *Inelastic Behavior of Solids* (edited by M. F. KANNINEN *et al.*), pp. 77–115. McGraw-Hill, New York.
ESHELBY, J. D. 1975 *J. Elasticity* **5,** 321.
FINE, M. E. 1964 *Introduction to Phase Transformations in Condensed Systems*. Macmillan, New York.
FOSDICK, R. L. and MACSITHIGH, G. 1983 *Archs ration. Mech. Analysis* **84,** 31.
GURTIN, M. E. 1983 *Archs ration. Mech. Analysis* **84,** 1.
HATORI, H. 1986 *Archs ration. Mech. Analysis* **92,** 247.

James, R. D.	1980	*Archs ration. Mech. Analysis* **73**, 125.
James, R. D.	1981	*Archs ration. Mech. Analysis* **77**, 143.
James, R. D.	1986	*J. Mech. Phys. Solids* **34**, 359.
Kestin, J.	1968	*A Course on Thermodynamics*, Vol. II. McGraw-Hill, New York.
Knowles, J. K.	1979	*J. Elasticity* **9**, 131.
Knowles, J. K. and Sternberg, E.	1978	*J. Elasticity* **8**, 329.
Lax, P. D.	1973	Hyperbolic systems of conservation laws and the mathematical theory of shock waves. No. 11, Regional Conference Series in Applied Mathematics, SIAM, Philadelphia.
Oleinik, O. A.	1959	*Uspekhii Matematicheskii Nauk* (*N.S.*) **14**, 165.
Porter, D. A. and Easterling, K. E.	1981	*Phase Transformations in Metals and Alloys*. Van Nostrand Reinhold, New York.
Rice, J. R.	1970	*ASME J. appl. Mech.* **37**, 728.
Rice, J. R.	1971	*J. Mech. Phys. Solids* **19**, 433.
Rice, J. R.	1975	In *Constitutive Equations in Plasticity* (edited by A. S. Argon), pp. 23–79. MIT Press, Cambridge, MA.
Rosakis, P.	1988	Ellipticity and deformations with discontinuous gradients in finite elastostatics, ONR Technical Report No. 4, Contract N00014-87-K-0117. California Institute of Technology, Pasadena, November, 1988.
Shearer, M.	1986	*Archs ration. Mech. Analysis* **93**, 45.
Silling, S.	1989	*J. Mech. Phys. Solids* **37**, 293.
Slemrod, M.	1983	*Archs ration. Mech. Analysis* **81**, 301.
Truesdell, C.	1969	*Rational Thermodynamics*. McGraw-Hill, New York.
Truesdell, C. and Noll, W.	1965	In *Handbuch der Physik*, III/3 (edited by S. Flugge). Springer, Berlin.
Yatomi, C. and Nishimura, N.	1983	*J. Elasticity* **13**, 311.

Arch. Rational Mech. Anal. 131 (1995) 67–100. © Springer-Verlag 1995

The Nature of Configurational Forces

MORTON E. GURTIN

Contents

1. Introduction

The standard forces associated with continua arise as a response to the motion of material points. That additional, *configurational*[1] forces may be needed to describe the internal structure of the material is clear from ESHELBY's work on lattice defects[2] and is at least intimated by GIBBS[3] in his discussion of multiphase equilibria. These studies are statical, based on variational arguments, with the configurational forces *defined* as derivatives of the energy. I take a different point of view. Although variational derivations may point the way toward a correct statement of basic laws, such derivations obscure the fundamental nature of balance laws in any general framework that includes dissipation. While I am not in favor of the capricious introduction of "fundamental physical laws", I do believe that *configurational forces should be viewed as basic primitive objects consistent with their own force balance*, rather than as variational constructs.[4]

My objective here is to demonstrate the role of configurational force balances in the study of dynamical phase transitions. In the standard theories of

(i) Stefan-type solidification,
(ii) interface motion neglecting bulk behavior,
(iii) solid-solid phase transitions

an *extra* interface condition — over and above those that follow from standard balance laws — is needed:

- for (i), the extra condition is the classical Stefan condition, temperature equals melting temperature, or a more general relation between temperature, curvature, and normal velocity;
- for (ii), the condition is the motion-by-curvature equation of BURKE & TURNBULL [1952] and MULLINS [1956], or a more general relation of that type;

[1] I use the adjective "*configurational*" to differentiate these forces from the standard forces, which I refer to as "*deformational*". In the past I used the term "accretive" rather than "configurational", but I now use "accretive" to describe the addition or removal of material.

[2] Cf. [1951, 1970]. ESHELBY [1951] remarks that the idea of a force on a lattice defect goes back to "an interesting paper" of BURTON [1892].

[3] GIBBS's discussion [1878, pp. 314–331] is paraphrased by CAHN [1980] as follows: "solid surfaces can have their physical area changed in two ways, either by creating or destroying surface without changing surface structure and properties per unit area, or by an elastic strain . . . along the surface keeping the number of surface lattice sites constant . . . ." The creation of surface involves configurational forces, while stretching the surface involves more standard deformational forces.

See also NOZIERES [1989, p. 26], who uses the term "chemical" rather than "configurational" and writes: "Such a concept of 'chemical stresses', although somewhat misleading, is often useful in asessing equilibrium shapes."

[4] It is difficult to imagine distinct force systems acting concurrently at each point of a body, which is perhaps why configurational forces have never been more than just variational constructs. Here I am reminded of opposition to the use of standard forces more than half a century after the publication of NEWTON's *Principia*: as TRUESDELL [1966] writes, "D'Alembert spoke of Newtonian forces as 'obscure and metaphysical beings, capable of nothing but spreading darkness over a science clear by itself.'"

- for (iii), (in the absence of interfacial stress) the condition is a kinetic relation of the type proposed by TRUSKINOVSKY [1987, 1991] and ABEYARATNE & KNOWLES [1990, 1991].

There are, I believe, three compelling arguments in support of configurational forces:

(i) Configurational forces provide a conceptual unification, *as each of these extra conditions is a consequence of the configurational force balance applied across the interface.*
(ii) Configurational forces lead to new results, an example being a weak formulation of the supercooled Stefan problem.
(iii) Configurational forces provide a valuable tool in the framing of new theories.[5]

The general configurational force balance upon which I base the theory is[6]

$$\int_{\partial R} \boldsymbol{Cm}\, da + \int_{R} \boldsymbol{f} da + \int_{\partial \mathcal{G}} \mathbf{C}\mathbf{v}\, ds + \int_{\mathcal{G}} \mathbf{e}\, da = 0, \tag{1.1}$$

with R a control volume that intersects the phase interface, $\boldsymbol{m}$ the outward unit normal to ∂R, $\mathcal{G}$ the portion of the interface in R, and $\mathbf{v}$ the outward unit normal to the boundary curve $\partial\mathcal{G}$ (Figure 1). The configurational fields appearing in (1.1) have the following interpretation: $\boldsymbol{C}$ is a bulk stress that acts in response to the exchange of material at the boundary of R; $\mathbf{C}$, a generalization of surface tension, is a stress within the interface that acts in response to increases in interfacial area as well as to changes in the orientation of the interface; $\mathbf{e}$ represents internal forces distributed over the interface; $\boldsymbol{f}$ represents internal forces distributed over the bulk volume. In the theories discussed here $\boldsymbol{f}$ is generally unimportant, but $\mathbf{e}$ is essential, as it represents dissipative forces associated with the kinetics of the interface.

Configurational forces are irrelevant when discussing defect-free single-phase materials, but their inclusion gives insight into the relation between $\boldsymbol{C}$, the bulk free energy Ψ, the standard bulk deformational-stress $\boldsymbol{S}$, and the deformation gradient $\boldsymbol{F}$. Here, to capture the mechanics associated with the addition and deletion of material points at the boundary of a portion of the body, *I use referential control volumes $R = R(t)$ whose boundaries evolve within the reference configuration.* Their use, which requires generalizations of the basic physical laws, leads to an important expression, $\boldsymbol{C} = \Psi\mathbf{1} - \boldsymbol{F}^T\boldsymbol{S}$, discovered by ESHELBY [1951, 1970]. My derivation of the Eshelby relation is accomplished without recourse to constitutive equations or to a variational principle;[7] the derivation is based on a version of the second law appropriate to a mechanical theory in conjunction with a requirement that this law be invariant under changes in the time-dependent parametrization of $\partial R(t)$.

---

[5] Cf. CERMELLI & GURTIN [1994].

[6] GURTIN [1994a, eq. (4.3)]. A less specific version appears in GURTIN [1988, eq. (3.2)] and GURTIN & STRUTHERS [1990, eq. (7.9)], where the balance is applied directly to the interface with the effects of $\boldsymbol{C}$ and $\mathbf{e}$ combined.

[7] And is hence applicable to theories, such as plasticity and viscoelasticity, for which memory effects render variational derivations inappropriate.

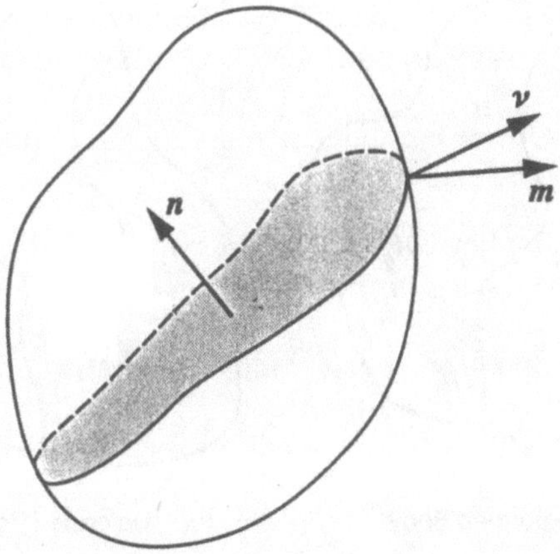

Figure 1. The portion $\mathscr{G}$ (shaded) of the interface contained in the control volume R; $\boldsymbol{m}$ is the outward unit normal to ∂R; $\boldsymbol{n}$ is the unit normal to the interface; $\boldsymbol{v}$ (tangent to the interface) is the outward unit normal to the boundary curve $\partial\mathscr{G}$.

The theories I discuss utilize the method of COLEMAN & NOLL [1963] to restrict constitutive equations. This procedure is based on the premise that the second law be satisfied in all conceivable processes; its rational application requires external fields (heat supplies, body forces, etc.) that ensure satisfaction of the underlying balance laws in all processes. Here, for convenience, I omit all mention of external fields, although the dissipation inequalities I use to reduce constitutive equations are the same as those that would occur were such fields present. I refer the reader to GURTIN [1988, 1993b], ANGENENT & GURTIN [1989], and GURTIN & STRUTHERS [1990] for statements fo the basic laws with external fields.

Notation. I use notation standard in continuum mechanics (cf. GURTIN [1981]). The body B is identified with the region of $\mathbb{R}^3$ it occupies in a fixed reference configuration; terms such as "referential volume" and "undeformed volume" are used interchangeably; X designates an arbitrary material point (point of B); fields Φ are described materially (as functions of (X, t)); $\Phi^\cdot$ denotes the material time-derivative of Φ (with respect to t holding X fixed), ∇ and div denote the material gradient and divergence (with respect to X holding t fixed).

2. Variational definition of configurational forces

To better understand the nature of configurational forces, I begin with a standard variational derivation for a coherent two-phase elastic solid, neglecting thermal and compositional variations as well as interfacial energy. I consider a body B whose phases α and β occupy closed complementary subregions B_α and B_β of B, with the interface $\mathscr{S} = B_\alpha \cap B_\beta$ a smooth, oriented surface whose continuous unit normal field $\boldsymbol{n}$ points outward from B_α (Figure 2). A *deformation* $\boldsymbol{y}$ of B is

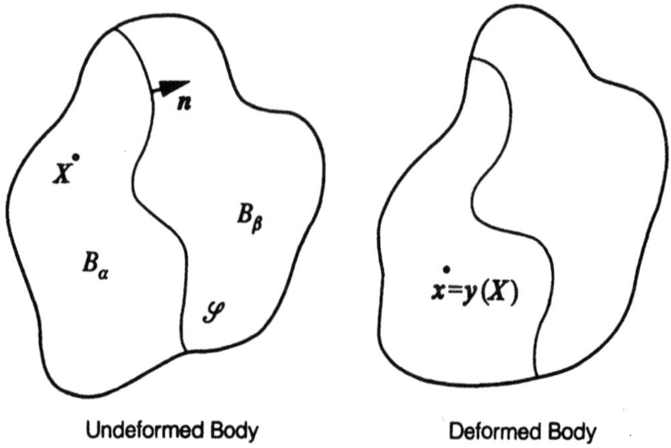

Figure 2. The regions B_α and B_β occupied by the phases α and β in the underformed body.

then a continuous function that assigns to each material point X in B a point $x = y(X)$ of space, has *deformation gradient*

$$F = \nabla y \tag{2.1}$$

continuous up to the interface from either side, and has $\det F > 0$.

I restrict attention to deformations y that obey a given boundary condition on ∂B. Then, for equilibrium, the position of the interface and the deformation minimize the total energy

$$E(\mathscr{S},y) = \int_{B_\alpha} \Psi \, dv + \int_{B_\beta} \Psi \, dv, \tag{2.2}$$

where the bulk energy $\psi(X)$, per unit volume, is given by constitutive equations

$$\Psi(X) = \Psi_\alpha(F(X)) \text{ in } B_\alpha, \quad \Psi(X) = \Psi_\beta(F(X)) \text{ in } B_\beta. \tag{2.3}$$

To formally compute the first variation $\delta E(\mathscr{S},y)$, which must vanish, the compatibility condition and the identity

$$[\delta y] = -(\delta\mathscr{S})[\mathbf{F}]n, \quad [fg] = \langle f \rangle [g] + \langle g \rangle [f] \tag{2.4}$$

are useful. Here δy (with $\delta y = \mathbf{0}$ on ∂B) is the variation of y; $\delta\mathscr{S}$, a scalar field, is the normal variation of $\mathscr{S}$; $[f]$ denotes the jump in a field f across the interface (limit from β minus that form α); and $\langle f \rangle$ denotes the average of the interfacial limits of f. (Less formally, by considering time-dependent departures $(\mathscr{S}(t), y(t))$ from $(\mathscr{S},y)$, the variations $\delta E(\mathscr{S},y)$, δy, and $\delta\mathscr{S}$ may be identified with $E(\mathscr{S},y)^{\cdot}$, $y^{\cdot}$, and the normal velocity V of $\mathscr{S}(t)$.) The requirement $\delta E(\mathscr{S},y) = 0$, (2.4), and the divergence theorem yield the expression

$$\int_{B_\alpha} \operatorname{div} S_\alpha \cdot \delta y \, dv + \int_{B_\beta} \operatorname{div} S_\beta \cdot \delta y \, dv + \int_{\mathscr{S}} \{[S]n \cdot \langle \delta y \rangle + ([\Psi] - \langle \mathrm{Sn} \rangle \cdot [Fn])\delta\mathscr{S}\} \, da = 0, \tag{2.5}$$

where $S(X)$ is the bulk stress, here defined as the derivative of the energ:

$$S = \partial_F \Psi_\alpha(F) \text{ in } B_\alpha, \quad S = \partial_F \Psi_\beta(F) \text{ in } B_\beta. \tag{2.6}$$

Since δy can be specified arbitrarily away from $\mathscr{S}$, and since $\langle \delta y \rangle$ and $\delta\mathscr{S}$ can be specified arbitrarily on $\mathscr{S}$, (2.5) yields the standard equilibrium equation

$$\operatorname{div} S = 0 \quad \text{in bulk} \tag{2.7}$$

(that is, in B_α and in B_β), the standard force balance

$$[S]n = 0 \quad \text{on the interface,} \tag{2.8}$$

and an additional condition[8]

$$[\Psi] = [Fn \cdot Sn] \quad \text{on the interface,} \tag{2.9}$$

often referred to as the Maxwell relation.

Since (2.9) cannot be derived from balance of force alone, this leads to the question of whether the Maxwell relation represents an additional "force balance". In fact it does. To see this, consider the "stress tensor"

$$C = \Psi 1 - F^T S \tag{2.10}$$

introduced by ESHELBY in his discussion of defects.[9] In terms of the Eshelby tensor, the Maxwell relation has the simple form $n \cdot [C]n = 0$. Further, the continuity of y across the interface implies that $[F]t = 0$ for any vector t tangent to the interface, so that (2.8) yields $t \cdot [C]n = 0$. Thus[10]

$$[C]n = 0 \quad \text{on the interface,} \tag{2.11}$$

implying continuity of the Eshelby traction across the interface. Further, a computation based on (2.6) and (2.7) yields the conclusion[11]

$$\operatorname{div} C = 0 \quad \text{in bulk,} \tag{2.12}$$

so that the force system corresponding to the Eshelby tensor satisfies a balance law; in fact, (2.11) and (2.12) together imply the more standard integral balance

$$\int_{\partial R} Cm \, da = 0 \tag{2.13}$$

for every subregion R of B, where m is the outward unit normal to ∂R.

---

[8] Cf. ESHELBY [1970], ROBIN [1974], LARCHÉ & CAHN [1978], GRINFELD [1981], JAMES [1981], GURTIN [1983].

[9] GRINFELD [1981] refers to C/ρ as a chemical-potential tensor.

[10] Cf. KAGANOVA & ROITBURD [1988].

[11] $\operatorname{div} C = 0 \Leftrightarrow \operatorname{div} S = 0$ in bulk, an equivalence not carried over to the interface conditions (2.8) and (2.11).

I henceforth use the term *deformational balance* for balances such as (2.7) and (2.8) involving the standard (Piola-Kirchhoff) stress $\boldsymbol{S}$, as opposed to the term *configurational balance*, which I reserve for balances of the form (2.13) involving the Eshelby tensor $\boldsymbol{C}$.

This analysis leads to the questions:

- Is there a formulation in which $\boldsymbol{C}$ is a primitive quantity, consistent with a force balance of the type (2.13), and in which the Eshelby relation (2.10) follows as a natural consequence?
- Aside from a possible better understanding of the underlying physics, does the introduction of configurational force lead to new results?

In what follows I attempt to answer these questions.

3. Basic ideas. Configurational forces in a single-phase material[12]

Configurational forces are irrelevant to the theory of single-phase continua without defects, but their study within that context provides essential information regarding their nature.

a. Referential control volumes that evolve with time

Let B be a single-phase body, and let $\boldsymbol{y}$ be a motion of B, so that $\boldsymbol{y}(X,t)$ is a deformation for each fixed t, with deformation gradient $\boldsymbol{F}(X,t)$ and material velocity $\boldsymbol{y}^{\cdot}(X,t)$ smooth functions.

As is standard, I formulate balance laws using referential control volumes (subregions of B). But as is not standard, I capture the mechanics associated with the addition and deletion of material points at the boundary of a portion of the body by using *referential control volumes whose boundaries evolve within the reference configuration.* Given such a control volume $R(t)$, I write $\mathscr{V}(X,t)$ for the normal velocity of $\partial R(t)$ in the direction of the outward unit normal $\boldsymbol{m}(X,t)$. Then for $\Phi(X,t)$ a smooth field,

$$\frac{d}{dt}\left\{\int_R \Phi\right\} = \int_R \Phi^{\cdot}\, dv + \int_{\partial R} \Phi \mathscr{V}\, da, \tag{3.1}$$

where

$$\frac{d}{dt}\left\{\int_R \Phi\, dv\right\} \quad \text{denotes} \quad \frac{d}{dt}\left\{\int_{R(t)} \Phi(X,t)\, dv(X)\right\}. \tag{3.2}$$

The evolving surface $\partial R(t)$ may be parametrized in a sufficiently small time interval and in a neighborhood of any of its points by a function of the form

[12] GURTIN [1994a].

$X = \hat{X}(u_1, u_2, t)$; the field

$$v(X,t) = \frac{\partial \hat{X}(u_1,u_2,t)}{\partial t} \tag{3.3}$$

then represents a velocity field for $\partial R(t)$ in that neighborhood. It is possible to use such parametrizations to construct a *velocity field* v for ∂R; that is, a smooth field $v(X,t)$ defined for all X on $\partial R(t)$ and *all* t in any (sufficiently small) time interval. A field v so constructed depends on the choice of local parametrizations, but its normal component is intrinsic:

$$v \cdot m = \mathcal{V}. \tag{3.4}$$

Under the motion y, $R(t)$ deforms to a region $\mathcal{R}(t) = y(R(t), t)$, and each local parametrization $X = \hat{X}(u_1, u_2, t)$ induces a corresponding local parametrization $x = \hat{x}(u_1, u_2, t) = y(\hat{X}(u_1, u_2, t), t)$ for $\partial\mathcal{R}(t)$; the corresponding velocity field

$$\bar{v}(X,t) = \frac{\partial \hat{x}(u_1,u_2,t)}{\partial t} \tag{3.5}$$

for $\partial\mathcal{R}(t)$ is related to v through the relation

$$\bar{v} = \dot{y} + Fv; \tag{3.6}$$

$\bar{v}$ is referred to as the *velocity field for $\partial\mathcal{R}$ induced by v* (Figure 3).

b. Classical deformational forces. Mechanical version of the second law

Let S denote the standard stress that arises in response to deformation, with S measured per unit referential area, and let ρ denote the reference density. Then,

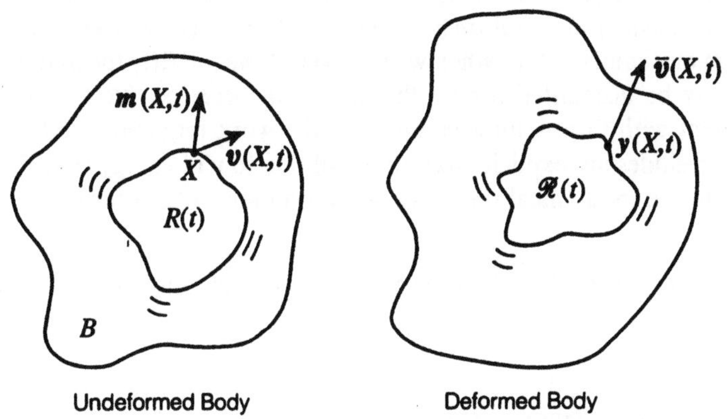

Figure 3. The time-dependent control volume $R(t)$, which deforms to $\mathcal{R}(t)$, with $v(X,t)$ a velocity field for $\partial R(t)$ and $\bar{v}(X,t)$ a corresponding velocity field for $\partial\mathcal{R}(t)$.

representing inertia through the body force

$$b = -\rho y^{\cdot\cdot}, \tag{3.7}$$

but neglecting other external forces, I write the standard balance laws for forces and moments in the form

$$\int_{\partial R} Sm\,da + \int_{R} b\,dv = 0, \quad \int_{\partial R} y \times Sm\,da + \int_{R} y \times b\,dv = 0 \tag{3.8}$$

for each referential control volume R, with m the outward unit normal to ∂R. Since R is arbitrary, a standard argument[13] yields the local relations

$$\operatorname{div} S + b = 0, \quad SF^T = FS^T. \tag{3.9}$$

In the absence of thermal and compositional effects the classical theory may be based on a "second law" that utilizes time-independent control volumes R and has the form[14]

$$\frac{d}{dt}\left\{\int_{R} \Psi\,dv\right\} \leqq \int_{\partial R} Sm \cdot y^{\cdot}\,da + \int_{R} b \cdot y^{\cdot}\,dv \tag{3.10}$$

with Ψ the *free energy*, per unit referential volume. For an *evolving* control volume $R(t)$ the standard generalization of (3.10) would include the term

$$\text{energy flow} = \int_{\partial R} \Psi \mathscr{V}\,da \tag{3.11}$$

on the right side. My discussion of configurational forces will be based on what I believe to be a more fundamental version of the second law.

c. Configurational stress. A version of the second law that accounts for accretion

I consider the dependence of $R(t)$ on t as representing the addition of material to — or the removal of material from — the boundary $\partial R(t)$. A standard precept of continuum mechanics is that when writing basic laws for $R(t)$ the material outside of $R(t)$ may be accounted for by the action of forces on $\partial R(t)$. I believe that — consistent with this — an accounting of the work required to add or remove material precludes an (explicit) accounting of the flow of energy into $R(t)$. In this spirit I base the mechanical theory on a "second law" of the form

$$\frac{d}{dt}\{\text{energy of } R(t)\} \leqq \{\text{rate at which work is performed on } R(t)\}. \tag{3.12}$$

---

[13] Cf. e.g., GURTIN [1981, p. 179].

[14] (3.10) (with Ψ the free energy) follows from standard statements of the first two laws (cf. GURTIN [1991]).

The view expressed above requires a careful treatment of the kinematics and mechanics of *accretion*, the term I use to describe the addition and removal of material. Kinematically, accretion is independent of the motion y, and since independent kinematical processes generally give rise to independent force systems, it seems reasonable to introduce additional forces that perform work during accretion. I therefore consider a tensor field C, the *configurational stress*, whose working accompanies the evolution of ∂R. More precisely, I choose local parametrizations $X = \hat{X}(u_1, u_2, t)$ for $\partial R(t)$, define a velocity field v for ∂R through (3.3), and assume that $Cm \cdot v$ represents the corresponding working, per unit area.

The working of the deformational stress S must also be taken into account. The motion of the deformed boundary $\partial\mathcal{R}(t) = y(\partial R(t), t)$ — for R independent of time — is described by the material velocity $y^\cdot$, and $Sm \cdot y^\cdot$ gives the required working. But when $R(t)$ depends on t, there is no intrinsic material description of $\partial\mathcal{R}(t)$, as material is continually being added and removed, and it would seem appropriate to use, as a velocity for $\partial\mathcal{R}(t)$, the derivative $\bar{v}(X, t)$ of $y(\hat{X}(u_1, u_2, t), t)$ with respect to t holding the surface parameters (u_1, u_2) fixed; for that reason I write the working of S in the form $Sm \cdot \bar{v}$.

I therefore take as an appropriate form of the second law the inequality

$$\frac{d}{dt}\left\{\int_R \Psi\, dv\right\} \leq \int_{\partial R} (Sm \cdot \bar{v} + Cm \cdot v)\, da + \int_R b \cdot y^\cdot\, dv \tag{3.13}$$

with v a velocity field for $\partial R = \partial R(t)$ and $\bar{v}$ the corresponding induced velocity field for $\partial\mathcal{R}$. (Note that the deformational working $Sm \cdot \bar{v}$ is a classical term $Sm \cdot y^\cdot$ plus a term $Sm \cdot Fv$ that accounts for the addition of deformed material to ∂R.) To ensure that the resulting theory be independent of local parametrizations, I require that (3.13) hold for any choice of the velocity field v for ∂R. This requirement of *invariance under reparametrization* has important consequences.

d. Bulk tension. Derivation of the Eshelby relation

Invariance of (3.13) under reparametrization is equivalent to invariance of the *working*

$$\mathcal{W}(R) = \int_{\partial R} (Sm \cdot \bar{v} + Cm \cdot v)\, da + \int_R b \cdot y^\cdot\, dv \tag{3.14}$$

$$= \int_{\partial R} \{Sm \cdot y^\cdot + (F^T Sm + Cm) \cdot v\}\, da + \int_R b \cdot y^\cdot\, dv. \tag{3.15}$$

Because of (3.4), changes in parametrization affect the tangential component of v, but leave the normal component unaltered. In fact, invariance of (3.15) under reparametrization is equivalent to the requirement that $(F^T Sm + Cm) \cdot t = 0$ on ∂R for all tangential vector fields t on ∂R; thus, since R is arbitrary, $(F^T S + C)m$ must be parallel to m for all m, so that

$$C + F^T S = \pi 1 \tag{3.16}$$

and, by (3.4), the working has the intrinsic form

$$\mathscr{W}(R) = \int_{\partial R} \boldsymbol{Sm} \cdot \boldsymbol{y}^{\cdot} da + \int_{R} \boldsymbol{b} \cdot \boldsymbol{y}^{\cdot} dv + \int_{\partial R} \pi \mathscr{V} da. \qquad (3.17)$$

The scalar field π is a *bulk tension* that works to increase the volume of R through the addition of material at its boundary. If we refer to the final term in (3.17) as the *configurational working*, then (3.17) may be written more suggestively as

$$\{\text{working}\} = \{\text{mechanical working}\} + \{\text{configurational working}\}. \qquad (3.18)$$

Note that the configurational working $\pi\mathscr{V}$ is not due solely to the action of the configurational stress $\boldsymbol{C}$; the deformational stress contributes also through the term $(\boldsymbol{Sm} \cdot \boldsymbol{Fm})\mathscr{V}$.

My next step is to relate π to the free energy Ψ. Using (3.1) and (3.17), I write the inequality (3.13) as

$$\int_{R} \Psi^{\cdot} dv \leqq \int_{\partial R} \{\boldsymbol{Sm} \cdot \boldsymbol{y}^{\cdot} + (\pi - \Psi)\mathscr{V}\} da + \int_{R} \boldsymbol{b} \cdot \boldsymbol{y}^{\cdot} dv. \qquad (3.19)$$

Given a time τ, it is possible to find a second referential control volume $R'(t)$ with $R'(\tau) = R(\tau)$, but with $\mathscr{V}'(\boldsymbol{X}, \tau)$, the normal velocity of $\partial R'(\tau)$, an *arbitrary* scalar field on $\partial R'(\tau)$; satisfaction of (3.19) for all such $\mathscr{V}'$ implies

$$\pi = \Psi. \qquad (3.20)$$

Bulk tension therefore coincides with bulk free-energy, a result analogous to the coincidence of surface tension and surface free-energy; but what is more important, (3.16) and (3.20) yield the Eshelby relation

$$\boldsymbol{C} = \Psi \boldsymbol{1} - \boldsymbol{F}^T \boldsymbol{S}. \qquad (3.21)$$

This derivation of the Eshelby relation was accomplished without recourse to constitutive equations or to a variational principle; the derivation was based on a version of the second law appropriate to referential control volumes whose boundaries evolve with time. This observation is not simply of pedagogical interest; it establishes the Eshelby relation as appropriate to theories, such as plasticity and viscoelasticity, for which memory effects render variational derivations inappropriate.

The result (3.16) is a consequence of the invariance of $\mathscr{W}(R)$ under reparametrization; it is independent of the particular form chosen for the second law and is hence more basic than (3.21). In fact, the identification of π with a "grand canonical potential" such as the free energy depends on whether or not there are associated transport processes; (3.16) is independent of such considerations.

Finally, in the notation of (3.11) and (3.18),

$$\{\text{configurational working}\} = \{\text{energy flow}\}, \qquad (3.22)$$

at least in this purely mechanical context, establishing consistency of the "second law" (3.13) with the more standard inequality (3.10) modified by (3.11).

e. The configurational force balance

In addition to the configurational stress $\boldsymbol{C}$, I allow for *internal configurational (body) forces* $\boldsymbol{f}$, which, being internal, contribute neither to the working (3.14) nor to the "second law" (3.13). As a second point of departure from the classical theory, I postulate a *configurational force balance*

$$\int_{\partial R} \boldsymbol{C}\boldsymbol{m}\, da + \int_{R} \boldsymbol{f}\, dv = \boldsymbol{0} \tag{3.23}$$

for each R, or equivalently,

$$\operatorname{div} \boldsymbol{C} + \boldsymbol{f} = \boldsymbol{0}. \tag{3.24}$$

$\boldsymbol{C}$ performs work when material is added to R through the motion of ∂R. Material is neither added nor removed from the interior of R. In fact, each material point $\boldsymbol{X}$ is constrained to a given position in the reference configuration for all time. Consequently, the force $\boldsymbol{f}$, which acts interior to R, performs no work (an observation consistent with its omission from (3.14)). I therefore consider $\boldsymbol{f}$ to be *indeterminate*,[15] in fact, as defined by the configurational balance (3.24).

Assume, for the moment, that the material is *elastic* and *homogeneous* with constitutive equations giving the stress $\boldsymbol{S}$ as the derivative of the free energy Ψ:

$$\Psi = \hat{\Psi}(\boldsymbol{F}), \quad \boldsymbol{S} = \partial_{\boldsymbol{F}} \hat{\Psi}(\boldsymbol{F}). \tag{3.25}$$

Then (3.21) yields an auxiliary relation giving $\boldsymbol{C}$ as a function of $\boldsymbol{F}$, and (3.9), (3.21), (3.24), and (3.25) yield $\boldsymbol{f} = \boldsymbol{F}^T\boldsymbol{b}$. If $\boldsymbol{b} = \boldsymbol{0}$ (equilibrium), then the internal configurational force vanishes. This is a direct consequence of *homogeneity*; for an *inhomogeneous* material with energy $\hat{\Psi}(\boldsymbol{F},\boldsymbol{X})$ and stress $\boldsymbol{S} = \partial_{\boldsymbol{F}} \hat{\Psi}(\boldsymbol{F},\boldsymbol{X})$,

$$\boldsymbol{f} = -\partial_{\boldsymbol{X}} \hat{\Psi} \tag{3.26}$$

and internal configurational forces are present (even though $\boldsymbol{b} = \boldsymbol{0}$). A one-dimensional cartoon giving an intuitive description of the configurational force system for homogeneous and inhomogeneous reference configurations is given in Figure 4.

Thus *for a single-phase elastic material the theory is equivalent to the classical theory* based on (3.9) with (3.24) and (3.21) considered as defining relations for $\boldsymbol{f}$ and $\boldsymbol{C}$; configurational forces play no role. On the other hand, for a two-phase system configurational forces play a pivotal role in the evolution of the *interface*, since it is at the interface that the material structure undergoes change.

In subsequent sections I demonstrate the role played by configurational forces in the study of evolving phase interfaces. But this does not seem the only circumstance in which this concept could be useful: configurational forces might form a basis for the systematic study of time-dependent defect structures such as dislocations and cracks.[16].

---

[15] That is, not specified constitutively (cf. TRUESDELL & NOLL [1965, p. 70]).

[16] An idea due to PAOLO PODIO-GUIDUGLI (private communication).

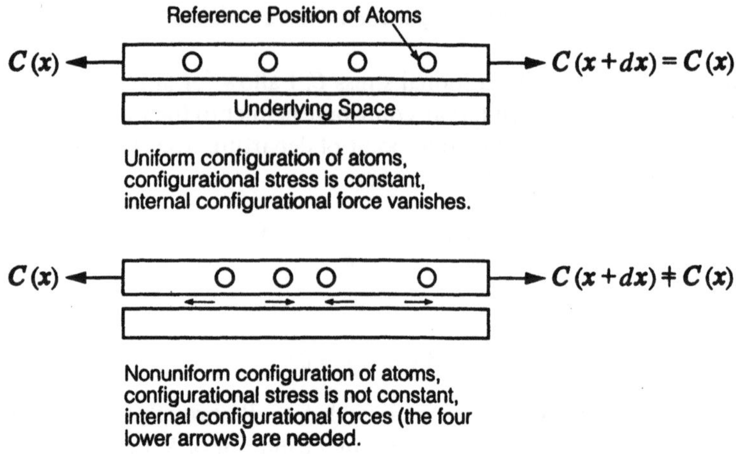

Figure 4. Cartoon showing why internal configurational forces respond to the inhomogeniety of the reference configuration.

4. Configurational forces for an evolving interface, neglecting bulk behavior

There are situations of physical interest in which the motion of a phase interface is effectively independent of deformational and transport processes in the bulk material;[17] here the underlying balance law is a configurational balance for the interface. I now turn toward a characterization of such situations.

a. Configurational force balance. Working

I consider two phases separated by a smoothly evolving surface $\mathscr{S}(t)$. I assume that $\mathscr{S}$ is oriented by a choice of unit normal field $\boldsymbol{n}$ and write V for the normal velocity, L for the curvature tensor, K for the total curvature (twice the mean curvature), and $\nabla_{\mathscr{S}}$ and $\mathrm{div}_{\mathscr{S}}$ for the surface gradient and surface divergence. (Cf. the Appendix for a discussion of evolving surfaces.)

I restrict attention to a configurational force system described by the fields:

- $\boldsymbol{C}$ *bulk stress*
- C *surface stress*
- $\boldsymbol{f}$ *internal bulk force*
- e *internal surface force*

$\boldsymbol{C}$ and $\boldsymbol{f}$ are as discussed in Section 3. The superficial vector field e represents internal forces distributed over the interface, while C, a superficial tensor field on

[17] Cf. the Introduction of TAYLOR, CAHN & HANDWERKER [1992].

$\mathscr{S}$, generalizes surface tension. Given a referential control volume $R(t)$, let

$$\mathscr{G}(t) = \mathscr{S}(t) \cap R(t)$$

denote the portion of the interface in $R(t)$ with $\boldsymbol{\nu}(X,t)$ (a tangential field on $\mathscr{S}(t)$) the outward unit normal to $\partial\mathscr{G}(t)$ (Figure 1); then $\mathbf{C}\boldsymbol{\nu}$ represents configurational forces within the interface applied to R across $\partial\mathscr{G}$.

The *configurational force balance* now takes the form

$$\int_{\partial R} \boldsymbol{C}\boldsymbol{m}\, da + \int_{R} \boldsymbol{f}\, da + \int_{\partial\mathscr{G}} \mathbf{C}\boldsymbol{\nu}\, ds + \int_{\mathscr{G}} \mathbf{e}\, da = 0, \tag{4.1}$$

with $R(t)$ a control volume and $\boldsymbol{m}$ the outward unit normal to ∂R. Shrinking R to the interface gives

$$\int_{\mathscr{G}} ([\boldsymbol{C}]\boldsymbol{n} + \mathbf{e})\, da + \int_{\partial\mathscr{G}} \mathbf{C}\boldsymbol{\nu}\, ds = \mathbf{0},$$

which, by virtue of the surface divergence theorem (A9), yields the *interfacial force balance*

$$[\boldsymbol{C}]\boldsymbol{n} + \mathbf{e} + \operatorname{div}_{\mathscr{S}}\mathbf{C} = \mathbf{0}. \tag{4.2}$$

On the other hand, restricting attention to R in (4.1) that do not interest the interface yields the bulk relation (3.24).

The motion of the curve $\partial\mathscr{G}(t)$ is characterized intrinsically by the velocity field $V\boldsymbol{n} + V_{\partial\mathscr{G}}\boldsymbol{\nu}$, where $V_{\partial\mathscr{G}}$, the tangential edge velocity of $\mathscr{G}$, is the velocity of $\partial\mathscr{G}$ in the direction of the normal $\boldsymbol{\nu}$. Alternatively, $\partial\mathscr{G}(t)$ may be parametrized locally by functions $\boldsymbol{X} = \boldsymbol{r}(u,t)$, which may be used to generate a *velocity field*

$$\boldsymbol{w}(X,t) = \frac{\partial \boldsymbol{r}(u,t)}{\partial t} \tag{4.3}$$

for $\partial\mathscr{G}(t)$. Then

$$\boldsymbol{w}\cdot\boldsymbol{n} = V, \quad \boldsymbol{w}\cdot\boldsymbol{\nu} = V_{\partial\mathscr{G}}, \tag{4.4}$$

but the component of $\boldsymbol{w}$ tangent to $\partial\mathscr{G}$ depends on the choice of local parametrizations.

Guided by the discussion leading to (3.14), I define the working $\mathscr{W}(R)$ by

$$\mathscr{W}(R) = \int_{\partial R} \boldsymbol{C}\boldsymbol{m}\cdot\boldsymbol{v}\, da + \int_{\partial\mathscr{G}} \mathbf{C}\boldsymbol{\nu}\cdot\boldsymbol{w}\, ds, \tag{4.5}$$

with the stipulation that *$\mathscr{W}(R)$ be independent of the particular local parametrizations $\boldsymbol{X} = \hat{\boldsymbol{X}}(u_1,u_2,t)$ and $\boldsymbol{X} = \boldsymbol{r}(u,t)$ used to determine the velocity fields $\boldsymbol{v}(X,t)$ and $\boldsymbol{w}(X,t)$ for $\partial R(t)$ and $\partial\mathscr{G}(t)$.* The argument leading to (3.16) then reduces $\boldsymbol{C}$ to a bulk tension:

$$\boldsymbol{C} = \pi\mathbf{1}. \tag{4.6}$$

Invariance under changes in parametrization for $\partial\mathcal{G}(t)$ has an equally stringent consequence. Because of (4.4), such changes effect the component of $\boldsymbol{w}$ tangential to $\partial\mathcal{G}$, but leave $\boldsymbol{w}$ otherwise unaltered. In fact, invariance of $\mathcal{W}(R)$ under changes in parametrization for $\partial\mathcal{G}$ is equivalent to the requirement that

$$\int_{\partial\mathcal{G}} \mathbf{C}\boldsymbol{v}\cdot\boldsymbol{t}\,ds = 0 \tag{4.7}$$

for every vector field $\boldsymbol{t}$ tangential to $\partial\mathcal{G}$. Since R and hence $\partial\mathcal{G}$ are arbitrary, it follows[18] that the tangential part of $\mathbf{C}$ is a surface tension:

$$\mathbf{C}_{\tan} = \sigma \boldsymbol{P} \tag{4.8}$$

(cf. (A3)) with σ the *surface tension*. Thus, by (A4),

$$\mathbf{C} = \sigma\boldsymbol{P} + \boldsymbol{n}\otimes\mathbf{c}, \tag{4.9}$$

where $\mathbf{c}$, the *surface shear*, is a tangential vector field that represents forces within $\mathcal{S}$ whose action is normal to $\mathcal{S}$.[19]

A computation, based on (A1), (A7), (A8), and (4.9), yields

$$\mathrm{div}_{\mathcal{S}}\mathbf{C} = (\sigma K + \mathrm{div}_{\mathcal{S}}\mathbf{c})\boldsymbol{n} + \nabla_{\mathcal{S}}\sigma - \mathbf{L}\mathbf{c}, \tag{4.10}$$

and since $\nabla_{\mathcal{S}}\sigma$ and $\mathbf{Lc}$ are tangential, (4.2) and (4.6) yield the *normal force balance*

$$\sigma K + \mathrm{div}_{\mathcal{S}}\mathbf{c} + [\pi] + e = 0 \tag{4.11}$$

with

$$e = \mathbf{e}\cdot\boldsymbol{n} \tag{4.12}$$

the *normal internal force*.

Next, by (3.4), (4.4), (4.5), (4.6), and (4.9),

$$\mathcal{W}(R) = \int_{\partial\mathcal{G}} (\sigma V_{\partial\mathcal{G}} + V\mathbf{c}\cdot\boldsymbol{v})\,ds + \int_{\partial R} \pi\mathcal{V}\,da, \tag{4.13}$$

and hence (A9), (A10), and (4.11) yield[20]

$$\mathcal{W}(R) = -\int_{\mathcal{G}} \{\sigma K V + \mathbf{c}\cdot\boldsymbol{n}^{\circ} + ([\pi] + e)V\}\,ds + \int_{\partial\mathcal{G}} \sigma V_{\partial\mathcal{G}}\,ds + \int_{\partial R} \pi\mathcal{V}^{\hat{}}\,da, \tag{4.14}$$

---

[18] GURTIN & STRUTHERS [1990, eq. (7.4)].

[19] GURTIN & STRUTHERS [1990, eq. (7.5)] (cf. GURTIN [1988]). For statics an essentially equivalent relation was derived by HERRING [1951], HOFFMAN & CAHN [1972], and CAHN & HOFFMAN [1974] by using variational arguments based on a constitutive equation $\psi = \hat{\psi}(\boldsymbol{n})$ for the interfacial energy, with σ and $\mathbf{c}$ defined by (4.19) and (4.22). The derivation given above is independent of constitutive equations.

[20] GURTIN [1988], GURTIN & STRUTHERS [1990].

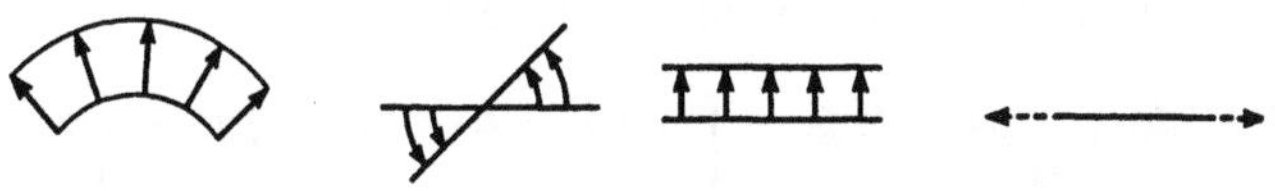

$-\sigma KV$ $\quad$ $\mathbf{c}\cdot\boldsymbol{n}^{\circ}$ $\quad$ $([\pi]+e)V$ $\quad$ $-\sigma V_{\partial\mathcal{G}}$

Figure 5. Contributions to the working at a phase interface.

where $(\ldots)^{\circ}$ denotes the time derivative following the normal trajectories of the interface. The right side of this relation catalogs the manner in which configurational forces perform work (Figure 5):

- The surface tension σ works to increase the area of $\mathcal{G}$: at internal points through $-\sigma KV$, at its boundary through $\sigma V_{\partial\mathcal{G}}$.
- The surface shear $\mathbf{c}$ works to change the orientation of $\mathcal{G}$.
- The jump $[\pi]$ in bulk tension and the normal internal force e work to advance the interface.
- The bulk tension works over ∂R to increase the volume of R.

Note that there is no expenditure of work associated with "tangential motion" of the interface (cf. the paragraph following (3.24)). Consistent with a "constraint" of this type, I leave as *indeterminate* the tangential component $\boldsymbol{Pe}$ of the internal force, an assumption that allows me to restrict attention to the normal balance (4.11).

b. The second law neglecting variations in temperature and composition

As before, I allow for a *bulk free-energy* Ψ, but, in accord with the physical assumptions underlying the current development, I assume that Ψ is constant in each phase and write

$$U = [\Psi]\ (= \text{constant}). \tag{4.15}$$

In addition, I now allow for an *interfacial free-energy* $\psi(X,t)$, per unit area, and write the second law in the form (3.12):

$$\frac{d}{dt}\left\{\int_R \Psi\, dv + \int_{\mathcal{G}} \psi\, ds\right\} \leqq \mathscr{W}(R) \tag{4.16}$$

for every evolving control volume $R(t)$, with $\mathcal{G}(t) = \mathscr{S}(t)\cap R(t)$.

The argument leading to (3.20) is valid here also. Further, by (4.15),

$$\frac{d}{dt}\left\{\int_R \Psi\, dv\right\} = -\int_{\mathcal{G}} UV\, da + \int_{\partial R} \Psi\mathscr{V}\, da, \tag{4.17}$$

so that, by (3.20), (4.14), and (A13), (4.16) becomes

$$\int_{\mathcal{G}}(\psi^\circ - \psi KV)\,da + \int_{\partial\mathcal{G}}\psi V_{\partial\mathcal{G}}\,ds \leqq -\int_{\mathcal{G}}\{\sigma KV + \mathbf{c}\cdot\boldsymbol{n}^\circ + Ve\}\,ds + \int_{\partial\mathcal{G}}\sigma V_{\partial\mathcal{G}}\,ds. \tag{4.18}$$

Given a time τ, it is possible to find a second referential control volume $R'(t)$ with $R'(\tau) = R(\tau)$, but with $V_{\partial\mathcal{G}'}(X,\tau)$, the normal velocity of $\partial\mathcal{G}'(\tau)$, an *arbitrary* scalar field on $\partial\mathcal{G}'(\tau)$; satisfaction of (4.18) for all such $V_{\partial\mathcal{G}'}$ implies

$$\sigma = \psi \tag{4.19}$$

and the surface tension and surface free-energy coincide.[21] Thus, since $\mathcal{G}$ is arbitrary, what remains is the *interfacial dissipation inequality*

$$\psi^\circ + \mathbf{c}\cdot\boldsymbol{n}^\circ + eV \leqq 0. \tag{4.20}$$

At this point it is worth noting the similarities between the bulk tension π and the surface tension σ:

- Bulk tension works to increase the volume of bulk material, surface tension works to increase the area of the interface.
- The configurational stresses $\boldsymbol{C} = \pi\mathbf{1}$ and $\mathbf{C}_{\tan} = \sigma\boldsymbol{P}$ have isotropic forms; these are not consequences of material symmetry, but follow instead from invariance under reparametrization.
- Both π and σ are related to energy: π to bulk free-energy, σ to interfacial free-energy.

c. *Constitutive equations*

Guided by the dissipation inequality (4.20), I consider constitutive equations for the interface giving the free energy ψ, the surface shear $\mathbf{c}$, and the normal internal force e as functions of the orientation $\boldsymbol{n}$ and the normal velocity V:

$$\psi = \hat{\psi}(\boldsymbol{n}, V), \quad \mathbf{c} = \hat{\mathbf{c}}(\boldsymbol{n}, V), \quad e = \hat{e}(\boldsymbol{n}, V). \tag{4.21}$$

I view the second law as a restriction on constitutive relations;[22] more precisely, granted (4.21), I require that the dissipation inequality (4.20) be satisfied in all motions of the interface:

$$\partial_V\hat{\psi}(\boldsymbol{n}, V)V^\circ + \{\partial_{\boldsymbol{n}}\hat{\psi}(\boldsymbol{n}, V) + \hat{\mathbf{c}}(\boldsymbol{n}, V)\}\cdot\boldsymbol{n}^\circ + \hat{e}(\boldsymbol{n}, V)V \leqq 0$$

---

[21] GURTIN & STRUTHERS [1990] (cf. GURTIN [1991]).

[22] This use of the second law is due to COLEMAN & NOLL [1963], who study single-phase thermoelastic materials; the extension to two-phase materials is given by GURTIN [1988] (cf. ANGENENT & GURTIN [1989], GURTIN [1993a]). An interesting feature of this procedure is that the local dissipation inequality generally suggests which fields should be given constitutive descriptions, a use of the second law that seems to lead — in all classical continuum theories — to the "correct" set of constitutive variables. This contrasts the standard formalism of studying balance laws to see where a lack of field equations may be compensated for by the introduction of constitutive relations.

in all such motions. It is possible to construct a motion of the interface for which $\boldsymbol{n}$, $\boldsymbol{n}^\circ$, V, and V° have arbitrarily assigned values at some given point and time, an observation that leads to the following constitutive restrictions: the free energy must be independent of the velocity V; the shear must be the negative of the derivative of the energy with respect to $\boldsymbol{n}$,

$$\mathsf{c} = -\partial_{\boldsymbol{n}}\hat{\psi}(\boldsymbol{n}); \tag{4.22}$$

the relation $e = \hat{e}(\boldsymbol{n}, V)$ for the normal internal force must have the form

$$e = -b(\boldsymbol{n}, V)V, \quad b(\boldsymbol{n}, V) \geqq 0, \tag{4.23}$$

with $b(\boldsymbol{n}, V)$ a constitutive quantity called the kinetic modulus.[23] These are the most general constitutive equations of the form (4.21) that are consistent with (4.20).

Anisotropy of the interface manifests itself in a nontrivial dependence of $\hat{\psi}(\boldsymbol{n})$ on $\boldsymbol{n}$; for an isotropic interface ψ is constant. An interesting consequence of (4.22) is that *for an anisotropic interface the surface shear cannot generally vanish*. This demonstrates the nonintuitive nature of configurational forces: the interface is presumed to be infinitesimally thin, yet it supports shear; thus the danger inherent in visualizing the interface as a membrane.

The surface shear must be balanced by couples exerted by the bulk material, although such couples, being indeterminate, need not be made explicit.[24] This furnishes an additional argument in support of the separate treatment of configurational forces when discussing deformation. If the variational treatment of Section 2, for an elastic material, is generalized to include an anisotropic interfacial energy, then the resulting Euler-Lagrange equations *in bulk* remain (2.7). Since these classical equations support neither bulk internal-couples nor bulk couple-stresses, a configurational system is needed to balance the couples induced by surface-shear.

d. Evolution equation for the interface

The evolution equation for the interface follows from the normal force-balance (4.11), (4.15), the tension-energy relations (3.20) and (4.19), the reduced constitutive relations (4.22) and (4.23), and the identity (A15):[25]

$$b(\boldsymbol{n}, V)V = \{\hat{\psi}(\boldsymbol{n})1 + \partial_{\boldsymbol{n}}\partial_{\boldsymbol{n}}\hat{\psi}(\boldsymbol{n})\}\cdot\mathsf{L} + U. \tag{4.24}$$

---

[23] For $b(\boldsymbol{n}, V)$ independent of V, $b(\boldsymbol{n})^{-1}$ is referred to as the mobility of the interface.

[24] GURTIN [1988, Remark 3.2].

[25] Proposed by UWAHA [1987, eq. (2)] in ($\mathbb{R}^2$ with $b = b(\boldsymbol{n})$) and independently by GURTIN [1988, eq. (8.3)]. Evolution according to (4.24) with $b = b(\boldsymbol{n})$ is studied by ANGENENT [1991], CHEN, GIGA & GOTO [1991], and SONER [1993]. The special case $V = -b(\boldsymbol{n})^{-1}U$ was introduced by FRANK [1958]. A formulation of (4.24) using a variational definition of the curvature term (TAYLOR [1992]) is given by TAYLOR, CAHN & HANDWERKER [1992], who give extensive references.

For an isotropic material with b independent of V and $U = 0$, (4.24), after a rescaling, reduces to the curve-shortening equation, $V = K$.[26] In (4.24) the derivatives must respect the constraint $|\boldsymbol{n}| = 1$. A simpler form of the equation follows if $\hat{\psi}(\boldsymbol{n})$ is extended from the unit sphere to $\mathbb{R}^3$ by defining $\tilde{\psi}(\boldsymbol{z}) = |\boldsymbol{z}|\hat{\psi}(\boldsymbol{z}/|\boldsymbol{z}|)$, for then the term $\{\ldots\}$ reduces to $\{\partial_{\boldsymbol{z}}\partial_{\boldsymbol{z}}\tilde{\psi}(\boldsymbol{z})\}$ at $\boldsymbol{z} = \boldsymbol{n}$. Finally, the analogous relation for evolution in $\mathbb{R}^2$, with $\boldsymbol{n}$ replaced by the counterclockwise angle φ from a fixed axis to $\boldsymbol{n}$, is[27]

$$b(\varphi, V)V = \{\hat{\psi}(\varphi) + \hat{\psi}''(\varphi)\}K + U. \tag{4.25}$$

The expression (4.24) represents the normal part of the configurational force balance (4.2). The indeterminancy of the tangential part of $\mathbf{e}$ renders the tangential part of (4.2) unimportant. Interestingly, a computation using $(A15)_1$, (4.6), (4.10), (4.19), and (4.22) shows the tangential balances to be satisfied automatically with $\boldsymbol{P}\mathbf{e} = \mathbf{0}$, so that $\mathbf{e} = e\boldsymbol{n}$.

For *nonsmooth* interfaces — which are possible when $\tilde{\psi}(\boldsymbol{z})$ is nonconvex — the evolution equation (4.24) is not, by itself, sufficient to describe the motion of the interface: the weaker form (4.1) must be used, for example across curves defined by a jump in the interface normal $\boldsymbol{n}$.[28]

5. Two-phase theory with deformation[29]

a. *Theory neglecting interfacial energy*

I now consider a coherent two-phase elastic solid, neglecting interfacial energy, interfacial stress, and, as before, thermal and compositional variations. The body B is presumed independent of time, but the subregions $B_\alpha(t)$ and $B_\beta(t)$ occupied by α and β as well as the interface $\mathscr{S}(t)$ depend on t. Coherency requires that motions $\boldsymbol{y}$ of B be continuous across the interface, with deformation gradient $\boldsymbol{F}$ and material velocity $\boldsymbol{y}^{\cdot}$ continuous up to the interface from either side, a constraint that yields the compatibility conditions

$$[\boldsymbol{y}^{\cdot}] = -V[\boldsymbol{F}]\boldsymbol{n}, \quad [\boldsymbol{F}]\boldsymbol{P} = \mathbf{0}, \tag{5.1}$$

with $\boldsymbol{n}$ the unit normal to $\mathscr{S}$ directed outward from B_α and V the normal velocity of $\mathscr{S}$.

---

[26] BURKE & TURNBULL [1952] and MULLINS [1956] introduced $V = K$ to study the motion of grain boundaries. Evolution according to this equation is discussed by many authors (cf. GURTIN [1993a, Footnote 12]).

[27] Derived using a thermomechanical argument by ANGENENT & GURTIN [1989], although the configurational force system is somewhat different (cf. GURTIN [1993a]).

[28] Cf. ANGENENT & GURTIN [1989] and GURTIN [1993a] for discussions of this issue in $\mathbb{R}^2$.

[29] GURTIN [1993b, 1994a].

While inertia is easily included, I neglect it to simplify the presentation. The basic balance laws are as discussed in Section 3 for single-phase materials, but with the interface accounted for by internal configurational forces with density $\mathbf{e}$ distributed over $\mathcal{S}$. The deformational and configurational balances thus take the form

$$\int_{\partial R} \boldsymbol{S}\boldsymbol{m}\,da = \mathbf{0}, \quad \int_{\partial R} \boldsymbol{y}\times\boldsymbol{S}\boldsymbol{m}\,da = \mathbf{0}, \tag{5.2}$$

$$\int_{\partial R} \boldsymbol{C}\boldsymbol{m}\,da + \int_{R} \boldsymbol{f}da + \int_{\mathcal{G}} \mathbf{e}\,da = \mathbf{0}, \tag{5.3}$$

where $R(t)$ is an arbitrary referential control volume with $\mathcal{G}(t) = \mathcal{S}(t) \cap R(t)$ the portion of the interface in $R(t)$.

The local force-balances now consist of bulk equations and interface conditions; the bulk equations remain (3.9) (with $\boldsymbol{b} = \mathbf{0}$) and (3.24), while the interface conditions are

$$[\boldsymbol{S}]\boldsymbol{n} = \mathbf{0}, \quad [\boldsymbol{C}]\boldsymbol{n} + \mathbf{e} = 0. \tag{5.4}$$

The working still has the form (3.14) (with $\boldsymbol{b} = \mathbf{0}$); there is no contribution of $\mathbf{e}$, as it acts internally. Restricting $R(t)$ to not intersect the interface leads to the relation (3.16) for the configurational stress. Further, since interfacial energy as well as variations in temperature and composition are neglected, the second law takes the form (3.13) and leads again to the bulk Eshelby relation (3.21).

To derive an expression for the second law at the interface, I restrict attention to time-independent control volumes that intersect the interface; granted this, (3.13) reduces to

$$\frac{d}{dt}\left\{\int_{R} \Psi\,dv\right\} \leqq \int_{\partial R} \boldsymbol{S}\boldsymbol{m}\cdot\boldsymbol{y}^{\cdot}\,da. \tag{5.5}$$

But

$$\frac{d}{dt}\left\{\int_{R} \Psi\,dv\right\} = -\int_{\mathcal{G}} [\Psi]\,V\,da + \int_{R} \Psi^{\cdot}\,dv. \tag{5.6}$$

Thus shrinking R to the interface yields the jump condition $[\Psi]V + [\boldsymbol{S}\boldsymbol{n}\cdot\boldsymbol{y}^{\cdot}] \geqq 0$, so that, by (5.1) and the first of (5.4), $([\Psi] - [\boldsymbol{F}\boldsymbol{n}\cdot\boldsymbol{S}\boldsymbol{n}])V \geqq 0$, or equivalently, by using (3.21) and the second of (5.4),

$$eV \leqq 0, \tag{5.7}$$

with $e = \mathbf{e}\cdot\boldsymbol{n}$ the normal internal force.

I consider bulk constitutive equations of the form (3.25) for each phase. (These are the most general constitutive equations — consistent with (5.5) — for ψ and $\boldsymbol{S}$ as functions of $\boldsymbol{F}$.) In addition, I consider a constitutive equation for the interface giving e as a function of V and a list z of fields such as $\boldsymbol{n}$ and the interfacial limits of

F; the requirement that (5.7) hold in all motions then yields the reduced constitutive relation

$$e = -bV, \quad b = b(z, V) \geqq 0, \tag{5.8}$$

with $b(z, V)$ a constitutive modulus.

The relations (3.21), (5.4), and (5.8) yield final interface conditions

$$[S]n = 0, \quad n \cdot [\Psi 1 - F^T S]n = bV, \tag{5.9}$$

which are replaced by

$$[S]n = -\rho[y^{\cdot}]V, \quad n \cdot [\Psi 1 - F^T S]n = -k + bV, \tag{5.10}$$

when inertia is included. Here $k = (\rho/2)[|Fn|^2]V^2$.

The kinetic relation $(5.10)_2$ was derived by Truskinovsky [1987, 1991] and Abeyaratne & Knowles [1990, 1991] on the basis of an argument that does not require the introduction of configurational forces. They note that the second law yields the inequality $([\Psi] - [Fn \cdot Sn] + k)V \geqq 0$, which they use to motivate a constitutive relation for the "driving traction" $f = -[\Psi] + [Fn \cdot Sn] - k$ of the form $f = -bV, b \geqq 0$. This argument is simpler than the one I have given, since it does not involve a configurational force balance, but I believe it to be unsound: k is a known function of F, n, and V, while Ψ and S are prescribed as functions of F through constitutive relations; therefore postulating an additional relation for $[\Psi] - [Fn \cdot Sn] + k$ seems both superfluous and arbitrary. I know of no other example from continuum mechanics in which fields specified by constitutive relations are presumed related by yet another such relation.

b. *Theory with interfacial energy and stress*

I now generalize the theory of Section 5a to include interfacial structure, but I merely sketch the analysis, as it is complicated. The inclusion of interfacial stress characterized by the superficial tensor fields

$\mathbf{S}$ *deformational surface stress*
$\mathbf{C}$ *configurational surface stress*

necessitates modification of (5.2) and (5.3) as follows: let ν denote the outward unit normal to $\partial\mathcal{G}$; then the terms

$$\int_{\partial\mathcal{G}} \mathbf{S}\nu \, ds, \quad \int_{\partial\mathcal{G}} y \times \mathbf{S}\nu \, ds, \quad \int_{\partial\mathcal{G}} \mathbf{C}\nu \, ds,$$

respectively, should be added to the left sides of the deformational force and moment balances and the left side of the configurational balance. This yields

$$[S]n + \mathrm{div}_{\mathcal{S}}\mathbf{S} = 0, \quad \mathbf{S}F^T = F\mathbf{S}^T, \quad [C]n + \mathbf{e} + \mathrm{div}_{\mathcal{S}}\mathbf{C} = 0, \tag{5.11}$$

with

$$\mathbf{F} = F_\alpha P = F_\beta P \tag{5.12}$$

the tangential deformation gradient (cf. $(5.1)_2$).

In addition to the bulk free-energy Ψ, there is an interfacial free-energy ψ, and the inequality representing the second law takes the form

$$\frac{d}{dt}\left\{\int_R \Psi\, dv + \int_{\mathscr{S}} \psi\, da\right\} \leqq \mathscr{W}(R) \tag{5.13}$$

for all $R(t)$, where $\mathscr{W}(R)$ now includes working of the surface forces:

$$\mathscr{W}(R) = \int_{\partial R} (S\boldsymbol{m}\cdot\bar{\boldsymbol{v}} + C\boldsymbol{m}\cdot\boldsymbol{v})\,da + \int_{\partial\mathscr{S}} (\mathbf{C}\boldsymbol{\nu}\cdot\boldsymbol{w} + \mathbf{S}\boldsymbol{\nu}\cdot\bar{\boldsymbol{w}})\,ds. \tag{5.14}$$

Here $\boldsymbol{v}$ is a velocity field for ∂R with $\bar{\boldsymbol{v}}$ the corresponding induced velocity field for $\partial\mathscr{R}$, while $\boldsymbol{w}$ is a velocity field for the curve $\partial\mathscr{S}(t)$ with $\bar{\boldsymbol{w}}$ the corresponding induced velocity field for the deformed curve $\boldsymbol{y}(\partial\mathscr{S}(t), t)$; that is, $\boldsymbol{w}(X, t)$ is computed using time-dependent parametrizations (4.3) for $\partial\mathscr{S}(t)$ and $\bar{\boldsymbol{w}} = \boldsymbol{y}^{\bullet} + \boldsymbol{F}\boldsymbol{w}$.

Invariance under reparametrization for $\partial R(t)$ then yields the Eshelby relation (3.21), while invariance under reparametrization for $\partial\mathscr{S}(t)$ yields an Eshelby relation for the interface:

$$\mathbf{C}_{\tan} = \psi\boldsymbol{P} - \mathbf{F}^T\mathbf{S}. \tag{5.15}$$

The resulting interfacial dissipation inequality, which generalizes (4.20), is

$$\psi^\circ + (\mathbf{c} + \mathbf{S}^T\boldsymbol{E}\boldsymbol{n})\cdot\boldsymbol{n}^\circ - \mathbf{S}\cdot\boldsymbol{E}^\circ + eV \leqq 0, \tag{5.16}$$

with $\mathbf{c}$ the configurational shear and $\boldsymbol{E}$ the average of the interfacial limits $\boldsymbol{F}_\alpha$ and $\boldsymbol{F}_\beta$ of $\boldsymbol{F}$:

$$\mathbf{c} = \mathbf{C}^T\boldsymbol{n}, \quad \boldsymbol{E} = \langle \boldsymbol{F} \rangle. \tag{5.17}$$

As constitutive equations for the interface I allow ψ, $\mathbf{S}$, $\mathbf{c}$, and e to depend on $\boldsymbol{n}$, V, $\boldsymbol{F}_\alpha$, and $\boldsymbol{F}_\beta$. A consequence of (5.16) is that ψ, $\mathbf{S}$, and $\mathbf{c}$ are independent of V and depend on $\boldsymbol{F}_\alpha$ and $\boldsymbol{F}_\beta$ through their average $\boldsymbol{E}$. In fact, the energy $\psi = \hat{\psi}(\boldsymbol{E}, \boldsymbol{n})$ determines $\mathbf{S}$ and $\mathbf{c}$ through the relations

$$\mathbf{S} = \partial_{\boldsymbol{E}}\hat{\psi}(\boldsymbol{E}, \boldsymbol{n}), \quad \mathbf{c} + \mathbf{S}^T\boldsymbol{E}\boldsymbol{n} = -\partial_{\boldsymbol{n}}\hat{\psi}(\boldsymbol{E}, \boldsymbol{n}), \tag{5.18}$$

$$e = -bV, \quad b = b(\boldsymbol{F}_\alpha, \boldsymbol{F}_\beta, \boldsymbol{n}, V) \geqq 0. \tag{5.19}$$

The final interface conditions consist of the compatibility condition (5.1) and the force balances[30]

$$[S]\boldsymbol{n} = -\mathrm{div}_{\mathscr{S}}\mathbf{S}, \quad \boldsymbol{n}\cdot[\Psi\mathbf{1} - \boldsymbol{F}^T S]\boldsymbol{n} + (\psi\boldsymbol{P} - \mathbf{F}^T S)\cdot\mathbf{L} + \mathrm{div}_{\mathscr{S}}\mathbf{c} = bV \tag{5.20}$$

[30] GURTIN & STRUTHERS [1990], GURTIN [1993b, 1994a]. See also LUSK [1992]. For statics $(5.20)_1$ was derived by GURTIN & MURDOCH [1974] from a force balance, while $(5.20)_2$ was derived by LEO & SEKERKA [1989] as an Euler-Lagrange equation for stable equilibria (cf. ALEXANDER & JOHNSON [1985], JOHNSON & ALEXANDER [1986], and FONSECA [1989]).

supplemented by the constitutive equations (5.18) and (5.19) as well as those for the bulk material. Note that $(5.20)_2$, which represents the normal configurational force balance, may be written in the form[31]

$$\boldsymbol{n}\cdot[\text{bulk Eshelby tensor}]\boldsymbol{n} + (\text{interfacial Eshelby tensor})\cdot\mathbf{L} + \operatorname{div}_{\mathcal{S}}\mathbf{c} = bV. \tag{5.21}$$

6. Solidification. The Stefan and Gibbs-Thompson conditions as consequences of the configurational force balance

To demonstrate the role of configurational forces in situations that are not purely mechanical, I turn now to two-phase heat flow, neglecting deformation. Paralleling (3.12), I write the first two laws for a control volume $R(t)$ as

$$\frac{d}{dt}\{\text{internal energy}\} = \{\text{heating}\} + \{\text{working}\},$$

$$\frac{d}{dt}\{\text{internal entropy}\} \geqq \{\text{entropy flux induced by heating}\}$$

in which the right sides include an accounting of the work and *heat* required to add or remove material, but make no explicit mention of flows of internal energy and internal entropy across ∂R.

a. Single-phase theory

To the classical fields

ε *internal energy*
η *internal entropy*
ϑ *absolute temperature*
$\boldsymbol{q}$ *heat flux vector*

I add three configurational fields

$\boldsymbol{C}$ *stress*
$\boldsymbol{f}$ *internal force*
Q *heating*

where Q is a scalar field, while $\boldsymbol{C}$ and $\boldsymbol{f}$ are as discussed previously, and I define the free energy through

$$\Psi = \varepsilon - \vartheta\eta. \tag{6.1}$$

[31] The insight afforded by the use of bulk and interfacial Eshelby tensors was pointed out to me by P. PODIO-GUIDUGLI (private communication).

For $R(t)$ an evolving control volume, with $\mathscr{V}$ the normal velocity of ∂R,

$$\int_{\partial R} Q\mathscr{V}\,da, \quad \int_{\partial R} \frac{Q}{\vartheta}\mathscr{V}\,da$$

represent flows of heat and entropy into R associated with the motion of ∂R.

The basic laws, for each evolving control volume $R(t)$, are *balance of energy*

$$\frac{d}{dt}\left\{\int_{R} \varepsilon\,dv\right\} = -\int_{\partial R} \boldsymbol{q}\cdot\boldsymbol{m}\,da + \int_{\partial R} Q\mathscr{V}\,da + \int_{\partial R} \boldsymbol{C}\boldsymbol{m}\cdot\boldsymbol{v}\,da, \tag{6.2}$$

growth of entropy

$$\frac{d}{dt}\left\{\int_{R} \eta\,dv\right\} \geqq -\int_{\partial R} \frac{\boldsymbol{q}}{\vartheta}\cdot\boldsymbol{m}\,da + \int_{\partial R} \frac{Q}{\vartheta}\mathscr{V}\,da, \tag{6.3}$$

and *balance of configurational forces*

$$\int_{\partial R} \boldsymbol{C}\boldsymbol{m}\,da + \int_{R} \boldsymbol{f}\,dv = 0, \tag{6.4}$$

with $\boldsymbol{m}$ the outward unit normal to ∂R, and with the stipulation that *balance of energy be independent of the particular local parametrizations used to determine the velocity field $\boldsymbol{v}$ for ∂R*. For R stationary (6.2) and (6.3) become

$$\frac{d}{dt}\left\{\int_{R} \varepsilon\,dv\right\} = -\int_{\partial R} \boldsymbol{q}\cdot\boldsymbol{m}\,da, \quad \frac{d}{dt}\left\{\int_{R} \eta\,dv\right\} \geqq -\int_{\partial R} \frac{\boldsymbol{q}}{\vartheta}\cdot\boldsymbol{m}\,da, \tag{6.5}$$

again demonstrating consistency with more standard ideas.

As before, invariance under changes in reparametrization for $\partial R(t)$ yields the conclusion that $\boldsymbol{C}$ is a pure tension π, as in (4.6); therefore, by (3.1) applied to ε in (6.2) and η in (6.3),

$$\varepsilon = \pi + Q, \quad \eta = \frac{Q}{\vartheta}, \tag{6.6}$$

relations that, when multiplied by $\mathscr{V}$, express balance of energy and entropy associated with the addition of material to R. A trivial but important corollary of these relations is that, once again, bulk tension and bulk free energy coincide: $\pi = \Psi$. Thus

$$\boldsymbol{C} = \Psi \boldsymbol{1}. \tag{6.7}$$

Finally, restricting attention to stationary control volumes, and using (6.1) and the energy balance to rewrite the entropy inequality yields the local relations

$$\varepsilon^{\cdot} = -\operatorname{div}\boldsymbol{q}, \tag{6.8}$$

$$\Psi^{\cdot} + \eta\vartheta^{\cdot} + \vartheta^{-1}\boldsymbol{q}\cdot\nabla\vartheta \leqq 0. \tag{6.9}$$

The standard constitutive equations consist of a relation between free energy and temperature, a relation giving the entropy as the negative of the derivative of the free energy with respect to temperature, and a Fourier law for heat conduction,[32]

$$\Psi = \hat{\Psi}(\vartheta), \quad \eta = -\hat{\Psi}'(\vartheta), \quad \boldsymbol{q} = -\boldsymbol{K}(\vartheta)\nabla\vartheta, \tag{6.10}$$

with $\boldsymbol{K}(\vartheta)$, the conductivity tensor, assumed positive definite. Granted these, the dissipation inequality (6.9) is satisfied identically. By (6.1) and (6.10), there is an auxiliary relation for the internal energy:

$$\varepsilon = \hat{\Psi}(\vartheta) - \vartheta\hat{\Psi}'(\vartheta). \tag{6.11}$$

The basic partial differential equation of the theory is given by balance of energy supplemented by $(6.10)_3$ and (6.11).

Note that, by the local form (3.24) of (6.4) and the equations (6.7) and (6.10),

$$\boldsymbol{f} = \eta\nabla\vartheta, \tag{6.12}$$

which I take as a defining relation for $\boldsymbol{f}$.

b. *The classical two-phase theory revisited. The Stefan condition as a consequence of the configurational force balance*

I now consider two phases, α and β, with $\Psi_\alpha(\vartheta)$ and $\Psi_\beta(\vartheta)$, and $\boldsymbol{K}_\alpha(\vartheta)$ and $\boldsymbol{K}_\beta(\vartheta)$ the corresponding free energies and conductivity tensors, and with resulting constitutive relations of the form (6.10). I assume further that there is a unique temperature, the *melting temperature* ϑ_M, at which the free energies of the individual phases coincide:

$$\Psi_\alpha(\vartheta_M) = \Psi_\beta(\vartheta_M). \tag{6.13}$$

I neglect interfacial structure, so that the basic laws remain (6.2)–(6.4). Further, I assume that the *temperature is continuous*, but allow the other fields to suffer jump discontinuities across the interface.

Balance of energy then yields the interfacial balance

$$[\varepsilon]V = [\boldsymbol{q}]\cdot\boldsymbol{n}, \tag{6.14}$$

which is the first of the classical interface conditions for the Stefan problem.

Next, the configurational balance (6.4) yields

$$[\boldsymbol{C}]\boldsymbol{n} = \boldsymbol{0}, \tag{6.15}$$

[32] These are consequences of (7.9) in conjunction with constitutive equations giving Ψ, η, and $\boldsymbol{q}$ as functions of θ and $\nabla\theta$ with $\boldsymbol{q}$ linear in $\nabla\theta$.

which, by (6.7), has the alternative form

$$[\Psi] = 0, \tag{6.16}$$

or, in view of the hypothesis ending in (6.13),

$$\vartheta = \vartheta_M \quad \text{on the interface.} \tag{6.17}$$

Thus *the classical Stefan condition equating the temperature at the interface to the melting temperature is equivalent to the configurational balance applied across the interface.*[33]

c. Weak form of the two-phase problem using the configurational balance

Not only does the configurational balance allow for a derivation of the classical Stefan condition, it allows for a weak formulation of the Stefan problem by replacing the condition $\vartheta = \vartheta_M$ on the interface, which is local and inappropriate to a weak formulation, with a partial differential equation. In particular, (6.5) and the configurational balance $\operatorname{div} \boldsymbol{C} = -\boldsymbol{f}$ with $\boldsymbol{C}$ given by (6.7) and $\boldsymbol{f}$ by (6.12) yield partial differential equations

$$\varepsilon^{\boldsymbol{\cdot}} = -\operatorname{div} \boldsymbol{q}, \quad \nabla\Psi = -\eta\nabla\vartheta, \quad \eta^{\boldsymbol{\cdot}} \geqq -\operatorname{div}\frac{\boldsymbol{q}}{\vartheta} \tag{6.18}$$

to be interpreted in a weak sense, for example in the sense of distributions. The distributional form of $(6.18)_1$ gives that partial differential equation classically in bulk and the balance (6.14) at the interface. The configurational balance $(6.18)_2$ is satisfied automatically in bulk; its only contribution is at the interface, where $\nabla\Psi$ is a distribution, as Ψ suffers a jump discontinuity ($\eta\nabla\vartheta$ does not contribute, as η and $\nabla\vartheta$ are bounded). In fact, $(6.18)_2$ formally yields (6.16) and hence the Stefan condition (6.17). Finally, $(6.18)_3$ is satisfied automatically in bulk as well as across the interface. To verify the latter assertion, note that $(6.18)_3$ yields $\vartheta[\eta]V \leqq [\boldsymbol{q}]\cdot\boldsymbol{n}$, or equivalently, by (6.1) and (6.14), $[\Psi]V \geqq 0$, an inequality satisfied by virtue of (6.16). It might therefore appear that the entropy inequality $(6.18)_3$ is superfluous, which is true when the interface moves smoothly, but there are situations involving large amounts of supercooling or superheating in which the interface moves "infinitely fast" resulting in an instantaneous change in phase for entire subregions of the body;[34] the entropy inequality is then needed to ensure that such instantaneous changes be consistent with the second law.[35]

[33] GURTIN [1988]. To this point the *assumption* of continuity of temperature across the interface is not needed; however, it is basic to Sections 6c and 6d.

[34] SHERMAN [1970], FASANO & PRIMICERIO [1977], GÖTZ & ZALTSMAN [1993], GURTIN [1994b].

[35] GURTIN [1994b].

d. The two-phase theory with surface structure. The Gibbs-Thompson condition as a consequence of the configurational balance[36]

I now generalize the theory to include surface structure by considering the basic laws for each evolving control volume $R(t)$ in the form

$$\frac{d}{dt}\left\{\int_R \varepsilon\, dv + \int_{\mathscr{S}} \bar{\varepsilon}\, da\right\} = -\int_{\partial R} \boldsymbol{q}\cdot\boldsymbol{m}\, da + \int_{\partial R} Q\mathscr{V}\, da + \int_{\partial\mathscr{S}} \bar{Q} V_{\partial\mathscr{S}}\, ds + \mathscr{W}(R), \tag{6.19}$$

$$\frac{d}{dt}\left\{\int_R \eta\, dv + \int_{\mathscr{S}} \bar{\eta}\, da\right\} \geqq -\int_{\partial R} \frac{\boldsymbol{q}}{\vartheta}\cdot\boldsymbol{m}\, da + \int_{\partial R} \frac{Q}{\vartheta}\mathscr{V}\, da + \int_{\partial\mathscr{S}} \frac{\bar{Q}}{\vartheta} V_{\partial\mathscr{S}}\, ds, \tag{6.20}$$

$$\int_{\partial R} \mathbf{C}\boldsymbol{m}\, da + \int_R \boldsymbol{f} da + \int_{\partial\mathscr{S}} \mathbf{C}\boldsymbol{v}\, ds + \int_{\mathscr{S}} \mathbf{e}\, da = \mathbf{0}, \tag{6.21}$$

where $\mathscr{W}(R)$ is given by (4.5), $\mathbf{C}$, $\mathbf{e}$ and Q are as discussed before, $\bar{\varepsilon}$ is the interfacial energy, $\bar{\eta}$ is the interfacial entropy, and $\bar{Q}$, a configurational heating, is an analog of Q in the sense that

$$\int_{\partial\mathscr{S}} \bar{Q} V_{\partial\mathscr{S}}\, ds, \quad \int_{\partial\mathscr{S}} \frac{\bar{Q}}{\vartheta} V_{\partial\mathscr{S}}\, ds$$

represent flows of heat and entropy into $\mathscr{S}$ induced by the motion of the boundary curve $\partial\mathscr{S}$.

The arguments used before then yield the bulk relations discussed in the paragraph containing (6.7), the interface relations

$$\bar{\eta} = \frac{\bar{Q}}{\vartheta}, \quad \sigma = \psi, \tag{6.22}$$

and the results (4.9) and (4.19) with the interfacial free-energy given by

$$\psi = \bar{\varepsilon} - \vartheta\bar{\eta}. \tag{6.23}$$

The configurational balance remains (4.11), but the remaining interface conditions are more complicated than before: balance of energy has the form

$$[\varepsilon]V = [\boldsymbol{q}]\cdot\boldsymbol{n} + \bar{\varepsilon}^{\circ} - \bar{\varepsilon}KV - \operatorname{div}_{\mathscr{S}}(V\mathbf{c}) \tag{6.24}$$

(cf. (A13), (4.13), (6.6), (6.22), (6.23)), an analogous inequality for the entropy follows from (6.20), and this inequality, (4.11), and (6.24) yield the interfacial dissipation inequality

$$\psi^{\circ} + \bar{\eta}\vartheta^{\circ} + \mathbf{c}\cdot\boldsymbol{n}^{\circ} + eV \leqq 0. \tag{6.25}$$

[36] This section, which represents a major improvement of my work [1988], is based on ideas introduced in GURTIN & STRUTHERS [1990] and GURTIN [1993, 1994a].

Guided by this inequality, I consider constitutive equations of the form (4.21), but with $(\vartheta, \boldsymbol{n}, V)$ as independent variables and with an additional constitutive equation, of the same form, for the entropy $\bar{\eta}$. The most general constitutive equations of this form consistent with the dissipation inequality (6.25) are

$$\psi = \hat{\psi}(\vartheta, \boldsymbol{n}), \quad \bar{\eta} = -\partial_\vartheta \hat{\psi}(\vartheta, \boldsymbol{n}), \quad \mathbf{c} = -\partial_{\boldsymbol{n}} \hat{\psi}(\vartheta, \boldsymbol{n}), \quad e = -b(\vartheta, \boldsymbol{n}, V)V, \tag{6.26}$$

with $b(\vartheta, \boldsymbol{n}, V) \geqq 0$. I henceforth assume that b is independent of V.

The resulting interface conditions consist of the energy balance (6.24) and the configurational balance (4.11) with $\pi = \Psi$, $\sigma = \psi$:

$$[\Psi] = -\psi K - \operatorname{div}_{\mathscr{S}} \mathbf{c} - e. \tag{6.27}$$

These interface conditions supplemented by the constitutive equations are the basic free-boundary conditions of the theory; the condition (6.27) replaces the classical Stefan condition.

The interface conditions (6.24) and (6.27) are complicated. In [1988] I formally derived an approximate theory appropriate to an interface whose free energy, internal energy, and kinetic coefficient are small, with the latter independent of V. Let

$$u = \frac{(\vartheta - \vartheta_M)}{\vartheta_M}, \quad l = \varepsilon_\beta(\vartheta_M) - \varepsilon_\alpha(\vartheta_M),$$
$$\psi_M(\boldsymbol{n}) = \hat{\psi}(\vartheta_M, \boldsymbol{n}), \quad b_M(\boldsymbol{n}) = b(\vartheta_M, \boldsymbol{n}), \tag{6.28}$$

where $\varepsilon_\alpha(\vartheta)$ and $\varepsilon_\beta(\vartheta)$ are the constitutive functions for the internal energy computed from the free energies $\Psi_\alpha(\vartheta)$ and $\Psi_\beta(\vartheta)$ via (6.11), so that l is the latent heat. Then the approximate interface conditions consist of an energy balance

$$lV = [\boldsymbol{q}] \cdot \boldsymbol{n} \tag{6.29}$$

and a generalized Stefan condition

$$lu = \{\psi_M(\boldsymbol{n})\mathbf{1} + \partial_{\boldsymbol{n}}\partial_{\boldsymbol{n}}\psi_M(\boldsymbol{n})\} \cdot \mathbf{L} - b_M(\boldsymbol{n})V, \tag{6.30}$$

which includes the effects of curvature and kinetics. For an isotropic material ψ_M and b_M are constants, which I write as ψ and b, and (6.30) reduces to[37]

$$lu = \psi K - bV, \tag{6.31}$$

which is the Gibbs-Thompson condition $lu = \psi K$ augmented by the term bV, which accounts for interface kinetics.

---

[37] $lu = -b(\boldsymbol{n})V$ was introduced by FRANK [1958] and used by CHERNOV [1963, 1964]; $lu = \psi K$ was introduced by MULLINS [1960] (in the context of mass transport) and used by MULLINS & SEKERA [1963, 1964]; $lu = \psi K - bV$ was used by VORONKOV [1964]. Cf. GURTIN [1993a, Footnote 84] for additional references.

Appendix on evolving surfaces

a. Surfaces

Let $\mathscr{S}$ be a smooth closed surface oriented by a choice of unit normal field $\boldsymbol{n}(X)$. The space $\boldsymbol{n}(X)^{\perp}$ of all vectors perpendicular to $\boldsymbol{n}(X)$ is the tangent space to $\mathscr{S}$ at X, and the tensor

$$P(X) = \mathbf{1} - \boldsymbol{n}(X) \otimes \boldsymbol{n}(X) \tag{A1}$$

projects vectors onto this tangent space.

In continuum mechanics tensors are generally linear transformations from $\mathbb{R}^3$ into itself, but of interest here are tensor fields T on $\mathscr{S}$ with the property that, at each X in $\mathscr{S}$, $\mathsf{T}(X)$ is a linear transformation from the tangent space at X into $\mathbb{R}^3$. These two notions of a tensor field may be reconciled by extending $\mathsf{T}(X)$ to vectors normal to $\mathscr{S}$ with the requirement that $\mathsf{T}(X)$ annihilate such vectors. Precisely, a *superficial tensor field* T on $\mathscr{S}$ is a function that associates with each X in $\mathscr{S}$ a linear transformation $\mathsf{T}(X)$ from $\mathbb{R}^3$ into $\mathbb{R}^3$ such that

$$\mathsf{T}\boldsymbol{n} = \mathbf{0}. \tag{A2}$$

T then admits a unique decomposition into *tangential* and *normal components* T_{tan} and $\mathbf{t}$, respectively:

$$\mathsf{T} = \mathsf{T}_{\text{tan}} + \boldsymbol{n} \otimes \mathbf{t}, \quad \mathsf{T}_{\text{tan}} = P\mathsf{T}, \quad \mathbf{t} = \mathsf{T}^T\boldsymbol{n}; \tag{A3}$$

given any vector field $\boldsymbol{v}$,

$$\mathsf{T}\boldsymbol{v} = \mathsf{T}_{\text{tan}}\boldsymbol{v} + (\mathbf{t}\cdot\boldsymbol{v})\boldsymbol{n} \tag{A4}$$

with $\mathsf{T}_{\text{tan}}\boldsymbol{v}$ a tangential vector field on $\mathscr{S}$. (Note that the *normal* component $\mathbf{t}$ of T is a *tangential vector field.*)

I define the surface gradient $\nabla_{\mathscr{S}}$ on $\mathscr{S}$ through the chain rule. Let $\varphi(X)$ be a smooth scalar field on $\mathscr{S}$ and $\boldsymbol{v}(X)$ a smooth vector field on $\mathscr{S}$. Then given any curve $z(t)$ on $\mathscr{S}$,

$$\varphi(z)^{\cdot} = \nabla_{\mathscr{S}}\varphi(z)\cdot z^{\cdot}, \quad \boldsymbol{v}(z)^{\cdot} = \nabla_{\mathscr{S}}\boldsymbol{v}(z)z^{\cdot} \tag{A5}$$

(which defines $\nabla_{\mathscr{S}}\boldsymbol{v}$ only on vectors tangent to $\mathscr{S}$, but, in accord with (A2), $\nabla_{\mathscr{S}}\boldsymbol{v}$ is extended by requiring that $(\nabla_{\mathscr{S}}\boldsymbol{v})\boldsymbol{n} = \mathbf{0}$). Then $\nabla_{\mathscr{S}}\varphi$ is a tangential vector field, while $\nabla_{\mathscr{S}}\boldsymbol{v}$ is a superficial tensor field. The surface divergence of $\boldsymbol{v}$ is defined by

$$\operatorname{div}_{\mathscr{S}}\boldsymbol{v} = \operatorname{tr}(\nabla_{\mathscr{S}}\boldsymbol{v}), \tag{A6}$$

while the surface divergence $\operatorname{div}_{\mathscr{S}}\mathsf{T}$ of a superficial tensor field is defined through the identity

$$\boldsymbol{a}\cdot\operatorname{div}_{\mathscr{S}}\mathsf{T} = \operatorname{div}_{\mathscr{S}}(\mathsf{T}^T\boldsymbol{a}) \tag{A7}$$

for every constant vector $\boldsymbol{a}$.

The *curvature tensor* $\mathbf{L}$ and *total curvature* K (twice the mean curvature) are defined by

$$\mathbf{L} = -\nabla_{\mathscr{S}}\boldsymbol{n}, \quad K = \operatorname{tr}\mathbf{L} = \mathbf{1}\cdot\mathbf{L} = -\operatorname{div}_{\mathscr{S}}\boldsymbol{n}. \tag{A8}$$

As is known, the curvature tensor is symmetric $\mathbf{L} = \mathbf{L}^T$ and (hence) tangential: $\mathbf{L}^T\boldsymbol{n} = \mathbf{0}$.

Let $\mathscr{G}$ denote a smooth subsurface of $\mathscr{S}$, and let $\boldsymbol{\nu}(X)$ denote the outward unit normal to the boundary curve $\partial\mathscr{G}$, so that $\boldsymbol{\nu}(X)$ is *tangent* to $\mathscr{S}$ at each $X \in \partial\mathscr{G}$. The surface divergence theorem then has the form

$$\int_{\partial\mathscr{G}} \boldsymbol{t}\cdot\boldsymbol{\nu}\, ds = \int_{\mathscr{G}} \operatorname{div}_{\mathscr{S}}\boldsymbol{t}\, da, \quad \int_{\partial\mathscr{G}} \mathbf{T}\boldsymbol{\nu}\, ds = \int_{\mathscr{G}} \operatorname{div}_{\mathscr{S}}\mathbf{T}\, da \tag{A9}$$

for $\boldsymbol{t}$ a tangential vector field and $\mathbf{T}$ a superficial tensor field.

b. Smoothly evolving surfaces

Now let $\mathscr{S}(t)$ depend smoothly on the time t. Let φ° denote the *normal time-derivative*[38] of a scalar, vector, or tensor field φ on $\mathscr{S}$. Then

$$\boldsymbol{n}^\circ = -\nabla_{\mathscr{S}}V, \tag{A10}$$

with V the normal velocity of $\mathscr{S}$.

Let $\mathscr{G}(t)$ denote a smoothly evolving subsurface of $\mathscr{S}(t)$ with $\boldsymbol{\nu}(x,t)$ the outward unit normal to $\partial\mathscr{G}(t)$. The motion of the curve $\partial\mathscr{G}(t)$ may be characterized intrinsically by the velocity field

$$V\boldsymbol{n} + V_{\partial\mathscr{G}}\boldsymbol{\nu}, \tag{A11}$$

where $V_{\partial\mathscr{G}}$, the *tangential edge velocity* of $\mathscr{G}$, is the velocity of $\partial\mathscr{G}$ in the direction of the normal $\boldsymbol{\nu}$.

For φ a superficial scalar field,

$$\frac{d}{dt}\int_{\mathscr{G}} \varphi\, da \quad \text{denotes} \quad \frac{d}{dt}\int_{\mathscr{G}(t)} \varphi(X,t)\, da(X). \tag{A12}$$

Then[39]

$$\frac{d}{dt}\int_{\mathscr{G}} \varphi\, da = \int_{\mathscr{G}} (\varphi^\circ - \varphi K V)\, da + \int_{\partial\mathscr{G}} \varphi V_{\partial\mathscr{G}}\, ds. \tag{A13}$$

[38] The derivative following the normal trajectories of the surface. Cf. GURTIN [1986, eq. (4.4)].

[39] Cf. PETRYK and MROZ [1986], GURTIN, STRUTHERS and WILLIAMS [1989], ESTRADA and KANWAL [1991], JARIC [1991].

c. *Functions of orientation*

Constitutive equations appropriate to a phase interface generally involve scalar functions $\varphi(\boldsymbol{n})$ and vector functions $\mathbf{f}(\boldsymbol{n})$ of the interface normal $\boldsymbol{n}$. The derivatives $\partial_{\boldsymbol{n}}\varphi(\boldsymbol{n})$ and $\partial_{\boldsymbol{n}}\mathbf{f}(\boldsymbol{n})$ are defined by the chain rule. Given any curve $\boldsymbol{n}(t)$ on the unit sphere,

$$\varphi(\boldsymbol{n})^{\cdot} = \{\partial_{\boldsymbol{n}}\varphi(\boldsymbol{n})\}\cdot\boldsymbol{n}^{\cdot}, \quad \mathbf{f}(\boldsymbol{n})^{\cdot} = \{\partial_{\boldsymbol{n}}\mathbf{f}(\boldsymbol{n})\}\boldsymbol{n}^{\cdot}; \tag{A14}$$

$\partial_{\boldsymbol{n}}\varphi(\boldsymbol{n})$ is tangent to the unit sphere, while $\partial_{\boldsymbol{n}}\mathbf{f}(\boldsymbol{n})$ is defined by (A14) only on vectors perpendicular to $\boldsymbol{n}$, but is extended by requiring that $\{\partial_{\boldsymbol{n}}\mathbf{f}(\boldsymbol{n})\}\boldsymbol{n} = \mathbf{0}$. Then for $\boldsymbol{n}$ the unit normal field on $\mathscr{S}$, a calculation using the chain rule and (A8) yields the identities

$$\nabla_{\mathscr{S}}\varphi(\boldsymbol{n}) = -\mathbf{L}\partial_{\boldsymbol{n}}\varphi(\boldsymbol{n}), \quad \nabla_{\mathscr{S}}\mathbf{f}(\boldsymbol{n}) = -\{\partial_{\boldsymbol{n}}\mathbf{f}(\boldsymbol{n})\}\mathbf{L}, \quad \operatorname{div}_{\mathscr{S}}\mathbf{f}(\boldsymbol{n}) = -\{\partial_{\boldsymbol{n}}\mathbf{f}(\boldsymbol{n})\}\cdot\mathbf{L}. \tag{A15}$$

Acknowledgement. This work was supported by the Army Research Office and the National Science Foundation. I am deeply grateful to PAOLO PODIO-GUIDUGLI for the many valuable discussions we have had concerning the material presented here; in particular, unpublished notes of his on variational descriptions of the Eshelby tensor were helpful to me in formulating the underlying ideas. I also acknowledge valuable discussions with METE' SONER.

References

[1878] GIBBS, J. W., On the equilibrium of hetrogeneous substances, *Trans. Connecticut Acad.* **3**, 108–248. Reprinted in: *The Scientific Papers of J. Willard Gibbs*, vol. 1, Dover, New York (1961).

[1892] BURTON, C. V., A theory concerning the constitution of matter, *Phil. Mag.* (Series 5) **33**, 191–203.

[1951] ESHELBY, J. D., The force on an elastic singularity, *Phil. Trans.* **A244**, 87–112.

[1951] HERRING, C., Surface tension as a motivation for sintering, *The Physics of Powder Metallurgy* (ed. W. E. KINGSTON), McGraw-Hill, New York.

[1952] BURKE, J. E. & D. TURNBULL, Recrystallization and grain growth, *Progress in Metal Physics* **3**, 220–292.

[1956] MULLINS, W. W., Two-dimensional motion of idealized grain boundaries, *J. Appl. Phys.*, **27**, 900–904.

[1958] FRANK, F. C., On the kinematic theory of crystal growth and dissolution processes, *Growth and Perfection of Crystals* (eds. R. H. DOREMUS, B. W. ROBERTS, D. TURNBULL), John Wiley, New York.

[1960] MULLINS, W. W., Grain boundary grooving by volume diffusion, *Trans. Met. Soc. AIME*, **218**, 354–361.

[1963] CHERNOV, A. A., Crystal growth forms and their kinetic stability [in Russian], *Kristallografiya* **8**, 87–93 (1963). English Transl. *Sov. Phys. Crystall.* **8**, 63–67 (1963).

[1963] COLEMAN, B. D. & W. NOLL, The thermodynamics of elastic materials with heat conduction and viscosity, *Arch. Rational Mech. Anal.* **13**, 167–178.

[1963] MULLINS, W. W. & R. F. SEKERKA, Morphological stability of a particle growing by diffusion or heat flow, *J. Appl. Phys.* **34**, 323–329.

[1964] CHERNOV, A. A., Application of the method of characteristics to the theory of the growth forms of crystals [in Russian], *Kristallografiya* **8**, 499–505. English Transl. *Sov. Phys. Crystall.* **8**, 401–405 (1964).

[1964] MULLINS, W. W. & R. F. SEKERKA, Stability of a planar interface during solidification of a dilute binary alloy, *J. Appl. Phys.* **35**, 444–451.

[1964] VORONKOV, V. V., Conditions for formation of mosaic structure on a crystallization front [in Russian], *Fizika Tverdogo Tela* **6**, 2984–2988. English Transl. *Sov. Phys. Solid State* **6**, 2378–2381 (1965).

[1965] TRUESDELL, C. A. & W. NOLL, The non-linear field theories of mechanics, *Handbuch der Physik*, **III/3** (ed. S. FLÜGGE), Springer-Verlag, Berlin.

[1966] TRUESDELL, C. A., Method and taste in natural philosophy, *Six Lectures on Modern Natural Philosophy*, Springer Verlag, New York.

[1970] ESHELBY, J. D., Energy relations and the energy-momentum tensor in continuum mechanics, *Inelastic Behavior of Solids* (ed. M. KANNINEN, W. ADLER, A. ROSENFIELD & R. JAFFEE), 77–115, McGraw-Hill, New York.

[1970] SHERMAN, B., A general one-phase Stefan problem, *Quart. Appl. Math.* **28**, 377–382 (1970).

[1972] HOFFMAN, D. W. & J. W. CAHN, A vector thermodynamics for anisotropic surfaces. 1. Fundamentals and applications to plane surface junctions, *Surface Sci.* **31**, 368–388.

[1974] CAHN, J. W. & D. W. HOFFMAN, A vector thermodynamics for anisotropic surfaces. 2. Curved and faceted surfaces, *Act. Metall.* **22**, 1205–1214.

[1974] GURTIN, M. E. & I. MURDOCH, A continuum theory of elastic material surfaces, *Arch. Rational Mech. Anal.*, **57**, 291–323.

[1974] ROBIN, P.-Y. F., Thermodynamic equilibrium across a coherent interface in a stressed crystal, *Am. Min.* **59**, 1286–1298.

[1977] FASANO, A. & M. PRIMICERIO, General free boundary problems for the heat equation, *J. Math. Anal. Appl., Part* 1, **57**, 694–723; Part 2, **58**, 202–231; Part 3, **59**, 1–14.

[1978] LARCHÉ, F. C. & J. W. CAHN, Thermochemical equilibrium of multiphase solids under stress, *Act. Metall.* **26**, 1579–1589.

[1980] CAHN, J. W., Surface stress and the chemical equilibrium of small crystals. 1. The case of the isotropic surface, *Act. Metall.* **28**, 1333–1338.

[1981] GRINFEL'D, M. A., On hetrogeneous equilibrium of nonlinear elastic phases and chemical potential tensors, *Lett. Appl. Engng. Sci.* **19**, 1031–1039.

[1981] GURTIN, M. E., *An Introduction to Continuum Mechanics*, Academic Press, New York.

[1981] JAMES, R., Finite deformation by mechanical twinning, *Arch. Rational Mech. Anal.* **77**, 143–176.

[1983] GURTIN, M. E., Two-phase deformations of elastic solids, *Arch. Rational Mech. Anal.* **84**, 1–29.

[1985] ALEXANDER, J. I. D. & W. C. JOHNSON, Thermomechanical equilibrium in solid-fluid systems with curved interfaces, *J. Appl. Phys.* **58**, 816–824.

[1986] GURTIN, M. E., On the two-phase Stefan problem with interfacial energy and entropy, *Arch. Rational Mech. Anal.* **96**, 199–241.

[1986] JOHNSON, W. C. & J. I. D. ALEXANDER, Interfacial conditions for thermomechanical equilibrium in two-phase crystals, *J. Appl. Phys.* **59**, 2735–2746.

[1986] PETRYK, H. & Z. MROZ, Time derivatives of integrals and functionals defined on varying volume and surface domains, *Arch. Mech.* **38**, 697–724.

[1987] TRUSKINOVSKY, L., Dynamics of nonequilibrium phase boundaries in a heat conducting nonlinear elastic medium, *J. Appl. Math. Mech.* (PMM), **51**, 777–784.

[1987] UWAHA, M., Asymptotic growth shapes developed from two-dimensional nuclei, *J. Crystal Growth* **80**, 84–90.

[1988] GURTIN, M. E., Multiphase thermomechanics with interfacial structure. 1. Heat conduction and the capillary balance law, *Arch. Rational Mech. Anal.*, **104**, 185–221.

[1988] KAGANOVA, I. M. & A. L. ROITBURD, Equilibrium between elastically-interacting phases [in Russian], *Zh. Eksp. Teor. Fiz.* **94**, 156–173. English Transl. *Sov. Phys. JETP* **67**, 1173–1183 (1988).

[1989] ANGENENT, S. & M. E. GURTIN, Multiphase thermomechanics with interfacial structure. 2. Evolution of an isothermal interface, *Arch. Rational Mech. Anal.* **108**, 323–391.

[1989] FONSECA, I., Interfacial energy and the Maxwell rule, *Arch. Rational Mech. Anal.*, **106**, 63–95.

[1989] GURTIN, M. E., A. STRUTHERS & W. O. WILLIAMS, A transport theorem for moving interfaces, *Quart. Appl. Math.* **47**, 773–777.

[1989] LEO, P. H. & R. F. SEKERKA, The effect of surface stress on crystal-melt and crystal-crystal equilibrium, *Act. Metall.* **37**, 3119–3138.

[1989] NOZIERES, P., *Shape and Growth of Crystals*, Notes of lectures given at the Beg-Rohu summer school.

[1990] ABEYARATNE, R. & J. K. KNOWLES, On the driving traction acting on a surface of strain discontinuity in a continuum, *J. Mech. Phys. Solids*, **38**, 345–360.

[1990] GURTIN, M. E. & A. STRUTHERS, Multiphase thermomechanics with interfacial structure. 3. Evolving phase boundaries in the presence of bulk deformation, *Arch. Rational Mech. Anal.* **112**, 97–160.

[1991] ABEYARATNE, R. & J. K. KNOWLES, Kinetic relations and the propagation of phase boundaries in elastic solids, *Arch. Rational Mech. Anal.*, **114**, 119–154.

[1991] ANGENENT, S. B., Parabolic equations for curves on surfaces, *Ann. Math.* (2) **132**, 451–483 (1990); **133**, 171–215.

[1991] CHEN, Y.-G., Y. GIGA & S. GOTO, Uniqueness and existence of viscosity solutions of generalized mean curvature flow equations, *J. Diff. Geom.* **33**, 749–786.

[1991] ESTRADA, R. & R. P. KANWAL, Non-classical derivation of the transport theorems for wave fronts, *J. Math. Anal. Appl.*, **159**, 290–297.

[1991] GURTIN, M. E., On thermodynamical laws for the motion of a phase interface, *Zeit. angew. Math. Phys.*, **42**, 370–388.

[1991] JARIC, J. P., On a transport theorem for moving interfaces, preprint.

[1991] TRUSKINOVSKY, L., Kinks versus shocks, in *Shock Induced Transitions and Phase Structures in General Media* (ed. R. FOSDICK, E. DUNN & M. SLEMROD), Springer-Verlag, Berlin (1991).

[1992] LUSK, M., Martensitic phase transitions with surface effects, Tech. Rept. 8, Div. Eng. Appl. Sci., Caltech.

[1992] TAYLOR, J., Mean curvature and weighted mean curvature, *Act. Metall.* **40**, 1475–1485.

[1992] TAYLOR, J., J. W. CAHN & C. A. HANDWERKER, Geometric models of crystal growth, *Act. Metall.* **40**, 1443–1474.

[1993] GÖTZ, I. G. & B. ZALTZMAN, Two-phase Stefan problem with supercooling, preprint.

[1993a] GURTIN, M. E., *Thermomechanics of Evolving Phase Boundaries in the Plane*, Oxford University Press, Oxford.

[1993b] GURTIN, M. E., The dynamics of solid-solid phase transitions. 1. Coherent interfaces, *Arch. Rational Mech. Anal.*, **123**, 305–335.

[1993] SONER, H. M., Motion of a set by the curvature of its boundary, *J. Diff. Eqs.* **101**, 313–372.

[1994] CERMELLI, P. & M. E. GURTIN, The dynamics of solid-solid phase transitions. 2. Incoherent interfaces, *Arch. Rational Mech. Anal.*, **127**, 41–99.

[1994a] GURTIN, M. E., The characterization of configurational forces, Addendum to GURTIN [1993b], *Arch. Rational Mech. Anal.*, **126**, 387–394.

[1994b] GURTIN, M. E., Thermodynamics and the supercritical Stefan equations with nucleations, *Quart. Appl. Math.* **52**, 133–155.

Department of Mathematics
Carnegie Mellon University
Pittsburgh, Pennsylvania 15213

(Accepted August 11, 1994)

III

Papers on Mathematics

Euro. Jnl of Applied Mathematics (1990), *vol.* 1, *pp.* 101–111

Solutions for the two-phase Stefan problem with the Gibbs–Thomson Law for the melting temperature

STEPHAN LUCKHAUS

Institut für Angewandte Mathematik, Universität Bonn, Wegelerstrasse 6, 5300 Bonn, West Germany

(*Received 2 February 1990*)

The coupling of the Stefan equation for the heat flow with the Gibbs–Thomson law relating the melting temperature to the mean curvature of the phase interface is considered. Solutions, global in time, are constructed which satisfy the natural *a priori* estimates. Mathematically the main difficulty is to prove a certain regularity in time for the temperature and the indicator function of the phase separately. A capacity type estimate is used to give an L_1 bound for fractional time derivatives.

1 Introduction

The equations considered couple the Stefan problem and the Gibbs–Thomson law on the free boundary; the latter says that the melting temperature of a solid at a curved phase boundary minus the melting temperature for a flat interface is proportional to the mean curvature.

Let $\Omega_t \subset \Omega$ be the region in a domain $\Omega \subset \mathbb{R}^3$ occupied by the liquid and set

$$\chi(t,x) = \begin{cases} 1 & \text{if} \quad x \in \Omega_t \\ -1 & \text{if} \quad x \notin \Omega_t \end{cases}.$$

Here u is the temperature of the material, and f a volume heat source. Assume for convenience the latent heat to be two, the specific heat and the thermal conductivity to be one; then the Stefan problem in its weak formulation is

$$\partial_t(u+\chi) - \Delta u = f \quad \text{in} \quad H_2^{-1}(\Omega) \tag{1}$$

(this is the continuity equation for the energy). If one chooses the origin such that the melting temperature for flat interfaces is 0 then the Gibbs–Thomson law reads: for almost all t there exists $\rho > 0$ such that for $x \in \partial\Omega_t \cap \Omega$, $\partial\Omega_t$ is a graph in $B_\rho(x) = \{y \mid |x-y| < \rho\}$, i.e. after a change of coordinates there exists g such that

$$\Omega_t \cap B_\rho(x) = \{(x', x_n) \mid x_n < g(x')\} \cap B_\rho(x)$$

and g and u satisfy:

$$-\alpha\nabla\left(\frac{\nabla g}{\sqrt{(1+|\nabla g|^2)}}\right)(x') = u(x', g(x')). \tag{2}$$

Boundary conditions have to be specified for (1) and (2). This will be done later, for the case that Ω is a container filled with the melting material. The aim of this article is to present a family of solutions of (1) and (2) for appropriate initial and boundary conditions, to discuss the high degree of nonuniqueness encountered, and to indicate a possible way of combining the solutions constructed here with the phenomenon of nucleation. The ultimate aim would be to find solutions which will converge as $\alpha \to 0$ to the unique weak solution of the Stefan problem without surface tension.

One of the main interests in studying the Gibbs–Thomson law for the melting temperature lies in the fact that the interface is 'stabilized', see e.g. the stability analysis in Chadam & Ortoleva (1983). Thus one hopes to avoid mushy regions with this law, whereas mushy regions will in general appear when the melting temperature is zero. Of course, in special cases (and in particular if $f = 0$) there exists a sharp interface of the ordinary Stefan problem, see Berger & Rogers (1984), Ei Ichi Hanzawa (1981) and Meirmanov (1980). The solutions constructed here for the system (1), (2) will always have a sharp interface, in contrast to the weak solutions previously constructed by Visintin, see Visintin (1984, 1988a, b). The mathematical tools which are used to get this result are regularity theory for minimal surfaces (see Almgren 1976, Giusti 1984) and a new estimate for a fractional time derivative of the temperature (see Lemmas 1 and 2 below).

2 Solutions of the problem with a 'natural' Lyapunov functional, and their properties

If one formally considers (1) and (2) in the case when, for example, $\partial\Omega_t$ is a C^2 interface moving with finite speed, and multiplies (1) by $u-u^0$, where u^0 is some prolongation of the boundary data, integrates over Ω and uses the relationship (2) to calculate $\int_\Omega \partial_t \chi u$ with the help of a partition of unity, then one arrives at

$$\partial_t \int\left(\frac{u^2}{2} - uu^0 - \chi u^0\right) + \int (u+\chi)\,\partial_t u^0 + \int \partial_t \chi u = -\int |\nabla u|^2 + \int \nabla u \nabla u^0 + \int f(u-u^0).$$

Also, specifying the boundary condition for g as $\partial_\nu g = 0$ on $\partial\Omega$

$$\int \partial_t \chi u = \alpha\,\partial_t \int |\nabla\chi| \left(\int |\nabla\chi| = 2.\ \text{Area}\ (\partial\Omega_t \cap \Omega)\right),$$

this boundary condition means that the surface tension between the material and the container wall is the same for the solid and the liquid.

Combining these one gets

$$\partial_t \int_\Omega \left[\frac{u^2}{2} + \alpha|\nabla\chi| - u^0(u+\chi)\right] \leqslant -\int_\Omega |\nabla u|^2 + \int_\Omega \nabla u \nabla u^0 + \int_\Omega f(u-u^0) - \int (u+\chi)\,\partial_t u^0.$$

The Lyapunov function which is used here,

$$\int_\Omega \frac{u^2}{2} + \alpha|\nabla\chi|,$$

is the same that Visintin and Gurtin introduced for this system.

Definition Let $Q = \Omega \times (0,T)$, $u^0 \in H_2^1(Q)$, $f \in L_2(Q)$, $e_0 \in L_2(Q)$. Suppose $u - u^0 \in L_2(0,T, H_2^1(\Omega))$, and Ω_t is such that, for almost all t, there exists $\rho = \rho(t)$ such that $\Omega_t \cap B_\rho(x) = \{(x',$

$x_n)|x_n < g_x(x')\} \cap B_\rho(x)$ for all x after a suitable rotation with $g_x \in C^1$. Suppose further that u, χ satisfy (1) in H^{-1}, and (2) for almost all t. Finally suppose that $(u-\chi)(t,.) \to_{t\to 0} e_0$ is in H^{-1}, and that for $(x', x_n) \in \partial\Omega$, and ν normal to $\partial\Omega$ in (x', x_n), $\nu.(\nabla g_x(x'), -1) = 0$.

Then if (3) holds for (u, χ) in an L_1 sense, the pair (u, χ) will be called a Lyapunov solution for the initial boundary value problem for (1) and (2).

Before going to the existence proof for Lyapunov solutions, we point out a few properties. They are most easily seen in a 'Gedankenexperiment' with Neumann boundary data (not treated here), but the discussion for suitable Dirichlet data is similar.

We start with $e_0 = u_0 + \chi_0$ where $u_0 = \text{const} < 0$, $\chi_0 \equiv -1$, and let $f \equiv a > 0$. Then we get a family of solutions for $t > 0$:

$$u = \begin{cases} u_0 + at & \text{for } t < t_0 \\ u_0 + at - 2 & \text{for } t > t_0 \end{cases}$$

$$\chi = \begin{cases} -1 & \text{for } t < t_0 \\ +1 & \text{for } t > t_0. \end{cases}$$

Only those solutions for which $at_0 + u_0 \geqslant 1$ are Lyapunov solutions, because for $at_0 + u_0 < 1$ the quantity $\int u^2/2$ increases by a positive jump at t_0. In other words the earliest time at which the phase can change is when u reaches half the latent heat. In particular, the Lyapunov solutions will not converge to the unique (see Brezis & Crandall 1979) solution of the Stefan problem as $\alpha \to 0$, i.e. to:

$$u = \begin{cases} u_0 + at & \text{for} \quad u_0 + at \leqslant 0 \\ 0 & \text{for} \quad 0 \leqslant u_0 + at \leqslant 2 \\ u_0 + at - 2 & \text{for} \quad u_0 + at \geqslant 2 \end{cases}$$

$$\chi = \begin{cases} -1 & \text{for} \quad u_0 + at \leqslant 0 \\ u_0 + at & \text{for} \quad 0 \leqslant u_0 + at \leqslant 2 \\ 1 & \text{for} \quad u_0 + at \geqslant 2. \end{cases}$$

On the other hand there cannot be a deterministic law treating melting and solidification symmetrically which makes the phase change by a jump in time before undercooling or overheating has reached half the latent heat, because by the continuity of the energy such a jump would take the state from one of critical undercooling to one of critical overheating and vice versa, so it would have to be reversed immediately.

A possible way out of the dilemma is the introduction of nucleation. The question as to how this might fit in with the Lyapunov solutions constructed here, is briefly and unfortunately only vaguely addressed in the concluding remarks.

3 The approximating Lyapunov solutions

The Lyapunov solutions are approximated by time discretization, that is, (1) is discretized implicitly in time and following an idea of A. Visintin the resulting equation is then used

5-2

to express u in terms of χ; χ is then determined as the solution of an appropriate minimization problem. The discretized problem is as follows:

$$\frac{u_h(t)-u_h(t-h)}{h}+\frac{\chi_h(t)-\chi_h(t-h)}{h}-\Delta u_h(t)=f(t) \quad (t=ih) \tag{1h}$$

$$u_h(t)-u^0(t)\in \mathring{H}^1_2(\Omega), \quad (u_h+\chi_h)(0)=e_0.$$

We define the operators K_h and K_h^0 by

$$K_h^0(v)=w \Leftrightarrow w-h\Delta w=v \quad \text{and} \quad w-u^0\in \mathring{H}^1_2(\Omega)$$

$$K_h(v)=w \Leftrightarrow w-h\Delta w=v \quad \text{and} \quad w\in \mathring{H}^1_2(\Omega).$$

Writing $e_h=u_h+\chi_h$, we have the equivalent formulation for $(1h)$:

$$u_h(t)=-K_h(\chi_h(t))+K_h^0(e_h(t-h)+hf(t)).$$

Define the functional $F_{A,h}(t,\chi)$ by

$$F_{A,h}(t,\chi)=\alpha\int_\Omega|\nabla\chi|+\frac{1}{2}\int_\Omega \chi K_h(\chi)-\frac{1}{2}\int_\Omega(-K_h(\chi)+K_h^0(e_h(t-h)$$

$$+hf(t))-u_h(t-h))^2+(\tfrac{1}{2}+A)\int_\Omega(\chi-\chi_h(t-h))\,K_h(\chi-\chi_h(t-h))$$

$$-\int_\Omega K_h^0(e_h(t-h)+hf(t))\,\chi.$$

Then $\chi_h(t)$ will be determined by $(2_{A,h})$ if one specifies $\chi_h(-h)=\chi_0$

$$F_{A,h}(t,\chi_h(t))=\min\{F_{A,h}(t,\chi)\,|\,\chi:\Omega\to\{1,-1\}\}. \tag{$2_{A,h}$}$$

Equation $(2_{A,h})$ can be solved since F is compact and bounded below for $u_h(t-h)\in L_2$. First let us derive the 'energy' estimate for u_h, χ_h, i.e. the equivalent of (3). Multiplying $(1h)$ by u_h-u^0 one gets

$$\frac{u_h^2(t)-u_h^2(t-h)}{2h}+\frac{|u_h(t)-u_h(t-h)|^2}{2h}-\frac{u^0(t)\,e_h(t)-u^0(t-h)\,e_h(t-h)}{h}$$

$$+\frac{\chi_h(t)-\chi_h(t-h)}{h}u_h(t)=\Delta u_h(t)(u_h(t)-u^0(t))+f(t)(u_h(t)$$

$$-u^0(t))-\frac{u^0(t)-u^0(t-h)}{h}e_h(t-h),$$

which can be abbreviated to

$$\partial_t^h\left(\frac{u_h^2}{2}-e_h u^0\right)+\partial_t^h\chi u_h(t)+\frac{h}{2}|\partial_t^h u|^2=\Delta u_h(u_h-u^0)+f(u_h-u^0)-\partial_t^h u^0 e_h.$$

On the other hand,

$$F_{A,h}(t,\chi_h(t))\leqslant F_{A,h}(t,\chi_h(t-h)),$$

i.e.

$$\alpha\int(|\nabla\chi_h(t)|-(\nabla\chi_h(t-h)))+\frac{1}{2}\int(\chi_h(t)\,K_h(\chi_h(t))-\chi_h(t-h)\,K_h(\chi_h(t-h)))$$

$$-\int K_h^0(e_h(t-h)+hf(t))(\chi_h(t)-\chi_h(t-h))$$

$$-\frac{1}{2}\int(u_h(t)-u_h(t-h))^2+(\tfrac{1}{2}+\Lambda)\int K_h(\chi_h(t)-\chi_h(t-h))(\chi_h(t)-\chi_h(t-h))$$

$$\leqslant 0.$$

The left-hand side is equal to

$$\alpha\int(|\nabla\chi_h(t)|-|\nabla\chi_h(t-h)|)+\int(K_h^0(e_h(t-h)+hf(t))$$

$$-(1+\Lambda)\,K_h(\chi_h(t)-\chi_h(t-h))).(\chi_h(t)-\chi_h(t-h))-\frac{1}{2}\int(uh\,u_h(t)(t)-u_h(t-h))^2$$

$$\geqslant\alpha\int(|\nabla\chi_h(t)|-|\nabla\chi_h(t-h)|)-\int(u_h(t)(\chi_h(t)-\chi_h(t-h))$$

$$-\frac{1}{2}\int(u_h(t)-u_h(t-h))^2.$$

Altogether this gives

$$\partial_t^h\int\left(\frac{u_h^2}{2}+\alpha|\nabla\chi_h|-u^0e_h\right)\leqslant-\int|\nabla u_h|^2+\int\nabla u_h\,\nabla u^0+\int f(u_h-u^0)-\int\partial_t^h\,u^0e_h(t-h). \quad (3h)$$

This leads to the estimate

$$\iint|\nabla u_h|^2+\sup_{0,T}\int\left(\frac{u_h^2}{2}+|\nabla\chi_h|\right)\leqslant\text{const.} \quad (4)$$

Next let us calculate the mean curvature of $\partial\Omega_{t,h}$, where $\Omega_{t,h}=\{x\,|\,\chi_h(t,x)>0\}$. For all comparison functions $\tilde{\chi}$, one has from $(2_{\Lambda,h})$, the binomial formula, and the pointwise estimate

$$K_h(\chi_h(t)+\tilde{\chi})(\chi_h(t)-\tilde{\chi})\geqslant-2|\chi_h(t)-\tilde{\chi}|=-2\chi_h(t)(\chi_h(t)-\tilde{\chi}),$$

the following integral estimate:

$$\alpha\int(|\nabla\chi_h(t)-|\nabla\tilde{\chi}|)-(2+2\Lambda)\int(\chi_h(t)+K_h(\chi_h(t-h)))(\chi_h(t)-\tilde{\chi})$$

$$+\int K_h(u_h(t)-u_h(t-h))(\chi_h(t)-\tilde{\chi})-\int K_h^0(u_h(t-h)+hf)(\chi_h(t)-\tilde{\chi})\leqslant 0.$$

Thus $\chi_h(t)$ solves the minimum problem for a functional of the form $G(\tilde{\chi}) = \int(|\nabla\tilde{\chi}| + \gamma\tilde{\chi})$ among $\tilde{\chi}: \Omega \to \{-1, 1\}$ where

$$|\gamma|_{L_{2n/(n-2)}} \leqslant \frac{c}{\alpha}(|u_h(t)|_{H_2^1} + |u_h(t-h)|_{H_2^1} + (\Lambda+1)).$$

For $n \leqslant 3$, the exponent $2n/(n-2)$ is larger than n. In this case, regularity theory for minimal surfaces (Giusti 1984), and quasi-minima for the area functional (Almgren 1976), give the existence of a ρ depending on Λ, α and the H_2^1 norm of u_h such that after a suitable rotation

$$\Omega_{t,h} \cap B_\rho(x) = \{(x', x_n) \mid g_{x,h}(x') > x_n\} \cap B_\rho(x)$$

and since $\frac{1}{4} < \{1 - n[2n/(n-2)]^{-1}\}$ for $n \leqslant 3$,

$$|g_{x,h}|_{C^{1,\frac{1}{4}}} < c = c\left(\Lambda, \frac{1}{\alpha}, |u_h(t-h)|_{H_2^1} + |u_h(t)|_{H_2^1}\right).$$

Using this regularity it is easy to calculate the first variation of F, which in local coordinates is

$$-\nabla.\left(\frac{\nabla g_{h,x}}{\surd(1+|\nabla g_{h,x}|^2)}\right)(x') = \frac{1}{\alpha}(K_h^0(u_h(t-h) + hf(t)) - K_h(u_h(t) - u_h(t-h))$$

$$+ (2+2\Lambda)(u_h(t) - K_h^0(u_h(t-h) + hf(t)))(x', g_{h,x}(x')). \quad (2h)'$$

The main point in proving the convergence of this scheme to a solution (1)–(3) is then to derive a 'continuity' estimate in time (it will in fact be an $H_1^{\frac{1}{2}}$ estimate) for u_h and χ_h separately from the energy estimate (4). That will give compactness for u_h and χ_h and it will also allow us to prove that for almost all t the right-hand side of $(2h)'$ converges to $u(t)$. This will be done in the next section.

4 A capacity-type estimate and the compactness of the sequence of approximating solutions

For convenience from now on we drop the h from the notation for the approximations. If one assumes that all the times occurring are multiples of h, there is no difference in the formulae. Note that equation (1) and estimate (4) give together

$$\sup_{0,T}\left(\int\frac{u^2}{2} + \frac{\alpha}{2}\int|\nabla\chi|\right) + \int_0^T\int_\Omega|\nabla u|^2 + \int_0^T|\partial_t(u+\chi)|_{H^{-1}}^2 < c. \quad (4)'$$

There is a difficulty in obtaining an estimate for a fractional time derivative of u and χ separately. The basic idea is that since χ has only discrete values and u is in H_2^1 for almost all times, jumps of u and χ cannot compensate each other, so, from the continuity of $u+\chi$ in H^{-1}, one should get estimates on $|\chi(t) - \chi(t-\tau)|_{L^1}$ and $|u(t) - u(t-\tau)|_{L^1}$ separately.

To make this idea work one has to convert it into a problem for *a priori* estimates. The following estimate is a kind of capacity estimate, giving a lower bound for the Dirichlet integral.

Lemma 1 *Let* $w \in H_2^1(\mathbb{R}^n)$, $\varphi: \mathbb{R}^n \to \{-2, 0, 2\}$, $n > 1$. *Then the estimate*

$$\int_{\mathbb{R}^n} |\varphi| \leqslant c\left(\int_{\mathbb{R}^n} |w+\varphi| + \left(\int_{\mathbb{R}^n} |\nabla w|^2\right)^{n/(2n-2)}\left(\int_{\mathbb{R}^n} |w+\varphi|\right)^{n/(2n-2)}\right)$$

holds with $c = c_n$.

Proof Let $a \wedge b = \min(a, b)$; then, by Sobolev, one has

$$\left(\int |(w-s_1)_+ \wedge s_2|^{n/(n-1)}\right)^{(n-1)/n} \leqslant c\int_{s_1 < w < s_2 + s_1} |\nabla w| \leqslant c|\{s_1 < w < s_1 + s_2\}|^{\frac{1}{2}} \int |\nabla w|^2|^{\frac{1}{2}};$$

with $s_1 = \frac{1}{2}$, $s_2 = 1$, and changing constants without changing notation,

$$\tfrac{1}{2}|\{w > \tfrac{3}{2}\}|^{(2n-2)/n} < c|\{\tfrac{1}{2} < w < \tfrac{3}{2}\}| \int |\nabla w|^2.$$

Using the same formula for $-w$ one arrives at

$$|\{|w| > \tfrac{3}{2}\}|^{(2n-2)/n} < c|\{\tfrac{1}{2} < |w| < \tfrac{3}{2}\}| \int |\nabla w|^2.$$

Now $|w| < \frac{3}{2}$, $\varphi \neq 0$ implies $|w+\varphi| > \frac{1}{2}$ and so one has

$$\frac{1}{2}\int |\varphi| < |\{|w| > \tfrac{3}{2}\}| + 2\int |w+\varphi|, \int |\varphi| < 4\int |w+\varphi| + c\left(|\{\tfrac{1}{2} < |w| < \tfrac{3}{2}\}| \int |\nabla w|^2\right)^{n/(2n-2)}.$$

But

$$|\{\tfrac{1}{2} < |w| < \tfrac{3}{2}\}| . \tfrac{1}{2} < \int |w+\varphi|$$

and so

$$\int |\varphi| < 4\int |w+\varphi| + c\left(\int |w+\varphi| \int |\nabla w|^2\right)^{n/(2n-2)}.$$

This lemma will be applied to $w = u(t) - u(t-\tau) - u^0(t) + u^0(t-\tau)$, $\varphi = \chi(t) - \chi(t-\tau)$. From (4)′, the fact that $\partial_t u^0 \in L^2$ and the Sobolev imbedding,

$$|w+\varphi|_{H_2^{-1}} < c\tau^{\frac{1}{2}}.$$

To get the L_1 estimate for $w+\varphi$ one uses standard interpolation. Define

$$\psi_\rho(x) = \rho^{-n}\psi\left(\frac{x}{\rho}\right) \quad \text{where } 0 \leqslant \psi \in L_0^\infty(\mathbb{R}^n), \quad \int \psi = 1.$$

Suppose $\partial\Omega$ is compact and Lipschitz. Then

$$\int |w+\varphi| \leqslant \int |(w+\varphi) * \psi_\rho - (w+\varphi)| + \int |(w+\varphi) * \psi_\rho|$$

$$\leqslant \left(\int_\Omega |\nabla w| + \int_\Omega |\nabla \varphi| + |\partial\Omega|\right). c\rho + |w+\varphi|_{H^{-1}(\Omega)} c|\Omega| \rho^{-1} + \int |w+\varphi| . \chi\{d(x, C\Omega) < \rho\}$$

$$\leqslant c_\Omega\left(1 + \sqrt{\left(\int |\nabla w|^2\right)} + \int |\nabla \varphi|\right)\rho + c_\Omega |w+\varphi|_{H^{-1}} \rho^{-1}.$$

Choosing $\rho = \tau^{\frac{1}{4}}(\int|\nabla w|^2)^{-\frac{1}{4}}$,

$$\int_\Omega |w+\varphi| \leqslant c_\Omega\left(1+\int|\nabla\varphi|+\tau^{-\frac{1}{2}}|w+\varphi|_{H^{-1}}\right)\left(\int|\nabla w|^2\right)^{\frac{1}{4}}\tau^{\frac{1}{4}}. \tag{6}$$

This estimate together with Lemma 1 gives the estimate of the fractional time derivatives needed.

Lemma 2 *Let* $\Omega \subset\subset \mathbb{R}^n$, $\partial\Omega \in C^{0,1}$,

$$u^0 \in H_2^1(Q), u \in L^\infty(0,\tau,L^2), u-u^0 \in L^2(0,T,\mathring{H}_2^1(\Omega)),$$

$$\chi \in L^\infty(0,T,BV(\Omega)) \quad \text{with} \quad \chi:\Omega\to\{-1,1\},$$

and

$$\partial_t(u+\chi) \in L^2(0,T,H^{-1}(\Omega));$$

then one has, with a constant c depending on all the norms above,

$$\int_\tau^T |\chi(t)-\chi(t-\tau)| < c\tau^{\delta_n}, \quad \frac{1}{\delta_n} = 13-\frac{8}{n}$$

$$\int_\tau^T |u(t)-u(t-\tau)| < c\tau^{\delta_n}.$$

Proof Set

$$w = (u(t)-u^0(t)-u(t-\tau)+u^0(t-\tau)),$$

$$\varphi = \chi(t)-\chi(t-\tau),$$

$$M_k = \left\{t \middle| \int_\Omega |\nabla w|^2(t) > K\right\}.$$

Then, in CM_k, by (6),

$$\int_\Omega |w+\varphi|(t) < c(K\tau)^{\frac{1}{4}}.$$

Also, by Lemma 1,

$$\int_\Omega |\varphi| < c((K\tau)^{\frac{1}{4}}+(K.(K\tau)^{\frac{1}{4}})^{n/(2n-2)})$$

$$\int_0^T\int_\Omega |\varphi| < c(|M_k|+(K\tau)^{\frac{1}{4}}+(K.(K\tau)^{\frac{1}{4}})^{n/(2n-2)})$$
$$< c\left(\frac{1}{K}+(K\tau)^{\frac{1}{4}}+(K.(K\tau)^{\frac{1}{4}})^{n/(2n-2)}\right).$$

Choosing $1/K = \tau^{\delta_n} = (K(K\tau)^{\frac{1}{4}})^{n/(2n-2)}$,

$$\iint|\varphi| < c\tau^{\delta_n}$$

and

$$\iint|u(t)-u(t-\tau)| < \iint|\varphi|+\iint|w+\varphi|+\iint|u^0(t)-u^0(t-\tau)| < c\tau^{\delta_n}+c\tau^{\frac{1}{5}}+c\tau$$

by (6). By Lemma 2 and estimate (6) one has compactness of u_h, χ_h in $L_1(Q)$ and one can assume convergence almost everywhere, to u, χ, solutions of equation (1). The next step is to obtain uniform estimates of g_h using the following. If $u_h \to u$ in measure and

$$\int_0^T \left(\int_\Omega |u_h|^{\bar{q}} \right)^{\bar{p}/\bar{q}} < \text{const.},$$

then

$$\int_0^T \left(\int_\Omega |u_h - u|^q \right)^{p/q} \to 0 \quad \text{for any} \quad p < \bar{p}, q < \bar{q}.$$

Proof Let $M_\epsilon = \{t \mid |\{|u_h - u|(t) > \epsilon\}| > \epsilon\}$. Then

$$\int_0^T \left(\int_\Omega |u_h - u|^q \right)^{p/q} < |M_\epsilon|^{1-(p/\bar{p})} 2 \sup_h \left(\int_0^T \left(\int_\Omega |u_h|^{\bar{q}} \right)^{\bar{p}/\bar{q}} \right)^{p/\bar{p}}$$

$$+ \epsilon^{(p/q)(p/\bar{q})} . 2 \sup_h \left(\int_0^T \left(\int |u_h|^{\bar{q}} \right)^{p/q} \right) + T\epsilon.$$

But $|M_\epsilon| \to_{h \to 0} 0$ and ϵ is arbitrary. Since $u \in L^2(0, T, H_2^1)$ one may use this with $\bar{p} = 2$, $\bar{q} = 2n/(n-2)$ to get, after selection of a subsequence,

Outside a set of t of measure zero for arbitrary

$$q < \frac{2n}{n-2}, \int |u_h(t)|^q \leqslant c(t, q) < \infty. \tag{7}$$

From (7) one gets that if $n \leqslant 3$, $\frac{1}{4} + \epsilon < 1 - (n/q) < (4-n)/2$,

$$|g_h|_{c^{1,\frac{1}{4}+\epsilon}(B_\rho(x))} \leqslant c(t) < \infty, \quad \rho \geqslant \rho(t) > 0,$$

so one may assume that for almost all t

$$|g_h - g|_{c^{1,\frac{1}{4}}(B_\rho(x))} \to 0, \quad \rho \geqslant \rho(t). \tag{8}$$

It remains to show convergence of the right-hand side of $(2h)'$. From the above and the trace theorem for H_2^1 functions one first gets, for $w \in L^2(0, T, H_2^1(\Omega))$,

$$\int_\Omega |w(t, x', g_h(t, x')) - w(t, x', g(t, x'))|^2 \leqslant \omega(t, h) \int |\nabla w(t)|^2 \tag{9}$$

with $\lim_{h \to 0} \omega(t, h) = 0$ for almost all t.

Now

$$\int |w(t, x', g(t, x')|^2 \leqslant c(t) \left(\delta \int_\Omega |\nabla w|^2 + \frac{1}{\delta} \int_\Omega w^2 \right)$$

$$\leqslant \tilde{c}(t) \left(\delta \int |\nabla w|^2 + \frac{1}{\delta} \left(\int |w| \right)^{4/(n+2)} \left(\int |w|^{2n/(n-2)} \right)^{(n-2)/(n+2)} \right)$$

$$\leqslant \bar{c}(t) \left(\delta \int |\nabla w|^2 + \frac{1}{\delta} \left(\int |w| \right)^{4/(n+2)} \left(\int |\nabla w|^2 \right)^{n/(n+2)} \right).$$

Choosing δ suitably one gets, with $c(t) = c|g|_{C^{0,1}}$,

$$\int |w(t,x',g(t,x')|^2 \leq c(t)\left(\int_\Omega |w|\right)^{2/(n+2)}\left(\int_\Omega |\nabla w|^2\right)^{(n+1)/(n+2)}. \tag{10}$$

Now choose a subset $\bar{M}_k$ of $(0,T)$ such that $|\bar{M}_k| \to_{k\to\infty} 0$.

$$c(t) < k, \quad \rho(t) > 1/k, \quad \text{for} \quad t \in \bar{M}_k,$$

$$\sup_{\bar{M}_k} \int_\Omega (h^2 f^2(t) + |u_h(t) - u(t)| + |u_h(t) - K_h^0(u_h(t-h))|) \underset{n\to 0}{\to} 0.$$

Then by (9) and (10)

$$(K_h^0(u_h(t-h) + hf(t)) + (2+2\Lambda)(u_h(t) - K_h^0(u_h(t-h) + hf))$$

$$- K_h(u_h(t) - u_h(t-h)))(.,g_h(.)) \to u(.,g(.)) \text{ in } L^2 \text{ for almost all } t \in M_k.$$

This together with (8) proves that (2) holds for almost all t, which means that u_h, χ_h converge to a Lyapunov solution of (1) and (2). The following theorem is thus proved.

Theorem *Suppose $u^0 \in H_2^1(Q)$, $Q = (0,T) \times \Omega$, $\Omega \subset\subset \mathbb{R}^3$ with Lipschitz boundary. Suppose $e_0 \in L_2(\Omega)$, $\chi_0: \Omega \to \{-1, 1\}$; then there exists a solution u, χ of equations* (1) *and* (2) *satisfying the 'energy' estimate* (3), *that is the limit, after selection of a subsequence, of the approximating solutions constructed by* (1_h), ($2_{\Lambda,h}$). *Here u has the boundary data u^0, $u+\chi$ has the initial data e_0.*

5 Nonuniqueness, the role of the penalization parameter Λ for the solution and the construction of non-Lyapunov solutions

The solutions constructed above are not unique: in general they differ, depending on the parameter Λ in ($2h$). It is easiest to see this in the example, again for simplicity with Neumann conditions, treated in §2. This is easy to calculate, as there is no space dependence. The solutions u, χ to which χ_h, u_h converge are then the solutions given in §2 with t_0 given by $(u_0 + at_0) = (1+2\Lambda)$, that is the first time when, in the limit $h \to 0$, $\chi \equiv -1$ will not be the absolute minimum for

$$-(u_0 + at_0) + (\tfrac{1}{2} + \Lambda)\int (\chi+1) K_h(\chi+1) + \frac{\alpha}{2}\int_\Omega |\nabla\chi|.$$

Thus one sees that the solutions differ for different Λ.

Another example is the melting of an ice block from two different point sources. The solutions will have two growing smooth regions of water in the ice. Without the Gibbs–Thomson law, these would grow until they touch each other, then forming the typical cusp singularity of the free boundary. The solutions constructed here do not form this cusp singularity. In order to get the two regions to touch, one would have to take Λ to infinity. The solutions constructed here will jump, forming a channel through the thin layer of ice separating the water when they have come close enough, since this will give a smaller value for $F_{\Lambda,h}(x)$.

For technical reasons, i.e. loss of sufficient compactness for g_h to ensure that the mean curvature equation is preserved in the limit it unfortunately does not seem to be possible to take Λ to infinity.

Note on the other hand that Λ penalizes only jumps; it does not seem to influence the smooth motion as the term

$$\int (\chi(t)-\chi(t-h))\, K_h(\chi(t)-\chi(t-h)) \text{ is of order } h^{\frac{3}{2}}$$

for a smooth boundary moving with finite speed.

Finally let us point out a way of constructing new solutions starting from the construction above. Suppose one lets the solution evolve between times t_i and t_{i+1}, and at the times t_{i+1}, where u, χ, have attained values $u^-(t_{i+1})$, $\chi^-(t_{i+1})$ one introduces arbitrary jumps from $\chi^-(t_{i+1})$ to $\chi^+(t_{i+1})$ which are compensated by defining

$$u^+(t_{i+1}) = u^-(t_{i+1}) + \chi^-(t_{i+1}) - \chi^+(t_{i+1}).$$

Then one lets the equation evolve again with these initial values at t_{i+1}. The result will be a solution, but in general not a Lyapunov solution, of the equation, since $u+\chi$ does not jump at t_{i+1}. A model for nucleation might consist of a random distribution of such jumps, leading to a distribution of solutions to the Stefan problem with the Gibbs–Thomson law.

One test for the validity of the assumed distribution of 'nucleation' jumps would be the convergence of the solutions to the unique weak solution of the Stefan problem as $\alpha \to 0$.

I would like to thank R. Gulliver, S. Howison, J. Ockendon and A. Visintin for fruitful discussions which helped me to understand some aspects of the problem, as I hope.

References

Almgren, F. J. 1976 Existence and regularity almost everywhere of elliptic variational problems with constraints. *Mem. AMS* **4**, 165.

Berger, A. E. & Rogers, J. C. W. 1984 Some properties of the nonlinear semigroup for the problem $u_t - \Delta f(u) = 0$. *Nonlinear Anal.* **8**, 909–939.

Brezis, H. & Crandall, M. G. 1979 Uniqueness of solutions of the initial value problem for $u_t - \Delta(\Phi(u)) = 0$. *J. Math. Pure Appl.* **58**, 153–163.

Chadam, J. & Ortoleva, P. 1983 The stabilizing effect of surface tension on the development of planar free boundaries in single phase Cauchy Stefan problems. *JIMA* **30**, 57–66.

Friedman, A. 1982 *Variational Principles and Free Boundary Problems.* Wiley.

Giusti, E. 1984 *Minimal Surfaces and Functions of Bounded Variation.* Birkhäuser.

Gurtin, M. 1986 On the two-phase Stefan problem with interfacial energy and entropy. *Arch. Rat. Mec. An.* **96**, 199–241.

Hanzawa, Ei Ichi 1981 Classical solution of the Stefan problem. *Tohoku Math. J.* **33**, 297–335.

Meirmanov, A. M. 1980 On a classical solution of the multidimensional Stefan problem for quasilinear parabolic equations. *Mat. Sb.* **112**, 170–192.

Rogers, J. C. W. 1983 The Stefan problem with surface tension. In *Free Boundary Problems, Theory and Applications I* (ed. A. Fasanao & M. Primicerio). Pitman.

Visintin, A. 1984 Stefan problem with surface tension. *Report of IAN of CNR.* Pavia.

Visintin, A. 1988a Surface tension effects in phase transitions. In *Material Instabilities in Continuums Mechanics* (ed. J. Ball). Clarendon.

Visintin, A. 1988b Stefan problem with surface tension. In *Mathematical Models for Phase Change Problems* (ed. J. F. Rodrigues). Birkhäuser.

J. DIFFERENTIAL GEOMETRY
33 (1991) 635-681

MOTION OF LEVEL SETS BY MEAN CURVATURE. I

L. C. EVANS & J. SPRUCK

Abstract

We construct a unique weak solution of the nonlinear PDE which asserts each level set evolves in time according to its mean curvature. This weak solution allows us then to define for any compact set Γ_0 a unique generalized motion by mean curvature, existing for all time. We investigate the various geometric properties and pathologies of this evolution.

1. Introduction

We set forth in this paper rigorous justification of a new approach for defining and then investigating the evolution of a hypersurface in $\mathbf{R}^n$ moving according to its mean curvature. This problem has been long studied using parametric methods of differential geometry (see, for instance, Gage [15], [16], Gage-Hamilton [17], Grayson [19], Huisken [23], Ecker-Huisken [10], etc.). In this classical setup, we are given at time 0 a smooth hypersurface Γ_0 which is, say, the connected boundary of a bounded open subset of $\mathbf{R}^n$. As time progresses we allow the surface to evolve, by moving each point at a velocity equal to $(n-1)$ times the mean curvature vector at that point. Assuming this evolution is smooth, we define thereby for each $t > 0$ a new hypersurface Γ_t. The primary problem is then to study geometric properties of $\{\Gamma_t\}_{t>0}$ in terms of Γ_0.

For the case $n = 2$ this program has been successfully carried out in great detail (see [17], [19]). For $n \geq 3$, however, it is fairly clear that even if Γ_0 is smooth, a smooth evolution as envisioned above cannot exist beyond some initial time interval. Imagine, for instance, Γ_0 to be the boundary of a "dumbbell" shaped region in $\mathbf{R}^3$, as illustrated in Figure 1 (next page).

In view of Grayson [20] and numerical calculations of Sethian [35], we expect that as time evolves, the surface will smoothly evolve (and shrink)

Received August 14, 1989. Both authors were supported by National Science Foundation grants DMS-86-10730, DMS-86-01531 (L. C. Evans), and DMS-8501952 (J. Spruck). The second author was also supported in part by Department of Energy grant DE-FG02-86ER250125.

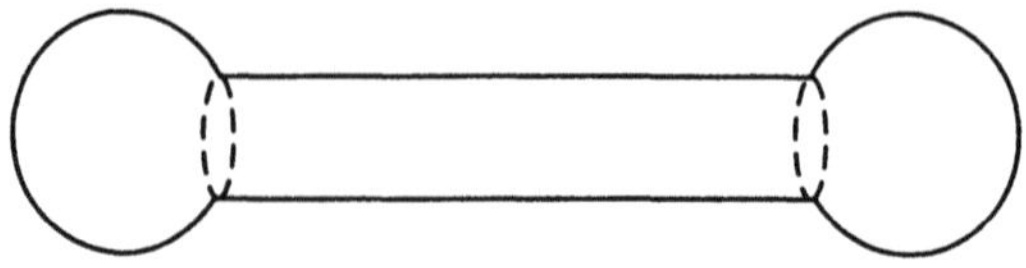

FIGURE 1

FIGURE 2

up until a critical time $t_* > 0$ when the two ends pinch off, as drawn in Figure 2.

After this time, the classical motion via mean curvature is undefined. In addition, if it were possible to define the subsequent motion in some reasonable way, we expect Γ_t for $t > t_*$ to comprise two pieces which pull apart at time t_*. If this were so, then Γ_t would have changed topological type. This possibility suggests inherent problems in the classical differential geometric approach of regarding Γ_0 as a parametrized surface: the parametrization will in general develop singularities.

What is needed is an alternative description of the evolution for all times $t > 0$, sufficiently general as to allow for the possible onset of singularities and attendant topological complications. To our knowledge there have been two different such undertakings, by Brakke [5] and by Osher-Sethian [33] (see also the note at the end of this section). Brakke [5] recasts the mean curvature motion problem (even in arbitrary codimension) into the setting of varifold theory from geometric measure theory (cf. Allard [2]). Brakke defines and then constructs an appropriate generalized varifold solution, which is defined for all time (although it may vanish after a finite time). He then deduces many geometric properties and under an additional density assumption establishes partial regularity. The principal drawback seems to be the lack of any uniqueness assertion.

A completely different viewpoint is to be found in the paper [33] by Osher and Sethian. Their approach, recast slightly, is this: given the initial hypersurface Γ_0 as above, select some continuous function $g: \mathbb{R}^n \to \mathbb{R}$ so that

$$\Gamma_0 = \{x \in \mathbb{R}^n | g(x) = 0\}. \tag{1.1}$$

Consider then the parabolic PDE

$$u_t = (\delta_{ij} - u_{x_i} u_{x_j} / |Du|^2) u_{x_i x_j} \quad \text{in } \mathbb{R}^n \times [0, \infty), \tag{1.2}$$

$$u = g \quad \text{on } \mathbb{R}^n \times \{t = 0\}, \tag{1.3}$$

for the unknown $u = u(x, t)$, $(x \in \mathbb{R}^n,\ t \geq 0)$. Now the PDE (1.2) says that each *level set of* u *evolves according to its mean curvature*, at least in regions where u is smooth and its spacial gradient Du does not vanish. Consequently, focusing our attention on the set $\{u = 0\}$, it seems reasonable in view of (1.1), (1.2) to *define*

$$\Gamma_t \equiv \{x \in \mathbb{R}^n | u(x, t) = 0\} \tag{1.4}$$

for each time $t > 0$. Osher and Sethian [33] and Sethian [35] introduce various techniques to study (1.2) and related PDE's numerically, thereby to track computationally the evolution of Γ_0 into Γ_t $(t \geq 0)$. (Notice by the way that our utilizing (1.1)–(1.3) amounts in the language of fluid mechanics to adopting an Eulerian viewpoint, as opposed to the Lagrangian, parametric viewpoint of classical differential geometry.)

Our purpose here is to provide theoretical justification for this approach. The undertaking is analytically subtle, principally because the mean curvature evolution equation (1.2) is nonlinear, degenerate, and indeed even undefined at points where $Du = 0$. In addition, it is not so clear that our definition (1.3) is independent of the choice of initial function g verifying (1.1). We will resolve these problems by introducing an appropriate definition of a weak solution for (1.2), inspired by the notion of so-called "viscosity solutions" of nonlinear PDE as in Evans [12], Crandall-Lions [9], Crandall-Evans-Lions [8], Lions [32], Jensen [25], and Ishii [24]. We then prove that there exists a unique weak solution of (1.2), and, further, that definition (1.3) is then independent of the choice of initial function g satisfying (1.1). We additionally check that $\{\Gamma_t\}_{t\geq 0}$ so defined agrees with the classical notion of motion via mean curvature, over any time interval for which the latter exists. Finally we employ the PDE (1.2) to deduce assorted geometric properties of $\{\Gamma_t\}_{t\geq 0}$.

The main theoretical advantage of (1.1)–(1.3) as compared with Brakke's varifold methods seems to us to be the following uniqueness assertion: the set Γ_t is unambiguously defined by (1.3) once we have a uniqueness assertion for the PDE (1.2). The primary disadvantage is that our techniques work only in codimension one.

In a companion paper [14] we give a new proof of short time existence for classical motion by mean curvature by studying the PDE solved by the distance function. We also hope to establish in a forthcoming paper a partial regularity theorem for $\{\Gamma_t\}_{t\geq 0}$.

Our paper is organized as follows. In §2 we motivate and introduce our definition of weak solution for (1.2) and in §3 we prove the uniqueness of a weak solution. §4 establishes the existence of a weak solution to (1.2). In §5 we verify the independence of the definition (1.3) on the choice of g. §6 contains a consistency check that the definition (1.3) agrees with the classical motion by mean curvature, if and so long as the latter exists. §§7 and 8 contain various geometric assertions, examples of pathologies, and conjectures.

After this work was completed, we learned of the recent paper of Chen, Giga, and Goto [7], which announces results very similar to ours, especially the existence of a unique weak solution of the PDE (1.2), (1.3). Their work includes as well generalizations to other geometric problems.

Another new paper concerning curvature and viscosity solutions is Trudinger [36].

2. Definition and elementary properties of weak solutions

2.1. Heuristics. We start with a formal derivation of the mean curvature evolution PDE (1.2). For this, suppose temporarily $u = u(x, t)$ is a smooth function whose spatial gradient $Du = (u_{x_1}, \cdots, u_{x_n})$ does not vanish in some open region O of $\mathbf{R}^n \times (0, \infty)$. Assume further that each level set of u smoothly evolves according to its mean curvature, as described in §1. We focus our attention onto any one such level set, and for definiteness consider the zero sets

$$\Gamma_t \equiv \{x \in \mathbf{R}^n | u(x, t) = 0\} \qquad (t \geq 0). \tag{2.1}$$

Let $\nu = \nu(x, t)$ be a smooth unit normal vector field to $\{\Gamma_t\}_{t\geq 0}$ in O. Then

$$-\frac{1}{n-1}\operatorname{div}(\nu)\nu$$

is the mean curvature vector field. Thus if we fix $t \geq 0, x \in \Gamma_t \cap O$, the point x evolves according to the nonautonomous ODE

$$\begin{cases} \dot{x}(s) = -[\operatorname{div}(\nu)\nu](x(s), s) & (s > t), \\ x(t) = x. \end{cases} \tag{2.2}$$

As $x(s) \in \Gamma_s$ $(s \geq t)$, we have $u(x(s), s) = 0$ $(s > t)$; and so

$$0 = \frac{d}{ds}u(x(s), s) = -[(Du \cdot \nu)\operatorname{div}(\nu)](x(s), s) + u_t(x(s), s).$$

Setting $s = t$, we discover

$$u_t = (Du \cdot \nu)\operatorname{div}(\nu) \quad \text{at } (x, t).$$

Choosing then

$$\nu \equiv \frac{Du}{|Du|} \tag{2.3}$$

it follows that

$$u_t = |Du| \operatorname{div}\left(\frac{Du}{|Du|}\right) = \left(\delta_{ij} - \frac{u_{x_i} u_{x_j}}{|Du|^2}\right) u_{x_i x_j} \quad \text{at } (x, t). \tag{2.4}$$

Similar reasoning demonstrates this PDE to hold throughout the region O.

Now, conversely, assume u is a smooth solution of (2.4) in some region O with Du nonvanishing. Fix $t > 0, x \in \Gamma_t \cap O$ and solve then the ODE (2.2), (2.3). Since u solves (2.4), we deduce as above

$$u(x(s), s) = 0 \qquad (s > t).$$

Consequently the zero sets, and similarly all the level sets, of u evolve in O according to their mean curvatures.

Since the motion of any level set thus depends only upon its own geometry, and not that of any other level set, our PDE (2.4) should be invariant under an arbitrary relabelling of these sets. Thus if $\Psi: \mathbf{R} \to \mathbf{R}$ is smooth, we expect that $v \equiv \Psi(u)$ will also be a solution of (2.4) in the region O. A direct calculation verifies this in the regions where $Dv \neq 0$. Hence we see that an *arbitrary function of a solution is still a solution*; this is in strong contrast to the situation for uniformly parabolic PDE's. Indeed, we may informally interpret (2.4) as being somehow "uniformly parabolic along each level set", but as being also "totally degenerate across different level sets".

2.2. Weak solutions. The foregoing heuristics done with, we turn now to the full mean curvature evolution equation:

$$u_t = (\delta_{ij} - u_{x_i} u_{x_j}/|Du|^2) u_{x_i x_j} \quad \text{in } \mathbf{R}^n \times (0, \infty), \tag{2.5}$$

$$u = g \quad \text{on } \mathbf{R}^n \times \{t = 0\}, \tag{2.6}$$

the function $g: \mathbf{R}^n \to \mathbf{R}$ being given. We want to define a notion of weak solution to (2.5). Since, however, the right-hand side of the PDE cannot be put into divergence form, we are not able to define a weak solution by means of formal integration by parts of derivatives onto a smooth test function (as for instance in Bombieri, De Giorgi, Giusti [4, §1]). We will instead follow Evans [12], Lions [32], Jensen [25], etc. and define our weak solution in terms of pointwise behavior with respect to a smooth test function. The primary difficulty will be to modify extant theory to cover the possibility that Du may vanish.

Definition 2.1. A function $u \in C(\mathbf{R}^n \times [0, \infty)) \cap L^\infty(\mathbf{R}^n \times [0, \infty))$ is a *weak subsolution* of (2.5) provided that if

$$u - \phi \text{ has a local maximum at a point } (x_0, t_0) \in \mathbf{R}^n \times (0, \infty) \tag{2.7}$$

for each $\phi \in C^\infty(\mathbf{R}^{n+1})$, then

$$\begin{cases} \phi_t \le (\delta_{ij} - \phi_{x_i}\phi_{x_j}/|D\phi|^2)\phi_{x_ix_j} & \text{at } (x_0, t_0) \\ \text{if } D\phi(x_0, t_0) \neq 0, \end{cases} \tag{2.8}$$

and

$$\begin{cases} \phi_t \le (\delta_{ij} - \eta_i\eta_j)\phi_{x_ix_j} \quad \text{at } (x_0, t_0) \\ \text{for some } \eta \in \mathbf{R}^n \text{ with } |\eta| \le 1, \text{ if } D\phi(x_0, t_0) = 0. \end{cases} \tag{2.9}$$

Definition 2.2. A function $u \in C(\mathbf{R}^n \times [0, \infty)) \cap L^\infty(\mathbf{R}^n \times [0, \infty))$ is a *weak supersolution* of (2.5) provided that if

$$u - \phi \text{ has a local maximum at a point } (x_0, t_0) \in \mathbf{R}^n \times (0, \infty) \tag{2.10}$$

for each $\phi \in C^\infty(\mathbf{R}^{n+1})$, then

$$\begin{cases} \phi_t \ge (\delta_{ij} - \phi_{x_i}\phi_{x_j}/|D\phi|^2)\phi_{x_ix_j} & \text{at } (x_0, t_0) \\ \text{if } D\phi(x_0, t_0) \neq 0, \end{cases} \tag{2.11}$$

and

$$\begin{cases} \phi_t \ge (\delta_{ij} - \eta_i\eta_j)\phi_{x_ix_j} \quad \text{at } (x_0, t_0) \\ \text{for some } \eta \in \mathbf{R}^n \text{ with } |\eta| \neq 1, \text{ if } D\phi(x_0, t_0) = 0. \end{cases} \tag{2.12}$$

Definition 2.3. A function $u \in C(\mathbf{R}^n \times [0, \infty)) \cap L^\infty(\mathbf{R}^n \times [0, \infty))$ is a *weak solution* of (2.5) provided u is both a weak subsolution and a weak supersolution.

As preliminary motivation for these definitions, suppose u is a smooth function on $\mathbf{R}^n \times (0, \infty)$ satisfying

$$u_t \le (\delta_{ij} - u_{x_i}u_{x_j}/|Du|^2)u_{x_ix_j}$$

wherever $Du \neq 0$. Our function u is thus a classical subsolution of (2.5) on $\{Du \neq 0\}$. Suppose now $Du(x_0, t_0) = 0$. Assume additionally there are points $(x_k, t_k) \to (x_0, t_0)$ for which $Du(x_k, t_k) \neq 0$ $(k = 1, 2, \dots)$. Then

$$u_t \le (\delta_{ij} - \eta_j^k\eta_j^k)u_{x_ix_j} \quad \text{at } (x_k, t_k)$$

for $\eta^k \equiv Du(x_k, t_k)/|Du(x_k, t_k)|$. Since $|\eta^k| = 1$ $(k = 1, 2, \dots)$ we may as necessary pass to a subsequence so that $\eta^k \to \eta$ in $\mathbf{R}^n$, $|\eta| = 1$. Passing to limits above, we find

$$u_t \leq (\delta_{ij} - \eta_i\eta_j)u_{x_ix_j} \quad \text{at } (x_0, t_0).$$

If, on the other hand, there do not exist such points $\{(x_k, t_k)\}_{k=1}^{\infty}$, then $Du = 0$ near (x_0, t_0), and so $D^2u = 0$ and u is a function of t only, near (x_0, t_0). Moving to the edge of the set $\{Du = 0\}$, we see that u is a nonincreasing function of t. Thus

$$u_t \leq (\delta_{ij} - \eta_i\eta_j)u_{x_ix_j} \quad \text{at } (x_0, t_0)$$

for any $\eta \in \mathbf{R}^n$.

Further motivation for our definition of weak solution, and, in particular, an explanation as to why we assume only $|\eta| \leq 1$ in (2.9), (2.12), will be found in §2.4.

2.3. An equivalent definition. It will be convenient to have at hand certain alternative definitions. We write $z = (x, t)$, $z_0 = (x_0, t_0)$ and below implicitly sum i, j from 1 to n.

Definition 2.4. A function $u \in C(\mathbf{R}^n \times [0, \infty)) \cap L^\infty(\mathbf{R}^n \times [0, \infty))$ is a weak subsolution of (2.5) if whenever $(x_0, t_0) \in \mathbf{R}^n \times (0, \infty)$ and

$$\text{(2.13)} \quad \begin{aligned} u(x, t) \leq\ & u(x_0, t_0) + p \cdot (x - x_0) + q(t - t_0) \\ & + \tfrac{1}{2}(z - z_0)^T R(z - z_0) + o(|z - z_0|^2) \quad \text{as } z \to z_0 \end{aligned}$$

for some $p \in \mathbf{R}^n$, $q \in \mathbf{R}$, $R = ((r_{ij})) \in S^{n+1\times n+1}$, then

$$\text{(2.14)} \qquad q \leq (\delta_{ij} - p_ip_j/|p|^2)r_{ij} \quad \text{if } p \neq 0$$

and

$$\text{(2.15)} \qquad q \leq (\delta_{ij} - \eta_i\eta_j)r_{ij} \quad \text{for some } |\eta| \leq 1, \text{ if } p = 0.$$

Definition 2.5. A function $u \in C(\mathbf{R}^n \times [0, \infty)) \cap L^\infty(\mathbf{R}^n \times [0, \infty))$ is a weak supersolution of (2.5) if whenever $(x_0, t_0) \in \mathbf{R}^n \times (0, \infty)$ and

$$\text{(2.16)} \quad \begin{aligned} u(x, t) \geq\ & u(x_0, t_0) + p \cdot (x - x_0) + q(t - t_0) \\ & + \tfrac{1}{2}(z - z_0)^T R(z - z_0) + o(|z - z_0|^2) \quad \text{as } z \to z_0 \end{aligned}$$

for some $p \in \mathbf{R}^n$, $q \in \mathbf{R}$, $R = ((r_{ij})) \in S^{n+1\times n+1}$, then

$$\text{(2.17)} \qquad q \geq (\delta_{ij} - p_ip_j/|p|^2)r_{ij} \quad \text{if } p \neq 0$$

and

$$q \geq (\delta_{ij} - \eta_i \eta_j) r_{ij} \quad \text{for some } |\eta| \leq 1, \text{ if } p = 0. \tag{2.18}$$

Theorem 2.6. *Definitions* 2.1 *and* 2.4 *are equivalent, and Definitions* 2.2 *and* 2.5 *are equivalent.*

This follows as in, for instance, Jensen [25], Ishii [24].

2.4. Properties of weak solutions.

Theorem 2.7. (i) *Assume* u_k *is a weak solution of* (2.5) *for* $k = 1, 2, \ldots$ *and* $u_k \to u$ *boundedly and locally uniformly on* $\mathbb{R}^n \times [0, \infty)$. *Then* u *is a weak solution.*

(ii) *An analogous assertion holds for weak subsolutions and supersolutions.*

Proof. 1. Choose $\phi \in C^\infty(\mathbb{R}^{n+1})$ and suppose first $u - \phi$ has a *strict* local maximum at some point $(x_0, t_0) \in \mathbb{R}^n \times (0, \infty)$. As $u^k \to u$ uniformly near (x_0, t_0),

(2.19) $u_k - \phi$ has a local maximum at a point (x_k, t_k) $(k = 1, 2, \ldots)$

with

$$(x_k, t_k) \to (x_0, t_0) \quad \text{as } k \to \infty. \tag{2.20}$$

Since u_k is a weak solution, we have either

$$\phi_t \leq (\delta_{ij} - \phi_{x_i}\phi_{x_i}/|D\phi|^2)\phi_{x_i x_j} \quad \text{at } (x_k, t_k) \tag{2.21}$$

if $D\phi(x_k, t_k) \neq 0$, or

$$\phi_t \leq (\delta_{ij} - \eta_i^k \eta_j^k)\phi_{x_i x_j} \quad \text{at } (x_k, t_k) \tag{2.22}$$

for some $\eta^k \in \mathbb{R}^n$ with $|\eta^k| \leq 1$, if $D\phi(x_k, t_k) = 0$.

2. Assume first $D\phi(x_0, t_0) \neq 0$. Then $D\phi(x_k, t_k) \neq 0$ for all large enough k. Hence we may pass to limits in the equalities (2.21) to discover

$$\phi_t \leq (\delta_{ij} - \phi_{x_i}\phi_{x_j}/|D\phi|^2) \quad \text{at } (x_0, t_0). \tag{2.23}$$

3. Next, suppose $D\phi(x_0, t_0) = 0$. We set

$$\xi^k \equiv \begin{cases} (D\phi/|D\phi|)(x_k, t_k) & \text{if } D\phi(x_k, t_k) \neq 0, \\ \eta^k & \text{if } D\phi(x_k, t_k) = 0. \end{cases} \tag{2.24}$$

Passing if necessary to a subsequence we may assume $\xi^k \to \eta$. Then $|\eta| \leq 1$. Utilizing now (2.22), we deduce as well

$$\phi_t \leq (\delta_{ij} - \eta_i \eta_j)\phi_{x_i x_j} \quad \text{at } (x_0, t_0). \tag{2.25}$$

4. If $u-\phi$ has only a local maximum at (x_0, t_0) we apply the above argument to

$$\psi(x, t) \equiv \phi(x, t) + |x - x_0|^4 + (t - t_0)^4,$$

so that $u - \psi$ has a strict local maximum at (x_0, t_0). Hence u is a weak subsolution. Similar reasoning verifies that u is a weak supersolution as well.

Theorem 2.8. *Assume u is a weak solution of* (2.5) *and $\Psi: \mathbf{R} \to \mathbf{R}$ is continuous. Then $v \equiv \Psi(u)$ is a weak solution.*

Proof. 1. Assume first Ψ is smooth, with

$$\Psi' > 0 \quad \text{on } \mathbf{R}. \tag{2.26}$$

Let $\phi \in C^\infty(\mathbf{R}^{n+1})$ and suppose $v - \phi$ has a local maximum at (x_0, t_0). Adding as necessary a constant to ϕ, we may assume

$$\begin{cases} v(x_0, t_0) = \phi(x_0, t_0), \\ v(x, t) \le \phi(x, t) \quad \text{for all } (x, t) \text{ near } (x_0, t_0). \end{cases} \tag{2.27}$$

In view of (2.26), $\Phi \equiv \Psi^{-1}$ is defined and smooth near $u(x_0, t_0)$, with

$$\Phi' > 0. \tag{2.28}$$

From (2.27) therefore we see

$$\begin{cases} u(x_0, t_0) = \psi(x_0, t_0), \\ u(x, t) \le \psi(x, t) \quad \text{for all } (x, t) \text{ near } (x_0, t_0), \end{cases} \tag{2.29}$$

where

$$\psi \equiv \Phi(\phi). \tag{2.30}$$

2. Since u is a weak solution we conclude

$$\psi_t \le (\delta_{ij} - \psi_{x_i}\psi_{x_j}/|D\psi|^2)\psi_{x_ix_j} \quad \text{at } (x_0, t_0) \tag{2.31}$$

if $D\psi(x_0, t_0) \ne 0$, and

$$\psi_t \le (\delta_{ij} - \eta_i\eta_j)\psi_{x_ix_j} \quad \text{at } (x_0, t_0) \tag{2.32}$$

for some $|\eta| \le 1$, if $D\psi(x_0, t_0) = 0$. Now $D\phi(x_0, t_0) = 0$ if and only if $D\psi(x_0, t_0) = 0$. Consequently (2.31) is obtained if $D\phi(x_0, t_0) \ne 0$; in which case we substitute (2.30) to compute

$$\Phi'\phi_t \le \left(\delta_{ij} - \frac{(\Phi')^2\phi_{x_i}\phi_{x_j}}{(\Phi')^2|D\phi|^2}\right)(\Phi'\phi_{x_ix_j} + \Phi''\phi_{x_i}\phi_{x_j}) \quad \text{at } (x_0, t_0).$$

Since $\Phi' > 0$, we simplify and obtain

(2.33) $$\phi_t \le (\delta_{ij} - \phi_{x_i}\phi_{x_j}/|D\phi|^2)\phi_{x_ix_j} \quad \text{at } (x_0, t_0).$$

Suppose on the other hand $D\phi(x_0, t_0) = 0$. Then (2.32) is valid for some $|\eta| \le 1$. We substitute (2.30) and compute

$$\Phi'\phi_t \le (\delta_{ij} - \eta_i\eta_j)(\Phi'\phi_{x_ix_j} + \Phi''\phi_{x_i}\phi_{x_j}) \quad \text{at } (x_0, t_0).$$

Since $D\phi = 0$, the term involving Φ'' is zero. Thus

(2.34) $$\phi_t \le (\delta_{ij} - \eta_i\eta_j)\phi_{x_ix_j} \quad \text{at } (x_0, t_0).$$

We similarly have the opposite inequalities to (2.33), (2.34) should $v - \phi$ have a local minimum at (x_0, t_0).

3. Now assume instead of (2.20) that

(2.35) $$\Psi' < 0 \quad \text{on } \mathbb{R}.$$

Then $\Phi' < 0$ on $\mathbb{R}$ as well. Thus (2.27) now implies

$$\begin{cases} u(x_0, t_0) = \psi(x_0, t_0), \\ u(x, t) \ge \psi(x, t) \quad \text{for all } (x, t) \text{ near } (x_0, t_0). \end{cases}$$

Since u is a weak solution either

$$\psi_t \ge (\delta_{ij} - \psi_{x_i}\psi_{x_j}/|D\psi|^2)\psi_{x_ix_j} \quad \text{at } (x_0, t_0)$$

if $D\psi(x_0, t_0) \ne 0$, or

$$\psi_t \ge (\delta_{ij} - \eta_i\eta_j)\psi_{x_ix_j} \quad \text{at } (x_0, t_0)$$

for some $|\eta| \le 1$,if $D\psi(x_0, t_0) = 0$. Since now $\Phi' < 0$, we deduce as above either (2.33) or (2.34).

4. We have so far shown that $v = \Psi(u)$ is a weak solution provided Ψ is smooth, with $\Psi' \ne 0$. Approximating and using Theorem 2.7 we draw the same conclusion if $\Psi' \ge 0$ or $\Psi' \le 0$ on $\mathbb{R}$.

5. Next assume Ψ is smooth and there exist finitely many points $-\infty = a_0 < a_1 < a_2 < \cdots < a_m < a_{m+1} = +\infty$ such that

(2.36) Ψ is monotone on the intervals (a_j, a_{j+1}) $(j = 0, \cdots, m)$

and

(2.37) Ψ is constant on the intervals $(a_j - \sigma, a_j + \sigma)$ $(j = 1, \cdots, m)$

for some $\sigma > 0$.

Suppose $v - \phi$ has a maximum at (x_0, t_0). Then

$$u(x_0, t_0) \in (a_j - \sigma/2, a_{j+1} + \sigma/2) \quad \text{for some } j \in \{0, \cdots, m\}.$$

As Ψ is monotone on $(a_j - \sigma, a_{j+1} + \sigma)$ and u is continuous, we can apply steps 1–4 in some neighborhood of (x_0, t_0) to deduce (2.33) or (2.34). The reverse inequalities are similarly obtained if $v - \phi$ has a minimum.

6. Finally suppose only that Ψ is continuous. We construct a sequence of smooth functions $\{\Psi^k\}_{k=1}^{\infty}$ each verifying the structural assumptions (2.36), (2.37) so that $\Psi^k \to \Psi$ uniformly on $[-\|u\|_{L^\infty}, \|u\|_{L^\infty}]$. Hence

$$v^k = \Psi^k(u) \to v \equiv \Psi(u)$$

bounded and uniformly. Then Theorem 2.7 asserts v to be a weak solution.

3. Uniqueness and comparison of weak solutions

3.1. Preliminaries. Our plan, as in Jensen [25] and Jensen-Lions-Souganidis [26], is to regularize using sup and inf convolutions, defined as follows. Assume $w: \mathbf{R}^n \times [0, \infty) \to \mathbf{R}$ is continuous and bounded. If $\epsilon > 0$, then we write

$$w^\epsilon(x; t) \equiv \sup_{\substack{y \in \mathbf{R}^n \\ s \in [0,\infty)}} \{w(y, s) - \epsilon^{-1}(|x-y|^2 + (t-s)^2)\}, \tag{3.1}$$

$$w_\epsilon(x, t) \equiv \inf_{\substack{y \in \mathbf{R}^n \\ s \in [0,\infty)}} \{w(y, s) + \epsilon^{-1}(|x-y|^2 + (t-s)^2)\}, \tag{3.2}$$

for $x \in \mathbf{R}^n$, $t \in [0, \infty)$. Note that since w is continuous and bounded, the "sup" and "inf" above can be replaced by "max" and "min".

Lemma 3.1 (Properties of sup and inf convolutions). *There exist constants A, B, C, depending only on $\|w\|_{L^\infty(R^n \times [0,\infty))}$, such that for $\epsilon > 0$ the following hold:*

(i) $w_\epsilon \le w \le w^\epsilon$ *on* $\mathbf{R}^n \times [0, \infty)$.

(ii) $\|w^\epsilon, w_\epsilon\|_{L^\infty(R^n \times [0,\infty))} \le A$.

(iii) *If* $y \in \mathbf{R}^n$, $s \in [0, \infty)$, *and* $w^\epsilon(x, t) = w(y, x) - \epsilon^{-1}(|x-y|^2 + (t-s)^2)$, *then*

$$|x - y|, |t - s| \le C\epsilon^{1/2} \equiv \sigma(\epsilon). \tag{3.3}$$

A similar assertion holds for w_ϵ.

(iv) $w^\epsilon, w_\epsilon \to w$ *as* $\epsilon \to 0^+$, *uniformly on compact subsets of* $\mathbf{R}^n \times [0, \infty)$.

(v) $\mathrm{Lip}(w^\epsilon), \mathrm{Lip}(w_\epsilon) \le B/\epsilon$.

(vi) *The mapping*

$$(x, t) \mapsto w^\epsilon(x, t) + \epsilon^{-1}(|x|^2 + t^2)$$

is convex, and the mapping

$$(x, t) \mapsto w_\epsilon(x, t) - \epsilon^{-1}(|x|^2 + t^2)$$

is concave.

(vii) *Assume* w *is a weak solution of* (2.5) *in* $\mathbf{R}^n \times (0, \infty)$. *Then* w^ϵ *is a weak subsolution on* $\mathbf{R}^n \times (\sigma(\epsilon), \infty)$. *Similarly, if* w *is a weak supersolution of* (2.5), w_ϵ *is a weak supersolution.*

(viii) *The function* w^ϵ *is twice differentiable a.e. and satisfies*

$$w_t^\epsilon \le (\delta_{ij} - w_{x_i}^\epsilon w_{x_j}^\epsilon / |Dw^\epsilon|^2) w_{x_i x_j}^\epsilon \tag{3.4}$$

at each point of twice differentiability in $\mathbf{R}^n \times (\sigma(\epsilon), \infty)$, *where* $Dw^\epsilon \neq 0$. *Similarly,* w_ϵ *is twice differentiable a.e. and satisfies*

$$w_{\epsilon t} \ge (\delta_{ij} - w_{\epsilon x_i} w_{\epsilon x_j} / |Dw_\epsilon|^2) w_{\epsilon x_i x_j} \tag{3.5}$$

at each point of twice differentiability in $\mathbf{R}^n \times (\sigma(\epsilon), \infty)$, *where* $Dw_\epsilon \neq 0$.

Proof. 1. Assertions (i) and (ii) are clear from the definitions, for $A = \|w\|_{L^\infty(R^n \times [0,\infty))}$. Statement (iii) follows from (ii), and then (iv) is a consequence of the uniform continuity of w on compact sets. In light of estimate (3.3) we have (v) as well.

2. For each $y \in \mathbf{R}^n$, $s \in [0, \infty)$, the mapping

$$(x, t) \mapsto w(y, s) - \epsilon^{-1}(|x - y|^2 + (t - s)^2) + \epsilon^{-1}(|x|^2 + t^2)$$

is affine. Consequently

$$\begin{aligned}(x, t) \mapsto & \sup_{\substack{y \in \mathbf{R}^n \\ s \in [0,\infty)}} [w(y, s) - \epsilon^{-1}(|x - y|^2 + (t - s)^2) + \epsilon^{-1}(|x|^2 + t^2)] \\ & = w^\epsilon(x, t) + \epsilon^{-1}(|x|^2 + t^2)\end{aligned}$$

is convex, and (v) is proved.

3. Assume $\phi \in C^\infty(\mathbf{R}^{n+1})$ and $w^\epsilon - \phi$ has a local maximum at a point (x_0, t_0), with $t_0 > \sigma(\epsilon)$. We then employ (3.3) to choose $(y_0, s_0) \in \mathbf{R}^n \times (0, \infty)$ so that

$$w^\epsilon(x_0, t_0) = w(y_0, s_0) - \epsilon^{-1}(|x_0 - y_0|^2 + (t_0 - s_0)^2).$$

Set

$$\psi(x, t) \equiv \phi(x + x_0 - y_0, t + t_0 - s_0). \tag{3.6}$$

Since $w^\epsilon - \phi$ has a local maximum at (x_0, t_0) we compute

$$\begin{aligned} w(y_0, s_0) - \epsilon^{-1}(|x_0 - y_0|^2 + (t_0 - s_0)^2) - \phi(x_0, t_0) \\ = w^\epsilon(x_0, t_0) - \phi(x_0, t_0) \geq w^\epsilon(x, t) - \phi(x, t) \\ \geq w(y, s) - \epsilon^{-1}(|x - y|^2 + (t - s)^2) - \phi(x, t) \end{aligned}$$

for all (x, t) near (x_0, t_0) and all $(y, s) \in \mathbb{R}^n \times [0, \infty)$. Fix (y, s) close to (y_0, s_0) and set $x = y + x_0 - y_0$, $t = s + t_0 - s_0$ as above, to discover

$$w(y_0, s_0) - \phi(x_0, t_0) \geq w(y, s) - \phi(y + x_0 - y_0, s + t_0 - s_0).$$

Using (3.6) we rewrite this as

$$w(y_0, s_0) - \psi(y_0, s_0) \geq w(y, s) - \psi(y, s)$$

for all (y, s) near (y_0, s_0). Hence $w - \phi$ has a local maximum at (y_0, s_0) and thus

$$\psi_t \leq (\delta_{ij} - \psi_{x_i}\psi_{x_i}/|D\psi|^2)\psi_{x_i x_j} \quad \text{at } (y_0, s_0)$$

if $D\psi(y_0, s_0) \neq 0$, and

$$\psi_t \leq (\delta_{ij} - \eta_i \eta_j)\psi_{x_j x_j} \quad \text{at } (y_0, s_0)$$

for some $|\eta| \leq 1$, if $D\psi(y_0, s_0) = 0$. Since

$$\begin{aligned} D\psi(y_0, s_0) = D\phi(x_0, t_0), \ \psi_t(y_0, s_0) = \phi_t(x_0, t_0), \\ D^2\phi(y_0, s_0) = D^2\phi(x_0, t_0), \end{aligned}$$

we immediately obtain

$$\phi_t \leq (\delta_{ij} - \phi_{x_i}\phi_{x_j}/|D\phi|^2)\phi_{x_i x_j} \quad \text{at } (x_0, t_0)$$

or

$$\phi_t \leq (\delta_{ij} - \eta_i \eta_j)\phi_{x_i x_j} \quad \text{at } (x_0, t_0)$$

according as $D\phi(x_0, t_0) = 0$ or not, and (vii) is proved.

4. Owing to (vi), $w^\epsilon(x, t) + \epsilon^{-1}(|x|^2 + t^2)$ is convex in (x, t) and so is twice differentiable a.e. according to a theorem of Alexandroff (see, e.g., Krylov [30, Appendix 2]). Thus w^ϵ is twice differentiable a.e. In view of (vii) and Theorem 2.6, (3.4) holds at points of twice differentiability, where $Dw^\epsilon \neq 0$. Hence (viii) is proved.

3.2. Comparison principle, uniqueness. We establish now a comparison assertion for weak solutions of our mean curvature evolution PDE. Many of the key technical devices in the proof are taken from Jensen [25] and Ishii [24].

Theorem 3.2. *Assume that u is a weak subsolution and v is a weak supersolution of* (2.5). *Suppose further*

$$u \le v \quad \text{on } \mathbb{R}^n \times \{t = 0\}.$$

Finally assume

$$\text{(3.8)} \qquad \begin{cases} u \text{ and } v \text{ are constant, with } u \le v, \\ \text{on } \mathbb{R}^n \times [0, \infty) \cap \{|x| + t \ge R\} \end{cases}$$

for some constant $R \ge 0$. Then

$$\text{(3.9)} \qquad u \le v \quad \text{on } \mathbb{R}^n \times [0, \infty).$$

In particular, a weak solution of (2.5), (2.6) *is unique.*

Proof. 1. Should (3.9) fail, then

$$\max_{(x,t)\in R^n\times[0,\infty)} (u - v) \equiv a > 0;$$

and so for $\alpha > 0$ small enough,

$$\text{(3.10)} \qquad \max_{(x,t)\in R^n\times[0,\infty)} (u - v - \alpha t) \ge a/2 > 0.$$

According to (3.8) we have

$$\text{(3.11)} \qquad u^\epsilon = u, \quad v_\epsilon = v \quad \text{on } \{|x| + t \ge 2R\}$$

for all small $\epsilon > 0$. Note further $u^\epsilon \to u$ and $v_\epsilon \to v$ uniformly. Consequently if we fix $\epsilon > 0$ small enough,

$$\text{(3.12)} \qquad \max_{(x,t)\in R^n\times[0,\infty)} (u^\epsilon - v_\epsilon - \alpha t) \ge a/4 > 0.$$

2. Given $\delta > 0$ define for $x, y \in \mathbb{R}^n$ and $t, t+s \in [0, \infty)$

$$\text{(3.13)} \quad \Phi(x, y, t, s) \equiv u^\epsilon(x+y, t+s) - v_\epsilon(x, t) - \alpha t - \delta^{-1}(|y|^4 + s^4).$$

Owing to (3.12) we see

$$\text{(3.14)} \qquad \max_{(x,t),(x+y,t+s)\in R^n\times[0,\infty)} \Phi \ge a/4 > 0.$$

Choose now $(x_1, t_1), (x_1 + y_1, t_1 + s_1) \in \mathbb{R}^n \times [0, \infty)$ so that

$$\text{(3.15)} \qquad \Phi(x_1, y_1, t_1, s_1) = \max_{(x,t),(x+y,t+s)\in R^n\times[0,\infty)} \Phi.$$

Note in view of (3.11), (3.13) and Lemma 3.1(ii) that such points exist. Since $\Phi(x_1, y_1, t_1, s_1) > 0$, (3.13) implies

$$\text{(3.16)} \qquad |y_1|, |s_1| \le C\delta^{1/4}.$$

3. We *claim* next that if $\epsilon, \delta > 0$ are fixed small enough, we have

$$t_1, t_1 + s > \sigma(\epsilon), \tag{3.17}$$

with $\sigma(\epsilon)$ defined in (3.3). Indeed if $t_1 \le \sigma(\epsilon)$, then

$$\begin{aligned} a/4 &\le \Phi(x_1, y_1, t_1, s_1) \\ &\le u^\epsilon(x_1 + y_1, t_1 + s_1) - v_\epsilon(x_1, t_1) \\ &= u(x_1 + y_1, t_1 + s_1) - v(x_1, t_1) + o(1) \quad \text{as } \epsilon \to 0 \\ &= u(x_1 + y_1, s_1) - v(x_1, 0) + o(1) \quad \text{as } \epsilon \to 0 \\ &= u(x_1, 0) - v(x_1, 0) + o(1) \quad \text{as } \epsilon, \delta \to 0 \\ &\le o(1) \quad \text{as } \epsilon, \delta \to 0, \end{aligned}$$

where we employed Lemma 3.1(ii), (3.16), (3.7), and the continuity of u, v. This is a contradiction for $\epsilon, \delta > 0$ small enough; whence $t_1 > \sigma(\epsilon)$. Owing to (3.16) we may as necessary adjust δ smaller to ensure (3.17). *Hereafter in the proof, $\alpha, \epsilon, \delta > 0$ are fixed.*

According to Lemma 3.1(vii),

(3.18) $\qquad u^\epsilon$ is a weak subsolution of (2.5) near $(x_1 + y_1, t_1 + s_1)$

and

(3.19) $\qquad v_\epsilon$ is a weak supersolution of (2.5) near (x_1, t_1).

4. We now demonstrate

$$y_1 \ne 0. \tag{3.20}$$

Assume for contradiction that in fact $y_1 = 0$. Then (3.13), (3.15) imply

$$\begin{aligned} u^\epsilon(x_1, t_1 + s_1) &- v_\epsilon(x, t_1) - \alpha t_1 - \delta^{-1} s_1^4 \\ &\ge u^\epsilon(x + y, t + s) - v_\epsilon(x, t) - \alpha t - \delta^{-1}(|y|^4 + s^4) \end{aligned}$$

for all $(x, t), (x + y, t + s) \in \mathbb{R}^n \times [0, \infty)$. Put $x = x_1$ and $t = t_1$ as above, and simplify to obtain the inequality

$$u^\epsilon(x_1 + y, t_1 + s) \le u^\epsilon(x_1, t_1 + s_1) + \delta^{-1}|y|^4 + \delta^{-1}(s^4 - s_1^4)$$

for $(x_1 + y, t_1 + s) \in \mathbb{R}^n \times [0, \infty)$. Set $r = s - s_1$ and rewrite to find

$$\begin{aligned} u^\epsilon(x_1 + y, t_1 + s_1 + r) \le u^\epsilon(x_1, t_1 + s_1) &+ 4s_1^3 r/\delta + 6s_1^2 r^2/\delta \\ &+ O(|r|^3 + |y|^4) \quad \text{as } (y, r) \to (0, 0). \end{aligned}$$

Since u^ϵ is a weak subsolution near $(x_1+y_1, t_1+s_1) = (x_1, t_1+s_1)$, we may invoke (2.13), (2.15) with $x_0 = x_1$, $t_0 = t_1 + s_1$, $p = 0$, $q = 4s_1^3/\delta$, $r_{n+1,n+1} = 12s_1^2/\delta$, $r_{ij} = 0$ otherwise. This gives

(3.22) $$4s_1^3/\delta \le 0.$$

Now go back and insert $y = x_1 - x$ and $s = t_1 + s_1 - t$ into (3.21). This yields after simplifications:

$$\begin{aligned} v_\epsilon(x, t) \ge v_\epsilon(x_1, t_1) &+ (4s_1^3/\delta - \alpha)(t - t_1) - 6s_1^2(t-t_1)^2/\delta \\ &+ O(|x - x_1|^4 + |t - t_1|^3) \quad \text{as } (x, t) \to (x_1, t_1). \end{aligned}$$

Now v_ϵ is a weak supersolution near (x_1, t_1). Thus (2.16), (2.18) with $x_0 = x_1$, $t_0 = t_1$, $p = 0$, $q = 4s_1^3/\delta - \alpha$, $r_{n+1,n+1} = -12s_1^2/\delta$, and $r_{ij} = 0$ otherwise, imply

(3.23) $$4s_1^3/\delta - \alpha \ge 0.$$

But now we have a contradiction with (3.22), since $\alpha > 0$. This establishes (3.20).

5. Note next that in general if $f: \mathbb{R}^m \to \mathbb{R}$ is convex, then so is the mapping $(w, z) \mapsto f(w + z)$ on $\mathbb{R}^{2m}$. Consequently Lemma 3.1(vi) asserts

$$(x, y, t, s) \mapsto u^\epsilon(x + y, t + s) + \epsilon^{-1}(|x - y|^2 + (t + s)^2)$$

is convex. As

$$(x, t) \mapsto -v_\epsilon(x, t) + \epsilon^{-1}(|x|^2 + t^2)$$

is convex as well, we see that

$$(x, y, t, s) \mapsto \Phi(x, y, t, s) + C(|x|^2 + |y|^2 + t^2 + s^2)$$

is convex near (x_1, y_1, t_1, s_1), for some sufficiently large constant $C = C(\epsilon, \delta)$. Since Φ additionally attains its maximum at (x_1, y_1, t_1, s_1) we may invoke Jensen [25]: there exist points $\{(x^k, y^k, t^k, s^k)\}_{k=1}^\infty$ such that

(3.24) $$(x^k, y^k, t^k, s^k) \to (x_1, y_1, t_1, s_1),$$

(3.25) $$\begin{gathered} \Phi,\ u^\epsilon \text{ and } v_\epsilon \text{ are each twice differentiable} \\ \text{at } (x^k, y^k, t^k, s^k)\ (k = 1, 2, \dots), \end{gathered}$$

(3.26) $$D_{x,y,t,s}\Phi(x^k, y^k, t^k, s^k) \to 0,$$

(3.27) $$D^2_{x,y,t,s}\Phi(x^k, y^k, t^k, s^k) \le o(1) I_{2n+2} \quad \text{as } k \to \infty.$$

6. Using (3.13), (3.25), we see

$$(3.28)\quad \begin{aligned} D_x\Phi(x^k, y^k, t^k, s^k) &= Du^\epsilon(x^k + y^k, t^k + s^k) - Dv_\epsilon(x^k, t^k) \\ &\equiv p^k - \overline{p}^k, \end{aligned}$$

$$\begin{aligned} D_y\Phi(x^k, y^k, t^k, s^k) &= Du^\epsilon(x^k + y^k, t^k + s^k) - 4|y^k|^2 y^k/\delta \\ &= p^k - 4|y^k|^2 y^k/\delta. \end{aligned}$$

Since $y^k \to y_1$, we deduce from (3.26) that

$$(3.30)\qquad p^k, \overline{p}^k \to 4|y_1|^2 y_1/\delta \equiv p \quad \text{in } \mathbf{R}^n.$$

Assertion (3.20) tells us $p \neq 0$ and so $p^k, \overline{p}^k \neq 0$ for large enough k.

Again employing (3.13), (3.26) we note

$$(3.31)\quad \begin{aligned} \Phi_t(x^k, y^k, t^k, s^k) &= u_t^\epsilon(x^k + y^k, t^k + s^k) - v_{\epsilon t}(x^k, t^k) - \alpha \\ &\equiv q^k - \overline{q}^k - \alpha. \end{aligned}$$

As u^ϵ and v_ϵ are Lipschitz, we may assume, upon passing to a subsequence and reindexing if necessary, that

$$(3.32)\qquad q^k \to q,\ \overline{q}^k \to \overline{q} \quad \text{in } \mathbf{R}.$$

Then (3.26) and (3.31) ensure

$$(3.33)\qquad q - \overline{q} = \alpha > 0.$$

7. Next, (3.13) and (3.25) imply

$$(3.34)\quad \begin{aligned} D_x^2\Phi(x^k, y^k, t^k, s^k) &= D^2u^\epsilon(x^k + y^k, t^k + s^k) - D^2v_\epsilon(x^k, t^k) \\ &\equiv R^k - \overline{R}^k. \end{aligned}$$

Now (3.27) forces

$$(3.35)\qquad R^k - \overline{R}^k \le \epsilon_k I_n,$$

where $\epsilon_k \to 0$. Furthermore, Lemma 3.1(vi) shows $R^k \ge -CI_n$ and $\overline{R}^k \le CI_n$, for $C = C(\epsilon)$. Thus

$$-CI_n \le R^k \le \overline{R}^k + \epsilon_k I_n \le CI_n.$$

We may consequently suppose, passing as necessary to subsequences, that

$$(3.36)\qquad R^k \to R,\ \overline{R}^k \to \overline{R} \quad \text{in } S^{n\times n},$$

with

$$R \leq \overline{R}. \tag{3.37}$$

8. Now recall (3.25) holds and $p^k \equiv Du^\epsilon(x^k + y^k, t^k + s^k)$, $\overline{p}^k \equiv Dv_\epsilon(x^k, t^k)$ are nonzero for large k. Since u^ϵ is a weak subsolution near $(x_1 + y_1, t_1 + s_1)$ and v_ϵ is a weak supersolution near (x_1, t_1), we thus have

$$q^k \leq (\delta_{ij} - p_i^k p_j^k/|p^k|^2)r_{ij}^k \quad \text{and} \quad \overline{q}^k \geq (\delta_{ij} - \overline{p}_i^k \overline{p}_j^k/|\overline{p}^k|^2)\overline{r}_{ij}^k$$

for all large k. We send k to infinity, recalling (3.30), (3.32), and (3.36) to obtain

$$q \leq (\delta_{ij} - p_i p_j/|p|^2)r_{ij} \quad \text{and} \quad \overline{q} \geq (\delta_{ij} - p_i p_j/|p|^2)\overline{r}_{ij},$$

and, by subtracting,

$$q - \overline{q} \leq (\delta_{ij} - p_i p_j/|p|^2)(r_{ij} - \overline{r}_{ij}).$$

Now the matrix $((\delta_{ij} - p_i p_j/|p|^2))$ is nonnegative and $R - \overline{R}$ is nonpositive, by (3.37). Consequently $q - \overline{q} \leq 0$, a contradiction to (3.33).

3.3. Contraction property.

Theorem 3.3. *Assume that u and v are weak solutions of* (2.5), *such that*

$$u \text{ and } v \text{ are constant on } \mathbb{R}^n \times [0, \infty) \cap \{|x| + t \geq R\} \tag{3.38}$$

for some constant $R > 0$. Then

$$\max_{0 \leq t < \infty} \|u(\cdot, t) - v(\cdot, t)\|_{L^\infty(R^n)} = \|u(\cdot, 0)0 - v(\cdot, 0)\|_{L^\infty(R^n)}. \tag{3.39}$$

Proof. Should (3.39) fail, we may assume

$$\max_{(x,t)\in R^n\times[0,\infty)} (u - v) \equiv a > \|u(\cdot, 0) - v(\cdot, 0)\|_{L^\infty(R^n)} \equiv b.$$

Then as in the proof of Theorem 3.2 as above, there exist $\alpha, \epsilon, \delta > 0$ such that $\max_{(x,t),(x+y,t+s)\in R^n\times[0,\infty)} \Phi > b$, where Φ is defined by (3.13). We find a point (x_1, y_1, t_1, s_1) satisfying (3.15) and check (3.17) is valid provided $\epsilon, \delta > 0$ are small enough. The rest of the proof follows from that for Theorem 3.2.

4. Existence of weak solutions

4.1. Approximation; geometric interpretation. We turn our attention now to constructing a weak solution of the initial value problem (2.5), (2.6). We will assume that

$$g \text{ is constant on } \{\mathbb{R}^n\} \cap \{|x| \geq S\} \tag{4.1}$$

for some constant $S > 0$ and additionally, for the moment at least, g is smooth.

Our intention is to approximate (2.5), (2.6) by the PDE

$$u_t^\epsilon = \left(\delta_{ij} - \frac{u_{x_i}^\epsilon u_{x_j}^\epsilon}{|Du^\epsilon|^2 + \epsilon^2}\right) u_{x_i x_j}^\epsilon \quad \text{in } \mathbf{R}^n \times (0, \infty), \tag{4.2}$$

$$u^\epsilon = g \quad \text{on } \mathbf{R}^n \times \{t = 0\}, \tag{4.3}$$

for $0 < \epsilon < 1$. (The superscript ϵ here and hereafter is only a label and does not mean the sup-convolution (3.1).)

We interpret (4.2), (4.3) geometrically as follows. Assuming for the moment $u^\epsilon = u^\epsilon(x, t)$ to be a smooth solution of (4.2), (4.3), write $y = (x, x_{n+1}) \in \mathbf{R}^{n+1}$ and define

$$v^\epsilon(y, t) \equiv u^\epsilon(x, t) - \epsilon x_{n+1}. \tag{4.4}$$

Then $|D_y v^\epsilon|^2 = |Du^\epsilon|^2 + \epsilon^2$, and thus our PDE (4.2) becomes

$$v_t^\epsilon = (\delta_{ij} - v_{y_i}^\epsilon v_{y_j}^\epsilon / |Dv^\epsilon|^2) v_{y_i y_j}^\epsilon \quad \text{in } \mathbf{R}^{n+1} \times [0, \infty), \tag{4.5}$$

$$v^\epsilon = g^\epsilon \quad \text{on } \mathbf{R}^{n+1} \times \{t = 0\}, \tag{4.6}$$

for $g^\epsilon(y) \equiv g(x) - \epsilon x_{n+1}$. As noted in §2, the PDE (4.5) says that each level set of v^ϵ evolves according to its mean curvature. This is, in particular, the case for the zero level sets

$$\Gamma_t^\epsilon \equiv \{y \in \mathbf{R}^{n+1} | v^\epsilon(y, t) = 0\}.$$

But according to (4.4) each Γ_t^ϵ is a graph:

$$\Gamma_t^\epsilon = \{y = (x, x_{n+1}) \in \mathbf{R}^{n+1} | x_{n+1} = \epsilon^{-1} u^\epsilon(x, t)\},$$

and Ecker and Huisken [10] have shown the evolution of an entire graph by mean curvature remains a smooth entire graph for all time.

Geometrically, if as in §1 we are given Γ_0 as the boundary of a smooth, bounded, simply connected open set U in $\mathbf{R}^n$, we select a smooth function g with $g = 0$ on Γ_0, $g < 0$ in U, $g > 0$ in $\mathbf{R}^n - \overline{U}$. Then $\Gamma_0^\epsilon \subset \mathbf{R}^{n+1}$ is the graph $\{x_{n+1} = \epsilon^{-1} g(x)\}$ as drawn in Figure 3 (next page).

For small ϵ, Γ_0^ϵ roughly approximates the cylinder $\Gamma_0 \times \mathbf{R}$. We may thus hope that for moderate $t > 0$ and small $\epsilon > 0$, the smooth graph Γ_t^ϵ will be close to the cylinder $\Gamma_t \times \mathbf{R}$, Γ_t denoting the evolution of Γ_0 via its mean curvature in $\mathbf{R}^n$ (see Figure 4).

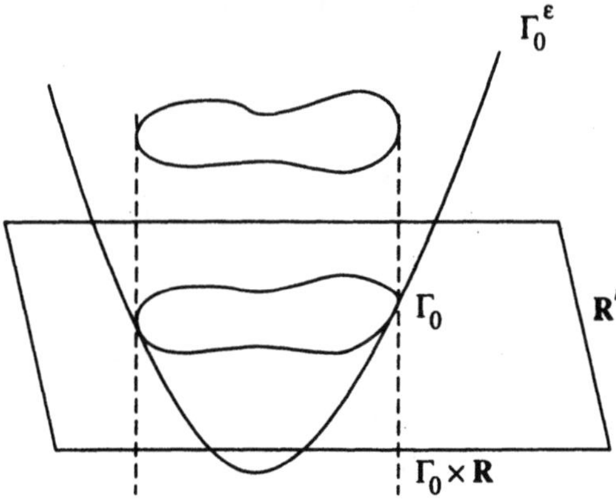

FIGURE 3

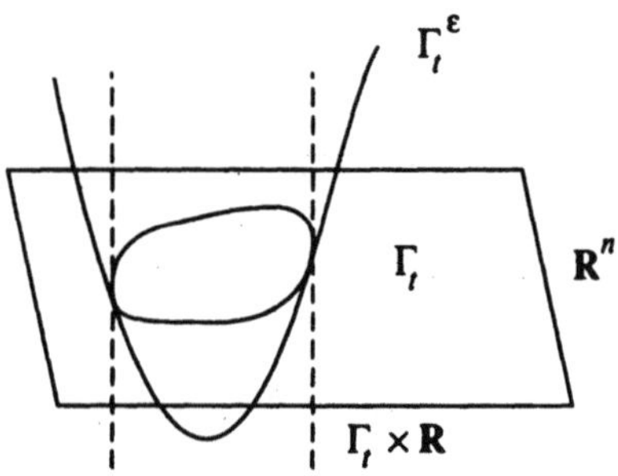

FIGURE 4

The idea then is that the complicated, possibly singular behavior of $\{\Gamma_t\}_{t\geq 0}$ in $\mathbf{R}^n$ will be approximated by the smooth evolution $\{\Gamma_t^\epsilon\}_{t\geq 0}$ in $\mathbf{R}^{n+1}$; in the sense that for a given $t > 0$, $\Gamma_t^\epsilon \approx \Gamma_t \times \mathbf{R}$ if $\epsilon > 0$ is very small. The illustrations provided make this expectation appear plausible, although there are a number of subtleties.

4.2. Solution of the approximate equations. We now investigate the approximations (4.2), (4.3) analytically.

Theorem 4.1. (i) *For each* $0 < \epsilon < 1$ *there exists a unique smooth, bounded solution* u^ϵ *of* (4.2), (4.3).

(ii) *Additionally,*

$$\sup_{0<\epsilon<1} \|u^\epsilon, Du^\epsilon, u_t^\epsilon\|_{L^\infty(R^n\times[0,\infty))} \leq C\|g\|_{C^{1,1}(R^n)}. \tag{4.7}$$

Proof. 1. For each $0 < \sigma < 1$, consider the PDE

$$u_t^{\epsilon,\sigma} = a_{ij}^{\epsilon,\sigma}(Du^{\epsilon,\sigma})u_{x_ix_j}^{\epsilon\sigma} \quad \text{in } \mathbf{R}^n \times [0, \infty), \tag{4.8}$$

$$u^{\epsilon,\sigma} = g \quad \text{on } \mathbf{R}^n \times \{t = 0\}, \tag{4.9}$$

for

$$a_{ij}^{\epsilon,\sigma}(p) \equiv (1+\sigma)\delta_{ij} - \frac{p_i p_j}{|p|^2+\epsilon^2} \qquad (p \in \mathbb{R}^n,\ 1 \le i, j \le n).$$

The smooth bounded coefficients $\{a_{ij}\}$ satisfy also the uniform ellipticity condition

$$\sigma|\xi|^2 \le a_{ij}^{\epsilon,\sigma}(p)\xi_i\xi_j \qquad (\xi \in \mathbb{R}^n)$$

for each $p \in \mathbb{R}^n$, and consequently classical PDE theory gives the existence of a unique smooth bounded solution $u^{\epsilon,\sigma}$ (see, e.g., Ladyzhenskaja, Solonnikov, and Ural′tseva [31]). By the maximum principle,

(4.10) $$\|u^{\epsilon,\sigma}\|_{L^\infty(R^n\times[0,\infty))} = \|g\|_{L^\infty(R^n)}.$$

2. Now differentiate (4.8) with respect to x_l:

$$u_{x_l t}^{\epsilon,\sigma} = a_{ij}^{\epsilon,\sigma}(Du^{\epsilon,\sigma})u_{x_l x_i x_j}^{\epsilon,\sigma} + a_{ij,p_k}^{\epsilon,\sigma}(Du^{\epsilon,\sigma})u_{x_l x_k}^{\epsilon,\sigma}u_{x_i x_j}^{\epsilon,\sigma}.$$

The maximum principle then implies

(4.11) $$\|Du^{\epsilon,\sigma}\|_{L^\infty(R^n\times[0,\infty))} = \|Dg\|_{L^\infty(R^n)}.$$

Similarly

(4.12) $$\|u_t^{\epsilon,\sigma}\|_{L^\infty(R^n\times[0,\infty))} = \|u_t^{\epsilon,\sigma}(\cdot, 0)\|_{L^\infty(R^n)} \le C\|D^2 g\|_{L^\infty(R^n)}.$$

3. Since

$$\left(1 - \frac{L^2}{L^2+\epsilon^2}\right)|\xi|^2 \le a_{ij}^{\epsilon,\sigma}(p)\xi_i\xi_j \qquad (\xi \in \mathbb{R}^n)$$

provided $|p| \le L$, we deduce from (4.10)–(4.12) and classical estimates that we have bounds, uniform in $0 < \sigma < 1$, on the derivatives of all orders of $\{u^{\epsilon,\sigma}\}_{0<\sigma<1}$. Consequently, uniqueness of the limit implies for each multi-index α,

$$D^\alpha u^{\epsilon,\sigma} \to D^\alpha u^\epsilon \quad \text{locally uniformly as } \sigma \to 0,$$

for a smooth function u^ϵ solving (4.2), (4.3). Estimate (4.7) follows from (4.10)–(4.12).

4.3. Passage to limits.

Theorem 4.2. *Assume* $g: \mathbb{R}^n \to \mathbb{R}$ *is continuous and satisfies* (4.1). *Then there exists a weak solution* u *of* (2.5), (2.6), *such that*

(4.13) $$u \text{ is constant on } \mathbb{R}^n \times [0,\infty) \cap \{|x| + t \ge R\}$$

for some $R > 0$, *depending only on the constant* S *from* (4.1).

Proof. 1. Suppose temporarily g is smooth. Employing estimate (4.7) we can extract a subsequence $\{u^{\epsilon_k}\}_{k=1}^\infty \subset \{u^\epsilon\}_{0<\epsilon\le 1}$ so that $\epsilon_k \to 0$ and

$u^{\epsilon_k} \to u$ locally uniformly in $\mathbb{R}^n \times [0, \infty)$, for some bounded, Lipschitz function u.

2. We assert now that u is a weak solution of (2.5), (2.6). For this, let $\phi \in C^\infty(\mathbb{R}^{n+1})$ and suppose $u - \phi$ has a *strict* local maximum at a point $(x_0, t_0) \in \mathbb{R}^n \times (0, \infty)$. As $u^{\epsilon_k} \to u$ uniformly near (x_0, t_0), $u^{\epsilon_k} - \phi$ has a local maximum at a point (x_k, t_k), with

$$(x_k, t_k) \to (x_0, t_0) \quad \text{as } k \to \infty. \tag{4.14}$$

Since u^{ϵ_k} and ϕ are smooth, we have

$$Du^{\epsilon_k} = D\phi, \quad u_t^{\epsilon_k} = \phi_t, \quad D^2 u^{\epsilon_k} \le D^2\phi \quad \text{at } (x_k, t_k).$$

Thus (4.2) implies

$$\phi_t - \left(\delta_{ij} - \frac{\phi_{x_i}\phi_{x_j}}{|D\phi|^2 + \epsilon_k^2}\right)\phi_{x_i x_j} \le 0 \quad \text{at } (x_k, t_k). \tag{4.15}$$

Suppose first $D\phi(x_0, t_0) \neq 0$. Then $D\phi(x_k, t_k) \neq 0$ for large k. We consequently may pass to limits in (4.15), recalling (4.14) to deduce

$$\phi_t \le (\delta_{ij} - \phi_{x_i}\phi_{x_j}/|D\phi|^2)\phi_{x_i x_j} \quad \text{at } (x_0, t_0). \tag{4.16}$$

Next, assume instead $D\phi(x_0, t_0) = 0$. Set

$$\eta^k \equiv \frac{D\phi(x_k, t_k)}{(|D\phi(x_k, t_k)|^2 + \epsilon_k^2)^{1/2}},$$

so that (4.15) becomes

$$\phi_t \le (\delta_{ij} - \eta_i^k \eta_j^k)\phi_{x_i x_j} \quad \text{at } (x_k, t_k). \tag{4.17}$$

Since $|\eta^k| \le 1$, we may assume, upon passing to a subsequence and reindexing if necessary, that $\eta^k \to \eta$ in $\mathbb{R}^n$ for some $|\eta| \le 1$. Sending k to infinity in (4.17), we discover

$$\phi_t \le (\delta_{ij} - \eta_i \eta_j)\phi_{x_i x_j} \quad \text{at } (x_0, t_0). \tag{4.18}$$

If $u-\phi$ has a local maximum, but not necessarily a strict local maximum at (x_0, t_0), we repeat the argument above with $\phi(x, t)$ replaced by

$$\tilde{\phi}(x, t) = \phi(x, t) + |x - x_0|^4 + (t - t_0)^4,$$

again to obtain (4.16) or (4.18).

Consequently, u is a weak subsolution. That u is a weak supersolution follows analogously.

3. Finally we verify u satisfies (4.13). Upon rescaling as necessary, we may as well assume

$$|g| \le 1 \quad \text{on } \mathbb{R}^n, \qquad g = 0 \quad \text{on } \mathbb{R}^n \cap \{|x| \ge 1\}. \tag{4.19}$$

Consider now the auxiliary function (cf. Brakke [5, p. 25])

$$v(x, t) \equiv \Psi(|x|^2/2 + (n-1)t) \qquad (x \in \mathbb{R}^n,\ t > 0), \tag{4.20}$$

for

$$\Psi(s) \equiv \begin{cases} 0 & (s \ge 2), \\ (s-2)^3 & (0 \le s \le 2). \end{cases}$$

Then for $\Psi \in C^2([0, \infty))$,

$$\Psi'(s) = \begin{cases} 0 & (s \ge 2), \\ 3(s-2)^2 & (0 \le s \le 2), \end{cases} \qquad \Psi''(s) = \begin{cases} 0 & (s \ge 2), \\ 6(s-2) & (0 \le s \le 2). \end{cases}$$

In particular,

$$|\Psi''(s)| \le C(\Psi'(s))^{1/2} \qquad (s \ge 0). \tag{4.21}$$

Now

$$\begin{aligned} v_t &- \left(\delta_{ij} - \frac{v_{x_i} v_{x_i}}{|Dv|^2 + \epsilon^2}\right) v_{x_i x_j} \\ &= (n-1)\Psi' - \left(\delta_{ij} - \frac{(\Psi')^2 x_i x_j}{(\Psi')^2 |x|^2 + \epsilon^2}\right) (\Psi' \delta_{ij} + \Psi'' x_i x_j) \\ &= \Psi' \left[(n-1) - \left(\delta_{ij} - \frac{(\Psi')^2 x_i x_j}{(\Psi'^2 |x|^2 + \epsilon^2}\right) \delta_{ij}\right] \\ &\quad - \Psi'' \left[\left(\delta_{ij} - \frac{(\Psi')^2 x_i x_j}{(\Psi')^2 |x|^2 + \epsilon^2}\right) x_i x_j\right] \\ &\equiv A + B. \end{aligned} \tag{4.22}$$

We further compute

$$A = -\Psi' \frac{\epsilon^2}{(\Psi')^2 |x|^2 + \epsilon^2} \le 0, \tag{4.23}$$

since $\Psi' \ge 0$. Moreover,

$$|B| = |\Psi''| \frac{\epsilon^2 |x|^2}{(\Psi')^2 |x|^2 + \epsilon^2}.$$

Now if $|\Psi'| \le \epsilon$, then

$$|B| \le |\Psi''|\,|x|^2 \le C|\Psi''| \quad (\text{since } \Psi'' = 0 \text{ if } |x| \ge 2) \\ \le C(\Psi')^{1/2} \quad (\text{by } (4.21)) \\ \le C\epsilon^{1/2}. \tag{4.24}$$

On the other hand if $|\Psi'| \ge \epsilon$, we have

$$|B| \le |\Psi''| \frac{\epsilon^2}{(\Psi')^2} \le \frac{C\epsilon^2}{|\Psi'|^{3/2}} \le C\epsilon^{1/2}. \tag{4.25}$$

Combining (4.22)–(4.25) yields

$$v_t - \left(\delta_{ij} - \frac{v_{x_i} v_{x_j}}{|Dx|^2 + \epsilon^2}\right) v_{x_i x_j} \le C\epsilon^{1/2}$$

and so

$$w_t^\epsilon \le \left(\delta_{ij} - \frac{w_{x_i}^\epsilon w_{x_j}^\epsilon}{|Dw^\epsilon|^2 + \epsilon^2}\right) w_{x_i x_j}^\epsilon \quad \text{in } \mathbf{R}^n \times (0, \infty) \tag{4.26}$$

for

$$w^\epsilon(x, t) \equiv v(x, t) - Ct\epsilon^{1/2}. \tag{4.27}$$

Now

$$w^\epsilon(x, 0) = \Psi(|x|^2/2) = 0 \quad \text{if } |x| \ge 2$$

and

$$w^\epsilon(x, 0) = \Psi(|x|^2/2) \le -1 \quad \text{if } |x| \le 1.$$

Consequently, we see from (4.19) that

$$w^\epsilon \le g \quad \text{on } \mathbf{R}^n \times \{t = 0\}. \tag{4.28}$$

Applying the maximum principle to (4.2), (4.3), (4.26), and (4.27), we deduce $w^\epsilon \le u^\epsilon$ in $\mathbf{R}^n \times [0, \infty)$ for each $0 < \epsilon < 1$. Sending $\epsilon = \epsilon_k$ to zero, we then have

$$\Psi(|x|^2/2 + (n-1)t) = v(x, t) \le u(x, t)$$

for all $x \in \mathbf{R}^n$, $t \ge 0$. Thus, $u \ge 0$ if $|x|^2/2 + (n+1)t \ge 2$. Similarly,

$$\tilde{w}_t^\epsilon \ge \left(\delta_{ij} - \frac{\tilde{w}_{x_i}^\epsilon \tilde{w}_{x_j}^\epsilon}{|D\tilde{w}^\epsilon|^2 + \epsilon^2}\right) \tilde{w}_{x_i x_j}^\epsilon \quad \text{in } \mathbf{R}^n \times (0, \infty), \tag{4.29}$$

$$\tilde{w}^\epsilon \ge g \quad \text{on } \mathbf{R}^n \times (0, \infty), \tag{4.30}$$

for $\tilde{w}^\epsilon \equiv -w^\epsilon$. As above we consequently deduce

$$u \le 0 \quad \text{if } |x|^2/2 + (n+1)t \ge 2.$$

Assertion (4.13) is proved.

4. According to the uniqueness assertion Theorem 3.2, in fact the full limit $\lim_{\epsilon \to 0} u^\epsilon = u$ exists. Note also from Theorem 3.3 that

$$\|u - \tilde{u}\|_{L^\infty(R^n \times [0,\infty))} = \|g - \tilde{g}\|_{L^\infty(R^n)} \tag{4.31}$$

if $\tilde{u}$ is the solution built as above for a smooth initial function $\tilde{g}$ satisfying (4.1).

Suppose at last g satisfies (4.1), but is only continuous. We select smooth $\{g^k\}_{k=1}^\infty$, satisfying (4.1) (for the same S) so that $g^k \to g$ uniformly on $\mathbb{R}^n$. Denote by u^k the solution of (2.5), (2.6) constructed above with initial function g^k. Utilizing (4.31) we see that the limit $\lim_{k\to\infty} u^k = u$ exists uniformly on $\mathbb{R}^n \times [0, \infty)$. According to Theorem 2.7 u is a weak solution of (2.5), (2.6).

5. Definition of the generalized evolution by mean curvature

We now make precise the definition of the motion $\{\Gamma_t\}_{t>0}$ for a given initial hypersurface Γ_0. In fact, let us assume now only that

$$\Gamma_0 \text{ is a compact subset of } \mathbb{R}^n. \tag{5.1}$$

Choose then any continuous function $g: \mathbb{R}^n \to \mathbb{R}$ satisfying

$$\Gamma_0 = \{x \in \mathbb{R}^n | g(x) = 0\} \tag{5.2}$$

and

$$g \text{ constant on } \mathbb{R}^n \cap \{|x| \ge S\} \tag{5.3}$$

for some $S > 0$. Utilizing Theorems 3.2 and 4.1, we see that there is a unique weak solution of the mean curvature evolution equation

$$u_t = (\delta_{ij} - u_{x_i} u_{x_j}/|Du|^2) u_{x_i x_j} \quad \text{in } \mathbb{R}^n \times (0, \infty), \tag{5.4}$$

$$u = g \quad \text{on } \mathbb{R}^n \times \{t = 0\}, \tag{5.5}$$

with

$$u \text{ constant on } \mathbb{R}^n \times [0, \infty) \cap \{|x| + t \ge R\} \tag{5.6}$$

for some $R > 0$.

Define then the compact set

$$\Gamma_t \equiv \{x \in \mathbf{R}^n \mid u(x, t) = 0\} \tag{5.7}$$

for each $t > 0$. We call $\{\Gamma_t\}_{t>0}$ the *generalized evolution by mean curvature* of the original compact set Γ_0.

We must first verify that $\{\Gamma_t\}_{t>0}$ is well defined.

Theorem 5.1. *Assume $\hat{g}: \mathbf{R}^n \to \mathbf{R}$ is continuous, with*

$$\Gamma_0 = \{x \in \mathbf{R}^n \mid \hat{g}(x) = 0\} \tag{5.8}$$

and

$$\hat{g} \text{ constant on } \mathbf{R}^n \cap \{|x| \geq S\}. \tag{5.9}$$

Suppose $\hat{u}$ is the unique weak solution of (5.4)–(5.6), *with $\hat{g}$ replacing g. Then*

$$\Gamma_t = \{x \in \mathbf{R}^n \mid \hat{u}(x, t) = 0\} \tag{5.10}$$

for each $t > 0$. Consequently our definition (5.7) *does not depend upon the particular choice of initial function g satisfying* (5.2), (5.3).

A related assertion for the level sets of solutions to homogeneous Hamilton-Jacobi PDE may be found in Evans-Souganidis [13, §7].

Proof. 1. First, we may as well assume $g \geq 0$ on $\mathbf{R}^n$ and thus $u \geq 0$ in $\mathbf{R}^n \times (0, \infty)$. Indeed, if g is negative somewhere, we can consider the PDE (5.4)–(5.6) with $|g|$ replacing g, the unique solution of which, owing to Theorems 2.8 and 3.2, is $|u|$. Our definition (5.7) is unchanged if we replace u by $|u|$. Similarly we may suppose $\hat{g}, \hat{u} \geq 0$. Set

$$\hat{\Gamma}_t = \{x \in \mathbf{R}^n \mid \hat{u}(x, t) = 0\} \qquad (t \geq 0).$$

2. For $k = 1, 2, \ldots$ write $E_0 = \emptyset$ and $E_k = \{x \in \mathbf{R}^n \mid g(x) \geq 1/k\}$, so that

$$E_1 \subset \cdots \subset E_k \subset E_{k+1} \subset \cdots, \qquad \mathbf{R}^n - \Gamma_0 = \bigcup_{k=1}^{\infty} E_k. \tag{5.11}$$

Define

$$a_k \equiv \max_{R^n - E_{k-1}} \hat{g} > 0 \qquad (k = 1, 2, \ldots). \tag{5.12}$$

Then $a_1 \geq a_2 \geq \cdots$ and $\lim_{k\to\infty} a_k = 0$, according to (5.8) and (5.11). Next define the continuous function $\Psi: [0, \infty) \to [0, \infty)$ satisfying

$$\begin{gathered} \Psi(0) = 0, \\ \Psi(1/k) = a_k \qquad (k = 1, 2, \ldots), \\ \Psi \text{ linear on } [1/(k+1), 1/k] \qquad (k = 1, 2, \ldots), \\ \Psi \text{ constant on } [1, \infty). \end{gathered}$$

3. Write $\tilde{g} = \Psi(g)$ and $\tilde{u} = \Psi(u)$. Then $\tilde{u}$ solves (5.4)–(5.6), with $\tilde{g}$ replacing g. Now $\tilde{g} = \hat{g} = 0$ on Γ_0. Furthermore, if $x \in E_k - E_{k-1}$, then

$$\tilde{g}(x) = \Psi(g(x)) \geq \Psi(1/k) = a_k \geq \hat{g}(x) \quad \text{by (5.12)}.$$

Thus $\tilde{g} \geq \hat{g}$ on $\mathbf{R}^n$. Consequently, Theorem 3.2 asserts

$$\tilde{u} = \Psi(u) \geq \hat{u} \geq 0 \quad \text{on } \mathbf{R}^n \times [0, \infty).$$

Thus if $x \in \Gamma_t$, then $\hat{u}(x, t) = 0$ and so $x \in \hat{\Gamma}_t$. Hence $\Gamma_t \subseteq \hat{\Gamma}_t$. The opposite inclusion is similarly proved, and therefore $\Gamma_t = \hat{\Gamma}_t$ for each $t > 0$. q.e.d.

In light of this theorem, we can regard the mappings $\Gamma_0 \mapsto \Gamma_t$ $(t \geq 0)$ as comprising a time-dependent evolution on the collection $\mathscr{K}$ of compact subsets of $\mathbf{R}^n$. Let us write

$$\mathscr{M}(t)\Gamma_0 \equiv \Gamma_t \qquad (t \geq 0) \tag{5.13}$$

explicitly to display the dependence of Γ_t on t and Γ_0. Then $\mathscr{M}(t): \mathscr{K} \to \mathscr{K}$ for each $t \geq 0$, and $\mathscr{M}(0)$ is the identity operator. We will call $\{\mathscr{M}(t)\}_{t\geq 0}$ the *mean-curvature semigroup on* $\mathscr{K}$.

To justify this terminology, let us verify the semigroup property.

Theorem 5.2. *We have*

$$\mathscr{M}(t+s) = \mathscr{M}(t)\mathscr{M}(s) \qquad (t, s \geq 0). \tag{5.14}$$

Proof. If $t, s > 0$ and $\Gamma_0 \in \mathscr{K}$, choose any continuous function g satisfying (5.2), (5.3). Let u be the corresponding unique weak solution of (5.4)–(5.6). Then

$$\mathscr{M}(t+s)\Gamma_0 = \Gamma_{t+s} = \{x \in \mathbf{R}^n | u(x, t+s) = 0\}, \tag{5.15}$$

$$\mathscr{M}(s)\Gamma_0 = \Gamma_s = \{x \in \mathbf{R}^n | u(x, s) = 0\}. \tag{5.16}$$

To compute $\mathscr{M}(t)\Gamma_s$ we select any continuous function $\hat{g}$ so that

$$\Gamma_s = \{x \in \mathbf{R}^n | \hat{g}(x) = 0\} \tag{5.17}$$

and $\hat{g}$ is constant outside some large ball. We then find the unique weak solution $\hat{u}$ of (5.4)–(5.6) (with $\hat{g}$ replacing g) and set

$$\mathscr{M}(t)\Gamma_s = \hat{\Gamma}_t = \{x \in \mathbf{R}^n | \hat{u}(x, t) = 0\}. \tag{5.18}$$

According to Theorem 5.1, this construction is independent of the particular choice of $\hat{g}$ satisfying (5.17). In particular, we may as well take $\hat{g}(x) = u(x, s)$ $(x \in \mathbf{R}^n)$. Owing then to the uniqueness of a weak solution to (5.4)–(5.6) we have

$$\hat{u}(x, t) = u(x, t+s) \qquad (x \in \mathbf{R}^n, \ t > 0).$$

Consequently (5.15) and (5.18) imply

$$\mathscr{M}(t+s)\Gamma_0 = \mathscr{M}(t)\mathscr{M}(s)\Gamma_0,$$

as required. This establishes (5.14). q.e.d.

Note that we make no assertions concerning continuity of the mapping $(t, \Gamma_0) \mapsto \mathscr{M}(t)\Gamma_0$.

6. Consistency with classical motion by mean curvature

We must now check that our generalized evolution by mean curvature agrees with the classical motion, if and so long as the latter exists. Let us therefore suppose for this section that Γ_0 is a smooth hypersurface, the connected boundary of a bounded open set $U \subset \mathbf{R}^n$. According to Hamilton [22], Gage-Hamilton [17], and Evans-Spruck [14], there exists a time $t_* > 0$ and a family $\{\Sigma_t\}_{0\le t<t_*}$ of smooth hypersurfaces evolving from $\Sigma_0 = \Gamma_0$ according to classical motion by mean curvature. In particular for each $0 \le t < t_*$, Σ_t is diffeomorphic to Γ_0, and is the boundary of an open set U_t diffeomorphic to $U_0 \equiv U$.

Theorem 6.1. *We have* $\Sigma_t = \Gamma_t$ $(0 \le t < t_*)$, *where* $\{\Gamma_t\}_{t\ge 0}$ *is the generalized evolution by mean curvature defined in* §5.

Proof. 1. Fix $0 < t_0 < t_*$, and define then for $0 \le t \le t_0$ the (signed) distance function

$$d(x, t) \equiv \begin{cases} -\operatorname{dist}(x, \Sigma_t) & \text{if } x \in U_t, \\ \operatorname{dist}(x, \Sigma_t) & \text{if } x \in \mathbf{R}^n \setminus \overline{U}_t. \end{cases}$$

As $\Sigma \equiv \bigcup_{0\le t\le t_0} \Sigma_t \times \{t\}$ is smooth, d is smooth in the regions

$$Q^+ \equiv \{(x, t) | 0 \le d(x, t) \le \delta_0, \ 0 \le t \le t_0\}$$

and

$$Q^- \equiv \{(x, t) | -\delta_0 \le d(x, t) \le 0, \ 0 \le t \le t_0\}$$

for $\delta_0 > 0$ sufficiently small.

2. Now if $\delta_0 > 0$ is small enough, for each point $(x, t) \in Q^+$ there exists a unique point $y \in \Sigma_t$ verifying $d(x, t) = |x - y|$. Consider now near (y, t) the smooth unit vector field $\nu \equiv Dd$ pointing from Σ into Q^+. Then

$$d_t(x, t) = (\operatorname{div} \nu)(y, t) \tag{6.1}$$

since $\{\Sigma_t\}_{0\le t\le t_*}$ is a classical evolution by mean curvature. Additionally, the eigenvalues of $D^2 d(x, t)$ are (see, e.g., Gilbarg-Trudinger [18, p. 355])

$$\left\{ \frac{-\kappa_1}{1-\kappa_1 d}, \dots, \frac{-\kappa_{n-1}}{1-\kappa_{n-1} d}, 0 \right\}, \tag{6.2}$$

$\kappa_1, \cdots, \kappa_{n-1}$ denoting the principal curvatures of Σ_t at the point y, calculated with respect to the unit normal field ν. Thus,

$$\Delta d(x, t) = -\sum_{i=1}^{n-1} \frac{\kappa_i}{1-\kappa_i d}. \tag{6.3}$$

However, $(\operatorname{div}\nu)(y, t) = -(\kappa_1 + \cdots + \kappa_{n-1})$, and so (6.1) and (6.3) imply

$$d_t - \Delta d = \left(\sum_{i=1}^{n-1} \frac{\kappa_i^2}{1-\kappa_i d}\right) d \quad \text{at } (x, t). \tag{6.4}$$

Since the quantity $\sum_{i=1}^{n-1} \kappa_i^2/(1-\kappa_i d)$ is uniformly bounded and $d \geq 0$ in Q^+, we deduce from (6.4) that

$$\underline{d} \equiv \alpha e^{-\lambda t} d \tag{6.5}$$

satisfies

$$\underline{d}_t - \Delta \underline{d} \leq 0 \quad \text{in } Q^+ \tag{6.6}$$

if $\lambda > 0$ is fixed large enough and $\alpha > 0$ (to be selected later). Furthermore, $|Dd|^2 = |\nu|^2 = 1$ and so $d_{x_i} d_{x_i x_j} = 0$ in Q^+ $(1 \leq j \leq n)$. The function $\underline{d}$ satisfies the same identity, whence (6.6) implies for each $\epsilon \geq 0$ that

$$\underline{d}_t - \left(\delta_{ij} - \frac{\underline{d}_{x_i} \underline{d}_{x_j}}{|D\underline{d}|^2 + \epsilon^2}\right) \underline{d}_{x_i x_j} \leq 0 \quad \text{in } Q^+. \tag{6.7}$$

We see therefore that $\underline{d}$ is a smooth subsolution of the approximate mean curvature evolution PDE (4.2) in Q^+.

3. Choose any Lipschitz function $g: \mathbf{R}^n \to \mathbf{R}^+$ so that $g(x) = \operatorname{dist}(x, \Sigma_0)$ near Σ_0, $\{g = 0\} = \Sigma_0$, and $g(x)$ is a positive constant for large $|x|$. For $0 < \epsilon < 1$ the approximating PDE (4.2), (4.3) then has a continuous solution u^ϵ, which is smooth in $\mathbf{R}^n \times (0, \infty)$. Additionally we have $u^\epsilon \to u$ locally uniformly, where

$$\Gamma_t = \{x \in \mathbf{R}^n | u(x, t) = 0\}, \qquad t \geq 0. \tag{6.8}$$

Now $u = g = \delta_0 > 0$ on $\{(x, 0) | \operatorname{dist}(x, \Sigma_0) = \operatorname{dist}(x, \Gamma_0) = \delta_0\}$; and, as u is continuous, we thus have

$$u \geq \delta_0/2 > 0 \quad \text{on } \{(x, t) | d(x, t) = \delta_0\} \tag{6.9}$$

for $0 \leq t \leq t_0$, provided $t_0 > 0$ is small enough. Hence (6.9) implies

$$u^\epsilon \geq \delta_0/4 \quad \text{on } \{(x, t) | d(x, t) = \delta_0\}$$

for $0 \le t \le t_0$, $0 < \epsilon \le \epsilon_0$, if $\epsilon_0 > 0$ is sufficiently small. Consequently there exists $0 < \alpha < 1$ so that

$$u^\epsilon \ge \underline{d} \quad \text{on } \{(x, t) | d(x, t) = \delta_0\} \tag{6.10}$$

for $0 \le t \le t_0$, $0 < \epsilon < \epsilon_0$, $\underline{d}$ defined by (6.5). Since $0 < \alpha < 1$, we have

$$u^\epsilon \ge \underline{d} \quad \text{on } \{(x, 0) | 0 \le d(x, 0) \le \delta_0\}. \tag{6.11}$$

Furthermore, $g \ge 0$ implies $u^\epsilon \ge 0$ and so

$$u^\epsilon \ge \underline{d} \quad \text{on } \{(x, t) | d(x, t) = 0\}. \tag{6.12}$$

4. Combining (6.10)–(6.12) we see that $u^\epsilon \ge \underline{d}$ on the parabolic boundary of Q^+. Since $\underline{d}$ solves (6.7) and u^ϵ solves (4.2), the maximum principle implies $u^\epsilon \ge \underline{d}$ in Q^+. Let $\epsilon \to 0$ to conclude

$$u > 0 \text{ in the interior of } Q^+. \tag{6.13}$$

A similar argument using instead $\underline{d} = -\alpha e^{-\lambda t} d$ shows

$$u > 0 \text{ in the interior of } Q^-. \tag{6.14}$$

Since $u > 0$ in $(R^n \setminus \{x | \operatorname{dist}(x, \Sigma_0) \le \delta_0\}) \times [0, t_0]$, we deduce from (6.13), (6.14), and (6.8) that

$$\Gamma_t \subseteq \Sigma_t = \{x | d(x, t) = 0\} \qquad (0 \le t \le t_0). \tag{6.15}$$

5. Now define a new function $\hat{g}: \mathbb{R}^n \to \mathbb{R}$ so that $\hat{g}(x) = d(x, 0)$ (the *signed* distance function to Σ_0) near $\Sigma_0 = \Gamma_0$, $\{\hat{g} = 0\} = \Sigma_0$, and $\hat{g}(x)$ is a positive constant for large $|x|$. Let $\hat{u}$ denote the unique weak solution of (2.5), (2.6), (4.13) for this new initial function $\hat{g}$. According to Theorem 5.1

$$\Gamma_t = \{x \in \mathbb{R}^n | \hat{u}(x, t) = 0\} \qquad (t \ge 0). \tag{6.16}$$

Since $\hat{g} < 0$ in U_0 we know by continuity that $\hat{u} < 0$ somewhere in U_t, provided $0 \le t \le t_0$ and t_0 is small. Similarly $\hat{u} > 0$ somewhere in $\mathbb{R}^n - \overline{U}_t$ for each $0 \le t \le t_0$. Fix any point $x_0 \in \Sigma_t$ and draw a smooth curve C in $\mathbb{R}^n$, intersecting Σ_t precisely at x_0 and connecting a point $x_1 \in U_t$, where $\hat{u}(x_1, t) < 0$, to a point $x_2 \in \mathbb{R}^n - \overline{U}_t$, where $\hat{u}(x_2, t) > 0$. As $\hat{u}$ is continuous, we must have $\hat{u}(x, t) = 0$ for some point x on the curve C. However (6.15) and (6.16) say that the set $\{x | \hat{u}(x, t) = 0\}$ lies in Σ_t. Thus $\hat{u}(x_0, t) = 0$. Since x_0 denotes any point on Σ_t we deduce from (6.15), (6.16) that

$$\Gamma_t = \Sigma_t \quad \text{if } 0 \le t \le t_0. \tag{6.17}$$

We have consequently demonstrated that the classical motion $\{\Sigma_t\}_{0\le t<t_*}$ and the generalized motion $\{\Gamma_t\}_{t\ge 0}$ agree at least on some short time interval $[0, t_0]$.

6. Write

$$s \equiv \sup_{0\le t<t_*} \{t | \Gamma_\tau = \Sigma_\tau \text{ for all } 0 \le \tau \le t\}$$

and suppose $s < t_*$. Then $\Gamma_t = \Sigma_t$ for all $0 \le t < s$, and so, applying the continuity of the solution u to (2.5) and (2.6) for g as above, we have $\Gamma_s \supseteq \Sigma_s$. On the other hand if $x \in \mathbb{R}^n - \Sigma_s$, there exists $r > 0$ so that $B(x, r) \subset \mathbb{R}^n - \Sigma_t$ for all $s - \epsilon \le t \le s$, $\epsilon > 0$ small enough. Using this we easily deduce $x \notin \Gamma_s$. Hence $\Gamma_s = \Sigma_s$. But then applying steps 1–5 we deduce $\Gamma_t = \Sigma_t$ for all $s \le t \le s + s_0 < t_*$, if $s_0 > 0$ is small enough. This contradicts the definition of s, and so in fact $s = t_*$. q.e.d.

Observe carefully that our argument in step 5 above improving (6.15) to (6.17) depends critically upon the possibility of finding an initial function $\hat{g}$ which changes sign above. Compare this with the geometric situation in Theorem 8.1 below.

7. Geometric properties of generalized evolution by mean curvature

We devote this section to establishing some elementary properties of the generalized evolution by mean curvature

$$\Gamma_0 \mapsto \mathcal{M}(t)\Gamma_0 \equiv \Gamma_t \qquad (t \ge 0) \tag{7.1}$$

for Γ_0 a compact subset of $\mathbb{R}^n$.

7.1. Localization and extinction. First of all, it is known that if Γ_0 is the sphere $\partial B(0, R)$, then

$$\Gamma_t = \begin{cases} \partial B(0, R(t)) & \text{if } 0 \le t < t^*, \\ \{0\} & \text{if } t = t^*, \\ \varnothing & \text{if } t > t^*, \end{cases} \tag{7.2}$$

where

$$R(t) \equiv (R^2 - 2(n-1)t)^{1/2} \quad \text{for } 0 \le t \le t^* \equiv R^2/2(n-1). \tag{7.3}$$

This assertion follows in our approach by noting $u(x, t) = \Psi(|x|^2 + 2(n-1)t)$ is a weak solution of (5.4), where $\Psi: \mathbb{R} \to \mathbb{R}$ is smooth with

$$\begin{cases} \Psi' \ge 0, \quad \Psi < 0 \text{ on } [0, R), \\ \Psi > 0 \text{ on } (R, 3R), \quad \Psi \equiv 1 \text{ on } [3R, \infty). \end{cases} \tag{7.4}$$

By making comparisons with the shrinking sphere (7.2) we derive now some elementary properties of the general motion (7.1) (cf. Brakke [5, pp. 29–30]).

Theorem 7.1. (a) *If* $\Gamma_0 \subset B(0, R)$, *then*

$$\Gamma_t = \varnothing \quad \textit{for } t > R^2/2(n-1). \tag{7.5}$$

(b) *We have*

$$\Gamma_t \subseteq \operatorname{conv}(\Gamma_0) \qquad (t \geq 0), \tag{7.6}$$

where $\operatorname{conv}(\Gamma_0)$ *denotes the convex hull of* Γ_0.

Proof. 1. Assume first $\Gamma_0 \subset B(0, R-\epsilon)$ for some $\epsilon > 0$. Let $g: \mathbf{R}^n \to \mathbf{R}$ be continuous, with $\Gamma_0 = \{g = 0\}$, $g = 1$ on $\mathbf{R}^n \cap \{|x| \geq 2R\}$. Set $\hat{g}(x) = \Psi(|x|^2)$, with Ψ satisfying (7.4) selected so that $\hat{g} \leq g$ on $\mathbf{R}^n$. Then

$$\hat{u} \leq u \quad \text{on } \mathbf{R}^n \times [0, \infty),$$

for $\hat{u}(x, t) = \Psi(|x|^2 + 2(n-1)t)$ and u the weak solution of (5.4)–(5.6). Thus $u > 0$, and so $\Gamma_t = \varnothing$, if $t > \frac{1}{2}R^2/(n-1)$.

In the general case, replace R by $R + \epsilon$ in this argument and send $\epsilon \to 0$.

2. Suppose $\Gamma_0 \subset \mathbf{R}^n_+ = \{x_n > 0\}$. Choose $R \gg 1$ so large that $\Gamma_0 \subset B(Re_n, R)$, for $e_n = (0, 0, \cdots, 0, 1)$. By the argument in step 1, we deduce $\Gamma_t \subset B(Re_n, R(t))$ for $0 \leq t \leq \frac{1}{2}R^2/(n-1)$, $R(t)$ defined as above. In particular, $\Gamma_t \subseteq \mathbf{R}^n_+$ for all $t \geq 0$. Replacing $\mathbf{R}^n_+$ in this argument by an open half-space containing Γ_0, we obtain (7.6).

7.2. Comparison of different sets moving by mean curvature.

Theorem 7.2. *Let* Γ_0 *and* $\hat{\Gamma}_0$ *be compact subsets of* $\mathbf{R}^n$, *and denote by* $\{\Gamma_t\}_{t\geq 0}$ *and* $\{\hat{\Gamma}_t\}_{t\geq 0}$ *the corresponding generalized motions by mean curvature. Suppose also*

$$\Gamma_0 \subseteq \hat{\Gamma}_0. \tag{7.7}$$

Then

$$\Gamma_t \subseteq \hat{\Gamma}_t \quad \textit{for each } t > 0. \tag{7.8}$$

We see therefore that if a compact set Γ_0 lies within another $\hat{\Gamma}_0$ at time zero, then the subsequent evolution Γ_t of Γ_0 lies within the subsequent evolution $\hat{\Gamma}_t$ of $\hat{\Gamma}_0$, for each $t > 0$. We will see in §8 that this assertion provides us with a useful tool for studying specific examples.

Proof. Choose continuous functions $g, \hat{g}: \mathbf{R}^n \to [0, \infty)$ so that $\Gamma_0 = \{g = 0\}$ and $\hat{\Gamma}_0 = \{\hat{g} = 0\}$, and g and $\hat{g}$ are constant on $\mathbf{R}^n \cap \{|x| \geq S\}$

for some $S > 0$. Replacing g by $g + \hat{g}$ if necessary, we may assume

$$\hat{g} \leq g \quad \text{on } \mathbb{R}^n. \tag{7.9}$$

Now let $\hat{u}, u$ denote the corresponding weak solutions of (5.4)–(5.6). Then (7.9) implies $0 \leq \hat{u} \leq u$ on $\mathbb{R}^n \times (0, \infty)$. Thus, $x \in \Gamma_t$ implies $x \in \hat{\Gamma}_t$, and so (7.8) is valid.

Theorem 7.3. *Assume Γ_0 and $\hat{\Gamma}_0$ are nonempty compact sets, and $\{\Gamma_t\}_{t\geq 0}$ and $\{\hat{\Gamma}_t\}_{t\geq 0}$ are the subsequent generalized motions by mean curvature. Then*

$$\operatorname{dist}(\Gamma_0, \hat{\Gamma}_0) \leq \operatorname{dist}(\Gamma_t, \hat{\Gamma}_t) \qquad (t \geq 0). \tag{7.10}$$

By definition, $\operatorname{dist}(\Gamma_t, \hat{\Gamma}_t) = +\infty$ *if* $\Gamma_t = \emptyset$, $\hat{\Gamma}_t = \emptyset$, *or both.*

Proof. 1. We may assume $\operatorname{dist}(\Gamma_0, \hat{\Gamma}_0) > 0$. Choose $g: \mathbb{R}^n \to \mathbb{R}$ so that

$$\begin{cases} \Gamma_0 = \{g = 0\}, \qquad \hat{\Gamma}_0 = \{g = 1\}, \\ g = 2 \quad \text{on } \mathbb{R}^n \cap \{|x| \geq S\} \text{ for some } S, \\ \operatorname{Lip}(g) = \operatorname{dist}(\Gamma_0, \hat{\Gamma}_0)^{-1}. \end{cases} \tag{7.11}$$

Then

$$\Gamma_t = \{u = 0\}, \qquad \hat{\Gamma}_t = \{u = 1\}, \tag{7.12}$$

with u denoting the corresponding weak solution of (5.4)–(5.6).

2. From the contraction property Theorem 3.3, we see that

$$\operatorname{Lip}(u(\cdot, t)) \leq \operatorname{Lip}(g) \qquad (t \geq 0). \tag{7.13}$$

If $\Gamma_t \neq \emptyset$ and $\hat{\Gamma}_t \neq \emptyset$, choose points $x \in \Gamma_t$, $\hat{x} \in \hat{\Gamma}_t$ so that

$$|x - \hat{x}| = \operatorname{dist}(\Gamma_t, \hat{\Gamma}_t).$$

Then using (7.11)–(7.13) we compute

$$1 = u(\hat{x}, t) - u(x, t) \leq \operatorname{Lip}(u)|x - \hat{x}| \leq \operatorname{dist}(\Gamma_0, \hat{\Gamma}_0)^{-1} \operatorname{dist}(\Gamma_t, \hat{\Gamma}_t).$$

This proves (7.10). q.e.d.

Inequality (7.10) implies in particular that two hypersurfaces evolving under generalized motion by mean curvature do not ever move closer to each other than they were initially. In particular, $\Gamma_t \cap \hat{\Gamma}_t = \emptyset$ for all $t > 0$ provided $\Gamma_0 \cap \hat{\Gamma}_0 = \emptyset$. Notice that this property is essential for our approach of representing the evolving surfaces as the level sets of a continuous function.

7.3. Positive mean curvature. Now let us assume that Γ_0 is a smooth connected hypersurface, the boundary of a bounded open set $U \subset \mathbb{R}^n$. We will suppose additionally that

$$\operatorname{div}(\nu) < 0 \quad \text{on } \Gamma_0, \tag{7.14}$$

ν denoting the inner unit normal vector field to Γ_0 (extended smoothly to some neighborhood of Γ_0). Inequality (7.14) says that Γ_0 has positive mean curvature with respect to the inner unit normal field. Consequently, if Γ_0 evolves according to mean curvature, we see from (2.2) that initially at least the motion is directed into U.

We show now that in fact Γ_t lies in U for all $t \geq 0$, and that Γ_t continues to have positive mean curvature, this last statement interpreted in an appropriate weak sense.

Expanding upon a suggestion of L. Caffarelli, our idea is to solve the mean curvature equation (5.4)–(5.6) by separating variables. Indeed we will show

$$u(x, t) \equiv v(x) - t \qquad (x \in U,\ t > 0), \tag{7.15}$$

where v is the (unique) weak solution of the stationary problem

$$-(\delta_{ij} - v_{x_i} v_{x_j}/|Dv|^2) v_{x_i x_j} = 1 \quad \text{in } U, \tag{7.16}$$

$$v = 0 \quad \text{on } \partial U = \Gamma_0. \tag{7.17}$$

We will further prove that

$$\Gamma_t = \{x \in U | v(x) = t\} \qquad (t \geq 0), \tag{7.18}$$

so that $\Gamma_t \subset U$ $(t \geq 0)$ and $\Gamma_t = \emptyset$ for $t > t^* \equiv \|v\|_{L^\infty(U)}$. Note also that in any open region where v is smooth and $|Dv| \neq 0$, we can rewrite (7.16) as

$$-\operatorname{div}(\nu) = 1/|Dv| > 0 \quad \text{for } \nu \equiv Dv/|Dv|.$$

As ν is the inward pointing unit normal field along $\Gamma_t \equiv \{v = t\}$, we informally interpret our PDE (7.16) as implying "Γ_t has positive mean curvature" for $0 \leq t < t^*$.

To carry out the foregoing program rigorously, let us first define $v \in C(\overline{U})$ to be a weak solution to (7.16) provided that if $u - \phi$ has a local maximum (minimum) at a point $x_0 \in U$ for each $\phi \in C^\infty(\mathbb{R}^n)$, then

$$-(\delta_{ij} - \phi_{x_i} \phi_{x_j}/|D\phi|^2)\phi_{x_i x_j} \leq (\geq) 1 \quad \text{at } x_0 \text{ if } D\phi(x_0) \neq 0 \tag{7.19}$$

and

$$\begin{cases} -(\delta_{ij}-\eta_i\eta_j)\epsilon_{x_ix_j} \le (\ge) 1 \quad \text{at } x_0 \text{ for some } \eta \in \mathbb{R}^n \\ \text{with } |\eta| \le 1, \text{ if } D\phi(x_0) = 0. \end{cases} \tag{7.20}$$

Theorem 7.4. *There exists a unique weak solution* v *of* (7.16), (7.17). *Furthermore, there are constants* $A, a > 0$ *so that*

$$\begin{gathered} a\,\mathrm{dist}(x, \Gamma_0) \le v(x) \le A\,\mathrm{dist}(x, \Gamma_0) \qquad (x \in \overline{U}), \\ |Dv(x)| \le A. \end{gathered} \tag{7.21}$$

Proof. 1. Similarly to §4, we approximate (7.16), (7.17) by the uniformly elliptic PDE

$$-\left(\delta_{ij} - \frac{v^\epsilon_{x_i} v^\epsilon_{x_j}}{|Dv^\epsilon|^2+\epsilon^2}\right) v^\epsilon_{x_ix_j} = 1 \quad \text{in } U, \tag{7.22}$$

$$v^\epsilon = 0 \quad \text{on } \partial U = \Gamma_0 \tag{7.23}$$

for $0 < \epsilon \le 1$. We will construct upper and lower barriers for (7.22), (7.33) of the form

$$w(x) = \lambda g(d(x)) \qquad (\lambda \in \mathbb{R},\ d(x) = \mathrm{dist}(x, \Gamma_0))$$

in a neighborhood $V \equiv \{0 < d(x) < 2\delta_0\}$ of Γ_0 in which d is smooth. Owing to the mean curvature condition (7.14), d satisfies

$$0 < b \le -\Delta d \le B, \quad d_{x_i} d_{x_j} d_{x_ix_j} \equiv 0 \tag{7.24}$$

in this region. We then use (7.24) to compute

$$\begin{aligned} Mw &\equiv \left(\delta_{ij} - \frac{w_{x_i} w_{x_j}}{|Dw|^2+\epsilon^2}\right) w_{x_ix_j} \\ &= \lambda(g'\Delta d + g'') - \frac{\lambda^3}{\lambda^2 g'^2+\epsilon^2} g'^2 g'' \\ &= \lambda g' \Delta d + \frac{\epsilon^2 \lambda g''}{\lambda^2 g'^2 + \epsilon^2}. \end{aligned} \tag{7.25}$$

Choosing $g(t) = \delta_0^2 - (t-\delta_0)^2$ we find from (7.24), (7.25) that $Mw \ge -c\lambda\delta_0 - 2\lambda > -1$ for λ sufficiently small. Since $w = 0$ on ∂V, $w < v^\epsilon$ in V by the maximum principle. In particular, $v^\epsilon(x) \ge a d(x)$ $(x \in U)$, where the constant a is independent of ϵ. To obtain the corresponding upper bound, we choose

$$g(t) = \log(2\delta_0/(2\delta_0 - t)).$$

Then $g(t)$ is convex on $[0, 2\delta_0)$, and satisfies

(7.26) $$g(0)=0, \quad g' \geq 1/(2\delta_0), \quad g''=g'^2, \quad g'(2\delta_0)=+\infty.$$

Again using (7.24)–(7.26), we find

$$Mw \leq -c\lambda + \epsilon^2/\lambda < -1$$

for λ sufficiently large. Since $\partial w/\partial \nu = +\infty$ on $\{d = 2\delta_0\}$, where ν denotes the exterior normal to V, we find that $v^\epsilon < w$ in V by a simple variant of the maximum principle. This gives the estimate

(7.27) $$v^\epsilon(x) \leq Ad(x) \qquad (x \in V).$$

To complete our preliminary estimates, we observe that (7.27) implies $|Dv^\epsilon| \leq A$ on Γ_0. By differentiating (7.22) with respect to x_l, we see that any derivative $v^\epsilon_{x_l}$ achieves its maximum and minimum on Γ_0. Thus $|Dv^\epsilon|$ is uniformly bounded in U and in particular $v^\epsilon \leq Ad$ in U.

2. As a consequence of step 1, we derived the uniform bounds

$$\sup_{0<\epsilon\leq 1} \|v^\epsilon\|_{C^{0,1}(U)} < \infty.$$

Hence we may extract a subsequence $\{v^{\epsilon_k}\}_{k=1}^\infty \subset \{v^\epsilon\}_{0<\epsilon\leq 1}$ so that $\epsilon_k \to 0$ and $v^{\epsilon_k} \to v$ uniformly on $\overline{U}$. As in the proof of Theorem 4.2, we verify that v is a weak solution of (7.16).

3. The uniqueness of this weak solution v will follow from the characterization of $\{\Gamma_t\}_{t\geq 0}$ below.

Theorem 7.5. *Let* $\{\Gamma_t\}_{t\geq 0}$ *denote the generalized evolution by mean curvature starting with* Γ_0. *Then* $\Gamma_t = \{x \in U | v(x) = t\}$ *for each* $t \geq 0$.

Proof. 1. Define $u(x, t) \equiv v(x) - t$ for $x \in U$, $t > 0$. It is then straightforward to verify that u is a weak solution of the mean curvature evolution equation

(7.28) $$u_t = (\delta_{ij} - u_{x_i}u_{x_j}/|Du|^2)u_{x_ix_j} \quad \text{in } U \times (0, \infty).$$

Set

(7.29) $$\hat{\Gamma}_t \equiv \{x \in U | v(x) = t\} = \{x \in U | u(x, t) = 0\} \qquad (t > 0).$$

2. Now let

$$\hat{u}(x, t) \equiv |u(x, t)| = |v(x) - t| \qquad (x \in U, \ t > 0).$$

In view of Theorem 2.8, $\hat{u}$ is a weak solution of

(7.30) $$\begin{cases} \hat{u}_t = (\delta_{ij} - \hat{u}_{x_i}\hat{u}_{x_j}/|D\hat{u}|^2)\hat{u}_{x_ix_j} & \text{in } U \times (0, \infty), \\ \hat{u} = t & \text{on } \partial U \times [0, \infty), \\ \hat{\mu} = v & \text{on } \overline{U} \times \{t = 0\}. \end{cases}$$

3. Choose any smooth function $g: \mathbb{R}^n \to \mathbb{R}$ so that

$$(7.31)\qquad \begin{cases} \Gamma_0 = \{g = 0\}, \quad g \geq 0, \quad Dg \neq 0 \quad \text{on } \Gamma_0, \\ g \text{ is constant on } \mathbb{R}^n \cap \{|x| \geq S\} \text{ for some } S > 0. \end{cases}$$

Let $w \geq 0$ be the unique weak solution of

$$(7.32)\qquad \begin{cases} w_t = (\delta_{ij} - w_{x_i} w_{x_j}/|Dw|^2) w_{x_i x_j} & \text{in } \mathbb{R}^n \times (0, \infty), \\ w = g \quad \text{on } \mathbb{R}^n \times \{t = 0\}, \end{cases}$$

so that

$$(7.33)\qquad \Gamma_t = \{x \in \mathbb{R}^n | w(x, t) = 0\} \qquad (t \geq 0).$$

According to our construction in §4, w is Lipschitz in t, and thus

$$|w(x, t)| \leq Ct \qquad (x \in \Gamma_0, \ t > 0)$$

for some constant C.

4. Employing now (7.21), we see that $\underline{w} \equiv \alpha w$ satisfies

$$\underline{w} \leq \begin{cases} v & \text{on } \overline{u} \times \{t = 0\}, \\ t & \text{on } \partial U \times (0, \infty) \end{cases}$$

if $\alpha > 0$ is sufficiently small.

Now the proof of our Comparison Theorem 3.2 can be modified to show from (7.30), (7.32) that $0 \leq \underline{w} \leq \hat{u}$ in $U \times [0, \infty)$. Thus $x \in \hat{\Gamma}_t$ implies $x \in \Gamma_t$, and so $\hat{\Gamma}_t \subseteq \Gamma_t$ for $t \geq 0$. Similarly, let us set

$$(7.34)\qquad \overline{w} \equiv \beta w$$

for some large constant β. Now (7.14) yields that if t_0 is sufficiently small, then

$$(7.35)\qquad \Gamma_t \subset U \qquad (0 \leq t \leq t_0).$$

Since $\partial U = \Gamma_0$, we may employ (7.35) and the semigroup property (5.14) to conclude $\Gamma_t \subset U$ $(t > 0)$. In particular, $w > 0$ on $\partial U \times (0, \infty)$. Consequently, for any $T > 0$ we may choose β so large that $\overline{w}$ defined by (7.34) satisfies $\hat{u} \leq \overline{w}$ on $U \times [0, T]$. Hence as above we find

$$\Gamma_t = \hat{\Gamma}_t \qquad (0 \leq t \leq T).$$

7.4. Convexity. We next recover certain assertions of Huisken [23], by suitably adapting various methods of Korevaar [29] and Kennington [27] for studying the convexity of solutions to nonlinear elliptic PDE. Kennington had previously proposed this method in [28] (see also the concluding remarks in Trudinger [36]).

Theorem 7.6. *Assume Γ_0 is the boundary of a smooth convex bounded open set U. Then there exists a time $t^* > 0$ such that Γ_t is the boundary of a convex, nonempty open set for $0 \le t < t^*$ and Γ_t is empty for $t > t^*$.*

Proof. 1. Because of §7.3 it suffices to consider the stationary PDE

$$(7.36) \qquad \begin{cases} -(\delta_{ij} - v_{x_i}v_{x_j}/|Dv|^2)v_{x_ix_j} = 1 & \text{in } U, \\ v = 0 \quad \text{on } \Gamma_0 = \partial U. \end{cases}$$

We will show that $\{x \in U | v(x) > t\}$ is convex for $0 \le t < t^*$, $t^* = \|v\|_{L^\infty}$. In fact, we will show that $\sqrt{v}$ is concave.

Formally, if $w = \sqrt{v}$ and v satisfies (7.36), then w solves

$$(7.37) \qquad -(\delta_{ij} - w_{x_i}w_{x_j}/|Dw|^2)w_{x_ix_j} = 1/2w \quad \text{in } U.$$

This suggests we consider approximations $w^\epsilon = \sqrt{v^\epsilon}$ satisfying

$$(7.38) \qquad \begin{aligned} Mw^\epsilon \equiv \left(\delta_{ij} - \frac{w^\epsilon_{x_i}w^\epsilon_{x_j}}{|Dw^\epsilon|^2 + \epsilon^2}\right) w^\epsilon_{x_ix_j} &= -\frac{1}{2w^\epsilon} \quad \text{in } U, \\ w^\epsilon &= 0 \qquad \text{on } \Gamma_0, \end{aligned}$$

$$(7.39) \qquad -\left(\delta_{ij} - \frac{v^\epsilon_{x_i}v^\epsilon_{x_j}}{|Dv^\epsilon|^2 + 4\epsilon^2 v^\epsilon}\right) v^\epsilon_{x_ix_j} = \frac{-2\epsilon^2|Dv^\epsilon|^2}{|Dv^\epsilon|^2 + 4\epsilon^2 v^\epsilon} + 1 \quad \text{in } U.$$

Because the convexity arguments are very sensitive to the form of the equation, we are forced into making a nice approximation w^ϵ to (7.37) and then making due with nastier approximations v^ϵ to (7.36).

2. We first demonstrate the existence of a solution $w^\epsilon \in C^2(U) \cap C^{1/2}(\overline{U})$ to (7.38). Consider therefore the PDE

$$(7.40) \qquad \begin{aligned} Ms^{\epsilon,\delta} \equiv \left(\delta_{ij} - \frac{w^{\epsilon,\delta}_{x_i}w^{\epsilon,\delta}_{x_j}}{|Dw^{\epsilon,\delta}|^2 + \epsilon^2}\right) w^{\epsilon,\delta}_{x_ix_j} &= -\frac{1}{2(w^{\epsilon,\delta} + \delta)} \quad \text{in } U, \\ w^{\epsilon,\delta} = 0 \quad \text{on } \Gamma_0, \end{aligned}$$

which has a unique smooth solution $w^{\epsilon,\delta} \ge 0$.

Choose a large ball $B(p, R)$ containing U with $\operatorname{dist}(p, U) \ge R/2$, and let $r \equiv |x - p|$. Set $w \equiv (2R - r)$. Then

$$Mw + \frac{1}{2(w+\delta)} = -\frac{(n-1)}{r} + \frac{1}{2(2R - r)} \le -\frac{(n-1)}{R} + \frac{1}{3R} < 0.$$

Since $w > 0$ on ∂U, $w > w^{\epsilon,\delta}$ in U by the maximum principle. Hence

$$(7.41) \qquad 0 \le w^{\epsilon,\delta} < 2R \quad \text{in } U,$$

with R independent of ϵ and δ.

Next, let $w \equiv \lambda\sqrt{d}$ in $V = \{0 < d(x) < \delta_0\}$. Using formula (7.25) of §7.3 (with $g(t) = \sqrt{t}$) we find

$$Mw + \frac{1}{2(w+\delta)} \leq -\frac{c\lambda}{\sqrt{d}} + \frac{1}{2(\lambda\sqrt{d}+\delta)} < 0$$

for λ sufficiently large. If in addition, we choose λ so that $\lambda\sqrt{\delta_0} \geq 2R$, then $w \geq w^{\epsilon,\delta}$ on ∂V, and thus $w \geq w^{\epsilon,\delta}$ on V by the maximum principle. In particular

$$0 \leq w^{\epsilon,\delta} \leq A\sqrt{d} \quad \text{in } U, \tag{7.42}$$

with A independent of ϵ and δ.

Estimate (7.42) implies that

$$|w^{\epsilon,\delta}(x) - w^{\epsilon,\delta}(y)| \leq C|x-y|^{1/2} \quad \text{if } x \in U,\ y \in \Gamma_0, \tag{7.43}$$

with C independent of ϵ and δ. We show that (7.43) holds for all $x, y \in U$ by the following well-known argument. Given $x, y \in U$ we set $\tau \equiv y - x$, $U_\tau \equiv \{z \in \mathbb{R}^n | z - \tau \in U\}$, and $w_\tau^{\epsilon,\delta}(z) \equiv w^{\epsilon,\delta}(z-\tau)$. Note that U_τ is open and nonempty since $y \in U_\tau$. On $U \cap U_\tau$, both $w^{\epsilon,\delta}$ and $w_\tau^{\epsilon,\delta}$ satisfy (7.40) and hence the difference $w = w^{\epsilon,\delta} - w_\tau^{\epsilon,\delta}$ satisfies a linear elliptic equation of the form $Lw + c(x)w = 0$ with $c(x) \geq 0$. Hence by the maximum principle,

$$|w(y)| \leq \max_{z \in \partial(U\cap U_\tau)} |w(z)| \qquad (y \in U \cap U_\tau).$$

Since for $z \in \partial(U \cap U_\tau)$ either $z \in \partial U$ or $z - \tau \in \partial U$, we have by (7.43) that

$$|w^{\epsilon,\delta}(y) - w^{\epsilon,\delta}(x)| = |w^{\epsilon,\delta}(y) - w_\tau^{\epsilon,\delta}(y)| \leq C|x-y|^{1/2}. \tag{7.44}$$

Finally, in order to pass to the limit for a sequence $\delta_k \searrow 0$, we need to establish some interior estimates for $w_k = w^{\epsilon,\delta_k}$. Let $W \subset\subset U$. Then we claim

$$\|w_k\|_{C^{2+\alpha}(W)} \leq M(\epsilon, \operatorname{dist}(W, \Gamma_0)) \tag{7.45}$$

with M independent of δ_k. By Schauder theory, (7.45) follows from an interior gradient estimate

$$\|Dw_k\|_{L^\infty(W)} \leq C(\epsilon, \operatorname{dist}(W, \Gamma_0)),$$

which in turn follows from Gilbarg-Trudinger [18, Theorem 15.5]. Therefore, we have established the existence of a (unique) solution w^ϵ of (7.38), and in addition the estimates

$$0 \le w^\epsilon \le A\sqrt{d}, \qquad 0 \le w^\epsilon \le 2R, \tag{7.46}$$
$$|w^\epsilon(x) - w^\epsilon(y)| \le C|x-y|^{1/2},$$

with A, C, R independent of ϵ.

3. Before we proceed to the proof of the concavity of w^ϵ, we shall need to establish the lower bound

$$w^\epsilon \ge ad \tag{7.47}$$

with a independent of ϵ.

Consider $w \equiv \lambda g(d)$ in $V = \{0 < d(x) < 2\delta_0\}$ with $g(t) = (\delta_0^2 - (t-\delta_0)^2)^{1/2}$. Then from formulas (7.24) and (7.25) we find

$$\begin{aligned} Mw &\ge -\frac{\lambda\delta_0}{g}\left(c + \frac{\epsilon^2\delta_0}{\lambda^2(d-\delta_0)^2 + \epsilon^2(\delta_0^2 - (d-\delta_0)^2)}\right) \\ &\ge -\frac{\lambda\delta_0}{g}\left(c + \frac{1}{\delta_0}\right) \quad \text{for } \lambda \ge \epsilon, \end{aligned}$$

and so

$$Mw + \frac{1}{2w} \ge -\frac{\lambda}{g}(\delta_0 c + 1) + \frac{1}{2\lambda g} \ge 0$$

for $\epsilon^2 \le \lambda^2 = 2(\delta_0 c + 1)$. With this choice, we see $w^\epsilon \ge w$ in $\{0 < d(x) < 2\delta_0\}$, and as in §7.3 the estimate (7.47) follows easily.

4. We can now show that w^ϵ is concave. For $x, y \in \overline{U}$ set $z = \lambda x + (1-\lambda)y$, $\lambda \in (0, 1)$ being fixed. The concavity function of w^ϵ is defined by

$$\mathscr{C}^\lambda(x, y) \equiv w^\epsilon(z) - \lambda w^\epsilon(x) - (1-\lambda)w^\epsilon(y) \qquad (x, y \in U).$$

The fundamental concavity maximum principle for $\mathscr{C}$ was established by Korevaar [29] for a large class of elliptic equations. The case at hand fails to satisfy Korevaar's condition. However, Kennington's improved concavity maximum principle [27, Theorem 3.1] does apply and so the infimum of $\mathscr{C}^\lambda$ is not attained on $U \times U$.

To complete the proof we must essentially show that w^ϵ is concave near Γ_0. Since $w^\epsilon = \sqrt{v^\epsilon}$ satisfies (7.38), (7.39), it is straightforward to see that $v^\epsilon \in C^{2+\alpha}(\overline{U})$ and $Dv^\epsilon \cdot \nu \ge a > 0$ for ν the interior normal to Γ_0. Using the strict convexity of U it is easy to check that

$$w^\epsilon_{ll} = \frac{1}{2\sqrt{v^\epsilon}} v^\epsilon_{ll} - \frac{(v^\epsilon_l)^2}{4(v^\epsilon)^{3/2}}$$

is strictly negative near Γ_0. It follows easily that $\mathscr{C}^\lambda \geq 0$ on $U \times U$ (for complete details, see Korevaar [29, Lemma 2.4] or Caffarelli-Spruck [6, Theorem 3.1]). This completes the proof that w^ϵ is concave.

5. Since w^ϵ is concave, it follows that $|Dw^\epsilon| \neq 0$ on each level set of w^ϵ below the maximum of w^ϵ. Hence all these level sets are smooth convex hypersurfaces.

We claim that these level sets have uniformly bounded principal curvatures. To see this, it suffices because of the convexity of these level sets to know that the mean curvature $\mathscr{H}$ with respect to the inward normal is uniformly bounded. But

$$\begin{aligned}\mathscr{H}|Dw^\epsilon| &= -\left(\delta_{ij} - \frac{w^\epsilon_{x_i} w^\epsilon_{x_j}}{|Dw^\epsilon|^2}\right) w^\epsilon_{x_i x_j} \\ &= \frac{1}{2w^\epsilon} + w^\epsilon_{x_i} w^\epsilon_{x_j} w^\epsilon_{x_i x_j}\left(\frac{1}{|Dw^\epsilon|^2} - \frac{1}{|Dw^\epsilon|^2 + \epsilon^2}\right).\end{aligned}$$

Since w^ϵ is concave we conclude that $0 \leq \mathscr{H} \leq 1/2w^\epsilon|Dw^\epsilon|$, and therefore $\mathscr{H}$ is uniformly bounded on each of the level sets below the maximum of w^ϵ.

6. We complete the proof of Theorem 7.6 by showing that $v^\epsilon \to v$ uniformly on $\overline{U}$, where v is the unique solution of (7.36) constructed in Theorem 7.4.

Since w^ϵ satisfies (7.38), v^ϵ satisfies

$$|v^\epsilon(x) - v^\epsilon(y)| \leq 4RC|x-y|^{1/2}, \qquad x, y \in U.$$

Hence, we may choose a sequence $\epsilon_k \to 0$ with $v^{\epsilon_k} \to v$ uniformly on $\overline{U}$. We assert that v is a weak solution of (7.36). As before, it suffices to consider $\phi \in C^\infty(\mathbb{R}^n)$ with $v - \phi$ having a *strict* local maximum at a point $x_0 \in U$. As $v^{\epsilon_k} \to v$ uniformly near x_0, $v^{\epsilon_k} - \phi$ has a local maximum at a point x_k, with $x_k \to x_0$ as $k \to \infty$.

Since v^{ϵ_k} and ϕ are smooth, we have

$$Dv^{\epsilon_k} = D\phi, \quad D^2 v^{\epsilon_k} \leq D^2\phi \quad \text{at } x_k.$$

Thus (7.39) implies

$$-\left(\delta_{ij} - \frac{\phi_{x_i}\phi_{x_j}}{|D\phi|^2 + 4\epsilon_k^2 v^{\epsilon_k}}\right)\phi_{x_i x_j} \leq -2\epsilon_k^2 \frac{|D\phi|^2}{|D\phi|^2 + 4\epsilon_k^2 v^{\epsilon_k}} + 1 \tag{7.48}$$

at x_k. Suppose first $D\phi(x_0) \neq 0$. Then $D\phi(x_k) \neq 0$ for large k. Consequently we may pass to the limit in (7.48) (since $0 \leq v^{\epsilon_k} \leq 4R^2$) to deduce

$$-\left(\delta_{ij}-\frac{\phi_{x_i}\phi_{x_j}}{|D\phi|^2}\right)\phi_{x_ix_j}\le 1 \quad \text{at } x_0 .$$

Next, assume instead $D\phi(x_0)=0$ and set

$$\eta^k\equiv\frac{D\phi(x_k)}{(|D\phi(x_k)|^2+4\epsilon_k^2 v^{\epsilon_k})^{1/2}},$$

so that (7.48) becomes

$$\text{(7.49)}\qquad -(\delta_{ij}-\eta_i^k\eta_j^k)\phi_{x_ix_j}\le -2\epsilon_k^2\frac{|D\phi|^2}{|D\phi|^2+4\epsilon_k^2 v_k^\epsilon}+1 \quad \text{at } x_k .$$

Since $|\eta^k|\le 1$, we may pass to a subsequence and reindex if necessary to ensure $\eta_k\to\eta$ in $\mathbb{R}^n$ for some $|\eta|\le 1$. Sending k to infinity in (7.49) we discover

$$\text{(7.50)}\qquad -(\delta_{ij}-\eta_i\eta_j)\phi_{x_ix_j}\le 1 \quad \text{at } x_0 .$$

Consequently v is a weak subsolution. Similarly, we find that v is a weak supersolution, and the proof of Theorem 7.6 is complete.

Remark 7.7. We have shown that if Γ_0 is smooth, then Γ_t is a $C^{1,1}$ convex hypersurface. In a subsequent paper, we will demonstrate that for arbitrary convex Γ_0, the surfaces $\{\Gamma_t\}_{t\ge 0}$ are actually smooth. Once this smoothness is demonstrated, it follows from the work of Huisken [23] that the $\{\Gamma_t\}_{t\ge 0}$ are strictly convex and shrink to a point.

8. Examples, pathologies, and conjectures

In this concluding section, we note various odd behavior allowed by our generalized mean curvature flow

$$\Gamma_0\mapsto\mathcal{M}(t)\Gamma_0=\Gamma_t \qquad (t\ge 0),$$

and set forth some related conjectures.

8.1. Instantaneous extinction. Suppose Σ_0 is the smooth, connected boundary of a bounded open subset $U\subset\mathbb{R}^n$, and let Γ_0 be a compact subset of Σ_0. If $\Gamma_0=\Sigma_0$, then we know from Theorem 6.1 that, at least for small times $t>0$, Γ_t is the classical evolution via mean curvature.

What happens if Γ_0 is a proper subset of Σ_0?

Theorem 8.1. *Assume that Γ_0 is compact, $\Gamma_0\subseteq\Sigma_0$, $\Gamma_0\ne\Sigma_0$. Then*

$$\text{(8.1)}\qquad \Gamma_t=\varnothing \quad \textit{for each } t>0 .$$

If we take Γ_0 to be, say, Σ_0 with a small disk D removed, we may informally regard (8.1) as asserting Γ_0 "pops" instantly. In this heuristic interpretation, we may think of Γ_0 as somehow having so much mean curvature concentrated along its boundary within Σ_0 that the hole then widens infinitely fast (see Figure 5).

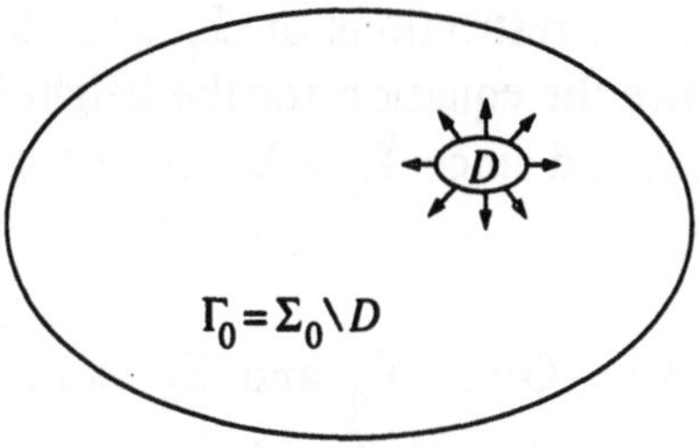

FIGURE 5

The proof of Theorem 8.1 will be given after the next assertion, of independent interest. Assume now that $\hat{\Sigma}_0$ is the smooth connected boundary of a bounded open set $\hat{U} \subset \mathbb{R}^n$ and that

$$\hat{\Sigma}_0 \subset \overline{U}, \tag{8.2}$$

with Σ_0 and U as above. Thus the surface $\hat{\Sigma}_0$ lies within the closed region $\overline{U}$ enveloped by Σ_0. Suppose further that

$$\hat{\Sigma}_0 \neq \Sigma_0. \tag{8.3}$$

Then choose a time $t_0 > 0$ so small that the classical evolutions $\{\Sigma_t\}$ and $\{\hat{\Sigma}_t\}$ starting at Σ_0 and $\hat{\Sigma}_0$, respectively, exist at least for times $0 \leq t \leq t_0$.

Theorem 8.2. *We have*

$$\Sigma_t \cap \hat{\Sigma}_t = \emptyset \quad \textit{for } 0 < t \leq t_0. \tag{8.4}$$

We are thus asserting that even if Σ_0 and $\hat{\Sigma}_0$ coincide except for a very small region (see Figure 6), then for any positive $t > 0$ the subsequent evolutions will have completely broken apart (as in Figure 7). The point

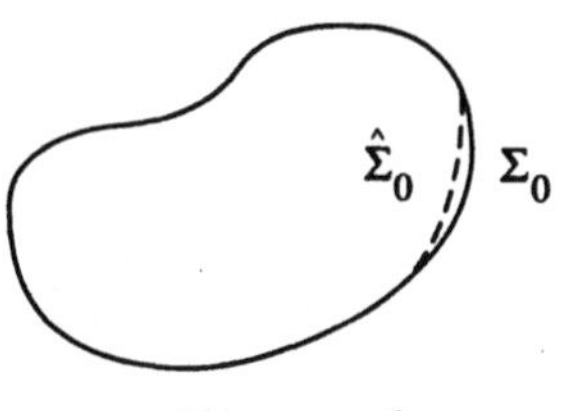

FIGURE 6

FIGURE 7

is that the PDE describing evolution by mean curvature is "uniformly parabolic along the surface" and thus admits infinite propagation speed for disturbances.

We will give the proof of Theorem 8.2 (as well as a new proof of the short time existence of classical mean curvature flow) in a separate paper [14]. Another proof follows by covering Σ_0 and $\widehat{\Sigma}_0$ by overlapping balls small enough so that the restrictions of Σ_t and $\widehat{\Sigma}_t$ to each ball can be written as graphs. Since the equation for the height function is uniformly parabolic for small t_0, and since $\widehat{\Sigma}_0 \neq \Sigma_0$, in at least one of the balls the surfaces Σ_t and $\widehat{\Sigma}_t$ must instantly separate. Thus in each ball the surfaces must also separate.

Proof of Theorem 8.1. Given Γ_0 and Σ_0 as in Theorem 8.1 we may choose a smooth, nearby surface $\widehat{\Sigma}_0$ to Σ_0 satisfying (8.2), (8.3), and $\Gamma_0 \subset \widehat{\Sigma}_0$. Then owing to Theorem 7.2 we have $\Gamma_t \subseteq \Sigma_t \cap \widehat{\Sigma}_t$ for small $t > 0$. Assertion (8.1) now follows from (8.4).

8.2. Development of an interior. The foregoing demonstrates that a "large" initial set Γ_0 can instantly vanish under the generalized mean curvature flow. An opposite and perhaps more surprising phenomenon is that the set Γ_t for $t > 0$ may develop an interior, even if Γ_0 had none.

The simplest example occurs if we take Γ_0 to be the union of the coordinate axes in the plane $\mathbb{R}^2$ (Figure 8). (Ignore for the moment that Γ_0 is not compact and so our theory in §5 is not really applicable.) To discover, heuristically at least, the subsequent evolution of Γ_0, consider instead the simpler figure as drawn in Figure 9. As for instance in Brakke [5, Figure 3] we expect this corner to evolve to the shape depicted in Figure 10 for times $t > 0$. Since Γ_0 is composed of four rotated copies of this corner, we expect from Theorem 7.2 that Γ_t will look like the shape in Figure 11. This assertion is at variance with Brakke [5, Figure 5]. Our Γ_t presumably contains the set shown in Figure 12, which he draws as one of the (nonunique!) evolutions for Γ_0. *We conjecture that our* Γ_t *contains all of the evolutions of* Γ_0 *allowed for by Brakke.*

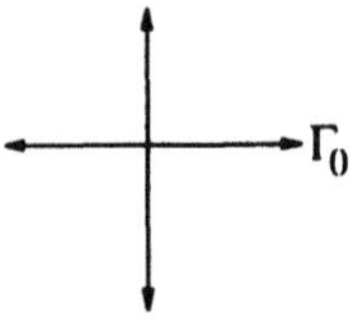

FIGURE 8

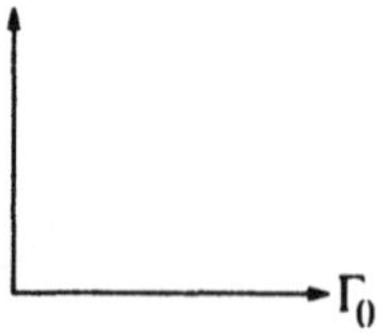

FIGURE 9

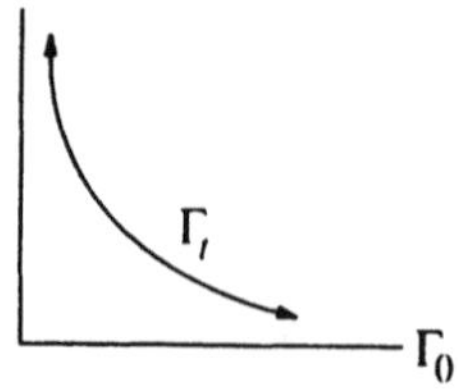

FIGURE 10

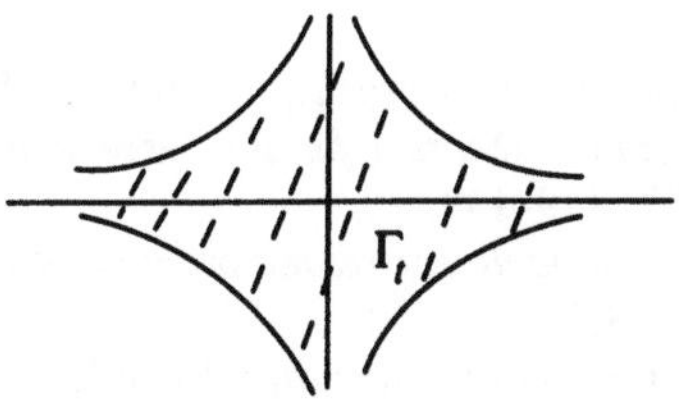

FIGURE 11

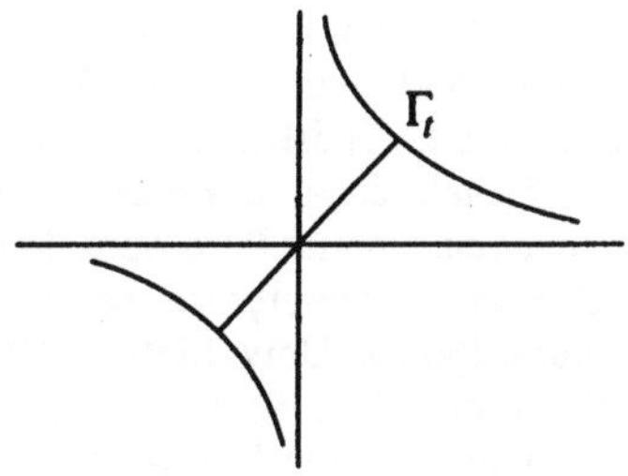

FIGURE 12

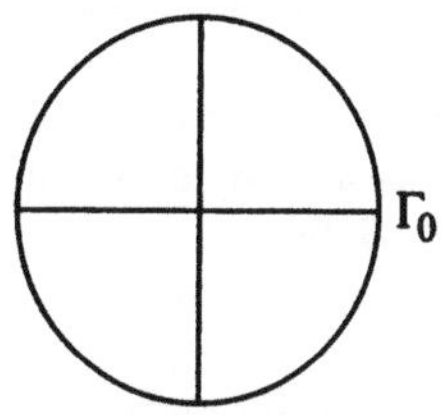

FIGURE 13

The discussion above can be modified to apply to various compact figures Γ_0, to which our theory does apply. We leave it to the reader to provide at least a heuristic proof that the set $\Gamma_0 \subset \mathbf{R}^2$ as drawn in Figure 13 will develop an interior.

Observe by the way that our approach regards a "figure eight" in R^2 as being embedded with a singularity at the crossing point. We in particular do not interpret this shape as an immersed circle, and consequently model its evolution completely differently than [1], [3], [11], etc.

We conjecture that if $\Gamma_0 = \Sigma_0$ is, as above, the boundary of a smooth open set, then Γ_t will never have an interior.

References

[1] U. Abresch & J. Langer, *The normalized curve shortening flow and homothetic solutions*, J. Differential Geometry **23** (1986) 175–196.

[2] W. Allard, *On the first variation of a varifold*, Ann. of Math. (2) **95** (1972) 417–491.

[3] S. Angenent, *Parabolic equations for curves on surfaces*. I, II, preprint.

[4] E. Bombieri, E. DeGiorgi & E. Giusti, *Minimal cones and the Bernstein problem*, Invent. Math. **7** (1969) 243–268.

[5] K. A. Brakke, *The motion of a surface by its mean curvature*, Princeton Univ. Press, Princeton, NJ, 1978.

[6] L. A. Caffarelli & J. Spruck, *Convexity properties of solutions to some classical variational problems*, Comm. Partial Differential Equations **7** (1982) 1337–1379.

[7] Y.-G. Chen, Y. Giga & S. Goto, *Uniqueness and existence of viscosity solutions of generalized mean curvature flow equations*, preprint, 1989.

[8] M. G. Crandall, L. C. Evans & P.-L. Lions, *Some properties of viscosity solutions of Hamilton-Jacobi equations*, Trans. Amer. Math. Soc. **282** (1984) 487–502.

[9] M. G. Crandall & P.-L. Lions, *Viscosity solutions of Hamilton-Jacobi equations*, Trans. Amer. Math. Soc. **277** (1983) 1–42.
[10] K. Ecker & G. Huisken, *Mean curvature evolution of entire graphs*, preprint, 1988.
[11] C. L. Epstein & M. I. Weinstein, *A stable manifold theorem for the curve shortening equation*, Comm. Pure Appl. Math. **40** (1987) 119–139.
[12] L. C. Evans, *A convergence theorem for solutions of nonlinear second order elliptic equations*, Indiana Univ. Math. J. **27** (1978) 875–887.
[13] L. C. Evans & P. E. Souganidis, *Differential games and representation formulas for solutions of Hamilton-Jacobi-Isaacs equations*, Indiana Univ. Math. J. **33** (1984) 773–797.
[14] L. C. Evans & J. Spruck, *Motion of level sets by mean curvature*. II, Trans. Amer. Math. Soc. (to appear).
[15] M. Gage, *An isoperimetric inequality with applications to curve shortening*, Duke Math. J. **50** (1983) 1225–1229.
[16] ——, *Curve shortening makes convex curves circular*, Invent. Math. **76** (1984) 357–364.
[17] M. Gage & R. S. Hamilton, *The heat equation shrinking convex plane curves*, J. Differential Geometry **23** (1986) 69–96.
[18] D. Gilbarg & N. S. Trudinger, *Elliptic partial differential equations of second order*, Springer, Berlin, 1983.
[19] M. Grayson, *The heat equation shrinks embedded plane curves to round points*, J. Differential Geometry **26** (1987) 285–314.
[20] ——, *A short note on the evolution of surfaces via mean curvature*, Duke Math. J. **58** (1989) 555–558.
[21] ——, *The shape of a figure-eight under the curve shortening flow*, Invent. Math. **96** (1989) 177–180.
[22] R. S. Hamilton, *Three manifolds with positive Ricci curvature*, J. Differential Geometry **17** (1982) 255–306.
[23] G. Huisken, *Flow by mean curvature of convex surfaces into spheres*, J. Differential Geometry **20** (1984) 237–266.
[24] H. Ishii, *On uniqueness and existence of viscosity solutions of fully nonlinear second order elliptic PDE's*, Comm. Pure Appl. Math. **42** (1989) 15–45.
[25] R. Jensen, *The maximum principle for viscosity solutions of fully nonlinear second order partial differential equations*, Arch. Rational Mech. Anal. **101** (1988) 1–27.
[26] R. Jensen, P.-L. Lions & P. E. Souganidis, *A uniqueness result for viscosity solutions of second order fully nonlinear partial differential equations*, Proc. Amer. Math. Soc. **102** (1988) 975–978.
[27] A. U. Kennington, *Power concavity and boundary value problems*, Indiana Univ. Math. J. **34** (1983) 687–704.
[28] ——, *Power concavity of solutions of Dirichlet problems*, Proc. Centre Math. Anal. Austral. Nat. Univ. **8** (1984) 133–136.
[29] N. J. Korevaar, *Convex solutions to nonlinear elliptic and parabolic boundary value problems*, Indiana Univ. Math. J. **32** (1983) 603–614.
[30] N. V. Krylov, *Nonlinear elliptic and parabolic equations of second order*, Reidel, Dordrecht, 1987.
[31] O. A. Ladyzhenskaja, V. A. Solonnikov & N. N. Ural′tseva, *Linear and quasilinear equations of parabolic type*, Amer. Math. Soc., Providence, RI, 1968.
[32] P.-L. Lions, *Optimal control of diffusion processes and Hamilton-Jacobi-Bellman equations*. I, Comm. Partial Differential Equations **8** (1983) 1101–1134.
[33] S. Osher & J. A. Sethian, *Fronts propagating with curvature dependent speed: algorithms based on Hamilton-Jacobi formulations*, J. Computational Phys. **79** (1988) 12–49.
[34] H. Parks & W. Ziemer, *Jacobi fields and regularity of functions of least gradient*, Ann. Scuola Norm. Sup. Pisa Cl. Sci. (4) **11** (1984) 505–527.

[35] J. A. Sethian, *Recent numerical algorithms for hypersurfaces moving with curvature-dependent speed: Hamilton-Jacobi equations and conservation laws*, J. Differential Geometry **31** (1990) 131–161.
[36] N. S. Trudinger, *The Dirichlet problem for the prescribed curvature equations*, preprint, 1989.

UNIVERSITY OF CALIFORNIA, BERKELEY

UNIVERSITY OF MASSACHUSETTS

J. DIFFERENTIAL GEOMETRY
33 (1991) 749-786

UNIQUENESS AND EXISTENCE OF VISCOSITY SOLUTIONS OF GENERALIZED MEAN CURVATURE FLOW EQUATIONS

YUN-GANG CHEN, YOSHIKAZU GIGA & SHUN'ICHI GOTO

1. Introduction

This paper treats degenerate parabolic equations of second order

$$u_t + F(\nabla u, \nabla^2 u) = 0 \tag{1.1}$$

related to differential geometry, where ∇ stands for spatial derivatives of $u = u(t, x)$ in $x \in \mathbb{R}^n$, and u_t represents the partial derivative of u in time t. We are especially interested in the case when (1.1) is regarded as an evolution equation for level surfaces of u. It turns out that (1.1) has such a property if F has a scaling invariance

$$F(\lambda p, \lambda X + \sigma p \otimes p) = \lambda F(p, X), \qquad \lambda > 0, \ \sigma \in \mathbb{R}, \tag{1.2}$$

for a nonzero $p \in \mathbb{R}^n$ and a real symmetric matrix X, where $\otimes$ denotes a tensor product of vectors in $\mathbb{R}^n$. We say (1.1) is *geometric* if F satisfies (1.2). A typical example is

$$u_t - |\nabla u| \operatorname{div}(\nabla u/|\nabla u|) = 0, \tag{1.3}$$

where ∇u is the (spatial) gradiant of u. Here $\nabla u/|\nabla u|$ is a unit normal to a level surface of u, so $\operatorname{div}(\nabla u/|\nabla u|)$ is its mean curvature unless ∇u vanishes on the surface. Since $u_t/|\nabla u|$ is a normal velocity of the level surface, (1.3) implies that a level surface of solution u of (1.3) moves by its mean curvature unless ∇u vanishes on the surface. We thus call (1.3) the *mean curvature flow equation* in this paper.

The motion of a closed (hyper)surface in $\mathbb{R}^n$ by its mean curvature has been studied by many authors [1], [3], [4], [8], [10], [12], [14], [15]. Such a motion is also important in the singular perturbation theory related to

Received August 1, 1989 and, in revised form, March 5, 1990. The first author is on leave from and was partially supported by Nankai Institute of Mathematics, Tianjin, China. The second author was partially supported by the Japan Ministry of Education, Science and Culture through grants No. 01740076 and 01540092 for scientific research.

phase transition phenomena (see [13], [23] and references therein). However, so far whole *unique* evolution families of surfaces were only constructed under geometric restrictions on initial surfaces such as convexity [10], [14], except $n = 2$ [3], [12]. When $n = 2$, M. Grayson [12] has shown that any embedded curve moved by its curvature never becomes singular unless it shrinks to a point. However, when $n \geq 3$ even embedded surfaces may develop singularities before it shrinks to a point. A barbell with a long, thin handle actually becomes singular in the middle in short time.

Our goal is to construct whole unique evolution families of surfaces even after the time when there appear singularities. Contrary to other authors (except [4]) we avoid parametrization and rather understand surfaces as level sets of solutions u of (1.3). We first study the initial value problem for a geometric, degenerate parabolic equation (1.1) with

$$u(0, x) = a(x) \in C_\alpha(\mathbb{R}^n) \tag{1.4}$$

for some constant α, where $C_\alpha(A)$ is the set of continuous functions a in A such that $a - \alpha$ is compactly supported in A. Recently P. L. Lions [22] introduced a class of weak solutions for degenerate elliptic equations of second order so that a comparison principle holds. Such solutions are called viscosity solutions, and a general theory is established by R. Jensen [19] and H. Ishii [17] (see also [6] for simplification). For a large class of geometric, degenerate parabolic equations including (1.3) we construct a unique global viscosity solution u_a in $C_\alpha([0, T] \times \mathbb{R}^n)$ of (1.1) and (1.4) for every $T > 0$. Since our $F(p, X)$ is singular at $p = 0$, as is observed in (1.3), even uniqueness of viscosity solutions does not follow directly from results in [17], [19]. We are forced to extend their theory to our situation. Existence of viscosity solutions is based on Perron's method discussed in [16], [17]. We construct viscosity sub- and supersolutions of (1.1) and (1.4) and obtain the viscosity solution u_a.

We now turn to the study of level surfaces of a viscosity solution u_a of (1.1) and (1.4). Let $\Gamma(t)$ be the γ-level surface of $u_a(t, \cdot)$ and let $D(t)$ be the set of $x \in \mathbb{R}^n$ such that $u_a(t, x) > \gamma$, where $\gamma > \alpha$. We call $u_a(t, \cdot)$ a *defining function* of $(\Gamma(t), D(t))$. When (1.1) is geometric and degenerate parabolic, using (1.2) we show that if u is a viscosity sub-(super)solution of (1.1), so is $\theta(u)$ provided that θ is continuous and nondecreasing. This is proved in §5 and the proof depends on Jensen's lemma on semiconvex functions [19]. By this property of (1.1) we prove that the family $(\Gamma(t), D(t))$ $(t \geq 0)$ is uniquely determined by $(\Gamma(0), D(0))$ and is independent of a choice of the defining functions a of $(\Gamma(0), D(0))$.

We call this evolution family $(\Gamma(t), D(t))$ $(t \geq 0)$ a *solution family* of (1.1) with initial data $(\Gamma(0), D(0))$. If $D(0)$ is a bounded open set and $\Gamma(0)$ $(\subset \mathbb{R}^n \backslash D(0))$ is a compact set containing the boundary $\partial D(0)$, then, evidently, there is a defining function $a \in C_\alpha(\mathbb{R}^n)$ of $(\Gamma(0), D(0))$. Existence of a viscosity solution u_a of (1.1) and (1.4) now yields a unique global solution family $(\Gamma(t), D(t))$ $(t \geq 0)$ of (1.1) for a given initial data $(\Gamma(0), D(0))$. In particular for the mean curvature flow equation (1.3) we construct a whole unique evolution family $\Gamma(t)$ moved by its mean curvature. Since $\Gamma(t)$ may not be regular, here its mean curvature is interpreted in some weak sense. By our comparison it is also proved that $(\Gamma(t), D(t))$ becomes empty in finite time provided that it is a solution family of (1.3) when $n \geq 2$. This extends a result of G. Huisken [14] because no geometric assumption of $\Gamma(0)$ is required in our approach. In [14] Huisken proved that $\Gamma(t)$ becomes extinct in finite time provided that $\Gamma(0)$ is a uniformly convex C^2 surface in $\mathbb{R}^n$ $(n \geq 3)$; for $n = 2$, see [10]. We also note that we need no regularity of $\Gamma(0)$. In [4] K. A. Brakke tried to construct a global evolution family $\Gamma(t)$ moved by its mean curvature by using geometric measure theory. However, his varifold solution is too weak to be regarded as an evolution of subsets in $\mathbb{R}^n$, and his solution may not be unique.

Our analysis works for a large class of geometric, degenerate parabolic equations (1.1) other than the mean curvature flow equation (1.3). Important examples generalizing (1.3) are

$$u_t - |\nabla u| \operatorname{div}(\nabla u / |\nabla u|) - \nu |\nabla u| = 0, \qquad \nu \in \mathbb{R}, \tag{1.5}$$

and its anisotropic version (cf. [13])

$$u_t - |\nabla u| \sum_{i=1}^{n} \frac{\partial}{\partial x_i}\left(\frac{\partial H}{\partial p_i}\left(\frac{\nabla u}{|\nabla u|}\right)\right) - \beta\left(\frac{\nabla u}{|\nabla u|}\right)|\nabla u| = 0, \tag{1.6}$$

where $H \in C^2(\mathbb{R}^n \backslash \{0\})$ is convex, positively homogeneous of degree one and β is continuous on a unit sphere in $\mathbb{R}^n$. In (1.5) one observes that the motion of $\Gamma(t)$ by constant speed ν is also considered as well as by the mean curvature. When $n = 2$, a global evolution family of curves $\Gamma(t)$ (even not embedded) moved by (1.6) with $\beta = 0$ is essentially constructed in [3] by a completely different method—parametrization of $\Gamma(t)$. In [3] strict convexity of H is also assumed. As is described in later sections our analysis works even when F depends on t. However, the case where F depends on x is not studied in the present paper.

This paper is organized as follows. §2 begins with the definition of viscosity sub- and supersolutions and treats its basic properties. Most of the results in this section are more or less known to specialists. However, the proof in our context is not explicitly written in the literature, and ideas of the proof are scattered in various literature, so we include the proof both for completeness and for the reader's convenience. In §3 we state a parabolic version of Ishii's lemma [17], [18], which is a key to establishing the comparison principle for viscosity solutions. Our results extend Proposition IV.1 in [18]. §4 establishes comparison results on viscosity sub- and supersolutions of degenerate parabolic equations, including (1.1) in a bounded domain even when $F = F(p, X)$ may not be continuous at $p = 0$. This extends the comparison results in [17], [18], where F is assumed to be continuous. §5 begins with the definition of geometric equations, and there we show that geometric, degenerate elliptic equations are invariant under an (orientation-preserving) change of a dependent variable in the viscosity sense, as is mentioned in the fourth paragraph of the Introduction. We use approximation and Jensen's lemma in [19] on semiconvex functions to show this fact. In §§6 and 7 we consider the initial-value problem of the geometric, degenerate parabolic equation (1.1) with (1.4). §6 establishes the unique global existence of viscosity solutions for a large class of (1.1) by Perron's method. Using results in §6, we construct a solution family $(\Gamma(t), D(t))$ of (1.1) for an arbitrary initial data $(\Gamma(0), D(0))$. The main body of this paper consists of §§4–7 preceded by preliminary §§2 and 3. This paper is written so that no previous knowledge of viscosity solutions in [16], [17], [18], [19], [22], [24] is required.

The results in this paper have been announced in [5].

After this work was completed, we were informed of a recent work of L. C. Evans and J. Spruck [9] closely related to ours. They also proved the existence of a unique viscosity solution and studied various properties of the level surfaces $\Gamma(t)$ of the solution, but only for the mean curvature flow equation (1.3). They showed that $\Gamma(t)$ is determined only by $\Gamma(0)$, which is not expected for general geometric equations having first-order terms such as (1.5) with $\nu \neq 0$. We also learned of works of S. Osher and J. A. Sethian [25] and Sethian [26] giving numerical algorithms for evolutions of surfaces with curvature-dependent speed. Their viewpoint of an evolution of surfaces is the same as ours. They regarded them as level surfaces of solutions of parabolic equations of second order.

We are grateful to Professor Hitoshi Ishii and Professor Robert Kohn for informing us of several recent works related to ours. We are also grateful to Professor L. Craig Evans for sending us his latest manuscript with J. Spruck [9] before its publication.

2. Basic properties of viscosity solutions

We recall the definition of a viscosity solution and collect some of its properties here. This section is almost paralleled to §2 in [17], although the situation is slightly different.

For a sequence of functions $h_k: L \to \mathbb{R}$ $(L \subset \mathbb{R}^d)$ $(k = 1, 2, \ldots)$ we associate its Γ^--*limit*

$$\lim_{k\to\infty}{}_{*}\, h_k : \overline{L} \to \widetilde{\mathbb{R}} = \mathbb{R} \cup \{\pm\infty\}$$

defined by

$$\lim_{k\to\infty}{}_{*}\, h_k(x) = \lim_{\substack{k\to\infty \\ \varepsilon\downarrow 0}} \inf_{l\geq k} \inf_{\substack{|x-y|<\varepsilon \\ y\in L}} h_l(y) \quad \text{for } x \in \overline{L},$$

where $\overline{L}$ denotes the closure of L in $\mathbb{R}^d$. When $h_k = h$ for all k, the Γ^--limit $\lim_{*k\to\infty} h_k$ is called the *lower semicontinuous* (*l.s.c.*) *relaxation* of h to $\overline{L}$ and is denoted by h_*. It is easy to see

$$h_*(x) = \lim_{\varepsilon\downarrow 0} \inf_{\substack{|x-y|<\varepsilon \\ y\in L}} h(y), \qquad x \in \overline{L}.$$

The *upper semicontinuous* (*u.s.c.*) *relaxation* of h to $\overline{L}$ is defined by $h^* = -(-h)_*$. The concept of Γ^--limit and the relaxations was introduced by E. De Giorgi [11] and it is important, for example in the calculus of variation.

For $A \subset \mathbb{R}^N$ we consider a dense subset W of $J(A) = A \times \mathbb{R} \times \mathbb{R}^N \times S^{N\times N}$, where $S^{N\times N}$ denotes the space of $N \times N$ real symmetric matrices. Let $E = E(y, s, q, Y)$ be a function from W to $\mathbb{R}$. Since W is dense in $J(A)$, the relaxations E_* and E^* are defined in $J(A)$ with value in $\widetilde{\mathbb{R}}$.

Definition 2.1. A function $u = u(y): A \to \mathbb{R}$ is called a *viscosity subsolution* (*supersolution*, respectively) of the equation

$$E(y, u, Du, D^2u) = 0 \quad \text{in } A \tag{2.1}$$

if $u^* < \infty$ ($u_* > -\infty$, resp.) in A, and for each pair $\phi \in C^2(A)$ and $\overline{y} \in A$ satisfying $\max_A(u^* - \phi) = (u^* - \phi)(\overline{y})$ ($\min_A(u_* - \phi) = (u_* - \phi)(\overline{y})$, resp.) it holds that

$$E_*(\overline{y}, u^*(\overline{y}), D\phi(\overline{y}), D^2\phi(\overline{y})) \leq 0$$
$$(E^*(\overline{y}, u_*(\overline{y}), D\phi(\overline{y}), D^2\phi(\overline{y})) \geq 0, \text{ resp.}).$$

Here $Du = (\partial u/\partial y_1, \cdots, \partial u/\partial y_N)$, and D^2u denotes the Hessian matrix of u. By $\phi \in C^2(A)$ we mean that ϕ has a C^2 extension $\tilde{\phi}$ in an open neighborhood of A.

When u has the second differential at $\bar{y} \in A$ it is easy to see that

$$E_*(\bar{y}, u(\bar{y}), Du(\bar{y}), D^2u(\bar{y})) \le 0$$

if u is a viscosity subsolution. A similar result holds for a viscosity supersolution. A function $u = u(y)$ is called a *viscosity solution* of (2.1) if it is both a viscosity sub- and supersolution of (2.1).

Our definition has a meaning for a wider class of E than in [7], [17], since we do not assume $W = J(A)$ here. We often suppress the word "viscosity", except in statements of theorems, since all solutions in this paper are considered in the viscosity sense.

In what follows we always assume that A is locally compact. We give below three basic properties of subsolutions.

Proposition 2.2. *Let S be a nonempty family of a subsolution of* (2.1) *and let u be a function defined on A by*

$$u(y) = \sup\{v(y); v \in S\} \quad \textit{for } y \in A.$$

Suppose $u^*(y) < \infty$ for $y \in A$. Then u is a subsolution of (2.2).

Proposition 2.3 (Existence). *Suppose that E is degenerate elliptic, i.e.,*

$$E(y, s, q, Y) \le E(y, s, q, Z) \quad \textit{in } W \textit{ if } Y \ge Z.$$

Let f and $g: A \to \mathbb{R}$ be respectively a sub- and supersolution of (2.1). *Suppose $f \le g$ in A. Then there exists a solution u of* (2.1) *satisfying $f \le u \le g$ in A.*

Proposition 2.4 (Stability). *Let $E, E_k: W \to \mathbb{R}$ and u_k be a subsolution of*

$$E_k(y, u_k, Du_k, D^2u_k) = 0 \quad \textit{in } A,$$

$k = 1, 2, \cdots$. *Assume $\liminf_{*k\to\infty} E_k \ge E_*$ and u_k converges to a function $u: A \to \mathbb{R}$ uniformly in each compact subset of A. Then, u is a subsolution of* (2.1) *with this E.*

When E is continuous in $W = J(A)$, Propositions 2.2 and 2.3 are studied in [17]. Existence of the solution (Proposition 2.3) is proved by Perron's method as in [16]. Although the proof is actually written for first order equations in [16], it still works in our situation with minor modifications. However, we give proofs both of Propositions 2.2 and 2.3 for the reader's convenience and for completeness because the proof for second order equations is not explicitly written in the literature.

We also give the proof of Proposition 2.4. The stability is often proved under stringent assumptions that E_k, $E: J(A) \to \mathbb{R}$ are continuous and that $E_k \to E$ uniformly on every compact subset of $J(A)$ (cf. [22]).

Proof of Proposition 2.2. First, we choose a function $\phi \in C^2(A)$ and a point $\bar{y} \in A$ such that

$$\max_A(u^* - \phi) = (u^* - \phi)(\bar{y}),$$

and then fix ϕ and $\bar{y}$. Here we can assume $(u^* - \phi)(\bar{y}) = 0$ without loss of generality, since the function $\phi(y)$ can be replaced by $\phi(y) + (u^* - \phi)(\bar{y})$. Putting $\psi(y) = \phi(y) + |y - \bar{y}|^4$, we see that $u^* - \psi$ attains its strict maximum in A at $\bar{y}$. Then

$$(u^* - \psi)(y) + |y - \bar{y}|^4 = (u^* - \phi)(y) \le (u^* - \phi)(\bar{y}) = (u^* - \psi)(\bar{y}) = 0$$

yields

$$(u^* - \psi)(y) \le -|y - \bar{y}|^4 \quad \text{for } y \in A. \tag{2.2}$$

By the definition of u^*, there is a sequence $\{x_k\} \subset A$ such that $x_k \to \bar{y}$ $(k \to \infty)$ and

$$\lim_{k\to\infty} a_k = (u^* - \phi)(\bar{y}) = 0$$

with $a_k = (u - \phi)(x_k)$. Since $u(y) = \sup\{v(y); v \in S\}$, there exists a sequence $\{v_k\} \subset S$ such that $v_k(x_k) > u(x_k) - 1/k$ $(k = 1, 2, \cdots)$. This implies

$$(v_k^* - \phi)(x_k) \ge (v_k - \phi)(x_k) > a_k - 1/k.$$

Since $v_k \le u$, by (2.2) we get

$$(v_k^* - \psi)(y) \le -|y - \bar{y}|^4 \quad \text{for } y \in A.$$

Since A is locally compact, there is a compact neighborhood of $\bar{y}$ which we denote by B. Because the function $(v_k^* - \phi)(x)$ is u.s.c. and has an upper bound, it attains its maximum in B at a point $y_k \in B$. We thus conclude

$$a_k - 1/k < (v_k^* - \psi)(x_k) \le (v_k^* - \psi)(y_k) \le -|y_k - \bar{y}|^4. \tag{2.3}$$

This implies $y_k \to \bar{y}$ $(k \to \infty)$. Indeed, if not, we would have

$$0 = \lim_{k\to\infty} (u - \phi)(x_k) \le \limsup_{k\to\infty}(1/k - |y_k - \bar{y}|^4) < 0,$$

which is a contradiction.

By (2.3) we can see $\lim_{k\to\infty}(v_k^* - \psi)(y_k) = 0$, and hence

$$\lim_{k\to\infty} v_k^*(y_k) = \lim_{k\to\infty} \psi(y_k) = \psi(\bar{y}) = u^*(\bar{y}).$$

Since v_k is a subsolution we see

$$E_*(y_k, v_k^*(y_k), D\psi(y_k), D^2\psi(y_k)) \le 0.$$

Since E_* is l.s.c. and $D\psi(\bar{y}) = D\phi(\bar{y})$, $D^2\psi(\bar{y}) = D^2\phi(\bar{y})$, this yields

$$E_*(\bar{y}, u^*(\bar{y}), D\phi(\bar{y}), D^2\phi(\bar{y})) \le 0,$$

which implies that u is a subsolution. q.e.d.

The above proof is due to Ishii [16], where he used $|y-\bar{y}|^2$ instead of $|y-\bar{y}|^4$. The modification term $|y-\bar{y}|^4$ is convenient to study the second order equations. The proof of Proposition 2.3 presented below is based on Perron's method and is essentially found in [16]. We begin with

Lemma 2.5. *Suppose that E is degenerate elliptic. Let $g: A \to \mathbb{R}$ be a supersolution of* (2.1). *Let S be the collection of all subsolutions v of* (2.1), *satisfying $v \le g$ in A. If $v \in S$ is not a supersolution of* (2.1), *then there is a function $w \in S$ and a point $z \in A$ such that $v(z) < w(z)$.*

Proof. Since v is not a supersolution, there exist $\phi \in C^2(A)$ and $\bar{y} \in A$ such that

$$\min_A (v_* - \phi) = (v_* - \phi)(\bar{y}) = 0$$

and

$$E^*(\bar{y}, v_*(\bar{y}), D\phi(\bar{y}), D^2\phi(\bar{y})) = E^*(\bar{y}, \phi(\bar{y}), D\phi(\bar{y}), D^2\phi(\bar{y})) < 0.$$

We may assume here

$$(v_* - \phi)(y) \ge |y-\bar{y}|^4 \quad \text{for } y \in A, \tag{2.4}$$

because the function ϕ can be modified as $\phi + |y-\bar{y}|^4$ if necessary. Evidently, we have $v_* \le g_*$ in A. We now obtain $v_*(\bar{y}) = \phi(\bar{y}) < g_*(\bar{y})$, since, otherwise, it would contradict the fact that g is a supersolution of (2.1). Considering that E^* is u.s.c. and $\phi \in C^2(A)$, for sufficiently small $\delta > 0$ we have

$$E^*(y, \phi(y) + \delta^4/2, D\phi(y), D^2\phi(y)) \le 0, \tag{2.5}$$

$$\phi(y) + \delta^4/2 \le g_*(y) \tag{2.6}$$

for $y \in B_{2\delta} \equiv B \cap \overline{B(\bar{y}, 2\delta)}$, where B is a compact neighborhood of $\bar{y}$, and $B(y, R)$ is the open ball of radius R centered at y. (Such B exists since A is locally compact.) Since E is degenerate elliptic, the inequality (2.5) indicates that, by $\phi \in C^2(A)$, the function $\phi(y)+\delta^4/2$ is a subsolution in $B_{2\delta}$. Furthermore, by (2.4) we have

$$v(y) \geq v_*(y) - \delta^4/2 \geq \phi(y) + \delta^4/2 \quad \text{on } B_{2\delta}\backslash B_\delta. \tag{2.7}$$

We now define $w(y)$ by

$$w(y) = \begin{cases} \max\{\phi(y)+\delta^4/2,\, v(y)\}, & y \in B_\delta, \\ v(y), & y \in A\backslash B_\delta. \end{cases}$$

It follows from (2.7) that

$$w(y) = \max\{\phi(y)+\delta^4/2,\, v(y)\} \quad \text{for } y \in B_{2\delta}.$$

According to Proposition 2.2, w is a subsolution of (2.1) over the whole of A and, thus, $w \in S$ since (2.6) holds.

However, we have

$$0 = (v_* - \phi)(\bar{y}) = \lim_{t\downarrow 0} \inf\{(v-\phi)(y)\,;\, y \in A \text{ and } |y-\bar{y}| \leq t\}.$$

This implies that there is a point $z \in B_\delta$ such that $v(z) - \phi(z) < \delta^4/2$, which yields $v(z) < w(z)$.

Proof of Proposition 2.3. We appeal to Perron's method. As in Lemma 2.5, we set $S = \{v\,;\, v \text{ is a subsolution of (2.1) and } v \leq g\}$. Since $f \in S$, we see $S \neq \varnothing$. We define

$$u(y) = \sup\{v(y)\,;\, v \in S\}.$$

By Proposition 2.2, u is a subsolution of (2.1), so $u \in S$ since $u \leq g$. Suppose that u were not a supersolution of (2.1). Then by Lemma 2.5 there would exist $w \in S$ such that $u(z) < w(z)$ for some $z \in A$. This is contrary to the definition of u. We thus conclude that u is a solution of (2.1) and $f \leq u \leq g$.

Proof of Proposition 2.4. Let $\phi \in C^2(A)$ and $\bar{y} \in A$ satisfy

$$\max_A (u^* - \phi) = (u^* - \phi)(\bar{y}).$$

Since A is locally compact and $u_k^* < \infty$, there is a compact neighborhood B of $\bar{y}$ on which u_k is bounded from above. Since u_k^* is u.s.c., one finds $y_k \in B$ $(k = 1, 2, \cdots)$ such that

$$\max_B (u_k^* - \phi) = (u_k^* - \phi)(y_k).$$

Here, we may assume $y_k \to \bar{y}$ $(k \to \infty)$ because $u_k \to u$ $(k \to \infty)$ uniformly on B.

Since $\lim_{*k\to\infty} E_k \ge E_*$ implies $\lim_{*k\to\infty} E_{k*} \ge E_*$, we have

$$
\begin{aligned}
&E_*(\bar{y}, u^*(\bar{y}), D\phi(\bar{y}), D^2\phi(\bar{y})) \\
&\le \lim_{k\to\infty}{}_* E_{k*}(\bar{y}, u^*(\bar{y}), D\phi(\bar{y}), D^2\phi(\bar{y})) \\
&= \lim_{\substack{k\to\infty \\ \varepsilon\downarrow 0}} \inf_{l\ge k} \inf_{\substack{|z-\bar{z}|<\varepsilon \\ z\in J(A)}} E_{l*}(y, s, q, Y),
\end{aligned} \tag{2.8}
$$

where $z = (y, s, q, Y)$ and $\bar{z} = (\bar{y}, u^*(\bar{y}), D\phi(\bar{y}), D^2\phi(\bar{y}))$. Since u_k converges to u uniformly on B, for each $\varepsilon > 0$ we see $|z_l - \bar{z}| < \varepsilon$ with $z_l = (y_l, u_l^*(y_l), D\phi(y_l), D^2\phi(y_l))$ for sufficiently large l. Hence, the right-hand side of (2.8) is dominated by

$$\liminf_{k\to\infty} E_{k*}(y_k, u_k^*(y_k), D\phi(y_k), D^2\phi(y_k)),$$

so we obtain the desired inequality

$$E_*(\bar{y}, u^*(\bar{y}), D\phi(\bar{y}), D^2\phi(\bar{y})) \le 0$$

which shows that u is a subsolution of (2.1), once we prove

$$E_{k*}(y_k, u_k^*(y_k), D\phi(y_k), D^2\phi(y_k)) \le 0 \quad \text{for all } k \ge 1. \tag{2.9}$$

Since u_k is a subsolution of $E_k(y, u_k, Du_k, D^2u_k) = 0$ in A and $\max_B(u_k^* - \phi) - (u_k^* - \phi)(y_k)$, we have indeed (2.9) so the proof is now complete.

3. Ishii's lemma on evolution equations

We are concerned with a special form of (2.1) called the evolution equation,

$$u_t + F(t, x, u, \nabla u, \nabla^2 u) = 0, \tag{3.1}$$

where ∇ stands for spatial derivatives. Our goal in this section is to prove a key lemma for our comparison theorem for (3.1) in §4. A similar result is first proved by Ishii [17] for (2.1) of nonevolution type. We state our main lemma.

Lemma 3.1. *Let Ω be a domain in $\mathbb{R}^n$ and $T > 0$. Let u be a locally bounded upper semicontinuous (u.s.c.) subsolution of*

$$u_t + F(t, x, u, \nabla u, \nabla^2 u) = 0 \quad \text{in } \Omega_T = (0, T] \times \Omega, \tag{3.2a}$$

and let v be a locally bounded lower semicontinuous (l.s.c.) supersolution of

$$v_t + G(t, x, v, \nabla v, \nabla^2 v) = 0 \quad \text{in } \Omega_T, \tag{3.2b}$$

where $F, G: J = \Omega_T \times \mathbb{R} \times \mathbb{R}^n \times S^{n\times n} \to \widetilde{\mathbb{R}}$ *are l.s.c. and u.s.c., respectively, and* $r \mapsto F(t, x, r, p, X)$ *and* $r \mapsto G(t, x, r, p, X)$ *are nondecreasing for all* $(t, x, r, p, X) \in J$. *Let* $\phi \in C^2((0, T] \times \Omega \times \Omega)$ *and* U *be a subdomain of* $\Omega \times \Omega$. *Let* $(\bar{t}, \bar{x}, \bar{y}) \in U_T$ *be a point such that*

$$u(\bar{t}, \bar{x}) - v(\bar{t}, \bar{y}) - \phi(\bar{t}, \bar{x}, \bar{y}) = \sup_{U_T}(u - v - \phi), \tag{3.3a}$$

where $U_T = (0, T] \times U$. *Suppose that*

(3.3b) $u(t, \cdot)$ *and* $v(t, \cdot)$ *are, for each* $t \in [0, T]$, *Lipschitz continuous on* $\overline{\Omega}$ *and there is a constant* $C > 0$, *independent of the time variable* t, *such that* $\nabla_x^2 u(t, x), -\nabla_y^2 v(t, y) \geq -CI$ *in* Ω_T *(in the sense of distributions).*

Then there are $X, Y \in S^{n\times n}$ *such that*

$$-CI \leq \begin{pmatrix} X & O \\ O & Y \end{pmatrix} \leq \nabla_{x,y}^2 \phi(\bar{t}, \bar{x}, \bar{y}), \tag{3.4a}$$

$$\begin{aligned} &\frac{\partial \phi}{\partial t}(\bar{t}, \bar{x}, \bar{y}) + F(\bar{t}, \bar{x}, u(\bar{t}, \bar{x}), \nabla_x \phi(\bar{t}, \bar{x}, \bar{y}), X) \\ &\quad - G(\bar{t}, \bar{y}, v(\bar{t}, \bar{y}), -\nabla_y \phi(\bar{t}, \bar{x}, \bar{y}), -Y) \leq 0, \end{aligned} \tag{3.4b}$$

where I *is the identity matrix.*

We first prove a weaker version of Lemma 3.1 which is essentially Proposition IV.1 of [18].

Proposition 3.2. *Suppose that* u, v, F, *and* G *are as in Lemma* 3.1. *If the maximum point* $(\bar{t}, \bar{x}, \bar{y})$ *is in* $\operatorname{int} U_T = (0, T) \times U$, *then the conclusion of Lemma* 3.1 *holds.*

The idea of proof is given in [18]. However the proof is not explicitly stated. We give it here for completeness and the reader's convenience. The crucial tool is Jensen's lemma on semiconvex functions.

Lemma 3.3 [19, Lemma 3.10]. *Let* D *be a bounded domain in* $\mathbb{R}^d$ *and let* w *be a Lipschitz continuous function in* $\overline{D}$. *For* $\delta > 0$, *define*

$$\begin{aligned} \mathcal{G}_\delta = \{z \in D;\ &\text{for some } p \in \mathbb{R}^d \text{ with } |p| \leq \delta, \\ &w(\zeta) \leq w(z) + p \cdot (\zeta - z) \text{ for all } \zeta \in D\}. \end{aligned}$$

Assume that there is a constant $K_0 > 0$ *such that* $D^2 w \geq -K_0 I$ *in* D *(in the sense of distributions). If* w *has an interior maximum* $(> \max_{\partial D} w)$, *then there are constants* $C_0 > 0$ *and* $\delta_0 > 0$ *such that* $\operatorname{meas}(\mathcal{G}_\delta) \geq C_0 \delta^d$ *for all* $\delta < \delta_0$.

As is observed in [17, Lemma 5.2], Lemma 3.3 yields the following result since a convex function is almost everywhere differentiable by Alexandroff's theorem [2].

Lemma 3.4. *Suppose that w is as in Lemma 3.3. Let $\overline{z} \in D$ be a maximum point of w. Then there are sequences $\{z_k\} \subset D$ $(k = 1, 2, \cdots)$ satisfying $z_k \to \overline{z}$ as $k \to \infty$ and $\{p_k\} \subset \mathbb{R}^d$ satisfying $|p_k| \le 1/k$ for all k such that the functions $z \mapsto w(z) + p_k \cdot z$ attain a maximum at z_k and have the second differential at z_k.*

To prove Proposition 3.2 we approximate u and v by its sup and inf convolutions as in [18], [20]. We recall properties of these convolutions.

Lemma 3.5 [21]. *Let D be a bounded domain in $\mathbb{R}^N$ and $f, g: D \to \mathbb{R}$ be bounded u.s.c. and l.s.c. functions, respectively. For $\varepsilon > 0$, we define*

$$f^\varepsilon(y) = \sup_{z \in D}\{f(z) - \varepsilon^{-1}|y - z|^2\} \quad \text{for } y \in \overline{D},$$

$$g_\varepsilon(y) = \inf_{z \in D}\{g(z) + \varepsilon^{-1}|y - z|^2\} \quad \text{for } y \in \overline{D},$$

and call f^ε, g_ε sup *and* inf *convolutions, respectively. Then f^ε (g_ε resp.) is Lipschitz continuous on $\overline{D}$ and semiconvex (semiconcave resp.) on D. More precisely, $f^\varepsilon(y) + \varepsilon^{-1}|y|^2$ is convex on D ($g_\varepsilon(y) - \varepsilon^{-1}|y|^2$ is concave on D resp.). In the definitions of* sup *(*inf *resp.) convolution the z's may be restricted by $|y - z| \le \lambda_0\sqrt{\varepsilon}$, where $\lambda_0 = (2\sup_D|f|)^{1/2}$ ($\lambda_0 = (2\sup_D|g|)^{1/2}$ resp.).*

Lemma 3.6 [18], [20]. *Suppose that f, g, and D are as in Lemma 3.5. Let $E: D \times \mathbb{R} \times \mathbb{R}^N \times S^{N\times N} \to \widetilde{\mathbb{R}}$ be an l.s.c. (u.s.c. resp.) function and $s \mapsto E(y, s, q, Y)$ be nondecreasing. If f (g resp.) is a sub- (super- resp.) solution of*

$$E(y, u, Du, D^2u) = 0 \quad \text{in } D,$$

then f^ε (g_ε resp.) is a sub- (super- resp.) solution of

$$E_\varepsilon(y, u, Du, D^2u) = 0 \quad \text{in } D^\varepsilon$$

$$E^\varepsilon(y, u, Du, D^2u) = 0 \quad \text{in } D^\varepsilon \text{ resp.}$$

Here $D^\varepsilon = \{y \in D; \mathrm{d}(y, \partial D) > \lambda_0\sqrt{\varepsilon}\}$ and

$$E_\varepsilon(y, s, q, Y) = \min\{E(z, s, q, Y);\ |y - z| \le \lambda_0\sqrt{\varepsilon},\ z \in D\}$$

$$E^\varepsilon(y, s, q, Y) = \max\{E(z, s, q, Y);\ |y - z| \le \lambda_0\sqrt{\varepsilon},\ z \in D\} \text{ resp.},$$

where $\mathrm{d}(y, \partial D)$ denotes the distance between y and ∂D.

These lemmas follow directly from definitions.

Proof of Proposition 3.2. Replacing ϕ by

$$(t, x, y) \mapsto \phi(t, x, y) + |t - \bar{t}|^2 + |x - \bar{x}|^4 + |y - \bar{y}|^4$$

if necessary, we may assume that $u(t, x) - v(t, y) - \phi(t, x, y)$ attains a strict maximum over $\overline{U_T}$ at $(\bar{t}, \bar{x}, \bar{y})$. Hence, we may assume that U_T is bounded and u, v are bounded there since they are locally bounded.

Let u^ε and v_ε be, respectively, sup and inf convolutions in time of u and v, i.e.,

$$u^\varepsilon(t, x) = \sup_{s\in[0,T]} \{u(s, x) - \varepsilon^{-1}(t-s)^2\},$$

$$v_\varepsilon(t, y) = \inf_{s\in[0,T]} \{v(s, y) + \varepsilon^{-1}(t-s)^2\}.$$

Since $u^\varepsilon - v_\varepsilon \downarrow u - v$ in $\overline{U}_T$, a similar argument to [17, p. 29, 2nd paragraph; see also Proof of Theorem 4.1 in this paper] implies that $u^\varepsilon(t, x) - v_\varepsilon(t, y) - \phi(t, x, y)$ also has an interior (strict) maximum over $\overline{U_T}$ if $\varepsilon > 0$ is sufficiently small and that the maximum point $(t^\varepsilon, x^\varepsilon, y^\varepsilon)$ tends to $(\bar{t}, \bar{x}, \bar{y})$ as $\varepsilon \downarrow 0$. By Lemma 3.5 we see u^ε and v_ε are Lipschitz continuous on $\overline{\Omega_T}$ and

$$D^2_{t,x} u^\varepsilon(t, x), -D^2_{t,y} v_\varepsilon(t, y) \geq -(C + 2\varepsilon^{-1})I \quad \text{in } \operatorname{int}\Omega_T$$

in the sense of distributions. We now apply Lemma 3.4 and observe that there are $(t^\varepsilon_k, x^\varepsilon_k, y^\varepsilon_k) \in \operatorname{int} U_T$ satisfying $(t^\varepsilon_k, x^\varepsilon_k, y^\varepsilon_k) \to (t^\varepsilon, x^\varepsilon, y^\varepsilon)$ as $k \to \infty$ and $(l^\varepsilon_k, p^\varepsilon_k, q^\varepsilon_k) \in \mathbf{R}^{2n+1}$ satisfying $|l^\varepsilon_k| + |p^\varepsilon_k| + |q^\varepsilon_k| \leq 1/k$ such that

$$(t, x, y) \mapsto u^\varepsilon(t, x) - v_\varepsilon(t, y) - \phi(t, x, y) - l^\varepsilon_k t - p^\varepsilon_k \cdot x + q^\varepsilon_k \cdot y$$

attains a maximum at $(t^\varepsilon_k, x^\varepsilon_k, y^\varepsilon_k)$ and has second differential at $(t^\varepsilon_k, x^\varepsilon_k, y^\varepsilon_k)$. These yield

$$D_{t,x,y}(u^\varepsilon - v_\varepsilon - \phi)(t^\varepsilon_k, x^\varepsilon_k, y^\varepsilon_k) = (l^\varepsilon_k, p^\varepsilon_k, -q^\varepsilon_k),$$

$$\nabla^2_{x,y}(u^\varepsilon - v_\varepsilon - \phi)(t^\varepsilon_k, x^\varepsilon_k, y^\varepsilon_k) \leq 0,$$

and by Lemma 3.6

$$\frac{\partial u^\varepsilon}{\partial t}(t^\varepsilon_k, x^\varepsilon_k) + F_\varepsilon(t^\varepsilon_k, x^\varepsilon_k, u^\varepsilon(t^\varepsilon_k, x^\varepsilon_k), \nabla_x u^\varepsilon(t^\varepsilon_k, x^\varepsilon_k), \nabla^2_x u^\varepsilon(t^\varepsilon_k, x^\varepsilon_k)) \leq 0,$$

$$\frac{\partial v_\varepsilon}{\partial t}(t^\varepsilon_k, y^\varepsilon_k) + G^\varepsilon(t^\varepsilon_k, y^\varepsilon_k, v_\varepsilon(t^\varepsilon_k, y^\varepsilon_k), \nabla_y v_\varepsilon(t^\varepsilon_k, y^\varepsilon_k), \nabla^2_y v_\varepsilon(t^\varepsilon_k, y^\varepsilon_k)) \geq 0.$$

Setting $X_k^\varepsilon = \nabla_x^2 u^\varepsilon(t_k^\varepsilon, x_k^\varepsilon)$ and $Y_k^\varepsilon = -\nabla_y^2 v_\varepsilon(t_k^\varepsilon, y_k^\varepsilon)$, we have

$$-CI \le \begin{pmatrix} X_k^\varepsilon & O \\ O & Y_k^\varepsilon \end{pmatrix} \le \nabla_{x,y}^2 \phi(t_k^\varepsilon, x_k^\varepsilon, y_k^\varepsilon), \tag{3.5a}$$

$$\begin{aligned} &\frac{\partial \phi}{\partial t}(t_k^\varepsilon, x_k^\varepsilon, y_k^\varepsilon) + l_k^\varepsilon \\ &+ F_\varepsilon(t_k^\varepsilon, x_k^\varepsilon, u^\varepsilon(t_k^\varepsilon, x_k^\varepsilon), \nabla_x \phi(t_k^\varepsilon, x_k^\varepsilon, y_k^\varepsilon) + p_k^\varepsilon, X_k^\varepsilon) \\ &- G^\varepsilon(t_k^\varepsilon, y_k^\varepsilon, v_\varepsilon(t_k^\varepsilon, y_k^\varepsilon), -\nabla_y \phi(t_k^\varepsilon, x_k^\varepsilon, y_k^\varepsilon) + q_k^\varepsilon, -Y_k^\varepsilon) \le 0. \end{aligned} \tag{3.5b}$$

Since $\nabla_{x,y}^2 \phi(t_k^\varepsilon, x_k^\varepsilon, y_k^\varepsilon)$ is bounded from above uniformly in ε and k, by compactness (see Lemma 5.3 in [17]) (3.5a) implies that there is an increasing sequence $\{k_j\}$, a decreasing sequence $\{\varepsilon_j\}$, and $X, Y \in S^{n\times n}$ such that $X_j = X_{k_j}^{\varepsilon_j} \to X$ and $Y_j = Y_{k_j}^{\varepsilon_j} \to Y$ as $j \to \infty$.

Let s_k^ε be a maximum point of $u(s, x_k^\varepsilon) - (t_k^\varepsilon - s)^2/\varepsilon$ over $[0, T]$. We observe that $|\bar{t} - s_k^\varepsilon| \le |\bar{t} - t_k^\varepsilon| + \lambda_0\sqrt{\varepsilon} \to 0$ as $\varepsilon \downarrow 0$, $k \to \infty$ and

$$u^\varepsilon(t_k^\varepsilon, x_k^\varepsilon) = u(s_k^\varepsilon, x_k^\varepsilon) - \varepsilon^{-1}(t_k^\varepsilon - s_k^\varepsilon)^2 \le u(s_k^\varepsilon, x_k^\varepsilon),$$

where $\lambda_0 = \max((2\sup|u|)^{1/2}, (2\sup|v|)^{1/2})$. Since u is u.s.c., we have

$$\limsup_{\substack{\varepsilon\downarrow 0 \\ k\to\infty}} u^\varepsilon(t_k^\varepsilon, x_k^\varepsilon) \le \limsup_{\substack{\varepsilon\downarrow 0 \\ k\to\infty}} u(s_k^\varepsilon, x_k^\varepsilon) \le u(\bar{t}, \bar{x}).$$

By definition of F_ε there is $\bar{s}_k^\varepsilon \in (0, T]$ satisfying $|t_k^\varepsilon - \bar{s}_k^\varepsilon| \le \lambda_0\sqrt{\varepsilon}$ such that

$$\begin{aligned} &F_\varepsilon(t_k^\varepsilon, x_k^\varepsilon, u^\varepsilon(t_k^\varepsilon, x_k^\varepsilon), \nabla_x \phi(t_k^\varepsilon, x_k^\varepsilon, y_k^\varepsilon) + p_k^\varepsilon, X_k^\varepsilon) \\ &\quad = F(\bar{s}_k^\varepsilon, x_k^\varepsilon, u^\varepsilon(t_k^\varepsilon, x_k^\varepsilon), \nabla_x \phi(t_k^\varepsilon, x_k^\varepsilon, y_k^\varepsilon) + p_k^\varepsilon, X_k^\varepsilon). \end{aligned}$$

Since F is l.s.c. and $r \mapsto F(t, x, r, p, X)$ is nondecreasing, we now have

$$\begin{aligned} &\liminf_{j\to\infty} F(\bar{s}_k^\varepsilon, x_k^\varepsilon, u^\varepsilon(t_k^\varepsilon, x_k^\varepsilon), \nabla_x \phi(t_k^\varepsilon, x_k^\varepsilon, y_k^\varepsilon) + p_k^\varepsilon, X_j) \\ &\quad \ge F(\bar{t}, \bar{x}, u(\bar{t}, \bar{x}), \nabla_x \phi(\bar{t}, \bar{x}, \bar{y}), X), \qquad \varepsilon = \varepsilon_j,\ k = k_j. \end{aligned}$$

Similarly,

$$\begin{aligned} &\limsup_{j\to\infty} G^\varepsilon(t_k^\varepsilon, y_k^\varepsilon, v_\varepsilon(t_k^\varepsilon, y_k^\varepsilon), -\nabla_y \phi(t_k^\varepsilon, x_k^\varepsilon, y_k^\varepsilon) + q_k^\varepsilon, -Y_j) \\ &\quad \le G(\bar{t}, \bar{y}, v(\bar{t}, \bar{y}), -\nabla_y \phi(\bar{t}, \bar{x}, \bar{y}), -Y), \qquad \varepsilon = \varepsilon_j,\ k = k_j. \end{aligned}$$

We now conclude from (3.5a) and (3.5b) that

$$-CI \le \begin{pmatrix} X & O \\ O & Y \end{pmatrix} \le \nabla_{x,y}^2 \phi(\bar{t}, \bar{x}, \bar{y})$$

and

$$\frac{\partial\phi}{\partial t}(\bar{t}, \bar{x}, \bar{y}) + F(\bar{t}, \bar{x}, u(\bar{t}, \bar{x}), \nabla_x\phi(\bar{t}, \bar{x}, \bar{y}), X) - G(\bar{t}, \bar{y}, v(\bar{t}, \bar{y}), -\nabla_y\phi(\bar{t}, \bar{x}, \bar{y}), -Y) \le 0,$$

which is the same as (3.4a) and (3.4b).

Remark 3.7. In Lemmas 3.1 and 3.6 the assumption that $r \mapsto F(t, x, r, p, X)$ is nondecreasing is for simplicity only. Assertions in both Lemmas 3.1 and 3.6 hold without this assumption.

Proof of Lemma 3.1. Again we may assume that

$$\Phi(t, x, y) = u(t, x) - v(t, y) - \phi(t, x, y)$$

attains its strict maximum over $\overline{U_T}$ at $(\bar{t}, \bar{x}, \bar{y})$. By Proposition 3.2 we may assume $\bar{t} = T$.

We first construct a "barrier" near T. Since Φ is u.s.c., there is a continuous nondecreasing function $m: [0, \infty) \to [0, \infty)$ with $m(0) = 0$ such that

$$\Phi(T, \bar{x}, \bar{y}) - \Phi(\bar{t}, \bar{x}, \bar{y}) \le m(T - t) \quad \text{for all } t,\ 0 < t \le T.$$

We take a continuous nondecreasing function $M: [0, \infty) \to [0, \infty)$ such that $M(0) = 0$ and $m(\sigma) < M(\sigma)$ for $\sigma > 0$ and that $M(\sigma)$ is C^2 for $\sigma > 0$. We now set

$$\begin{aligned} \phi_\alpha(t, x, y) &= \phi(t, x, y) - \alpha M(T - t), \\ \Phi_\alpha(t, x, y) &= u(t, x) - v(t, y) - \phi_\alpha(t, x, y), \end{aligned} \qquad \text{for } 0 \le \alpha \le 1.$$

Since Φ_α is u.s.c., the set Σ_α of all maximum points of Φ_α over U_T is closed. We set

$$t_\alpha = \sup\{t;\ (t, x, y) \in \Sigma_\alpha \text{ for some } (x, y) \in U\}.$$

Since Σ_α is closed, there is $(x_\alpha, y_\alpha) \in U$ such that $(t_\alpha, x_\alpha, y_\alpha) \in \Sigma_\alpha$. We set

$$\Lambda = \{\alpha \in [0, 1];\ t_\alpha < T\}$$

and observe that Λ is an open set in $[0, 1]$. Since

$$\begin{aligned} \Phi_1(T, \bar{x}, \bar{y}) = \Phi(T, \bar{x}, \bar{y}) &< \Phi(t, \bar{x}, \bar{y}) + M(T - t) \\ &= \Phi_1(t, \bar{x}, \bar{y}) \quad \text{for all } t < T \end{aligned}$$

by the definition of M, we see $1 \in \Lambda$. Since $\bar{t} = T$, we see $0 \notin \Lambda$ which implies $\Lambda \ne [0, 1]$. Let $\tilde{\beta}$ be a boundary point of $\Lambda^c = \{\beta \in [0, 1];\ t_\beta = T\}$. If $t_\alpha = T$, then

$$\Phi_\alpha(T, x, y) = \Phi(T, x, y)$$

holds, so we have $(x_\alpha, y_\alpha) = (\overline{x}, \overline{y})$. We thus observe that there is a sequence $\{\alpha_j\}$ in Λ such that $\alpha_j \to \tilde{\beta}$ as $j \to \infty$ and

$$(t_{\alpha_j}, x_{\alpha_j}, y_{\alpha_j}) = (t_j, x_j, y_j) \to (T, \overline{x}, \overline{y}) \quad \text{as } j \to \infty.$$

By $t_j < T$ and Proposition 3.2 there are $X_j, Y_j \in S^{n\times n}$ such that

$$\text{(3.6a)} \qquad -CI \le \begin{pmatrix} X_j & O \\ O & Y_j \end{pmatrix} \le \nabla^2_{x,y}\phi_{\alpha_j}(t_j, x_j, y_j),$$

$$\text{(3.6b)} \quad \frac{\partial \phi_{\alpha_j}}{\partial t}(t_j, x_j, y_j) + F(t_j, x_j, u(t_j, x_j), \nabla_x\phi_{\alpha_j}(t_j, x_j, y_j), X_j) \\ - G(t_j, y_j, v(t_j, y_j), -\nabla_y\phi_{\alpha_j}(t_j, x_j, y_j), -Y_j) \le 0.$$

Since

$$\nabla^2_{x,y}\phi_{\alpha_j}(t_j, x_j, y_j) = \nabla^2_{x,y}\phi(t_j, x_j, y_j)$$

is bounded from above uniformly in j, (3.6a) implies there is a sequence (still denoted X_j, Y_j) and $X, Y \in S^{n\times n}$ such that $X_j \to X$ and $Y_j \to Y$ as $j \to \infty$. We also see

$$\frac{\partial \phi_{\alpha_j}}{\partial t} = \frac{\partial \phi}{\partial t} + \alpha M'(T-t) \ge \frac{\partial \phi}{\partial t}$$

since $\nabla\phi_{\alpha_j} = \nabla\phi$ and M is nondecreasing. Since F and G are, respectively, l.s.c. and u.s.c., letting $j \to \infty$ in (3.6a, b) yields (3.4a,b) with $\bar{t} = T$ provided that

$$\text{(3.7)} \qquad \lim_{j\to\infty} u(t_j, x_j) = u(T, \overline{x}), \qquad \lim_{j\to\infty} v(t_j, y_j) = v(T, \overline{y}).$$

It remains to prove (3.7). Suppose that (3.7) were false. Since u and v are u.s.c. and l.s.c., respectively, then either

$$\liminf_{j\to\infty} u(t_j, x_j) < u(T, \overline{x}) \quad \text{or} \quad \limsup_{j\to\infty} v(t_j, y_j) > v(T, \overline{y})$$

will hold, This would imply

$$\liminf_{j\to\infty} \Phi_{\alpha_j}(t_j, x_j, y_j) < \Phi(T, \overline{x}, \overline{y}).$$

Since Φ_{α_j} attains its maximum at (t_j, x_j, y_j), we see

$$\Phi_{\alpha_j}(T, \overline{x}, \overline{y}) = \Phi(T, \overline{x}, \overline{y}) \le \Phi_{\alpha_j}(t_j, x_j, y_j),$$

which leads to a contradiction.

4. Parabolic comparison theorem

We consider an evolution equation

$$u_t + F(t, u, \nabla u, \nabla^2 u) = 0 \tag{4.1}$$

of degenerate parabolic type, i.e., F is degenerate elliptic. In other words

$$F(t, r, p, X+Y) \le F(t, r, p, X) \quad \text{for all } Y \ge O,\ Y \in S^{n\times n}$$

holds where F is defined. Our goal in this section is to show a comparison theorem on viscosity solutions of (4.1) in a bounded domain, even if $F(t, r, p, X)$ may not be continuous at $p = 0$. A comparison theorem on viscosity solutions was first proved by P. L. Lions [22] for some special degenerate elliptic equations. Later his results were extended in [19], [17] for general degenerate elliptic equations $E(u(y), Du(y), D^2u(y)) = 0$, where E is assumed to be continuous in its variables (see also [6], [20]). A parabolic comparison theorem is also discussed in [24], [18], where $F = F(t, r, p, X)$ is still assumed to be continuous in p.

Let Ω be a bounded domain in $\mathbb{R}^n$ and $T > 0$. We assume that F satisfies

(4.2a) $F\colon J_0 = (0, T] \times \mathbb{R} \times (\mathbb{R}^n \setminus \{0\}) \times S^{n\times n} \to \mathbb{R}$ is continuous,

(4.2b) F is degenerate elliptic on J_0,

(4.2c) for all $M > 0$, there is a constant $c_0 = c_0(n, T, M)$ such that $r \mapsto F(t, r, p, X) + c_0 r$ is nondecreasing for all $(t, r, p, X) \in J_0$ with $|r| \le M$,

(4.2d) $-\infty < F_*(t, r, 0, O) = F^*(t, r, 0, O) < \infty$ for all $t \in [0, T]$, $r \in \mathbb{R}$.

We now state our main comparison result.

Theorem 4.1. *Let u and v be, respectively, sub- and supersolutions of* (4.1) *in Ω_T. Suppose that F satisfies* (4.2a)–(4.2d). *If $u^* \le v_*$ on $\partial_p\Omega_T = \{0\} \times \Omega \cup [0, T] \times \partial\Omega$, then $u^* \le v_*$ on Ω_T.*

Remark 4.2. We may assume that (4.1) has a form

$$u_t + u + F(t, u, \nabla u, \nabla^2 u) = 0 \tag{4.3}$$

with

(4.2c′) $r \mapsto F(t, r, p, X)$ is nondecreasing for all $(t, r, p, X) \in J_0$

if we replace u (v resp.) by $e^{\lambda t}u$ ($e^{\lambda t}v$ resp.) with sufficiently large λ.

The crucial step to prove Theorem 4.1 is the following property on viscosity solutions.

Lemma 4.3. *Let u and v be, respectively, u.s.c. sub- and l.s.c. supersolutions of* (4.3) *in* Ω_T, *where* F *satisfies* (4.2a), (4.2b), (4.2c′), *and* (4.2d). *Assume that* u *and* v *are locally bounded in* Ω_T. *Let* $\phi \in C^2(\mathbb{R}^n)$ *such that*

(4.4a) $$\nabla\phi(x) = 0 \quad \textit{if and only if} \quad x = 0,$$

(4.4b) $$\phi(x) = \phi(-x),$$

(4.4c) *there is a constant* $C_1 > 0$ *such that* $|\phi(x) - \phi(0)| \le C_1|x|^2$ *for sufficiently small* $|x|$

(*this actually follows from* (4.4a) *and* $\phi \in C^2(\mathbb{R}^n)$),

(4.4d) $$\phi(0) \le \phi(x) \quad \textit{for all } x \in \mathbb{R}^n.$$

Define $\psi(x, y) = \phi(x - y)$ *and* $Q_T = (0, T] \times \Omega \times \Omega$. *Assume that*

(4.5a) $u(t, \cdot)$ *and* $v(t, \cdot)$ *are Lipschitz continuous on* $\overline{\Omega}$ *for each* $t \in [0, T]$ *and there is a constant* $C_2 > 0$ (*independent of* t) *such that* $\nabla_x^2 u(t, x), -\nabla_y^2 v(t, y) \ge -C_2 I$ *on* Ω_T (*in the sense of distributions*),

(4.5b) $(\bar{t}, \bar{x}, \bar{y}) \in Q_T$ *is a point such that*
$$u(\bar{t}, \bar{x}) - v(\bar{t}, \bar{y}) - \psi(\bar{x}, \bar{y}) = \sup_{Q_T}(u - v - \psi) > \max_{\partial_p Q_T}(u - v - \psi),$$

where $\partial_p Q_T = \{0\} \times \Omega \times \Omega \cup [0, T] \times \partial(\Omega \times \Omega)$. *Then* $u(\bar{t}, \bar{x}) - v(\bar{t}, \bar{y}) \le 0$.

Proof. We divide the proof into several cases depending on $\bar{x}$ and $\bar{y}$. We first discuss the case $\bar{x} \ne \bar{y}$.

Case 1 ($\bar{x} \ne \bar{y}$). Since $\bar{x} \ne \bar{y}$ implies $\nabla\phi(\bar{x} - \bar{y}) \ne 0$, the proof parallels the argument found in [17, pp. 29–30]. By Lemma 3.1 and $\partial\psi/\partial t = 0$, there are $X, Y \in S^{n\times n}$ such that

(4.6a) $$-C_2 I \le \begin{pmatrix} X & O \\ O & Y \end{pmatrix} \le \nabla_{x,y}^2 \psi(\bar{x}, \bar{y}) = \begin{pmatrix} \nabla^2\phi(\bar{x} - \bar{y}) & -\nabla^2\phi(\bar{x} - \bar{y}) \\ -\nabla^2\phi(\bar{x} - \bar{y}) & \nabla^2\phi(\bar{x} - \bar{y}) \end{pmatrix},$$

(4.6b) $$u(\bar{t},\bar{x}) + F_*(\bar{t},u(\bar{t},\bar{x}),\nabla_x\psi(\bar{x},\bar{y}),X) - v(\bar{t},\bar{y}) - F^*(\bar{t},v(\bar{t},\bar{y}),-\nabla_y\psi(\bar{x},\bar{y}),-Y) \le 0.$$

The second inequality in (4.6a) is equivalent to

$$Xp\cdot p + Yq\cdot q \le \nabla^2\phi(\bar{x}-\bar{y})p\cdot p - 2\nabla^2\phi(\bar{x}-\bar{y})p\cdot q + \nabla^2\phi(\bar{x}-\bar{y})q\cdot q \quad \text{for all } p, q \in \mathbb{R}^n.$$

This yields $X \le -Y$ by taking $p = q$. Since $F_* = F^* = F$ on J_0, and

$$\nabla_x\psi(x,y) = -\nabla_y\psi(x,y) = \nabla\phi(x-y)$$

and $\nabla\phi(\bar{x}-\bar{y}) \neq 0$, applying (4.2b) and (4.2c′) in (4.6b) with $X \le -Y$ we obtain

$$u(\bar{t},\bar{x}) - v(\bar{t},\bar{y}) \le 0.$$

Case 2 $(\bar{x}=\bar{y})$. We set

$$\tilde{\psi}(t,x,y) = \psi(x,y) + (\bar{t}-t)^2,$$

and observe that $w - \tilde{\psi}$ attains its maximum over Q_T at $(\bar{t},\bar{x},\bar{x})$, where

$$w(t,x,y) = u(t,x) - v(t,y).$$

For $\eta \in \mathbb{R}^n$ we set

$$\Phi_\eta(t,x,y) = w(t,x,y) - \phi(x-y-\eta) - (\bar{t}-t)^2$$

and divide the situation into the following Case 2a and its negation Case 2b.

Case 2a. For some $\kappa > 0$ there is $(t_\eta, x_\eta, y_\eta) \in Q_{\bar{t}}$ with $x_\eta - y_\eta = \eta$ such that

(4.7) $$\Phi_\eta(t_\eta,x_\eta,y_\eta) = \sup\{\Phi_\eta(t,x,y);\, x,y\in\Omega,\ |x-y|<\kappa,\ t\in(0,\bar{t}]\}$$

for all $\eta \in \mathbb{R}^n$ with $|\eta| < \kappa$.

We discuss Case 2a. We set

$$f(\eta) = \sup\{w(t_\eta,x,y) - (\bar{t}-t_\eta)^2;\, x,y\in\Omega,\ x-y=\eta\}.$$

By (4.7) we have

$$w(t,x,y) - (\bar{t}-t)^2 \le \phi(x-y-\eta) + w(t_\eta,x_\eta,y_\eta) - (\bar{t}-t_\eta)^2 - \phi(0)$$

for all $x, y \in \Omega$ with $|x-y| < \kappa$, $t \in (0,T]$. This, in particular, yields

$$f(\xi) \le \phi(\xi-\eta) + f(\eta) - \phi(0)$$

by taking $x - y = \xi$ and $t = t_\xi$. From (4.4b) it now follows that

$$-\phi(\xi - \eta) + \phi(0) \leq f(\xi) - f(\eta) \leq \phi(\xi - \eta) - \phi(0)$$

which together with (4.4c) implies that f is differentiable in η, $|\eta| < \kappa$, and that ∇f identically equals zero. Hence, $f(\eta)$ is a constant for $|\eta| < \kappa$. This implies

$$\sup_{|x-y|<\kappa} \{w(t_\eta, x, y) - (\bar{t} - t_\eta)^2\} = \sup_{x\in\Omega} w(\bar{t}, x, x). \tag{4.8}$$

We shall show that

$$\sup_{|x-y|<\kappa,\, t\in(0,\bar{t}]} \{w(t, x, y) - (\bar{t} - t)^2\} = w(\bar{t}, \bar{x}, \bar{x}). \tag{4.9}$$

Property (4.7) with $\eta = 0$ yields

$$\sup_{|x-y|<\kappa,\, t\in(0,\bar{t}]} \{w(t, x, y) - (\bar{t} - t)^2 - \phi(x - y)\} = w(\bar{t}, \bar{x}, \bar{x}) - \phi(0),$$

from which we deduce

$$w(\bar{t}, \bar{x}, \bar{x}) - \phi(0) \geq w(\bar{t}, x, x) - \phi(0) \quad \text{for all } x \in \Omega.$$

We thus conclude

$$\sup_{x\in\Omega} w(\bar{t}, x, x) = w(\bar{t}, \bar{x}, \bar{x}). \tag{4.10}$$

Observe that

$$\sup_{x-y=\eta,\, t\in(0,\bar{t}]} \{w(t, x, y) - (\bar{t} - t)^2\} \leq w(t_\eta, x_\eta, y_\eta) - (\bar{t} - t_\eta)^2, \quad |\eta| < \kappa, \tag{4.11}$$

by taking $x - y = \eta$ in (4.7). Applying (4.8) and (4.10) to (4.11) yields

$$\sup_{|x-y|<\kappa,\, t\in(0,\bar{t}]} \{w(t, x, y) - (\bar{t} - t)^2\} \leq w(\bar{t}, \bar{x}, \bar{x}).$$

The converse inequality is trivial, so we now obtain (4.9).

By (4.9) we now apply Lemma 3.1 with $\phi = (\bar{t} - t)^2$ and

$$U = \{(x, y) \in \Omega \times \Omega;\ |x - y| < \kappa\},$$

and observe that there are $X, Y \in S^{n\times n}$ such that

$$\begin{pmatrix} X & O \\ O & Y \end{pmatrix} \leq O, \tag{4.12a}$$

$$\begin{aligned} u(\bar{t}, \bar{x}) + F_*(\bar{t}, u(\bar{t}, \bar{x}), 0, X) - v(\bar{t}, \bar{x})& \\ - F^*(\bar{t}, v(\bar{t}, \bar{x}), 0, -Y) \leq 0.& \end{aligned} \tag{4.12b}$$

The estimate (4.12a) yields $X \leq O \leq -Y$. By (4.2b) we now have

$$F_*(t, r, 0, X) \geq F_*(t, r, 0, O), \qquad F^*(t, r, 0, -Y) \leq F^*(t, r, 0, O).$$

Applying these with (4.2c′) and (4.2d) to (4.12b) gives

$$u(\bar{t}, \bar{x}) - v(\bar{t}, \bar{x}) \leq 0,$$

and the proof of Case 2a is now complete. The assumption (4.2d) is only invoked here.

It remains to discuss the following Case 2b. In this case there is a sequence $\{\eta_i\}$ in $\mathbb{R}^n$ with $|\eta_i| < 1/i$ and $(t_i, x_i, y_i) \in Q_{\bar{t}}$ with $x_i - y_i \neq \eta_i$ $(i = 1, 2, \ldots)$ such that (4.7) holds. Since $\nabla\phi(x_i - y_i - \eta_i) \neq 0$, the same argument in Case 1 shows that

$$u(t_i, x_i) - v(t_i, y_i) \leq 0. \tag{4.13}$$

There is a convergent subsequence (still denoted $\{(t_i, x_i, y_i)\}$) in Q_T. Since the maximum of Φ_0 is not attained for $t < \bar{t}$, its limit is expressed as $(\bar{t}, \bar{\bar{x}}, \bar{\bar{y}}) \in Q_T$ by (4.5b). The limit $(\bar{t}, \bar{\bar{x}}, \bar{\bar{y}})$ is a maximum point of Φ_0 since Φ_0 is u.s.c. Letting $i \to \infty$ in (4.13) yields

$$u(\bar{t}, \bar{\bar{x}}) - v(\bar{t}, \bar{\bar{y}}) \leq 0 \tag{4.14}$$

provided that

$$\lim_{i\to\infty} u(t_i, x_i) = u(\bar{t}, \bar{\bar{x}}), \qquad \lim_{i\to\infty} v(t_i, y_i) = v(\bar{t}, \bar{\bar{y}}). \tag{4.15}$$

By (4.4d) it follows from (4.14) that

$$u(\bar{t}, \bar{x}) - v(\bar{t}, \bar{x}) \leq 0.$$

It remains to prove (4.15) to complete the proof of Case 2. The idea is similar to the proof of (3.7). Suppose that (4.15) were false. Then we would have

$$\liminf_{i\to\infty} \Phi_{\eta_i}(t_i, x_i, y_i) < \Phi_0(\bar{t}, \bar{\bar{x}}, \bar{\bar{y}}), \tag{4.16}$$

since u and v are u.s.c. and l.s.c., respectively. However, since Φ_{η_i} attains its maximum at (t_i, x_i, y_i) and Φ_0 attains its maximum at $(\bar{t}, \bar{x}, \bar{x})$ and $(\bar{t}, \bar{\bar{x}}, \bar{\bar{y}})$, we see

$$\begin{aligned}\Phi_{\eta_i}(t_i, x_i, y_i) &\geq \Phi_{\eta_i}(\bar{t}, \bar{x}, \bar{y}) = w(\bar{t}, \bar{x}, \bar{x}) - \phi(\eta_i)\\ &\geq \Phi_0(\bar{t}, \bar{\bar{x}}, \bar{\bar{y}}) - C_1/i^2\end{aligned}$$

by (4.4b) and (4.4c). This leads to a contradiction to (4.16), so we obtain (4.15). q.e.d.

Lemma 4.3 plays a crucial role in the proof of Theorem 4.1. The remaining part of the proof of Theorem 4.1 parallels the argument in [17, pp. 29–30] as is described below.

Proof of Theorem 4.1. By Remark 4.2 we may assume that the equation has the form (4.3). By the definition of sub- (super- resp.) solution and the boundedness of Ω_T, replacing u (v resp.) by

$$\{\max(u(t,x),-M)\}^* \quad (\{\min(v(t,x),M)\}_* \text{ resp.})$$

for sufficiently large M we may assume that u (v resp.) is bounded u.s.c. (l.s.c. resp.) on $\overline{\Omega_T} = \Omega_T \cup \partial_p\Omega_T$. The assumption $u^* \le v_*$ implies there is a modulus function m (i.e., $m\colon [0,\infty) \to [0,\infty)$ is continuous, nondecreasing, concave and $m(0)=0$) such that

$$u(t,x) - v(t,y) \le m(|x-y|) \quad \text{on } \partial_p Q_T,$$

where

$$Q_T = (0,T] \times \Omega \times \Omega \quad \text{and} \quad \partial_p Q_T = \{0\} \times \Omega \times \Omega \cup [0,T] \times \partial(\Omega \times \Omega).$$

We choose $\{a_\lambda\}_{\lambda\in\Lambda}$ and $\{b_\lambda\}_{\lambda\in\Lambda}$ as positive numbers, with Λ a suitable index set, such that

$$m(\sigma) = \inf_{\lambda\in\Lambda}(a_\lambda\sigma + b_\lambda).$$

For fixed $\lambda \in \Lambda$ and $\delta > 0$ define ϕ, ψ by

$$\phi(z) = a_\lambda(|z|^2 + \delta)^{1/2} + b_\lambda \quad \text{and} \quad \psi(x,y) = \phi(x-y).$$

We set $w(t,x,y) = u(t,x) - v(t,y)$ and want to prove $w \le \psi$ on Q_T, which obviously ensures our assertion. To do this, we suppose that $\sup_{Q_T}(w-\psi) > 0$, and then get a contradiction. Set

$$w^\varepsilon(t,x,y) = u^\varepsilon(t,x) - v_\varepsilon(t,y) \quad \text{for } \varepsilon > 0,$$

where u^ε (v_ε resp.) is the sup (inf resp.) convolution in x (y resp.). We know that $w^\varepsilon \downarrow w$ on $\overline{Q_T}$ as $\varepsilon \downarrow 0$, and w^ε and w are u.s.c. on $\overline{Q_T}$. As $\varepsilon \downarrow 0$ we see

$$\left\{(w^\varepsilon - \phi)(t,x,y) - \max_{\partial_p Q_T}(w-\phi)\right\}^+ \downarrow 0 \quad \text{for } (t,x,y) \in \partial_p Q_T,$$

where $f(z)^+ = \max(f(z),0)$. By Dini's theorem this convergence is uniform on $\partial_p Q_T$. Note also that $\max_{\partial_p Q_T}(w-\phi) < 0 < \sup_{Q_T}(w-\phi)$. Thus we find that $w^\varepsilon - \phi$ attains a maximum over Q_T at a point $(\bar t, \bar x, \bar y)$ of Q_T^σ if $\varepsilon > 0$ and $\sigma > 0$ are sufficiently small, where $Q_T^\sigma = (\sigma, T] \times \Omega^\sigma \times \Omega^\sigma$ and $\Omega^\sigma = \{x \in \Omega;\ \mathrm{d}(x,\partial\Omega) > \sigma\}$. We get, for sufficiently small

$\sigma > 0$,

$$(w^\varepsilon - \psi)(\bar{t}, \bar{x}, \bar{y}) = \sup_{Q_T^\sigma}(w^\varepsilon - \psi) > \max_{\partial_p Q_T^\sigma}(w^\varepsilon - \psi).$$

Since u^ε (v_ε resp.) is a sub- (super- resp.) solution of (4.3) in Ω_T^σ by Lemma 3.6, and since $u^\varepsilon(t, \cdot)$ ($v_\varepsilon(t, \cdot)$ resp.) by Lipschitz continuous on $\overline{\Omega}$ for each $t \in [0, T]$ and

$$\nabla_x^2 u^\varepsilon(t, x), -\nabla_y^2 v^\varepsilon(t, y) \geq -2\varepsilon^{-1} I \quad \text{in } \Omega_T \text{ (in the sense of distributions)}$$

by Lemma 3.5, applying Lemma 4.3 with u^ε and v_ε yields $w^\varepsilon(\bar{t}, \bar{x}, \bar{y}) \leq 0$. Hence, we have

$$w^\varepsilon(t, x, y) - \psi(x, y) \leq 0 \quad \text{in } Q_T,$$

which is a contradiction. Therefore, $\sup_{Q_T}(w - \phi) \leq 0$, i.e.,

$$u(t, x) - v(t, y) \leq a_\lambda(|x - y|^2 + \delta)^{1/2} + b_\lambda \quad \text{on } Q_T.$$

Letting $\delta \to 0$ and taking the infimum for $\lambda \in A$, we find $u(t, x) - v(t, y) \leq m(|x - y|)$ on Q_T holds. We now conclude that $u \leq v$ on Ω_T and this completes the proof. q.e.d.

As in [17], Theorem 4.1 yields uniqueness and existence of solutions by Perron's method.

Theorem 4.4. *Suppose Ω is a bounded domain in $\mathbb{R}^n$, and F is as in Theorem* 4.1. *For given data $g \in C(\partial_p \Omega_T)$ there is at most one viscosity solution u of the initial boundary value problem of* (4.1) *in Ω_T, with $u^* = u_* = g$ on $\partial_p \Omega_T$.*

This follows directly from Theorem 4.1 by comparison.

Theorem 4.5. *Suppose Ω is a bounded domain in $\mathbb{R}^n$, and F is as in Theorem* 4.1. *Suppose that there is a subsolution f and a supersolution g of* (4.1) *in Ω_T satisfying $f \leq g$ on Ω_T and $f_* = g^*$ on $\partial_p \Omega_T$. Then there is a viscosity solution u of* (4.1) *satisfying $u \in C(\overline{\Omega_T})$ and $f \leq u \leq g$ on $\overline{\Omega_T}$, where $\overline{\Omega_T} = \partial_p \Omega_T \cup \Omega_T$.*

Proof. By Proposition 2.3 there is a viscosity solution u of (4.1) in Ω_T satisfying $f \leq u \leq g$ on Ω_T. Assumption $f_* = g^*$ on $\partial_p \Omega_T$ implies $u^* \leq u_*$ on $\partial_p \Omega_T$, which leads to $u^* \leq u_*$ on Ω_T by Theorem 4.1. Hence, we have $u^* = u_*$, i.e., $u \in C(\overline{\Omega_T})$.

Remark 4.6. Our method also yields a comparison theorem for elliptic equations $u + F(u, \nabla u, \nabla^2 u) = 0$. In fact Theorem 3.1 in [17] can be extended even when F is not continuous at $p = 0$ provided (4.2d) holds.

5. Geometric parabolic equations

In this section we prove that geometric, degenerate parabolic equations are invariant under an (orientation-preserving) change of a dependent variable even in the viscosity sense. Our main tools are an approximation of solutions by sup (inf) convolution in Lemma 3.5, Jensen's lemma (Lemmas 3.3 and 3.4), and stability (Proposition 2.4). We begin by discussing geometric equations of nonevolution type.

Definition 5.1. Let $E\colon W \to \mathbb{R}$, where $W \subset J(\mathbb{R}^N) = \mathbb{R}^N \times \mathbb{R} \times \mathbb{R}^N \times S^{N\times N}$. We say the equation $E = 0$ is *geometric* in W if E satisfies

(5.1a) $$E = E(y, s, q, Y) \text{ is independent of } s,$$

and for $\lambda > 0$ and $\mu \in \mathbb{R}$ there is $C_i = C_i(\lambda, \mu) > 0$ $(i = 1, 2)$ such that

(5.1b) $$C_1 E(y, q, Y) \leq E(y, \lambda q, \lambda Y + \mu q \otimes q) \leq C_2 E(y, q, Y)$$

holds whenever each term is well defined. Here $\otimes$ denotes a tensor product of vectors in $\mathbb{R}^N$. It is easy to see that the equations $E_* = 0$ and $E^* = 0$ are geometric in $\overline{W}$ if $E = 0$ is geometric in W.

The following is our main result for geometric, degenerate elliptic equations in this section.

Theorem 5.2. *Let A be an open set in $\mathbb{R}^N$, and W be a dense subset of $J(A)$. Let u be a locally bounded viscosity sub- (super- resp.) solution of*

(5.2) $$E(y, u, Du, D^2u) = 0 \quad \text{in } A.$$

If E is degenerate elliptic and $E = 0$ is geometric, then $\theta(u)$ is a viscosity sub- (super- resp.) solution whenever $\theta\colon \mathbb{R} \to \mathbb{R}$ is a continuous nondecreasing function.

We first prove Theorem 5.2 assuming more regularity on θ and u so that Jensen's lemma is applicable.

Lemma 5.3. *Let A, W, u, and E be as in Theorem 5.2. Suppose that A is bounded and that u is semiconvex (concave resp.) in A and Lipschitz continuous on $\overline{A}$. If θ is an increasing function in $C^2(\mathbb{R})$ with $\theta' > 0$, then $\theta(u)$ is a sub- (super- resp.) solution of (5.2).*

Proof. We only prove the case where u is a subsolution since the other case can be proved similarly. Since u is semiconvex in A and Lipschitz continuous on a compact set $\overline{A}$, we see

$$D^2(\theta \circ u) = (\theta'' \circ u) Du \otimes Du + (\theta' \circ u) D^2 u \geq -CI \quad \text{in } A$$

(in the sense of distribution) with some C independent of y, where

$(f \circ g)(y) = f(g(y))$. This implies that $\theta \circ u$ is also semiconvex in A. We also observe that $\theta \circ u$ is Lipschitz continuous in $\overline{A}$.

Suppose that $\phi \in C^2(A)$ and $\overline{y} \in A$ satisfy

$$\max_A(\theta \circ u - \phi) = \theta(u(\overline{y})) - \phi(\overline{y}).$$

(Here we may assume $\phi \in C^2(\overline{A})$ by modifying ϕ away from $\overline{y}$.) Applying Lemma 3.4 to $\theta \circ u - \phi$, we see there are sequences $\{y_k\}$ in A and $\{q_k\}$ in $\mathbb{R}^N$ such that

$$(5.3) \qquad y_k \to \overline{y} \quad \text{and} \quad q_k \to 0 \quad \text{as } k \to \infty,$$

$$(5.4) \qquad \max_A(\theta \circ u - \phi - q_k \cdot y) = \theta(u(y_k)) - \phi(y_k) - q_k \cdot y_k,$$

$$(5.5) \qquad \theta \circ u \text{ has second differential at } y_k.$$

Thus

$$(5.6) \quad E_*(y_k, D(\theta \circ u)(y_k), D^2(\theta \circ u)(y_k)) \le C_2 E_*(y_k, Du(y_k), D^2u(y_k)),$$

since $E_* = 0$ is geometric in $J(A)$, and (5.5) implies

$$D(\theta \circ u)(y_k) = \theta'(u(y_k))Du(y_k),$$
$$D^2(\theta \circ u)(y_k) = \theta''(u(y_k))Du(y_k) \otimes Du(y_k) + \theta'(u(y_k))D^2u(y_k)$$

with $\theta'(u(y_k)) > 0$. (5.6) yields

$$(5.7) \qquad E_*(y_k, D(\theta \circ u)(y_k), D^2(\theta \circ u)(y_k)) \le 0,$$

since u is a subsolution of (5.2) and the right-hand side of (5.6) is nonpositive; notice that $u(y)$ has second differential at $y = y_k$ by (5.5) since $\theta \in C^2(\mathbb{R})$ with $\theta' > 0$ implies that the inverse $\theta^{-1} \in C^2(\mathbb{R})$.

By (5.4) we have

$$D(\theta \circ u)(y_k) = D\phi(y_k) + q_k, \qquad D^2(\theta \circ u)(y_k) \le D^2\phi(y_k).$$

Since E_* is degenerate elliptic, (5.7) gives

$$(5.8) \qquad E_*(y_k, D\phi(y_k) + q_k, D^2\phi(y_k)) \le 0.$$

Since E_* is l.s.c., letting $k \to \infty$ in (5.8) and noting (5.3), we obtain

$$E_*(\overline{y}, D\phi(\overline{y}), D^2\phi(\overline{y})) \le 0,$$

which shows that $\theta \circ u = \theta(u)$ is a subsolution of (5.2) in A. q.e.d.

We need to approximate $\theta \in C(\mathbb{R})$ by C^2 functions.

Lemma 5.4. *Suppose that $\theta:\mathbb{R} \to \mathbb{R}$ is a continuous nondecreasing function. Then there is a sequence $\{\theta_k\}$ in $C^2(\mathbb{R})$ of increasing functions with $\theta_k' > 0$ such that $\theta_k \to \theta$ uniformly in $\mathbb{R}$ as $k \to \infty$.*

Proof. 1°. We approximate θ by nondecreasing piecewise linear functions. For an integer j and positive integer k we set

$$a_j^{(k)} = \sup\{t;\, \theta(t) \le j/k\}.$$

Since θ is a continuous nondecreasing function, we see the sequence $\{a_j^{(k)}\}_{j=-\infty}^{\infty}$ in $\tilde{\mathbb{R}} = \mathbb{R} \cup \{\pm\infty\}$ has no accumulation points in $\mathbb{R}$ and $a_j^{(k)} < a_{j+1}^{(k)}$ (unless both are infinite with the same sign). We now define a continuous nondecreasing piecewise linear function θ_k such that it is linear except at $a_j^{(k)}$ and agrees with θ at $a_j^{(k)} \in \mathbb{R}$. It is easy to see that $\theta_k \to \theta$ uniformly in $\mathbb{R}$ as $k \to \infty$.

2°. We approximate a nondecreasing piecewise linear function θ_k by a nondecreasing C^2 function. This is easy because all we need is to mollify θ_k near nondifferentiable points $a_j^{(k)}$. We still denote C^2 approximation of θ_k by θ_k.

3°. We approximate the nondecreasing C^2 function θ_k by a C^2 function whose derivative is always positive. Let $\beta \in C^2(\mathbb{R})$ be a bounded C^2 function with $\beta' > 0$. If we set

$$\overline{\theta}_k(t) = \theta_k(t) + \beta(t)/k,$$

then this $\overline{\theta}_k$ converges to θ uniformly in $\mathbb{R}$ as $k \to \infty$. Since $\overline{\theta}_k$ is a C^2 increasing function with $\overline{\theta}_k' > 0$, this completes the proof.

Proof of Theorem 5.2. We may assume that A is bounded and u is bounded in A since our problem is local. Let u^ε be the sup convolution of u^* in Lemma 3.5. By Lemmas 3.5 and 3.6 we see u^ε is semiconvex in A and Lipschitz continuous on $\overline{A}$, and is a subsolution of

$$E_\varepsilon(y, u^\varepsilon, Du^\varepsilon, D^2u^\varepsilon) = 0 \quad \text{in } A^\varepsilon. \tag{5.9}$$

Let θ_k be an approximation of θ in Lemma 5.4. By Lemma 5.3 we observe that $\theta_k(u^\varepsilon)$ is also a subsolution of (5.9). Since the convergence $u^\varepsilon \downarrow u^*$ is monotone and u^* is u.s.c., applying Dini's theorem we see u^ε converges to u^* uniformly in $\overline{A}$. This implies that $\theta_k(u^\varepsilon)$ converges to $\theta(u^*)$ uniformly in $\overline{A}$. We now apply the stability Proposition 2.4 and conclude that $\theta(u^*)$ is a subsolution of (5.2) since $\lim_{*\varepsilon\downarrow 0} E_\varepsilon \ge E_*$. Since $\theta(u)^* = \theta(u^*)$, it follows that $\theta(u)$ is a subsolution of (5.2). q.e.d.

We now give a version of Theorem 5.2 for a parabolic equation

$$E(y, u, Du, D^2u) = u_t + F(t, x, u, \nabla u, \nabla^2 u) = 0, \tag{5.10}$$

where $y = (t, x)$, $D = (\partial_t, \nabla)$, and $F: (0, T] \times W \to \mathbb{R}$ with $W \subset J(\mathbb{R}^n)$. It turns out that the word "geometric" is consistently used in both the Introduction and Definition 5.2.

Proposition 5.5. *Suppose that* E *is expressed as in* (5.10). *Then the equation* $E = 0$ *is geometric if (and only if)* F *satisfies*

(5.11a) $\qquad F = F(t, x, r, p, X)$ *is independent of* r,

and

(5.11b) $$F(t, x, \lambda p, \lambda X + \sigma p \otimes p) = \lambda F(t, x, p, X)$$

for $\lambda > 0$ *and* $\sigma \in \mathbb{R}$, *whenever each term is well defined.*

If F *satisfies* (5.11a) *and* (5.11b), *we say* F *is geometric as in the Introduction.*

The proof is straightforward from the definitions and is thus omitted.

Theorem 5.6 (Parabolic version). *Let* Ω *be an open set in* $\mathbb{R}^n$ *and* $T > 0$. *Let* W *be a dense subset of* $J(\Omega) = \Omega \times \mathbb{R} \times \mathbb{R}^n \times S^{n \times n}$. *Suppose that* F *is degenerate elliptic and geometric in* $(0, T] \times W$. *If* u *is a locally bounded viscosity sub- (super-) solution of* (5.10) *in* $\Omega_T = (0, T] \times \Omega$, *then so is* $\theta(u)$ *whenever* $\theta: \mathbb{R} \to \mathbb{R}$ *is a continuous nondecreasing function.*

Proof. Let E be as in (5.10). By assumptions of F we see E is degenerate elliptic and $E = 0$ is geometric. Applying Theorem 5.2 with $A = \operatorname{int} \Omega_T$ shows that $\theta(u)$ is a sub- (super- resp.) solution of (5.10) in $\operatorname{int} \Omega_T = (0, T) \times \Omega$. This together with the next lemma implies that $\theta(u)$ is a sub- (super- resp.) solution of (5.10) in Ω_T.

Lemma 5.7. *If* u *is a subsolution in* $(0, T) \times \Omega$ *of* (5.10), *then* u *is a subsolution in* $\Omega_T = (0, T] \times \Omega$ *of* (5.10).

Proof. The proof is similar to that of Lemma 3.1 (admitting Proposition 3.2). We may assume that $u^* - \phi$ attains its maximum over Ω_T at $(T, \overline{x})$, $\overline{x} \in \Omega$, where $\phi \in C^2(\Omega_T)$. As in the proof of Lemma 3.1 we shift ϕ by ϕ_α, and find a sequence $(t_\alpha, x_\alpha) \in \operatorname{int} \Omega_T$ such that $u^* - \phi_\alpha$ attains its maximum at (t_α, x_α) and that $(t_\alpha, x_\alpha) \to (T, \overline{x})$. Since u is a subsolution of (5.10) in $\operatorname{int} \Omega_T$, passing to the limit in

$$E_*(y_\alpha, u^*(y_\alpha), D\phi(y_\alpha), D^2\phi(y_\alpha)) \le 0, \qquad y_\alpha = (t_\alpha, x_\alpha),$$

yields

$$E_*(\overline{y}, u^*(\overline{y}), D\phi(\overline{y}), D^2\phi(\overline{y})) \le 0, \qquad \overline{y} = (T, \overline{x}).$$

Remark 5.8. In [24, Proposition 2.2] there is a proof of Lemma 5.7. Our proof is different from that in [24].

We conclude this section by listing examples of geometric, degenerate parabolic equations. We shall suppress the word "degenerate" (because all geometric equations are degenerate).

Example 5.9. The mean curvature flow equation as well as its generalization

$$u_t - |\nabla u| \operatorname{div}(\nabla u/|\nabla u|) - \nu|\nabla u| = 0, \qquad \nu \in \mathbb{R}, \tag{5.12}$$

is a geometric parabolic equation. Indeed, (5.12) is expressed as

$$u_t + F(\nabla u, \nabla^2 u) = 0 \tag{5.13}$$

with

$$F(p, X) = -\operatorname{trace}((I - \overline{p} \otimes \overline{p})X) - \nu|p|, \qquad \overline{p} = p/|p|.$$

A calculation shows

$$\begin{aligned} F(\lambda p, \lambda X + \sigma p \otimes p) &= -\operatorname{trace}((I - \overline{p} \otimes \overline{p})(\lambda X + \sigma p \otimes p)) - \lambda\nu|p| \\ &= \lambda F(p, X) - \sigma \operatorname{trace}((I - \overline{p} \otimes \overline{p})p \otimes p). \end{aligned}$$

Since $(\overline{p} \otimes \overline{p})(p \otimes p) = p \otimes p$, the last term disappears so F satisfies (5.11a,b). By Proposition 5.5 we see (5.12) is geometric. Since

$$-\operatorname{trace}((I - \overline{p} \otimes \overline{p})Y) \leq 0 \quad \text{for } Y \geq O,$$

we see F is degenerate elliptic, i.e.,

$$F(p, X + Y) \leq F(p, X) \quad \text{for } Y \geq O.$$

We thus conclude (5.12) is a geometric parabolic equation.

Example 5.10 (An anisotropic version of (5.12)). We consider

$$u_t - |\nabla u| \sum_{i=1}^{n} \frac{\partial}{\partial x_i}\left(\frac{\partial H}{\partial p_i}\left(\frac{\nabla u}{|\nabla u|}\right)\right) - \beta\left(\frac{\nabla u}{|\nabla u|}\right)|\nabla u| = 0, \tag{5.14}$$

where $H \in C^2(\mathbb{R}^n \backslash \{0\})$ is convex and positively homogeneous of degree 1, i.e.,

$$H(\lambda p) = \lambda H(p) \quad \text{for } \lambda > 0,\ p \in \mathbb{R}^n \backslash \{0\}.$$

If $H(p) = |p|$ and $\beta = \nu$, (5.14) is the same as (5.12). (5.14) is a geometric parabolic equation. Indeed, (5.14) is expressed as (5.13) with

$$F(p, X) = -\operatorname{trace}(A(\overline{p})(I - \overline{P} \otimes \overline{p})X) - \beta(\overline{p})|p| \tag{5.15}$$

and $n \times n$ matrix

$$A(\overline{p}) = \left(\frac{\partial^2 H}{\partial p_i \partial p_j}(\overline{p})\right), \qquad \overline{p} = \frac{p}{|p|}.$$

As in Example 5.9 we easily observe that F is geometric. The convexity of H yields $A(\overline{p}) \geq 0$ which implies that F is degenerate elliptic. We thus conclude that (5.14) is a geometric parabolic equation. We remark that (5.15) is also expressed as

$$(5.16)\qquad F(p, X) = -\operatorname{trace}(A(\overline{p})X) - \beta(\overline{p})|p|.$$

Indeed, the homogeneity of H implies

$$H = \sum_{i=1}^{n} \frac{\partial H}{\partial p_i} p_i.$$

Differentiating this identity in p_j, we easily see $A(p)p \otimes p = 0$.

The anisotropic version of (5.12) is important in studying anisotropic phase transition phenomena such as crystal growth. We refer to [13] for its background.

6. Existence and uniqueness of solutions

In this section, as applications of results in §4 we construct a unique continuous viscosity solution of the initial value problem for a geometric parabolic equation

$$(6.1)\qquad u_t + F(t, \nabla u, \nabla^2 u) = 0 \quad \text{in } \mathbf{R}^n_T = (0, T] \times \mathbf{R}^n,$$

$$(6.2)\qquad u(0, x) = a(x)$$

for $a \in C_\alpha(\mathbf{R}^n)$, i.e., $a - \alpha$ is continuous with compact support in $\mathbf{R}^n$, where $\alpha \in \mathbf{R}$. We also establish a comparison theorem as well as uniqueness of solutions. Since $\mathbf{R}^n_T$ is unbounded, we shall reduce our problem to the case where the domain is bounded by using "barriers" so that the results in §4 are applicable. To show the existence of solutions we construct sub- and supersolutions of the initial value problem (6.1)–(6.2). This leads to the existence of solutions of (6.1)–(6.2) by Perron's method as in Theorem 4.5.

We begin by constructing sub- and supersolutions of (6.1). In what follows we shall always assume that $F = F(t, x, p, X)$ is continuous and degenerate elliptic in $(0, T] \times \mathbf{R}^n \times (\mathbf{R}^n \backslash \{0\}) \times S^{n\times n}$.

Lemma 6.1. *Suppose that* F *is geometric and that*

$$(6.3_-)\qquad F_*(t, x, p, -I) \le c_-(|p|),$$

$$(6.3_+)\qquad F^*(t, x, p, I) \ge -c_+(|p|)$$

for some $c_\pm(\sigma) \in C^1[0, \infty)$ *and* $c_\pm(\sigma) \ge c_0 > 0$ *with some constant* c_0. *We set*

$$(6.4)\quad u^\pm(t, x) = \pm(t + w_\pm(\rho)), \quad \rho = |x|, \text{ with } w_\pm(\rho) = \int_0^\rho \frac{\sigma}{c_\pm(\sigma)} d\sigma.$$

Then u^- (u^+ *resp.*) *is a* C^2 *sub- (super- resp.) solution of* (6.1) *in* $\mathbf{R} \times \mathbf{R}^n$.

Proof. We only show that u^- is a subsolution of (6.1) when (6.3_) holds, since the proof for u^+ is parallel. By definition (6.4), $u = u^- \in C^2(\mathbb{R} \times \mathbb{R}^n)$. Since F_* is geometric and

$$\nabla w_-(\rho) = w_-'(\rho)\nabla\rho,$$
$$\nabla^2 w_-(\rho) = w_-''(\rho)\nabla\rho \otimes \nabla\rho + w_-'(\rho)\nabla^2\rho,$$
$$\nabla^2\rho = \rho^{-1}(I - \nabla\rho \otimes \nabla\rho)$$

with $w_-'(\rho) > 0$, a calculation yields

$$\begin{aligned} F_*(t, x, \nabla u, \nabla^2 u) &= F_*(t, x, -w_-'(\rho)\nabla\rho, -w_-'(\rho)\rho^{-1}I) \\ &= w_-'(\rho)\rho^{-1}F_*(t, x, -\rho\nabla\rho, -I), \end{aligned}$$

which together with (6.4) gives

$$u_t + F_*(t, x, \nabla u, \nabla^2 u) = -1 + F_*(t, x, -x, -I)/c_-(\rho),$$

since $\rho\nabla\rho = x$. Thus the assumption (6.3_) implies

$$u_t + F_*(t, x, \nabla u, \nabla^2 u) \le 0,$$

which means that $u = u_-$ is a subsolution of (6.1).

Example 6.2. If $F(t, x, p, X)$ is of degree one in X and independent of t and x, then F has a form

$$F(p, X) = -\operatorname{trace}(A(p)X) + B(p).$$

If F is geometric and continuous in $p \in \mathbb{R}^n\backslash\{0\}$, we see easily that F satisfies $(6.3_\pm)$ by taking

$$c_\pm(\rho) = \sup_{|p|=1} \operatorname{trace} A(p) + \rho \sup_{|p|=1} |B(p)|.$$

In particular, (5.12) and (5.14) in Examples 5.9 and 5.10 satisfy $(6.3_\pm)$. When (6.1) is the mean curvature flow equation (5.12) with $\nu = 0$,

$$F(p, X) = -\operatorname{trace}((I - \overline{p} \otimes \overline{p})X), \qquad \overline{p} = p/|p|,$$

we have

$$u^\pm(t, x) = \pm(t + |x|^2/2(n-1)).$$

Moreover, $u^\pm$ is a *solution* of (5.12) with $\nu = 0$.

Lemma 6.3. *Suppose that F is geometric and that* (6.3_) $((6.3_+)$ *resp.) holds. For* $u^\pm$ *we set*

$$U_{\xi h}^\pm(t, x) = h(u^\pm(t, x - \xi)), \qquad \xi \in \mathbb{R}^n,$$

where $u^{\pm}$ is defined in Lemma 6.1, and h is a continuous nondecreasing function in $\mathbb{R}$. Then $U_{\xi h}^{-}$ ($U_{\xi h}^{+}$ resp.) is a sub- (super- resp.) solution of (6.1) in $\mathbb{R} \times \mathbb{R}^n$.

Proof. By a translation of the variable x, Lemma 6.1 implies that $u^{-}(t, x-\xi)$ is a subsolution of (6.1). Since F is geometric, by Theorem 5.6

$$U_{\xi h}^{-} = h(u^{-}(t, x-\xi))$$

is again a subsolution of (6.1). The proof for $U_{\xi h}^{+}$ is parallel so is omitted.

Proposition 6.4. *Suppose that F is geometric and that (6.3_) ((6.3$_+$) resp.) holds. Then for every $a \in C(\mathbb{R}^n)$ there is a lower (upper resp.) semi-continuous sub- (super- resp.) solution v^{-} (v^{+} resp.) of (6.1) in $[0, \infty) \times \mathbb{R}^n$ satisfying*

$$v^{-} \le a \le v^{+} \quad \text{for } t \ge 0 \qquad \text{and} \qquad v^{\pm} = a \quad \text{at } t = 0.$$

Proof. We construct only a subsolution v^{-} of (6.1)–(6.2) by using $U_{\xi h}^{-}$ since a supersolution v^{+} of (6.1)–(6.2) is constructed parallelly from $U_{\xi h}^{+}$. Since $u^{-}(t, x)$ is decreasing in $|x|$ and t, for each $\xi \in \mathbb{R}^n$ the continuity of a guarantees that there is a continuous nondecreasing function $h = h_\xi : \mathbb{R} \to \mathbb{R}$ with $h(0) = a(\xi)$ such that $U_{\xi h}^{-}(t, x) \le a(x)$ for $t \ge 0$. Since $U_{\xi h}^{-}$ is a subsolution of (6.1) in $\mathbb{R} \times \mathbb{R}^n$, by Proposition 2.2 the function

$$v^{-}(t, x) = \sup\{U_{\xi h}^{-}(t, x);\, h = h_\xi,\ \xi \in \mathbb{R}^n\} \le a(x)$$

is again a subsolution of (6.1) in $[0, \infty) \times \mathbb{R}^n$. Since $h_\xi(0) = a(\xi)$ and $U_{\xi h}^{-}(0, x) \le a(x)$ with $h = h_\xi$, we observe that $v^{-}(0, x) = a(x)$ so v^{-} satisfies (6.2). The continuity of $U_{\xi h}^{-}$ implies that v^{-} is l.s.c.

We next introduce "barriers" to handle the unbounded domain $\mathbb{R}^n$.

Lemma 6.5. *For $\omega \ge 0$ we set*

$$\psi^{\pm}(t, x) = \begin{cases} \mp (|x| - \omega t)^4 & \text{if } |x| > \omega t, \\ 0 & \text{otherwise.} \end{cases} \tag{6.5}$$

Suppose that F is degenerate elliptic and satisfies

$$F_*(t, x, p, O) \le \nu_- |p|, \tag{6.6\_}$$

$$F^*(t, x, p, O) \ge -\nu_+ |p| \tag{6.6$_+$}$$

with $\nu_\pm \ge 0$ independent of t, x, and p. Then ψ^{-} (ψ^{+} resp.) is a C^2 sub- (super- resp.) solution of (6.1) in $\mathbb{R} \times \mathbb{R}^n$ provided that $\omega \ge \nu_-$ ($\omega \ge \nu_+$ resp.).

Proof. Again we only show that ψ^- is a subsolution of (6.1). Clearly, ψ^- is C^2. Since $\psi = \psi^-$ is convex and F is degenerate elliptic, by (6.6_) we obtain

$$F_*(t, x, \nabla\psi, \nabla^2\psi) \le F_*(t, x, \nabla\psi, O) \le \nu_-|\nabla\psi|.$$

A calculation shows that $\psi = \psi^-$ solves $\psi_t + \omega|\nabla\psi| = 0$. If $\omega \ge \nu_-$, this yields

$$\psi_t + F_*(t, x, \nabla\psi, \nabla^2\psi) \le \psi_t + \nu_-|\nabla\psi| \le 0.$$

Proposition 6.6. *Suppose that F is geometric and degenerate elliptic and satisfies* (6.3_) ((6.3$_+$) *resp.*). *Then F satisfies* (6.6_) ((6.6$_+$) *resp.*) *with $\nu_\pm = c_\pm(1)$.*

Proof. If F is degenerate elliptic, (6.3_) implies $F(t, x, p, O) \le c_-(|p|)$. Since F is geometric, we see

$$F(t, x, p, O) = |p|F(t, x, \bar{p}, O) \le |p|c_-(1), \qquad \bar{p} = p/|p|,$$

which yields (6.6_) with $\nu_- = c_-(1)$. The estimate (6.6$_+$) is parallelly proved. q.e.d.

We now state the uniqueness and comparison of viscosity solutions of (6.1)–(6.2) in $\mathbf{R}^n_T$. By $f \in C_\alpha(A)$ we mean that $f \in C(A)$ and $f - \alpha$ has a compact support in A, where $\alpha \in \mathbf{R}$.

Theorem 6.7 (Uniqueness and comparison). *Let $T > 0$. Assume that $F(t, p, X)$ is continuous in $(0, T] \times (\mathbf{R}^n\backslash\{0\}) \times S^{n\times n}$ and is geometric and degenerate elliptic, and that F satisfies* (6.6$_\pm$) *and*

$$-\infty < F_*(t, 0, O) = F^*(t, 0, O) < \infty. \tag{6.7}$$

Then for $a \in C_\alpha(\mathbf{R}^n)$ there is at most one viscosity solution $u_a \in C_\alpha([0, T] \times \mathbf{R}^n)$ of (6.1) – (6.2). *Moreover, if $b \ge a$ with $b \in C_\beta(\mathbf{R}^n)$ for some $\beta \ge \alpha$, then $u_b \ge u_a$ in $\mathbf{R}^n_T$.*

Proof. We may assume $\alpha = 0 \le \beta$. For $\psi^\pm$ in (6.5) we set

$$f_R = \min(\psi^- - R^4, 0), \qquad g_R = \max(\psi^+ + R^4, \beta), \tag{6.8}$$

where $\omega \ge \nu_\pm$ and $R > 0$. We take R large enough so that $f_R \le a(x) \le b(x) \le g_R$ holds at $t = 0$. By Lemma 6.5, ψ^- and ψ^+ are, respectively, sub- and supersolutions of (6.1) in $\mathbf{R}^n_T$. Since F is geometric and the functions

$$\theta_-(\sigma) = \min(\sigma - R^4, 0), \qquad \theta_+(\sigma) = \max(\sigma + R^4, \beta)$$

are continuous nondecreasing functions, by Theorem 5.6 we conclude that $f_R = \theta_-(\psi^-)$ and $g_R = \theta_+(\psi^+)$ are, respectively, sub- and supersolutions of (6.1) in $\mathbb{R}^n_T$. Take R_1 such that u_a, $u_b - \beta$, f_R, $g_R - \beta$ are supported in $[0, T] \times B(R_1)$, where $B(R_1)$ denotes the open ball of radius R_1 centered at the origin. Applying the comparison Theorem 4.1 with $\Omega = B(R_1)$ yields $u_b \geq u_a$. This in particular deduces the uniqueness of u_a for a given $a \in C_\alpha(\mathbb{R}^n)$.

Theorem 6.8 (Global existence). *Let $T > 0$. Assume that $F(t, p, X)$ is continuous in $(0, T] \times (\mathbb{R}^n \backslash \{0\}) \times S^{n\times n}$ and is geometric and degenerate elliptic, and that F satisfies (6.3±) and (6.7). Then for $a \in C_\alpha(\mathbb{R}^n)$ there is a (unique) viscosity solution $u_a \in C_\alpha([0, T] \times \mathbb{R}^n)$ of (6.1)–(6.2).*

Proof. We may assume $\alpha = 0$. Since $(6.3_\pm)$ implies $(6.6_\pm)$ by Proposition 6.6, f_R and g_R in (6.8) with $\beta = 0$ are, respectively, sub- and supersolution of (6.1) in $\mathbb{R}^n_T$. We take R large so that $f_R \leq a(x) \leq g_R$ at $t = 0$. Let $v^\pm$ be sub- and supersolutions of (6.1)–(6.2) constructed in Proposition 6.4, and set

$$f = \max(v^-, f_R), \qquad g = \min(v^+, g_R).$$

Then, by Proposition 2.2, f and g are, respectively, sub- and supersolutions of (6.1)–(6.2) in $\mathbb{R}^n_T$ and are supported in $[0, T] \times B(R_1)$ for sufficiently large R_1. Since f and g are, respectively, lower and upper semicontinuous, we now apply the existence Theorem 4.5 with $\Omega = B(R_1)$ to get a solution u_a of (6.1)–(6.2) satisfying $f \leq u_a \leq g$ supported in $[0, T] \times B(R_1)$. This u_a solves (6.1)–(6.2) in $\mathbb{R}^n_T$ by extending zero outside $B(R_1)$, and satisfies $u_a \in C_\alpha([0, T] \times \mathbb{R}^n)$.

Remark 6.9. Condition (6.7) follows from $(6.6_\pm)$ if $(t, X) \mapsto F(t, p, X)$ is equicontinuous for small p. In particular, if $F(t, p, X)$ is of degree one in X and independent of t as in Example 6.2, all assumptions on F in Theorems 6.7 and 6.8 are fulfilled provided that F is geometric and degenerate elliptic. We thus observe that our Theorems 6.7 and 6.8 are applicable to equations in Examples 5.9 and 5.10. In [5] Theorem 6.7 and a weaker version of Theorem 6.8 are stated.

7. Evolution of level surfaces

We now study the γ-level set $\Gamma(t)$ of the solution u_a of (6.1)–(6.2) in Theorems 6.7 and 6.8. Our goal is to show that the γ-level set

$$\Gamma(t) = \{x \in \mathbb{R}^n ; u_a(t, x) = \gamma\} \tag{7.1}$$

and the open set surrounded by $\Gamma(t)$,

$$D(t) = \{x \in \mathbb{R}^n ; u_a(t, x) > \gamma\}, \qquad \gamma > \alpha, \tag{7.2}$$

are uniquely determined by $(\Gamma(0), D(0))$ and do not depend on a choice of the defining functions a of $(\Gamma(0), D(0))$. In other words, (6.1) can be regarded as an evolution equation of $(\Gamma(t), D(t))$. By the existence Theorem 6.8 we see that there is a unique global evolution family $(\Gamma(t), D(t))$ ($t \geq 0$) for (6.1) with initial data $(\Gamma(0), D(0))$. No regularity of $\Gamma(0)$ is assumed. It turns out that all we need to obtain $(\Gamma(t), D(t))$ is for $D(0)$ to be a bounded open set and for $\Gamma(0)$ ($\subset \mathbf{R}^n \backslash D(0)$) to be a compact set containing $\partial D(0)$. In particular, when (6.1) is the mean curvature flow equation (1.3) we construct a whole unique evolution family $\Gamma(t)$ moved by its mean curvature. Since $\Gamma(t)$ may be singular, the mean curvature here is understood in some weak sense. By the comparison Theorem 4.1 we shall also show that $\Gamma(t)$ becomes extinct in finite time provided that $n \geq 2$. This extends a result of Huisken [14] where he proved this fact when $\Gamma(0)$ is a uniformly convex C^2 hypersurface in $\mathbf{R}^n$ ($n \geq 3$) (see [10] for $n = 2$).

Theorem 7.1 (Uniqueness). *Suppose that F and u_a are as in Theorem 6.7 with $a \in C_\alpha(\mathbf{R}^n)$. Let $\Gamma(t)$ and $D(t)$ be defined by* (7.1)–(7.2). *If $\gamma > \alpha$, then the evolution family $(\Gamma(t), D(t))$ for $t \geq 0$ is uniquely determined by $(\Gamma(0), D(0))$ and is independent of a, α, and γ. We call $(\Gamma(t), D(t))$ a solution family of* (6.1) *with initial data $(\Gamma(0), D(0))$.*

To prove Theorem 7.1 we prepare an elementary lemma on the comparison of continuous functions.

Lemma 7.2. *Suppose that $a, b \in C(\overline{D})$ are positive in D and vanish on ∂D, where D is an open set in $\mathbf{R}^n$. Then there exists a continuous (strictly) increasing function $\theta: \mathbf{R} \to \mathbf{R}$ such that $a(x) \leq \theta(b(x))$ in D with $\theta(0) = 0$ provided that either*

(i) *D is bounded, or*

(ii) *D is an exterior domain, i.e., $\mathbf{R}^n \backslash D$ is a nonempty compact set and $a(x), b(x) \in C_\alpha(\overline{D})$ for some $\alpha > 0$.*

Proof. We prove only that case (i) holds. The proof of case (ii) is similar so is omitted. We set

$$a_1(r) = \sup\{a(x); x \in D, \mathrm{d}(x, \partial D) \leq r\},$$
$$b_1(r) = \inf\{b(x); x \in D, \mathrm{d}(x, \partial D) \geq r\}$$

for r, $0 \leq r \leq R$, where $R = \sup\{\mathrm{d}(x, \partial D); x \in D\}$ and $\mathrm{d}(x, \partial D)$ denotes the distance between x and ∂D. Since a_1 and b_1 are nondecreasing continuous functions with $a_1(0) = b_1(0) = 0$ and $a_1(r) > 0$, $b_1(r) > 0$ for $0 < r \leq R$, we see that

$$\overline{a}(r) = a_1(r) + r, \qquad \overline{b}(r) = b_1(r)r/R$$

are both continuous increasing functions on $[0, R]$ and satisfy

$$(7.3) \qquad \overline{a}(r) > 0, \quad \overline{b}(r) > 0 \quad \text{for } 0 < r \le R, \ \overline{a}(0) = \overline{b}(0) = 0.$$

By the definition of $\overline{a}$ and $\overline{b}$ we observe that

$$(7.4) \qquad 0 \le a(x) \le \overline{a}(r), \quad 0 \le \overline{b}(r) \le b(x) \quad \text{with } r = \mathrm{d}(x, \partial D).$$

Since $\overline{b}$ is continuous and increasing, so is its inverse function $\overline{b}^{-1}$. We now set $\theta = \overline{a} \circ \overline{b}^{-1}$ and find θ is continuous and increasing in $[0, \overline{b}(R)]$. We extend θ outside $[0, \overline{b}(R)]$ so that θ is a continuous increasing function on $\mathbb{R}$. Evidently, by (7.3) we obtain $\theta(0) = 0$. Since $\overline{a}(r) = \theta(\overline{b}(r))$, $0 \le r \le R$, it follows from (7.4) that

$$a(x) \le \overline{a}(r) = \theta(\overline{b}(r)) \le \theta(b(x)) \quad \text{for all } x \in \overline{D}.$$

Proof of Theorem 7.1. Suppose $b \in C_\beta(\mathbb{R}^n)$ and $u_b \in C_\beta([0, T] \times \mathbb{R}^n)$ solves (6.1) with initial data b. We set

$$\Gamma'(t) = \{x \in \mathbb{R}^n ; u_b(x, t) = \gamma'\},$$
$$D'(t) = \{x \in \mathbb{R}^n ; u_b(x, t) > \gamma'\}, \qquad \gamma' > \beta.$$

It suffices to prove

$$(7.5) \qquad (\Gamma'(t), D'(t)) = (\Gamma(t), D(t)) \quad \text{for all } t \ge 0$$

provided that

$$(7.6) \qquad (\Gamma'(0), D'(0)) = (\Gamma(0), D(0)).$$

Since F is geometric, applying Theorem 5.6 we may assume $\gamma' = \gamma = 0$ and $\alpha = \beta < 0$ by a translation and dilation of the dependent variable. By (7.6) we apply Lemma 7.2 and conclude that there are continuous increasing functions $\theta_1, \theta_2 \colon \mathbb{R} \to \mathbb{R}$ such that

$$a(x) \le \theta_1(b(x)), \quad b(x) \le \theta_2(a(x)) \quad \text{for } x \in \mathbb{R}^n$$

with $\theta_1(0) = \theta_2(0) = \gamma = 0$. Since F is geometric, by Theorem 5.6 $\theta_1(u_b)$ and $\theta_2(u_a)$ are solutions of (6.1) with initial data $\theta_1(b)$ and $\theta_2(a)$ respectively. Our comparison Theorem 6.7 now yields

$$u_a \le \theta_1(u_b), \quad u_b \le \theta_2(u_a) \quad \text{in } \mathbb{R}^n_T.$$

This implies (7.5) which completes the proof.

Theorem 7.3 (Existence). *Suppose that F is as in Theorem 6.8 for all $T > 0$, and that $D(0)$ is a bounded open set and $\Gamma(0)$ ($\subset \mathbb{R}^n \backslash D(0)$) is a compact set containing $\partial D(0)$. Then there exists a unique solution family $(\Gamma(t), D(t))$ for all $t \ge 0$ of (6.1) with initial data $(\Gamma(0), D(0))$.*

Proof. By the assumptions on $(\Gamma(0), D(0))$, there is a defining function $a \in C_\alpha(\mathbb{R}^n)$, $\alpha < 0$ such that

$$\Gamma(0) = \{x \in \mathbb{R}^n ; a(x) = 0\}, \qquad D(0) = \{x \in \mathbb{R}^n ; a(x) > 0\}.$$

Such $a(x)$ is constructed, for example, by

$$a(x) = \begin{cases} \mathrm{d}(x, \Gamma(0)) & \text{if } x \in D(0), \\ \max(-\mathrm{d}(x, \Gamma(0)), \alpha) & \text{if } x \notin D(0). \end{cases}$$

By Theorem 6.8 there is a unique solution u_a of (6.1)–(6.2). We now find a solution family $(\Gamma(t), D(t))$ of (6.1) defined by (7.1)–(7.2) with $\gamma = 0$. Its uniqueness follows from Theorem 7.1.

Corollary 7.4. *Let F and $(\Gamma(0), D(0))$ be as in Theorem 7.3, and $(\Gamma(t), D(t))$ be the solution family of (6.1) with initial data $(\Gamma(0), D(0))$. If F satisfies*

$$F^*(t, p, -I) \geq c > 0 \tag{7.7}$$

with some constant c independent of $t > 0$ and $p \in \mathbb{R}^n \backslash \{0\}$, then $\Gamma(t)$ and $D(t)$ become empty in finite time.

Proof. As in the proof of Lemma 6.1, a calculation shows that

$$g(t, x) = -(t + \rho^2/2c), \qquad \rho = |x|, \tag{7.8}$$

is a *supersolution* of (6.1) since (7.7) holds. For $(\Gamma(0), D(0))$ we take a defining function $a \in C_\alpha(\mathbb{R}^n)$ with $\alpha < 0$ as in the proof of Theorem 7.3. If $M > 0$ is sufficiently large, then $a(x) \leq h(g(0, x))$, where $h(\tau) = \max(\tau + M, \alpha)$. Since h is continuous and nondecreasing, applying Theorem 5.6 we observe that $h(g(t, x))$ is a supersolution of (6.1) in $\mathbb{R}^n_T$ for every $T > 0$. Since $u_a \in C_\alpha([0, T] \times \mathbb{R}^n)$, definition (7.8) implies that both $u_a - \alpha$ and $h(g) - \alpha$ are supported in $[0, T] \times B(R)$ for sufficiently large R. We now apply the comparison Theorem 4.1 with $\Omega = B(R)$, and find $u_a \leq h(g)$ in $\mathbb{R}^n_T$ for all $T > 0$. Definition (7.8) shows that $h(g) - \alpha = 0$ for sufficiently large t, say $t > T'$. This implies that $u_a(x, t) \leq \alpha$ for $t > T'$. In particular, $\Gamma(t)$ and $D(t)$ become empty for $t > T'$.

Corollary 7.5. *Let $(\Gamma(0), D(0))$ be as in Theorem 7.3, and $(\Gamma(t), D(t))$ be the solution family of (5.14) with initial data $(\Gamma(0), D(0))$. Assume that H and β in (5.14) satisfy*

$$\inf_{|p|=1} \operatorname{trace} A(p) > 0 \quad \text{and} \quad \sup_{|p|=1} \beta(p) \leq 0, \tag{7.9}$$

where $A = \nabla^2 H$. Then $\Gamma(t)$ and $D(t)$ become extinct in finite time. In *particular, if* (5.14) *is the mean curvature flow equation* (1.3), *then* $\Gamma(t)$ *and* $D(t)$ *become extinct in finite time provided that* $n \geq 2$.

Proof. By (5.16), assumption (7.9) implies (7.7). We also observe that (7.9) holds for the mean curvature equation when $n \geq 2$. Thus applying Corollary 7.4 completes the proof.

References

[1] U. Abresch & J. Langer, *The normalized curve shortening flow and homothetic solutions*, J. Differential Geometry **23** (1986) 175–196.

[2] A. D. Alexandroff, *Almost everywhere existence of the second differential of a convex function and some properties of convex surfaces connected with it*, Leningrad Gos. Univ. Uchen. Zap. Ser. Mat. Nauk **6** (1939) 3–35. (Russian)

[3] S. Angenent, *Parabolic equations for curves on surfaces.* II: *Intersections, blow up and generalized solutions*, CMS-Technical Summary Report #89-24, Univ. of Wisconsin.

[4] K. A. Brakke, *The motion of a surface by its mean curvature*, Princeton Univ. Press, Princeton, NJ, 1978.

[5] Y.-G. Chen, Y. Giga & S. Goto, *Uniqueness and existence of viscosity solutions of generalized mean curvature flow equations*, Proc. Japan Acad. Ser. A **65** (1989) 207–210.

[6] M. G. Crandall, *Semidifferentials, quadratic forms and fully nonlinear elliptic equations of second order*, Ann. Inst. H. Poincaré Anal. Non Linéare **6** (1989) 419–435.

[7] P. Dupuis & H. Ishii, *On oblique derivative problems for fully nonlinear second order elliptic PDE's on nonsmooth domain*, preprint.

[8] C. Epstein & M. Weinstein, *A stable manifold theorem for the curve shortening equation*, Comm. Pure Appl. Math. **40** (1987) 119–139.

[9] L. C. Evans & J. Spruck, *Motion of level sets by mean curvature.*, I, this issue, pp. 635–681.

[10] M. Gage & R. Hamilton, *The heat equation shrinking of convex plane curves*, J. Differential Geometry **23** (1986) 69–96.

[11] E. De Giorgi, *Convergence problems for functionals and operators*, Proc. Internat. Meeting on Recent Methods in Nonlinear Analysis (E. De Georgi et al., eds.), Pitagora Editrice, Bologna, 1979, 131–188.

[12] M. Grayson, *The heat equation shrinks embedded plane curves to round points*, J. Differential Geometry **26** (1987) 285–314.

[13] M. Gurtin, *Towards a nonequilibrium thermodynamics of two-phase materials*, Arch. Rational Mech. Anal. **100** (1988) 275–312.

[14] G. Huisken, *Flow by mean curvature of convex surfaces into spheres*, J. Differential Geometry **20** (1984) 237–266.

[15] ——, *Asymptotic behavior for singularities of the mean curvature flow*, J. Differential Geometry **31** (1990) 285–299.

[16] H. Ishii, *Perron's method for Hamilton-Jacobi equations*, Duke Math. J. **55** (1987) 369–384.

[17] ——, *On uniqueness and existence of viscosity solutions of fully nonlinear second order elliptic PDE's*, Comm. Pure Appl. Math. **42** (1989) 15–45.

[18] H. Ishii & P. L. Lions, *Viscosity solutions of fully nonlinear second-order elliptic partial differential equations*, J. Differential Equations **83** (1990) 26–78.

[19] R. Jensen, *The maximum principle for viscosity solutions of fully nonlinear second-order partial differential equations*, Arch. Rational Mech. Anal. **101** (1988) 1–27.

[20] R. Jensen, P. L. Lions & P. E. Souganidis, *A uniqueness result for viscosity solutions of second order fully nonlinear partial differential equations*, Proc. Amer. Math. Soc. **102** (1988) 975–978.

[21] J. M. Lasry & P. L. Lions, *A remark on regularization in Hilbert spaces*, Israel J. Math. **55** (1986) 257–266.

[22] P. L. Lions, *Optimal control of diffusion processes and Hamilton-Jacobi-Bellman equations, Part* II: *Viscosity solutions and uniqueness*, Comm. Partial Differential Equations **8** (1983) 1229–1276.

[23] P. de Mottoni & M. Schatzman, *Evolution géometric d'interfaces*, C.R. Acad. Sci. Paris Sér. I Math. **309** (1989) 453–458.

[24] D. Nunziante, *Uniqueness of viscosity solutions of fully nonlinear second order parabolic equations with noncontinuous time-dependence*, Differential Integral Equations **3** (1990) 77–91.

[25] S. Osher & J. A. Sethian, *Fronts propagating with curvature dependent speed: Algorithms based on Hamilton-Jacobi formulations*, J. Computational Phys. **79** (1988) 12–49.

[26] J. A. Sethian, *Numerical algorithms for propagating interfaces: Hamilton-Jacobi equations and conservation laws*, J. Differential Geometry **31** (1990) 131–161.

HOKKAIDO UNIVERSITY, JAPAN

Arch. Rational Mech. Anal. 131 (1995) 139–197. © Springer-Verlag 1995

Convergence of the Phase-Field Equations to the Mullins-Sekerka Problem with Kinetic Undercooling

H. METE SONER

Dedicated to Mort Gurtin on the occasion of his sixtieth birthday

Communicated by the Editor

Abstract

I prove that the solutions of the phase-field equations, on a subsequence, converge to a weak solution of the Mullins-Sekerka problem with kinetic undercooling. The method is based on energy estimates, a monotonicity formula, and the equipartition of the energy at each time. I also show that for almost all t, the limiting interface is $(d-1)$-rectifiable with a square-integrable mean-curvature vector.

1. Introduction

Phase-field equations for solidification were introduced by CAGINALP [7, 8], COLLINS & LEVINE [15], FIX [19] and LANGER [24] to treat phenomena not covered by the classical Stefan problem. These equations, for the temperature (deviation) θ and the phase field φ, consist of a heat equation

$$c\theta_t + l\varphi_t = k\Delta\theta \tag{1.1}$$

and a Ginzburg-Landau equation

$$\beta\varphi_t = \lambda\Delta\varphi - \nu W'(\varphi) + l\theta \tag{1.2}$$

where c, l, k, β, λ and ν are positive constants and W is a double-well potential whose wells, of equal depth, correspond to the solid and liquid phases.

Recently thermodynamically consistent models have been developed in FRIED & GURTIN [20], PENROSE & FIFE [27], WANG et al. [33] and in references therein; in particular, [20], [27], and [33] allow the latent heat l to depend on the order parameter φ.

The main goal here is to rigorously study the global-time asymptotics of (1.1) and (1.2) in the limit $\varepsilon \downarrow 0$ for

$$c, k = 1, \quad \beta = \lambda = \varepsilon, \quad \nu = \frac{1}{\varepsilon}, \quad l = l(\varphi), \tag{1.3}$$

and for specificity with

$$W(\varphi) = \tfrac{1}{2}(1 - \varphi^2)^2, \quad l = 1 - \varphi^2, \tag{1.4}$$

a choice that is essentially the same as Model II in [33]. My analysis can be modified to analyze any smooth function l vanishing at the minimizers of W; i.e., any l of the form

$$l(\varphi) = (1 - \varphi^2)H(\varphi),$$

where $H \geqq 0$ is an arbitrary smooth function. A specific choice of H corresponds to Model I in [33], while the particular choice $l = W$ would simplify some of the analysis (cf. Remark 4.1 below). Observe that, granted (1.4), the nonlinearity $-\nu W'(\varphi) + l\theta$ in (1.2) vanishes at ± 1 for any value of θ.

The formal analyses of [7, 9, 15, 19] at least indicate that solutions of the Ginzburg-Landau equation (1.2) form a sharp interface whose normal velocity depends linearly on the mean curvature and the temperature of the interface. To describe this result precisely, let $(\theta^\varepsilon, \varphi^\varepsilon)$ be the solution of the phase-field equations with parameters consistent with (1.3) and assume that $(\theta^\varepsilon, \varphi^\varepsilon)$ converges to (θ, φ). Since the two minima of W are ± 1, it is easy to prove that $|\varphi| = 1$ almost everywhere. Let $\Gamma(t)$ be the interface separating the two regions

$$\Omega(t) = \{x : \varphi(t, x) = -1\}$$

and $\{\varphi = 1\}$. Then, formally, (θ, Ω) is a solution of the heat equation

$$\theta_t - \Delta\theta = -(h(\varphi))_t = \tfrac{4}{3}(\chi_{\Omega(t)})_t, \quad h(\varphi) = \varphi - \tfrac{1}{3}\varphi^3 \tag{1.5}$$

everywhere, coupled with the geometric equation

$$\vec{V} = H - \theta n \tag{1.6}$$

at the interface $\Gamma(t)$, where χ_Ω is the indicator of the set Ω, and where $\vec{V}$, n and H are, respectively, the normal velocity vector, the outward unit vector, and the mean-curvature vector of the interface $\Gamma(t)$. A derivation of these sharp interface equations from thermodynamics as well as an exhaustive list of earlier references are given in GURTIN's book [21, Chapter 3]. In 1964, MULLINS & SEKERKA [26] studied the linear stability of a related system of equations obtained by replacing (1.6) by the Gibbs-Thompson condition: $\theta = -K$. They showed that planar interfaces are unstable under some perturbations, thus explaining the dendritic growth observed in solidification. I refer to equations (1.5), (1.6) as the *Mullins-Sekerka problem* with kinetic undercooling.

My chief result is that, in the limit, θ and Ω constitute a weak solution of the Mullins-Sekerka problem with kinetic undercooling. This result is global in time; I do not assume the existence of a solution of (1.5), (1.6). Therefore I also provide an existence result for this limit problem, extending a previous result of CHEN & REITICH [12] for local-time existence. To the best of my knowledge, the only other global results are due to LUCKHAUS [25] and ALMGREN & WANG [3]. They proved the global existence of weak solutions for the heat equation (1.5) coupled with the Gibbs-Thompson condition: $\theta = -K$.

There are two essential difficulties in the analysis of (1.5), (1.6): a solution (θ, Ω) of (1.5), (1.6) can start out smooth and yet, in finite time, the boundary of Ω may develop geometric singularities, and θ may blow up pointwise (see the example in the Appendix). These difficulties also complicate the analysis of convergence. Since θ is unbounded, θ^ε does not converge to θ uniformly. For that reason I cannot use results of [4] concerning the convergence of (1.2) with a given continuous temperature field. The asymptotics of the Cahn-Allen equation, which is (1.2) with $l = 0$, is studied in [18] via sub- and supersolutions constructed from the weak solutions of the mean-curvature flow; unfortunately the approach of [18] is not directly applicable to the phase-field equations, as they do not have a maximum principle and there is no a priori weak theory for the limit equations.

I overcome these difficulties by utilizing the energy estimates in §2.2, and a monotonicity result in §5. The latter is an extension of the monotonicity formula of CHEN & STRUWE [13], which originates from STRUWE's formula for parabolic flow of harmonic maps [32], and a later result of ILMANEN [23] for the Cahn-Allen equation, which originates form HUISKEN's formula for smooth mean-curvature flows [22]. My main observation is that the geometric equation (1.6) is not simply a perturbation of the mean-curvature flow, and therefore the monotonicity should involve the mathematical energy

$$\int_{\mathbb{R}^d} \frac{\varepsilon}{2} |\nabla\varphi^\varepsilon|^2 + \frac{1}{\varepsilon} W(\varphi^\varepsilon) + \frac{1}{2}(\theta^\varepsilon)^2 \, dx$$

related to the system (1.5) and (1.6). The main technical difficulty is then to show that the discrepancy measure

$$\xi^\varepsilon(t; A) = \int_A \frac{\varepsilon}{2} |\nabla\varphi^\varepsilon|^2 - \frac{1}{\varepsilon} W(\varphi^\varepsilon) \, dx \tag{1.7}$$

has non-positive limiting value. For the Cahn-Allen equation, $\xi^\varepsilon \leqq 0$ follows easily from the maximum principle. For the phase-field equations, however, it follows from a series of estimates obtained in §4. In later sections, following ILMANEN [23], I prove that the weak* limit of ξ^ε is indeed equal to zero.

I close this introduction with a brief survey of related results. Equations (1.1), (1.2) with $c = \beta = 0$, $l = 1$ form the Cahn-Hilliard equation. Recently the convergence of the Cahn-Hilliard equation to the Hele-Shaw problem was proved by ALIKAKOS, BATES & CHEN [2] using a spectral estimate of CHEN [11]. In contrast to this paper, they assume the existence of a smooth solution to the limiting problem. Briefly, their method is to construct approximate solutions for the "ε problem" that are close to the smooth solution of the limit problem. They then use the spectral estimates to bound the error terms. Also, STOTH [30] studied the asymptotic limit of the phase-field equations with radial symmetry. Independently, a radially symmetric problem in an annular domain with one interface was studied in [10]. Asymptotics of the Cahn-Allen equation, obtained by setting l to zero in (1.2), have been studied extensively. An exhaustive list of references related to the Cahn-Allen equation can be found in my paper [29].

This paper is organized as follows. In the next section I outline the background and state the main results. In §3, several elementary estimates are obtained. A gradient estimate is proved in §4; this estimate implies that ξ^ε is non-positive in the limit. In §5, I derive a monotonicity result which I use in § 6 to prove a clearing-out lemma. I then establish the equipartition of energy in §7. In that section, I also show that the Hausdorff dimension of the interface is $d-1$. I complete the proof in Section 8. In the appendix, for a simple radially symmetric example studied jointly with ILMANEN, I prove the pointwise blowup of the temperature.

2. Preliminaries

The following notation is used throughout the paper. $C_c^\infty(A \to B)$ denotes the set of all compactly supported, smooth functions on A, with values in B. $\mathscr{D}'(A)$ denotes the set of all distributions defined on A. For a measure space (A, μ) and for $p \in [1, \infty]$, $L^p(A; d\mu)$ denotes the set of all functions that are p-integrable with respect to the measure μ. When μ is the Lebesgue measure, we use the notation $L^p(A)$. For $T < \infty$ and $p \in [1, \infty]$, $\|\cdot\|_{p,T}$ denotes the norm in $L^p((0,T) \times \mathscr{R}^d)$. For $R > 0$ and $x \in \mathscr{R}^d$,

$$B_R = \{y \in \mathscr{R}^d : |y| \leqq R\}, \quad B_R(x) = \{y \in \mathscr{R}^d : |y - x| \leqq R\}.$$

For two $d \times d$ matrices M and N,

$$M:N = \sum_{i,j=1}^{d} M_{ij} N_{ij}.$$

S^{d-1} denotes the set of all unit vectors in $\mathscr{R}^d$. For $p \in \mathscr{R}^d$, $p \otimes p$ denotes the $d \times d$ matrix with entries $p_i p_j$.

For a Radon measure μ on $\mathscr{R}^d$ and a continuous bounded function ψ,

$$\mu(\psi) := \int_{\mathscr{R}^d} \psi(x) d\mu(x).$$

$\mathscr{H}^k$ denotes the k-dimensional Hausdorff measure (cf. [28]). Finally

$$Q = (0, \infty) \times \mathscr{R}^d$$

and for $(\tau, \xi) \in Q$, $G(\tau, \xi)$ is the heat kernel:

$$G(\tau, \xi) = (4\pi\tau)^{-d/2} \exp\left(\frac{|\xi|^2}{4\tau}\right).$$

2.1. Equations

For a scalar u, set

$$W(u) = \tfrac{1}{2}(u^2-1)^2,$$

$$h(u) = u - \tfrac{1}{3}u^3, \quad g(u) = h'(u) = (1-u^2) = \sqrt{2W(u)}.$$

The heat equation (1.1) and the order-parameter equation (1.2) — with these functions and with parameters as in (1.3), (1.4) — take the form,

$$\varphi_t^\varepsilon - \Delta\varphi^\varepsilon + \frac{1}{\varepsilon^2}W'(\varphi^\varepsilon) - \frac{1}{\varepsilon}g(\varphi^\varepsilon)\theta^\varepsilon = 0 \quad \text{in } (0,\infty)\times\mathcal{R}^d, \tag{OPE}$$

$$\theta_t^\varepsilon - \Delta\theta^\varepsilon + g(\varphi^\varepsilon)\varphi_t^\varepsilon = 0 \quad \text{in } (0,\infty)\times\mathcal{R}^d. \tag{HE}$$

For $\varepsilon > 0$, let $(\varphi^\varepsilon, \theta^\varepsilon)$ be the unique, smooth, bounded solution of the phase-field equations satisfying the initial data

$$\varphi^\varepsilon(x,0) = \varphi_0^\varepsilon(x), \quad \theta^\varepsilon(x,0) = \theta_0^\varepsilon(x), \quad x \in \mathcal{R}^d. \tag{IC}$$

We assume that

$$|\varphi_0^\varepsilon(x)| \leqq 1 \quad \forall x \in \mathcal{R}^d. \tag{A1}$$

Then since $W'(\pm 1) = g(\pm 1) = 0$, by the maximum principle,

$$|\varphi^\varepsilon(t,x)| < 1 \quad \forall (t,x) \in (0,\infty)\times\mathcal{R}^d.$$

For a real number τ, $q(\tau) = \tanh(\tau)$ satisfies

$$q'' = W'(q), \quad q' = \sqrt{2W(q)} = g(q),$$

and q is the standing wave associated with the reaction diffusion equation with nonlinearity W'. Since $|\varphi^\varepsilon| < 1$, we may define z^ε by

$$\varphi^\varepsilon(t,x) = q\left(\frac{z^\varepsilon(t,x)}{\varepsilon}\right) \Leftrightarrow z^\varepsilon = \varepsilon q^{-1}(\varphi^\varepsilon).$$

Then z^ε satisfies

$$z_t^\varepsilon - \Delta z^\varepsilon - \theta^\varepsilon + \frac{2\varphi^\varepsilon}{\varepsilon}(|\nabla z^\varepsilon|^2 - 1) = 0 \tag{ZE}$$

(observe that $g(\varphi^\varepsilon) = g(q(z^\varepsilon/\varepsilon)) = q'(z^\varepsilon/\varepsilon)$ and $q'' = 2\varphi^\varepsilon q'$).

2.2. Energy

For a Borel subset $A \subset \mathcal{R}^d$, define

$$\mu^\varepsilon(t;A) = \int_A \tfrac{\varepsilon}{2}|\nabla\varphi^\varepsilon|^2 + \tfrac{1}{\varepsilon}W(\varphi^\varepsilon)\,dx,$$

$$\hat{\mu}^\varepsilon(t;A) = \mu^\varepsilon(t;A) + \int_A \tfrac{1}{2}(\theta^\varepsilon)^2\,dx.$$

In terms of z^ε,

$$\mu^\varepsilon(t; dx) = \frac{1}{2\varepsilon}\left(q'\left(\frac{z^\varepsilon(t,x)}{\varepsilon}\right)\right)^2 [|\nabla z^\varepsilon(t,x)|^2 + 1]\, dx,$$

and the discrepancy measure ξ^ε (cf. (1.7)) is given by

$$\xi^\varepsilon(t; dx) = \frac{1}{2\varepsilon}\left(q'\left(\frac{z^\varepsilon(t,x)}{\varepsilon}\right)\right)^2 [|\nabla z^\varepsilon(t,x)|^2 - 1]\, dx. \tag{2.1}$$

By differentiation and integration by parts we obtain

$$\frac{d}{dt}\hat{\mu}^\varepsilon(t; \mathscr{R}^d) = -\int_{\mathscr{R}^d} \varepsilon(\varphi_t^\varepsilon)^2 + |\nabla\theta^\varepsilon|^2\, dx.$$

We assume that the initial data satisfy

$$\hat{\mu}^\varepsilon(0; \mathscr{R}^d) \leqq C_1^*, \quad \varepsilon > 0. \tag{A2}$$

Then

$$\hat{\mu}^\varepsilon(t; \mathscr{R}^d) + \int_0^t \int_{\mathscr{R}^d} \varepsilon(\varphi_t^\varepsilon)^2 + |\nabla\theta^\varepsilon|^2\, dx\, dt \leqq C_1^*, \quad \varepsilon, t \geqq 0. \tag{2.2}$$

(Assumption (A2) can be relaxed as in [29]). We now localize this estimate. Let ψ be any *positive*, smooth, compactly supported function. Then

$$\frac{d}{dt}\hat{\mu}^\varepsilon(t;\cdot)(\psi) = -\int_{\mathscr{R}^d} \varepsilon\psi\left[\left(\varphi_t^\varepsilon + \frac{\nabla\varphi^\varepsilon\cdot\nabla\psi}{2\psi}\right)^2 + \left|\nabla\theta^\varepsilon + \frac{\theta^\varepsilon\nabla\psi}{2\psi}\right|^2\right]$$

$$+ \int_{\mathscr{R}^d} \frac{\varepsilon}{4\psi}(\nabla\varphi^\varepsilon\cdot\nabla\psi)^2 + \frac{(\theta^\varepsilon)^2}{4\psi}|\nabla\psi|^2$$

$$\leqq \left[\sup_x \frac{|\nabla\psi(x)|^2}{2\psi(x)}\right] \int_{\{\psi>0\}} \frac{\varepsilon}{2}|\nabla\varphi^\varepsilon|^2 + \frac{1}{2}(\theta^\varepsilon)^2$$

$$\leqq \|D^2\psi\|_\infty\, \hat{\mu}^\varepsilon(t; \{\psi > 0\}).$$

Here we have used the fact that, for any positive C^2 function,

$$\frac{|\nabla\psi(x)|^2}{2\psi(x)} \leqq \|D^2\psi\|_\infty.$$

Hence there is a constant $C(\psi)$, independent of ε, such that the map

$$t \mapsto \int \psi(x)\hat{\mu}^\varepsilon(t; dx) - C(\psi)t \tag{2.3}$$

is non-increasing.

We close this subsection by obtaining a similar local energy identity for the classical solutions of the limit equation. Suppose that $(\theta(t,x), \Omega(t))$ is a classical solution of the Mullins-Sekerka problem (1.5), (1.6). Let $\vec{V}$, H, and n be, respectively, the normal velocity vector, the mean-curvature vector and the outward normal of the interface $\Gamma(t) = \partial\Omega(t)$. Let $\mu(t;\cdot)$ be equal to $\frac{4}{3}$ times the surface measure of $\partial\Omega(t)$. Then (1.5) is equivalent to

$$(\theta_t - \Delta\theta)\,dx = \vec{V}\cdot n\mu(t; dx).$$

For a smooth, compactly supported $\psi(x)$, we have

$$\frac{d}{dt}\hat{\mu}(t;\cdot)(\psi) = \int [-\vec{V}\cdot H\psi + \vec{V}\cdot \nabla\psi]\mu(t; dx) + \int \theta\theta_t\psi\,dx$$

$$= \int [\vec{V}\cdot(-H + \theta n)\psi + \vec{V}\cdot\nabla\psi]\mu(t; dx) + \int \tfrac{1}{2}\theta^2\Delta\psi - |\nabla\psi|^2\psi\,dx.$$

Since $\vec{V} = H - \theta n$ by (1.6), it follows that

$$\frac{d}{dt}\hat{\mu}(t;\cdot)(\psi) = \int [-|\vec{V}|^2\psi + \vec{V}\cdot\nabla\psi]\mu(t; dx) + \int \tfrac{1}{2}\theta^2\Delta\psi - |\nabla\theta|^2\psi\,dx. \quad (2.4)$$

Observe that this identity is very similar to that used by BRAKKE to develop a weak theory for mean-curvature flows (cf. [5], [23, §1]).

2.3. *Subsequence*

The energy estimate (2.2) yields

$$\sup_{\varepsilon, t>0} \|\theta^\varepsilon(t,\cdot)\|_2 < \infty.$$

Hence there are a subsequence, denoted by ε, and an L^2 function θ such that

$$\theta^\varepsilon \rightharpoonup \theta \quad \text{in weak } L^2((0,T)\times\mathscr{R}^d),$$

for every $T > 0$. We will show that this convergence is, in fact, in the strong topology (see Proposition 3.4 below). Moreover, by the arguments of BRONSARD & KOHN [6], this sequence can be chosen so that

$$h(\varphi^\varepsilon) \to h(\varphi) = \tfrac{2}{3}\varphi \quad \textit{in } L^1_{\text{loc}}((0,\infty)\times\mathscr{R}^d), \quad \varphi^\varepsilon \to \varphi \quad \text{a.e.},$$

where φ is a function of bounded variation, and $|\varphi(t,x)| = 1$, for almost every (t,x). Since $(h')^2 = 2W$, (2.2) implies that, for $0 \leqq s < t$ and $\varepsilon > 0$,

$$\|h(\varphi^\varepsilon(t,\cdot)) - h(\varphi^\varepsilon(s,\cdot))\|_1 \leqq \int_s^t \int |h'(\varphi^\varepsilon(r,x))\varphi^\varepsilon_t(r,x)|\,dx\,dr$$

$$\leqq \left(\int_s^t \int \tfrac{\varepsilon}{2}(\varphi^\varepsilon_t(r,x))^2\,dx\,dr\right)^{1/2} \left(\int_s^t \mu^\varepsilon(r;\mathscr{R}^d)\,dr\right)^{1/2}$$

$$\leqq C_1^*\sqrt{t-s}. \quad (2.5)$$

Hence

$$h(\varphi^\varepsilon(t;\cdot)) \to h(\varphi(t;\cdot)) \quad \text{in } L^1_{\text{loc}}(\mathscr{R}^d), \quad (2.6)$$

uniformly in the variable t.

The energy estimate implies that, for each $t \geqq 0$, the sequence $\{\hat{\mu}^\varepsilon(t,\cdot)\}_{\varepsilon>0}$ is precompact in the weak* topology of Radon measures. By a diagonalization argument we construct a sequence, denoted by ε again, such that as $\varepsilon \downarrow 0$, $\hat{\mu}^\varepsilon(t,\cdot)$ is weak* convergent for all rational $t \geqq 0$. Then, by the monotonicity estimate (2.3), we construct a further sequence so that $\{\hat{\mu}^\varepsilon(t,\cdot)\}_{\varepsilon>0}$ is convergent for all $t \geqq 0$. Therefore there are a subsequence, denoted by ε, and a family of Radon measures $\hat{\mu}(t,\cdot)$ that satisfy

$$\lim_{\varepsilon \downarrow 0} \hat{\mu}^\varepsilon(t,\cdot) \rightharpoonup \hat{\mu}^\varepsilon(t,\cdot) \quad \forall t \geqq 0$$

in the weak* topology of Radon measures. See [23, §5.4] for further details of this argument. Now for a Borel subset $B \subset [0, \infty) \times \mathscr{R}^d$, define

$$\hat{\mu}(B) = \int\int_B \hat{\mu}(t; dx)\, dt, \quad \mu(B) = \hat{\mu}(B) - \tfrac{1}{2}\int\int_B \theta^2\, dx\, dt.$$

Then the strong convergence of θ^ε to θ (cf. Proposition 3.4 below) implies that $\mu \geqq 0$ and

$$\mu^\varepsilon(t,\cdot) \rightharpoonup \mu(t,\cdot) \quad \forall t \geqq 0.$$

Since the interface condition (1.6) involves not only the mean-curvature vector, which is independent of orientation, but also the normal vector, we introduce yet another measure, m^ε, that keeps track of the normal direction. For $(t, x, n) \in [0, \infty) \times \mathscr{R}^d \times S^{d-1}$, define

$$v^\varepsilon(t,x) = \begin{cases} \dfrac{\nabla\varphi^\varepsilon(t,x)}{|\nabla\varphi^\varepsilon(t,x)|} & \text{if } |\nabla\varphi^\varepsilon(t,x)| \neq 0, \\ v_0 & \text{if } \nabla\varphi^\varepsilon(t,x) = 0, \end{cases}$$

$$dm^\varepsilon(t,x,n) = dt\mu^\varepsilon(t; dx)\delta_{\{v^\varepsilon(t,x)\}}(dn),$$

where $v_0 \in S^{d-1}$ is arbitrary and $\delta_{\{v^\varepsilon\}}$ is the Dirac measure located at v^ε. Observe that m^ε is independent of the choice of v_0.

Since S^{d-1} is compact, there is a further sequence, denoted by ε, such that dm^ε is weak* convergent. By a slicing argument (cf. [16, Theorem 10, page 14]) we conclude that there exist probability measures $N(t,x,\cdot)$ on S^{d-1} such that as ε tends to zero,

$$dm^\varepsilon \rightharpoonup dm = dt\mu(t; dx)N(t, x; dn).$$

Finally define $\bar{m}^\varepsilon$ by

$$d\bar{m}^\varepsilon = -z_t^\varepsilon dm^\varepsilon.$$

In §8 we show that there are a subsequence, denoted by ε, and

$$v \in L^2((0,T) \times \mathscr{R}^d \times S^{d-1}; dm) \quad \forall T > 0$$

such that

$$\bar{m}^\varepsilon \rightharpoonup \bar{m}, \quad d\bar{m} = v(t,x,n)\, dm.$$

2.4. Initial data and assumptions

In addition to (A1), (A2) we assume that

$$\| \nabla z_0^\varepsilon \|_\infty \leqq 1, \tag{A3}$$

$$\sup_{0<\varepsilon\leqq 1} \varepsilon \| D^2 z_0^\varepsilon \|_\infty < \infty, \tag{A4}$$

$$\sup_{0<\varepsilon\leqq 1} \sup_{\substack{x\in\mathscr{R}^d \\ R>0}} \frac{\mu^\varepsilon(0; B_R(x))}{R^{d-1}} < \infty, \tag{A5}$$

where $B_R(x)$ is the sphere centered at x with radius R. We also assume that

$$\sup_{0<\varepsilon\leqq 1} \{ \| \theta_0^\varepsilon \|_1 + \| \theta_0^\varepsilon \|_\infty + \sqrt{\varepsilon} \| \nabla \theta_0^\varepsilon \|_\infty \} < \infty, \tag{A6}$$

$$\sup_{0<\varepsilon\leqq 1} \{ \varepsilon^3 \| D^3 \varphi_0^\varepsilon \|_\infty + \varepsilon^2 \| D^2 \theta_0^\varepsilon \|_\infty \} < \infty. \tag{A7}$$

Since

$$\| \theta_0^\varepsilon \|_p^p \leqq \| \theta_0^\varepsilon \|_1 \| \theta_0^\varepsilon \|_\infty^{p-1} \quad \text{for } 1 \leqq p < \infty,$$

observe that (A6) implies that

$$\sup_{0<\varepsilon\leqq 1} \| \theta_0^\varepsilon \|_p = K(p) < \infty. \tag{2.7}$$

Finally we assume that there is $\theta_0 \in L^2(\mathscr{R}^d)$ such that

$$\theta_0^\varepsilon \to \theta_0 \quad \text{in } L^2\text{-strong}. \tag{A8}$$

While (A1)–(A8) may seem restrictive, in fact, they are merely technical assumptions which are consistent with approximations to any smooth initial data. Indeed, if $\theta_0^\varepsilon = \theta_0$ is a smooth, compactly supported function, then θ_0 satisfies (A6), (A7) and (A8) trivially. Suppose that Γ_0 is a bounded, smooth hypersurface in $\mathscr{R}^d$. Let $d(x)$ be the signed distance of x to Γ_0 and let $\hat{d}$ be an appropriate modification of d outside of a tubular neighborhood of Γ_0 such that all derivaties of $\hat{d}$ up to order three are bounded and such that $2|\hat{d}| \geqq |d|$. Then $z_0^\varepsilon = \hat{d}$ satisfies (A1)–(A7).

Finally we note that the term $\sqrt{\varepsilon}$ appearing in (A6) is not essential. Indeed if (A6) holds with ε^ν for some $\nu \geqq \frac{1}{2}$, then we can prove the same results with minor changes.

2.5. Varifolds, rectifiable measures, etc.

In this subsection, we recall several definitions and results from geometric measure theory ([28], [23, §1]).

Following [23, §1.7], we call a Radon measure μ on $\mathscr{R}^d$ *k-rectifiable*, if there are a $\mathscr{H}^k$-measurable, locally k-rectifiable set $X \subset \mathscr{R}^d$ and

$$f \in L^1_{\text{loc}}(X; d\mathscr{H}^k \llcorner X) \quad (\mathscr{H}^k \llcorner X(A) = \mathscr{H}^k(A \cap X))$$

such that

$$\mu(A) = \int_{A \cap X} f(x) d\mathscr{H}^k(x) \quad \text{for any Borel set } A.$$

When μ is k-rectifiable, for μ-almost every x, the *measure-theoretic tangent plane*

$$T_x\mu = \lim_{\lambda \downarrow 0} \mu_{x,\lambda}, \quad (\mu_{x,\lambda}(A) = \lambda^{-k}\mu(x + \lambda A))$$

exists and is a positive multiple of $\mathscr{H}^k$ restricted to a k-plane. With an abuse of notation, we use $T_x\mu$ to denote this k-plane.

A general *k-varifold* is a Radon measure on $\mathscr{R}^d \times G_k(\mathscr{R}^d)$, where $G_k(\mathscr{R}^d)$ is the Grassman manifold of unoriented k-planes in $\mathscr{R}^d$. The mass measure $\| V \|$ is defined by

$$\| V \|(A) = V(A \times G_k(\mathscr{R}^d)).$$

For every k-rectifiable Randon measure μ, there is a corresponding (rectifiable) k-varifold V_μ defined by

$$dV_\mu(x, S) = d\mu(x) d\delta_{\{T_x\mu\}}(S),$$

where $\delta_{\{T_x\mu\}}$ is the Dirac measure located at $T_x\mu$. Note that $\| V_\mu \| = \mu$. We say that a k-rectifiable Radon measure μ has a generalized *mean-curvature vector*

$$H \in L^1_{\text{loc}}(\mathscr{R}^d \to \mathscr{R}^d; d\mu)$$

if for any smooth, compactly supported vector field $Y(x)$,

$$\int \text{tr}(\nabla Y(x) P(x))\, d\mu = - \int Y(x) \cdot H(x)\, d\mu, \tag{2.8}$$

where $P(x)$ is the projection on the tangent plane $T_x\mu$. In the terminology of geometric measure theory, the left-hand side of (2.8) is the first variation δV_μ of the varifold V_μ [1].

2.6. Main results

First we recall the convergence results stated in §2.3.

Theorem 2.1 (Convergence). *There are a sequence, denoted by* ε, *functions* $\theta \in L^2_{\text{loc}}((0, \infty) \times \mathscr{R}^d)$, $v \in L^2_{\text{loc}}((0, \infty) \times \mathscr{R}^d \times S^{d-1}; dm)$, *non-negative Radon measures* $\{\mu(t;\cdot), \hat{\mu}(t;\cdot)\}_{t \geq 0}$ *and probability measures* $\{N(t, x; \cdot)\}_{(t,x) \in (0,\infty) \times \mathscr{R}^d}$ *such that, as* ε *tends to zero,*

$$\theta^\varepsilon(t, \cdot) \to \theta(t, \cdot) \quad \textit{strongly in } L^2_{\text{loc}}(\mathscr{R}^d) \quad \forall t \geqq 0,$$

$$h(\varphi^\varepsilon) \to h(\varphi) \quad \textit{strongly in } L^1_{\text{loc}}((0, \infty) \times \mathscr{R}^d).$$

Moreover, $|\varphi| = 1$, $\varphi^\varepsilon(t, x) \to \varphi(t, x)$ *for almost every* $(t, x) \in (0, \infty) \times \mathscr{R}^d$, *and*

$$h(\varphi^\varepsilon(t, \cdot)) \to h(\varphi(t, \cdot)) \quad \textit{strongly in } L^1_{\text{loc}}(\mathscr{R}^d),$$

uniformly for $t \geqq 0$. *In addition, the following convergence results in the weak* topology of Radon measures, are valid:*

$$\mu^\varepsilon(t,\cdot) \rightharpoonup \mu(t,\cdot) \quad \forall t \geqq 0,$$

$$\hat{\mu}^\varepsilon(t,\cdot) \rightharpoonup \hat{\mu}(t,\cdot) \quad \forall t \geqq 0 \quad (\hat{\mu}(t;dx) = \mu(t;dx) + \tfrac{1}{2}\theta^2\,dx),$$

$$m^\varepsilon \rightharpoonup m, \quad dm(t,x,n) = \mu(t;dx)\,dt\,N(t,x;dn),$$

$$\bar{m}^\varepsilon \rightharpoonup \bar{m}, \quad d\bar{m}^\varepsilon = -z_t^\varepsilon dm^\varepsilon, \quad d\bar{m} = v(t,x,n)\,dm.$$

For every $T > 0$, *the functions* θ *and* v *satisfy*

$$\sup_{t \geqq 0} \|\theta(t,\cdot)\|_2 + \|\nabla\theta\|_{2,T} < \infty, \tag{2.9}$$

$$\|v\|_{L^2((0,T)\times\mathscr{R}^d\times S^{d-1};\,dm)} < \infty.$$

The strong convergence of θ^ε is proved in Proposition 3.4, and the convergence of $\bar{m}^\varepsilon$ and the integrability of v are proved in §8. The remaining assertions were established in §2.3.

Set $d\mu(t,x) = \mu(t;dx)\,dt$. Let Γ be the support of μ and Γ_t be the t cross section of Γ. In the terminology of §2.5, we have the following regularity result.

Theorem 2.2 (Regularity). *For almost every* $t \geqq 0$, $\mu(t,\cdot)$ *is* $(d-1)$*-rectifiable and has a generalized mean-curvature vector* $H(t,x)$. *Moreover for every* $T > 0$,

$$|H| \in L^2((0,T)\times\mathscr{R}^d; d\mu),$$

$$\sup_{t \leqq T} \mathscr{H}^{d-1}(\Gamma_t) < \infty,$$

$$\theta \in L^1_{\text{loc}}((0,\infty)\times\mathscr{R}^d; d\mu),$$

and the support of the probability measure $N(t,x;\cdot)$ *is orthogonal to* $T_x\mu(t;\cdot)$ *for* μ*-almost every* (t,x). *In particular,*

$$\iiint \nabla Y(t,x) : (I - n\otimes n)\,dm = -\iint Y(t,x)\cdot H(t,x)\,d\mu$$

for all $Y \in C_c^\infty((0,\infty)\times\mathscr{R}^d \to \mathscr{R}^d)$.

The estimate of Hausdorff measure is proved in Proposition 7.2. The existence and the square-integrability of H, the integrability of θ with respect to μ, and the orthogonality of N are all proved in §8. The final assertion of the theorem follows from the orthogonality of N and the defining property of H.

The next result states that the limit of $(\theta^\varepsilon, \varphi^\varepsilon)$ weakly satisfies (1.5), (1.6). However the lack of regularity of the limit functions necessitates the use of measures μ and m.

Let θ, φ, μ, $\hat{\mu}$, N, and v be as in Theorem 2.1. Recall that $Q = (0,\infty)\times\mathscr{R}^d$.

Theorem 2.3 (Limit Equations). *For any* $\psi \in C_c^\infty(Q \to \mathscr{R})$ *and* $Y \in C_c^\infty(Q \to \mathscr{R}^d)$,

$$\iint -(\psi_t + \Delta\psi)\theta \, dx\, dt = \iiint v\psi \, dm \tag{2.10}$$

$$= \iint \psi_t h(\varphi)\, dx\, dt, \tag{2.11}$$

$$\iiint Y\cdot n(\theta + v)\, dm = -\iiint \nabla Y : (I - n\otimes n)\, dm$$

$$= \iint Y\cdot H\, d\mu, \tag{2.12}$$

$$\iiint Y\cdot n\, dm = \iint Y\cdot \nabla(h(\varphi))\, dx\, dt.$$

For any $0 \leqq s \leqq t$ *and* $\phi \in C_c^\infty(Q \to [0,\infty))$, *the following Brakke-type inequality holds*:

$$\hat{\mu}(\phi)(t) - \hat{\mu}(\phi)(s) \leqq \int_s^t \iint (-v^2\phi - vn\cdot\nabla\phi)\, dm$$

$$+ \int_s^t \int (\tfrac{1}{2}\theta^2\Delta\phi - |\nabla\theta|^2\phi)\, dx\, dt. \tag{2.13}$$

This theorem is proved in §8.

The system (2.10)–(2.13) constitutes a weak formulation of the Mullins-Sekerka equations (1.5) and (1.6). Indeed, set

$$V(t,x) = \int v(t,x,n) N(t,x;dn), \quad \vec{V}(t,x) = \int nv(t,x,n)N(t,x;dn),$$

$$\Omega(t) = \{x : \varphi(t,x) = -1\}.$$

Then (2.11) yields

$$\theta_t - \Delta\theta + (h(\varphi))_t = 0 \quad \text{in } \mathscr{D}'(Q).$$

Since $(h(\varphi))_t = -(\frac{4}{3})(\chi_\Omega)_t$, by (2.10) and (2.11), V is formally equal to $\frac{4}{3}$ times the normal velocity of the interface $\partial\Omega$. Suppose that $N(t,x;\cdot)$ is a Dirac measure located at $n(t,x) \in S^{d-1}$. Then the orthogonality of N to the tangent plane implies that $n(t,x)$ is orthogonal to $\partial\Omega$, and (2.12) is equivalent to

$$\vec{V} = Vn = H - \theta n \quad \text{on } \partial\Omega.$$

The equation (2.12) is therefore a weak formulation of (1.6) provided that V is the normal velocity of $\mu(t;\cdot)$. Indeed, the inequality (2.13) provides a weak formulation of this statement; compare (2.13) to (2.4).

Remarks on regularity and uniqueness

1. Suppose that $\Gamma(t)$ is smooth. Then does the weak formulation proved in Theorem 2.3 imply that $\Gamma(t)$ and θ satisfy the Mullins-Sekerka problem classically? Alternatively, suppose that there is a smooth solution of the Mullins-Sekerka problem in $(0,T)\times\mathscr{R}^d$. Then does this classical solution agree with the limit functions constructed in this paper?

2. An attendant modification of [23, §9] together with the results of this paper imply that for any compactly supported smooth function $\phi(x)$ and for $t > 0$, if

$$d := \limsup_{s \to t} \frac{1}{s-t} \int \phi(x)(\hat{\mu}(s; dx) - \hat{\mu}(t; dx)) > -\infty,$$

then $\mu(t, \cdot)$ restricted to $\{\phi > 0\}$ is $(d-1)$-rectifiable with a generalized mean-curvature vector $H(t, \cdot)$. Moreover,

$$\begin{aligned} d \leqq & -\textstyle\iint \phi(x)|H(t,x) + \theta(t,x)n|^2 \mu(t; dx) N(t,x; dn) \\ & -\textstyle\int \phi(x) |\nabla\theta(t,x)|^2 \, dx + \frac{1}{2} \int \Delta\phi(x) \theta^2(t,x) \, dx \\ & -\textstyle\iint D\phi(x) \cdot [H(t,x) + n\theta(t,x)] \mu(t; dx) N(t,x; dn). \end{aligned}$$

It seems that this inequality together with some of the formulae proved in Theorem 2.3 provide a Brakke-type weak-formulation of the Mullins-Sekerka problem. Further analysis of these equations may yield a generalization of a partial-regularity result of Brakke [5].

3. Simple examples indicate that $N(t, x; dn)$ may not be a Dirac measure at some points (t, x). This corresponds to interface "piling-up" at such points. An interesting question is to estimate the dimension of these points at which $N(t, x; dn)$ is not a Dirac measure. Since the heat equation (1.5) does not have any external forcing term, we expect this set to be of lower dimension.

4. A related question is whether the equation

$$[v(t, x, n) + \theta(t, x)]n = H(t, x) \tag{2.14}$$

holds for dm almost every (t, x, n). The equation (2.12) implies (2.14) only after integration with respect to $N(t, x; dn)$. Radially symmetric examples indicate that (2.14) may be true.

Suppose that (2.14) holds. Then, formally, if $\theta(t, x) \neq 0$ and $N(t, x, \cdot)$ is not a Dirac measure, then $v(t, x, n)$ is different for each n in the support of $N(t, x; \cdot)$. Therefore formally $N(s, x, \cdot)$ would become a Dirac measure for $s > t$ and s near t.

3. Elementary Estimates

In this section we obtain several elementary estimates by using the heat kernel

$$G(\tau, \xi) = (4\pi\tau)^{-d/2} \exp\left(-\frac{|\xi|^2}{4\tau}\right), \quad (\tau, \xi) \in (0, \infty) \times \mathscr{R}^d.$$

Since $h' = g$, the heat equation (HE) and integration by parts yield

$$\theta^\varepsilon(t, x) = A^\varepsilon(t, x) + B^\varepsilon(t, x), \tag{3.1}$$

where

$$A^\varepsilon(t,x) = (G(t,\cdot) * [\theta_0^\varepsilon + H^\varepsilon(0,\cdot) - H^\varepsilon(t,\cdot)])(x),$$

$$B^\varepsilon(t,x) = \int_0^t (G_\tau(\tau,\cdot) * (H^\varepsilon(t,\cdot) - H^\varepsilon(t-\tau,\cdot)))(x)\,d\tau,$$

$$H^\varepsilon(t,x) = h(\varphi^\varepsilon(t,x)),$$

and * denotes convolution in the x-variable.

All the constants in this and later sections depend on T but we generally suppress this dependence in our notation.

Constants independent of ε are denoted by K; this constant may change from one line to the next.

Lemma 3.1. *There is a constant K such that*

$$\varepsilon \| \nabla\varphi^\varepsilon \|_{\infty,T} \leqq K, \tag{3.2}$$

$$\| \theta^\varepsilon \|_{\infty,T} \leqq K[1 + |\ln \varepsilon|]. \tag{3.3}$$

Proof.

1. Fix $T > 0$ and set

$$m^\varepsilon(T) = \varepsilon \| \nabla\varphi^\varepsilon \|_{\infty,T}, \quad n^\varepsilon(T) = \| \theta^\varepsilon \|_{\infty,T},$$

$$f^\varepsilon = -\frac{1}{\varepsilon^2} W'(\varphi^\varepsilon) + \frac{1}{\varepsilon} g(\varphi^\varepsilon)\theta^\varepsilon.$$

Then

$$\| f^\varepsilon \|_{\infty,T} \leqq \frac{1}{\varepsilon^2}[2 + \varepsilon n^\varepsilon(T)]$$

and the order-parameter equation (OPE) may be rewritten as

$$\varphi_t^\varepsilon - \Delta\varphi^\varepsilon = f^\varepsilon.$$

2. Fix $(t,x) \in [0,T] \times \mathcal{R}^d$. Then for any $\sigma \in (0,t]$, (OPE) yields

$$\nabla\varphi^\varepsilon(t,x) = (\nabla G(\sigma,\cdot) * \varphi^\varepsilon(t-\sigma,\cdot))(x) + \int_0^\sigma (\nabla G(\tau,\cdot) * f^\varepsilon(t-\tau,\cdot))(x)\,d\tau.$$

Observe that for any $\tau > 0$,

$$\sqrt{\tau}\, \| \nabla G(\tau,\cdot) \|_1 = \frac{(\pi)^{-d/2}}{2} \int_{\mathcal{R}^d} |y| e^{-|y|^2}\,dy = K.$$

Therefore

$$\left| \int_0^\sigma [\nabla G(\tau,\cdot) * f^\varepsilon(t-\tau,\cdot)](x)\,dt \right| \leqq \frac{2K(2 + \varepsilon n^\varepsilon(T))}{\varepsilon^2}\sqrt{\sigma},$$

$$|(\nabla G(\sigma,\cdot) * \varphi^\varepsilon(t-\sigma,\cdot))(x)| \leqq \frac{K}{\sqrt{\sigma}}.$$

Also, if $\sigma = t$, then

$$|(\nabla G(\sigma,\cdot) * \varphi^\varepsilon(t-\sigma,\cdot))(x)| = |(G(t,\cdot) * \nabla\varphi_0^\varepsilon(x)| \leqq \|\nabla\varphi_0^\varepsilon\|_\infty .$$

3. Now use the foregoing inequalities with $\sigma = \varepsilon^2 \wedge t$; the result is

$$|\nabla\varphi^\varepsilon(t,x)| \leqq \begin{cases} \dfrac{K}{\varepsilon}[5 + 2\varepsilon n^\varepsilon(T)] & \text{if } t \geqq \varepsilon^2, \\ \dfrac{K}{\varepsilon}[\varepsilon\|\nabla\varphi_0^\varepsilon\|_\infty + 2(2+\varepsilon n^\varepsilon(T))] & \text{if } t \leqq \varepsilon^2. \end{cases}$$

By (A3), $\varepsilon\|\nabla\varphi_0^\varepsilon\|_\infty \leqq K$; we therefore conclude that

$$m^\varepsilon(T) \leqq K[1 + \varepsilon n^\varepsilon(T)]. \tag{3.4}$$

4. Let $A^\varepsilon, B^\varepsilon$ be as in (3.1). Then

$$|A^\varepsilon(t,x)| \leqq \|\theta_0^\varepsilon\|_\infty + 2.$$

For $\sigma \in (0, t]$,

$$\left|\int_\sigma^t [G_\tau(\tau,\cdot) * (H^\varepsilon(t,\cdot) - H^\varepsilon(t-\tau,\cdot))](x)\,d\tau\right| \leqq 2\int_\sigma^t \|G_\tau(\tau,\cdot)\|_1\,d\tau.$$

Observe that

$$\tau\|G_\tau(\tau,\cdot)\|_1 \leqq \frac{(\pi)^{-d/2}}{2}\int_{\mathscr{R}^d}\left(\frac{d}{2} + |y|^2\right)e^{-|y|^2}\,dy \leqq K.$$

Hence

$$\left|\int_\sigma^t [G_\tau(\tau,\cdot) * (H^\varepsilon(t,\cdot) - H^\varepsilon(t-\tau,\cdot))](x)\,d\tau\right| \leqq K\ln\left(\frac{t}{\sigma}\right).$$

Since $\Delta G = G_\tau$ and $\nabla H = g\nabla\varphi^\varepsilon$, integrating by parts we obtain

$$\begin{aligned} &\left|\int_0^\sigma [G_\tau(\tau,\cdot) * (H^\varepsilon(t,\cdot) - H^\varepsilon(t-\tau,\cdot))](x)\,d\tau\right| \\ &\quad = \left|\int_0^\sigma\int_{\mathscr{R}^d} \nabla G(\tau, x-y)\cdot[\nabla H(t,y) - \nabla H(t-\tau,y)]\,dy\,d\tau\right| \\ &\quad \leqq \int_0^\sigma \|\nabla G(\tau,\cdot)\|_1(\|\nabla\varphi^\varepsilon(t-\tau,\cdot)\|_\infty + \|\nabla\varphi^\varepsilon(t,\cdot)\|_\infty)\,d\tau \\ &\quad \leqq \frac{K}{\varepsilon} m^\varepsilon(T)\sqrt{\sigma}. \end{aligned}$$

5. Estimates obtained in Step 4 and (3.1) yield

$$|\theta^\varepsilon(t,x)| \leqq \|\theta_0^\varepsilon\|_\infty + 2 + K\ln\left(\frac{t}{\sigma}\right) + \frac{K\sqrt{\sigma}}{\varepsilon} m^\varepsilon(T).$$

Choose $\sigma = \varepsilon^2 \wedge t$; then

$$n^\varepsilon(T) \leqq \|\theta_0^\varepsilon\|_\infty + 2 + K \ln T + K |\ln \varepsilon| + K m^\varepsilon(T), \tag{3.5}$$

and hence (3.4) and (A6) yield

$$n^\varepsilon(T) \leqq K_0(1 + |\ln \varepsilon| + \varepsilon n^\varepsilon(T)).$$

Hence (3.2) and (3.3) hold for all $\varepsilon > 0$ satisfying

$$2K_0\varepsilon \leqq 1 \Leftrightarrow \varepsilon \leqq \varepsilon_0 = (\tfrac{1}{2}K_0).$$

For $1 \geqq \varepsilon \geqq \varepsilon_0$, (3.2) and (3.3) can be proved easily. □

Remark 3.1. The family θ^ε is not necessarily uniformly bounded in ε. Indeed consider the Mullins-Sekerka problem with radial symmetry and one interface. If the radius R_0 of the initial interface is sufficiently small and the initial temperature θ_0 is sufficiently large, then the radius $R(t)$ of the interface becomes zero in a finite time T. Since the phase-field equations with radial symmetry are known to approximate the Mullins-Sekerka problem [30], this example, which is discussed in the Appendix, shows that θ^ε is not uniformly bounded in ε.

Next we use the techniques developed in this section to obtain uniform bounds for $\varepsilon^2|D^2\varphi^\varepsilon|$ and $\varepsilon|\nabla\theta^\varepsilon|$.

Lemma 3.2.

$$\sup_{0<\varepsilon\leqq 1} \{\varepsilon^2[\|D^2\varphi^\varepsilon\|_{\infty,T} + \|\varphi_t^\varepsilon\|_{\infty,T}] + \varepsilon\|\nabla\theta^\varepsilon\|_{\infty,T}\} < \infty. \tag{3.6}$$

Proof.
1. Differentiate the (OPE) to obtain

$$\varphi_{x_j t}^\varepsilon - \Delta\varphi_{x_j}^\varepsilon = F_j^\varepsilon,$$

$$F_j^\varepsilon = -\frac{1}{\varepsilon^2} W''(\varphi^\varepsilon)\varphi_{x_j}^\varepsilon + \frac{1}{\varepsilon} g'(\varphi^\varepsilon)\varphi_{x_j}^\varepsilon\theta^\varepsilon + \frac{1}{\varepsilon} g(\varphi^\varepsilon)\theta_{x_j}^\varepsilon.$$

Using (3.2) and (3.3) we conclude that

$$\|F_j^\varepsilon\|_{\infty,T} \leqq K\left[\frac{1}{\varepsilon^3} + \frac{1}{\varepsilon}\|\nabla\theta^\varepsilon\|_{\infty,T}\right]$$

for some constant K. Set

$$\bar{m}^\varepsilon(T) = \varepsilon^2\|D^2\varphi^\varepsilon\|_{\infty,T}, \quad \bar{n}^\varepsilon(T) = \varepsilon\|\nabla\theta^\varepsilon\|_{\infty,T}.$$

Then we use (A4) and (3.2) as in Step 2 of the proof of Lemma 3.1; the result is

$$\bar{m}^\varepsilon(T) \leqq K[1 + \varepsilon\bar{n}^\varepsilon(T)]; \tag{3.7}$$

hence (3.3), (3.7), and (OPE) yield

$$\varepsilon^2\|\varphi_t^\varepsilon\|_{\infty,T} \leqq K[1 + \varepsilon\bar{n}^\varepsilon(T)]. \tag{3.8}$$

2. Let A^ε and B^ε be as in (3.1). Then,

$$\nabla A^\varepsilon(t,x) = (G(t,\cdot) * \nabla(\theta_0^\varepsilon + H^\varepsilon(0,\cdot) - H^\varepsilon(t,\cdot)))(x),$$

$$\nabla B^\varepsilon(t,x) = \int_0^t (\nabla G_\tau(\tau,\cdot) * [H^\varepsilon(t,\cdot) - H^\varepsilon(t-\tau,\cdot)])(x)\,d\tau.$$

Fix $(t,x) \in [0,T] \times \mathscr{R}^d$. In view of (A6) and (3.2),

$$|\nabla A^\varepsilon(t,x)| \leqq \|\nabla\theta_0^\varepsilon\|_\infty + 2\|\nabla\varphi^\varepsilon\|_{\infty,T} \leqq \frac{K}{\varepsilon}.$$

Also for $\sigma \in (0, t \wedge 1]$,

$$\left|\int_\sigma^t [G_\tau(\tau,\cdot) * (H^\varepsilon(t,\cdot) - H^\varepsilon(t-\tau,\cdot))](x)\,d\tau\right|$$

$$\leqq 2\int_\sigma^t \|\nabla G_\tau(\tau,\cdot)\|_1\,d\tau \leqq \int_\sigma^t K\tau^{-3/2}\,d\tau \leqq K\left(\frac{1}{\sqrt{\sigma}} - \frac{1}{\sqrt{t}}\right).$$

3. Integrating by parts in the τ-variable, we obtain

$$\left|\int_0^\sigma [G_\tau(\tau,\cdot) * (H^\varepsilon(t,\cdot) - H^\varepsilon(t-\tau,\cdot)]](x)\,d\tau\right|$$

$$\leqq |\nabla G(\sigma,\cdot) * (H^\varepsilon(t,\cdot) - H^\varepsilon(t-\sigma,\cdot)]](x)| + \left|\int_0^\sigma (\nabla G(\tau,\cdot) * (H_t^\varepsilon(t-\tau,\cdot))(x)\,d\tau\right|$$

$$\leqq \frac{K}{\sqrt{\sigma}} \|H^\varepsilon(t,\cdot) - H^\varepsilon(t-\sigma,\cdot)\|_\infty + \int_0^\sigma \|\nabla G(\tau,\cdot)\|_1 \|H_t^\varepsilon\|_{\infty,T}\,d\tau.$$

$$\leqq K\sqrt{\sigma}\,\|H_t^\varepsilon\|_{\infty,T} \leqq K\sqrt{\sigma}\,\|\varphi_t^\varepsilon\|_{\infty,T}.$$

4. Combine Steps 2 and 3, and choose $\sigma = \varepsilon^2 \wedge t$; then

$$|\nabla\theta^\varepsilon(t,x)| \leqq \tfrac{K}{\varepsilon}(1 + \varepsilon^2\|\varphi_t^\varepsilon\|_{\infty,T}).$$

As in the last step of the previous lemma, this estimate together with (3.7) and (3.8) imply (3.6) for sufficiently small $\varepsilon \leqq \varepsilon_0$. But for $\varepsilon \geqq \varepsilon_0$, (3.6) holds trivially. □

Assumption (A7) and the arguments of Lemma 3.2 yield

$$\sup_{0<\varepsilon\leqq 1} \{\varepsilon^3\|D^3\varphi^\varepsilon\|_{\infty,T} + \varepsilon^2\|D^2\theta^\varepsilon\|_{\infty,T} + \varepsilon^2\|\theta_t^\varepsilon\|_{\infty,T}\} < \infty. \tag{3.9}$$

Lemma 3.3. *For* $1 \leqq p < \infty$ *and* $T \geqq 0$,

$$\sup_{0<\varepsilon\leqq 1,\, t\leqq T} \|\theta^\varepsilon(t,\cdot)\|_{L^p(\mathscr{R}^d)} < \infty. \tag{3.10}$$

Proof.
1. Recall that, by (2.5),

$$\|H^\varepsilon(t_1,\cdot) - H^\varepsilon(t_0,\cdot)\|_1 \leqq C_1^*\sqrt{t_1 - t_0}.$$

Since $|H^\varepsilon| \leqq 1$,

$$\|H^\varepsilon(t_1,\cdot) - H^\varepsilon(t_0,\cdot)\|_p^p \leqq 2^p C_1^* \sqrt{t_1 - t_0}.$$

2. Let A^ε and B^ε be as in (3.1). Then, by (2.7) and the previous step,

$$\|A^\varepsilon\|_p \leqq \|\theta_0^\varepsilon\|_p + \|H^\varepsilon(t,\cdot) - H^\varepsilon(0,\cdot)\|_p \leqq K(1 + t^{1/2p}).$$

3. By Step 1,

$$\|B^\varepsilon\|_p \leqq \int_0^t \|G_\tau(\tau,\cdot) * (H^\varepsilon(t,\cdot) - H^\varepsilon(t-\tau,\cdot))\|_p \, d\tau$$

$$\leqq \int_0^t \|G_\tau(\tau,\cdot)\|_1 \|H^\varepsilon(t,\cdot) - H^\varepsilon(t-\tau,\cdot)\|_p \, d\tau$$

$$\leqq K \int_0^t \tau^{-1+1/2p} \, d\tau. \quad \square$$

We close this section by proving the strong convergence of the sequence θ^ε.

Proposition 3.4. *For every $t \geqq 0$, $\theta^\varepsilon(t,\cdot)$ converges to $\theta(t,\cdot)$ strongly in $L^2_{\mathrm{loc}}(\mathscr{R}^d)$. In particular,*

$$\mu^\varepsilon(t;\cdot) \rightharpoonup \mu(t;\cdot) \quad \forall t \geqq 0,$$

in the weak topology of Radon measures.*

Proof.
1. Let $\bar\theta^\varepsilon = \theta^\varepsilon - \hat\theta^\varepsilon$ with $\hat\theta^\varepsilon$ the unique solution of

$$\hat\theta_t^\varepsilon - \Delta\hat\theta^\varepsilon = 0 \quad \text{in } (0,\infty) \times \mathscr{R}^d$$

with initial data $\hat\theta^\varepsilon(0,x) = \theta_0^\varepsilon(x)$. Then (A8) implies that $\hat\theta^\varepsilon(t,\cdot)$ converges to

$$\hat\theta(t,\cdot) = G(t,\cdot) * \theta_0$$

strongly in $L^2(\mathscr{R}^d)$.

2. By integration by parts, $\bar\theta^\varepsilon = \bar\theta^{\varepsilon,1} + \bar\theta^{\varepsilon,2}$, where

$$\bar\theta^{\varepsilon,1}(t,\cdot) = G(t,\cdot) * [H^\varepsilon(0,\cdot) - H^\varepsilon(t,\cdot)],$$

$$\bar\theta^{\varepsilon,2}(t,;) = \int_0^t G_\tau(\tau,\cdot) * [H^\varepsilon(t,\cdot) - H^\varepsilon(t-\tau,\cdot)] \, d\tau,$$

and $H(t,x) = h(\varphi(t,x))$. Clearly (2.6) implies that for every $t \geqq 0$, $\bar\theta^{\varepsilon,1}(t,\cdot)$ converges to

$$\bar\theta^1(t,\cdot) = G(t,\cdot) * [H(0,\cdot) - H(t,\cdot)]$$

strongly in $L^2_{\mathrm{loc}}(\mathscr{R}^d)$.

3. For $t, \sigma > 0$, set $\delta = \min\{\sigma, t\}$. Then

$$\int_\delta^t G_\tau(\tau,\cdot) * [H^\varepsilon(t,\cdot) - H^\varepsilon(t-\tau,\cdot)]\, d\tau$$

converges to

$$\int_\delta^t G_\tau(\tau,\cdot) * [H(t,\cdot) - H(t-\tau,\cdot)]\, d\tau$$

strongly in $L^2_{\text{loc}}(\mathscr{R}^d)$. And by (2.5),

$$\begin{aligned}\left\|\int_0^\delta (G_\tau(\tau,\cdot) * [H^\varepsilon(t,\cdot) - H^\varepsilon(t-\tau,\cdot)]\, d\tau\right\|_2 &\leqq \int_0^\delta \|G_\tau(\tau,\cdot)\|_1 \|H^\varepsilon(t,\cdot) - H^\varepsilon(t-\tau,\cdot)\|_2\, d\tau,\\ &\leqq K \int_0^\delta \tfrac{1}{\tau} \|H^\varepsilon(t,\cdot) - H^\varepsilon(t-\tau,\cdot)\|_1^{1/2}\, d\tau,\\ &\leqq K(\delta)^{1/4} \leqq K\sigma^{1/4}.\end{aligned}$$

A similar argument shows that

$$\left\|\int_0^\delta G_\tau(\tau,\cdot) * [H(t,\cdot) - H(t-\tau,\cdot)]\, d\tau\right\|_2 \leqq K\sigma^{1/4}.$$

Therefore, for all $\sigma, t, R > 0$,

$$\limsup_{\varepsilon\to 0} \|\theta^\varepsilon(t,\cdot) - \theta(t,\cdot)\|_{L^2(B_R)} \leqq K\sigma^{1/4},$$

where

$$\theta(t,x) = \hat{\theta}(t,x) + G(t,\cdot) * [H(0,\cdot) - H(t,\cdot)])(x) + \int_0^t G_\tau(\tau,\cdot) * [H(t,\cdot) - H(t-\tau,\cdot)]\, d\tau. \quad \square$$

An elementary argument, very similar to the proof of Proposition 3.4, shows that the map

$$t \mapsto \|\theta^\varepsilon(t,\cdot)\|_2$$

is uniformly Hölder continuous in $\varepsilon \in (0, 1]$, with exponent $\frac{1}{4}$. This fact will not be used in our analysis.

4. A Gradient Estimate

The main result of this section is

Theorem 4.1. *For $T > 0$, there exists a constant $K^* = K^*(T)$ satisfying*

$$|\nabla z^\varepsilon(t,x)|^2 \leqq 1 + \sqrt{\varepsilon} K^*(1 + |z^\varepsilon(t,x)|), \tag{4.1}$$

for all $(t,x) \in [0,T] \times \mathscr{R}^d$, and $0 < \varepsilon \leqq 1$.

This estimate is an essential ingredient of the monotonicity result that is proved in §5. In particular, (4.1) implies that the weak* limit of the discrepancy measure ξ^ε introduced earlier (cf. (1.7), (2.1)) is nonpositive. Later we show that this limit is zero (see Proposition 7.3.).

The proof of this estimate, which is tangential to the main thrust of this paper, will be completed in several steps. Before I start the lengthy analysis, I briefly explain the main idea. Set

$$w^\varepsilon = |\nabla z^\varepsilon|^2.$$

In view of the equation (ZE),

$$w_t^\varepsilon + \mathscr{L}_t^\varepsilon w^\varepsilon + R^\varepsilon(t, x, w^\varepsilon) - 2\nabla\theta^\varepsilon \cdot \nabla z^\varepsilon \leqq 0, \tag{4.2}$$

where for $\psi \in C^2(\mathscr{R}^d)$,

$$\mathscr{L}_t^\varepsilon \psi(x) = -\Delta\psi(x) + \frac{4\varphi^\varepsilon(t, x)}{\varepsilon} \nabla z^\varepsilon(t, x) \cdot \nabla\psi(x),$$

$$R^\varepsilon(t, x, r) = \frac{4}{\varepsilon^2} q'\left(\frac{z^\varepsilon(t, x)}{\varepsilon}\right) r(r-1), \quad r \geqq 0.$$

In [29, §8], I obtained pointwise estimates for a differential inequality obtained by setting the last term involving $\nabla\theta^\varepsilon$ in (4.2) to zero. Here we start by using the technique developed in [29]. Using (3.6), we first obtain the crude estimate that

$$|2\nabla\theta^\varepsilon \cdot \nabla z^\varepsilon| \leqq 2\|\nabla\theta^\varepsilon\|_{\infty, T} w^\varepsilon \leqq \frac{K}{\varepsilon} w^\varepsilon$$

for $w^\varepsilon \geqq 1$. Then the proof of Proposition 8.1 in [29] yields that w^ε is uniformly bounded in ε.

Our next step is to obtain a uniform bound for $\varepsilon|z_t^\varepsilon|$ (see (4.8)). Using these estimates, we shall obtain a bound for $|\nabla\theta^\varepsilon|$, which is slightly better than (3.6). Finally, we shall use this new estimate of $|\nabla\theta^\varepsilon|$ in (4.2) together with an argument similar to the ones used in [29] to obtain (4.1).

Remark 4.1. In the phase-field equations if, in contrast to the choice we actually made, we take

$$g = 2W = (q')^2,$$

then w satisfies

$$w_t^\varepsilon + \mathscr{L}_t^\varepsilon w^\varepsilon + R^\varepsilon(t, x, w^\varepsilon) - 2q'\nabla\theta^\varepsilon \cdot \nabla z^\varepsilon - \frac{2}{\varepsilon} q''\theta^\varepsilon w^\varepsilon \leqq 0,$$

and the proof of the estimate (4.1) simplifies greatly. Indeed an attendant modification of the proof of Proposition 4.1 below yields this estimate.

As in §3, we fix $T > 0$ and denote by K all constants depending only on T.

Proposition 4.2. *There is a $K = K(T)$ satisfying*

$$|\nabla z^\varepsilon(t,x)|^2 \leqq K(1 + |z^\varepsilon(t,x)|), \quad (t,x) \in [0,T] \times \mathscr{R}^d. \tag{4.3}$$

Proof.
1. Fix $T > 0$ and set

$$K_0 = 2 \sup_{0<\varepsilon\leqq 1} \varepsilon \| \nabla\theta^\varepsilon \|_{\infty,T},$$

$$\hat{\mathscr{L}}\psi = \mathscr{L}_t^\varepsilon\psi - 2|\nabla\theta^\varepsilon(t,x)|\psi.$$

Then

$$w_t^\varepsilon + \hat{\mathscr{L}}w^\varepsilon + R^\varepsilon(t,x,w^\varepsilon) \leqq 0. \tag{4.4}$$

In the next several steps we construct a "supersolution" to (4.4).

2. Let $z_0 > 0$ be the point that satisfies

$$q'\left(\frac{z_0}{\varepsilon}\right) = \varepsilon^{1/4} \Rightarrow z_0 = \varepsilon(q')^{-1}(\varepsilon^{1/4}).$$

Then z_0 behaves like $\varepsilon|\ln\varepsilon|$ as ε tends to zero. Indeed,

$$\lim_{\varepsilon\to 0} \frac{z_0}{\varepsilon|\ln\varepsilon|} = \frac{1}{8}.$$

Now define

$$h_\varepsilon(r) = \begin{cases} \frac{1}{2}C_\varepsilon r^2 + 1, & |r| \leqq z_0, \\ (K_0 + 1)[|r| - z_0] + h_\varepsilon(z_0), & |r| > z_0, \end{cases}$$

where

$$C_\varepsilon = \frac{K_0 + 1}{z_0},$$

so that h_ε is continuously differentiable with Lipschitz derivatives. Finally we set

$$W = 1 + h_\varepsilon(z^\varepsilon).$$

3. By (ZE) and a direct calculation, we obtain

$$\begin{aligned} I &= W_t + \hat{\mathscr{L}}W + R^\varepsilon(t,x,W) \\ &\geqq h_\varepsilon'(z^\varepsilon)[z_t^\varepsilon - \Delta z^\varepsilon + \tfrac{4}{\varepsilon}\varphi^\varepsilon w^\varepsilon] - h_\varepsilon''(z^\varepsilon)w^\varepsilon - \frac{K_0}{\varepsilon}W + \frac{4}{\varepsilon^2}q'\left(\frac{z^\varepsilon}{\varepsilon}\right)h_\varepsilon(z^\varepsilon)W \\ &\geqq \frac{2}{\varepsilon}\varphi^\varepsilon h_\varepsilon'(z^\varepsilon)(w^\varepsilon + 1) + \frac{4}{\varepsilon^2}q'\left(\frac{z^\varepsilon}{\varepsilon}\right)h_\varepsilon(z^\varepsilon)W - \frac{K_0}{\varepsilon}W - h_\varepsilon''(z^\varepsilon)w^\varepsilon + h_\varepsilon'(z^\varepsilon)\theta^\varepsilon. \end{aligned}$$

Observe that $h_\varepsilon \geqq 1$, $|h_\varepsilon'| \leqq K_0 + 1$ and

$$\| h_\varepsilon'' \|_\infty = C_\varepsilon, \quad \lim_{\varepsilon\to 0} \varepsilon C_\varepsilon = 0. \tag{4.5}$$

Hence

$$I \geqq \frac{2}{\varepsilon}\varphi^\varepsilon h'_\varepsilon(w^\varepsilon + 1) + \frac{4}{\varepsilon^2}q' W - \frac{K_0}{\varepsilon}W - C_\varepsilon w^\varepsilon - (K_0 + 1)\|\theta^\varepsilon\|_{\infty,T}.$$

4. Suppose that

$$|z^\varepsilon(t,x)| \leqq z_0.$$

(The opposite case will be discussed in the next step). Then

$$q'\left(\frac{z^\varepsilon}{\varepsilon}\right) \geqq q'\left(\frac{z_0}{\varepsilon}\right) = \varepsilon^{1/4}.$$

Since $\varphi^\varepsilon h'_\varepsilon \geqq 0$, $W \geqq 1$, by (3.3), we obtain

$$\begin{aligned} I &\geqq \frac{4}{\varepsilon^2}q' W - \frac{K_0}{\varepsilon}W - C_\varepsilon w^\varepsilon - (K_0 + 1)\|\theta^\varepsilon\|_{\infty,T} \\ &\geqq \left(\frac{4}{\varepsilon^{7/4}} - \frac{K_0}{\varepsilon}\right)W - C_\varepsilon w^\varepsilon - (K_0 + 1)K(1 + |\ln\varepsilon|) \\ &\geqq C_\varepsilon(W - w^\varepsilon) \end{aligned}$$

for sufficiently small $\varepsilon > 0$.
5. Suppose that $|z^\varepsilon(t,x)| \geqq z_0$. Then

$$h'_\varepsilon(|z^\varepsilon(t,x)|) = K_0 + 1,$$

$$|\varphi^\varepsilon(x,t)| = q\left(\frac{|z^\varepsilon(t,x)|}{\varepsilon}\right) \geqq q\left(\frac{z_0}{\varepsilon}\right) \geqq \frac{1}{2}$$

for sufficiently small $\varepsilon > 0$. Therefore

$$\varphi^\varepsilon(t,x)h'_\varepsilon(z^\varepsilon(t,x)) = |\varphi^\varepsilon(t,x)|h'_\varepsilon(|z^\varepsilon(t,x)|) \geqq \tfrac{1}{2}[K_0 + 1].$$

Since $q' \geqq 0$,

$$\begin{aligned} I &\geqq \frac{2\varphi^\varepsilon}{\varepsilon}h'_\varepsilon(z^\varepsilon)(w^\varepsilon + 1) - \frac{K_0}{\varepsilon}W - C_\varepsilon w^\varepsilon - (K_0 + 1)\|\theta^\varepsilon\|_{\infty,T} \\ &\geqq \left(\frac{K_0 + 1}{\varepsilon} - C_\varepsilon\right)w^\varepsilon - \frac{K_0}{\varepsilon}W + (K_0 + 1)\left[\frac{1}{\varepsilon} - \|\theta^\varepsilon\|_{\infty,T}\right], \end{aligned}$$

and (3.3) and (4.5) imply that $I \geqq 0$ on $\{w^\varepsilon \geqq W\}$.
6. In Steps 3, 4 and 5, we proved that for every $T > 0$ there is an $\varepsilon_0 = \varepsilon_0(T) > 0$ that satisfies

$$W_t + \hat{\mathscr{L}}W + R^\varepsilon(t,x,W) \geqq C_\varepsilon(W - w^\varepsilon)$$

on $(0, T)\times\mathscr{R}^d\cap\{w^\varepsilon \geqq W\}$ for all $0<\varepsilon\leqq\varepsilon_0(T)$. Also in Step 1, we showed that

$$w_t^\varepsilon + \hat{\mathscr{L}}w^\varepsilon + R^\varepsilon(t, x, w^\varepsilon) \leqq 0 \quad \text{in } (0, T)\times\mathscr{R}^d.$$

Since $W \geqq 1 \geqq w^\varepsilon(0, x)$, by the maximum principle we conclude that $W \geqq w^\varepsilon$ on $(0, T)\times\mathscr{R}^d$. See the proof of Proposition 4.2 in [29, §8] for an application of the maximum principle in a very similar situation.

7. Since

$$h_\varepsilon(z) \leqq (K_0 + 1)|z| + 1,$$

we conclude that

$$w^\varepsilon \leqq W = 1 + h_\varepsilon(z^\varepsilon) \leqq 1 + (K_0 + 1)(|z^\varepsilon| + 1). \quad \square$$

Our next step is a crude estimate of $|z_t^\varepsilon|$. We obtain a better estimate in Lemma 4.4.

Lemma 4.3. *For* $0 < \varepsilon \leqq 1$,

$$|z_t^\varepsilon(t, x)| \leqq \frac{K}{\varepsilon^2}, \quad (t, x) \in [0, T]\times\mathscr{R}^d. \tag{4.6}$$

Proof.

1. For $\alpha > 0$, set

$$\Omega = \left\{(t, x) \in [0, T)\times\mathscr{R}^d : \left|\frac{z^\varepsilon(t, x)}{\varepsilon}\right| > \alpha\right\}.$$

By (3.6),

$$|z_t^\varepsilon| = \varepsilon\left(q'\left(\frac{z^\varepsilon}{\varepsilon}\right)\right)^{-1}|\varphi_t^\varepsilon| \leqq \frac{K}{\varepsilon}\left(q'\left(\frac{z^\varepsilon}{\varepsilon}\right)\right)^{-1}.$$

Hence, on the complement of Ω,

$$|z_t^\varepsilon| \leqq \frac{K}{\varepsilon q'(\alpha)}.$$

2. Set $v = z_t^\varepsilon$. Differentiate (ZE) to obtain

$$v_t - \Delta v + \frac{4\varphi^\varepsilon}{\varepsilon}\nabla z^\varepsilon\cdot\nabla v + \frac{2}{\varepsilon^2}q'\left(\frac{z^\varepsilon}{\varepsilon}\right)(|\nabla z^\varepsilon|^2 - 1)v = \theta_t^\varepsilon. \tag{4.7}$$

3. For $K_1 > 0$, let

$$V = \frac{K_1}{\varepsilon}\left(1 + \frac{|z^\varepsilon|}{\varepsilon}\right).$$

We show that for appropriately chosen K_1 and α, the function V is a supersolution of (4.7) in Ω. Indeed in Ω,

$$I = V_t - \Delta V + \frac{4\varphi^\varepsilon}{\varepsilon} \nabla z^\varepsilon \cdot \nabla V + \frac{2}{\varepsilon^2} q'\left(\frac{z^\varepsilon}{\varepsilon}\right)(|\nabla z^\varepsilon|^2 - 1)V$$

$$\geqq \frac{K_1 z^\varepsilon}{\varepsilon^2 |z^\varepsilon|}\left[z_t^\varepsilon - \Delta z^\varepsilon + \frac{4\varphi^\varepsilon}{\varepsilon}|\nabla z^\varepsilon|^2\right] - \frac{2K_1}{\varepsilon^3} q'\left(\frac{z^\varepsilon}{\varepsilon}\right)\left[\frac{|z^\varepsilon|}{\varepsilon} + 1\right]$$

$$\geqq \frac{K_1}{\varepsilon^3}\left\{2|\varphi^\varepsilon|(1 + |\nabla z^\varepsilon|^2) - \varepsilon \|\theta^\varepsilon\|_{\infty, T} - 2 \sup_{r \geqq \alpha} q'(r)(r+1)\right\}$$

$$\geqq \frac{K_1}{\varepsilon^3}\left\{2q(\alpha) - \varepsilon \|\theta^\varepsilon\|_{\infty, T} - 2 \sup_{r \geqq \alpha} q'(r)(r+1)\right\}.$$

Since $q'(r)$ is exponentially small for large values of r, (3.3) and (3.9) imply that there are constants K_1 and α such that

$$I \geqq \|\theta_t^\varepsilon\|_{\infty, T} \quad \text{in } \Omega$$

for all sufficiently small ε. By redefining K_1, if necessary, we may assume that

$$\inf_{\partial\Omega} V = \frac{K_1}{\varepsilon}(1 + \alpha) \geqq \frac{K}{\varepsilon q'(\alpha)} = \sup_{\Omega^c} |z_t^\varepsilon|.$$

4. We proved that there is an $\varepsilon_0 > 0$ such that for, $0 < \varepsilon \leqq \varepsilon_0$, V is a supersolution of (4.7) in Ω. Moreover $V \geqq v$ on $\partial\Omega$. Therefore, by the maximum principle,

$$V \geqq v \quad \text{in } \Omega, \varepsilon \leqq \varepsilon_0.$$

Hence, by Step 1,

$$v = z_t^\varepsilon \leqq \frac{K}{\varepsilon}\left(1 + \frac{|z^\varepsilon|}{\varepsilon}\right)$$

for all $0 < \varepsilon \leqq \varepsilon_0$. For $\varepsilon_0 \leqq \varepsilon < 1$ this last estimate is easier to prove. These arguments also yield the same bound for $-z_t^\varepsilon$.

5. Set

$$\hat{\Omega} = \{|z^\varepsilon| \geqq 1\}, \quad \hat{V} = \frac{K}{\varepsilon^2}(1 + t).$$

Then on $[0, T] \times \mathscr{R}^d \cap \hat{\Omega}$,

$$\hat{V}_t - \Delta \hat{V} + \frac{4\varphi^\varepsilon}{\varepsilon} \nabla z^\varepsilon \cdot \nabla \hat{V} + \frac{2}{\varepsilon^2} q'\left(\frac{z^\varepsilon}{\varepsilon}\right)(|\nabla z^\varepsilon|^2 - 1)\hat{V}$$

$$\geqq \frac{K}{\varepsilon^2} - \frac{K(1+T)}{\varepsilon^4} q'\left(\frac{1}{\varepsilon}\right) \geqq \|\theta_t^\varepsilon\|_{\infty, T}$$

for sufficiently small ε. Also, by Step 4, $\hat{V} \geqq |v|$ on $\partial\hat{\Omega}$. Hence (4.6) follows from the maximum principle. □

Next we improve (4.6).

Lemma 4.4.

$$|z_t^\varepsilon(t,x)| + |D^2 z^\varepsilon(t,x)| \leqq \frac{K}{\varepsilon}(1+|z^\varepsilon(t,x)|), \quad (t,x)\in[0,T]\times\mathscr{R}^d. \tag{4.8}$$

Proof. Fix $T>0$. All the constants in this proof depend in T. Set

$$k^\varepsilon = \sup\left\{\frac{\varepsilon|z^\varepsilon_{x_i x_j}(t,x)|}{1+|z^\varepsilon(t,x)|} : (t,x)\in[0,T]\times\mathscr{R}^d,\ i,j=1,\ldots,d\right\}.$$

1. In view of (4.3), for any $t \leqq T$ and $x, y\in\mathscr{R}^d$,

$$(|z^\varepsilon(t,y)|+1) \leqq e^{K|x-y|}(|z^\varepsilon(t,x)|+1).$$

Also (4.6) implies that there is a K^* such that for all $\tau\in[0,t]$,

$$\begin{aligned}(|z^\varepsilon(t-\tau,y)+1) &\leqq \left(1+\frac{K^*\tau}{\varepsilon^2}\right)(|z^\varepsilon(t,y)|+1)\\ &\leqq \left(1+\frac{K^*\tau}{\varepsilon^2}\right)e^{K^*|x-y|}(|z^\varepsilon(t,x)|+1).\end{aligned} \tag{4.9}$$

2. Fix $(t_0,x_0)\in[0,T]\times\mathscr{R}^d$. For any $h\in(0,t_0\wedge 1]$, (ZE) yields

$$z^\varepsilon_{x_i x_j}(t_0,x_0) = a+b+c,$$

where

$$a = (G_{x_i}(h,\cdot) * z^\varepsilon_{x_j}(t_0-h,\cdot))(x_0),$$

$$b = \int_0^h (G_{x_i}(\tau,\cdot) * \theta^\varepsilon_{x_j}(t_0-\tau,\cdot))(x_0)\,d\tau,$$

$$c = \int_0^h (G_{x_i}(\tau,\cdot) * F^\varepsilon_{x_j}(t_0-\tau,\cdot))(x_0)\,d\tau,$$

$$F^\varepsilon = \frac{2\varphi^\varepsilon}{\varepsilon}(1-|\nabla z^\varepsilon|^2).$$

3. If $h=t_0$, by (A4), we obtain

$$|a| \leqq \|G(h,\cdot)\|_1 \|D^2 z_0^\varepsilon\|_\infty \leqq \frac{K}{\varepsilon}.$$

When $h<t_0$, (4.3) and (4.9) imply that

$$\begin{aligned}|a| &\leqq \int_{\mathscr{R}^d} K|\nabla G(h,x_0-y)|[1+|z^\varepsilon(t_0-h,y)|]^{1/2}\,dy\\ &\leqq K\left(1+\frac{K^*h}{\varepsilon^2}\right)^{1/2}[1+|z^\varepsilon(t_0,x_0)|]^{1/2}\int_{\mathscr{R}^d} e^{\frac{1}{2}K^*|w|}|\nabla G(h,w)|\,dw\\ &\leqq \frac{K}{\sqrt{h}}\left(1+\frac{K^*h}{\varepsilon^2}\right)^{1/2}(1+|z^\varepsilon(t_0,x_0)|)^{1/2}.\end{aligned}$$

4. By (3.6),

$$|b| \leqq \frac{K}{\varepsilon}\sqrt{h}.$$

5. Differentiate F^ε to obtain

$$\nabla F^\varepsilon = \frac{2}{\varepsilon^2} q'\left(\frac{z^\varepsilon}{\varepsilon}\right) \nabla z^\varepsilon (1 - |\nabla z^\varepsilon|^2) - \frac{4\varphi^\varepsilon}{\varepsilon} D^2 z^\varepsilon \nabla z^\varepsilon.$$

The definition of k^ε, (4.3) and (4.9) yield

$$\begin{aligned} |F^\varepsilon_{x_j}(t_0 - \tau, y)| &\leqq \frac{K}{\varepsilon^2}\left[\sup_r q'(r)(1 + \varepsilon r)^{3/2} + (1 + |z^\varepsilon(t_0 - \tau, y)|)^{3/2} k^\varepsilon\right] \\ &\leqq \frac{K}{\varepsilon^2}\left[1 + k^\varepsilon\left(1 + \frac{K^*\tau}{\varepsilon^2}\right)^{3/2} (1 + |z^\varepsilon(t_0, x_0)|)^{3/2} \exp\left(\frac{3}{2} K^* |x_0 - y|\right)\right]. \end{aligned}$$

Therefore

$$|c| \leqq C^* \frac{\sqrt{h}}{\varepsilon^2}\left[1 + k^\varepsilon\left(1 + \frac{K^* h}{\varepsilon^2}\right)^{3/2} (1 + |z^\varepsilon(t_0, x_0)|)^{3/2}\right]$$

for some C^*, and without loss of generality we may assume that $C^* \geqq 1$.
6. Choose

$$h = \min\{t_0, \varepsilon^2 [4(1 + K^*)^3 (1 + |z^\varepsilon(t_0, x_0)|)(C^*)^2]^{-1}\}.$$

Since $h \leqq \varepsilon^2$, $(1 + K^* h/\varepsilon^2) \leqq (1 + K^*)$ and therefore

$$|c| \leqq \frac{1}{2\varepsilon}(1 + |z^\varepsilon(t_0, x_0)|) k^\varepsilon + \frac{C^*}{\varepsilon},$$

and, by Step 3,

$$|a| \leqq \frac{K}{\varepsilon}(1 + K^*)^2 C^* (1 + |z^\varepsilon(t_0, x_0)|).$$

Therefore

$$|z^\varepsilon_{x_i x_j}(t_0, x_0)| \leqq \frac{1}{\varepsilon}(1 + |z^\varepsilon(t_0, x_0)|)[K + \tfrac{1}{2} k^\varepsilon].$$

The inequality (4.8) follows from this estimate and (ZE). □

We continue by improving the estimate for $|\nabla \theta^\varepsilon|$.

Lemma 4.5. *For every* $(t, x) \in [0, T] \times \mathscr{R}^d$,

$$|\nabla \theta^\varepsilon(t, x)| \leqq \frac{K}{\sqrt{\varepsilon[\varepsilon + (|z^\varepsilon(t, x)| \wedge 1)]}} \tag{4.10}$$

for some constant $K = K(T)$.

Proof. Fix $(t_0, x_0) \in [0, T] \times \mathscr{R}^d$ and set

$$p^\varepsilon = \frac{|z^\varepsilon(t_0, x_0)| \wedge 1}{\varepsilon}.$$

If $p^\varepsilon \leqq 1$, (4.10) at (t_0, x_0) follows from (3.6); so we may assume that $p^\varepsilon \geqq 1$.
1. For $\varepsilon, \lambda > 0$, set

$$O_{\varepsilon,\lambda} = [t_0 - \varepsilon^2 \lambda p^\varepsilon, t_0 + \varepsilon^2 \lambda p^\varepsilon] \times B^\varepsilon, \quad B^\varepsilon = \{|x - x_0| \leqq \varepsilon \lambda p^\varepsilon\}.$$

We assert that there exists $\lambda = \lambda(T) > 0$ satisfying

$$|z^\varepsilon(t, x)| \geqq \frac{\varepsilon}{2} p^\varepsilon \quad \forall (t, x) \in O_{\varepsilon,\lambda}.$$

Use (4.3) and (4.8) to construct a constant $K = K(T)$ such that

$$|z^\varepsilon(s, y)| + 1 \leqq (|z^\varepsilon(t, x)| + 1) \exp K\left(\frac{|t - s|}{\varepsilon} + |x - y|\right)$$

for all $s, t \leqq T$. Now suppose that

$$z^\varepsilon(t_0 + \varepsilon^2 \tau, x_0 + \varepsilon y) = \tfrac{1}{2}\varepsilon p^\varepsilon \quad \text{for some } (\tau, y) \in \mathscr{R}^{d+1}.$$

We use the previous estimate with $(s, y) = (t_0, x_0)$ and $(t, x) = (t_0 + \varepsilon^2 \tau, x_0 + \varepsilon y)$ to obtain

$$1 + \varepsilon p^\varepsilon \leqq 1 + |z^\varepsilon(t_0, x_0)| \leqq \left(1 + \frac{\varepsilon}{2} p^\varepsilon\right) e^{\varepsilon K(|\tau| + |y|)}.$$

Since $\varepsilon p^\varepsilon \leqq 1$,

$$\varepsilon K(|\tau| + |y|) \geqq \ln\left(\frac{1 + \varepsilon p^\varepsilon}{1 + \frac{\varepsilon}{2} p^\varepsilon}\right) \geqq \ln\left(1 + \frac{\varepsilon}{4} p^\varepsilon\right) \geqq \frac{\varepsilon}{8} p^\varepsilon.$$

Hence, for $\lambda = 1/8K$,

$$|z^\varepsilon(t, x)| \geqq \tfrac{1}{2}\varepsilon p^\varepsilon \quad \forall (t, x) \in O_{\varepsilon,\lambda}.$$

2. Set

$$\sigma = \min\{t_0, \varepsilon^2 \lambda p^\varepsilon\}.$$

By integrating by parts and by (3.1), we obtain

$$\nabla\theta^\varepsilon(t_0, x_0) = a + b + c$$

where

$$a = (\nabla G(t_0, \cdot) * [\theta_0^\varepsilon + H^\varepsilon(0, \cdot) - H^\varepsilon(t_0, \cdot)])(x_0),$$

$$b = \int_\sigma^{t_0} (\nabla G_\tau(\tau, \cdot) * [H^\varepsilon(t_0, \cdot) - H^\varepsilon(t_0 - \tau, \cdot)])(x_0)\, d\tau,$$

$$c = \int_0^\sigma (\nabla G_\tau(\tau, \cdot) * [H^\varepsilon(t_0, \cdot) - H^\varepsilon(t_0 - \tau, \cdot)])(x_0)\, d\tau.$$

3. Since $\varepsilon p^\varepsilon \leqq 1$, (A6) yields

$$|(\nabla G(t_0,\cdot) * \theta_0^\varepsilon)(x_0)| = |(G(t_0,\cdot) * \nabla \theta_0^\varepsilon)(x_0)| \leqq \frac{K}{\sqrt{\varepsilon}} \leqq \frac{K}{\varepsilon\sqrt{p^\varepsilon}}.$$

Also, when $t_0 \geqq \varepsilon^2 \lambda p^\varepsilon$,

$$|(\nabla G(t_0,\cdot) * [H^\varepsilon(0,\cdot) - H^\varepsilon(t_0,\cdot)])(x_0)| \leqq \frac{K}{\sqrt{t_0}} \leqq \frac{K}{\varepsilon\sqrt{\lambda p^\varepsilon}}.$$

However, if $t_0 \leqq \varepsilon^2 \lambda p^\varepsilon$, then

$$\{0\} \times B^\varepsilon \subset O_{\varepsilon,\lambda}$$

so that by (3.2) and Step 1,

$$|\nabla H^\varepsilon(0, y)| = |g(\varphi^\varepsilon(0, y))\, \nabla\varphi^\varepsilon(0, y)| \leqq \frac{K}{\varepsilon} q'\left(\frac{z^\varepsilon(0, y)}{\varepsilon}\right)$$

$$\leqq \frac{K}{\varepsilon} q'\left(\frac{p^\varepsilon}{2}\right) \quad \forall y \in B^\varepsilon.$$

Hence,

$$\begin{aligned} |(\nabla G(t_0,\cdot) * H^\varepsilon(0,\cdot))(x_0)| &= |(G(t_0,\cdot) * \nabla H^\varepsilon(0,\cdot))(x_0)| \\ &\leqq \int_{B^\varepsilon} G(t_0, x_0 - y) q'\left(\frac{p^\varepsilon}{2}\right) \frac{K}{\varepsilon}\, dy + \int_{\mathscr{R}^d - B^\varepsilon} G(t_0, x_0 - y) \frac{K}{\varepsilon}\, dy \\ &\leqq \frac{K}{\varepsilon}\left[q'\left(\frac{p^\varepsilon}{2}\right) + \int G(1, w) \chi_{\{\sqrt{40}}{}^{|w| \geqq \varepsilon\lambda p^\varepsilon\}}\, dw\right] \\ &\leqq \frac{K}{\varepsilon}\left[q'\left(\frac{p^\varepsilon}{2}\right) + G\left(1, \frac{\varepsilon\lambda p^\varepsilon}{\sqrt{4t_0}}\right)\right]. \end{aligned}$$

Since $t_0 \leqq \varepsilon^2 \lambda p^\varepsilon$ and $p^\varepsilon \geqq 1$,

$$|(\nabla G(t_0,\cdot) * H^\varepsilon(0,\cdot))(x_0)| \leqq \frac{K}{\varepsilon\sqrt{p^\varepsilon}}.$$

Indeed, we can estimate this quantity by a function decaying faster than the square root, but this sharper estimate does not improve the final estimate.

Next, we estimate $|(\nabla G(t_0,\cdot) * H^\varepsilon(t_0,\cdot))(x_0)|$ exactly the same way to obtain

$$|a| \leqq \frac{K}{\varepsilon\sqrt{p^\varepsilon}}.$$

4. Since $\| \nabla G_\tau(\tau,\cdot)\|_1 \leqq K\tau^{-3/2}$,

$$|b| \leqq K\left(\frac{1}{\sqrt{\sigma}} - \frac{1}{\sqrt{t_0}}\right) \leqq \frac{K}{\varepsilon\sqrt{p^\varepsilon}}.$$

5. By integration by parts in the variable t,

$$|c| \leqq \left| \int_0^\sigma (\nabla G(\tau,\cdot) * g(\varphi^\varepsilon(t_0-\tau,\cdot))\varphi_t^\varepsilon(t_0-\tau,\cdot))(x_0)\,d\tau \right| + |(\nabla G(\sigma,\cdot) * [H^\varepsilon(t_0,\cdot) - H^\varepsilon(t_0-\sigma,\cdot)])(x_0))|.$$

Since $\sigma \leqq \varepsilon^2 \lambda p^\varepsilon$,

$$[t_0-\sigma, t_0] \times B^\varepsilon \subset O_{\varepsilon,\lambda}.$$

Therefore, for any $y \in B^\varepsilon$, $\tau \in [0,\sigma]$, (3.6) yields

$$|g^\varepsilon(\varphi^\varepsilon(t_0-\tau,y))\varphi_t^\varepsilon(t_0-\tau,y)| \leqq |\varphi_t^\varepsilon| q'\left(\frac{p^\varepsilon}{2}\right) \leqq \frac{K}{\varepsilon^2} q'\left(\frac{p^\varepsilon}{2}\right).$$

As in Step 3,

$$\left| \int_0^\sigma (\nabla G(\tau,\cdot) * g(\varphi^\varepsilon(t_0-\tau,\cdot))\varphi_t^\varepsilon(t_0-\tau,\cdot))(x_0)\,d\tau \right|$$

$$\leqq \int_0^\sigma \int_{B^\varepsilon} |\nabla G(\tau, x_0-y)| \frac{K}{\varepsilon^2} q'\left(\frac{p^\varepsilon}{2}\right) d\tau + \int_0^\sigma \int_{\mathscr{R}^d - B^\varepsilon} |\nabla G(\tau, x_0-y)| \frac{K}{\varepsilon^2}\, d\tau$$

$$\leqq \frac{K\sqrt{\sigma}}{\varepsilon^2}\left[q'\left(\frac{p^\varepsilon}{2}\right) + \left| \nabla G\left(1, \frac{\varepsilon\lambda p^\varepsilon}{\sqrt{4\sigma}}\right)\right| \right] \leqq \frac{K}{\varepsilon\sqrt{p^\varepsilon}}.$$

Also, if $\sigma = \varepsilon^2\lambda p^\varepsilon$,

$$|(\nabla G(\sigma,\cdot) * [H^\varepsilon(t_0,\cdot) - H^\varepsilon(t_0-\sigma,\cdot)])(x_0)| \leqq K \|\nabla G(\sigma,\cdot)\|_1 \leqq \frac{K}{\sqrt{\sigma}} \leqq \frac{K}{\varepsilon\sqrt{p^\varepsilon}}.$$

If $\sigma = t_0$, then by step 3,

$$|(\nabla G(\sigma,\cdot) * [H^\varepsilon(t_0,\cdot) - H^\varepsilon(t_0-\sigma,\cdot)])(x_0)|$$

$$= |(G(t_0,\cdot) * [\nabla H^\varepsilon(t_0,\cdot) - \nabla H^\varepsilon(0,\cdot)])(x_0)| \leqq \frac{K}{\varepsilon\sqrt{p^\varepsilon}}.$$

6. Finally, we combine Steps 3, 4 and 5 to conclude that

$$|\nabla\theta^\varepsilon(t_0,x_0)| \leqq \frac{K}{\varepsilon\sqrt{p^\varepsilon}}. \quad \square$$

We are now in a position to prove Theorem 4.1.

Proof of Theorem 4.1. This proof is very similar to the proof of Proposition 4.2.
1. Let z_0 be as in Proposition 4.3, i.e.,

$$q'\left(\frac{z_0}{\varepsilon}\right) = \varepsilon^{1/4}.$$

Set

$$K_\varepsilon = (z_0)^{-3/2}.$$

Since $q'(r)$ decays exponentially,

$$\lim_{\varepsilon\to 0} \frac{z_0}{\varepsilon|\ln \varepsilon|} = \frac{1}{8}, \quad \lim_{\varepsilon\to 0} (\varepsilon|\ln \varepsilon|)^{3/2} K_\varepsilon < \infty. \tag{4.11}$$

2. For a real number r, set

$$f_\varepsilon(r) = \begin{cases} \frac{1}{2}K_\varepsilon r^2 + 1, & |r| \leqq z_0, \\ 2[\sqrt{r} - \sqrt{z_0}] + f_\varepsilon(z_0), & |r| \in [z_0, 1], \\ |r| - 1 + f_\varepsilon(1), & |r| \geqq 1. \end{cases}$$

Observe that f_ε is continuously differentiable with Lipschitz continuous derivatives.
3. For $K^* > 1$ define

$$W = 1 + \sqrt{\varepsilon} K^* f_\varepsilon(z^\varepsilon).$$

In the next three steps, we show that for K^* large enough, W is a "supersolution" of (4.4).

Let $\hat{\mathscr{L}}$, R^ε and w^ε be as in Proposition 4.2. Then

$$\begin{aligned} I &= W_t + \hat{\mathscr{L}} W + R^\varepsilon(t, x, W) \\ &= \sqrt{\varepsilon} K^* f_\varepsilon' \left[z_t^\varepsilon - \Delta z^\varepsilon + \frac{4\varphi^\varepsilon}{\varepsilon} w^\varepsilon \right] - \sqrt{\varepsilon} K^* f_\varepsilon'' w^\varepsilon \\ &\quad + \frac{4}{\varepsilon^2} q'\left(\frac{z^\varepsilon}{\varepsilon}\right) \sqrt{\varepsilon} K^* f_\varepsilon W - 2|\nabla\theta^\varepsilon(t,x)|\, W \\ &\geqq \sqrt{\varepsilon} K^* \left\{ f_\varepsilon' \left[\frac{2\varphi^\varepsilon}{\varepsilon}(w^\varepsilon + 1) + \theta^\varepsilon \right] - f_\varepsilon'' w^\varepsilon + \frac{4}{\varepsilon^2} q'\left(\frac{z^\varepsilon}{\varepsilon}\right) W \right\} - 2|\nabla\theta^\varepsilon| W. \end{aligned}$$

4. We split the estimate of I into the three cases:

$$\text{(a) } |z^\varepsilon| \leqq z_0, \quad \text{(b) } |z^\varepsilon| \in [z_0, 1], \quad \text{(c) } |z^\varepsilon| \geqq 1,$$

and start with case a. Since $z_0 = \varepsilon (q')^{-1}(\varepsilon^{1/4})$,

$$q'\left(\frac{z^\varepsilon}{\varepsilon}\right) \geqq q'\left(\frac{z_0}{\varepsilon}\right) = \varepsilon^{1/4}.$$

By (4.11),

$$\sqrt{\varepsilon}|f_\varepsilon'(z^\varepsilon)| = \sqrt{\varepsilon} K_\varepsilon |z^\varepsilon| \leqq \sqrt{\varepsilon} z_0 K_\varepsilon \leqq \frac{K}{\sqrt{|\ln \varepsilon|}}$$

for some constant K. Since $f_\varepsilon' \varphi^\varepsilon \geqq 0$,

$$I \geqq K^* \left[-\frac{K}{\sqrt{|\ln \varepsilon|}} \|\theta^\varepsilon\|_{\infty,T} - \sqrt{\varepsilon} K_\varepsilon w^\varepsilon + 4\varepsilon^{-5/4} W \right] - 2\|\nabla\theta^\varepsilon\|_{\infty,T} W.$$

Using (3.3), (3.6), (4.11), and the fact that $W \geqq 1$, we construct $\varepsilon_0 = \varepsilon_0(T) > 0$ such that

$$I \geqq \sqrt{\varepsilon} K^* K_\varepsilon (W - w^\varepsilon), \quad \varepsilon \leqq \varepsilon_0, t \leqq T$$

for any $K^* \geqq 1$.
5. Suppose that $|z^\varepsilon| \geqq 1$. Then for sufficiently small ε,

$$f'_\varepsilon \varphi^\varepsilon = |\varphi^\varepsilon| \geqq \tfrac{1}{2}.$$

Moreover, $f''_\varepsilon(z^\varepsilon) = 0$ and, by (4.10),

$$|\nabla \theta^\varepsilon(t, x)| \leqq \frac{\hat{K}}{\sqrt{\varepsilon}}.$$

Since $q' \geqq 0$,

$$I \geqq \frac{K^*}{\sqrt{\varepsilon}}(w^\varepsilon + 1) - \sqrt{\varepsilon} K^* \|\theta^\varepsilon\|_{\infty, T} - \frac{2\hat{K}}{\sqrt{\varepsilon}} W.$$

So if $K^* \geqq 2\hat{K}$, (3.3) implies that $I \geqq 0$ on $\{w^\varepsilon \geqq W\}$ for all sufficiently small ε.
6. Finally we consider the case $|z^\varepsilon| \in [z_0, 1]$. Then, for sufficiently small $\varepsilon > 0$,

$$f'_\varepsilon(z^\varepsilon)\varphi^\varepsilon = f'_\varepsilon(|z^\varepsilon|)|\varphi^\varepsilon| \geqq \frac{1}{2} f'_\varepsilon(|z^\varepsilon|) = \frac{1}{2\sqrt{|z^\varepsilon|}}.$$

Moreover, by the construction of f_ε,

$$\sqrt{\varepsilon} f'_\varepsilon(|z^\varepsilon|) \leqq 1, \quad f''_\varepsilon(z^\varepsilon) \leqq 0 \Rightarrow -f''_\varepsilon w^\varepsilon \geqq 0.$$

Since $\varepsilon \leqq z_0 \leqq |z^\varepsilon| \leqq 1$, by (4.10),

$$|\nabla \theta^\varepsilon(t, x)| \leqq \frac{\hat{K}}{\sqrt{\varepsilon |z^\varepsilon|}},$$

$$I \geqq \frac{K^*}{\sqrt{\varepsilon^* z^\varepsilon|}}(w^\varepsilon + 1) - K^* \|\theta^\varepsilon\|_{\infty, T} - \frac{2\hat{K}}{\sqrt{\varepsilon |z^\varepsilon|}} W.$$

Hence, there exists a constant $\varepsilon_0 > 0$ such that, on $\{w^\varepsilon \geqq W\}$, $I \geqq 0$ for all $K^* \geqq 2\hat{K}$ and $\varepsilon \in (0, \varepsilon_0]$.
7. Steps 4, 5 and 6 yield

$$I \geqq \sqrt{\varepsilon} K^* K_\varepsilon (W - w^\varepsilon) \quad \text{on } \{w^\varepsilon \geqq W\}$$

for $K^* \geqq 2\hat{K}$ and $\varepsilon \in (0, \varepsilon_0]$. By (A3), $W(0, x) \geqq 1 \geqq |\nabla z_0^\varepsilon|^2$ and therefore the maximum principle implies that $W \geqq w^\varepsilon$ for $\varepsilon \leqq \varepsilon_0$ (see [29, §8] for the details of an application of the maximum principle in a similar situation). Since by construction

$$f_\varepsilon(r) \leqq |r| - f_\varepsilon(1) \leqq |r| + 4,$$

this proves (4.1) for all $\varepsilon \leqq \varepsilon_0$. For $\varepsilon \geqq \varepsilon_0$, (4.1) follows from (4.3). □

The following lemma will be useful in the next section.

Lemma 4.6. *Suppose that there is a bounded, open set $O \subset (0, \infty) \times \mathscr{R}^d$ for which*

$$\beta = \liminf_{\varepsilon \to 0} \inf_{(s,y) \in \bar{O}} |\varphi^\varepsilon(s, y)| > 0.$$

Then for every $(s, y) \in O$,

$$\liminf_{(s',y') \to (s,y) \varepsilon \to 0} |z^\varepsilon(s', y')| \geqq \inf\{|\hat{y} - y| : (s, \hat{y}) \notin O\}.$$

Proof.
1. Since $\bar{O}$ is compact, there is an $\varepsilon_0 > 0$ satisfying

$$|\varphi^\varepsilon(s, y)| \geqq \tfrac{1}{2}\beta \quad \forall (s, y) \in \bar{O}, \ \varepsilon \leqq \varepsilon_0.$$

Since φ^ε is continuous and $h(\varphi^\varepsilon)$ is convergent in L^1_{loc}, either

$$\varphi^\varepsilon(s, y) \geqq \tfrac{1}{2}\beta \quad \forall (s, y) \in \bar{O}, \ \varepsilon \leqq \varepsilon_0 \tag{4.12}$$

or

$$\varphi^\varepsilon(s, y)| \leqq -\tfrac{1}{2}\beta \quad \forall (s, y) \in \bar{O}, \ \varepsilon \leqq \varepsilon_0. \tag{4.13}$$

2. Multiply (ZE) (of Section 2.1) by ε to get

$$2\varphi^\varepsilon(|\nabla z^\varepsilon|^2 - 1) = \varepsilon(-z^\varepsilon_t + \Delta z^\varepsilon) + \varepsilon\theta^\varepsilon. \tag{4.14}$$

In view of (3.3),

$$\lim_{\varepsilon \to 0} \varepsilon \|\theta^\varepsilon\|_{\infty, T} = 0.$$

Set

$$z^*(t, x) = \limsup_{\varepsilon \to 0, (s,y) \to (t,x)} z^\varepsilon(s, y), \quad z_*(t, x) = \liminf_{\varepsilon \to 0, (s,y) \to (t,x)} z^\varepsilon(s, y).$$

Then, as ε approaches zero, (4.14) yields

$$|Dz_*| - 1 \geqq 0 \quad \text{in } O \text{ if (4.12) holds,} \tag{4.15}$$

$$|Dz^*| - 1 \leqq 0 \quad \text{in } O \text{ if (4.13) holds.}$$

These inequalities are to be understood in the viscosity sense [14]; the details of this argument are given in [29, Lemma 4.1].

For $(s, y) \in \bar{O}$ set

$$d(s, y) = \inf\{|y' - y| : (s, y') \notin O\} \text{ if (4.12) holds,}$$

$$d(s, y) = -\inf\{|y' - y| : (s, y') \in O\} \text{ if (4.13) holds.}$$

Then $d(s, y)$ satisfies (4.15) in the viscosity sense and the comparison theorems for the eikonal equation [14] imply that

$$z_*(s, y) \geqq d(s, y) \quad \text{if (4.12) holds,}$$

$$z^*(s, y) \leqq d(s, y) \quad \text{if (4.13) holds.} \quad \square$$

5. Monotonicity Formula

In this section we obtain an extension of the monotonicity formula of CHEN & STRUWE [13], which originates from STRUWE's formula for parabolic flow of harmonic maps [32], and a later result of ILMANEN [23] for the Cahn-Allen equation, which originates from HUISKEN's formula for smooth mean-curvature flows [22].

For $x, x_0 \in \mathscr{R}^d$, $0 \leqq t < t_0$, let

$$\rho(t, x; t_0, x_0) = (4\pi(t_0 - t))^{1/2} G(t_0 - \tau, x_0 - x)$$
$$= (4\pi(t_0 - t))^{-(d-1)/2} \exp\left(-\frac{|x - x_0|^2}{4(t_0 - t)}\right).$$

Then

$$\nabla_x \rho = -\frac{(x - x_0)}{2(t_0 - t)} \rho,$$
$$\rho_t = \left[\frac{d-1}{2(t_0 - t)} - \frac{|x - x_0|^2}{4(t_0 - t)^2}\right] \rho,$$
$$D_x^2 \rho = \left[-\frac{I}{2(t_0 - t)} + \frac{(x - x_0) \otimes (x - x_0)}{4(t_0 - t)^2}\right] \rho,$$

where I is the identity matrix and $\otimes$ is the tensor product. For $t \geqq 0$ and any Borel set $A \subset \mathscr{R}^d$, let $\mu^\varepsilon(t; A)$, $\hat{\mu}^\varepsilon(t; A)$ and ξ^ε be as in §2.2 and (1.7), and set

$$\alpha^\varepsilon(t; t_0, x_0) = \int_{\mathscr{R}^d} \rho(t, x; t_0, x_0) \hat{\mu}^\varepsilon(t; dx).$$

Theorem 5.1. *There is a constant C_d, depending only on the dimension, such that*

$$\frac{d}{dt} \alpha^\varepsilon(t; t_0, x_0) \leqq \frac{1}{2(t_0 - t)} \int_{\mathscr{R}^d} \rho(t, x; t_0, x_0) \xi^\varepsilon(t; dx) + \frac{C_d}{\sqrt{t_0 - t}}. \tag{5.1}$$

Proof. Fix (t_0, x_0). We suppress the dependence on (t_0, x_0) in our notation.

1. By (OPE) and (HE),

$$\frac{d}{dt} \alpha^\varepsilon(t) = \int \rho_t \cdot \hat{\mu}^\varepsilon + \int \rho \left(\varepsilon \nabla \varphi^\varepsilon \cdot \nabla \varphi_t^\varepsilon + \frac{1}{\varepsilon} W'(\varphi^\varepsilon) \varphi_t^\varepsilon + \theta^\varepsilon \theta_t^\varepsilon\right)$$
$$= \int \rho_t \hat{\mu}^\varepsilon - \int \varepsilon \nabla \rho \cdot \nabla \varphi^\varepsilon \varphi_t^\varepsilon$$
$$+ \int \left[\varepsilon \varphi_t^\varepsilon \left(-\Delta \varphi^\varepsilon + \frac{1}{\varepsilon^2} W'(\varphi^\varepsilon)\right) + \theta^\varepsilon (\Delta \theta^\varepsilon - g(\varphi^\varepsilon) \varphi_t^\varepsilon)\right] \rho$$
$$= \int \rho_t \hat{\mu}^\varepsilon - \int \varepsilon \nabla \rho \cdot \nabla \varphi^\varepsilon \varphi_t^\varepsilon - \int \varepsilon (\varphi_t^\varepsilon)^2 \rho + \int \theta^\varepsilon \Delta \theta^\varepsilon \rho =$$

$$= \int \rho_t \hat{\mu}^\varepsilon + \varepsilon \int \left[\nabla\rho \cdot \nabla\varphi^\varepsilon \varphi_t^\varepsilon + \frac{(\nabla\rho \cdot \nabla\varphi^\varepsilon)^2}{\rho} \right]$$

$$- \varepsilon \int \left(\varphi_t^\varepsilon + \frac{\nabla\rho \cdot \nabla\varphi^\varepsilon}{\rho} \right)^2 \rho - \int |\nabla\theta^\varepsilon|^2 \rho + \int \Delta\rho \frac{1}{2} (\theta^\varepsilon)^2.$$

Since

$$\rho_t + \Delta\rho = -\frac{1}{2(t_0 - t)} \rho, \quad \hat{\mu}^\varepsilon = \mu^\varepsilon + \frac{1}{2} (\theta^\varepsilon)^2,$$

it follows that

$$\frac{d}{dt} \alpha^\varepsilon(t) = -\varepsilon \int \left(\varphi_t^\varepsilon + \frac{\nabla\rho \cdot \nabla\varphi^\varepsilon}{\rho} \right)^2 \rho$$

$$+ \int \left[\rho_t \mu^\varepsilon + \nabla\rho \cdot \nabla\varphi^\varepsilon \left(\varepsilon \Delta\varphi^\varepsilon - \frac{1}{\varepsilon} W'(\varphi^\varepsilon) \right) + \varepsilon \frac{(\nabla\rho \cdot \nabla\varphi^\varepsilon)^2}{\rho} \right]$$

$$- \int \rho \left[|\nabla\theta^\varepsilon|^2 + \frac{1}{4(t_0 - t)} (\theta^\varepsilon)^2 \right] + \int \nabla\rho \cdot \nabla\varphi^\varepsilon g(\varphi^\varepsilon) \theta^\varepsilon.$$

2. Let $\nu = \nu^\varepsilon$ be as in §2.3, i.e.,

$$\nu = \frac{\nabla\varphi^\varepsilon}{|\nabla\varphi^\varepsilon|}$$

for $|\nabla\varphi^\varepsilon| \neq 0$. Set

$$\hat{T} = \varepsilon \left(\nu \otimes \nu - \frac{1}{2} I \right) |\nabla\varphi^\varepsilon|^2 - \frac{1}{\varepsilon} W(\varphi^\varepsilon) I.$$

Then

$$\hat{T} = \varepsilon \nabla\varphi^\varepsilon \otimes \nabla\varphi^\varepsilon - \left(\frac{\varepsilon}{2} |\nabla\varphi^\varepsilon|^2 + \frac{1}{\varepsilon} W(\varphi^\varepsilon) \right) I,$$

$$\sum_{i=1}^{d} \frac{\partial}{\partial x_i} \hat{T}_{ij} = \varphi_{x_j}^\varepsilon \left(\varepsilon \Delta\varphi^\varepsilon - \frac{1}{\varepsilon} W'(\varphi^\varepsilon) \right).$$

Since $\xi^\varepsilon + \mu^\varepsilon = \varepsilon |\nabla\varphi^\varepsilon|^2 \, dx$,

$$\hat{T} \, dx = (\nu \otimes \nu) \xi^\varepsilon - (I - \nu \otimes \nu) \mu^\varepsilon.$$

Let k be the second term appearing in the expression at the end of Step 1:

$$k = \int \left[\rho_t \mu^\varepsilon + \nabla\rho \cdot \nabla\varphi^\varepsilon \left(\varepsilon \Delta\varphi^\varepsilon - \frac{1}{\varepsilon} W'(\varphi^\varepsilon) \right) + \varepsilon \frac{(\nabla\rho \cdot \nabla\varphi^\varepsilon)^2}{\rho} \right].$$

Integration by parts and the identity $\xi^\varepsilon + \mu^\varepsilon = \varepsilon |\nabla\varphi^\varepsilon|^2 \, dx$ yield

$$k = \int \rho_t \mu^\varepsilon - D^2\rho : \hat{T} \, dx + \frac{(\nabla\rho \cdot \nu)^2}{\rho} (\xi^\varepsilon + \mu^\varepsilon)$$

$$= \int \left[\rho_t + D^2\rho : (I - \nu \otimes \nu) + \frac{(\nabla\rho \cdot \nu)^2}{\rho} \right] \mu^\varepsilon + \int \left[\frac{(\nabla\rho \cdot \nu)^2}{\rho} - D^2\rho : \nu \otimes \nu \right] \xi^\varepsilon,$$

where $M:N = \operatorname{trace} MN$, for symmetric matrices M, N. Explicit formulae for the derivatives of ρ imply that, for any unit vector v,

$$\rho_t + D^2\rho:(I - v\otimes v) + \frac{(\nabla\rho\cdot v)^2}{\rho} = 0,$$

$$\frac{(\nabla\rho\cdot v)^2}{\rho} - D^2\rho:v\otimes v = \frac{\rho}{2(t_0 - t)}.$$

Hence

$$k = \int \frac{\rho}{2(t_0 - t)}\,\xi^\varepsilon(t; dx).$$

3. Recall that $H^\varepsilon = h(\varphi^\varepsilon)$ and $\nabla H^\varepsilon = \nabla\varphi^\varepsilon g(\varphi^\varepsilon)$. By an integration by parts,

$$\int \nabla\rho\cdot\nabla\varphi^\varepsilon g(\varphi^\varepsilon)\theta^\varepsilon = -\int [\Delta\rho H^\varepsilon\theta^\varepsilon + H^\varepsilon\nabla\rho\cdot\nabla\theta^\varepsilon].$$

In view of Steps 1 and 2,

$$\frac{d}{dt}\alpha^\varepsilon(t) \leqq \frac{1}{2(t_0 - t)}\int \rho\,\xi^\varepsilon(t; dx) + I + J,$$

where

$$I = -\int [\rho|\nabla\theta^\varepsilon|^2 + H^\varepsilon\nabla\rho\cdot\nabla\theta^\varepsilon],$$

$$J = -\int\left[\frac{\rho}{4(t_0 - t)}(\theta^\varepsilon)^2 + \Delta\rho H^\varepsilon\theta^\varepsilon\right].$$

4. Since $|\varphi^\varepsilon| < 1$, $|H^\varepsilon| \leqq \frac{2}{3} < 1$. Hence

$$\begin{aligned} I &= -\int \rho\left|\nabla\theta^\varepsilon + H^\varepsilon\frac{\nabla\rho}{2\rho}\right|^2 + \frac{1}{4}\int\frac{|\nabla\rho|^2}{\rho}|H^\varepsilon|^2 \\ &\leqq \frac{1}{4}\int \frac{|x - x_0|^2}{4(t_0 - t)^2}\rho(t, x; t_0, x_0)\; dx \\ &= \frac{1}{2\sqrt{t_0 - t}}\int_{\mathbb{R}^d} (\pi)^{-(d-1)/2}|y|^2 e^{-|y|^2}\,dy. \end{aligned}$$

5. To estimate J, we observe that

$$\begin{aligned} |\Delta\rho H^\varepsilon\theta^\varepsilon| &\leqq \left[\frac{d-1}{2(t_0 - t)} + \frac{|x - x_0|^2}{4(t_0 - t)^2}\right]|H^\varepsilon||\theta^\varepsilon|\rho \\ &= \frac{\rho}{2(t_0 - t)}\left[d - 1 + \frac{|x - x_0|^2}{2(t_0 - t)}\right]|H^\varepsilon||\theta^\varepsilon|. \end{aligned}$$

Since

$$2|H^\varepsilon||\theta^\varepsilon| \leqq \left[d - 1 + \frac{|x_0 - x|^2}{2(t_0 - t)}\right]^{-1}(\theta^\varepsilon)^2 + \left[d - 1 + \frac{|x_0 - x|^2}{2(t_0 - t)}\right]|H^\varepsilon|^2,$$

we have

$$|\Delta\rho H^\varepsilon\theta^\varepsilon| \leqq \frac{(\theta^\varepsilon)^2\rho}{4(t_0-t)} + C(t,x)\rho,$$

where

$$C(t,x) = \frac{1}{4(t_0-t)}\left[d-1+\frac{|x_0-x|^2}{2(t_0-t)}\right]^2.$$

Hence

$$J \leqq \int C(t,x)\rho dx$$

$$= \frac{1}{2\sqrt{t_0-t}}\int_{\mathscr{R}^d}(\pi)^{-(d-1)/2}(d-1+2|y|^2)e^{-|y|^2}dy.$$

6. Combining the previous steps, we obtain (5.1) with

$$C_d = \tfrac{1}{2}(\pi)^{(1-d)/2}\int_{\mathscr{R}^d}[|y|^2+(d-1+2|y|^2)^2]e^{-|y|^2}dy. \quad \square$$

Remark 5.1. Suppose that in the heat equation (1.1), $\lambda = c\varepsilon$ with some constant $c > 0$. Then (HE) in §2.1 takes the form

$$\theta_t^\varepsilon - c\Delta\theta^\varepsilon + g(\varphi^\varepsilon)\varphi_t^\varepsilon = 0.$$

This change does not affect the results of the preceeding sections. However, for $c \neq 1$, the monotonicity formula (5.1) has to be modified: For any $\beta > 0$, there is a constant $C_{d,\beta}$ such that

$$\frac{d}{dt}\alpha^\varepsilon \leqq \frac{1}{2(t_0-t)}\int \rho\, d\xi^\varepsilon + \frac{C_{d,\beta}}{(t_0-t)^{1/2+\beta}}.$$

Since the new error term $E(t) = (t_0-t)^{-1/2-\beta}$ is integrable over $(0, t_0)$, the main results of this paper and their proof remain unchanged.

The proof of the modified monotonicity result is very similar to that of (5.1). Indeed, in Step 1 we now have an additional error term

$$L := \frac{1-c}{2}\int \rho_t(\theta_\varepsilon)^2\,dx.$$

For any $p > 1$, Hölder's inequality and (3.10) yield

$$|L| \leqq \frac{|1-c|}{c}\,\|\rho_t\|_q\,\|(\theta_\varepsilon(t,\cdot))^2\|_p \leqq K_{p,c}(t_0-t)^{-1/2-d/2p}.$$

For a given $\beta > 0$, choose $p = d/2\beta$. $\square$

The monotonicity formula and the gradient estimate (4.1) yield

Corollary 5.2. *For any $T > 0$, there exists a constant $K = K(T)$ such that, for any $x_0 \in \mathscr{R}^d$ and $0 \leqq t \leqq r < t_0 \leqq T$,*

$$\alpha^\varepsilon(r; t_0, x_0) \leqq \alpha^\varepsilon(t; t_0, x_0)\left(\frac{t_0-t}{t_0-r}\right)^{K\sqrt{\varepsilon}} + K\int_t^r\left(\frac{t_0-\tau}{t_0-r}\right)^{K\sqrt{\varepsilon}}\frac{d\tau}{\sqrt{t_0-\tau}}. \quad (5.2)$$

Moreover, as ε tends to zero,

$$\alpha(r; t_0, x_0) \leqq \alpha(t; t_0; x_0) + C_d[\sqrt{t_0 - t} - \sqrt{t_0 - r}]. \tag{5.3}$$

Proof. Since

$$\xi^\varepsilon(t; dx) = \frac{1}{2\varepsilon}\left(q'\left(\frac{z^\varepsilon}{\varepsilon}\right)\right)^2 (|\nabla z^\varepsilon|^2 - 1),$$

(4.1) and (5.1) yield

$$\frac{d}{dt}\alpha^\varepsilon(t) \leqq \frac{1}{2(t_0 - t)} \int \rho \frac{1}{2\varepsilon}\left(q'\left(\frac{z^\varepsilon}{\varepsilon}\right)\right)^2 K^* \sqrt{\varepsilon}(1 + |z^\varepsilon|) + \frac{C_d}{\sqrt{t_0 - t}}.$$

Observe that

$$\frac{1}{2\varepsilon}\left(q'\left(\frac{z^\varepsilon}{\varepsilon}\right)\right)^2 dx \leqq \frac{1}{2\varepsilon}\left(q'\left(\frac{z^\varepsilon}{\varepsilon}\right)\right)^2 [1 + |\nabla z^\varepsilon|^2]\, dx = \mu^\varepsilon(t; dx),$$

$$\left(q'\left(\frac{z^\varepsilon}{\varepsilon}\right)\right)^2 |z^\varepsilon|^2 \leqq \varepsilon^2 \left(\sup_{r \geq 0} q'(r) r\right)^2 \leqq 4\varepsilon^2.$$

Hence

$$\frac{1}{2\varepsilon}\left(q'\left(\frac{z^\varepsilon}{\varepsilon}\right)\right)^2 (1 + |z^\varepsilon|)\, dx \leqq \frac{1}{2\varepsilon}\left(q'\left(\frac{z^\varepsilon}{\varepsilon}\right)\right)^2 \left(\frac{3}{2} + \frac{1}{2}|z^\varepsilon|^2\right) dx \leqq \frac{3}{2}\mu^\varepsilon(t; dx) + \varepsilon\, dx$$

and consequently

$$\begin{aligned}
\frac{d}{dt}\alpha^\varepsilon(t) &\leqq \frac{K\sqrt{\varepsilon}}{(t_0 - t)} \int \rho(\mu^\varepsilon + \varepsilon\, dx) + \frac{C_d}{\sqrt{t_0 - t}} \\
&\leqq \frac{K\sqrt{\varepsilon}}{(t_0 - t)} \alpha^\varepsilon(t) + \frac{K\varepsilon\sqrt{\varepsilon}}{t_0 - t} \int \rho\, dx + \frac{C_d}{\sqrt{t_0 - t}} \\
&\leqq \frac{K\sqrt{\varepsilon}}{(t_0 - t)} \alpha^\varepsilon(t) + \frac{K}{\sqrt{t_0 - t}}.
\end{aligned}$$

Now an application of Gronwall's inequality yields (5.2). □

6. Clearing-Out

In this section we follow the proof of [29, Theorem 5.1] to prove an extension of the clearing-out lemma established in [23, 29].

Theorem 6.1. *For every $T > 0$, there are positive constants η, $t^* > 0$, depending on T, such that if*

$$\int \rho(t, x; t_0, x_0)\mu(t; dx) \leqq \eta \tag{6.1}$$

for some t, t_0, x_0 satisfying

$$(t_0 - t^*) \leqq t < t_0 \leqq T, \tag{6.2}$$

then there exists a neighborhood O of (t_0, x_0) such that

$$\lim_{(s', y') \to (s, y) \varepsilon \to 0} |z^\varepsilon(s', y')| > 0, \quad \forall (s, y) \in O. \tag{6.3}$$

In particular,

$$(t_0, x_0) \notin \overline{\bigcup_{t \geqq 0} \{t\} \times \operatorname{spt} \mu(t, \cdot)}. \tag{6.4}$$

Proof. Fix t, t_0, x_0. Suppose that (6.1), (6.2) hold with some η, t^* that will be chosen later.
1. Hölder's inequality yields

$$\begin{aligned}\int \rho(t, x; t_0, x_0)(\theta^\varepsilon(t, x))^2\, dx &\leqq \|\rho(t, \cdot; t_0, x_0)\|_{p'} \|(\theta^\varepsilon(t, \cdot))^2\|_p \\ &= (4\pi(t_0 - t))^{1/2} \|G(t_0 - t, \cdot)\|_{p'} \|\theta^\varepsilon(t, \cdot))^2\|_p\end{aligned}$$

for any $1 \leqq p \leqq \infty$, where p' is the conjugate of p: $\frac{1}{p} + \frac{1}{p'} = 1$. Since

$$\|G(\tau, \cdot)\|_{p'} \leqq K(p) \tau^{-d/2p},$$

we have

$$\int \rho(t, x; t_0, x_0)(\theta^\varepsilon(t, x))^2\, dx \leqq \hat{K}(p)(t_0 - t)^{1/2(1 - d/p)} \|(\theta^\varepsilon(t, \cdot))^2\|_p.$$

Choose $p = d + 1$ and use (3.10) to obtain

$$\int \rho(t, x; t_0, x_0)(\theta^\varepsilon(t, x))^2\, dx \leqq K^*(t_0 - t)^\gamma, \quad 0 < \varepsilon \leqq 1,$$

for some constants K^* and $\gamma > 0$.
2. The continuity of ρ, the convergence of μ^ε to μ and (2.2) imply that there are a constant $\varepsilon_0 > 0$ and a neighborhood U of (t_0, x_0) such that for all $\varepsilon \leqq \varepsilon_0$ and $(s, y) \in U$,

$$t + \varepsilon^2 < \frac{t + t_0}{2} < s, \tag{6.5}$$

$$\int \rho(t, x; s, y) \mu^\varepsilon(t; dx) \leqq 2\eta.$$

Step 1 yields

$$\begin{aligned}\alpha^\varepsilon(t; s, y) &= \int \rho(t, x; s, y)[\mu^\varepsilon(t; dx) + \tfrac{1}{2}(\theta^\varepsilon(t, x))^2\, dx] \\ &\leqq 2\eta + \tfrac{1}{2} K^*(s - t)^\gamma, \quad (s, y) \in U, \ \varepsilon \leqq \varepsilon_0.\end{aligned}$$

Note that ε_0 may depend on η, t and U.

3. Since $s-\varepsilon^2 > t$, we may use (5.2) with $(t_0, x_0) = (s, y)$ and $r = s - \varepsilon^2$ to obtain

$$\alpha^\varepsilon(s-\varepsilon^2; s, y) \leqq \left(\frac{s-t}{\varepsilon^2}\right)^{K\sqrt{\varepsilon}} \alpha^\varepsilon(t; s, y) + K \int_t^{s-\varepsilon^2} \left(\frac{s-\tau}{\varepsilon^2}\right)^{K\sqrt{\varepsilon}} \frac{d\tau}{\sqrt{s-\tau}}$$

$$\leqq \left(\frac{s-t}{\varepsilon^2}\right)^{K\sqrt{\varepsilon}} \left[\alpha^\varepsilon(t; s, y) + K \int_t^{s-\varepsilon^2} \frac{d\tau}{\sqrt{s-\tau}} d\tau\right]$$

$$\leqq \left(\frac{s-t}{\varepsilon^2}\right)^{K\sqrt{\varepsilon}} [\alpha^\varepsilon(t; s, y) + 2K\sqrt{s-t}].$$

As ε approaches to zero, $\varepsilon^{-2K\sqrt{\varepsilon}}$ converges to 1 and therefore, by (6.5), there is a constant $0 < \hat{\varepsilon}_0 \leqq \varepsilon_0$ that satisfies

$$\left(\frac{s-t}{\varepsilon^2}\right)^{K\sqrt{\varepsilon}} \leqq 2, \quad \varepsilon \leqq \hat{\varepsilon}_0, \ (s, y) \in U.$$

Then, by Step 2,

$$\alpha^\varepsilon(s-\varepsilon^2; s, y) \leqq 4\eta + K^*(s-t)^\gamma + 4K\sqrt{s-t}$$

for all $(s, y) \in U$ and $\varepsilon \leqq \hat{\varepsilon}_0$. Set

$$\hat{U} = U \cap (t_0 - t^*, t_0 + t^*) \times \mathscr{R}^d,$$

so that, for any $(s, y) \in \hat{U}$ and t, t_0 satisfying (6.2), $(s-t) \leqq 2t^*$, and consequently for an appropriately chosen $t^* = t^*(\eta)$,

$$K^*(s-t)^\gamma + 4K\sqrt{s-} \leqq K^*(2t^*)^\gamma + 4K\sqrt{2t^*} \leqq \eta.$$

Therefore

$$\alpha(s-\varepsilon^2, s, y) \leqq 5\eta, \quad (s, y) \in \hat{U}, \ \varepsilon \leqq \hat{\varepsilon}_0.$$

Recall that this estimate is obtained under the assumption that (6.1) holds with t, t_0 satisfying (6.2) with $t^* = t^*(\eta)$ and that we have not chosen η yet.

4. Let $B_\varepsilon(y)$ be the sphere centered at y with radius ε. For any $x \in B_\varepsilon(y)$,

$$\rho(s-\varepsilon^2, x; s, y) = (4\pi\varepsilon^2)^{-(d-1)/2} \exp\left(-\frac{|x-y|^2}{4\varepsilon^2}\right)$$

$$\geqq [(4\pi)^{-(d-1)/2} e^{-1/4}] \varepsilon^{-(d-1)} = (K_* \varepsilon^{d-1})^{-1}$$

for some constant K_*. Therefore

$$\hat{\mu}^\varepsilon(s-\varepsilon^2; B_\varepsilon(y)) \leqq \left[\min_{x \in B_\varepsilon(y)} \rho(s-\varepsilon^2, x; s, y)\right]^{-1} \alpha^\varepsilon(s-\varepsilon^2; s, y)$$
$$\leqq 5K_* \eta \varepsilon^{d-1} \quad \forall (s, y) \in \hat{U}, \ \varepsilon \leqq \hat{\varepsilon}_0. \tag{6.6}$$

5. Define

$$\beta = \liminf_{\varepsilon \to 0} \inf_{(s,y) \in \hat{U}} |\varphi^\varepsilon(s, y)|.$$

In this step we show that for a carefully chosen η, we have $\beta \geqq \frac{7}{8}$.

Suppose that $\beta < \frac{7}{8}$. Then there are $\varepsilon_n \to 0$ and $(s_n, y_n) \in \hat{U}$ satisfying

$$|\varphi^{\varepsilon_n}(s_n - \varepsilon_n^2, y_n)| < \tfrac{7}{8} \Rightarrow |z^{\varepsilon_n}(s_n - \varepsilon_n^2, y_n)| < \varepsilon_n q^{-1}(\tfrac{7}{8}).$$

Using (4.1) we construct K_0 and n_0, independent of η, such that

$$|z^{\varepsilon_n}(s_n - \varepsilon_n^2, x)| < \varepsilon_n[q^{-1}(\tfrac{7}{8}) + K_0] \quad \forall x \in B_{\varepsilon_n}(y_n), \quad n \geqq n_0,$$

and therefore

$$W(\varphi^{\varepsilon_n}(s_n - \varepsilon_n^2, x)) > W(q(q^{-1}(\tfrac{7}{8}) + K_0)) \quad \forall x \in B_{\varepsilon_n}(y_n), \quad n \geqq n_0.$$

Hence, for $n \geqq n_0$,

$$\mu^{\varepsilon_n}(s_n - \varepsilon_n^2; B_{\varepsilon_n}(y_n)) \geqq \int_{B_{\varepsilon_n}(y_n)} \frac{1}{\varepsilon_n} W(\varphi^{\varepsilon_n}(\varepsilon_n - \varepsilon_n^2, x))\,dx$$

$$> w_d W(q(q^{-1}(\tfrac{7}{8}) + K_0))(\varepsilon_n)^{d-1},$$

where w_d is the volume of the d-dimensional unit sphere. Now choose

$$\eta = \frac{w_d}{5K_*} W(q(q^{-1}(\tfrac{7}{8}) + K_0)), \tag{6.7}$$

where K_* is the constant appearing in (6.6). With this choice of η, (6.7) contradicts (6.6). Hence $\beta \geqq \frac{7}{8}$.

In the foregoing discussion we have established the following: If (6.1) and (6.2) hold for some t, t_0 with η as in (6.7) and t^* as in Step 3, then there exists a neighborhood $\hat{U}$ of (t_0, x_0) such that

$$\beta = \liminf_{\varepsilon \to 0} \inf_{(s,y) \in \hat{U}} |\varphi^\varepsilon(s, y)| \geqq \tfrac{7}{8}.$$

Now, by Lemma 4.6, (6.3) holds on any open set O satisfying $\bar{O} \subset \hat{U}$ and

$$\liminf_{\varepsilon \to 0} \inf_{(s,y) \in \bar{O}} |z^\varepsilon(s, y)| > 0.$$

Then (4.1) yields

$$\mu^\varepsilon(\bar{O}) = \int\int_{\bar{O}} \tfrac{1}{2\varepsilon}(q'(\tfrac{z^\varepsilon}{\varepsilon}))^2(|\nabla z^\varepsilon|^2 + 1)\,dx\,dt$$

$$\leqq \int\int_{\bar{O}} \tfrac{1}{2\varepsilon}(q'(\tfrac{z^\varepsilon}{\varepsilon}))^2(2 + K\sqrt{\varepsilon}(1 + |z^\varepsilon|))\,dx\,dt,$$

and therefore $\mu^\varepsilon(\bar{O})$ converges to zero as ε tends to zero, proving (6.4). □

7. Dimension of Γ and Equipartition of Energy

Let $\Lambda(t) \subset \mathscr{R}^d$ be the support of $\mu(t;\cdot)$ and $\Gamma \subset \bar{Q}$ be the support of $d\mu = \mu(t; dx)\,dt$. Then

$$\Gamma \subset \overline{\bigcup_{t \geq 0} \{t\} \times \Lambda(t)}.$$

Suppose that $(t_0, x_0) \notin \Gamma$. Then there is a neighborhood U of (t_0, x_0) such that $U \cap \Gamma = \emptyset$. Therefore

$$\lim_{t \uparrow t_0} \int \rho(t, x; t_0, x_0)\mu(t; dx) = 0$$

and, by Theorem 6.1, (t_0, x_0) satisfy (6.4). Hence

$$\Gamma = \overline{\bigcup_{t \geq 0} \{t\} \times \Lambda(t)}.$$

Let Γ_t be the t-section of Γ. In this section we first estimate the Hausdorff dimension of Γ_t (cf. [17]). Then we show that the discrepancy measure ξ^ε, defined in (1.7), converges to zero, hence proving the equipartition of energy. Our arguments closely follow Sections 6, 7 and 8 in [23].

The next theorem follows from [34, Theorem 5.12.4.].

Theorem 7.1. *Let μ be a positive Borel measure satisfying*

$$M(\mu) = \sup_{x \in \mathscr{R}^d,\, R > 0} \frac{\mu(B_R(x))}{R^{d-1}} < \infty.$$

Then there is a constant K_d, depending only on the dimension d but not on μ, such that

$$\left| \int \varphi(x)\mu(dx) \right| \leq K_d M(\mu) \| \nabla\varphi \|_1 \quad \forall \varphi \in C_c^\infty(\mathscr{R}^d).$$

We continue with an estimate of the dimension of the interface.

Proposition 7.2. *For every $T > 0$ there is $K(T) > 0$ such that*

$$\mu^\varepsilon(r; B_R(x)) \leq K(T) R^{d-1}, \tag{7.1}$$

$$\mathscr{H}^{d-1}(\Gamma_r) \leq K(T), \tag{7.2}$$

for all $0 < \varepsilon \leq 1$, $R > 0$ and $0 \leq r \leq T$.

Proof.

1. Theorem 7.1, (A5) and (A6) imply that

$$\begin{aligned}
\alpha^\varepsilon(0; t_0, x_0) &= \int \rho(0, x; t_0, x_0)\left[\mu^\varepsilon(0; dx) + \tfrac{1}{2}(\theta_0^\varepsilon(x))^2\,dx\right] \\
&\leq K\left[\| \nabla_x \rho(0, \cdot; t_0, x_0) \|_1 + \| \rho(0, \cdot; t_0, x_0) \|_1\right] \\
&\leq K(\sqrt{t_0} + 1)
\end{aligned}$$

for some constant K, independent of ε.

2. If $R \geqq (C_1^*)^{1/(d-1)} = R_0$, then the energy estimate (2.2) yields

$$\mu^\varepsilon(r; B_R(x)) \leqq \hat{\mu}^\varepsilon(r; R^d) \leqq C_1^* \leqq R^{d-1}.$$

Hence (7.1) holds for all $r \geqq 0$ and $R \geqq R_0$ with constant $K(T) = 1$.
3. Fix $0 \leqq r \leqq T$, $\varepsilon \leqq R \leqq R_0$ and $x_0 \in \mathscr{R}^d$. Then for $t_0 > r$,

$$\begin{aligned}\mu^\varepsilon(r; B_R(x_0)) &\leqq \hat{\mu}^\varepsilon(r; B_R(x_0)) \\ &\leqq \left[\inf_{x \in B_R(x_0)} \rho(r, x; t_0, x_0)\right]^{-1} \alpha^\varepsilon(r; t_0, x_0) \qquad (7.3) \\ &= (4\pi(t_0 - r))^{(d-1)/2} \exp\left(\frac{R^2}{4(t_0 - r)}\right) \alpha^\varepsilon(r; t_0, x_0).\end{aligned}$$

Choose $t_0 = r + R^2$ so that $t_0 \leqq T + R_0^2 = T_*$ and, by Step 1 and (5.2),

$$\begin{aligned}\alpha^\varepsilon(r; t_0, x_0) &\leqq \alpha^\varepsilon(0; t_0, x_0)\left(\frac{t_0}{R^2}\right)^{K\sqrt{\varepsilon}} + K\int_0^r \left(\frac{t_0 - \tau}{R^2}\right)^{K\sqrt{\varepsilon}} \frac{d\tau}{\sqrt{t_0 - \tau}} \\ &\leqq K\left(\frac{t_0}{R^2}\right)^{K\sqrt{\varepsilon}} (\sqrt{t_0} + 1).\end{aligned}$$

Since $R \geqq \varepsilon$, there is a constant $K = K(T)$ satisfying

$$\alpha^\varepsilon(r; r + R^2, x_0) \leqq K, \quad 0 < \varepsilon \leqq 1, \quad r \leqq T.$$

Then (7.3) implies that

$$\mu^\varepsilon(r; B_R(x_0)) \leqq (4\pi)^{(d-1)/2} e^{1/4} K R^{d-1}, \quad \varepsilon \leqq R \leqq R_0;$$

hence (7.1) holds for all $R \geqq \varepsilon$.
4. In this step we study the case $0 < R \leqq \varepsilon$. The inequality (3.2) yields

$$\mu^\varepsilon(r; B_R(x_0)) = \int_{B_R(x_0)} \frac{\varepsilon}{2} |\nabla \varphi^\varepsilon|^2 + \frac{1}{\varepsilon} W(\varphi^\varepsilon) \leqq \frac{K}{\varepsilon} |B_R(x_0)| = \frac{K}{\varepsilon} R^d$$

for $0 \leqq r \leqq T$. Since $R \leqq \varepsilon$, $R^d \varepsilon^{-1} \leqq R^{d-1}$, this completes the proof of (7.1) for all R.
5. The inequality (7.2) follows from Theorem 6.1, (7.1) and an application of the Besicovitch covering theorem (see the proof of [23, §6.3]). □

In the remainder of this section we prove that ξ^ε converges to zero. Our proof is a direct modification of Sections 7 and 8 in [23].

Let η be as in Theorem 6.1. Define

$$Z^- = \left\{(t, x) \in \Gamma \cap [0, T] \times \mathscr{R}^d : \sup_{s \downarrow t} \int \rho(t, x; s, y) \mu(s; dy) < \eta\right\}.$$

Then Section 7 in [23] implies that for any $\delta > 0$,

$$\mathscr{H}^{d-2+\delta}(Z_t^-) = 0 \quad \text{for almost every } t \in [0, T]. \qquad (7.4)$$

Let ξ^ε be as in Section 5. For a Borel set $A \subset [0, T] \times \mathscr{R}^d$ define

$$\xi^\varepsilon(A) = \int_A \xi^\varepsilon(t; dx)\, dt.$$

Since $|\xi^\varepsilon| \leqq \mu^\varepsilon$, by passing to a further subsequence we assume that ξ^ε converges to a Borel measure ξ in the weak* topology of Radon measures.

Proposition 7.3. $\xi = 0$.

Proof.
1. For $\varepsilon > 0$ and any Borel set $A \subset \bar{Q}$, let

$$v^\varepsilon(A) = \int_A \frac{1}{2\varepsilon}\left(q'\left(\frac{z^\varepsilon}{\varepsilon}\right)\right)^2 (|\nabla z^\varepsilon|^2 - 1)^+ \, dx\, dt,$$

$$\lambda^\varepsilon(A) = \int_A \frac{1}{2\varepsilon}\left(q'\left(\frac{z^\varepsilon}{\varepsilon}\right)\right)^2 (|\nabla z^\varepsilon|^2 - 1)^- \, dx\, dt$$

where for any real number b, $(b)^+ = \max\{b, 0\}$, $b^- = \max\{-b, 0\}$. Then

$$\xi^\varepsilon = v^\varepsilon - \lambda^\varepsilon.$$

2. Equation (4.1) and the proof of Corollary 5.2 imply that

$$v^\varepsilon(A) \leqq K\sqrt{\varepsilon}[\mu^\varepsilon(A) + \varepsilon|A|], \quad 0 < \varepsilon \leqq 1 \tag{7.5}$$

for any $A \subset [0, T] \times \mathscr{R}^d$. Hence v^ε converges to zero and λ^ε converges to $-\xi$.
3. Fix $(s, y) \in [0, \infty) \times \mathscr{R}^d$ and $0 < \sigma \leqq s$. Integrate (5.1) on $[0, s - \sigma]$. Using (7.5) and the exponential decay of ρ, we let ε go to zero to obtain

$$\alpha(s - \sigma; s, y) - \alpha(0; s, y) \leqq - \int_0^{s-\sigma} \int_{\mathscr{R}^d} \frac{1}{2(s-t)} \rho(t, x; s, y)\, d\lambda(t, x) + 2C_d(\sqrt{s} - \sqrt{\sigma}).$$

This inequality and Step 1 of Proposition 7.2 yield

$$\int_0^{s-\sigma} \int_{\mathscr{R}^d} \frac{1}{2(s-t)} \rho(t, x; s, y)\, d\lambda(t, x) \leqq K(\sqrt{s} + 1).$$

Fix $T > 0$ and integrate this inequality against $\mu(s; dy)\, ds$ and then use (2.2); the result is

$$\int_0^{T+1} \int_{\mathscr{R}^d} \int_0^{s-\sigma} \int_{\mathscr{R}^d} \frac{1}{2(s-t)} \rho(t, x; s, y)\, d\lambda(t, x) \mu(s; dx)\, ds$$

$$\leqq \int_0^{T+1} \int_{\mathscr{R}^d} K(\sqrt{s} + 1) \mu(s; dy)\, ds \leqq \hat{C}(T)$$

for some constant $\hat{C}(T)$ depending on T.

4. Fubini's theorem and the monotone convergence theorem enable us to send σ to zero to obtain

$$\int_0^{T+1}\int_{\mathscr{R}^d}\int_t^{T+1}\int_{\mathscr{R}^d}\frac{1}{2(s-t)}\rho(t,x;s,y)\mu(s;dy)\,ds\,d\lambda(t,x)\leqq\hat{C}(T).$$

Hence

$$\int_t^{t+1}\frac{1}{2(s-t)}\int_{\mathscr{R}^d}\rho(t,x;s,y)\mu(s;dy)\,ds\leqq C(x,t)<\infty \tag{7.6}$$

for λ almost every $(t,x)\in[0,T]\times\mathscr{R}^d$.

5. Fix (t,x) such that (7.6) holds. For $s\in(t,t+1]$ define

$$\beta=\ln(s-t),\quad h(s)=\int_{\mathscr{R}^d}\rho(t,x;s,y)\mu(s;dy).$$

Then (7.6) implies that

$$\int_{-\infty}^{0} h(t+e^{\beta})\,d\beta<\infty. \tag{7.7}$$

We wish to prove that

$$\lim_{s\downarrow t} h(s)=0.$$

Clearly (7.7) implies that $h(t+e^{\beta})$ converges to zero on a subsequence. We now use the monotonicity of h to prove convergence on the whole sequence.

6. Following [23], for $\gamma\in(0,1]$ we choose a decreasing sequence $\beta_i\to-\infty$ such that

$$|\beta_{i+1}-\beta_i|\leqq\gamma,\quad h(t+e^{\beta_i})\leqq\gamma.$$

Then, for any $\beta\in[\beta_i,\beta_{i-1})$,

$$\begin{aligned}h(t+e^{\beta})&=\int\rho(t,x;t+e^{\beta},y)\mu(t+e^{\beta};dy)\\&=\int\rho(t+e^{\beta},x;t+2e^{\beta},y)\mu(t+e^{\beta};dy)\\&=\alpha(t+e^{\beta};t+2e^{\beta},x).\end{aligned}$$

Use (5.3) to obtain

$$\begin{aligned}h(t+e^{\beta})&\leqq\alpha(t+e^{\beta_i};t+2e^{\beta},x)+C_d[\sqrt{2e^{\beta}-e^{\beta_i}}-\sqrt{e^{\beta}}]\\&\leqq\alpha(t+e^{\beta_i};t+2e^{\beta},x)+C_d\sqrt{2e^{\beta}},\end{aligned} \tag{7.8}$$

and the preceding identity with $\beta=\beta_i$ yields

$$\gamma\geqq h(t+e^{\beta_i})=\alpha(t+e^{\beta_i};t+2e^{\beta_i},x). \tag{7.9}$$

7. We assert that for any $\delta > 0$ there is a $\gamma(\delta, T) > 0$ satisfying

$$\alpha(t_0; t_0 + R_1, x) \leqq (1+\delta)\alpha(t_0; t_0 + R_0, x) + \delta \tag{7.10}$$

for all $0 \leqq t_0 \leqq T+1$, $x \in \mathscr{R}^d$ and $0 \leqq R_0 \leqq R_1 \leqq (\gamma(\delta)+1)R_0$. This result follows from (7.1) and it is stated in [23, Lemma 3.4(iv)]. We postpone the elementary proof of (7.10) to the next step and complete the proof of the Proposition.

Set

$$t_0 = t + e^{\beta_i}, \quad R_1 = 2e^{\beta} - e^{\beta_i}, \quad R_0 = e^{\beta_i}$$

so that

$$\frac{R_1}{R_0} = \sqrt{2e^{\beta-\beta_i} - 1} \leqq \sqrt{2[e^{\beta-\beta_i} - 1] + 1} \leqq 1 + K\gamma$$

for some constant K. So if $K\gamma \leqq \gamma(\delta)$, then (7.10) holds and, by (7.8) and (7.9),

$$\begin{aligned} h(t + e^{\beta}) &\leqq \alpha(t + e^{\beta_i}; t + 2e^{\beta}, x) + C_d\sqrt{2e^{\beta}} \\ &\leqq (1+\delta)\alpha(t + e^{\beta_i}; t + 2e^{\beta_i}, x) + \delta + C_d\sqrt{2e^{\beta}} \\ &= (1+\delta)h(t + e^{\beta_i}) + \delta + C_d\sqrt{2e^{\beta}} \\ &\leqq (1+\delta)\gamma + \delta + C_d\sqrt{2e^{\beta}} \end{aligned}$$

for all $\delta > 0$ and $0 < \gamma \leqq \gamma_0(\delta)$. Now pass to the limit $i \to \infty$, $\gamma \to 0$ and then $\delta \to 0$, to obtain

$$\lim_{s \downarrow t} h(s) = 0$$

for every (t, x) satisfying (7.6). Recall that (7.6) holds for λ-almost every (t, x). On the other hand, (7.4) and (7.1) imply that

$$\limsup_{s \downarrow t} h(s) \geqq \eta > 0$$

for μ-almost every (t, x). Since $\lambda = -\xi$ is absolutely continuous with respect to μ, we conclude that $\lambda = -\xi = 0$.

8. In this step we establish (7.10). Recall that

$$\alpha(t_0; t_0 + \tau, x_0) = \int \left(\frac{1}{4\pi\tau}\right)^{(d-1)/2} e^{-|x_0 - y|^2/4\tau} \mu(t_0; dy).$$

Without loss of generality we take $x_0 = 0$. Set $\mu(dy) = \mu(t_0; dy)$,

$$f(\tau) = \int \left(\frac{1}{4\pi\tau}\right)^{(d-1)/2} e^{-|y|^2/4\tau} \mu(dy).$$

Then, for any $0 < \alpha < 1$,

$$f\left(\frac{\tau}{1-\alpha}\right) \leqq \int \left(\frac{1}{4\pi\tau}\right)^{(d-1)/2} e^{-|y|^2/4\tau(1-\alpha)} \mu(dy).$$

Furthermore, for $0 < \delta$ and $\alpha \leqq \frac{1}{2}$,

$$I = f\left(\frac{\tau}{1-\alpha}\right) - (1+\delta) f(\tau)$$
$$\leqq \int \left(\frac{1}{4\pi\tau}\right)^{(d-1)/2} e^{-|y|^2/4\tau}\left(e^{\alpha(|y|^2)/4\tau} - (1+\delta)\right)$$
$$\leqq \int_{|y| \geqq \Lambda} \left(\frac{1}{4\pi\tau}\right)^{(d-1)/2} e^{-|y|^2/4\tau} \mu(dy),$$

where

$$\Lambda = \sqrt{\frac{4\tau}{\alpha} \ln(1+\delta)}\,.$$

Since by (7.1), $\mu(\{|y| \leqq R\}) \leqq KR^{d-1}$, and since the integrand is radially symmetric, an integration by parts yields

$$I \leqq \int_{\Lambda}^{\infty} \left(\frac{1}{(4\pi\tau)}\right)^{(d-1)/2} \frac{R}{4\tau} e^{-R^2/4\tau} KR^{d-1}\, dR.$$

By a change of variables,

$$I \leqq K \int_{\Lambda/2\sqrt{2\tau}}^{\infty} |\xi|^d e^{-|\xi|^2} d\xi \leqq K \exp\left(-\frac{\ln(1+\delta)}{2\alpha}\right) \leqq K \exp\left(-\frac{1}{2\alpha}\right) \leqq \delta$$

provided α is sufficiently small. □

8. Passage to the Limit

In this section we complete the proofs of Theorems 2.1, 2.2, and 2.3. We start with the following lemma.

Lemma 8.1. *For any $T > 0$ and $\alpha \geqq 0$ there are constants $K(T, \alpha)$ and $K(T)$ such that, for any Borel set $B \subset \mathscr{R}^d$,*

$$\sup_{0<\varepsilon\leqq 1} \int_0^T \int_B (1 + |z^\varepsilon(t,x)|)^\alpha \mu^\varepsilon(t; dx)\, dt \leqq K(T,\alpha)(1 + |B|), \tag{8.1}$$

$$\int_0^T \int_B (z_t^\varepsilon(t,x))^2 \mu^\varepsilon(t; dx)\, dt \leqq K(T)(1 + \sqrt{\varepsilon}|B|). \tag{8.2}$$

Proof. Set

$$\Omega = \{(t,x) \in [0,T] \times B; |z^\varepsilon(t,x)| \leqq 1\}.$$

Then the energy estimate (2.2) yields

$$\int_0^T \int_B (1 + |z^\varepsilon(t,x)|)^\alpha \mu^\varepsilon(t;dx)\,dt$$
$$\leqq 2^\alpha \int_\Omega \mu^\varepsilon(t;dx)\,dt + \int_{\Omega^c} (1+|z^\varepsilon|)^\alpha \mu^\varepsilon(t;dx)\,dt$$
$$\leqq (2^\alpha + 1) C_1^* T + \int_{\Omega^c} [(1+|z^\varepsilon|)^\alpha - 1]\mu^\varepsilon(t;dx)\,dt,$$

where Ω^c denote the complement of Ω. For any $r \geqq 1$,

$$(1+r)^\alpha - 1 \leqq \alpha r \max\{1, (1+r)^{\alpha-1}\} \leqq \alpha r (1+r)^\alpha.$$

Since $|z^\varepsilon| > 1$ on Ω^c, this inequality and (4.3) imply that, on Ω^c,

$$[(1+|z^\varepsilon|)^\alpha - 1]\mu^\varepsilon(t;dx) = [(1+|z^\varepsilon|)^\alpha - 1]\frac{1}{2\varepsilon}\left(q'\left(\frac{z^\varepsilon}{\varepsilon}\right)\right)^2 (|\nabla z^\varepsilon|^2 + 1)$$
$$\leqq K\alpha \frac{|z^\varepsilon|}{\varepsilon}(1+|z^\varepsilon|)^{\alpha+1}\left(q'\left(\frac{z^\varepsilon}{\varepsilon}\right)\right)^2$$
$$\leqq K\alpha \sup_{0<\varepsilon\leqq 1} \sup_{r \geqq 1} \frac{r}{\varepsilon}(1+r)^{\alpha+1}\left(q'\left(\frac{r}{\varepsilon}\right)\right)^2$$
$$= K\alpha \sup_{0<\varepsilon\leqq 1} \sup_{\bar r > \varepsilon} \bar r (1+\varepsilon \bar r)^{\alpha+1} (q'(\bar r))^2$$
$$= K\alpha \sup_{\bar r > 0} \bar r (1+\bar r)^{\alpha+1} (q'(\bar r))^2 = C^*(\alpha) < \infty.$$

This proves (8.1). To prove (8.2), first recall that, by (2.2),

$$\int_0^T \int_{\mathbb{R}^d} \varepsilon(\varphi_t^\varepsilon)^2\,dx\,dt = \int_0^T \int_{\mathbb{R}^d} \frac{1}{\varepsilon}(z_t^\varepsilon)^2 (q')^2\,dx\,dt \leqq C_1^*,$$

where q' is evaluated at $(z^\varepsilon/\varepsilon)$. Hence, by (2.2) and (4.1),

$$\int_0^T \int_B (z_t^\varepsilon)^2 \mu^\varepsilon(t;dx)\,dt = \int_0^t \int_B \frac{1}{2\varepsilon}(z_t^\varepsilon)^2 (q')^2 (1 + |\nabla z^\varepsilon|^2)\,dx\,dt$$
$$\leqq C_1^* + K\sqrt{\varepsilon}\left[\int_0^T \int_B \frac{1}{2\varepsilon}(z_t^\varepsilon)^2 (q')^2 (1+|z^\varepsilon|)\right] dx\,dt$$
$$\leqq C_1^* + K\sqrt{\varepsilon}\left[\int_0^T \int_{\mathbb{R}^d} \frac{\varepsilon}{2}(\varphi_t^\varepsilon)^2\,dx\,dt + \int_0^T \int_B \frac{1}{2\varepsilon}(z_t^\varepsilon)^2 (q')^2 |z^\varepsilon|\,dx\,dt\right]$$

By (4.6) and (2.2),

$$\int_0^T\int_B \frac{1}{2\varepsilon}(z_t^\varepsilon)^2(q')^2|z^\varepsilon|\,dx\,dt$$

$$\leqq \int_0^T\int_B \frac{1}{2\varepsilon}(z_t^\varepsilon)^2(q')^2[1+|z^\varepsilon|\chi_{|z^\varepsilon|\geqq 1}]\,dx\,dt$$

$$\leqq \int_0^T\int_{\mathscr{R}^d} \varepsilon(\varphi_t^\varepsilon)^2 + \int_0^T\int_B \frac{K}{\varepsilon^5}(q')^2|z^\varepsilon|\chi_{|z^\varepsilon|\geqq 1}\,dx\,dt$$

$$\leqq C_1^* + \int_0^T\int_B \frac{K}{\varepsilon^5}\sup_{r\geqq 1} rq'\left(\frac{r}{\varepsilon}\right)dx\,dt \leqq K[1+|B|T]. \quad \square$$

Proof of Theorem 2.1. The only assertion left to prove is the convergence of $\bar{m}^\varepsilon$, where

$$d\bar{m}^\varepsilon = -z_t^\varepsilon\,dm^\varepsilon.$$

The L^2 estimate (8.2) implies that

$$\sup_{0<\varepsilon\leqq 1} |\bar{m}^\varepsilon|([0,T]\times B_R\times S^{d-1}) < \infty \quad \text{for } R, T > 0.$$

Hence on a subsequence, denoted by $\varepsilon, \bar{m}^\varepsilon$ converges to a Radon measure $\bar{m}$. Moreover, (8.2) implies that $\bar{m}$ is absolutely continuous with respect to m; let v denote the corresponding Radon-Nikodym derivative, so that, by (8.2),

$$v \in L^2((0,T)\times\mathscr{R}^d\times S^{d-1}; dm). \quad \square$$

Proof of Theorem 2.2. We first prove the existence of the mean-curvature vector H. Following [23, §9.3], let $V^\varepsilon(t;\cdot)$ be the varifold (cf. [28])

$$V^\varepsilon(t; dx\,dS) = \delta_{\{(v^\varepsilon)^\perp\}}(dS)\mu^\varepsilon(t; dx)$$

so that $(V^\varepsilon(t;\cdot))^{(x)}$ is supported at $(v^\varepsilon(t,x))^\perp$ and

$$\|V^\varepsilon(t;\cdot)\| = \mu^\varepsilon(t;\cdot).$$

1. In this step we show that

$$\sup_{0<\varepsilon\leqq 1}\int_0^T\int \varepsilon\left[-\Delta\varphi^\varepsilon + \frac{1}{\varepsilon^2}W'(\varphi^\varepsilon)\right]^2 dx\,dt < \infty. \tag{8.3}$$

Since $g^2 = 2W$, (OPE) implies that

$$\varepsilon\left[-\nabla\varphi^\varepsilon + \frac{1}{\varepsilon^2}W'(\varphi^\varepsilon)\right]^2 = \varepsilon\left[-\varphi_t^\varepsilon + \frac{1}{\varepsilon}g(\varphi^\varepsilon)\theta^\varepsilon\right]^2 \leqq 2\varepsilon(\varphi_t^\varepsilon)^2 + \frac{4}{\varepsilon}W(\varphi^\varepsilon)(\theta^\varepsilon)^2,$$

while Theorem 7.1 and (7.1) yield

$$\tfrac{1}{\varepsilon}\int W(\varphi^\varepsilon)(\theta^\varepsilon)^2\,dx \leqq \int (\theta^\varepsilon)^2\,d\mu^\varepsilon \leqq K\,\|\nabla(\theta^\varepsilon)^2\|_1 \leqq K\int(\theta^\varepsilon)^2 + |\nabla\theta^\varepsilon|^2.$$

Hence

$$\int_0^T\int \varepsilon\left[-\Delta\varphi^\varepsilon + \frac{1}{\varepsilon^2}W'(\varphi^\varepsilon)\right]^2 dx\,dt$$

$$\leqq \int_0^T\int [2\varepsilon(\varphi_t^\varepsilon)^2 + K|\nabla\theta^\varepsilon|^2 + K(\theta^\varepsilon)^2]\,dx\,dt,$$

and (8.3) follows from (2.2).

2. For any smooth vector field $Y(x)$, the definition of V^ε and the definition of the first variation (cf. [28]) imply that

$$\delta V^\varepsilon(t;\cdot)(Y) = \int \nabla Y : S V^\varepsilon(t;dx\,dS) = \int \nabla Y:(I - \nu^\varepsilon\otimes\nu^\varepsilon)\mu^\varepsilon(t;dx).$$

Let $\hat{T}$ be as in Step 2 of Theorem 5.1. Recall that

$$(I-\nu^\varepsilon\otimes\nu^\varepsilon)\mu^\varepsilon = (\nu^\varepsilon\otimes\nu^\varepsilon)\xi^\varepsilon - \hat{T}\,dx,$$

$$\operatorname{div}\hat{T} = -\varepsilon\,\nabla\varphi^\varepsilon\left[-\Delta\varphi^\varepsilon + \frac{1}{\varepsilon^2}W'(\varphi^\varepsilon)\right].$$

As in Theorem 5.1,

$$\begin{aligned}\delta V^\varepsilon(t;\cdot)(Y) &= \int \nabla Y:[(\nu^\varepsilon\otimes\nu^\varepsilon)\,d\xi^\varepsilon - \hat{T}\,dx]\\ &= \int Y\cdot\operatorname{div}\hat{T}\,dx + \nabla Y:(\nu^\varepsilon\otimes\nu^\varepsilon)\,d\xi^\varepsilon\\ &= -\int \varepsilon Y\cdot\nabla\varphi^\varepsilon\left[-\Delta\varphi^\varepsilon + \frac{1}{\varepsilon^2}W'(\varphi^\varepsilon)\right]dx + \int \nabla Y:\nu^\varepsilon\otimes\nu^\varepsilon\,d\xi^\varepsilon.\end{aligned}$$

Hence

$$|\delta V^\varepsilon(t;\cdot)(Y)| \leqq \left(\int \varepsilon|Y|^2|\nabla\varphi^\varepsilon|^2\,dx\right)^{1/2}\left(\int\varepsilon\left[-\Delta\varphi^\varepsilon + \frac{1}{\varepsilon^2}W(\varphi^\varepsilon)\right]^2\right)^{1/2} + \int|\nabla Y|\,d|\xi^\varepsilon|.$$

3. In view of (8.3), Proposition 7.3 and (2.2), for every $T > 0$ there is a constant $K(T)$ satisfying

$$\limsup_{\varepsilon\downarrow 0}\int_0^T |\delta V^\varepsilon(t;\cdot)(Y)|\,dt \leqq K(T)\,\|Y\|_\infty$$

for all $Y\in C_c^\infty(\mathscr{R}^d\to\mathscr{R}^d)$. Choose a further subsequence $\varepsilon_n\downarrow 0$ such that the Radon measures $dV^{\varepsilon_n}(t;\cdot)\,dt$ on $\mathscr{R}^d\times G_{d-1}(\mathscr{R}^d)\times[0,\infty)$ are convergent in the weak* topology. By a slicing argument [16, Theorem 10, page 14], we conclude that there are varifolds $\tilde{V}(t;\cdot)$ that satisfy

$$dV^{\varepsilon_n}(t;\cdot)\,dt \rightharpoonup d\tilde{V}(t;\cdot)\,dt.$$

4. By definition,

$$\int_0^T |\delta\tilde{V}(t;\cdot)(Y)|\,dt = \sup_{|h(t)|\leq 1} \int_0^T \left(\int Y(x):S\,\tilde{V}(t;dx\,dS)\right) h(t)\,dt,$$

so that Step 3 yields

$$\int_0^T |\delta\tilde{V}(t,\cdot)(Y)|\,dt \leq K(T)\,\|Y\|_\infty \quad \forall Y \in C_c^\infty(\mathscr{R}^d \to \mathscr{R}^d).$$

Since $C_c^\infty(\mathscr{R}^d \to \mathscr{R}^d)$ is separable,

$$K(t) := \sup_{|Y|\leq 1} |\delta V^\varepsilon(t,\cdot)(Y)| < \infty$$

for almost every $t \geq 0$.
5. Let $t \geq 0$ be a point with $K(t) < \infty$. Then (7.2) and ALLARD's theorem of rectifiability [1, 5.5(2)] imply that $\|\tilde{V}(t;\cdot)\|$ is $d-1$ rectifiable. Moreover, by the definition of the varifolds $V^\varepsilon(t;\cdot)$,

$$\|V^\varepsilon(t;\cdot) = \mu^\varepsilon(t;\cdot) \;\Rightarrow\; \|\tilde{V}(t,\cdot)\| = \mu(t;\cdot).$$

Since a $(d-1)$-rectifiable varifold is uniquely determined by its mass measure,

$$\tilde{V}(t;\cdot) = V_{\mu(t;\cdot)}.$$

Hence $dV^\varepsilon(t;\cdot)\,dt$ converges on the entire original sequence ε, and more importantly, $\mu(t;\cdot)$ is $(d-1)$-rectifiable.

We have also proved that

$$|\delta V_{\mu(t;\cdot)}(Y)| \leq \liminf_{\varepsilon\to 0} \left(\int \varepsilon\left[-\Delta\varphi^\varepsilon + \frac{1}{\varepsilon^2} W'(\varphi^\varepsilon)\right]^2 dx\right)^{1/2} \left(\int |Y|^2 \mu(t;dx)\right)^{1/2}.$$

Hence, for almost every $t \geq 0$, $\mu(t,\cdot)$ has a generalized mean-curvature vector $H(t,x)$ and

$$\int |H(t,x)|^2 \mu(t;dx) \leq \liminf_{\varepsilon\to 0} \int \varepsilon\left[-\Delta\varphi^\varepsilon + \frac{1}{\varepsilon^2} W'(\varphi^\varepsilon)\right]^2 dx.$$

Step 1 implies that $H \in L^2((0,T)\times\mathscr{R}^d; d\mu)$ and in Step 2 we have established that, for any $Y \in C_c^\infty(\mathscr{R}^d \to \mathscr{R}^d)$,

$$\iint Y\cdot H(t,x)\,dt\,\mu(t;dx) = -\int \delta\tilde{V}(t;\cdot)(Y)\,dt = -\lim_{\varepsilon\to 0} \int \delta V^\varepsilon(t;\cdot)(Y)\,dt \quad (8.4)$$

$$= \lim_{\varepsilon\to 0} \iint \varepsilon Y\cdot\nabla\varphi^\varepsilon \left[-\Delta\varphi^\varepsilon + \frac{1}{\varepsilon^2} W'(\varphi^\varepsilon)\right] dx\,dt.$$

6. In this step we show that $\theta \in L^1_{\text{loc}}(d\mu)$. In view of (2.9), there exists a sequence of smooth functions θ_k satisfying

$$\lim_{k\to\infty} \|\theta_k - \theta\|_{2,T} = 0, \quad \sup_k \|\nabla\theta_k\|_{2,T} < \infty \tag{8.5}$$

for every $T > 0$. Fix $\lambda > 0$ and $T > 0$. Then, by Theorem 7.1, (7.1) and (2.2),

$$\begin{aligned}\int_0^T \int_{\mathscr{R}^d} |\theta_k - \theta_l| \, dt\mu(t; dx) &\leq \int_0^T \int_{\mathscr{R}^d} \left[\frac{\lambda}{2}|\theta_k - \theta_l|^2 + \frac{1}{2\lambda}\right] dt\, \mu(t; dx)\\ &\leq K\left[\frac{\lambda}{2}\|\nabla(|\theta_k - \theta_l|^2)\|_{1,T} + \frac{T}{\lambda}\right]\\ &\leq K\left[\lambda\|\theta_k - \theta_l\|_{2,T}\|\nabla(\theta_k - \theta_l)\|_{2,T} + \frac{T}{\lambda}\right],\end{aligned}$$

which converges to zero as $k, l \to \infty$, since we can take $\lambda \to \infty$. Hence $\theta \in L^1_{\text{loc}}(d\mu)$ and

$$\iiint \theta(t, x) n \cdot Y(t, x) \, dm(t, x, n) = \lim_{k\to\infty} \iiint \theta_k n \cdot Y \, dm.$$

7. For $n \in S^{d-1}$, let $Pn \in G_{d-1}(\mathscr{R}^d)$ be the $(d-1)$-dimensional, unoriented plane orthogonal to n so that $P: S^{d-1} \to G_{d-1}(\mathscr{R}^d)$ is a surjective map. Then, by definition,

$$\frac{dV^\varepsilon(t; \cdot)}{\mu^\varepsilon(t; dx)} = \frac{dm^\varepsilon}{\mu^\varepsilon(t; dx)\, dt} \circ P^{-1}.$$

By the weak* convergence of these measures,

$$\delta_{T_x\mu(t;\cdot)} = \frac{dV_{\mu(t;\cdot)}(t; \cdot)}{\mu(t; dx)} = \frac{dm}{\mu(t; dx)\, dt} \circ P^{-1} = N(t, x; \cdot) \circ P^{-1}.$$

Therefore the support of $N(t, x; \cdot)$ is orthogonal to $T_x\mu(t; \cdot)$ for $d\mu$-almost all (t, x). □

Proof of Theorem 2.3.

1. Let $\psi(t, x)$ be a smooth compactly supported function. Then the action of the distribution $\theta_t - \Delta\theta$ on ψ is given by

$$I(\psi) = -\iint (\psi_t + \Delta\psi)\theta \, dx\, dt = \lim_{\varepsilon\to 0} \iint (\theta^\varepsilon_t - \Delta\theta^\varepsilon)\psi \, dx\, dt,$$

so that, by (HE),

$$\begin{aligned}I(\psi) &= -\lim_{\varepsilon\to 0} \iint g(\varphi^\varepsilon)\varphi^\varepsilon_t \psi \, dx\, dt\\ &= -\lim_{\varepsilon\to 0} \iint \frac{1}{\varepsilon}\left(q'\left(\frac{z^\varepsilon}{\varepsilon}\right)\right)^2 z^\varepsilon_t \psi \, dx\, dt\\ &= \lim_{\varepsilon\to 0} \iint \psi \, d\bar{m}^\varepsilon + \lim_{\varepsilon\to 0} \iint \psi z^\varepsilon_t \, d\xi^\varepsilon\\ &= \iiint v(t, x, n)\psi(t, x) \, dm + \lim_{\varepsilon\to 0} \iint \psi z^\varepsilon_t \, d\xi^\varepsilon.\end{aligned}$$

We assert that the second term in the last expression is equal to zero. Indeed, since $|\xi^\varepsilon| \leqq \mu^\varepsilon$, the Cauchy-Schwarz inequality yields

$$\left|\iint \psi z_t^\varepsilon \, d\xi^\varepsilon\right| \leqq \left(\iint |\psi|^2 d|\xi^\varepsilon|\right)^{1/2} \left(\iint_{\operatorname{spt}\psi} |z_t^\varepsilon|^2 \, d\mu^\varepsilon\right)^{1/2}.$$

By Proposition 7.3 and (8.2), the right-hand side of the previous expression converges to zero as ε tends to zero; hence (2.10) holds. Equation (2.11) follows after an integration by parts in the variable t.

2. Let Y be a compactly supported, smooth vector field. The definitions of $\bar{m}^\varepsilon$ and $v(t, x, n)$ yield

$$\begin{aligned}
L(Y) &:= \iiint v(t,x,n) n \cdot Y(t,x) \, dm = \iiint n \cdot Y(t,x) \, d\bar{m} \\
&= -\lim_{\varepsilon \to 0} \iint z_t^\varepsilon v^\varepsilon \cdot Y \mu^\varepsilon(t; dx) \, dt \\
&= -\lim_{\varepsilon \to 0} \iint z_t^\varepsilon v^\varepsilon \cdot Y \frac{1}{\varepsilon}\left(q'\left(\frac{z^\varepsilon}{\varepsilon}\right)\right)^2 dx\, dt - \lim_{\varepsilon \to 0} \iint z_t^\varepsilon v^\varepsilon \cdot Y \, d\xi^\varepsilon \, dt.
\end{aligned}$$

As in Step 1, the second term in the above expression is zero. Next we use the identities

$$z_t^\varepsilon q'\left(\frac{z^\varepsilon}{\varepsilon}\right) = \varepsilon \varphi_t^\varepsilon, \quad v^\varepsilon q'\left(\frac{z^\varepsilon}{\varepsilon}\right) = \frac{\varepsilon \nabla \varphi^\varepsilon}{|\nabla z^\varepsilon|}, \quad g(\varphi^\varepsilon) = \sqrt{2W(\varphi^\varepsilon)} = q',$$

together with (OPE) and (8.4) to obtain

$$\begin{aligned}
L(Y) &= -\lim_{\varepsilon \to 0} \iint \varepsilon \varphi_t^\varepsilon \nabla \varphi^\varepsilon \cdot Y \frac{1}{|\nabla z^\varepsilon|} dx\, dt \\
&= -\lim_{\varepsilon \to 0} \iint q'\left(\frac{z^\varepsilon}{\varepsilon}\right) \nabla \varphi^\varepsilon \cdot Y \frac{1}{|\nabla z^\varepsilon|} \theta^\varepsilon \, dx\, dt \\
&\quad + \lim_{\varepsilon \to 0} \iint \varepsilon Y \cdot \nabla \varphi^\varepsilon \left[-\Delta \varphi^\varepsilon + \frac{1}{\varepsilon^2} W'(\varphi^\varepsilon)\right] \frac{1}{|\nabla z^\varepsilon|} dx\, dt \\
&= -\lim_{\varepsilon \to 0} \iint \frac{1}{\varepsilon}\left(q'\left(\frac{z^\varepsilon}{\varepsilon}\right)\right)^2 v^\varepsilon \cdot Y \theta^\varepsilon \, dx\, dt \\
&\quad + \lim_{\varepsilon \to 0} \iint \varepsilon Y \cdot \nabla \varphi^\varepsilon \left[-\Delta \varphi^\varepsilon + \frac{1}{\varepsilon^2} W'(\varphi^\varepsilon)\right] dx\, dt \\
&\quad + \lim_{\varepsilon \to 0} \iint \varepsilon Y \cdot \nabla \varphi^\varepsilon \left[-\Delta \varphi^\varepsilon + \frac{1}{\varepsilon^2} W'(\varphi^\varepsilon)\right] \left(\frac{1 - |\nabla z^\varepsilon|}{|\nabla z^\varepsilon|}\right) dx\, dt \\
&:= \lim_{\varepsilon \to 0} I^\varepsilon + \iint H \cdot Y \, dt \mu(t; dx) + \lim_{\varepsilon \to 0} E^\varepsilon. \qquad (8.6)
\end{aligned}$$

3. In this step we show that E^ε converges to zero. By (8.3),

$$|E^\varepsilon| \leqq \|Y\|_\infty \left(\iint\limits_{\operatorname{spt} Y} \varepsilon \left[-\Delta\varphi^\varepsilon + \frac{1}{\varepsilon^2} W'(\varphi^\varepsilon) \right]^2 \right)^{1/2} \left(\iint\limits_{\operatorname{spt} Y} \varepsilon |\nabla\varphi^\varepsilon|^2 \left(\frac{1 - |\nabla z^\varepsilon|}{|\nabla z^\varepsilon|} \right)^2 \right)^{1/2} .$$

$$\leqq K(Y) \left(\iint\limits_{\operatorname{spt} Y} \frac{1}{\varepsilon} \left(q'\left(\frac{z^\varepsilon}{\varepsilon}\right)\right)^2 (1 - |\nabla z^\varepsilon|)^2 \right)^{1/2},$$

while (4.1) yields

$$\begin{aligned}(1 - |\nabla z^\varepsilon|)^2 &= (1 - |\nabla z^\varepsilon|)^2 \chi_{\{|\nabla z^\varepsilon| \geqq 1\}} + (1 - |\nabla z^\varepsilon|)^2 \chi_{\{|\nabla z^\varepsilon| < 1\}} \\ &\leqq K\varepsilon(1 + |z^\varepsilon|)^2 + |1 - |\nabla z^\varepsilon|^2|.\end{aligned}$$

Hence

$$E^\varepsilon \leqq K(Y) \left(\iint\limits_{\operatorname{spt} Y} K\varepsilon(1 + |z^\varepsilon|)^2 \, d\mu^\varepsilon + d|\xi^\varepsilon| \right)^{1/2},$$

and by Proposition 7.3 and (8.1), the limit of E^ε is zero.
4. Let θ_k be as in (8.5). In Step 6 of the previous proof we have shown that $\theta \in L^1_{\text{loc}}(d\mu)$ and

$$I(\theta) := -\iiint Y \cdot n\theta \, dm = -\lim_{k\to\infty} \iiint \theta_k n \cdot Y \, dm .$$

Then

$$\begin{aligned} I(\theta) - I^\varepsilon = &\iint \frac{1}{\varepsilon} (q')^2 v^\varepsilon \cdot Y(\theta^\varepsilon - \theta_k) \, dx \, dt \\ &+ \iiint \left(\frac{1}{\varepsilon} (q')^2 v^\varepsilon \cdot Y \, dx \, dt - n \cdot Y \, dm^\varepsilon \right) \theta_k \\ &+ \iiint n \cdot Y\theta_k (dm^\varepsilon - dm) + \iiint n \cdot Y(\theta_k - \theta) \, dm. \end{aligned}$$

Since

$$\frac{1}{\varepsilon} (q')^2 v^\varepsilon \cdot Y \, dx \, dt - n \cdot Y \, dm^\varepsilon = v^\varepsilon \cdot Y \, d\xi^\varepsilon,$$

Proposition 7.3 and the convergence of m^ε to m yields

$$\begin{aligned}\limsup_{\varepsilon\to 0} |I^\varepsilon - I(\theta)| \leqq \limsup_{\varepsilon\to 0} &\left| \iint \frac{1}{\varepsilon} (q')^2 v^\varepsilon \cdot Y(\theta^\varepsilon - \theta_k) \, dx \, dt \right| \\ &+ \left| \iiint n \cdot Y(\theta_k - \theta) \, dm \right| .\end{aligned}$$

Recall that θ^ε converges to θ strongly in L^2_{loc} (cf. Proposition 3.4). So as in Step 6 of the previous proof,

$$\limsup_{\varepsilon\to 0}\left|\iint \frac{1}{\varepsilon}(q')^2 v^\varepsilon\cdot Y(\theta^\varepsilon-\theta_k)\,dx\,dt\right|$$

$$\leqq \|Y\|_\infty \limsup_{\varepsilon\to 0}\iint_{\mathrm{spt}\,Y}|\theta^\varepsilon-\theta_k|\,dt\,\mu^\varepsilon(t;dx)$$

$$\leqq K\|Y\|_\infty\left[\lambda\|\theta-\theta_k\|_{2,T}+\frac{T}{\lambda}\right],$$

for any $\lambda>0$. Finally, let k and then λ go to infinity to show that I^ε converges to $I(\theta)$.
5. Combining the previous steps we conclude that

$$\iiint Y\cdot n(v+\theta)\,dm=\iint H\cdot Y\,d\mu$$

for any smooth vector field Y proving (2.12).
6. The computations of §2.3 imply that, for any $\phi\in C_c^\infty(\mathscr{R}^d\to[0,\infty))$,

$$\frac{d}{dt}\int\phi(x)\hat{\mu}^\varepsilon(t;dx)=-\int\phi[\varepsilon(\varphi_t^\varepsilon)^2+|\nabla\theta^\varepsilon|^2]\,dx+\tfrac{1}{2}\int\Delta\phi(\theta^\varepsilon)^2\,dx$$

$$-\varepsilon\int\nabla\phi\cdot\nabla\varphi^\varepsilon\varphi_t^\varepsilon\,dx.$$

By Step 2,

$$\iiint_s^t Y\cdot nv\,dm=-\lim\iint_s^t \varepsilon\varphi_t^\varepsilon\nabla\varphi^\varepsilon\cdot Y\frac{1}{|\nabla z^\varepsilon|}\,dx\,dr.$$

We now proceed as in Step 3 to obtain

$$\iiint_s^t Y\cdot nv\,dm=-\lim\iint_s^t\varepsilon\varphi_t^\varepsilon\nabla\varphi^\varepsilon\cdot Y\,dx\,dr.$$

Hence

$$\hat{\mu}(\phi)(t)-\hat{\mu}(\phi)(s)\leqq-\liminf\iint_s^t\phi(\varphi_t^\varepsilon)^2\,dx\,dr$$

$$+\iiint_s^t vn\cdot\nabla\phi\,dm+\iint_s^t(\tfrac{1}{2}\theta^2\Delta\phi-|\nabla\theta|^2\phi)\,dx\,dr.$$

Since ϕ is compactly supported, following the proof of the estimate (8.2) we find that

$$\liminf\iint_s^t\phi(\varphi_t^\varepsilon)^2\,dx\,dr=\liminf\iint_s^t\phi(z_t^\varepsilon)^2\mu^\varepsilon(r;dx)\,dr.$$

For any $w \in C_c^\infty(Q \times S^{d-1} \to \mathscr{R})$,

$$0 \leqq \liminf \int_s^t\!\!\int\!\!\int \phi(z_i^\varepsilon + w)^2 \, dm^\varepsilon$$

$$= \liminf \int_s^t\!\!\int \phi(z_i^\varepsilon)^2 \mu^\varepsilon(r; dx)\, dr + \int_s^t\!\!\int\!\!\int \phi(w^2 - 2wv)\, dm;$$

hence

$$\int_s^t\!\!\int\!\!\int \phi v^2 \, dm = \int_s^t\!\!\int\!\!\int \phi[(w-v)^2 - (w^2 - 2wv)]\, dm$$

$$\leqq \liminf \int_s^t\!\!\int \phi(\varphi_i^\varepsilon)^2 \, dx\, dr + \int_s^t\!\!\int\!\!\int \phi(w-v)^2 \, dm.$$

Since $v \in L^2(dm)$,

$$\int_s^t\!\!\int\!\!\int \phi v^2 \, dm \leqq \liminf \int_s^t\!\!\int \phi(\varphi_i^\varepsilon)^2 \, dx\, dr,$$

which proves (2.13). □

9. Appendix

In this section we study a simple radially symmetric solution of the Mullins-Sekerka problem. We show that if the initial radius of the interface is sufficiently small, then the temperature is *not* a bounded function.

The Mullins-Sekerka problem (1.5), (1.6) with radial symmetry and one interface, takes the form

$$\theta_t - \Delta\theta = \tfrac{4}{3}(\chi_{\{|x| \leqq R(t)\}})_t$$

while the radius $R(t)$ of the interface is a solution of

$$R'(t) = -\frac{1}{R(t)} - \theta(t, R(t)), \quad t \in (0, T_{\text{ext}}),$$

where the extinction (or melting) time T_{ext} is defined to be the first time at which $R(T_{\text{ext}}) = 0$, if there is such a time; otherwise $T_{\text{ext}} = \infty$. Let $(R(t), \theta(t, r))$ be the solution of this problem with initial data

$$\theta(0, \cdot) \equiv 0, \quad R(0) = R_0.$$

The following result was obtained in collaboration with T. Ilmanen.

Theorem 9.1. *For all sufficiently small R_0, the extinction time $T_{\text{ext}} < \infty$ and*

$$\lim_{\delta \downarrow 0} \frac{\theta(T_{\text{ext}} - \delta, 0)}{|\ln(\delta)|} < 0.$$

Proof.

1. Let G be the heat kernel. Then

$$\theta(t, x) = \tfrac{4}{3}\int_0^t R'(s)\left[\int_{\{|y|=R(s)\}} G(t-s, x-y)\, d\mathscr{H}^{d-1}(y)\right] ds$$

and there is a constant $C(d)$ such that for any $\tau, \rho > 0$ and $x \in \mathscr{R}^d$,

$$\int_{\{|y|=\rho\}} G(\tau, x-y)\, d\mathscr{H}^{d-1}(y) \leqq \frac{3C(d)}{8\sqrt{\tau}}.$$

Set

$$\bar{\theta}(t) := \sup\{|\theta(s, x)| : (s, x) \in [0, t] \times \mathscr{R}^d\},$$

$$\bar{K}(t) := \sup\left\{\frac{1}{R(s)} : s \in [0, t]\right\}$$

so that

$$|\theta(t, x)| \leqq \frac{4}{3}\int_0^t \sup\{|R'(s)| : s \in [0, t]\}\frac{3C(d)}{8\sqrt{t-s}}\, ds$$

$$\leqq C(d)\sqrt{t}[\bar{K}(t) + \bar{\theta}(t)].$$

Hence

$$\bar{\theta}(t) \leqq C(d)\sqrt{t}[\bar{K}(t) + \bar{\theta}(t)].$$

2. Set

$$t^* = \inf\{t \in [0, T_{ext}] : R'(t) = 0\},$$

or $t^* = \infty$ if this set is empty. Since $\theta_0 \equiv 0$, it follows that $t^* > 0$ and

$$R'(t) < 0, \quad \bar{K}(t) = \frac{1}{R(t)} \quad \forall t \in [0, t^*].$$

Also, if $t^* < T_{ext}$, then

$$\bar{\theta}(t^*) \geqq -\theta(t^*, R(t^*)) = \frac{1}{R(t^*)} = \bar{K}(t^*).$$

3. Set

$$t_0 = \min\left\{T_{ext}, \frac{1}{(3C(d))^2}\right\}$$

so that, for $s \in [0, t_0]$,

$$\bar{\theta}(t) \leqq \frac{1}{3}(\bar{K}(t) + \bar{\theta}(t)) \;\Rightarrow\; \bar{\theta}(t) \leqq \frac{1}{2}\bar{K}(t) = \frac{1}{2R(t)}.$$

Therefore $t_0 < t^*$ and

$$R'(s) \leqq -\frac{1}{2R(s)} \quad \forall t \in [0, t_0]. \tag{9.1}$$

Since $R(t_0) \geqq 0$, the preceeding differential inequality yields

$$R(s) \geqq \sqrt{t_0 - s} \quad \forall s \in [0, t_0].$$

Suppose that

$$R_0 < \frac{1}{3C(d)};$$

then

$$\sqrt{t_0} \leqq R_0 < \frac{1}{3C(d)},$$

and we conclude that $T_{ext} = t_0 < \infty$. Also, since $\theta \leqq 0$, we have

$$R' \geqq -\frac{1}{R(t)}; \tag{9.2}$$

hence

$$\sqrt{T_{ext} - s} \leqq R(s) \leqq \sqrt{2(T_{ext} - s)} \quad \forall s \in [0, T_{ext}]. \tag{9.3}$$

4. Set

$$\Lambda(\zeta) := \int_{\{|y|=\zeta\}} G(1, y)\, d\mathscr{H}^{d-1}(y).$$

For $\delta > 0$, by (9.1) and (9.3),

$$\begin{aligned} \theta(T_{ext} - \delta, 0) &\leqq -\frac{2}{3} \int_0^{T_{ext}-\delta} \frac{1}{R(s)} \left[\int_{\{|y|=R(s)\}} G(T_{ext} - \delta - s, y)\, d\mathscr{H}^{d-1}(y) \right] ds \\ &= -\frac{2}{3} \int_0^{T_{ext}-\delta} \frac{1}{R(s)} \frac{1}{\sqrt{T_{ext} - \delta - s}} \Lambda\left(\frac{R(s)}{\sqrt{T_{ext} - \delta - s}} \right) ds \\ &\leqq -\frac{2}{3\sqrt{2}} \int_0^{T_{ext}-\delta} \frac{1}{\sqrt{\tau}\sqrt{\tau + \delta}} \Lambda\left(\frac{R(T_{ext} - \delta - \tau)}{\sqrt{\tau}} \right) d\tau. \end{aligned}$$

For $\tau \geqq \delta$,

$$1 \leqq \sqrt{\frac{\delta + \tau}{\tau}} \leqq \frac{R(T_{ext} - \delta - \tau)}{\sqrt{\tau}} \leqq \sqrt{\frac{2(\delta + \tau)}{\tau}} \leqq 2\sqrt{2}.$$

Hence, for $\tau \geqq \delta$,

$$\Lambda\left(\frac{R(T_{ext} - \delta - \tau)}{\sqrt{\tau}} \right) \geqq \lambda_0 > 0,$$

where λ_0 is an appropriate constant. Therefore

$$\theta(T_{ext} - \delta, 0) \leqq -\frac{2\lambda_0}{3\sqrt{2}} \int_\delta^{T_{ext}-\delta} \frac{1}{\sqrt{\tau + \delta}\sqrt{\tau}}\, d\tau \leqq -\frac{\lambda_0}{3} \int_\delta^{T_{ext}-\delta} \frac{1}{\tau}\, d\tau. \quad \square$$

Acknowledgements. Part of this work was done during my visit, supported by SFB256, to the University of Bonn. I thank TOM ILMANEN and STEFAN LUCKHAUS for valuable discussions during that visit. I also thank TOM ILMANEN and MORT GURTIN for their careful reading of an earlier version of this paper, and for several suggestions and corrections which led to substantial changes. Finally I express my deepest gratitude to MORT GURTIN for his support, encouragement and numerous conversations.

This work was partially supported by the Army Research Office and the National Science Foundation through the Center for Nonlinear Analysis and under the National Science Foundation grant DMS-9200801.

References

1. W. ALLARD, On the first variation of a varifold, *Ann. of Math.* **95** (1972), 417–491.
2. N. D. ALIKAKOS, P. W. BATES & X. CHEN, Convergence of the Cahn-Hilliard equation to the Hele-Shaw model, *Arch. Rational Mech. Anal.* **128** (1994), 165–205.
3. F. ALMGREN & L. WANG, Mathematical existence of crystal growth with Gibbs-Thompson curvature effects (1993), preprint.
4. G. BARLES, H. M. SONER, & P. E. SOUGANIDIS, Front propagation and phase field theory, *SIAM. J. Cont. Opt.*, March 1993, issue dedicated to W. H. FLEMING, 439–469.
5. K. A. BRAKKE, *The Motion of a Surface by its Mean Curvature*, Princeton University Press (1978).
6. L. BRONSARD & R. KOHN, Motion by mean curvature as the singular limit of Ginzburg-Landau model, *J. Diff. Eqs.* **90** (1991), 211–237.
7. G. CAGINALP, Surface tension and supercooling in solidification theory, *Lecture Notes in Physics*, **216** (1984), 216–226.
8. G. CAGINALP, An analysis of a phase field model of a free boundary, *Arch. Rational Mech. Anal.* **92** (1986), 205–245.
9. G. CAGINALP, Stefan & Hele-Shaw type models as asymptotic limits of the phase-field equations, *Physical Review A*, **39** (1989), 5887–5896.
10. G. CAGINALP & X. CHEN, Phase field equations in the singular limit of sharp interface equations (1992), preprint.
11. X. CHEN, Spectrums for the Allen-Cahn, Cahn-Hilliard and phase-field equations for generic interfaces (1993), preprint.
12. X. CHEN & F. REITICH, Local existence and uniqueness of solutons of the Stefan problem with surface tension and kinetic undercooling, *J. Math. Anal. Appl.* **164** (1992), 350–362.
13. Y. CHEN & M. STRUWE, Existence and partial regularity results for the heat flow for harmonic maps, *Math. Z.* **201** (1989), 83–103.
14. M. G. CRANDALL, H. ISHII & P.-L. LIONS, User's guide to viscosity solutions of second order partial differential equations, *Bull. Amer. Math. Soc.* **27** (1992), 1–67.
15. J. B. COLLINS & H. LEVINE, Diffuse interface model of diffusion-limited crystal growth, *Phys. Rev. B*, **31** (1985), 6119–6122.
16. L. C. EVANS, Weak Convergence Methods for Nonlinear Partial Differential Equations, *CBMS Regional Series in Mathematics* **74** (1990).
17. L. C. EVANS & R. GARIEPY, *Measure Theory and Fine Properties of Functions*, CRC Press (1992).
18. L. C. EVANS, H. M. SONER & P. E. SOUGANIDIS, Phase transitions and generalized motion by mean curvature, *Comm. Pure Appl. Math.* **45** (1992), 1097–1123.

19. G. FIX, Phase field methods for free boundary problems, in *Free boundary problems: theory and applications*, edited by B. FASANO & M. PRIMICERIO, Pitman (1983), 580–589.
20. E. FRIED & M. GURTIN, Continuum theory of thermally induced phase transitions based on an order parameter, *Physica D* **68** (1993), 326–343.
21. M. E. GURTIN, *Thermodynamics of Evolving Phase Boundaries in the Plane*, Oxford University Press (1993).
22. G. HUISKEN, Asymptotic behavior for singularities of the mean curvature flow, *J. Diff. Geometry* **31** (1990), 285–299.
23. T. ILMANEN, Convergence of the Allen-Cahn equation to the Brakke's motion by mean curvature, *J. Diff. Geometry* **38** (1993), 417–461.
24. J. S. LANGER, unpublished notes (1978).
25. S. LUCKHAUS, Solutions of the two phase Stefan problem with the Gibbs-Thompson law for the melting temperature, *European J. Appl. Math.* **1** (1990), 101–111.
26. W. W. MULLINS & R. F. SEKERKA, Stability of a planar interface during solidification of a dilute binary alloy, *J. Appl. Phys.* **35** (1964), 444–451.
27. O. PENROSE & P. FIFE, Theormodynamically consistent models for the kinetics of phase transitions, *Physica D* **43** (1990), North-Holland, 44–62.
28. L. SIMON, *Lectures on Geometric Measure Theory*, Centre for Mathematical Analysis, Australian National University (1984).
29. H. M. SONER, Ginzburg-Landau equation and motion by mean curvature, I: convergence, II: development of the initial interface, *J. Geometric Analysis* (1993), to appear.
30. B. STOTH, The Stefan problem with the Gibbs-Thompson law as singular limit of phase-field equations in the radial case, *European J. Appl. Math.* (1992), to appear.
31. B. STOTH, Convergence of the Cahn-Hilliard equation to the Mullins-Sekerka problem in spherical symmetry, preprint.
32. M. STRUWE, On the evolution of harmonic maps in higher dimensions, *J. Diff. Geometry* **28** (1988), 485–502.
33. S. L. WANG, R. F. SEKERKA, A. A. WHEELER, B. T. MURRAY, S. R. CORIELL, R. J. BRAUN & G. B. MCFADDEN, Thermodynamically-consistent phase-field models for solidification, *Physica D* **69** (1993), 189–200.
34. W. ZIEMER, *Weakly Differentiable Functions*, Springer-Verlag (1989).

Department of Mathematics
Carnegie Mellon University
Pittsburgh, PA 15213-3890

(Accepted January 15, 1995)

Papers Reprinted

Papers on Materials Science

1951 *Herring, C.*, Surface Tension as a Motivation for Sintering. In: *The Physics of Powder Metallurgy* (W.E. Kingston, ed.), pp. 143-179. McGraw-Hill, New York 1951

1956 *Mullins, W. W.* , Two-Dimensional Motion of Idealized Grain Boundaries. J. Appl. Phys. **27,** 900-904
Reprinted with permission from the American Institute of Physics, Woodbury, NY, USA

1963 *Mullins, W. W.* and *R. F. Sekerka,* Morphological Stability of a Particle Growing by Diffusion or Heat Flow. J. Appl. Phys. **34,** 323-324
Reprinted with permission from the American Institute of Physics, Woodbury, NY, USA

1970 *Eshelby, J. D.,* Energy Relations and the Energy-Momentum Tensor in Continuum Mechanics. In: *Inelastic Behavior of Solids* (M. Kanninen, W. Adler, A. Rosenfeld, and R. Jaffe, eds.). McGraw-Hill, New York 1970, pp. 77-115
Reprinted with kind permission from McGraw-Hill, New York, USA

1985 *Larché, F. C.* and *J. W. Cahn,* The Interactions of Composition and Stress in Crystalline Solids. Acta Metall. **33,** 331-357
Reprinted with kind permission from Elsevier Science Ltd, The Boulevard, Langford Lane, Kidlington OX5 1GB, UK

Papers on Continuum Mechanics

1988 *Gurtin, M. E.,* Multiphase Thermomechanics with Interfacial Structure. 1. Heat Conduction and the Capillary Balance Law. Arch. Rational Mech. Anal. **103,** 195-221
Springer Berlin Heidelberg

1989 *Leo, P. H.* and *R. F. Sekerka,* The Effect of Surface Stress on Crystal-Melt and Crystal-Crystal Equilibrium. Acta Metall. **37,** 3119-3138
Reprinted with kind permission from Elsevier Science Ltd, The Boulevard, Langford Lane, Kidlington OX5 1GB, UK

1989 *Angenent, S. B.* and *M. E. Gurtin,* Multiphase Thermomechanics with Interfacial Structure. 2. Evolution of an Isothermal Interface. Arch. Rational Mech. Anal. **108,** 323-391
Springer Berlin Heidelberg

1990 *Abeyaratne, R.* and *J. K. Knowles,* On the Driving Traction Acting on a Surface of Strain Discontinuity in a Continuum. J. Mech. Phys. Solids **38,** 345-360
Reprinted with kind permission from Elsevier Science Ltd, Pergamon Imprint, Oxford, England

1995 *Gurtin, M. E.,* The Nature of Configurational Forces. Arch. Rational Mech. Anal. **131,** pp. 67-100
Springer Berlin Heidelberg

Papers on Mathematics

1990 *Luckhaus, S.,* Solutions for the Two-Phase Stefan Problem with the Gibbs–Thomson Law for the Melting Temperature. Euro. J. Appl. Math., **1,** 101-111
Reprinted with kind permission from Cambridge University Press, Cambridge, United Kingdom

1991 *Evans, L. C.* and *J. Spruck,* Motion of Level Set by Mean Curvature. J. Diff. Geom. **33,** 635-681
Reprinted with kind permission from the editor-in-chief, Lehigh University, Bethlehem, PA, USA

1991 *Chen, Y.-G., Y. Giga,* and *S. Goto,* Uniqueness and Existence of Viscosity Solutions of Generalized Mean Curvature Flow Equations. J. Diff. Geom. **33,** 749-786
Reprinted with kind permission from the editor-in-chief, Lehigh University, Bethlehem, PA, USA

1995 *Soner, H. M.,* Convergence of the Phase Field Equations to the Mullins-Sekerka Problem with Kinetic Undercooling. Arch. Rational Mech. Anal. **131,** pp. 139-137
Springer Berlin Heidelberg

MIX
Papier aus verantwortungsvollen Quellen
Paper from responsible sources
FSC® C105338

FSC
www.fsc.org

If you have any concerns about our products,
you can contact us on
ProductSafety@springernature.com

In case Publisher is established outside the EU,
the EU authorized representative is:
Springer Nature Customer Service Center GmbH
Europaplatz 3, 69115 Heidelberg, Germany

Printed by Libri Plureos GmbH
in Hamburg, Germany